Fluid Mechanics and Hydraulic Machines

S.S. RATTAN
Department of Mechanical Engineering,
Regional Engineering College,
Kurukshetra.

KHANNA BOOK PUBLISHING CO. (P) LTD.
Publisher of Engineering and Computer Books
4C/4344, Ansari Road, Darya Ganj, New Delhi-110002
Phone : 011-23244447-48 **Mobile:** +91-9910909320
E-mail : contact@khannabooks.com
Website : www.khannabooks.com

Dear Readers,

To prevent the piracy, this book is secured with HIGH SECURITY HOLOGRAM on the front title cover. In case you don't find the hologram on the front cover title, please write us to at contact@khannabooks.com or whatsapp us at +91-99109 09320 and avail special gift voucher for yourself.

Specimen of Hologram on front Cover title:

Moreover, there is a SPECIAL DISCOUNT COUPON for you with EVERY HOLOGRAM.

How to avail this SPECIAL DISCOUNT:

Step 1: Scratch the hologram
Step 2: Under the scratch area, your "coupon code" is available
Step 3: Logon to www.khannabooks.com
Step 4: Use your "coupon code" in the shopping cart and get your copy at a special discount
Step 5: Enjoy your reading!

Price: ₹ 548.00

ISBN: 978-81-87522-46-1

Fluid Mechanics and Hydraulic Machines *by* S.S. Rattan

Edition: 2025

To view complete list of books, Please scan the QR Code:

Published by:
Khanna Book Publishing Co. (P) Ltd.
CIN: U22110DL1998PTC095547

Visit us at: **www.khannabooks.com**
Write us at: **contact@khannabooks.com**

Printed in Bharat.

To

Neena, Ravneet and Jasmeet

PREFACE

Writing a book on a subject on which there alerady exists innumerable books in the market could not be termed as a wise decision in general. However since the publication of my book on *Theory of Machines*, my perception has changed. The overwhelming success of that book convinced me that a scope for improvement always exists. It could be in the manner of treatment on presentation of the subject matter or in the technique of solving the numerical problems. The success of that book motivated me to venture into writing this book. Also, my readers always prompted me to write another book. Naturally, while writing, I was mainly guided by the same objectives in mind *i.e.* compactness, lucidity, simplicity, comprehensiveness and completeness.

The book has been written to present a straightforward and modern approach to the subject of *Fluid Mechanics and Hydraulic Machines*. The endeavor is to present the subject matter in the most comprehensive and usable form. The derivation of fundamental relations has been kept as simple as possible. The emphasis throughout is on physical concepts which have been evolved from simple and readily understood principles. The simplest mathematical methods have been employed in preference to those which may be more elegant but less obvious. The theory is reinforced with illustrations and suitably worked out examples. This should also help students develop an orderly approach to problem solving.

The SI system is used throughout the book as it has been adopted by most of the universities by now. The contents of the book have been chosen keeping in view the syllabi of most of the teaching institutions. The data for the solved examples and the problems at the end of chapters have been taken from the examinations of these institutions with suitable modifications wherever necessary.

Though the aim throughout has been simplicity of presentation, yet extra care has been taken to make some sections more intelligible where it is felt that the students usually face difficulties in the comprehension. Due importance is given to the topics on Hydraulic Machines and the readers will find that the portion is not just a supplement of a book on Fluid Mechanics. There are detailed descriptions of these machines along with a number of solved and unsolved problems.

In preparing the manuscript, I relied heavily on the works of renowned authors whose writings are considered classics in the field and I am indebted to them. I also acknowledge the help of my many colleagues who helped me in one form or the other in preparing this treatise.

A creation by a human being can never be perfect. A number of mistakes might have crept in. I will be highly grateful to the readers for their uninhibited comments and pointing out the errors in the text.

I am also thankful to my publishers for taking keen interest in the publication of the book. Had they not taken the inititative, the book would have never seen the light of the day.

Last but not the least I am also indebted to my wife and children for sparing me to do this arduous task of preparing the manuscript. But for their sacrifice, I would not have been able to complete it in the most satisfying way. They happily bore the hardships during the period.

S.S. Rattan

CONTENTS

SYMBOLS

a	area, acceleration
A	area, width
b	width
B	width, centre of buoyancy
c	chord length
C	constant, coefficient, velocity of pressure wave
C_D^*	skin friction drag, local coefficient of drag
d	diameter
D	diameter
e	eccentricity, internal energy, slope of energy line
E	modulus of elasticity, energy, specific energy
f	force, function, coefficient of friction, frictional resistance
F	force
g	acceleration due to gravity
G	centre of gravity
h	pressure head, height, depth
H	height, depth, head
i	slope of bed
I	moment of inertia
k	constant roughness height, ratio
K	constant, bulk modulus of elasticity, conveyance, Bazin's constant
K_u	speed ratio
K_f	flow ratio
l	length
L	length, stroke
m	mass, hydraulic mean depth, hydraulic radius, jet ratio
M	mass, momentum
n	ratio
N	speed, Kutter's constant
p	pressure
P	power, perimeter
q	discharge, heat
Q	discharge
r	radius, crank length
R	raidius, reaction force, resultant, universal gas constant
s	span, slip
S	specific gravity (relative density),
t	time

T	time, torque, temperature, thickness
u	velocity, tangential velocity
u_*	shear velocity
v	velocity
V	volume, velocity
w	specific weight, weight, velocity
W	weight, work
x	depth, displacement
y	displacement, depth
z	displacement, potential energy, datum head, depth
Z	number of buckets or blades
θ	angle, momentum thickness
α	angle, angular acceleration, kinetic energy correction factor
β	angle, momentum correction factor
γ	shear strain, ratio of specific heats
η	efficiency
ϕ	velocity potential function
μ	dynamic viscosity
ν	kinematic viscosity
ε	linear strain
ρ	mass density
π	dimensionles group, 3.1416
σ	surface tension
δ	increment of a quantity, boundary layer thickness
δ'	thickness of boundary sublayer
δ^*	displacement thickness
δ^{**}	energy thickness
τ	shear stress
Γ	circulation
ω	angular velocity
ψ	stream function

Subscripts

a	approach, atmospheric, axial
c	coupling
B	buoyant
c	contraction, vena-contracta, circumferential, critical
d	discharge, draft tube, delivery
D	drag

e	elnargement
f	friction
g	through centre of gravity, generator
h	horizontal, hydraulic
i	inlet, impeller
L	loss, liquid, lift
m	mercury, mean, model, mechanical
max	maximum
n	normal
o	initial value, overall, outlet
p	prototype, pressure
P	pump
r	radial, ratio, relative, resultant
s	specific, suction
t	turbulent, tangential, total
th	theoretical
T	turbine
u	tangential, unit
v	vertical, velocity, volumetric, vapour
w	water, whirl
x	x-component
y	y-component
z	z-component
θ	angular direction

Non-dimensional groups

Re	Reynolds number
Fr	Froude number
We	Weber number
Eu	Euler number
M	Mach number
σ	cavitation number
C_p	pressure coefficient

Chapter 1

FLUID PRESSURE

1.1 INTRODUCTION

Fluids are substances which can flow and offer no resistance to change of shape. **Fluid mechanics** is that branch of science which deals with the behaviour of fluids when they are subjected to changes of pressure, frictional resistance, flow through various types of orifices and nozzles and production of power etc.

Fluids may be divided into three types : **Liquids, vapours** and **gases.**

Liquids offer great resistance to compression and are not much affected by change of temperature. They are usually treated as incompressible in fluid mechanics, thus ignoring the small amount of compression due to its volumetric elasticity. No liquid can exist as a liquid at zero pressure. All known liquids vaporise at various pressures above zero depending upon the temperature. A liquid can withstand a small amount of tension due to molecular attraction between its particles and which causes a shear resistance between two adjacent layers. This phenomenon is known as **viscosity**.

The portion of the subject that deals with water is called **Hydraulics** which will comprise a main part of the book.

Vapours and gases are compressible fluids. Their density is changed with the change in pressure and temperature. Vapours differ from the pure gases in the sense that the vapours are also subjected to condensation as well as evaporation of the moisture which may be in suspension. Thus a vapour may be called a gas whose pressure and temperature are very near to the liquid state.

A hypothetical fluid having no viscosity and surface tension and is incompressible is known as **ideal fluid**. In true sense, no ideal fluid exists in nature. The concept of ideal fluid helps in simplifying the mathematical analysis. The practical fluids *ie* fluids having viscosity, surface tension and compressibility are known as **real fluids**.

That branch of fluid mechanics which deals with the study of fluids at rest is known as **fluid statics.**

1.2 PROPERTIES OF FLUIDS

The main properties of fluids are :

1. Density or Mass Density

Density or **mass density** is the ratio of the mass of a fluid to its volume.

$$\text{Mass density, } \rho = \frac{\text{Mass of fluid}}{\text{Volume of fluid}}$$

Density of liquids may be considered constant.

Units are kg/m^3. Density of water is 1000 kg/m^3.

2. Specific Weight (Weight Density)

Specific weight or the **weight density** is the ratio between the weight of a fluid to its volume.

$$\text{Specific weight, } w = \frac{\text{Weight of fluid}}{\text{Volume of fluid}} = \frac{\text{Mass of fluid} \times g}{\text{Volume of fluid}} = \rho g$$

Units of specific weight are N/m^3. Specific weight of water is $9{\cdot}81 \times 1000$ N/m^3 or 9·81 kN/m^3.

3. Specific Volume

The volume of a fluid per unit mass is called its **specific volume.**

$$\text{Specific volume} = \frac{\text{Volume of fluid}}{\text{Mass of fluid}} = \frac{1}{\rho}$$

Units are m^3/kg.

4. Specific Gravity (Relative Density)

Specific gravity which is also called **relative density** is the ratio of specific weight of a fluid to that of a standard fluid (for liquids, the standard fluid is water at 4°C and for gases it is air).

$$S = \frac{\text{Density of fluid}}{\text{Density of water or air}}$$

It has no units.

5. Viscosity (Dynamic Viscosity)

It is the property of a fluid that offers resistance against shearing of one layer of fluid over another adjacent layer. It is caused due to the cohesion and interaction between particles.

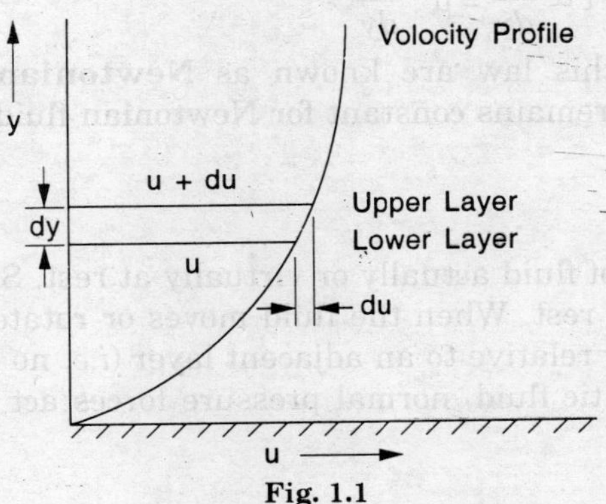

Fig. 1.1

In Fig. 1.1, the two layers are at a distance dy apart. The upper layer moves at a higher velocity than the lower one. Thus the upper layer causes a shear stress on the lower layer whereas the lower layer on the upper one. The shear stress is proportional to the rate of change of velocity with respect to y *i.e.*

$$\text{shear stress, } \tau \propto \frac{du}{dy}$$

$$= \mu \cdot \frac{du}{dy}$$

where μ is the constant of proportionality and is known as the **coefficient of dynamic viscosity** or only viscosity.

$$\text{Thus} \qquad \mu = \frac{\tau}{du/dy}$$

$$= \frac{\text{Shear stress}}{\text{Rate of shear strain}}$$

= Shear stress required to provide unit rate of shear strain

Units can be determined as under :

$$\mu = \frac{\tau}{du/dy} = \frac{\text{Force/Area}}{\text{Change of velocity/Change of distance}} = \frac{\text{N/m}^2}{\text{(m/s)/m}} = \text{N . s/m}^2$$

6. Kinematic Viscosity

The ratio of dynamic viscosity to the density of a fluid is known as its **kinematic viscosity.**

$$\nu = \frac{\text{Viscosity}}{\text{Density}}$$

$$\text{Units :} \qquad \frac{\text{N . s/m}^2}{\text{kg/m}^3} = \frac{(\text{kg . m/s}^2)\,.\,(\text{s/m}^2)}{\text{kg/m}^3} = \text{m}^2\text{/s}$$

7. Newton's Law of Viscosity

It states that the shear stress on a fluid element layer is directly proportional to the rate of shear strain.

i.e.,
$$\tau \propto \frac{du}{dy} = \mu \, . \, \frac{du}{dy}$$

The fluids that obey this law are known as **Newtonian fluids** and the others as **non-Newtonian fluids.** μ remains constant for Newtonian fluids.

1.3 FLUID STATICS

Fluid statics is the study of fluid actually or virtually at rest. Still fluid in a container is an example of fluid actually at rest. When the fluid moves or rotates as a solid mass and there is no movement of one layer relative to an adjacent layer (*i.e.* no shear stresses act), the fluid is virtually at rest. In a static fluid, normal pressure forces act on all free bodies and there are no shear stresses.

1.4 FLUID PRESSURE

A fluid contained in a vessel exerts force at all the points on the sides and the bottom of the container. The force per unit area is known as **pressure**. Pressure (or intensity of pressure) is a scalar quantity. This implies that the pressure acts equally in all directions at a point in a fluid. Areas and forces are vector quantities. Thus pressure force is a vector quantity with magnitude equal to the product of the pressure intensity and the magnitude of the area and the direction is normal to the area.

Let dA = an infinitesimal small area in a large mass of fluid

and dF = force acting on the area dA in the normal direction

Then intensity of pressure (or simply pressure) is given by,

$$p = \frac{dF}{dA}$$

However, if the force is uniformly distributed over the whole area A,

$$p = \frac{F}{A} \qquad ...(1.1)$$

Fluid pressure is also known as **static** or **hydrostatic pressure** to make it more specific.

Units of pressure are N/m^2.

1.5 PASCAL'S LAW

Pascal's law states that in a static fluid, the pressure or the intensity of pressure at a point is the same in all directions. This can be proved by considering a small wedge-shaped free body of unit length (perpendicular to the paper) at a point (x, y) in a fluid at rest (Fig. 1.2).

There cannot be any shear forces, so only the normal surface forces and gravity forces act on the body.

Let p_x = pressure acting on the face AB

p_y = pressure acting on the face AC

p_s = pressure acting on the face BC

w = specific weight of the fluid

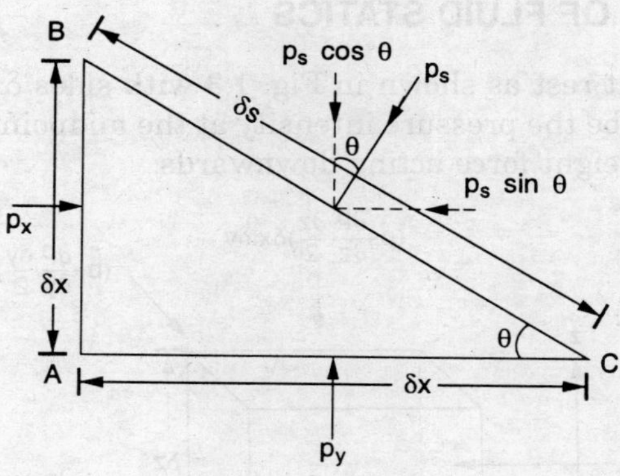

Fig. 1.2

Forces acting on the body are,

$$F_{AB} = p_x \,.\, (\delta y \,.\, 1)$$

$$F_{AC} = p_y \,.\, (\delta x \,.\, 1)$$

$$F_{BC} = p_s \,.\, (\delta s \,.\, 1)$$

Weight of the fluid element,

$$W = \frac{\delta x \,.\, \delta y}{2} \,.\, 1 \,.\, w$$

For equilibrium, ΣF_x and ΣF_y should be equal to zero,

i.e. $\Sigma F_x = p_x \,.\, \delta y - p_s \,.\, \delta s \,.\, \sin\theta = 0$

or $p_x \,.\, \delta y - p_s \,.\, \delta y = 0$

or $\delta y\,(p_x - p_s) = 0$

As $\delta y \neq 0, \therefore\ p_x = p_s$. ...(i)

Also, $\Sigma F_y = p_y \,.\, \delta x - p_s \,.\, \delta s \,.\, \cos\theta - \dfrac{\delta x \,.\, \delta y \,.\, w}{2}$

or $p_y \,.\, \delta x - p_s \,.\, \delta x - \dfrac{\delta x \,.\, \delta y \,.\, w}{2} = 0$

Third term on the LHS being the product of two very small quantities can be neglected.

$\therefore$ $\delta x\,(p_y - p_s) = 0$

As $\delta x \neq 0,\ p_y = p_s$ ...(ii)

From (i) and (ii),

$$p_x = p_y = p_s$$

As θ is an arbitrary angle, pressure is same in all the directions at a point in a static fluid.

1.6 GENERAL EQUATION OF FLUID STATICS

Consider an element of fluid at rest as shown in Fig. 1.3 with sides δx, δy and δz in x, y and z directions respectively. Let p be the pressure intensity at the midpoint C. The only body force acting on the element is the weight force acting downwards.

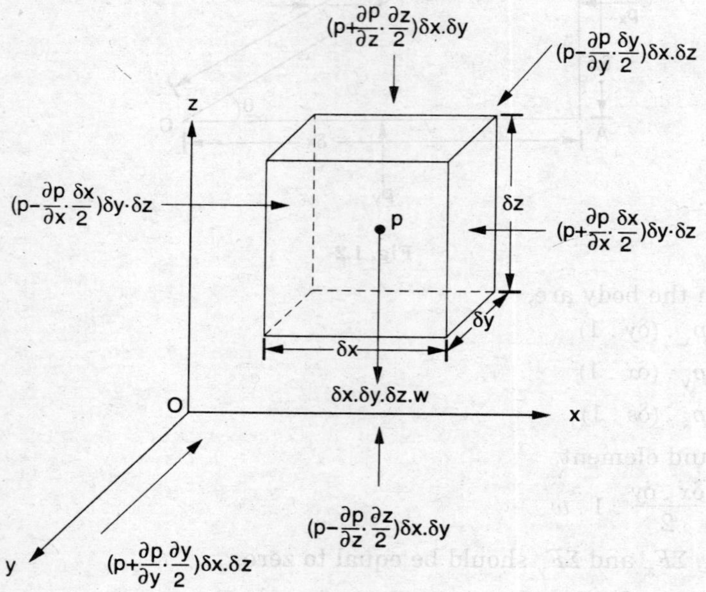

Fig. 1.3

Weight of the body = $\delta x \,.\, \delta y \,.\, \delta z \,.\, w$ in the downward direction

The forces acting on the fluid element in the x-, y- and z-directions are,

Force in the x-direction, $\delta F_x = \left[p - \dfrac{\partial p}{\partial x} \cdot \dfrac{\delta x}{2}\right] \delta y \,.\, \delta z - \left[p + \dfrac{\partial p}{\partial x} \cdot \dfrac{\delta x}{2}\right] \delta y \,.\, \delta z$

$$= \frac{\partial p}{\partial x} \cdot \partial x \, \partial y \, \partial z$$

Similarly,

Force in the y-direction, $\delta F_y = \left[p - \dfrac{\partial p}{\partial y} \cdot \dfrac{\delta y}{2}\right] \delta x \,.\, \delta z - \left[p + \dfrac{\partial p}{\partial y} \cdot \dfrac{\delta y}{2}\right] \delta x \,.\, \delta z$

$$= -\frac{\partial p}{\partial y} \cdot \partial x \, \partial y \, \partial z$$

Force in the z-direction, $\delta F_z = \left[p - \dfrac{\partial p}{\partial z} \cdot \dfrac{\delta z}{2}\right] \delta x \,.\, \delta y - \left[p + \dfrac{\partial p}{\partial z} \cdot \dfrac{\delta z}{2}\right] \delta x \,.\, \delta y - \delta x \,.\, \delta y \,.\, \delta z \,.\, w$

$$= -\frac{\partial p}{\partial z} \cdot \partial x \, \partial y \, \partial z - \delta x \,.\, \delta y \,.\, \delta z \,.\, w$$

$$= \left(\frac{\partial p}{\partial z} + w\right) \delta x \,.\, \delta y \,.\, \delta z$$

Since the fluid is at rest, the force vectors in the x-, y- and z-directions must be zero. *i.e.*

$$\delta F_x = 0 \text{ or } \frac{\partial p}{\partial x} = 0 \text{ i.e. no pressure variation in } x\text{-direction}$$

$$\delta F_y = 0 \text{ or } \frac{\partial p}{\partial y} = 0 \text{ i.e. no pressure variation in } y\text{-direction}$$

$$\delta F_z = 0 \text{ or } \frac{\partial p}{\partial z} + w = 0$$

or

$$\frac{\partial p}{\partial z} = -w$$

As the pressure changes only in the z-direction, the partial derivative may be replaced with exact derivative.

i.e.

$$\frac{dp}{dz} = -w \qquad ...(1.2)$$

Minus sign indicates that the pressure decreases in the positive z-direction or the pressure intensity increases with depth from the free surface of fluid.

For an incompressible fluid w is constant. Integrating equation 1.2 for such a fluid and considering two points 1 and 2 at different heights,

$$\int_1^2 dp = -w \int_1^2 dz$$

which gives,

$$p_2 - p_1 = w(z_1 - z_2)$$

This equation gives the pressure difference between two points at different heights of a liquid in a container.

If h is taken as the depth of liquid from the free surface where the pressure is zero, then we can also write

$$p = wh \qquad ...(1.3)$$

where p indicates the pressure at a depth h from the free surface. This also implies that the pressure is same at all the points on a horizontal level.

1.7 PRESSURE HEAD OF A FLUID

The pressure at any point in a fluid depends on the depth of that point from the free surface or the height of the free surface above that point. It is sometimes convenient to express a fluid pressure by the height of the free surface which can cause that pressure, *i.e.*

$$h = p/w \qquad ...(1.4)$$

where h is the height of the free surface of fluid and is known as the **static head.** Thus the intensity of pressure of a fluid may be expressed either as a force/area or as an equivalent static head. When the pressure is expressed in terms of static head, it is known as the **pressure head** at the point.

In case the pressure at a point in a fluid at rest is known (let it be p_0), the pressure at any point below it at a depth h will be,

$$p = p_0 + wh$$

A change in p_0 will cause an equal change in pressure at all other points.

The equation 1.4 may be derived by considering a vertical column of liquid of finite height h and area a such that its upper surface touches the free surface of liquid (Fig. 1.4). For equilibrium of the liquid column, its weight must be balanced by a vertical upward pressure force at the base equal to the weight of the liquid column.

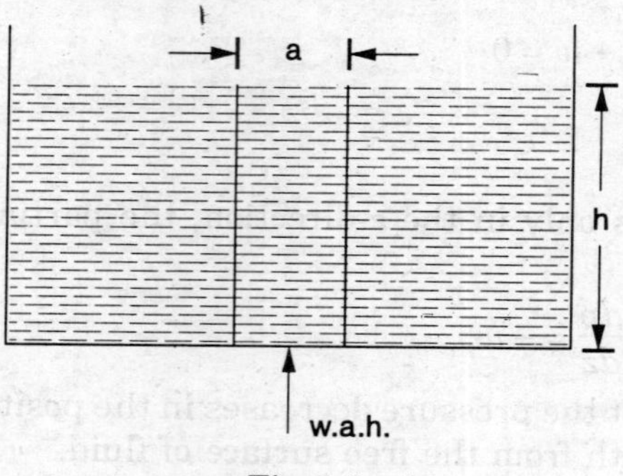

Fig. 1.4

Weight of the liquid column $= wah$

Thus upward pressure force on the base $= wah$

or upward pressure on the base $=$ Force/area

$= wah/a$

$= wh$

Thus pressure at any horizontal surface in a liquid at rest is simply due to the head or height of the free surface of liquid above the horizontal level. In Fig. 1.5, though the weight of liquid in each part differs, the pressures at points A, B, C, D and E lying on the same horizontal level are same because the vertical height of the liquid column is same as they are interconnected. Also, the pressure at point F is the same as at points A or B etc. This implies that pressure at a point is given by equation $p = wh$ irrespective of the fact whether the point has actual water column or not above it.

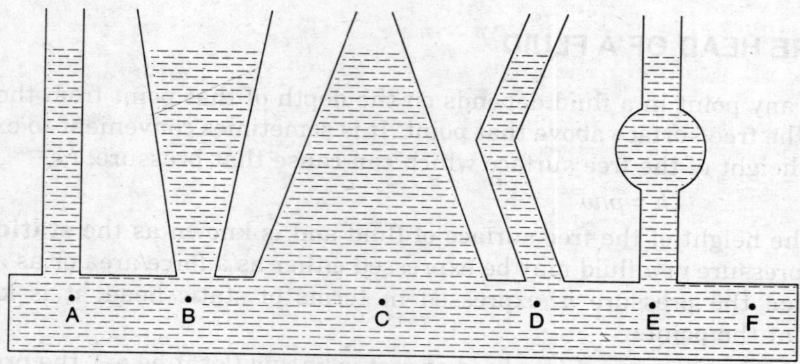

Fig. 1.5

Consider another vessel of the shape as shown in Fig. 1.6. The pressure at point B is wh. Since the same pressure must exist in a horizontal plane, the pressure wh will be acting on the bottom surface throughout and the total pressure force will be wAh. This is equal to the weight of liquid corresponding to CDEF. This is known as the **hydrostatic paradox.** It states that the resultant force on the bottom of a container may be several times greater than the water it contains. However, this is possible only in the cases where the walls of the container curve inwards so that there is an upward thrust. The net resultant force on the inside of the container will simply be the weight of the liquid contained in it.

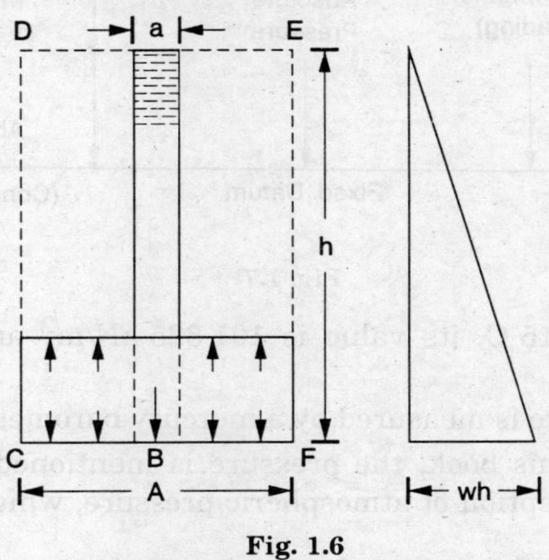

Fig. 1.6

1.8 ABSOLUTE, GAUGE AND VACUUM PRESSURES

Usually, pressures are expressed in two ways by assuming different datums in the two cases. In the first case, the absolute zero is adopted as the datum whereas in the second, the pressure above the local atmospheric pressure is considered.

Absolute Pressure

It is the pressure difference between its value and the absolute zero or complete vacuum.

Gauge Pressure

When the pressure is expressed as a difference between its value and the local atmospheric pressure, it is known as **gauge pressure**. This is the pressure which is measured by the pressure measuring instruments.

Vacuum pressure

It is defined as the pressure below the atmospheric pressure.

Fig. 1.7 defines the absolute, gauge and the vacuum pressures.

Mathematically,

Absolute pressure = Atmospheric pressure + Gauge pressure

= Atmospheric pressure − Vacuum pressure

The value of the local atmospheric pressure varies with temperature and the elevation of

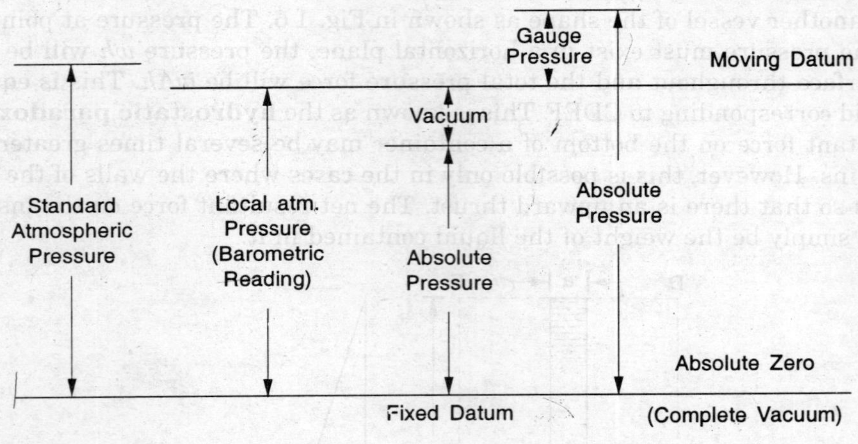

Fig. 1.7

the place. At sea level and 15°C, its value is 101·325 kN/m^2 and is known as **Standard atmospheric pressure.**

Local atmospheric pressure is measured by a mercury barometer or an aneroid barometer.

Throughout the text of this book, the pressure is mentioned as gauge pressure unless stated otherwise with the exception of atmospheric pressure, which is an absolute pressure.

1.9 MEASUREMENT OF PRESSURE

The main devices used to measure the pressure are the manometers and the mechanical gauges. Manometers are the devices which employ liquid columns to determine difference in pressure whereas the mechanical gauges use springs or dead weights to balance the fluid columns to know the pressure. The manometers are also divided into simple manometers and differential manometers and have been discussed in the following sections.

1.10 SIMPLE MANOMETERS

A simple manometer consists of a glass tube, the one end of which is connected to the point where the pressure is to be measured and the other end is open to the atmosphere. Following types of simple manometers are commonly used :

1. Piezometer
2. *U*-tube manometer
3. Single column manometer

1. Piezometer

It is the simplest form of a manometer. It consists of a glass tube having one of its ends fixed to the gauge point where the pressure is to be measured and the other end raised vertically to some height without the danger of overflowing. The liquid from the pipe or the tank will rise in the pipe as shown in Fig. 1.8 upto a height indicating the pressure at the connecting

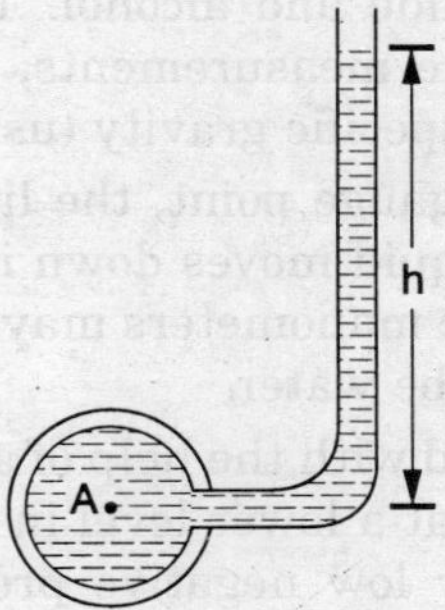

Fig. 1.8

point. Pressure so measured indicates the gauge pressure *i.e.* the pressure of the liquid above the atmospheric pressure as one end of the tube is under the atmospheric pressure.

i.e. $p_A = w \cdot h$

Piezometers can be used for low pressure liquids only. Otherwise, the length of the tubes will be quite large which is not feasible. The rise of the liquid in the piezometer can be measured above any desired point and the head measured will correspond to that point. Thus tubes may be inserted at the top, on the side or on the bottom of the vessel. The size of the piezometer glass tube should be between 12 mm and 25 mm to avoid the effect of the capillary action on the height of the liquid column.

A piezometer tube cannot be used for measuring negative pressure, since in that case the air enters the pipe through the tube.

2. *U*-Tube Manometer

As mentioned earlier, piezometers can only be used for low pressures of the liquids. To measure high pressure, *U*-tube manometers are used. A *U*-tube instead of a simple vertical tube is fixed to the requisite point of pressure. The other end remains open to the atmosphere as usual (Fig. 1.9a).

A liquid of specific gravity higher than the specific gravity of the liquid whose pressure is to be measured, is used in the *U*-tube. Moreover, the liquid used in this way should be immiscible with the main liquid in the pipe or the container. Liquids frequently used are

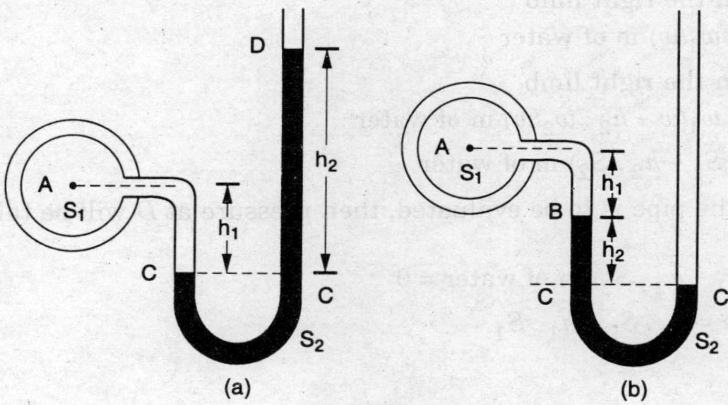

Fig. 1.9

mercury, salt solution, oil, carbon disulphide and alcohol. The choice will depend upon the pressure to be measured. For low pressure measurements, liquids with low specific gravity and for high pressures, liquids with high specific gravity (usually mercury) may be used.

When the *U*-tube is connected to the gauge point, the liquid from the pipe or the vessel enters the *U*-tube and the manometeric liquid moves down in the limb connected to the pipe or vessel and rises in the open limb. *U*-tube manometers may be used to measure the pressure of a gas. The liquid used in this case may be water.

Negative pressure can also be measured with the help of a *U*-tube manometer. In that case the liquid in the open limb will be resting at a lower level in comparison to the liquid level in the pipe or the container (Fig. 1.9b). For low negative pressure same liquid can be used whereas for high pressures, a different liquid has to be used.

A manometer works on the principle of pressure height relationship *i.e.* the pressure in a mass of liquid at rest increases in the downward direction and decreases in the upward direction and on a horizontal plane the liquid pressure is constant.

The following general procedure may be adopted while solving manometer problems :

1. Start at one end. Write the pressure at that end in an appropriate unit or in an appropriate symbol if it is unknown.
2. Add or subtract the change in pressure from one meniscus to the next. If the next meniscus is lower, the pressure is to be added and for higher meniscus, the pressure is to be subtracted, keeping the units same.
3. The procedure is continued till the other end of the gauge is reached.
4. Equate the expression to the pressure at that point.

Refer Fig. 1.9a. Let the liquid in the pipe and the left limb of manometer be lighter of specific gravity S_1 and specific weight w_1 and that in the right limb be heavier of specific gravity S_2 and specific weight w_2. If the pressure head of liquid in the pipe is equivalent to h m of water, Then

Pressure at point C in the left limb

$$= h \text{ m of water} + h_1 \text{ m of lighter liquid}$$

$$= h \text{ m of water} + h_1 \,.\, w_1/w \text{ m of water}$$

$$= (h + h_1 \,.\, w_1/w) \text{ m of water}$$

Pressure at point C in the right limb

$$= (h + h_1 \,.\, w_1/w) \text{ m of water}$$

Pressure at point D in the right limb

$$= (h + h_1 \,.\, w_1/w - h_2 \,.\, w_2/w) \text{ m of water}$$

$$= (h + h_1 \,.\, S_1 - h_2 \,.\, S_2) \text{ m of water}$$

If gauge pressure in the pipe is to be evaluated, then pressure at D will be taken as zero. Thus

$$(h + h_1 \,.\, S_1 - h_2 \,.\, S_2) \text{ m of water} = 0$$

or

$$h = h_2 \,.\, S_2 - h_1 \cdot S_1 \quad \text{...(1.5)}$$

An alternate way can be by equating the liquid pressures at points $C–C$ in the left as well as in the right limb. *i.e.*

$$h + h_1 . S_1 = h_2 . S_2$$

or $$h = h_2 . S_2 - h_1 . S_1$$

For Fig. 1.9b,

$$h + h_1 . S_1 + h_2 . S_2 = 0$$

or $$h = -(h_1 . S_1 + h_2 . S_2) \quad ...(1.6)$$

3. Single Column Manometer

U-tube manometers just discussed require to take the readings at more than one point to know the pressure in the pipe. This is due to lowering of liquid level in one limb and rise in the other. In a single column manometer, one of the two limbs is made so large in area (about 100 times) that the variation of level in this limb is safely neglected (Fig. 1.10). Thus reading of the liquid only in the narrow limb is to be taken which directly gives the fluid pressure in the pipe. The narrow limb of the manometer may be made vertical or inclined. The inclined form is preferred for the measurement of low pressures. As there remains no necessity of taking the reading in the larger limb, it is usually made of non-transparent material.

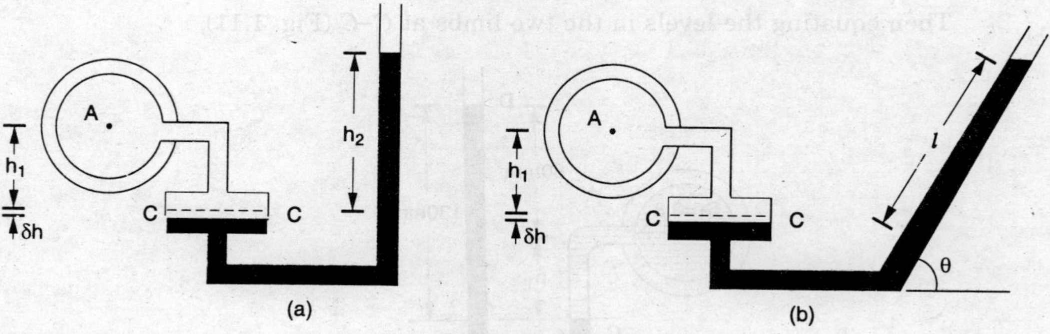

Fig. 1.10

Let h = pressure head in the pipe (m of water)

δh = fall of heavy liquid in the reservoir

h_1 = height of centre of pipe above $C–C$

h_2 = rise of heavy liquid in the narrow pipe

S_1 = specific gravity of liquid (light) in the pipe

S_2 = specific gravity of heavy liquid

A = cross-sectional area of the reservoir

a = cross-sectional area of the right limb of manometer

When the manometer is attached to the pipe,

Decrease in the volume of heavy liquid in the reservoir

= Increase in the volume of heavy liquid in the narrow limb

i.e. $A \cdot \delta h = a \cdot h_2$

or $\delta h = h_2 \cdot a/A$

Equating the pressures at the datum C–C,

$$h + (h_1 + \delta h) \cdot S_1 = (h_2 + \delta h) \cdot S_2$$

or $$h = h_2 \cdot S_2 - h_1 \cdot S_1 + \delta h\,(S_2 - S_1)$$

or $$h = h_2 \cdot S_2 - h_1 \cdot S_1 + h_2 \cdot (a/A)\,(S_2 - S_1) \quad \text{...(1.7)}$$

If the area a is very small as compared to A, then the ratio a/A may be neglected and

$$h = h_2 \cdot S_2 - h_1 \cdot S_1 \quad \text{...(1.8)}$$

In case of inclined column manometer, let l be the length of the heavy liquid moved in the right limb above C–C. Then

$$h = l \sin\theta \cdot S_2 - h_1\ S_1 \quad \text{...}(h_2 = l \sin\theta) \text{ ...(1.9)}$$

Example 1.1 The left end of a U-tube manometer is connected to a pipe through which an oil of specific gravity 0·8 is flowing and the right limb is open to the atmosphere. The level of mercury in the right limb is 80 mm above the centre line of the pipe. Determine the absolute pressure of the oil in the pipe if the difference of mercury levels in the two limbs is 130 mm.

Solution Let the pressure head in the pipe be h m of water.

Then equating the levels in the two limbs at C–C (Fig. 1.11),

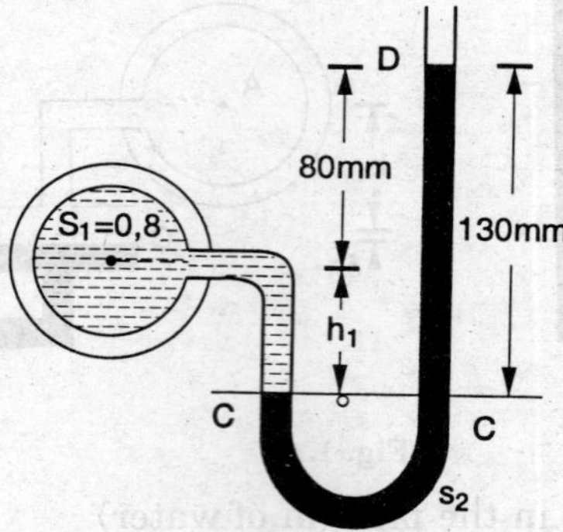

Fig. 1.11

$$h + h_1 \cdot S_1 = h_2 \cdot S_2$$

$$h + (0{\cdot}13 - 0{\cdot}08) \times 0{\cdot}8 = 0{\cdot}13 \times 13{\cdot}6$$

$$h = 1{\cdot}728 \text{ m}$$

$$p = wh = 9{\cdot}81 \times 1{\cdot}728 = 16{\cdot}95 \text{ kN/m}^2$$

Absolute pressure $= p_{atm} + p_{gauge}$

$= 100 + 16{\cdot}95$

$= 116{\cdot}95 \text{ kN/m}^2$

Example 1.2 A simple U-tube manometer is connected to a pipe at one end. The pipe contains a liquid of specific gravity 0·85 and has vacuum pressure. The other end is exposed to atmosphere. If the difference of mercury level in the two limbs is 300 mm and the centre of pipe is 200 mm above the level of mercury in the left limb, find the vacuum pressure in the pipe.

Solution Let the pressure head in the pipe be h m of water.

Then equating the levels in the two limbs at C–C,

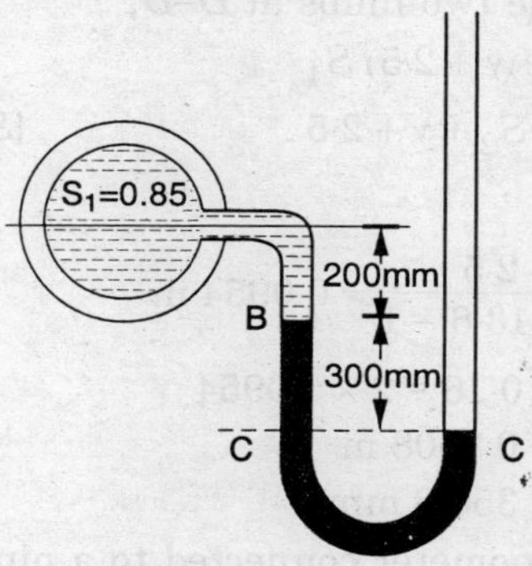

Fig. 1.12

$$h + h_1 . S_1 + h_2 . S_2 = 0$$

$$h + 0{\cdot}2 \times 0{\cdot}85 + 0{\cdot}3 \times 13{\cdot}6 = 0$$

$$h = -\,4{\cdot}25 \text{ m}$$

$$p_{\text{vacuum}} = wh = 9{\cdot}81 \times 4{\cdot}25 = 41{\cdot}7 \text{ kN/m}^2$$

Example 1.3 A *U*-tube manometer is connected to a conical vessel as shown in Fig. 1.13a. The reading of the manometer shown is when the vessel is empty. Determine the reading of the manometer when the vessel is completely filled with water.

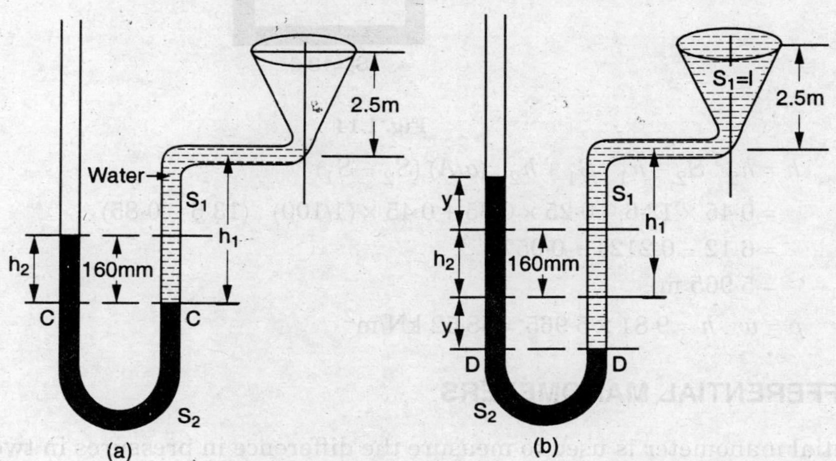

Fig. 1.13

Solution Equating the levels in the two limbs at C–C,

$$h_1 \,.\, S_1 = h_2 \,.\, S_2$$

$$h_1 = h_2 \,.\, S_2 \qquad (S_1 = 1 \text{ for water}) \ldots\text{(i)}$$

When the vessel is full of water, let the level of mercury rise by a height y m in the left limb and fall by y m in the right limb (Fig. 1.13b).

Equating the levels in the two limbs at $D–D$,

$$(h_2 + 2y)\, S_2 = (h_1 + y + 2{\cdot}5)\, S_1$$

$$h_2 \,.\, S_2 + 2y \,.\, S_2 = h_2 \,.\, S_2 + y + 2{\cdot}5 \qquad [S_1 = 1 \text{ and } h_1 = h_2\, S_2 \text{ from (i)}]$$

$$y\,(2S_2 - 1) = 2{\cdot}5$$

$$y = \frac{2{\cdot}5}{2S_2 - 1} = \frac{2{\cdot}5}{2 \times 13{\cdot}6 - 1} = 0{\cdot}0954 \text{ m}$$

Reading of manometer $= 0{\cdot}16 + 2 \times 0{\cdot}0954$

$= 0{\cdot}3508$ m

$= 350{\cdot}8$ mm

Example 1.4 A single column manometer connected to a pipe contains a liquid of specific gravity 0·85 (Fig. 1.14). The area of the reservoir is 100 times the area of the tube. Determine the pressure in the pipe.

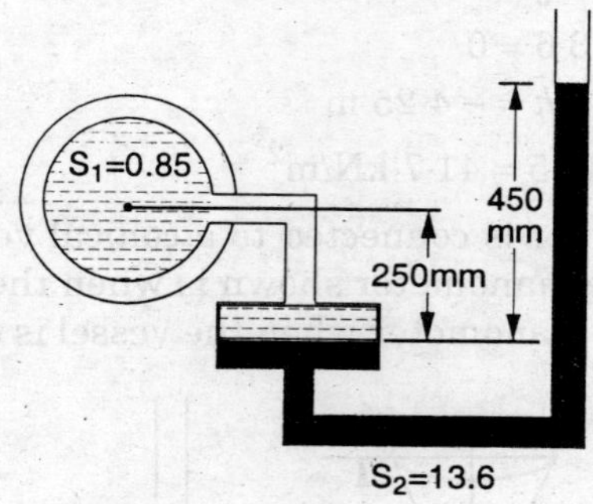

Fig. 1.14

Solution $h = h_2 \,.\, S_2 - h_1 \,.\, S_1 + h_2 \,.\, (a/A)\,(S_2 - S_1)$ (Eq. 1.7)

$= 0{\cdot}45 \times 13{\cdot}6 - 0{\cdot}25 \times 0{\cdot}85 + 0{\cdot}45 \times (1/100)\,.\,(13{\cdot}6 - 0{\cdot}85)$

$= 6{\cdot}12 - 0{\cdot}2124 + 0{\cdot}0574$

$= 5{\cdot}965$ m

$p = w \,.\, h = 9{\cdot}81 \times 5{\cdot}965 = 58{\cdot}52$ kN/m^2

1.11 DIFFERENTIAL MANOMETERS

A differential manometer is used to measure the difference in pressures in two pipes or at two points in a single pipe. It consists of a U-tube containing a heavy liquid. Its two ends are connected to the two points of which the pressure difference is to be found. There are two main types of differential manometers :

1. *U*-tube differential manometer
2. Inverted *U*-tube differential manometer

1. *U*-Tube Differential Manometer

The connected pipes may be at the same level or they may be at different levels.

When the pipes are at different levels (Fig. 1.15a),

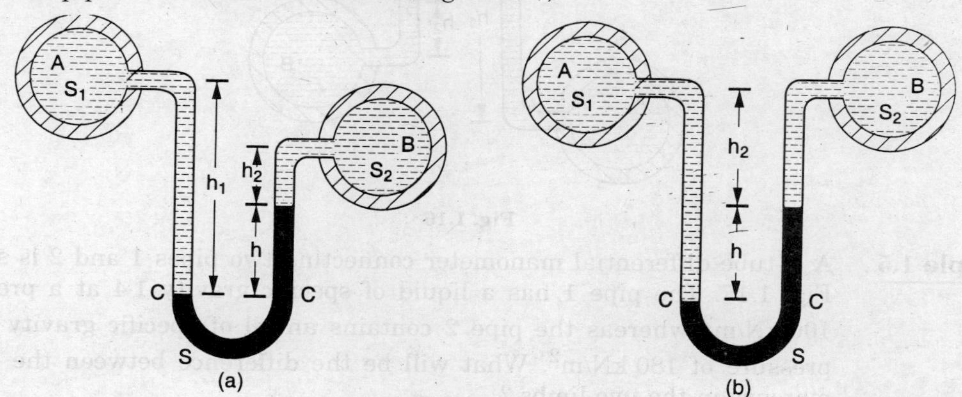

Fig. 1.15

Equating the pressures at *C–C*,

$$h_A + h_1 \,.\, S_1 = h_B + h_2 \,.\, S_2 + h \,.\, S$$

or
$$h_A - h_B = h_2 \,.\, S_2 - h_1 \,.\, S_1 + h \,.\, S \qquad ...(1.10)$$

When the pipes are at the same level (Fig. 1.15b),

$$h_A + (h_2 + h) \,.\, S_1 = h_B + h_2 \,.\, S_2 + h \,.\, S$$

or
$$h_A - h_B = h_2 \,.\, S_2 - h_2 \,.\, S_1 + h\,(S - S_1) \qquad ...(1.11)$$

If the two pipes contain the same liquid, then $S_1 = S_2$

$$h_A - h_B = h_2 \,.\, S_1 - h_2 \,.\, S_1 + h\,(S - S_1)$$
$$= h\,(S - S_1) \qquad ...(1.12)$$

2. Inverted *U*-Tube Differential Manometer

An inverted *U*-tube differential manometer is, usually, used to measure the difference of low pressures and where higher accuracy is desired. It consists of an inverted *U*-tube containing a light liquid. The two ends are connected to the two pipes as usual (Fig. 1.16).

Assuming that the pressures in the pipes *A* and *B* are h_A and h_B m of water column respectively. Equating the pressures at *C–C*,

$$h_A - h_1 \,.\, S_1 + h \,.\, S + h_2 \,.\, S_2 = h_B$$

or
$$h_A - h_B = h_1 \,.\, S_1 - h_2 \,.\, S_2 - h \,.\, S \qquad ...(1.13)$$

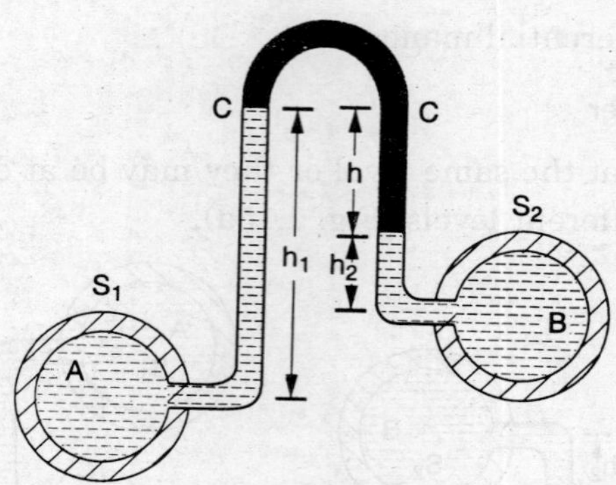

Fig. 1.16

Example 1.5 A *U*-tube differential manometer connecting two pipes 1 and 2 is shown in Fig. 1.17. The pipe 1 has a liquid of specific gravity 1·4 at a pressure of 100 kN/m^2 whereas the pipe 2 contains an oil of specific gravity 0·9 at a pressure of 180 kN/m^2. What will be the difference between the levels of mercury in the two limbs ?

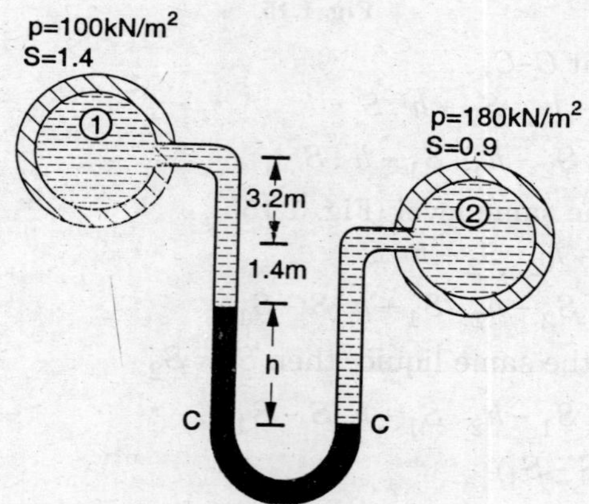

Fig. 1.17

Solution Let the difference between the levels of mercury in the two limbs be h.

Equating the pressure heads in the two limbs at C–C,

$$\frac{100}{9 \cdot 81} + (3 \cdot 2 + 1 \cdot 4) \times 1 \cdot 4 + 13 \cdot 6\, h = \frac{180}{9 \cdot 81} + (1 \cdot 4 + h) \times 0 \cdot 9$$

$$12 \cdot 7\, h = 8 \cdot 155 + 1 \cdot 26 - 6 \cdot 44$$

$$= 2 \cdot 975$$

$$h = 0 \cdot 234 \text{ m}$$

Example 1.6 Fig. 1.18 shows a differential manometer connected at two points A and B. Determine the absolute pressure at A if the air pressure at B is 120 kN/m^2.

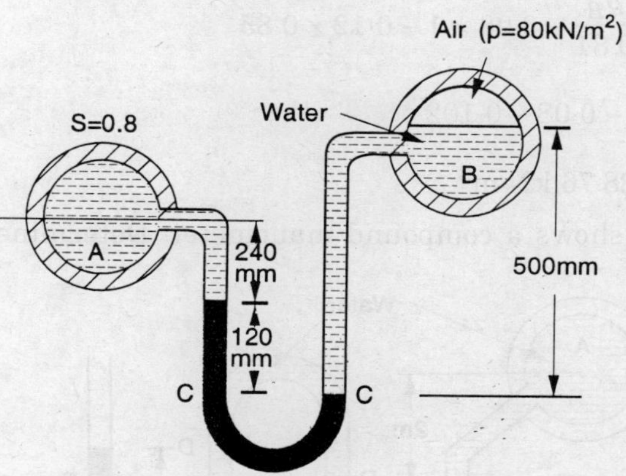

Fig. 1.18

Solution Let the pressure at A be p_A.

Equating the pressure heads in the two limbs at C–C,

$$\frac{p_A}{9{\cdot}81} + 0{\cdot}24 \times 0{\cdot}8 + 0{\cdot}12 \times 13{\cdot}6 = \frac{80}{9{\cdot}81} + 0{\cdot}5 \times 1$$

$$\frac{p_A}{9{\cdot}81} + 0{\cdot}192 + 1{\cdot}632 = \frac{80}{9{\cdot}81} + 0{\cdot}5$$

$$p_A = 67{\cdot}01 \text{ kN/m}^2$$

Example 1.7 An inverted differential manometer contains an oil of specific gravity 0·85 and is connected to two different pipes carrying water under pressure (Fig. 1.19). Determine the pressure in the pipe B if the pressure in the pipe A is 3 m of water.

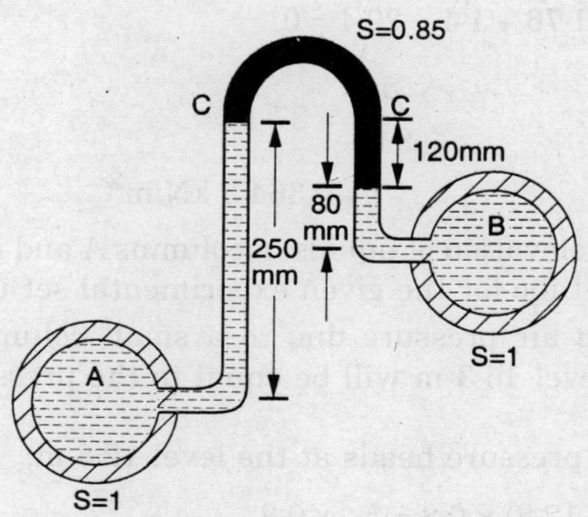

Fig. 1.19

Solution Let the pressure at B be p_B.

Equating the pressure heads in the two limbs at C–C,

$$3 - 0{\cdot}25 \times 1 = \frac{p_B}{9{\cdot}81} - 0{\cdot}08 \times 1 - 0{\cdot}12 \times 0{\cdot}85$$

$$2{\cdot}75 \times \frac{p_B}{9{\cdot}81} - 0{\cdot}08 - 0{\cdot}102$$

$$p_A = 28{\cdot}76 \text{ kN/m}^2$$

Example 1.8 Fig. 1.20 shows a compound manometer. Determine the gauge pressure at *A*.

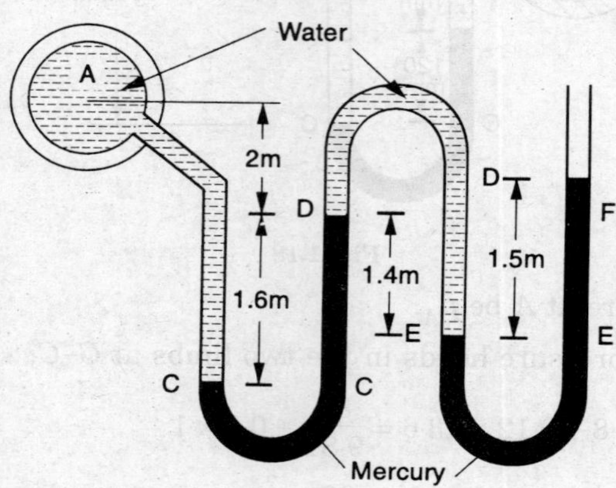

Fig. 1.20

Solution Let the pressure at *A* be p_A.

$$\frac{p_A}{9{\cdot}81} + (2 + 1{\cdot}6) \times 1 - 1{\cdot}6 \times 13{\cdot}6 + 1{\cdot}4 \times 1 - 1{\cdot}5 \times 13{\cdot}6 = 0$$

(at *C–C*) (at *D–D*) (at *E–E*) (at *F*)

$$\frac{p_A}{9{\cdot}81} + 3{\cdot}6 - 21{\cdot}76 + 1{\cdot}4 - 20{\cdot}4 = 0$$

$$\frac{p_A}{9{\cdot}81} = 37{\cdot}16$$

$$p_A = 364{\cdot}5 \text{ kN/m}^2$$

Example 1.9 Find the elevation of liquids in columns *A* and *B* and the deflection of mercury in the *U*-tube for the given experimental set up shown in Fig. 1.21.

Solution Assuming the air pressure due to a small column of air to be negligible, the pressure at level 16·3 m will be equal to the pressure indicated by the pressure gauge.

Equating the pressure heads at the level 12·8 m,

$$\frac{-20}{9{\cdot}81} + (16{\cdot}3 - 12{\cdot}8) \times 0{\cdot}8 = h_A \times 0{\cdot}8$$

$$h_A = \frac{2{\cdot}8 - 2{\cdot}039}{0{\cdot}8}$$

$$= 0{\cdot}952 \text{ m}$$

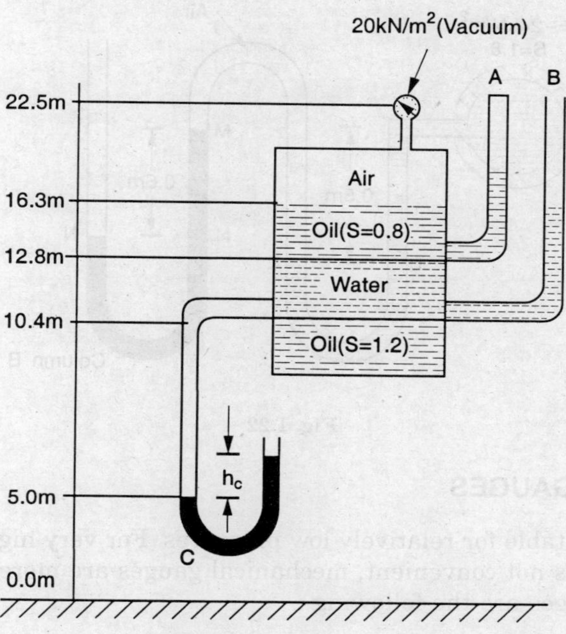

Fig. 1.21

Thus elevation of oil in column $A = (12{\cdot}8 + 0{\cdot}952) = 13{\cdot}752$ m

Equating the pressure heads at the level 10·4 m,

$$0{\cdot}952 \times 0{\cdot}8 + (12{\cdot}8 - 10{\cdot}4) \times 1 = h_B$$

$$h_B = 0{\cdot}762 + 2{\cdot}4$$

$$= 3{\cdot}162 \text{ m}$$

Elevation of oil in column $B = 10{\cdot}4 + 3{\cdot}162 = 13{\cdot}562$ m

Equating the pressure heads at the level 5 m,

$$[3{\cdot}162 + (10{\cdot}4 - 5)] \times 1 = h_C \times 13{\cdot}6$$

$$h_C = 0{\cdot}63 \text{ m}$$

Elevation of mercury in column $C = 5{\cdot}0 + 0{\cdot}63 = 5{\cdot}63$ m

Example 1.10 Determine the specific gravity of the liquid of column B for the manometer shown in Fig. 1.22.

Solution Let the specific gravity of liquid in column B be S_B. The pressure due to air column between L and M can be neglected and thus pressures at L and M are assumed equal.

Equating the pressure heads at levels in the two limbs at N–N.

$$\frac{-24}{9{\cdot}81} + 0{\cdot}5 \times 1{\cdot}8 + 0{\cdot}6 \times S_B = 0$$

$$S_B = 2{\cdot}577$$

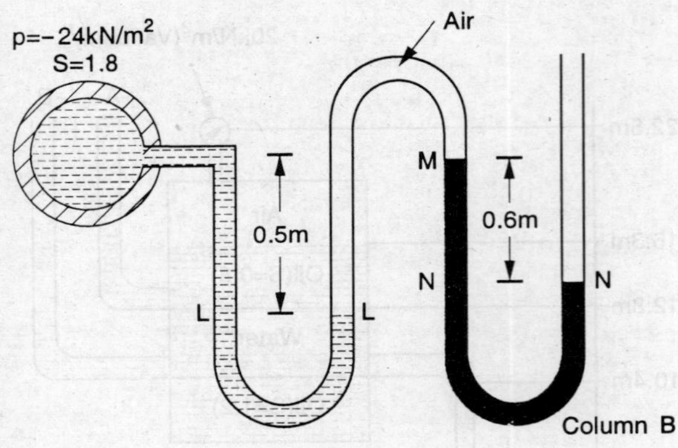

Fig. 1.22

1.12 MECHANICAL GAUGES

The manometers are suitable for relatively low pressures. For very high pressures and where the use of manometers is not convenient, mechanical gauges are more suitable. The common types of mechanical gauges are the following :

Bourden Tube Pressure Gauge

A bourden tube pressure gauge may be used to measure high as well as low pressures. It consists of a curved hollow metal tube usually made of bronze or nickle metal and of elliptical cross-section as shown in Fig. 1.23. When this gauge tube is connected to the fluid whose pressure is to be measured, the fluid flows into the tube changing the curvature of the tube

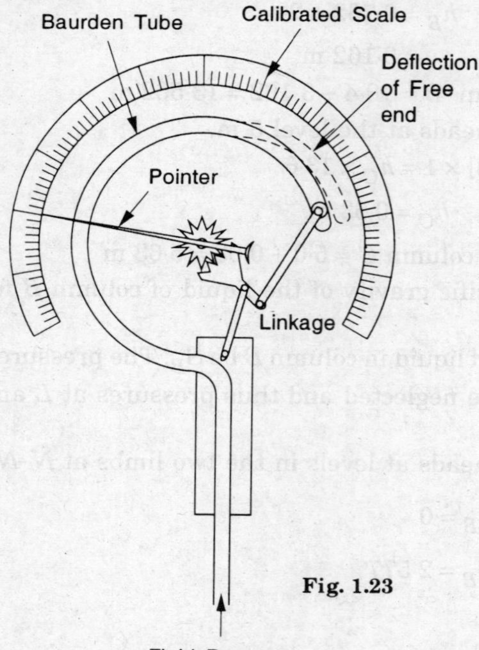

Fig. 1.23

according to the pressure. The moving end of the tube rotates a pointer on a dial through a linkage system. The pointer moves over a calibrated scale directly giving the pressure of the fluid. A pressure gauge can also be made to measure the vacuum pressure. A gauge used to measure the pressures both above and below the atmospheric pressure is known as compound bourden tube.

Diaphragm Pressure Gauge

A diaphragm gauge consists of a corrugated metallic diaphragm instead of a bourden tube (Fig. 1.24). When the gauge is connected to the fluid, the fluid under pressure deforms the diaphragm. The elastic deformation of the diaphragm rotates a pointer on a calibrated scale with the help of a pinion arrangement. Thus the reading of the pointer directly gives the pressure of the fluid.

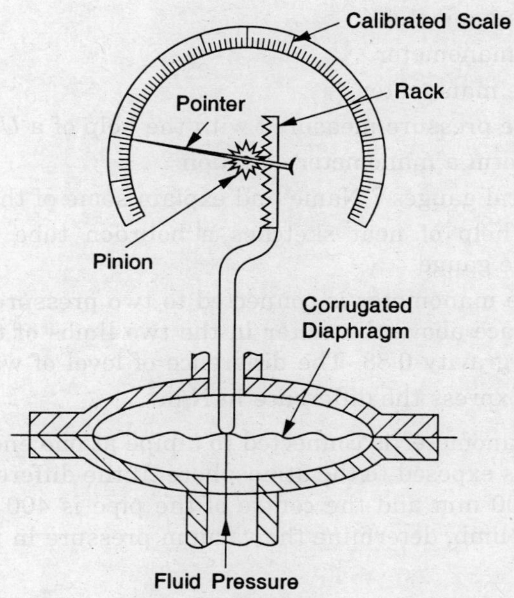

Fig. 1.24

PROBLEMS

1. Define the following terms :
   (a) Fluid mechanics
   (b) Fluid statics
   (c) Ideal and real fluids
2. What are fluids ? How are they classified ?
3. Define the following properties of fluids :
   Density, specific weight, specific gravity, specific volume.
4. What is viscosity ? State Newton's law of viscosity. What is a Newtonian fluid ?

5. What do you mean by kinematic viscosity and dynamic viscosity ? What are their units ?
6. Define the terms pressure intensity and pressure head. Deduce an expression for the pressure intensity at a point in a fluid.
7. State and prove Pascal's law.
8. Define the terms :

   Absolute pressure, gauge pressure and vacuum pressure.
9. What are manometers ? How are they classified ?
10. Distinguish between simple and differential type of manometers.
11. Explain the following with the help of neat sketches :
    (a) Piezometer tube
    (b) Single column manometer
    (c) Double column manometer
    (d) Inverted *U*-tube manometer
12. How is the negative pressure measured with the help of a *U*-tube manometer ?
13. State how do you form a manometer equation ?
14. What are mechanical gauges ? Name and explain some of them.
15. Explain with the help of neat sketches a bourdon tube pressure gauge and a diaphragm pressure gauge.
16. An inverted *U*-tube manometer is connected to two pressure points in a horizontal water pipe. The space above the water in the two limbs of the manometer is filled with oil of specific gravity 0·88. The difference of level of water column in the two limbs is 400 mm. Express the difference in N/m^2. (470·1 N/m^2)
17. A simple *U*-tube manometer is connected to a pipe at one end which contains water and the other end is exposed to the atmosphere. If the difference of mercury level in the two limbs is 600 mm and the centre of the pipe is 400 mm above the level of mercury in the left limb, determine the vacuum pressure in the pipe. (86·4 kN/m^2 absolute)
18. A simple manometer is used to measure the pressure of water flowing in a pipeline, the left limb of which is connected to the pipe and the right limb is open to the atmosphere. The centre of the pipe is in level with that of mercury in the right limb. If the difference of mercury level in the two limbs is 120 mm, find the pressure in the pipe. (1·512 m of weater, 14·83 kN/m^2)
19. A differential manometer containing mercury is connected at two points of a pipe carrying oil of specific gravity 0·9. The difference in the levels of mercury in the two limbs is 120 mm, the level being higher in the right limb. Determine the difference in pressure at the two points. (15·2 kN/m^2)
20. A *U*-tube mercury differential manometer is connected to the inlet and the throat of a venturimeter placed with its axis horizontal in a pipeline. Find the difference in pressure between the inlet and the throat if the manometer reading is 400 mm and water flows through the pipeline. (5·04 m of water)

21. Find the gauge readings A and B in Fig. 1.25 if the local atmospheric pressure is 740 mm of mercury. Assume the vapour pressure of mercury and pressure due to short column of air to be negligible. (11·235 kN/m^2 vacuum, 11·49 kN/m^2)

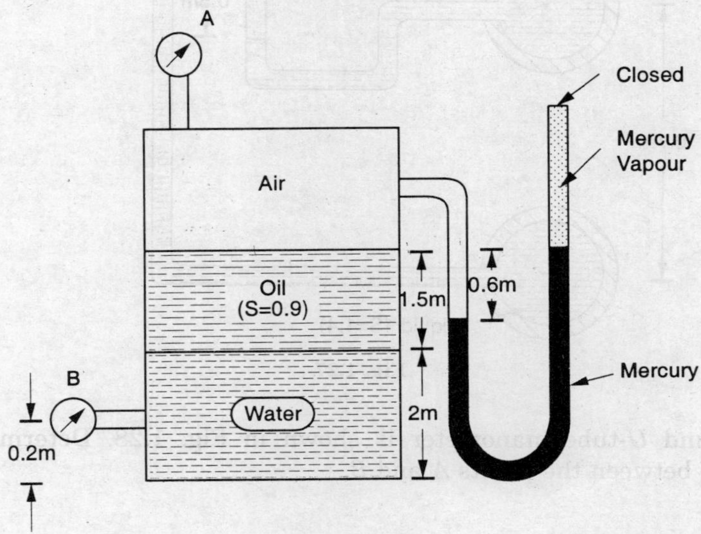

Fig. 1.25

22. A micrometer has a ratio of reservoir to limb area as 60. It is used to measure the pressure in a pipe containing a liquid of specific gravity 0·9. Determine the pressure in the pipe for the manometer reading shown in Fig. 1.26. (12·94 kN/m^2)

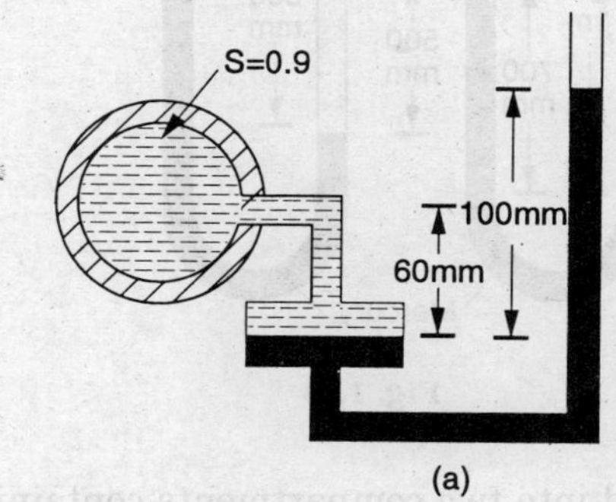

Fig. 1.26

23. Determine the pressure difference between A and B (Fig. 1.27). (12·31 kN/m^2)

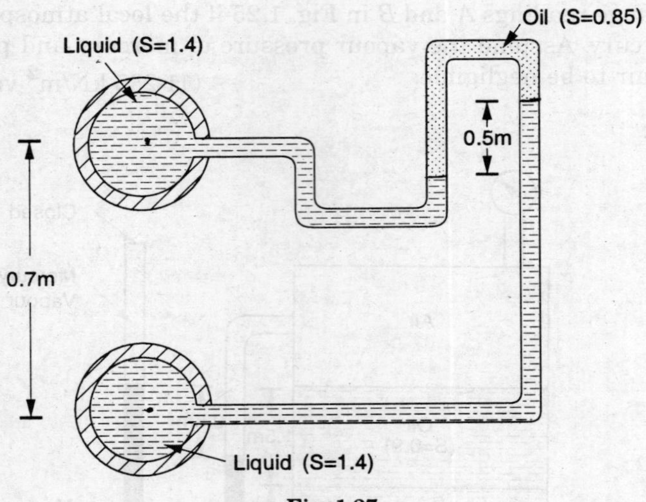

Fig. 1.27

24. A compound *U*-tube manometer is shown in Fig. 1.28. Determine the pressure difference between the points *A* and *B*. (162·4 kN/m^2)

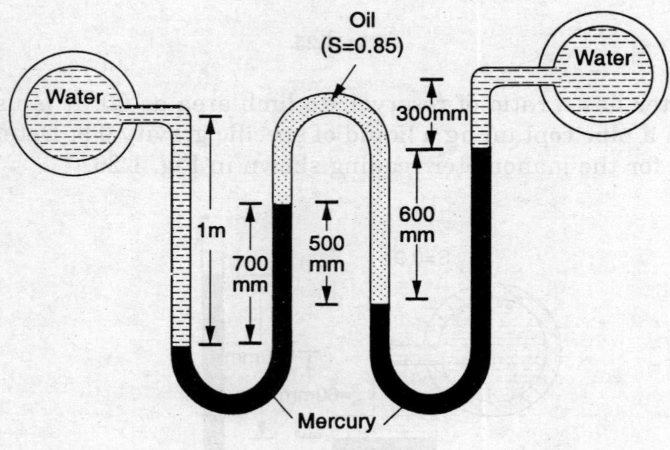

Fig. 1.28

25. A closed vessel is divided into two compartments containing oil and water as shown in Fig. 1.29. Find the value of *h*. (8·645 m)

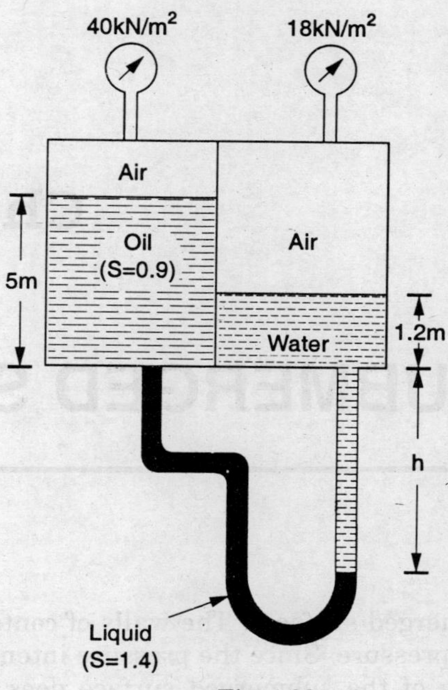

Fig. 1.29

Chapter 2

SUBMERGED SURFACES

2.1 INTRODUCTION

Fluids exert pressure on all submerged surfaces. The walls of containers, pipes or dams etc. experience forces due to the fluid pressure. Since the pressure intensity at a point varies with the depth of the fluid, the whole of the submerged surface does not experience the same intensity of pressure. **Total pressure** on the surface is the net force exerted by the fluid on the whole surface. As the lower portions of the submerged surfaces are subjected to higher pressures, the point of application of total pressure (known as **centre of pressure**) lies below the centre of gravity of the surface.

Submerged surfaces may be classified as :

1. Horizontal plane surface
2. Vertical plane surface
3. Inclined plane surface
4. Curved surface

2.2 HORIZONTAL PLANE SURFACE

Fig. 2.1 shows a horizontal plane surface submerged in a liquid.

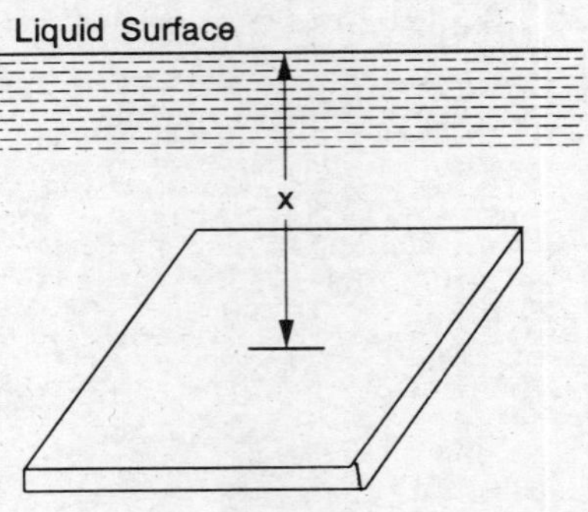

Fig. 2.1

Total Pressure

Let A = surface area

x = depth of the plane surface

w = specific weight of the liquid.

As all the points on the surface area are at the same depth from the liquid surface, the entire surface is subjected to a uniform pressure intensity of $p = wx$.

Total pressure (force), F = Pressure intensity × Area

$= wx \,.\, A$

$= wAx$ ...(2.1)

Centre of Pressure

Centre of pressure (CP) will be at a depth of x as all the elementary areas are subjected to same intensity of pressure.

2.3 VERTICAL PLANE SURFACE

The total pressure and the centre of pressure are found as follows :

Total Pressure

Let a vertical plane surface be submerged in a liquid (Fig. 2.2). In this case, as the pressure varies with depth, divide the whole surface into a number of horizontal parallel elementary strips. Then, pressure on an elementary strip may be considered to be the same.

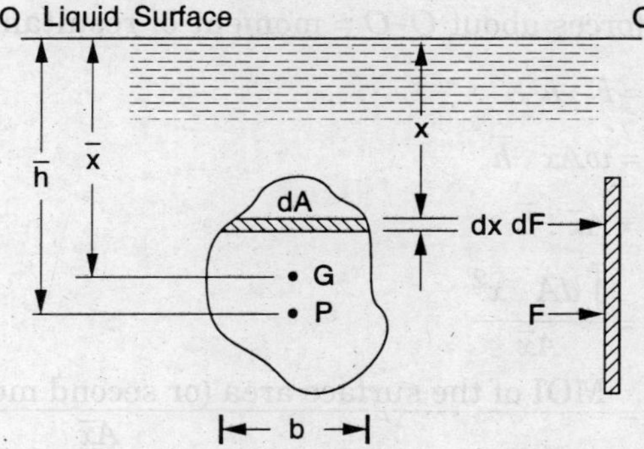

Fig. 2.2

Let dx = thickness of the elementary strip

x = depth of the elementary strip from free surface of liquid

b = breadth of the strip

p = intensity of pressure on the strip = wx

dA = area of the strip = $b \,.\, dx$

Then dF = total pressure on the strip

= pressure intensity × area

$= wx \,.\, dA$

Total pressure force,

$$F = \int dF$$

$$= \int wx \,.\, dA$$

$$= w \int dA \,.\, x$$

$$= w \int \text{Elementary area} \times \text{Its distance from free surface of liquid}$$

$$= w \int \text{moment of elementary area about the free surface}$$

$$= w \times \text{moment of whole area about the free surface}$$

$$= w \times \text{Surface area} \times \text{Depth of its } CG \text{ from the free surface}$$

$$= wA\bar{x} \qquad \text{...(2.2)}$$

where $\bar{x}$ is the depth of CG of the submerged surface from the liquid surface.

Centre of Pressure

In this case, intensity of pressure is not same for all the elementary areas. A lower elementary area is subjected to a higher intensity of pressure than that of an upper one. Thus the resultant of all elementary pressures act on a point lower than the CG of the area. The point of application of the total pressure force can be calculated by applying the principle of moments which states that the moments of differential forces about an axis is equal to the moment of the resultant force about the same axis (Refer Fig. 2.2).

Let $\quad \bar{h}$ = depth of centre of pressure from the free surface of liquid

Let O–O denote the free surface of liquid. Then

$\int$ Moment of elementary forces about O–O = moment of resultant force about O–O.

or

$$\int dF \,.\, x = F \,.\, \bar{h}$$

$$\int (wx \,.\, dA) \,.\, x = wA\bar{x} \,.\, \bar{h}$$

$$\int x^2 \,.\, dA = A\bar{x} \,.\, \bar{h}$$

$$\bar{h} = \frac{\int dA \,.\, x^2}{A\bar{x}}$$

$$= \frac{\text{MOI of the surface area (or second moment of area) about } O\text{–}O}{A\bar{x}}$$

$$= \frac{I_0}{A\bar{x}}$$

But $I_0 = I_G + A\bar{x}^2$ from the theorem of parallel axis.

where I_G = moment of inertia of the area about an axis through the CG and parallel to O–O.

$$\therefore \qquad \bar{h} = \frac{I_G + A\bar{x}^2}{A\bar{x}}$$

$$= \frac{I_G}{A\bar{x}} + \bar{x} \qquad \text{...(2.3)}$$

If a rectangular vertical area is immersed in a liquid such that an edge of the area coincides with the liquid surface,

$$\bar{h} = \frac{bh^3/12}{bh \cdot h/2} + \frac{h}{2} = \frac{2}{3} h \qquad ...(2.4)$$

2.4 INCLINED PLANE SURFACE

On an inclined surface, the total pressure and the centre of pressure are found as in the previous case.

Total Pressure

Let an inclined plane surface be submerged in liquid as shown in Fig. 2.3

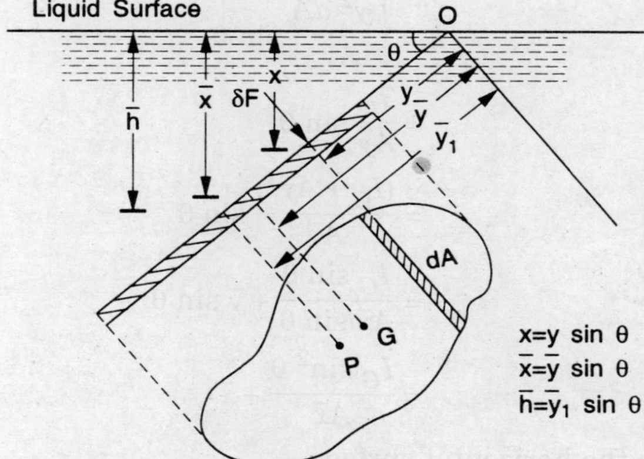

Fig. 2.3

Dividing the inclined surface into number of elementary areas as before, and considering an elementary strip of thickness dy at a distance y from O. (a point on liquid surface where the immersed surface will meet, if produced).

Let θ be the angle of the inclined surface with the free surface of water.

$$dF = p \cdot dA$$
$$= \int w \cdot x \cdot dA$$
$$= w\,(y \sin\theta) \cdot dA$$

Total pressure force,

$$F = \int dF$$
$$= \int w \cdot y \sin\theta \cdot dA$$
$$= w \sin\theta \int dA \cdot y$$
$$= w \sin\theta \int \text{Elementary area} \times \text{Its distance from } O$$
$$= w \sin\theta \times \text{Whole area} \times \text{Distance of its } CG \text{ from } O$$
$$= w \sin\theta \cdot A \cdot \bar{y}$$

$$= w\,A\,\bar{y}\sin\theta$$

$$= wA\bar{x} \qquad (\bar{x} = \bar{y}\sin\theta) \qquad ...(2.5)$$

where $\bar{x}$ is the depth of centre of gravity of the submerged area from the free surface.

Centre of Pressure

$\int$ Moment of elementary forces about O = Moment of resultant force about O.

$$\int dF \,.\, y = F\,\bar{y}_1$$

$$\int (wy\sin\theta \,.\, dA) \,.\, y = (wA\bar{y}\sin\theta)\,\frac{\bar{h}}{\sin\theta}$$

$$\int y^2\, dA = (A\bar{y})\,\frac{\bar{h}}{\sin\theta}$$

$$\bar{h} = \frac{\int y^2\, dA}{A\bar{y}}\sin\theta$$

$$= \frac{I_0}{A\bar{y}}\sin\theta$$

$$= \frac{(I_G + A\bar{y}^2)}{A\bar{y}}\sin\theta$$

$$= \frac{I_G\sin\theta}{A\bar{x}/\sin\theta} + \bar{y}\sin\theta$$

$$= \frac{I_G\sin^2\theta}{A\bar{x}} + \bar{x} \qquad ...(2.6)$$

when $\theta = 0°$, $\bar{h} = \bar{x}$ for the horizontal surface

when $\theta = 90°$, $\bar{h} = \bar{x} + \frac{I_G}{A\bar{x}}$ for vertical surface

If a rectangular inclined area of inclined length l is immersed in a liquid such that an edge of the area coincides with the liquid surface and makes an angle θ,

$$\bar{h} = \frac{(bl^3/12)\sin^2\theta}{bl\,.\,(l/2)\sin\theta} + \frac{l}{2}\sin\theta$$

$$= \frac{l}{6}\sin\theta + \frac{l}{2}\sin\theta$$

$$= \frac{2}{3}\,l\sin\theta$$

$$= \frac{2}{3}\,h \qquad ...(2.7)$$

Thus, centre of pressure lies at $\frac{2}{3}h$ from the liquid surface whether the surface is vertical or inclined.

2.5 CURVED SURFACE

In case of curved surface submerged in liquid, the direction of dF changes from point to point as it is normal to the surface at all points. Total force on the area can be easily known if horizontal and vertical components of the area are considered.

Total Pressure

Consider a strip of elementary area dA at a depth x from the free surface of liquid (Fig. 2.4).

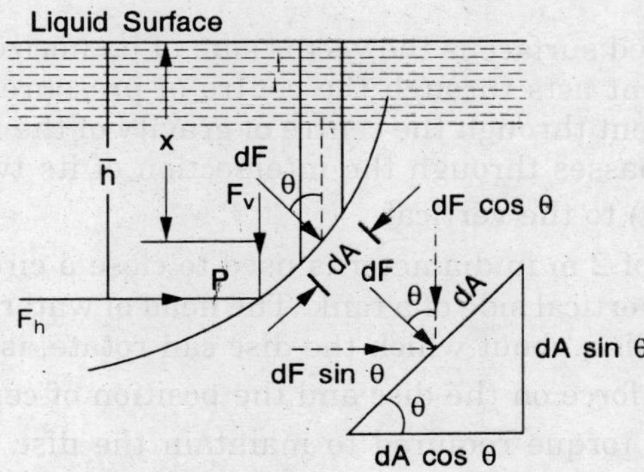

Fig. 2.4

Let dF_h = horinzontal component of force dF

dF_v = vertical component of force dF

Then total horizontal force on the curved surface,

$$F_h = \int dF_h$$

$$= \int dF \sin\theta$$

$$= \int w \,.\, dA \,.\, x \,.\, \sin\theta$$

$dA \,.\, \sin\theta$ can be written as $(dA)_V$, the vertical projected area.

$$\therefore \quad F_h = \int wx\,(dA)_V$$

$$= w \int (dA)_V \,.\, x$$

$$= wA_V \,.\, \bar{x} \qquad \ldots(2.8)$$

where $\bar{x}$ is the depth of CG of A_V from the free surface of liquid.

Similarly,

$$F_v = \int dF_v$$

$$= \int dF \cos\theta$$

$$= \int wxdA \,.\, \cos\theta$$

$$= w \int (dA)_H \,.\, x \qquad [(dA)_H = \text{Horizontal projected area}]$$

$$= w \int \text{Volume of liquid above the area } (dA)_H$$

= w × Volume of liquid above the whole curved surface

= Weight of liquid above the curved surface

Then, $$F = \sqrt{F_h^2 + F_v^2} \qquad ...(2.9)$$

As the curved surface immersed in a static fluid is in equilibrium, the force components on the outer surface are the same in magnitude as on the inner surface, but are opposite in direction, so, the vertical component on the outer surface is also found by assuming weight of water contained in the curved surface, though actually it may not contain any water at all.

Centre of Pressure

Total pressure on a curved surface is the vector sum of its horizontal and vertical components. The horizontal component acts through the centre of pressure of the vertical projected area and the vertical component through the centre of gravity of the liquid supported on the curved surface. The resultant passes through the intersection of its two components and acts at an angle of $\alpha = \tan^{-1} (F_h/F_v)$ to the vertical.

Example 2.1 A disc of 2 m in diameter is used to close a circular opening of the same size in the vertical side of a tank. The head of water above the horizontal diameter of the disc, about which the disc can rotate, is 3 m. Determine

(i) the force on the disc and the position of centre of pressure.

(ii) the torque required to maintain the disc in equilibrium in the vertical position.

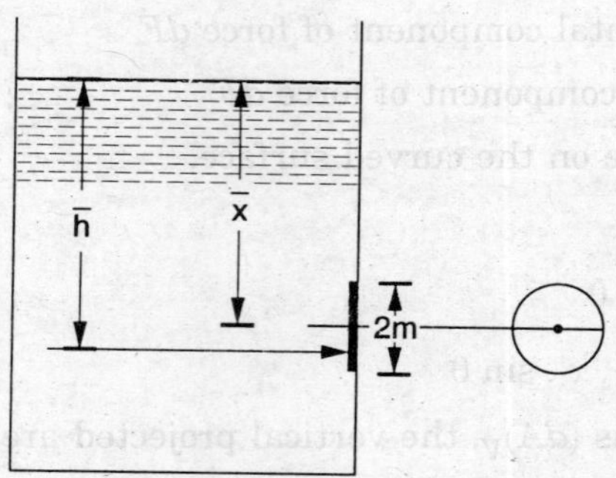

Fig. 2.5

Solution $A = \frac{\pi}{4} \times 2^2 = 3{\cdot}142 \text{ m}^2$ $\qquad \bar{x} = 3$ m

(i) $F = wA\bar{x} = 9{\cdot}81 \times 3{\cdot}142 \times 3 = 92{\cdot}46$ kN

$$\bar{h} = \frac{I_G}{A\bar{x}} + \bar{x} = \frac{\frac{\pi}{64} d^4}{\left(\frac{\pi}{4} d^2\right) \bar{x}} + \bar{x} = \frac{d^2}{16\,\bar{x}} + \bar{x} = \frac{2^2}{16 \times 3} + 3 = 3{\cdot}0833$$

(ii) Torque required = $F \, . \, (\bar{h} - \bar{x}) = 92{\cdot}46 \times (3{\cdot}0833 - 3) = 7{\cdot}705$ kN . m

Example 2.2 A pipe line 4 m in diameter contains a gate valve. The pressure at the centre of the pipe is 200 kN/m^2. The pipe is filled with an oil of specific gravity 0·9. What will be the force exerted by the oil upon the gate ? Also locate the postion of the centre of pressure.

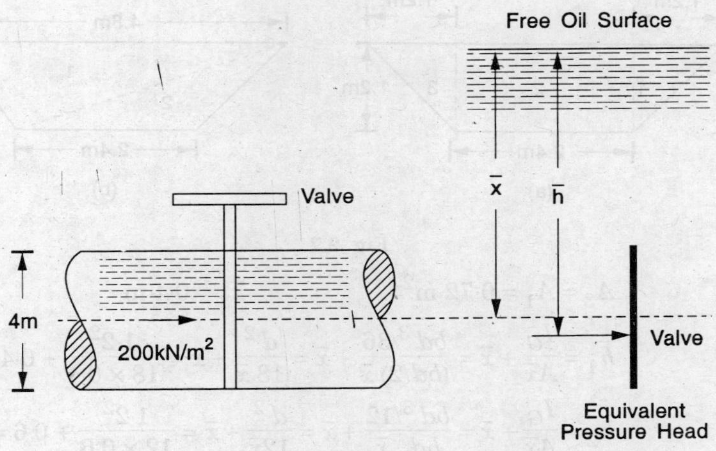

Fig. 2.6

Solution $p = 200 \text{ kN/m}^2$ $\quad S = 0{\cdot}9$

$d = 4$ m

$$A = \frac{\pi}{4} d^2 = \frac{\pi}{4} \times 4^2 = 4\pi \text{ m}^2$$

$$F = \text{Pressure} \times \text{Area} = 200 \times 4\pi = 2513{\cdot}3 \text{ kN}$$

$$\rho_0 = S \times \rho_w = 0{\cdot}9 \times 1000 \text{ kg/m}^3 \qquad \left(\because S = \frac{\rho_0}{\rho_w}\right)$$

$$w = \rho_0 \times g = (0{\cdot}9 \times 1000 \times 9{\cdot}81) \text{ N/m}^3$$

$$= 0{\cdot}9 \times 9{\cdot}81 \text{ kN/m}^3$$

$$\bar{x} = \frac{p}{w} = \frac{200}{0{\cdot}9 \times 9{\cdot}81} = 22{\cdot}653 \text{ m}$$

$$\bar{h} = \frac{I_G}{A\bar{x}} + \bar{x} = \frac{\frac{\pi}{64} d^4}{\left(\frac{\pi}{4} d^2\right) \bar{x}} + \bar{x} = \frac{d^2}{16\,\bar{x}} + \bar{x} = \frac{4^2}{16 \times 22{\cdot}653} + 22{\cdot}653$$

$$= 0{\cdot}044 + 22{\cdot}653 = 22{\cdot}697 \text{ m}$$

or centre of pressure lies $(22{\cdot}697 - 22{\cdot}653) = 0{\cdot}044$ m or 44 mm below the centre of pipe.

Example 2.3 A trapezoidal channel 2·4 m wide at the bottom and 1·2 m deep has side slopes 1 : 1. Find the total pressure and the centre of pressure on the vertical gate closing the channel when it is full of water.

Solution The problem can be solved by dividing the whole area into sub areas. Let whole of the area be divided into three sub areas as shown in Fig. 2.7a.

$$A_1 = \frac{1{\cdot}2 \times 1{\cdot}2}{2} = 0{\cdot}72 \text{ m}^2 \qquad \bar{x}_1 = 1{\cdot}2/3 = 0{\cdot}4 \text{ m}$$

$$A_2 = 2{\cdot}4 \times 1{\cdot}2 = 2{\cdot}88 \text{ m}^2 \qquad x_2 = 1{\cdot}2/2 = 0{\cdot}6 \text{ m}$$

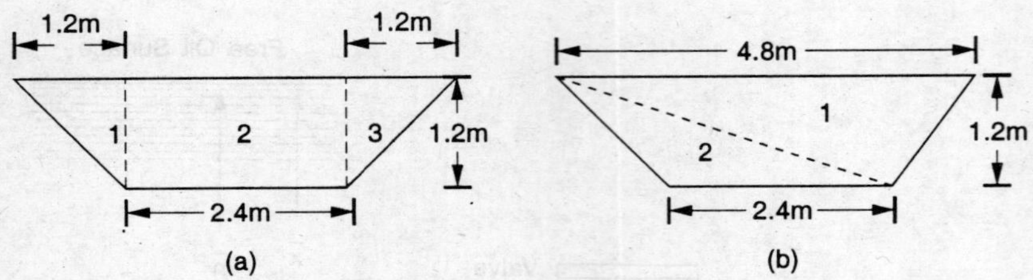

Fig. 2.7

$$A_3 = A_1 = 0{\cdot}72 \text{ m}^2 \qquad \bar{x}_3 = 0{\cdot}4 \text{ m}$$

$$\bar{h}_1 = \frac{I_G}{A\bar{x}} + \bar{x} = \frac{bd^3/36}{(bd/2)\,\bar{x}} + \bar{x} = \frac{d^2}{18\,\bar{x}} + \bar{x} = \frac{1{\cdot}2^2}{18 \times 0{\cdot}4} + 0{\cdot}4 = 0{\cdot}2 + 0{\cdot}4 = 0{\cdot}6 \text{ m}$$

$$\bar{h}_2 = \frac{I_G}{A\bar{x}} + \bar{x} = \frac{bd^3/12}{bd\,.\,\bar{x}} + \bar{x} = \frac{d^2}{12\bar{x}} + \bar{x} = \frac{1{\cdot}2^2}{12 \times 0{\cdot}6} + 0{\cdot}6 = 0{\cdot}2 + 0{\cdot}6 = 0{\cdot}8 \text{ m}$$

$$\bar{h}_3 = \bar{h}_1 = 0{\cdot}6 \text{ m}$$

$$F_1 = wA\bar{x} = 9{\cdot}81 \times 0{\cdot}72 \times 0{\cdot}4 = 2{\cdot}825 \text{ kN} = F_3$$

$$F_2 = 9{\cdot}81 \times 2{\cdot}88 \times 0{\cdot}6 = 16{\cdot}952 \text{ kN}$$

$$\therefore \quad F = F_1 + F_2 + F_3 = 2{\cdot}825 + 16{\cdot}952 + 2{\cdot}825 = 22{\cdot}602 \text{ kN}$$

Centre of pressure is found by taking moments about the upper surface.

$$F\bar{h} = F\bar{h}_1 + F\bar{h}_2 + F\bar{h}_3$$

$$22{\cdot}602 \times \bar{h} = 2{\cdot}825 \times 0{\cdot}6 + 16{\cdot}952 \times 0{\cdot}8 + 2{\cdot}825 \times 0{\cdot}6$$

$$\bar{h} = 0{\cdot}75 \text{ m}$$

The whole area could also be divided into two sub areas as shown in Fig. 2.7b.

$$A_1 = \frac{4{\cdot}8 \times 1{\cdot}2}{2} = 2{\cdot}88 \text{ m}^2 \qquad \bar{x}_1 = 0{\cdot}4 \text{ m}$$

$$A_2 = \frac{2{\cdot}4 \times 1{\cdot}2}{2} = 1{\cdot}44 \text{ m}^2 \qquad \bar{x}_2 = 0{\cdot}8 \text{ m}$$

$$\bar{h}_1 = \frac{4{\cdot}8 \times 1{\cdot}2^3/36}{2{\cdot}88 \times 0{\cdot}4} + 0{\cdot}4 = 0{\cdot}6 \text{ m} \quad \text{or} \quad \frac{1{\cdot}2^2}{18 \times 0{\cdot}4} + 0{\cdot}4 = 0{\cdot}6 \text{ m}$$

$$\bar{h}_1 = \frac{2{\cdot}4 \times 1{\cdot}2^3/36}{1{\cdot}44 \times 0{\cdot}8} + 0{\cdot}8 = 0{\cdot}9 \text{ m} \quad \text{or} \quad \frac{1{\cdot}2^2}{18 \times 0{\cdot}8} + 0{\cdot}8 = 0{\cdot}9 \text{ m}$$

$$F = 9{\cdot}81 \times 2{\cdot}88 \times 0{\cdot}4 + 981 \times 1{\cdot}44 \times 0{\cdot}8$$

$$= 11{\cdot}301 + 11{\cdot}301$$

$$= 22{\cdot}602 \text{ kN}$$

$$22{\cdot}602 \times \bar{h} = 11{\cdot}301 \times 0{\cdot}6 + 11{\cdot}301 \times 0{\cdot}9$$

$$\bar{h} = 0{\cdot}75 \text{ m}$$

Example 2.4 A tank contains water upto a height of 300 mm above the base. An immiscible liquid of specific gravity 0·9 is filled above the water upto a height of 600 mm. Determine the total pressure and the position of centre of pressure on one side of the tank the width of which is 2 m.

Solution In case of vertically immersed rectangular sections having more than one fluid, the calculations become easier by drawing pressure diagrams as shown in Fig. 2.8b.

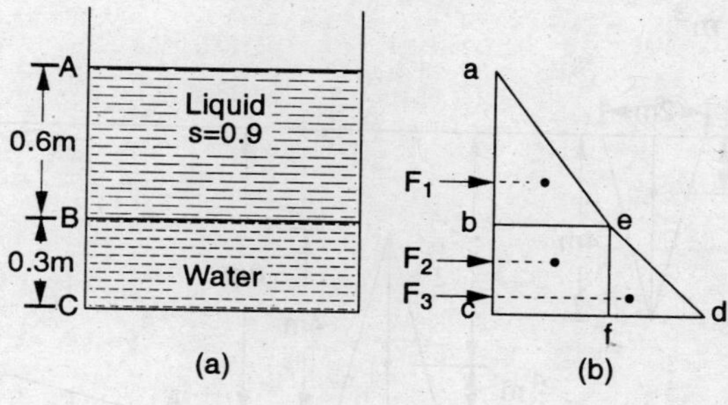

Fig. 2.8

$$p_A = 0$$

$$p_B = wh = 9{\cdot}81 \times 0{\cdot}9 \times 0{\cdot}6 = 5{\cdot}3 \text{ kN/m}^2 = be$$

$$p_C = p_B + (9{\cdot}81 \times 0{\cdot}3) = 5{\cdot}30 + 2{\cdot}94 = 8{\cdot}24 \text{ kN/m}^2 = cf + fd$$

$$F_1 = \text{Surface area} \times \text{Mean pressure}$$

$$= (0{\cdot}6 \times 2) \times \frac{5{\cdot}3}{2} = 3{\cdot}18 \text{ kN}$$

$$F_2 = (0{\cdot}3 \times 2) \times 5{\cdot}3 = 3{\cdot}18 \text{ kN}$$

$$F_3 = (0{\cdot}3 \times 2) \times \frac{2{\cdot}94}{2} = 0{\cdot}88 \text{ kN}$$

Total force, $F = 3{\cdot}18 + 3{\cdot}18 + 0{\cdot}88 = 7{\cdot}24$ kN

Centre of pressure :

$$\bar{h}_1 = \frac{2}{3} ab = \frac{2}{3} \times 0{\cdot}6 = 0{\cdot}4 \text{ m}$$

$$\bar{h}_2 = ab + \frac{bc}{2} = 0{\cdot}6 + \frac{0{\cdot}3}{2} = {\cdot}75 \text{ m}$$

$$\bar{h}_3 = ab + \frac{2}{3} ef = 0{\cdot}6 + \frac{2 \times 0{\cdot}3}{3} = 0{\cdot}8 \text{ m}$$

Taking moments of all the forces about A,

$$F \times \bar{h} = F_1 h_1 + F_2 h_2 + F_3 h_3$$

$$7{\cdot}24 \times \bar{h} = 3{\cdot}18 \times 0{\cdot}4 + 3{\cdot}18 \times 0{\cdot}75 + 0{\cdot}88 \times 0{\cdot}8$$

$$\bar{h} = 0{\cdot}6023 \text{ m}$$

Example 2.5 An isosceles triangular plate of 2 m base and 4 m height is immersed vertically in an oil of specific gravity 0·9. Determine the total pressure on the plate and the centre of pressure if

(i) The base of the plate coincides with the free surface of oil

(ii) The base of the plate is 8 m below the water surface (apex above the base)

(iii) The axis of symmetry passing through the apex being horizontal and 8 m below the water surface.

Solution $h = 4$ m $\qquad S = 0{\cdot}9$

$$A = \frac{2 \times 4}{2} = 4 \text{ m}^2$$

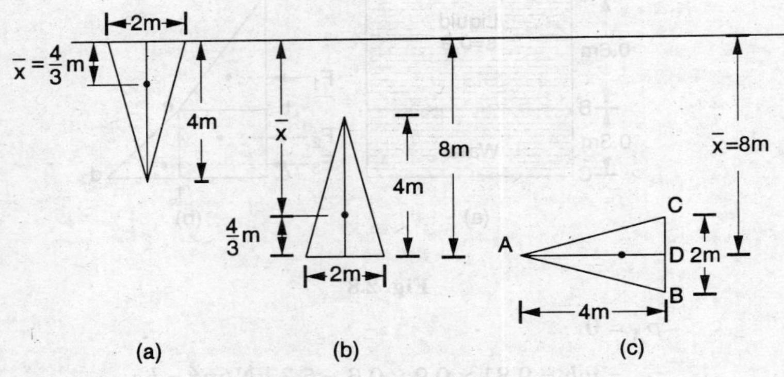

Fig. 2.9

(i) $\bar{x} = \frac{4}{3}$ m

$$F = wA\bar{x}$$

$$= (0{\cdot}9 \times 9{\cdot}81) \times 4 \times \frac{4}{3}$$

$$= 47{\cdot}088 \text{ kN}$$

$$\bar{h} = \frac{I_G}{A\bar{x}} + \bar{x} = \frac{bh^3/36}{(bh/2)\bar{x}} + \bar{x} = \frac{h^2}{18\,\bar{x}} + \bar{x} = \frac{4^2 \times 3}{18 \times (4/3)} + \frac{4}{3} = 2 \text{ m}$$

(ii) $\bar{x} = \left(8 - \frac{4}{3}\right) = \frac{20}{3}$ m

$$F = (0{\cdot}9 \times 9{\cdot}81) \times 4 \times \frac{20}{3}$$

$$= 235{\cdot}4 \text{ kN}$$

$$\bar{h} = \frac{h^2}{18\,\bar{x}} + \bar{x} = \frac{4^2 \times 3}{18 \times 20} + \frac{20}{3} = 6{\cdot}8 \text{ m}$$

(iii) $\bar{x} = 8$ m

$$F = (0{\cdot}9 \times 9{\cdot}81) \times 4 \times 8$$

$$= 282{\cdot}5 \text{ kN}$$

I = Moment of inertia of ABD about AD + Moment of inertia of ACD about AD

$$= \frac{4 \times 1^3}{12} + \frac{4 \times 1^3}{12}$$

$$= \frac{2}{3} \text{ m}^4$$

$$\overline{h} = \frac{I_G}{A\overline{x}} + \overline{x} = \frac{2/3}{4 \times 8} + 8 = 8{\cdot}0208 \text{ m}$$

Example 2.6 A rectangular gate 4 m × 2 m is hinged at its base and inclined at an angle of 60° with the horizontal (Fig. 2.10). A weight of 40 kN acting at an angle of 90° keeps the upper end of the gate in position. Neglecting the weight of the gate, determine the level of water when the gate begins to fall.

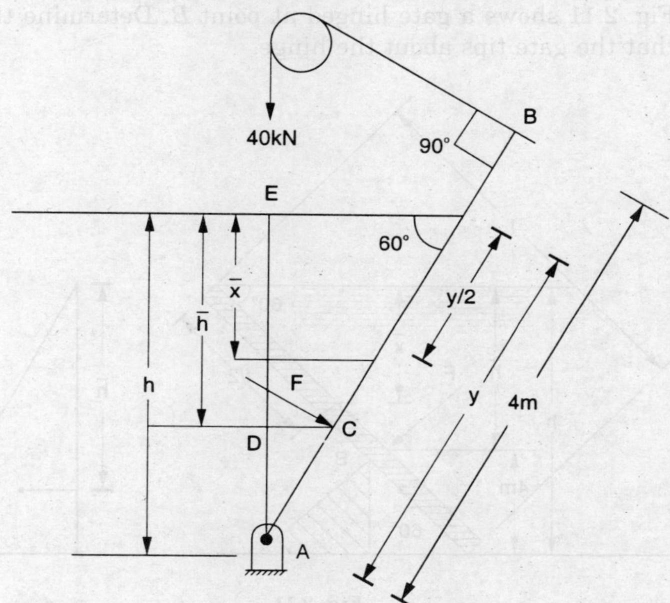

Fig. 2.10

Solution

$$\overline{x} = \frac{y}{2} \sin 60° = 0{\cdot}433y$$

$$F = wA\overline{x}$$

$$= 9{\cdot}81 \times 2y \times 0{\cdot}433y$$

$$= 8{\cdot}496y$$

$$\overline{h} = \frac{I_G \sin^2 \theta}{A\overline{x}} + \overline{x} = \frac{by^3 \sin^2 \theta}{12by\overline{x}} + \overline{x} = \frac{y^2 \sin^2 \theta}{12\overline{x}} + \overline{x}$$

$$= \frac{y^2 \sin^2 60°}{12 \times 0{\cdot}433y} + 0{\cdot}433y$$

$$= 0{\cdot}144y + 0{\cdot}433y$$

$$= 0{\cdot}577y \qquad \left(\text{or } \overline{h} = \frac{2}{3}h = \frac{2}{3} y \sin 60° = 0{\cdot}577y\right)$$

Now, $AC = \dfrac{AD}{\sin 60°} = \dfrac{h - \overline{h}}{\sin 60} = \dfrac{0{\cdot}866y - 0{\cdot}577y}{\sin 60} = 0{\cdot}334y$

Taking moments of forces about A,

$$F \times AC = 50 \times AB$$

$$8{\cdot}496y^2 \times 0{\cdot}334y = 40 \times 4$$

or $$y^3 = 56{\cdot}38$$

$$y = 3{\cdot}83 \text{ m}$$

$$h = 3{\cdot}83 \sin 60^\circ$$

$$= 3{\cdot}32 \text{ m}$$

Example 2.7 Fig. 2.11 shows a gate hinged at point *B*. Determine the height of water so that the gate tips about the hinge.

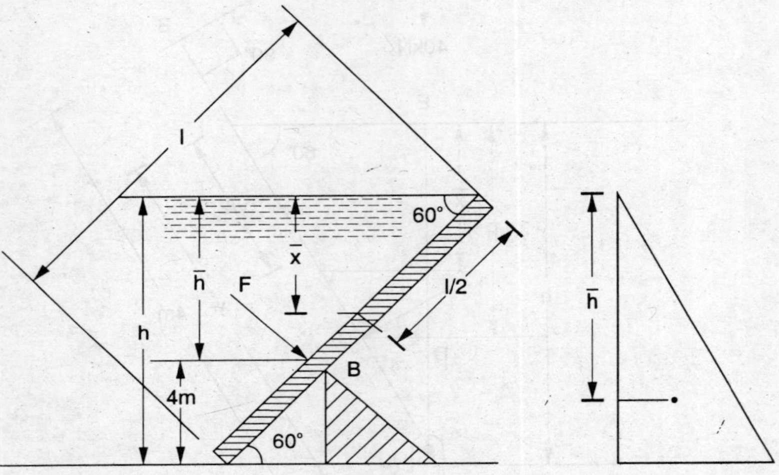

Fig. 2.11

Solution Considering the width of the gate as unity, tipping of the gate about *B* would be possible only if the centre of pressure of the resultant pressure force lies above point *B*.

$$\bar{x} = \frac{l}{2} \sin 60^\circ$$

$$\bar{h} = \frac{I_G \sin^2 \theta}{A\bar{x}} + \bar{x}$$

$$h - 4 = \frac{bl^3 \sin^2 60^\circ}{12 \, . \, (b \, . \, l) \cdot \frac{l}{2} \sin 60^\circ} + \frac{l}{2} \sin 60^\circ$$

$$l \sin 60^\circ - 4 = \frac{l \sin 60^\circ}{6} + \frac{l}{2} \sin 60^\circ$$

$$0{\cdot}866l - 4 = 0{\cdot}1443l + 0{\cdot}433l$$

$$0{\cdot}2887l = 4$$

$$l = 13{\cdot}85$$

$$h = l \sin 60^\circ$$

$$= 12 \text{ m}$$

Alternately,

$$\bar{h} = \frac{2}{3} h$$

or

$$(h - 4) = \frac{2}{3} h$$

$$3h - 12 = 2h$$

$$h = 12 \text{ m}$$

Example 2.8 A sliding gate 2·4 m wide and 1·2 m high weighs 20 kN. It lies in a vertical plane and has a coefficient of friction 0·1 between itself and the guides. If the upper edge of the gate is 5 m from the free surface of water, determine the vertical force required to raise the gate.

Solution

$$\bar{x} = 5 + \frac{1 \cdot 2}{2} = 5 \cdot 6 \text{ m}$$

$$F = wA\bar{x}$$

$$= 9 \cdot 81 \times 2 \cdot 4 \times 1 \cdot 2 \times 5 \cdot 6$$

$$= 158 \cdot 2 \text{ kN}$$

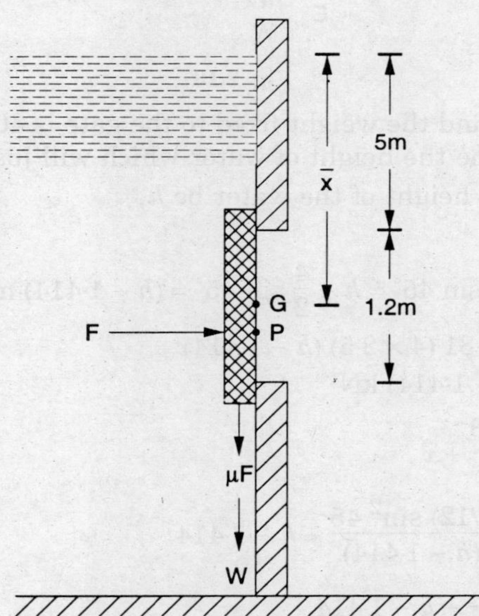

Fig. 2.12

i.e. normal force on the gate = 158·2 kN

Friction force = $\mu R_n = \mu F$

$$= 0 \cdot 1 \times 1582$$

$$= 15 \cdot 82 \text{ kN}$$

Vertical force required to raise the gate = 20 + 15·82 = 35·82 kN

Example 2.9 A rectangular sluice gate *AB*, 4 m long and 2·5 m wide is hinged at *A* as shown in Fig. 2.13. It is kept closed by a weight fixed to the gate. The total weight

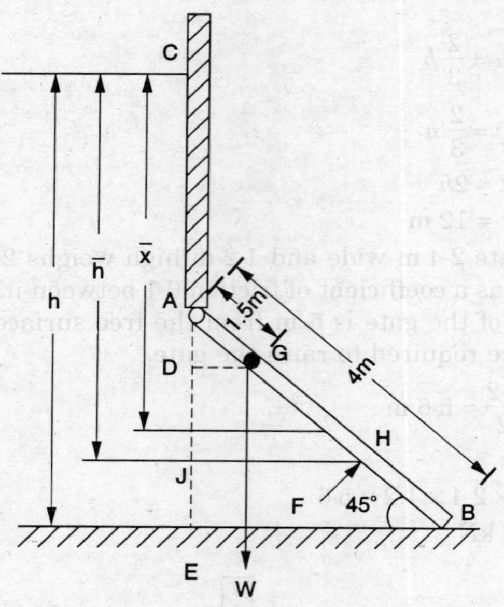

Fig. 2.13

of the gate and the weight fixed to the gate is 400 kN and acts through point *G*. Determine the height of water which will just make the gate to open.

Solution Let the required height of the water be h,

$$AB = 4 \text{ m}$$

$$\bar{x} = h - \frac{AB}{2} \sin 45° = h - \frac{4}{2} \sin 45° = (h - 1{\cdot}414) \text{ m}$$

$$F = wA\bar{x} = 9{\cdot}81\ (4 \times 2{\cdot}5)\ (h - 1{\cdot}414)$$

$$= 98{\cdot}1\ (h - 1{\cdot}414) \text{ kN}$$

$$\bar{h} = \frac{I_G \sin^2 \theta}{A\bar{x}} + \bar{x}$$

$$= \frac{(2{\cdot}5 \times 4^3/12) \sin^2 45}{2{\cdot}5 \times 4\ (h - 1{\cdot}414)} + h - 1{\cdot}414$$

$$= \frac{0{\cdot}667}{h - 1{\cdot}414} + h - 1{\cdot}414$$

Taking moments about hinge point A,

$$400 \times DG = F \times AH$$

$$400 \times 15 \sin 45 = 98{\cdot}1\ (h - 1{\cdot}414) \frac{AJ}{\sin 45°}$$

$$AJ = \frac{3{\cdot}058}{h - 1{\cdot}414}$$

Also $AJ = h - AC$

$$= h - (CE - AE)$$

$$= \bar{h} - (h - 4 \sin 45^\circ)$$

$$= \left[\frac{0.667}{h - 1.414} + h - 1.414\right] - (h - 2.828)$$

$$= \frac{0.667}{h - 1.414} + 1.414$$

Comparing the two values of AJ,

$$\frac{0.667}{h - 1.414} + 1.414 = \frac{3.058}{h - 1.414}$$

$$0.667 + 1.414\, h - 2 - 3.058$$

$$1.414\, h = 4.391$$

$$h = 3.105 \text{ m}$$

Example 2.10 A circular plate of 8 m diameter with a circular hole of 2 m diameter is immersed in water at an angle of 30° to the horizontal as shown in Fig. 2.14. The bottom edge of the plate is 8 m below the free water surface. Determine

(i) The total pressure on the plate

(ii) The depth of the centre of pressure.

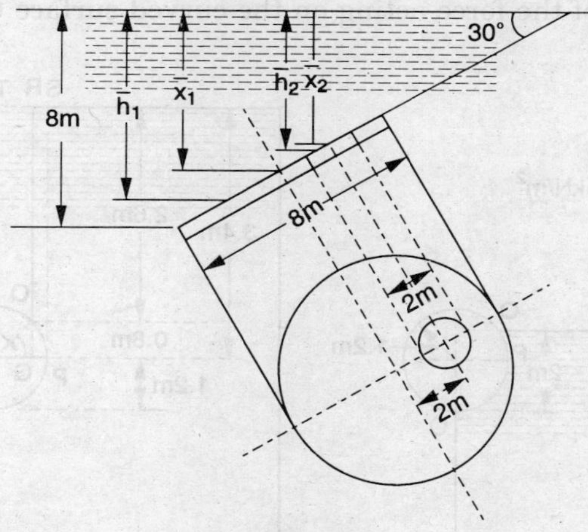

Fig. 2.14

Solution From the figure,

$$\bar{x}_1 = 8 - 4 \sin 30^\circ = 6 \text{ m}$$

$$\bar{x}_2 = 8 - 6 \sin 30^\circ = 5 \text{ m}$$

(i) Force on the solid plate,

$$F_1 = wA\bar{x} = 9.81 \times \frac{\pi}{4} \times 8^2 \times 6 = 2958.6 \text{ kN}$$

$$\bar{h}_1 = \frac{\left(\frac{\pi}{64} d^4\right) \sin^2 30^\circ}{\frac{\pi}{4} d^2 \bar{x}} + \bar{x} = \frac{d^2 \sin^2 30^\circ}{16\, \bar{x}} + \bar{x} = \frac{8^2 \sin^2 30^\circ}{16 \times 6} + 6 = 6.1667 \text{ m}$$

Force on the area of hole,

$$F_2 = wA\bar{x} = 9{\cdot}81 \times \frac{\pi}{4}(2)^2 \times 5 = 154{\cdot}1 \text{ kN}$$

$$\bar{h}_2 = \frac{2^2 \sin^2 30^\circ}{16 \times 5} + 5 = 5{\cdot}0125 \text{ m}$$

Force on the plate $= F_1 - F_2$

$= 2958{\cdot}6 - 154{\cdot}1$

$= 2804{\cdot}5$ kN

(ii) $$\bar{h} = \frac{F_1 \times \bar{h}_1 - F_2 \times \bar{h}_2}{F}$$

$$= \frac{2958{\cdot}6 \times 6{\cdot}1667 - 154{\cdot}1 \times 5{\cdot}0125}{2804{\cdot}5}$$

$= 6{\cdot}23$ m

Example 2.11 Fig. 2.15a shows the cross-section of a tank. The tank is 2·5 m long and is full of water. An empty cylinder of radius 1·2 m lies along the length of the tank on one of its corners. Determine the horizontal and the vertical components of the force acting on the curved surface *OPQ* of the cylinder.

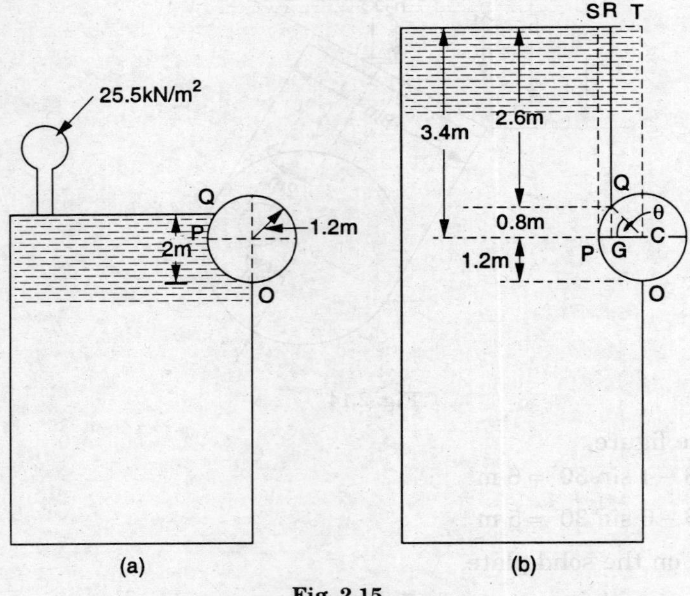

Fig. 2.15

Solution $h = \dfrac{p}{w} = \dfrac{25{\cdot}5}{9{\cdot}81} = 2{\cdot}6$ m

Free water surface is at a height of 2·6 m from the top of the tank. The equivalent free surface of water is shown in Fig. 2.15b.

Horizontal force,

$$F_h = wA\bar{x} = 9{\cdot}81 \times (2 \times 2{\cdot}5) \times \left(2{\cdot}6 + \frac{2}{2}\right)$$
$$= 176{\cdot}58 \text{ kN}$$

F_v = Net upward force

= Upward force on surface OP – Downward force on surface PQ

= Weight of water contained in $OPSTCO$ – Weight of water contained in $PSRQP$

$$\text{Area } OPSTCO = \frac{\pi}{4} r^2 + PC \,.\, PS$$
$$= \frac{\pi}{4}(1{\cdot}2)^2 + 1{\cdot}2 \times 3{\cdot}4$$
$$= 5{\cdot}21 \text{ m}^2$$

Area $PSRQP$ = area $PGRS$ – area $PGQP$

$= PG \,.\, PS$ – (Area CPQ – CGQ)

$$= PG \,.\, PS - \left(\text{Area } CPQ - \frac{CG \,.\, GQ}{2}\right)$$

$$\sin\theta = \frac{GQ}{CQ} = \frac{0{\cdot}8}{1{\cdot}2} = 0{\cdot}667$$
$$\theta = 41{\cdot}8^\circ$$
$$PG = PC - GC = 1{\cdot}2 - CQ\cos 41{\cdot}8^\circ = 1{\cdot}2 - 1{\cdot}2\cos 41{\cdot}8^\circ = 0{\cdot}305$$

$$\therefore \quad PSRQP = 0{\cdot}305 \times 3{\cdot}4 - \left[\frac{\pi(1{\cdot}2)^2 \times 41{\cdot}8}{360} - \frac{1{\cdot}2\cos 41{\cdot}8^\circ \times 0{\cdot}8}{2}\right]$$
$$= 1{\cdot}0385 - (0{\cdot}5253 - 0{\cdot}3578)$$
$$= 0{\cdot}871 \text{ m}^2$$
$$F_v = 9{\cdot}81\,(5{\cdot}21 - 0{\cdot}871) \times 2{\cdot}5 = 106{\cdot}4 \text{ kN}$$

Example 2.12 A cylinder of 6 m diameter and 3 m long has water on its both sides as shown in Fig. 2.16. If the weight of the cylinder is 250 kN, find the magnitude of the horizontal and vertical components of the force that will keep the cylinder just touching the floor. Also find the magnitude and the location of the resultant force.

Solution

$$F_{h1} = 9{\cdot}81 \times (6 \times 3) \times \frac{6}{2} = 529{\cdot}74 \text{ kN}$$

$$F_{h2} = 9{\cdot}81 \times (3 \times 3) \times \frac{3}{2} = 132{\cdot}44 \text{ kN}$$

Thus net horizontal force acting on the cylinder = 529·74 – 132·44 = 397·3 kN. This force acts towards right. Therefore, 397·3 kN force acting towards left is required to keep the cylinder in position.

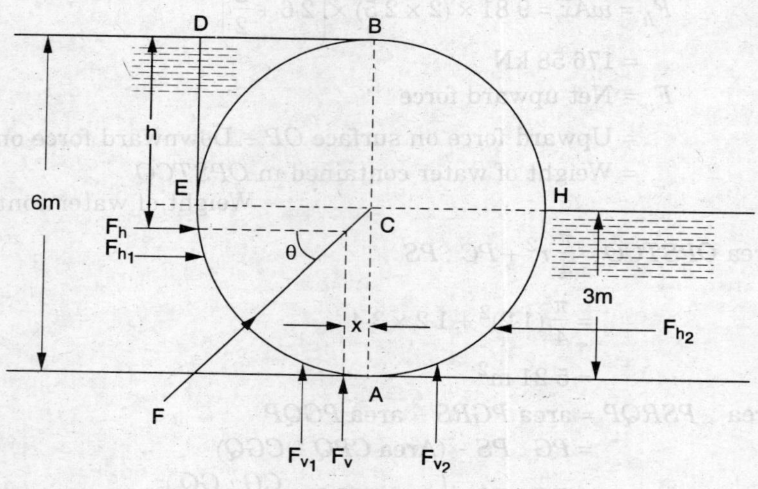

Fig. 2.16

F_{v1} = Upward force acting on surface AE – Downward force acting on surface BE

= Weight of water in area $AEDBCA$ – Weight of water in area $EDBE$

= Weight of water in area $AEBCA$

$= w\left(\dfrac{\pi r^2}{2}\right) \times 3$

$= 9{\cdot}81 \times \dfrac{\pi \times 3^2}{2} \times 3$

= 416·06 kN upwards

F_{v2} = Weight of water in area $ACHA$

$= w\left(\dfrac{\pi r^2}{4}\right) \times 3$

$= 9{\cdot}81 \times \dfrac{\pi \times 3^2}{4} \times 3$

= 208·03 kN upwards

F_v = 416·06 + 208·03 = 624·09 kN

Resultant, $F = \sqrt{(397{\cdot}3)^2 + (624{\cdot}09)^2}$

$= \sqrt{157847{\cdot}3 + 389488{\cdot}3}$

= 739·82 kN

$\tan\theta = \dfrac{F_y}{F_x} = \dfrac{624{\cdot}09}{397{\cdot}3} = 1{\cdot}5708$

θ = 57·52°

Location :

F_{h1} acts at $\dfrac{2}{3} \times 6 = 4$ m from top or 2 m from A

F_{h2} acts at $\frac{2}{3} \times 3 = 2$ m from top or 1 m from A

Taking moments about A,

$$F_h \times h = F_{h1} \times 2 + F_{h2} \times 1$$

$$397{\cdot}3 \times h = 529{\cdot}74 \times 2 - 132{\cdot}44 \times 1$$

$$h = 2{\cdot}333 \text{ m from } A$$

F_{v1} acts at $\frac{4r}{3\pi}$ from vertical diameter

i.e. $\frac{4 \times 3}{3\pi} = 1{\cdot}273$ m from AB (towards left)

F_{v2} acts at $\frac{4 \times 3}{3\pi} = 1{\cdot}273$ m from AB (towards right)

$$F_v \times x = F_{v1} \times 1{\cdot}273 - F_{v2} \times 1{\cdot}273$$

$$624{\cdot}09 \times x = 416{\cdot}06 \times 1{\cdot}273 - 208{\cdot}03 \times 1{\cdot}273$$

$$= (416{\cdot}06 - 208{\cdot}03) \times 1{\cdot}273$$

$$x = 0{\cdot}424 \text{ m from } AB \text{ (towards left)}$$

Since the weight of the cylinder is 250 kN, the additional force required to keep the cylinder in position will be

$F - 250 = 624{\cdot}09 - 250 = 374{\cdot}09$ kN.

Example 2.13 Fig. 2.17 shows a radial gate AB of radius 2 m. The width of the gate is also 2 m. Determine the magnitude, direction and the point of application of the resultant force acting on it.

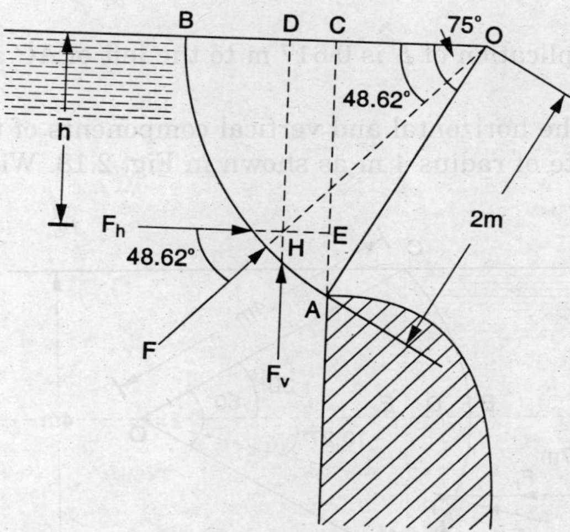

Fig. 2.17

Solution $AC = 2 \sin 75^\circ = 1{\cdot}932$ m

$$F_h = 9{\cdot}81 \times (1{\cdot}932 \times 2) \times \frac{1{\cdot}932}{2}$$

$$= 36{\cdot}61 \text{ kN}$$

$$\bar{h} = 1{\cdot}932 \times \frac{2}{3} = 1{\cdot}288 \text{ m}$$

$$F_v = \text{Weight of water in } ABC$$

$$= w \times \text{Area } ABC \times \text{Width}$$

$$= w \times (\text{Area } ABO = \text{Area } ACO) \times \text{Width}$$

$$= 9{\cdot}81 \times \left[\pi\,(2)^2 \times \frac{75}{360} - \frac{1{\cdot}932 \times 2\cos 75^\circ}{2}\right] \times 2$$

$$= 9{\cdot}81\,(2{\cdot}618 - 0{\cdot}5) \times 2$$

$$= 9{\cdot}81 \times 2{\cdot}118 \times 2$$

$$= 41{\cdot}55 \text{ kN}$$

$$F = \sqrt{36{\cdot}61^2 + 41{\cdot}55^2}$$

$$= 55{\cdot}38 \text{ kN}$$

$$\tan\theta = \frac{F_v}{F_h} = \frac{41{\cdot}55}{36{\cdot}61} = 1{\cdot}135$$

$$\theta = 48{\cdot}62^\circ$$

As the gate is radial, the resultant force (normal) must pass through *O*.

$$CD = OD - OC = \frac{DP}{\tan 48{\cdot}62^\circ} - 2\cos 75^\circ$$

$$= \frac{1{\cdot}288}{1{\cdot}135} - 2\cos 75^\circ$$

$$= 1{\cdot}135 - 2\cos 75^\circ$$

$$= 0{\cdot}617 \text{ m}$$

Thus point of application of *F* is 0·617 m to the left of *AC* and 1·288 m below the water level *OB*.

Example 2.14 **Determine the horizontal and vertical components of the pressure acting on a tainter gate of radius 4 m as shown in Fig. 2.18. Width of the gate is 3 m.**

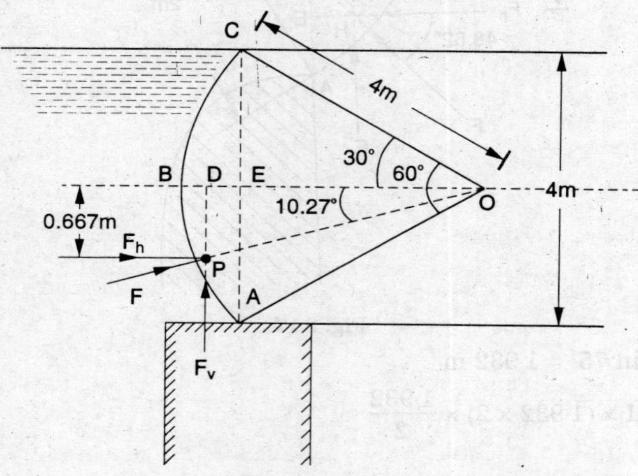

Fig. 2.18

Solution OAC is an equilateral triangle,

$$AC = OA = CO = 4 \text{ m}$$

$$F_h = wA\bar{x}$$

$$= w \times (AC \times \text{Width}) \frac{AC}{2}$$

$$= w \times (2EC \times \text{Width}) \,.\, EC$$

$$= 9{\cdot}81 \times [(2 \times 4 \sin 30°) \times 3] \times 4 \sin 30°$$

$$= 9{\cdot}81 \times 12 \times 2$$

$$= 235{\cdot}44 \text{ kN}$$

$$\bar{h} = \frac{4}{3} = 1{\cdot}333 \text{ m above the bottom edge}$$

or $(2 - 1{\cdot}333 = 0{\cdot}667 \text{ m})$ below OB

or $DP = 0{\cdot}667$ m

$$F_v = \text{Weight of water contained in portion } ABC$$

$$= w\left[\pi r^2 \times \frac{60}{360} - 2\left(\frac{4 \sin 30 \times 4 \cos 30}{2}\right)\right] \times 3$$

$$= w\left[\pi \times 4^2 \times \frac{60}{360} - 6{\cdot}93\right] \times 3$$

$$= 9{\cdot}81 \times (8{\cdot}378 - 6{\cdot}93) \times 3$$

$$= 42{\cdot}66 \text{ kN}$$

$$\tan\theta = \frac{42{\cdot}66}{235{\cdot}44} = 0{\cdot}181,\ \theta = 10{\cdot}27°$$

As the gate is radial, the resultant force (normal) must pass through O.

$$DE = DO - EO$$

$$= DP \cot 10{\cdot}27° - 4 \cos 30°$$

$$= 0{\cdot}667 \cot 10{\cdot}27° - 4 \cos 30°$$

$$= 3{\cdot}681 - 3{\cdot}464$$

$$= 0{\cdot}217 \text{ m}$$

Thus point of application of F is 0·217 m to the left of AC and 1·333 m above the bottom of the tainter.

Example 2.15 Fig. 2.19 shows a plate $ABDEF$ immersed vertically in water. Determine the total force and the location of centre of pressure for the plate.

Solution Force on area $ABEF = wAx_1$

$$= 9{\cdot}81 \times (3 \times 3) \times \left(\frac{3}{2}\right) = 132{\cdot}44 \text{ kN}$$

$$\text{Force on area } BDE = wA\bar{x}_2 = 9{\cdot}81 \times \left(\frac{3 \times 3}{2}\right) \times \left(3 + \frac{3}{3}\right) = 176{\cdot}58 \text{ kN}$$

Total pressure force $= 132{\cdot}44 + 176{\cdot}58$

$= 309{\cdot}02$ kN

Centre of pressure of area $ABEF$, $\bar{h}_1 = 3 \times \frac{2}{3} = 2$ m below AE

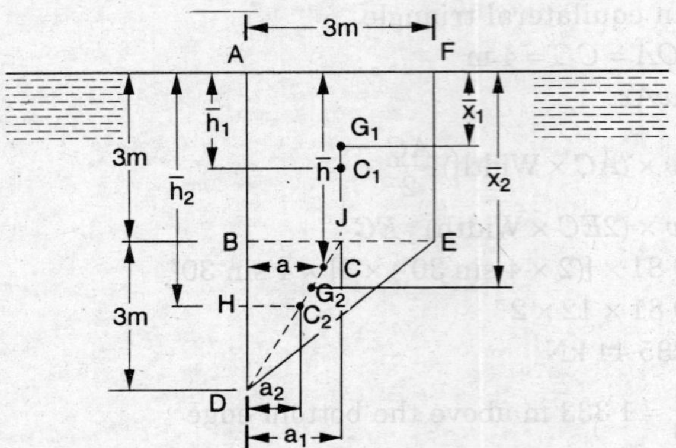

Fig. 2.19

Centre of pressure of area BDE, $\bar{h}_2 = \dfrac{I_G}{A\bar{x}} + \bar{x}_2$

$$= \frac{bh^3/36}{(bh/2)\,\bar{x}} + \bar{x}_2$$

$$= \frac{h^2}{18\bar{x}_2} + \bar{x}_2$$

$$= \frac{3^2}{18 \times 4} + 4$$

$$= 4{\cdot}125 \text{ m}$$

Centre of pressure for the total force is found by taking moments about the water surface

$$309{\cdot}02 \times h = 132{\cdot}44 \times 2 + 176{\cdot}58 \times 4{\cdot}125$$

$$h = 3{\cdot}214 \text{ m}$$

Horizontal location of the centre of pressure can be obtained by taking moments about ABD.

Distance of centre of pressure for force on $ABEF$ from ABD = 3/2 = 1·5 m. Distance of centre of pressure for force on BDE from ABD is found from the fact that for any elementary horizontal strip at BE, the centre of pressure is at the mid point (J) of BE and at the lowest point, it is at D. Its variation with the depth is along the line JD. However, as the centre of pressure for the area is to be at a depth of 4·125 m from the free surface or (6 – 4·125 = 1·875 m) from the bottom point D, its distance from ABC can be found from the similar triangles BJD and HC_2D,

$$\frac{HC_2}{BJ} = \frac{DH}{DB}$$

$$HC_2 = BJ \cdot \frac{DH}{DB} = 1{\cdot}5 \times \frac{1{\cdot}875}{3} = 0{\cdot}9375 \text{ m}$$

$$\therefore \quad 309{\cdot}02 \times a = 132{\cdot}44 \times 1{\cdot}5 + 176{\cdot}58 \times 0{\cdot}9375$$

$$a = 1{\cdot}179 \text{ m}$$

Thus the centre of pressure is 3·214 m below the water surface and 1·179 m from *ABD*.

2.6 DAMS

A dam is a structure to store water and to provide head to a power house. Consider a dam as shown in Fig. 2.20.

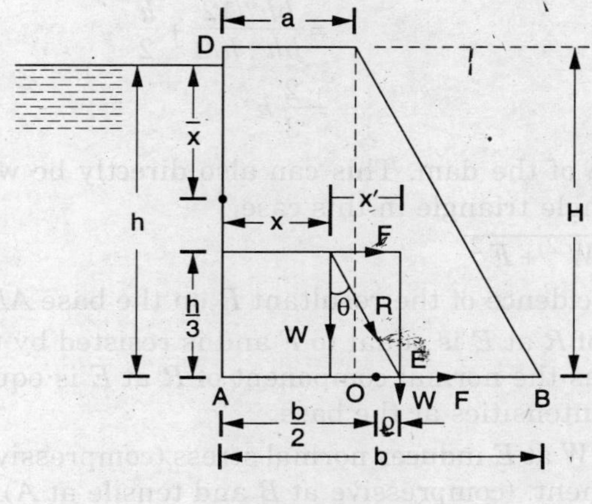

Fig. 2.20

There are two main forces acting on the dam, static water pressure and the weight of the dam.

Let h = height of water on the upstream side

H = height of dam

a = top width of dam

b = bottom width of dam

W = weight of the masonry dam acting vertically downwards through CG of the dam

$$= \frac{\rho\,(a+b)\,.\,H}{2}$$

x = horizontal distance of the line of action of weight W from point A.

Taking moments of the rectangular and triangular areas about vertical face AD,

$$a\,.\,H\,.\,\frac{a}{2} + \frac{(b-a)\,.\,H}{2}\left[a + \frac{(b-a)}{3}\right] = \left[aH + \frac{(b-a)\,H}{2}\right].\,x$$

$$\frac{a^2}{2} + \frac{ab}{2} - \frac{a^2}{2} + \frac{b^2}{6} + \frac{a^2}{6} - \frac{2ab}{6} = \frac{a+b}{2}\,.\,x$$

$$3ab + b^2 + a^2 - 2ab = 3(a+b)\,.\,x$$

$$x = \frac{a^2 + b^2 + ab}{3(a+b)} \qquad \text{...(2.10)}$$

Considering one meter length of the dam,

Horizontal thrust of water on the dam, $F = wA\bar{x}$

$$= wAh/2$$
$$= w\,(h\,.\,1)\,.\,h/2 \qquad ...(2.11)$$
$$= wh^2/2$$

Centre of pressure is given by,

$$\bar{h} = \frac{I_G}{A\bar{x}} + \bar{x}$$
$$= \frac{bh^3/12}{bh\,.\,h/2} + \frac{h}{2}$$
$$= \frac{2}{3}h \qquad \text{from the water surface}$$

or at $h/3$ from the base of the dam. This can also directly be written from the pressure diagram which is a right angle triangle in this case.

Resultant thrust, $R = \sqrt{W^2 + F^2}$

Let E be the point of incidence of the resultant R on the base AB.

Horizontal component of R at E is equal to F and is resisted by the frictional force at the base against sliding whereas the normal component of R at E is equal to W, balanced by the resultant of normal stress intensities at the base.

The vertical component W at E induces normal stress (compressive) at the base due to load W, and due to bending moment. (compressive at B and tensile at A).

The bending moment = Load × Eccentricity

$$= W \times e$$

Normal stress due to axial load, $p_0 = \dfrac{W}{A} = \dfrac{W}{b \times 1} = \dfrac{W}{b}$ (Compressive)

Normal stress due to BM, $p_b = \dfrac{My}{I} = \dfrac{W\,.\,e\,.\,y}{\frac{1}{12}\,.\,1\,.\,b^3} = \dfrac{12Wey}{b^3}$

Eccentricity e is given by

$$e = AE - AO$$
$$= AE - \frac{b}{2}$$
$$= (x + x') - \frac{b}{2}$$
$$= x + \frac{h}{3}\tan\theta - \frac{b}{2}$$
$$= x + \frac{h}{3}\frac{F}{W} - \frac{b}{3} \qquad ...(2.12)$$

Maximum bending stress will be at the two ends of the base.

Compressive stress at outer edge $B = p_0 + p_b$

$$= \frac{W}{b} + \frac{6We}{b^2} \qquad ...(y = b/2)$$

$$= \frac{W}{b}\left(1 + \frac{6e}{b}\right) \qquad \text{...(2.13)}$$

Compressive stress at inner edge $A = p_0 - p_b$

$$= \frac{W}{b}\left(1 - \frac{6e}{b}\right) \qquad \text{...(2.14)}$$

For safety of the dam, tension in the base is to be avoided,

i.e. the term $\left(1 - \frac{6e}{b}\right)$ must be positive.

or $$\frac{6e}{b} < 1 \quad \text{or} \quad e < \frac{b}{6} \qquad \text{...(2.15)}$$

i.e. the point E must lie within $b/6$ on either side of midpoint of the base or the resultant R must act within middle third of the base. Also to prevent sliding of the dam, frictional force should be more than the horizontal force.

i.e. $$\mu W > F \qquad \text{...(2.16)}$$

Example 2.16 A concrete dam trapezoidal in section is 12 m high. Its top width is 2 m and the bottom width 8 m. The face exposed to water has a slope of 1 horizontal to 12 vertical. The water level in the dam coincides with the top of the dam. The weight of the masonary is 25 kN/m^3 and the coefficient of friction between the bottom of the dam and the soil is 0·6. Check the stability of the dam.

Solution $F = wA\bar{x} = 9{\cdot}81 \times (12 \times 1) \times \frac{12}{2} = 706{\cdot}32$ kN

Considering 1 m length of the dam,

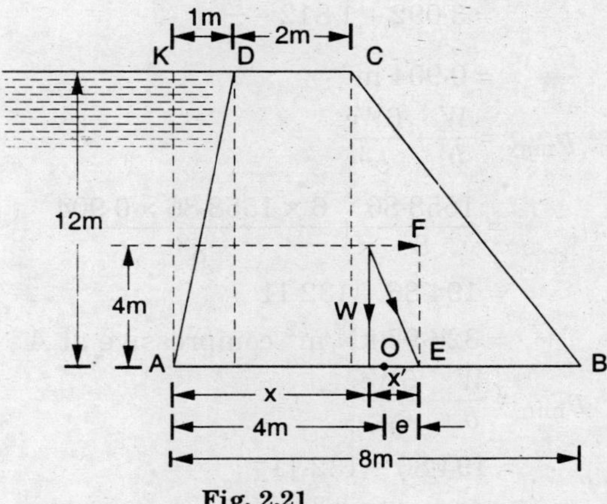

Fig. 2.21

Weight of water acting as weight on the dam *i.e.* weight of water contained in area ADK,

$$F_w = \frac{12 \times 1}{2} \times 9{\cdot}81 = 58{\cdot}86 \text{ kN}$$

$$W = \left[\frac{8+2}{2} \times 12 \times 1\right] \times 25 = 1500 \text{ kN}$$

Total vertical force $W' = W + F_w$

$$= 1500 + 58{\cdot}86$$

$$= 1558{\cdot}86 \text{ kN}$$

To find centre of gravity of the combined vertical force, take moments of the forces about A,

$$1558{\cdot}86 \times x = \left[58{\cdot}86 \times \frac{1}{3}\right] + \left[\left(\frac{12 \times 1}{2} \times 25\right) \times \frac{2}{3}\right]$$

$$+ \left[(12 \times 2 \times 25) \times \left(1 + \frac{2}{2}\right)\right] + \left[\left(\frac{5 \times 12}{2} \times 25\right) \times \left(3 + \frac{5}{3}\right)\right]$$

$$= 19{\cdot}6 + 100 + 1200 + 1633{\cdot}3$$

$$= 4819{\cdot}6$$

$$x = 3{\cdot}092 \text{ m}$$

Now, $$\frac{x'}{12/3} = \frac{F}{W}$$

$$= \frac{706{\cdot}32}{1558{\cdot}86}$$

$$x' = 1{\cdot}812 \text{ m}$$

$$e = x + x' - \frac{b}{2}$$

$$= 3{\cdot}092 + 1{\cdot}812 - \frac{8}{2}$$

$$= 0{\cdot}904 \text{ m}$$

$$p_{\max} = \frac{W}{b} + \frac{6We}{b^2}$$

$$= \frac{1558{\cdot}86}{8} + \frac{6 \times 1558{\cdot}86 \times 0{\cdot}904}{8^2}$$

$$= 194{\cdot}86 + 132{\cdot}11$$

$$= 326{\cdot}97 \text{ kN/m}^2 \text{ compressive at } A$$

$$p_{\min} = \frac{W}{b} - \frac{6We}{b^2}$$

$$= 194{\cdot}86 - 132{\cdot}11$$

$$= 62{\cdot}75 \text{ kN/m}^2 \text{ compressive at } B$$

Checking of stability

1. Since $e = 0{\cdot}904$ lies between $b/6 = 8/6 = 1{\cdot}333$ m, no tension is present in the base as observed above.
2. Frictional force = $\mu W = 0{\cdot}6 \times 1558{\cdot}86 = 935{\cdot}3$ kN

which is more than F (= 706·32 kN). Therefore, dam is safe from sliding.

3. Clockwise moments about $B = F \times h/3 = 706{\cdot}32 \times 4 = 2825{\cdot}3$ kN . m

Counter-clockwise moments about $B = W' \times (b - x)$

$$= 1558{\cdot}86 \times (8 - 3{\cdot}092)$$
$$= 7650{\cdot}9 \text{ kN . m}$$

Since the moments due to the weight of the dam are more than overturning moments, the dam is safe against overturning.

$$\text{Factor of safety} = \frac{7650{\cdot}9}{2825{\cdot}3} = 2{\cdot}71$$

Example 2.17 The curved face of a dam is shaped according to the relation $y = x^2/16$ as shown in Fig. 2.22. The height of water retained by the dam is 16 m. Determine the magnitude and direction of the resultant water pressure acting on the curved face for a unit width of the dam.

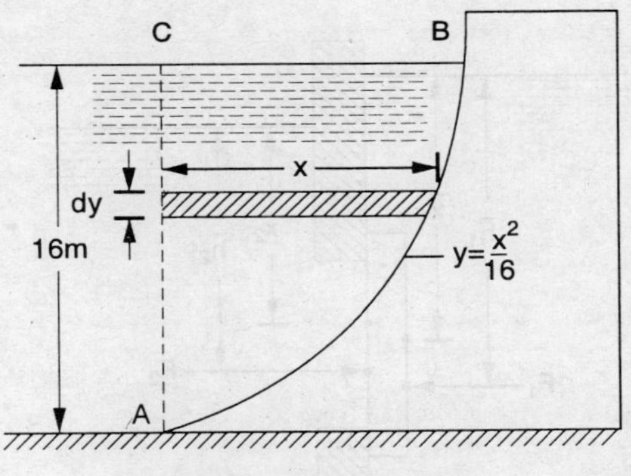

Fig. 2.22

Solution $y = \dfrac{x^2}{16}$ or $x = 4\sqrt{y}$

$$F_h = 9{\cdot}81 \times (16 \times 1) \times (16/2) = 1255{\cdot}68 \text{ kN}$$

F_v = Weight of water contained in area ABC

$= w \times$ Volume of water contained in ABC

$$= w\left[\int_0^{16} x \,.\, dy\right] \times 1$$

$$= w\int_0^{16} 4\sqrt{y} \,.\, dy$$

$$= 4w\left[\frac{y^{3/2}}{3/2}\right]$$

$$= 9{\cdot}81 \times 4 \times \frac{2}{3} \times (16)^{3/2}$$

$$= 1674{\cdot}24 \text{ kN}$$

$$\text{Resultant water pressure} = \sqrt{1255{\cdot}68^2 + 1674{\cdot}24^2}$$
$$= 2092{\cdot}18 \text{ kN}$$

$$\text{Direction of resultant force} = \tan\theta = \frac{1674{\cdot}24}{1255{\cdot}61} = 1{\cdot}333$$
$$\theta = 53{\cdot}13^\circ$$

2.7 GATES

To control the flow of water in rivers, dams or pipes etc. gates are provided.

Sluice Gates

A sluice gate is used to control the flow of water through the openings in a structure. It moves in vertical direction with the help of rollers that travel on vertical rails. A sluice gate may be acted upon by water pressure on one side or on both sides. Fig. 2.23 shows a sluice gate acted upon by water pressure on both sides.

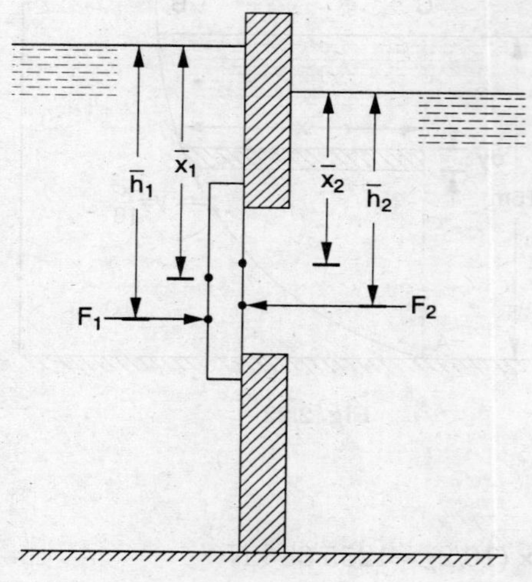

Fig. 2.23

Let x_1 = height of water on the upstream side

x_2 = height of water on the downstream side

$F_1 = wa\bar{x}_1$

$F_2 = wa\bar{x}_2$

Resultant force, $F = F_1 - F_2$

The two forces act at the centres of pressure on the two sides respectively. Position of the resultant force may be determined by taking the moments of forces about the upper or lower edge of the gate.

Lock Gates

Lock gates are used to change the water level in canals to facilitate navigation. Whenever there is a vertical fall in a river or canal, it becomes necessary to raise or lower the water level so that the boat can be transferred from one side to the other. In such a case, two pairs of lock gates are used and a chamber is made in-between (Fig. 2.24).

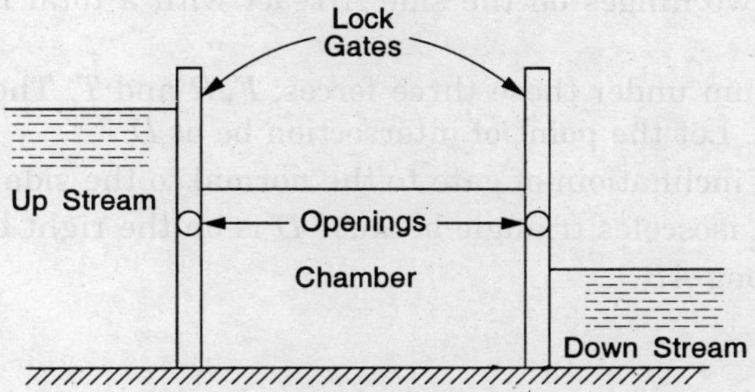

Fig. 2.24

When a boat is to be transferred from the upper side to the lower side, water through the openings in the upper lock gate is allowed to pass whereas the openings in the lower lock gate are closed. When the water level rises and becomes equal to the upstream level, the boat is taken in the chamber by opening the pair of upper lock gate. Now the openings in the lower lock gate are opened and that of the upper lock gate are closed. Water level slowly lowers and when it becomes equal to the water level on the downstream side, the boat is moved out of the chamber.

To transfer the boat from the downstream side to the upstream side, the procedure is repeated in the reverse order.

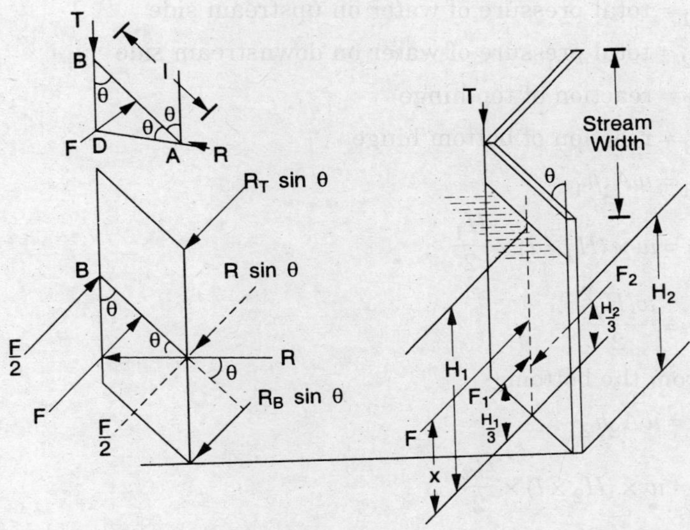

Fig. 2.25

A pair of lock gates is shown in Fig. 2.25. Each gate is supported by two hinges, one at the top and the other at the bottom end of the gate.

The water pressure acts with a resultant force F at the centre of pressure of each gate. The centre of pressure is at the midpoint of AB and $H_1/3$ from the bottom of the gate. The other gate of the pair acts on the gate AB with a reaction force T normal to the surface of contact of the two gates. Two hinges on the side A react with a total force R, the direction being unknown.

The gate is in equilibrium under these three forces; F, R and T. Therefore, all the forces must intersect at one point. Let the point of intersection be at D.

Let θ = angle of inclinatioin of gate to the normal to the side of the canal.

The triangle ADB is an isosceles triangle because D is on the right bisector of AB.

Resolving the forces along AB,

$$R \cos \theta = T \cos \theta$$

or

$$R = T$$

Resolving the forces normal to AB,

$$F = R \sin \theta + T \sin \theta$$

$$= (R + T) \sin \theta$$

$$= (R + R) \sin \theta \qquad ....(R = T)$$

$$= 2R \sin \theta$$

$$R = \frac{F}{2 \sin \theta} \qquad ...(2.17)$$

Further,

Let H_1 = height of water on upstream side

H_2 = height of water on downstream side

F_1 = total pressure of water on upstream side

F_2 = total pressure of water on downstream side

R_T = reaction of top hinge

R_B = reaction of bottom hinge

$$F_1 = wA_1\bar{h}_1$$

$$= w \times (H_1 \times l) \times \frac{H_1}{2}$$

$$= \frac{wlH_1^2}{2}$$

acts at $H_1/3$ from the bottom.

and

$$F_2 = wA_2\bar{h}_2$$

$$= w \times (H_2 \times l) \times \frac{H_2}{2}$$

$$= \frac{wlH_2^2}{2}$$

acts at $H_2/3$ from the bottom.

Resultant $F = F_1 - F_2$

Let x be the height of the point of application of F from the bottom, then

$$F \cdot x = F_1 \cdot \frac{H_1}{3} - F_2 \cdot \frac{H_2}{3}$$

Also $$R = R_T + R_B = \frac{F}{2 \sin \theta}$$

or $$\frac{F}{2} = R_T \sin \theta + R_B \sin \theta \qquad ...(2.18)$$

Half of the water pressure is taken by the reaction of the hinge edge of the gate whereas the rest half is borne by the reaction of the other gate.

Example 2.18 Each gate of a lock is 8 m high and is supported on one side by two hinges, each 1 m from the top and the bottom. When the gates are closed, they make an angle of 120°. The width of the lock is 5·2 m. The water levels are 6 m and 3 m on the upstream and the downstream sides respectively. Determine

(i) Resultant water force on each gate

(ii) Magnitude of the reaction between the gates

(iii) Force on each hinge.

Solution $$l = \frac{5{\cdot}2/2}{\sin 60^\circ} = 3 \text{ m}$$

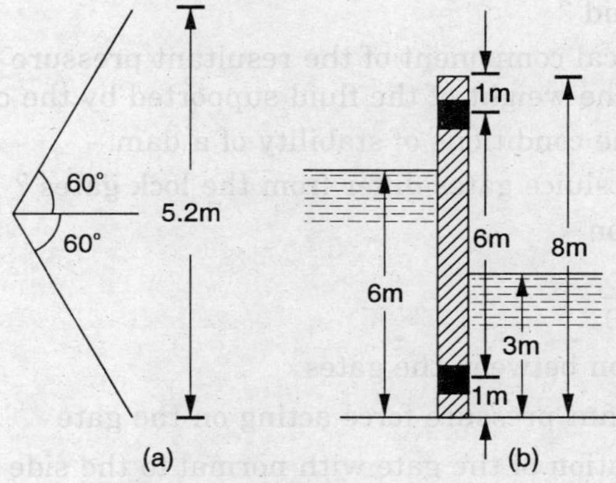

Fig. 2.26

(i) $$F_1 = 9{\cdot}81 \times (3 \times 6) \times \frac{6}{2} = 529{\cdot}74 \text{ at 2 m from bottom}$$

$$F_2 = 9{\cdot}81 \times (3 \times 3) \times \frac{3}{2} = 132{\cdot}44 \text{ at 1 m from bottom}$$

$$F = F_1 - F_2 = 529{\cdot}74 - 132{\cdot}44 = 397{\cdot}3 \text{ kN}$$

(ii) Taking moments about bottom of gate,

$$397{\cdot}3 \times x = 529{\cdot}74 \times 2 - 132{\cdot}44 \times 1$$

$$x = 2{\cdot}333 \text{ m}$$

$$R = \frac{F}{2 \sin \theta} = \frac{397{\cdot}3}{2 \sin 30} = 397{\cdot}3 \text{ kN}$$

$$R_T + R_B = 397{\cdot}3 \text{ kN} = R$$

(iii) Taking moments about bottom hinge

$$R_T \times 6 + R_B \times 0 = 397{\cdot}3 \times (2{\cdot}333 - 1)$$

$$R_T = 154{\cdot}48 \text{ kN}$$

$$R_B = 397{\cdot}3 - 154{\cdot}5 = 242{\cdot}8 \text{ kN}$$

PROBLEMS

1. Define the terms : (a) total pressure (b) centre of pressure.
2. Derive an expression for the total pressure acting on a vertical plane surface immersed in a liquid at rest. Locate the position of centre of pressure also.
3. Show that the centre of pressure on any plane surface immersed in liquid is always below its centroid.
4. Deduce expressions for the total pressure and depth of centre of pressure in case of an inclined plane surface submerged in a fluid.
5. How do you find the total pressure and the resultant pressure on a curved surface submerged in a liquid ?
6. Show that the vertical component of the resultant pressure on an immersed curved surface is equal to the weight of the fluid supported by the curved surface.
7. State and deduce the conditions of stability of a dam.
8. In what way do the sluice gates differ from the lock gates ?
9. Derive the expression

$$R = \frac{F}{2 \sin \theta}$$

where R = reaction between the gates

F = resultant pressure force acting on the gate

θ = inclination of the gate with normal to the side of canal

10. An annular ring of internal and external diameters as 2 m and 1 m respectively is vertically immersed in water such that the centre of the ring is 2 m below the surface of water. Determine the total force and the position of centre of pressure for the ring below the water surface. (46·1 kN, 2·135 m)
11. A circular plate 3 m in diameter is immersed in water such that its least and greatest depths below the water surface are 1 m and 2 m respectively. Determine the total pressure on one surface of the plate and the position of the centre of pressure. (104 kN, 1·542 m)

12. A square lamina of 600 mm side is immersed vertically in water such that one of the diagonols is vertical. The upper corner is 600 mm below the surface of water. Determine the force on one side and the location of centre of pressure.

(3·61 kN, 1·0536 m)

13. Determine the moment of hydrostatic force for a gate defined by $y = 2x^2$ as shown in Fig. 2.27. (149·4 kN . m)

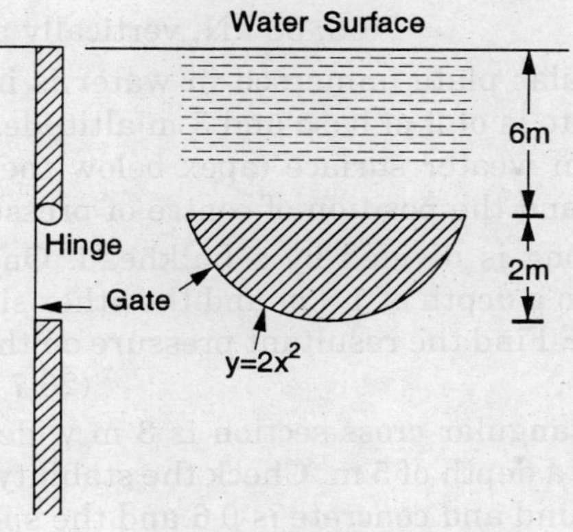

Fig. 2.27

14. A box tank has a square base of side 500 mm and is 1 m in height. It is filled with a liquid having specific gravity 2 upto a height of 500 mm above the base and the remainder space is filled with water which is immiscible with the other liquid. Determine the magnitude and the line of action of the resultant force exerted on one side of the box tank. (3·606 kN, 700 mm below the water surface)

15. A door in a tank is in the form of a quadrant of a cylinder of 2 m radius and 2·4 m wide (Fig. 2.28). Determine the resultant force on the door and the vertical height of resultant on the gate. (248·1 kN, 1·3 m)

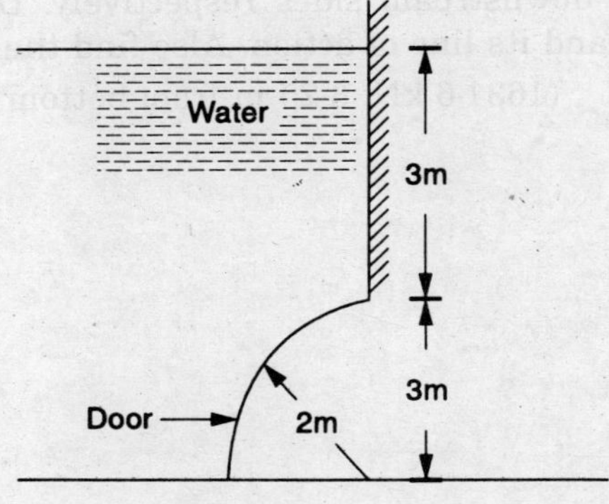

Fig. 2.28

16. One of the vertical sides of a square vessel has a hemispherical projection of diameter 1·2 m. Determine the horizontal and vertical forces acting on the projection if the vessel contains water upto a height of 3 m above the centre of hemisphere.
(35·5 kN, 133·1 kN)

17. A hemisphere bulge of diameter 1·6 m is provided at the bottom of a tank. Find the magnitude and direction of the resultant force if the depth of water above the horizontal floor of the tank is 4 m.
(68, 38 kN, vertically at the centre of hemisphere)

18. The plane of a triangular plate immersed in water is inclined at 30° with the free water surface. The plate is of 2 m base and 3 m altitude. The base is parallel to and at a depth of 2 m from weater surface (apex below the base). Determine the total pressure on the plate and the position of centre of pressure. (73·58 kN, 2·55 m)

19. A storage tank 3 m long is divided by a bulkhead. On one side, petrol of specific gravity 0·78 is stored to a depth of 1·8 m and the other side has oil of specific gravity 0·88 to a depth of 0·9 m. Find the resultant pressure on the bulkhead and the position at which it acts. (26·7 kN, 718 mm from the base)

20. A concrete wall of rectangular cross-section is 3 m wide and 7 m high. One face of the wall holds water to a depth of 5 m. Check the stability of the wall if the coefficient of friction between ground and concrete is 0·6 and the specific weight of the concrete is 20 kN/m^2. (No tension at base, no overturning, no sliding)

21. A concrete dam of trapezoidal section is 8 m wide at the base and 3 m wide at the top. The face exposed to water is vertical and is 16 m high. The water level in the dam coincides with the top of the dam. The density of the concrete is 2400 kg/m^3. Determine the resultant thrust on the base per m length of the dam and the point where it intersects the base. (2422·7 kN, 6·17 m)

22. Each gate of a lock is 9 m high and is supported on one side by two hinges, one placed at 0·8 m and the other at 6 m from the bottom of lock. The lock is 8 m wide and the gates make an angle of 120° when closed. The water levels are 9 m and 3 m on the upstream and the downstream sides respectively. Determine the resultant water force on each gate and its line of action. Also find the force on each hinge.
(1631·6 kN, 3·25 m from bottom, R_t = 768·7 kN, R_b = 862·9 kN)

Chapter 3

BUOYANCY

3.1 INTRODUCTION

We know, a body submerged or floating in a liquid experiences a vertical uplift, known as *buoyant force*. Its magnitude is given by the familiar Archimedes' principle of buoyancy *i.e.* when a body is fully or partially immersed in a liquid, it is buoyed up by a force equal to the weight of fluid it displaces.

Consider a submerged body as shown in Fig. 3.1. The resultant horizontal pressure on the body is to be zero due to the fact that on any horizontal prism, the pressures on the two sides are the same and opposite in direction.

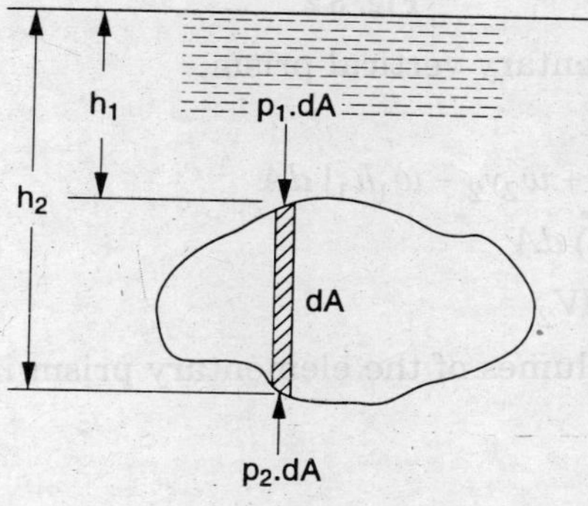

Fig. 3.1

To find the vertical lift, divide the body into a number of vertical prisms. On the two ends of a vertical prism of cross section dA, pressure forces are p_1dA and p_2dA, the latter being greater than the former. The difference of the two forces is the net vertical upward buoyant force,

i.e. $$dF_B = p_2dA - p_1dA$$

$$= wh_2dA - wh_1dA$$

$$= wdA\,(h_2 - h_1)$$
$$= wdV$$

where dV is the volume of the vertical prism. On integrating this expression, the buoyant force F_B on the entire submerged body may be obtained.

$$F_B = \int dF_B$$
$$= w \int dV$$
$$= wV \qquad ...(3.1)$$

= Weight of liquid displaced by the submerged body of volume V

The buoyant force acts through the centroid of the displaced volume and is known as *centre of buoyancy.*

Now, consider a body floating at the surface of separation of two fluids. Let w_1 and w_2 be specific weights of the two fluids (Fig. 3.2).

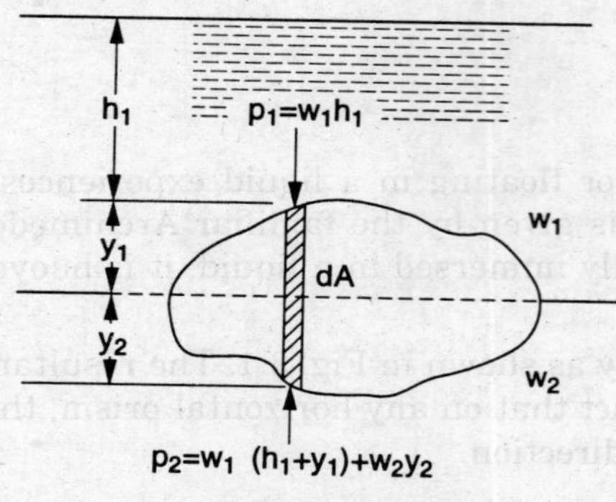

Fig. 3.2

The buoyant force on an elementary vertical prism,

$$dF_B = p_2 dA - p_1 dA$$
$$= [w_1 (h_1 + y_1) + w_2 y_2 - w_1 h_1]\, dA$$
$$= (w_1 y_1 + w_2 y_2)\, dA$$
$$= w_1 dV_1 + w_2 dV_2$$

where dV_1 and dV_2 are the volumes of the elementary prism in the two fluids.

Total buoyant force,

$$F_B = \int dF_B$$
$$= \int w_1 dV_1 + \int w_2 dV_2$$
$$= w_1 V_1 + w_2 V_2 \qquad ...(3.2)$$

When a body floats at the free surface of a liquid, specific weight of air becomes negligible as compared to the specific weight of the liquid. Thus if fluid 1 is air in the above case,

$$F_B = w_1 V_1 + w_2 V_2$$
$$= w_2 V_2$$

Thus buoyant force is equal to the weight of the fluid displaced. Also as the floating body is statically in equilibrium, therefore, for vertical equilibrium of the body, the buoyant force must balance the weight of the body,

or $$F_B = W \quad ...(3.3)$$

Example 3.1 A wooden block 3 m long, 1m wide and 0·5 m in height floats in water. If the specific gravity of the block is 0·6, determine the volume of concrete of specific weight 24 kN/m^3 which may be put on the block that will immerse the

(i) block completely in water

(ii) block and the concrete completely in water.

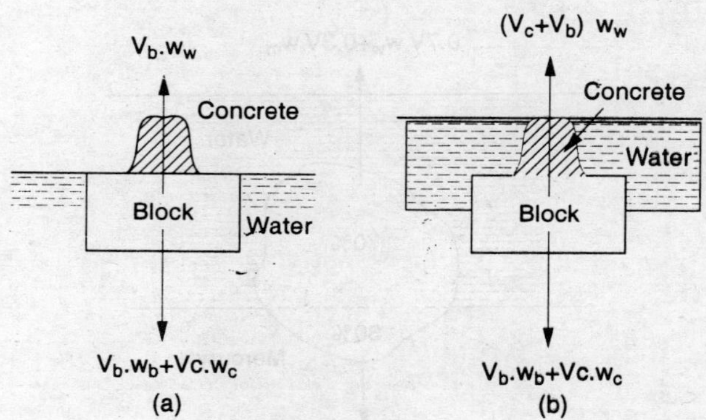

Fig. 3.3

Solution W_b = Volume × Specific weight of block

$= (3 \times 1 \times 0{\cdot}5) \times (0{\cdot}6 \times 9{\cdot}81)$

$= 8{\cdot}829$ kN

(i) When block is completely immersed (Fig. 3.3a),

Volume of water displaced = Volume of block

$= 3 \times 1 \times 0{\cdot}5$

$= 1{\cdot}5 \text{ m}^3$

Let V_c be the volume of concrete.

Downward force = Upward force

Weight of concrete and block = Weight of water displaced

$$V_c \times 24 + 8{\cdot}829 = 1{\cdot}5 \times 9{\cdot}81$$

$$V_c = 0{\cdot}245 \text{ m}^3$$

(ii) When block and the concrete are completely immersed (Fig. 3.3b),

Volume of water displaced = Volume of block + Volume of concrete

$= (1{\cdot}5 + V_c) \text{ m}^3$

Weight of concrete and block = Weight of water displaced

$$V_c \times 24 + 8{\cdot}829 = (V_c + 1{\cdot}5) \times 9{\cdot}81$$
$$= 9{\cdot}81V_c + 14{\cdot}72$$

or
$$14{\cdot}19V_c = 5{\cdot}89$$
$$V_c = 0{\cdot}415 \text{ m}^3$$

Example 3.2 A metallic body floats at the interface of mercury and water such that 30% of its volume is submerged in mercury and 70% in water. Determine the specific weight of the metallic body. Specific gravity of mercury is 13·6.

Solution Let V be the volume of the metallic body.

For equilibrium (Fig. 3.4),

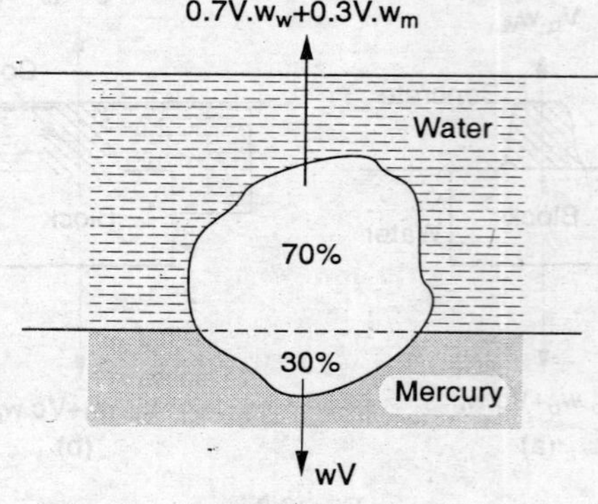

Fig. 3.4

Weight of the body = Weight of water displaced + Weight of mercury displaced

$$wV = 0{\cdot}7 \text{ V} \times 9{\cdot}81 + 0{\cdot}3 \text{ V} \times 13{\cdot}6 \times 9{\cdot}81$$
$$w = (0{\cdot}7 + 0{\cdot}3 \times 13{\cdot}6) \times 9{\cdot}81$$
$$w = 46{\cdot}89 \text{ kN/m}^3$$

Example 3.3 A wooden log 1 m in diameter and 8 m long floats in water. The specific gravity of the log is 0·7. Determine the depth of the log in water.

Solution Weight of log = Weight of water displaced

= Volume of water displaced × 9·81

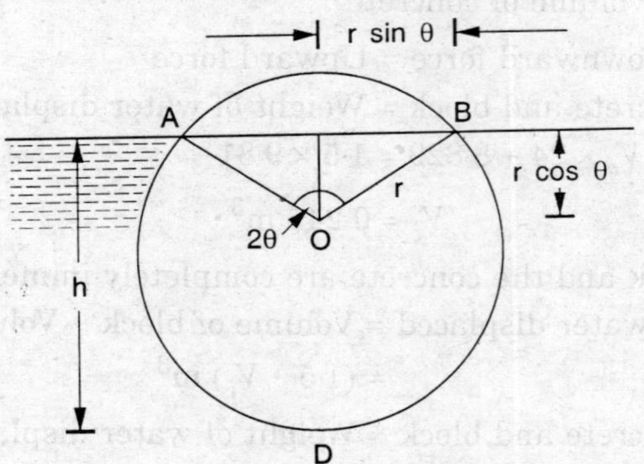

Fig. 3.5

$$(0{\cdot}7 \times 9{\cdot}81) \times (\pi/4) \times 1^2 \times 8 = \text{Volume of water displaced} \times 9{\cdot}81$$

$$\text{Volume of water displaced} = \frac{(0{\cdot}7 \times 9{\cdot}81) \times (\pi/4) \times 1^2 \times 8}{9{\cdot}81}$$

$$= 4{\cdot}398 \text{ m}^3$$

i.e. Area $ABDA \times$ Length of log $= 4{\cdot}398$ m^3

or Area $ABDA = 4{\cdot}398/8 = 0{\cdot}55$ m^2

Area $AOBDA$ + Area $AOBA = 0{\cdot}55$ m^2

$$\pi r^2 \left[\frac{360 - 2\theta}{360}\right] + \left[\frac{1}{2} \times r \sin\theta \,.\, r\cos\theta\right] \times 2 = 0{\cdot}55$$

$$\pi \times 0{\cdot}5^2 \left[1 - \frac{\theta}{180}\right] + (0{\cdot}5)^2 \,.\, \sin\theta \,.\, \cos\theta = 0{\cdot}55$$

$$0{\cdot}25\pi - \frac{0{\cdot}25\pi\,\theta}{180} + 0{\cdot}25 \sin\theta\cos\theta = 0{\cdot}55$$

or $$\theta - 57{\cdot}296 \sin\theta\cos\theta = 53{\cdot}95$$

By trial and error,

$$\theta = 71{\cdot}5^\circ$$

$$h = 0{\cdot}5 + 0{\cdot}5\cos 71{\cdot}5^\circ$$

$$= 0{\cdot}659 \text{ m}$$

Example 3.4 A float valve is used to regulate the flow of oil of specific gravity 0·9 in a cistern. The float is of spherical shape with a diameter of 300 mm. *AOB* is a weightless link mounted on a frictionless hinge at *O* and angle *AOB* is 135°. The *AOB* carries the float on one end and a valve at the other end which regulates the flow of oil through the pipe. The length of *OA* is 100 mm and the distance between the hinge and the float is 1 m. When there is no flow, *AO* is vertical. The force required to press the valve on to the seat to completely stop the flow into the cistern is 80 N. If the flow of oil is stopped when the free surface of oil in the cistern is 700 mm below the hinge, determine the weight of the float.

Solution Refer Fig. 3.6.

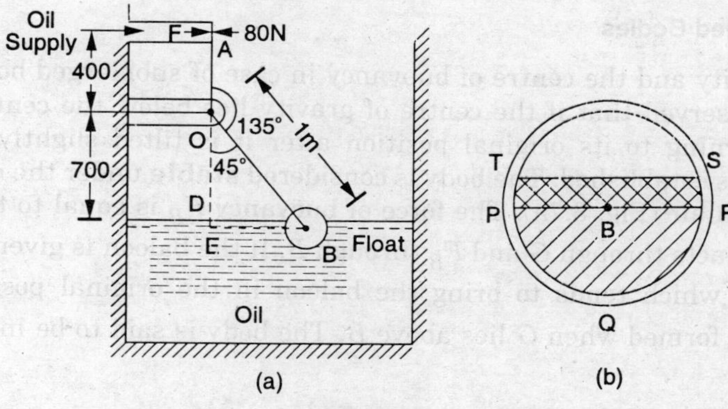

Fig. 3.6

$\angle BOE = 180 - 135 = 45$

On stopping the flow of oil, the centre of the float will be at B.

i.e. at a depth of $1000 \times \cos 45^\circ$

or $OE = 707{\cdot}11$ mm

$DE = OE - OD = 707{\cdot}11 - 700 = 7{\cdot}11 \text{ mm} = 0{\cdot}00711 \text{ m}$

Buoyant force = Weight of oil displaced

= Volume of oil displaced × Specific weight

= [Volume of semi sphere (PQR) + Volume contained in the area $PRST$] . w

$$= w\left[\frac{1}{2}\left(\frac{4}{3} \times \pi r^3\right) + \pi r^2 . DE\right]$$

$$= 0{\cdot}9 \times 9{\cdot}81 \times \left[\frac{1}{2}\left(\frac{4}{3} \times \pi \times 0{\cdot}15^3\right) + \pi \times 0{\cdot}15^2 \times 0{\cdot}00711\right]$$

$$= 8{\cdot}829 \times (0{\cdot}007069 + 0{\cdot}0005026)$$

$$= 0{\cdot}0668 \text{ kN or } 66{\cdot}8 \text{ N}$$

The buoyant force and the weight (force) of the float passes through the same vertical line through B.

Let the weight of the float be W.

Then net upward force on the float = Weight of float − Buoyant force

$= (66{\cdot}8 - W)$

Taking moments about hinge O,

$$F \times 0{\cdot}4 = (66{\cdot}8 - W) \times BE$$

$$80 \times 0{\cdot}4 = (66{\cdot}8 - W) . (1 \times \cos 45^\circ)$$

$$W = 45{\cdot}32 \text{ N}$$

3.2 STABILITY OF SUBMERGED AND FLOATING BODIES

Stability of a submerged or a floating body depends upon its capability to come back to its original position when tilted slightly. Further, the stability is determined by the relative positions of the centre of gravity of the body and the centre of buoyancy.

Stability of Submerged Bodies

The centre of gravity and the centre of buoyancy in case of submerged bodies always remain fixed. It can be observed that if the centre of gravity lies below the centre of buoyancy, the body will be returning to its original position after it is tilted slightly. This is because a rightening couple is established. The body is considered **stable** under the conditions. Consider a baloon floating in air (Fig. 3.7a). The force of buoyancy F_B is equal to the weight of baloon for equilibrium. W acts through G and F_B through B. If the baloon is given a slight tilt, W and F_B form a couple which tends to bring the baloon in the original positioin. Such type of restoring couple is formed when G lies above B. The body is said to be in stable equilibrium.

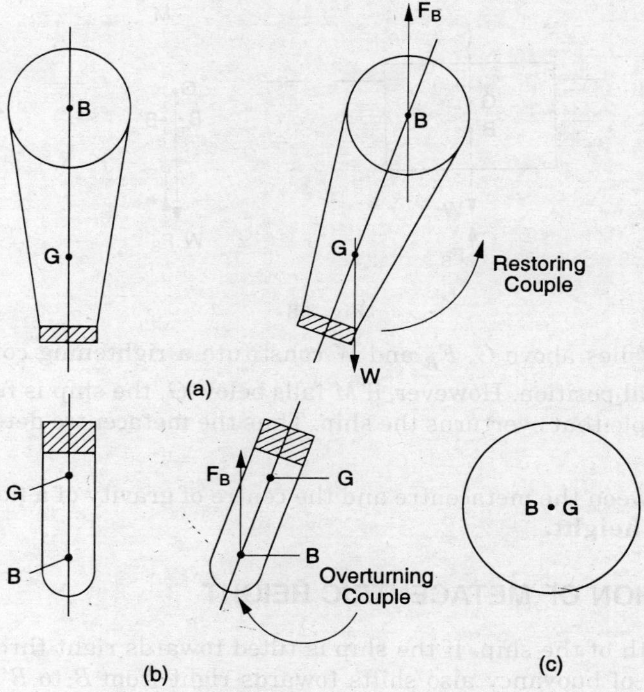

Fig. 3.7

However, if the centre of gravity of the body lies above the centre of buoyancy, the couple produced further tends to rotate the body away from its original position. The body is called **unstable.** Take the case of a test tube fitted with a heavy lead stopper and immersed in a liquid (Fig. 3.7b). On giving a small tilt to the tube, an overturning couple is formed which tends to further tilt the tube. This is the unstable condition with G above B.

The body is called in **neutral equilibrium** if the two centres coincide. Under this, the body assumes a new position after it is tilted slightly. A homogeneous sphere submerged in a liquid is an example of neutral equilibrium in which the sphere takes a new position on tilting (Fig. 3.7c).

Stability of Floating Bodies

Stability of floating bodies differs from the stability of the submerged bodies in that a body may be in stable equilibrium even if its centre of gravity lies above the centre of buoyancy. Instead, if a new point called the **metacentre** lies above the centre of gravity of the body, the body is in stable equilibrium.

Consider the stability of a ship. Assume that its centre of buoyancy shifts from position B to B' when it is tilted through a small angle θ. (Fig. 3.8). B is the centroid of the displaced liquid when the ship is in the upright position and B', the new centroid of the submerged volume after the ship is tilted. The point of intersection of a vertical line through B' and the line joining B and G is called **metacentre** M.

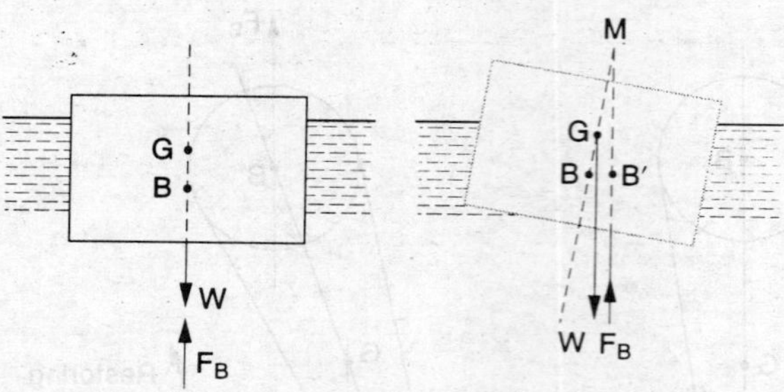

Fig. 3.8

Now, so far as M lies above G, F_B and W constitute a rightening couple and the ship is brought to the original position. However, if M falls below G, the ship is further tilted because F_B and W form a couple that overturns the ship. Thus the metacentre determines the stability of a floating ship.

The distance between the metacentre and the centre of gravity of a floating body is known as the **metacentric height.**

3.3 DETERMINATION OF METACENTRIC HEIGHT

Consider a unit length of the ship. If the ship is tilted towards right through a small angle θ (Fig. 3.9), the centre of buoyancy also shifts towards right from B to B'. This is because an

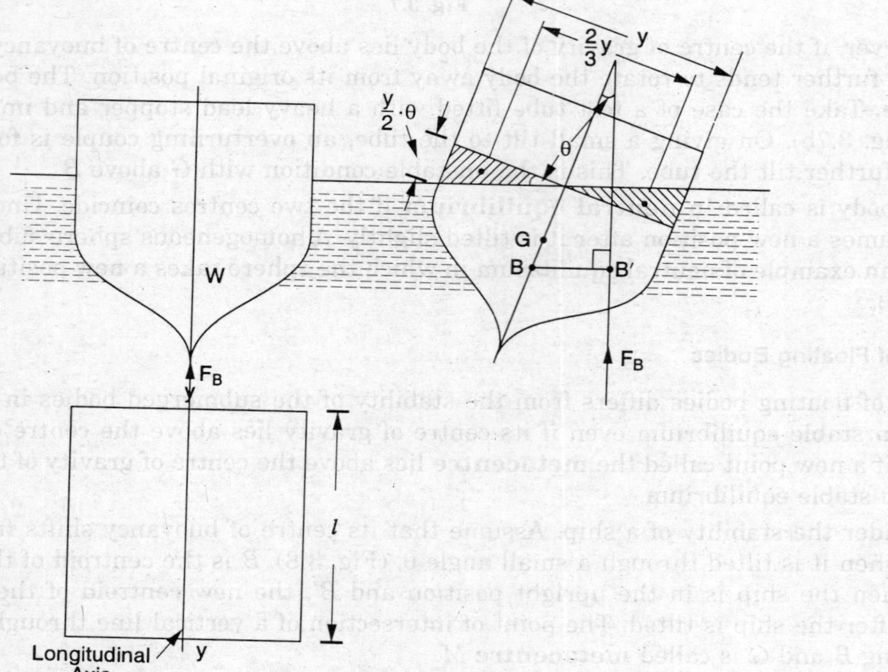

Fig. 3.9

additional buoyant force dF_B acts on the right half as the volume of the displaced liquid is increased and also due to reduction in the buoyant force on the left half by an equal amount due to decrease in the displaced volume. The upward force due to the left hand wedge and the downward force due to the right hand wedge constitute a couple which must be equal to the moment due to shifting of centre of buoyancy from B to B'.

Thus

The rightening couple due to shifting of centre of buoyancy

= Moments of the couple formed by forces due to two wedges

$$\text{The rightening couple} = F_B \times BB'$$

$$= \text{Force of buoyancy} \times MB \sin\theta$$

$$= \text{Weight of displaced liquid} \times MB \sin\theta$$

$$= wV \,.\, MB \,.\, \theta$$

where V is the volume of liquid displaced by the ship and remains constant during the tilting of the ship and also $\sin\theta \approx \theta$ for small angles of tilt.

Thus $\quad wV \times MB \times \theta = dF_B \times$ Distance between the centroids of the wedges

$$= w \,.\, V_w \times \frac{2}{3} y \qquad \ldots(V_w = \text{Volume of a wedge})$$

$$= \frac{2}{3} w \, \frac{1}{2}\left(l \,.\, \frac{y}{2} \,.\, \frac{y}{2} \,.\, \theta\right) y$$

$$= \frac{1}{12} w l y^3 \,.\, \theta$$

$$V \,.\, MB = \frac{l y^3}{12}$$

= Moment of inertia of the liquid line area $(l \times y)$ about its longitudinal axis $y - y$

or $$V \,.\, MB = I$$

or $$MB = \frac{I}{V} \qquad \ldots(3.4)$$

Metacentric height $GM = MB - BG$

$$= \frac{I}{V} - BG \qquad \ldots(3.5)$$

Example 3.5 A wooden block $3 \times 2 \times 1$ m in size and of specific gravity 0·8 floats in water. Determine its metacentric height.

Solution Refer Fig. 3.10

Weight of block = Weight of water displaced

$$w_b \,.\, AH = w_w Ah$$

$$h = \frac{w_b}{w} H$$

= Specific gravity of block × Height of block

= 0·8 × 1

= 0·8 m

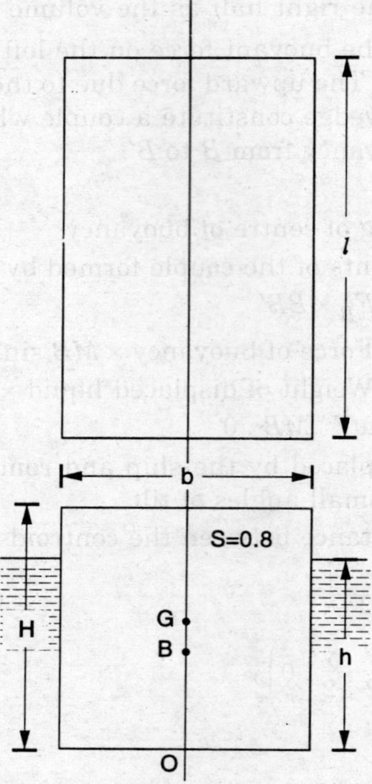

Fig. 3.10

$$GM = \frac{I}{V} - BG$$

$$= \frac{lb^3/12}{lbh} - (OG - OB)$$

$$= \frac{b^2}{12h} - \left(\frac{H}{2} - \frac{h}{2}\right)$$

$$= \frac{2^2}{12 \times 0{\cdot}8} - \left(\frac{1}{2} - \frac{0{\cdot}8}{2}\right)$$

$$= 0{\cdot}417 - 0{\cdot}1$$

$$= 0{\cdot}317 \text{ m}$$

Example 3.6 Determine the metacentric height of a solid cylinder 5 m in diameter and 4 m high floating in water with its axis vertical. The specific gravity of the material of the cylinder is 0·7. State whether the equilibrium of the cylinder is stable or unstable.

Solution Depth of cylinder in water (Fig. 3.11),

h = Specific gravity of cylinder × Height of cylinder

= 0·7 × 4 = 2·8 m

$$GM = \frac{I}{V} - BG$$

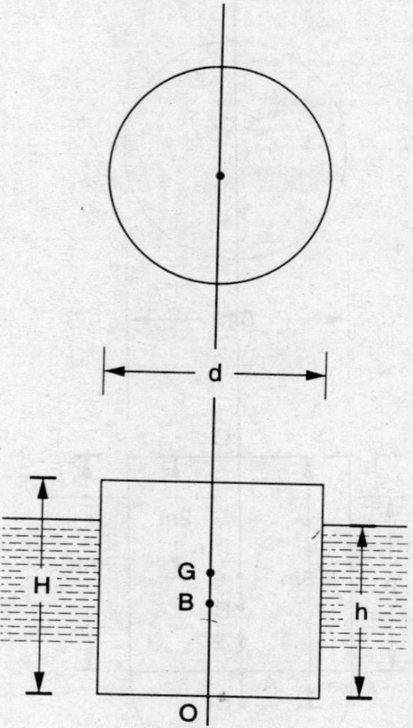

Fig. 3.11

$$= \frac{(\pi/64)\, d^4}{(\pi/4)\, d^2 .\, h} - (OG - OB)$$

$$= \frac{d^2}{16h} - \left(\frac{H}{2} - \frac{h}{2}\right)$$

$$= \frac{5^2}{16 \times 2{\cdot}8} - \left(\frac{4}{2} - \frac{2{\cdot}8}{2}\right)$$

$$= 0{\cdot}558 - 0{\cdot}6$$

$$= -\,0{\cdot}042 \text{ m}$$

-ve sign means that M is below G.

Therefore, unstable.

Example 3.7 A floating body is made up of two portions. The upper cylinderical portion is 5 m in diameter and 3 m deep. The lower portion is curved one. The body displaces 98·1 kN of water and its centre of gravity is 2 m below the top. The curved portion displaces a volume of 2000 litres of water and its centre of buoyancy lies at 3·5 m below the top of the body. Determine the metacentric height of the body.

Solution Total volume of water displaced $= \frac{98{\cdot}1}{9{\cdot}81} = 10 \text{ m}^3$

Volume displaced by curved portion = 2000 litres

$= 2 \text{ m}^3$

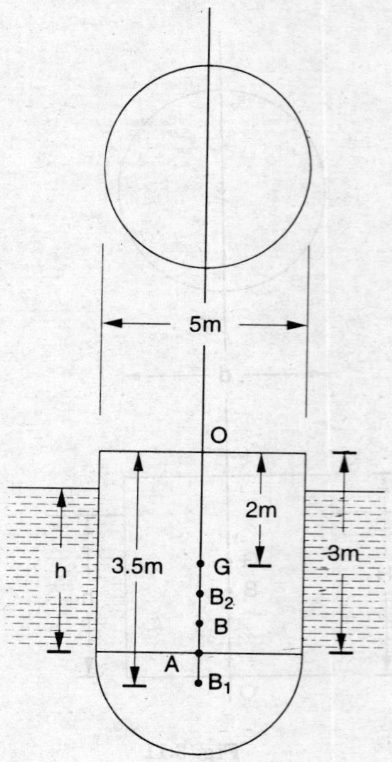

Fig. 3.12

= Total volume displaced by body – Volume displaced by curved portion
= 10 – 2
= 8 m^3

Let B be the centre of pressure for the whole body and B_1 and B_2 be the centres of pressure for the curved and the cylinderical portions respectively.

$$OB_1 = 3{\cdot}5 \text{ m}$$

h = Depth of cylinderical portion immersed in weater

$$= \frac{\text{Volume of cylinderical portion immersed in water}}{\text{Area of cross-section}}$$

$$= \frac{8}{(\pi/4) \times 5^2} = 0{\cdot}4074 \text{ m}$$

$$OB_2 = OA - AB_2 = OA - \frac{h}{2} = 3 - \frac{0{\cdot}4074}{2} = 2{\cdot}796 \text{ m}$$

$$OB = \frac{2 \times 3{\cdot}5 + 8 \times 2{\cdot}796}{10} = 2{\cdot}937 \text{ m}$$

$$GM = \frac{I}{V} - BG$$

$$= \frac{(\pi/64)\, d^4}{V} - (OB - OG)$$

$$= \frac{(\pi/64) \times 5^4}{10} - (2{\cdot}937 - 2)$$
$$= 3{\cdot}068 - 0{\cdot}937$$
$$= 2{\cdot}131 \text{ m}$$

Example 3.8 A solid cylinder is 150 mm in diameter and 600 mm long. The base of the cylinder is 20 mm thick and made of a material of specific gravity 5. The specific gravity of the remaining portion of the cylinder is 0·5. State, if it can float vertically in water.

Solution Let G be the centre of gravity of the cylinder with a as the cross-sectional area. Then taking moments about point O (Fig. 3.13),

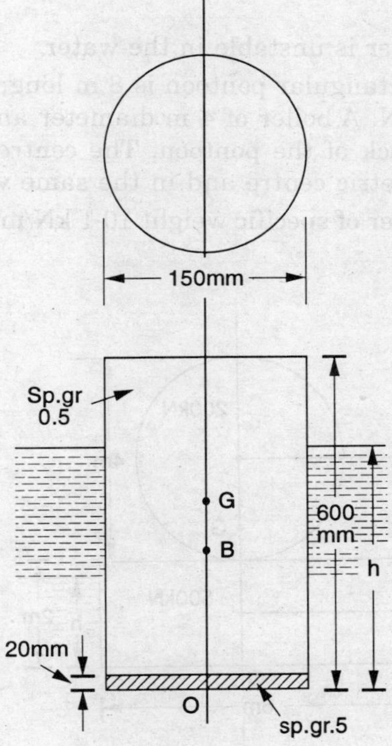

Fig. 3.13

$$[(a \times 0{\cdot}58 \times 0{\cdot}5) + (a \times 0{\cdot}02 \times 5)]\, OG$$
$$= (a \times 0{\cdot}58 \times 0{\cdot}5)\left(\frac{0{\cdot}58}{2} + 0{\cdot}02\right) + (a \times 0{\cdot}02 \times 5)\left(\frac{0{\cdot}02}{2}\right)$$
$$OG = 0{\cdot}233 \text{ m}$$

Weight of cylinder = Weight of water displaced

$$a \times 0{\cdot}58 \times 9{\cdot}81 \times 0{\cdot}5 + a \times 0{\cdot}02 \times 9{\cdot}81 \times 5 = a \times h \times 9{\cdot}81$$
$$h = 0{\cdot}58 \times 0{\cdot}5 + 0{\cdot}02 \times 5$$
$$= 0{\cdot}29 + 0{\cdot}1$$
$$= 0{\cdot}39 \text{ m}$$

$$GM = \frac{I}{V} - BG$$

$$= \frac{(\pi/64)\, d^4}{(\pi/4)\, d^2 \,.\, h} - (OG - OB)$$

$$= \frac{d^2}{16h} - \left(OG - \frac{h}{2}\right)$$

$$= \frac{0{\cdot}15^2}{16 \times 0{\cdot}39} - \left(0{\cdot}233 - \frac{0{\cdot}39}{2}\right)$$

$$= 0{\cdot}00361 - 0{\cdot}038$$

$$= -\,0{\cdot}0344 \text{ m}$$

i.e. M lies below G.

Therefore, the cylinder is unstable in the water.

Example 3.9 The size of a rectangular pontoon is 8 m long, 6 m broad and 2 m deep and it weighs 500 kN. A boiler of 4 m diameter and weighing 200 kN is carried on the upper deck of the pontoon. The centre of gravity of each lies at its respective geometric centre and in the same vertical line. If the whole unit floats in sea water of specific weight 10·1 kN/m^3, find the metacentric height.

Solution (Fig. 3.14)

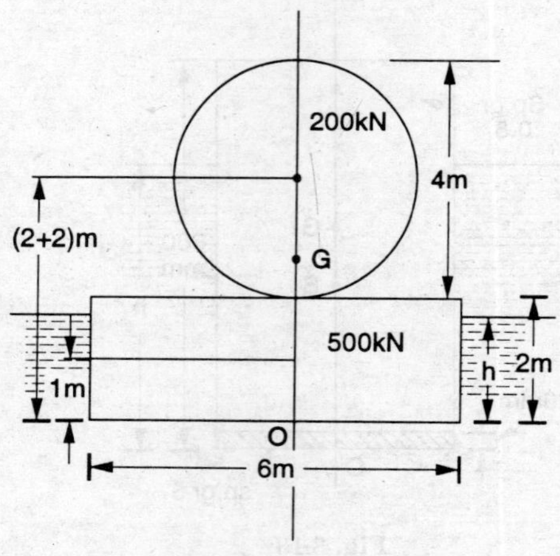

Fig. 3.14

$$GM = \frac{I}{V} - BG$$

Volume of water displaced,

$$V = \frac{\text{Weight of water displaced}}{\text{Specific weight}}$$

$$= \frac{500 + 200}{10{\cdot}1}$$

$$= 69{\cdot}31 \text{ m}^3$$

Height of portion immersed in water,

$$h = \frac{V}{\text{Area of Cross-section}} = \frac{69{\cdot}31}{8 \times 6} = 1{\cdot}444 \text{ m}$$

To find the combined centre of gravity of the pontoon and the boiler, take moments about point O,

$$500 \times 1 + 200 \times (2 + 2) = 700 \times OG$$

$$OG = 1{\cdot}857 \text{ m}$$

$$GM = \frac{I}{V} - BG$$

$$= \frac{8 \times 6^3/12}{69{\cdot}31} - (OG - OB)$$

$$= 2{\cdot}078 - \left(1{\cdot}857 - \frac{1{\cdot}444}{2}\right)$$

$$= 2{\cdot}078 - 1{\cdot}135$$

$$= 0{\cdot}943 \text{ m}$$

Example 3.10 A cylinder of specific gravity 0·64 is required to float in an oil of specific gravity 0·96 with its longitudinal axis vertical. Find the maximum value of *L*/*D* (Length to diameter) ratio of the cylinder.

Solution (Fig. 3.15)

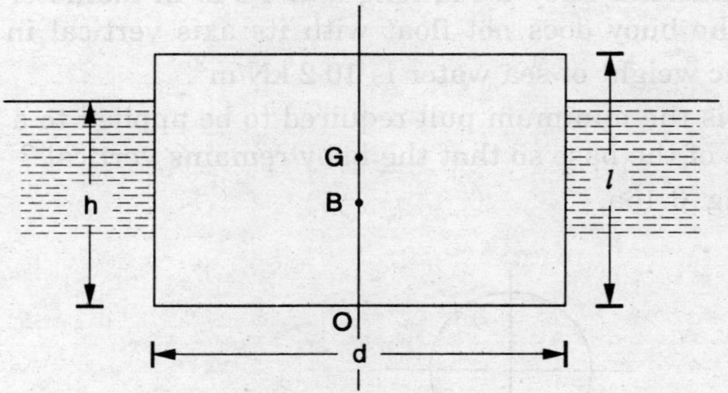

Fig. 3.15

Weight of cylinder = Weight of oil displaced

Volume of cylinder × w_c = Volume of oil displaced × w_o

$$\frac{\pi}{4} d^2 l \times S_c = \frac{\pi}{4} d^2 h \times S_o$$

$$h = \frac{S_c}{S_0} l = \frac{0{\cdot}64}{0{\cdot}96} l = \frac{2}{3} l$$

For the cylinder to float with its longitudinal axis vertical,

$$GM > 0$$

$$\frac{I}{V} - BG > 0$$

$$\frac{\frac{\pi}{64}d^4}{\frac{\pi}{4}d^2 \cdot h} - (OG - OB) > 0$$

$$\frac{d^2}{16h} - \left(\frac{l}{2} - \frac{h}{2}\right) > 0$$

$$\frac{d^2}{16 \times (2/3)\,l} - \left(\frac{l}{2} - \frac{2l}{3 \times 2}\right) > 0 \qquad \left(h = \frac{2}{3}l\right)$$

$$\frac{3}{32}\frac{d^2}{l} - \frac{l}{6} > 0$$

$$\frac{3}{32}\frac{d^2}{l} > \frac{l}{6}$$

$$\frac{d^2}{l^2} > \frac{16}{9}$$

$$\frac{l^2}{d^2} < \frac{9}{16}$$

or $$\frac{l}{d} < \frac{3}{4}$$

Example 3.11 A cylinderical buoy 2·4 m long and 1·6 m in diameter weighs 20 kN. Show that the buoy does not float with its axis vertical in the sea water if the specific weight of sea water is 10·2 kN/m^3.

What is the minimum pull required to be applied to a chain attached to the centre of the base so that the buoy remains vertical ?

Solution (i) Refer Fig. 3.16a,

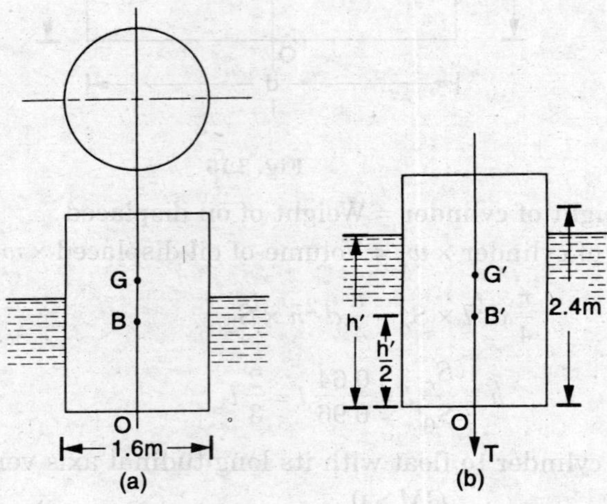

Fig. 3.16

$H = 2{\cdot}4$ m $\quad d = 1{\cdot}6$ m

$W = 20$ kN $\quad w = 10{\cdot}2$ kN/m^3

Volume of water displaced,

$$V = \frac{W}{w} = \frac{20}{10{\cdot}2} = 1{\cdot}961 \text{ m}^3$$

Depth of immersion,

$$h = \frac{\text{Volume}}{\text{Area}} = \frac{1{\cdot}961}{(\pi/4) \times 1{\cdot}6^2} = 0{\cdot}975 \text{ m}$$

$$GM = \frac{I}{V} - BG$$

$$= \frac{\frac{\pi}{64} \times 1{\cdot}6^4}{1{\cdot}961} - (OG - OB)$$

$$= \frac{0{\cdot}322}{1{\cdot}961} - \left(\frac{2{\cdot}4}{4} - \frac{0{\cdot}975}{2}\right)$$

$$= 0{\cdot}064 - 0{\cdot}712$$

$$= -\,0{\cdot}648$$

This shows that M lies below G. Therefore, it is unstable.

(ii) Let T be the pull (Fig. 3.16b),

$$W' = A \times h' \times w_s$$

$$20 + T = \frac{\pi}{4} \times 1{\cdot}6^2 \times h' \times 10{\cdot}2$$

$$h' = \frac{20 + T}{20{\cdot}51}$$

The combined centre of gravity due to the weight of the cylinder and the tension T in the chain can be found by taking moments about the point O,

$$OG' = \frac{20 \times 1{\cdot}2 + T \times 0}{20 + T} = \frac{24}{20 + T}$$

$$GM = \frac{I}{V} - B'G'$$

$$= \frac{\frac{\pi}{64} d^4}{\frac{\pi}{4} d^2 h'} - (OG' - OB')$$

$$= \frac{d^2}{16} \frac{1}{h'} - \left[OG' - \frac{1}{2} h'\right]$$

$$= \frac{1{\cdot}6^2}{16} \cdot \frac{20{\cdot}51}{20 + T} - \left[\frac{24}{20 + T} - \frac{1}{2}\left(\frac{20 + T}{20{\cdot}51}\right)\right]$$

$$= \frac{3{\cdot}28}{(20 + T)} - \left(\frac{24}{20 + T} - \frac{20 + T}{41{\cdot}02}\right)$$

To keep the buoy vertical, $GM > 0$,

$$\frac{134{\cdot}6 - 984{\cdot}5 + (20 + T)^2}{41{\cdot}02\,(20 + T)} > 0$$

or $\quad (20 + T)^2 - 849{\cdot}9 > 0$

or $\quad 20 + T > \sqrt{849{\cdot}9}$

$$> 29{\cdot}15$$

$$T > 9{\cdot}15 \text{ kN}$$

Example 3.12 A cone of specific gravity S_c floats in a liquid of specific gravity S_l with its apex downwards. If 2α is the apex angle of the cone, H the vertical height and h the depth of immersioin, show that for stable equilibrium

(i) $\sec^2\alpha > \dfrac{H}{h}$

(ii) $\sec^2\alpha > \left(\dfrac{S_l}{S_c}\right)^{1/3}$

(iii) Find the value of the least apex angle for stable equilibrium if the specific gravities of the cone and the liquid are 0·6 and 0·9 respectively.

Solution Fig. 3.17,

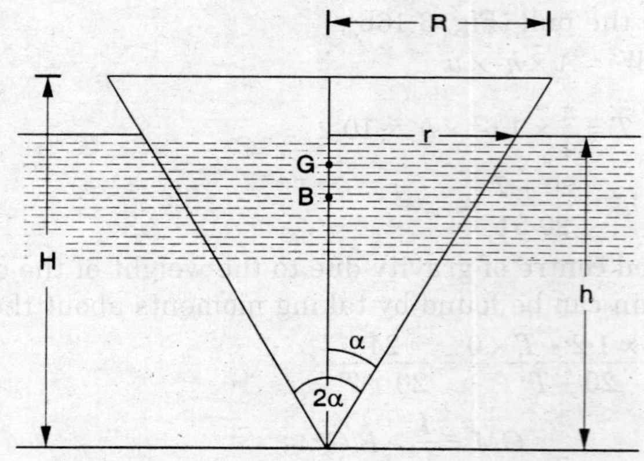

Fig. 3.17

Let R = radius of the cone

r = radius of the cone at the water surface

For stable equilibrium, $GM > 0$

$$\frac{I}{V} - BG > 0$$

$$\frac{\dfrac{\pi}{4}\,.\,r^4}{\dfrac{1}{3}\,\pi r^2\,.\,h} - (OG - OB) > 0$$

$$\frac{3}{4}\frac{r^2}{h} - \left(\frac{3}{4}H - \frac{3}{4}h\right) > 0$$

$$\frac{3}{4}\frac{(h \tan \alpha)^2}{h} - \frac{3}{4}(H - h) > 0$$

$$h \tan^2\alpha - H + h > 0$$

$$h (\tan^2\alpha + 1) > 0$$

$$h \sec^2\alpha > H$$

$$\sec^2\alpha > \frac{H}{h} \qquad \text{...(i)}$$

Weight of cone = Weight of liquid displaced

$$w_c \times \frac{1}{3}\pi R^2 H = w_l \times \frac{1}{3}\pi r^2 h$$

$$\frac{(w_c/w_w)}{(w_l/w_w)} . R^2 H = r^2 h$$

$$\frac{S_c}{S_l}(H^2 \tan^2\alpha) H = (h \tan^2\alpha) h$$

$$\left(\frac{H}{h}\right)^3 = \frac{S_l}{S_c}$$

$$\frac{H}{h} = \left(\frac{S_l}{S_c}\right)^{1/3}$$

From (i), $\sec^2\alpha > \frac{H}{h}$

or $\sec^2\alpha > \left(\frac{S_l}{S_c}\right)^{1/3}$

$$\sec^2\alpha > \left(\frac{0{\cdot}9}{0{\cdot}6}\right)^{1/3}$$

$$\sec^2\alpha > 1{\cdot}145$$

$$\sec \alpha > 1{\cdot}07$$

$$\cos \alpha > \frac{1}{1{\cdot}07} = 0{\cdot}935$$

$$\alpha > 20{\cdot}83°$$

$$2\alpha > 41{\cdot}66°$$

Least apex angle = 41·66°

Example 3.13 A cone of radius R, height H and specific gravity S floats in water with its apex downwards. Show that for stable equilibrium of the cone

$$H < \sqrt{\frac{R^2 S^{1/3}}{1 - S^{1/3}}}$$

Solution From Example 3.12,

$$\sec^2\alpha > \frac{1}{S^{1/3}} \qquad (S_l = 1, S_c = S)$$

or $$1 + \tan^2\alpha > \frac{1}{S^{1/3}}$$

or $$\tan^2\alpha > \left(\frac{1}{S^{1/3}} - 1\right)$$

$$> \frac{1 - S^{1/3}}{S^{1/3}}$$

$$\frac{R^2}{H^2} > \frac{1 - S^{1/3}}{S^{1/3}}$$

$$\frac{R^2 S^{1/3}}{1 - S^{1/3}} > H^2$$

or $$H < \sqrt{\frac{R^2 S^{1/3}}{1 - S^{1/3}}}$$

3.4 EXPERIMENTAL METHOD FOR METACENTRIC HEIGHT

Let W = weight of the floating ship including w

w = weight placed on the centre of deck of the ship.

Let G and B, be the centre of gravity and centre of buoyancy respectively of the ship as shown in Fig. 3.18. Now, the weight w is moved across the deck of the ship through a distance x. Let G' be the new position of the centre of gravity due to the shifting of this weight. Also let centre of buoyancy shift to B' due to the change in the geometrical form of the fluid displaced. The ship is tilted through an angle θ which can be measured by means of a plumb line and a horizontal scale.

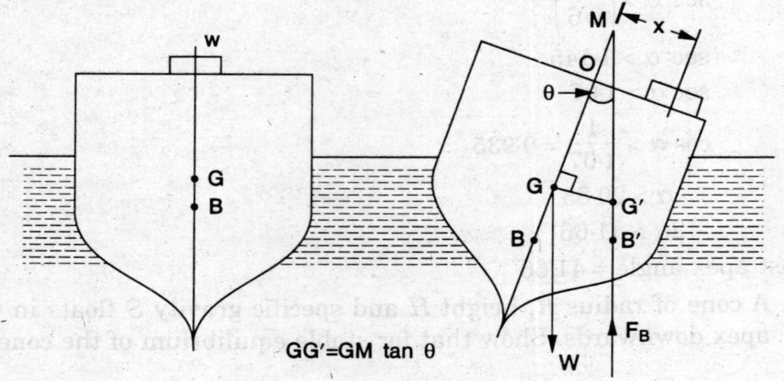

Fig. 3.18

Under equilibrium, the moment due to the movement of weight w on the deck must be equal to the moment of the total weight W due to shift of G,

$$w \,.\, x = W \,.\, GG'$$
$$= W \,.\, GM \,.\, \tan\theta$$

or $$GM = \frac{w \,.\, x}{W \,.\, \tan\theta} \approx \frac{w \,.\, x}{W \,.\, \theta} \quad \text{(For small angles, } \tan\theta \approx \theta) \quad ...(3.6)$$

Example 3.14 The weight of a ship 80 m long and 9 m broad is 18 MN. On moving a weight of 250 kN across the deck through a distance of 5 m, the ship is tilted through 5°. The moment of inertia of the ship at water line about its longitudinal axis is 70% of the moment of inertia of the curcumscribing rectangle. If the centre of buoyancy is 2·5 m below the water line, find the metacentric height and the position of centre of gravity of the ship. Specific weight of sea water is 10·2 kN/m^3.

Solution $W = 18$ MN $= 18000$ kN $\quad w = 250$ kN

$l = 80$ m $\quad b = 9$ m

$x = 5$ m $\quad \theta = 5°$

$OB = 2{\cdot}5$ m

$$GM = \frac{w \,.\, x}{W \,.\, \tan\theta}$$
$$= \frac{250 \times 5}{18000 \times \tan 5°}$$
$$= 0{\cdot}794 \text{ m}$$

Refer Fig. 3.19,

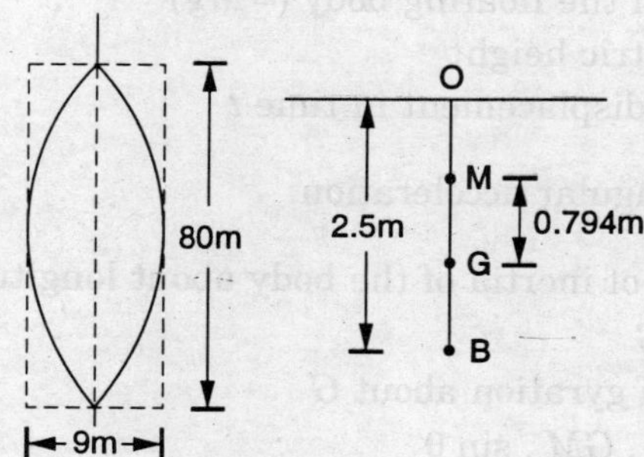

Fig. 3.19

$$I = 0{\cdot}7 \times \left(\frac{lb^3}{12}\right) = 0{\cdot}7 \times \frac{80 \times 9^3}{12} = 3402 \text{ m}^4$$

and V = Volume of water displaced by ship

$$= \frac{\text{Weight of the ship}}{\text{Specific weight of water}} = \frac{18000}{10{\cdot}2} = 1764{\cdot}7 \text{ m}^3$$

$$GM = \frac{I}{V} - BG$$

$$0{\cdot}794 = \frac{3402}{1764{\cdot}7} - (OB - OG)$$
$$= 1{\cdot}928 - (2{\cdot}5 - OG)$$
$$OG = 1{\cdot}366 \text{ m}$$

3.5 OSCILLATION OF A FLOATING BODY

If the overturning couple (which tilts a body through a small angle θ) is suddenly removed, a floating body may oscillate about its metacentre similar to the motion of a simple pendulum (Fig. 3.20).

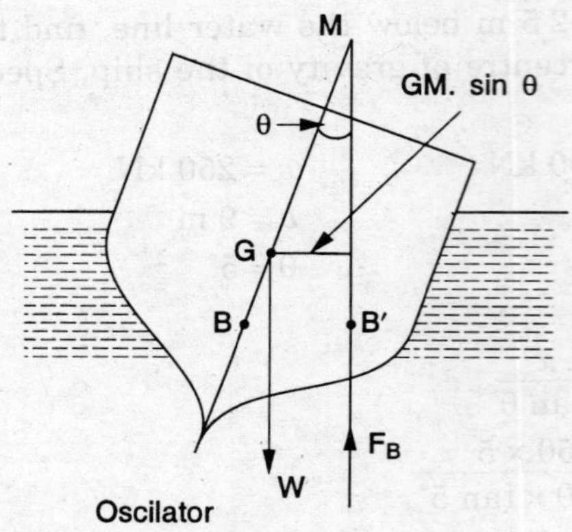

Fig. 3.20

Let W = weight of the floating body (= Mg)

GM = metacentric height

θ = angle of displacement in time t

$\alpha = \dfrac{d^2\theta}{dt^2}$ = angular acceleration

I_g = moment of inertia of the body about longitudinal axis through centre of gravity G

k = radius of gyration about G

Rightening moment = $W \,.\, GM \,.\, \sin\theta$

Inertia torque = $-I_g \,.\, \alpha$

In the expression for moment of inertia, axis of oscillation is assumed to pass through G. This is approximately valid if metacentric height is small. Negative sign in the expression for inertia torque indicates that the restoring couple tries to decrease the angle θ.

The Rightening moment = Inertia torque

or $$W \,.\, GM \,.\, \sin\theta = -I_g \,.\, \alpha$$

or $$W \,.\, GM \,.\, \theta = -\frac{W}{g} k^2 \times \frac{d^2\theta}{dt^2}$$ (if θ is small, $\sin\theta \approx \theta$)

or $$\frac{d^2\theta}{dt^2} + \frac{GM \cdot g \cdot \theta}{k^2} = 0 \qquad ...(3.7)$$

and its solution is,

$$\theta = A \sin \omega t + B \cos \omega t$$

where A and B are the constants of integration. The time period T can be found as follows :

Let at $t = 0, \quad \theta = 0°$

$$0 = 0 + B$$

or $B = 0$

Also, when $t = T/2$, $\theta = 0°$

$$0 = A \sin \frac{\omega T}{2} + 0$$

But A cannot be zero.

$$\therefore \quad \sin \frac{\omega T}{2} = 0$$

or $$\omega \frac{T}{2} = 0 \text{ or } \pi$$

If $T = 0$, there is no oscillation, thus T cannot be zero.

$$\therefore \quad \sqrt{\frac{GM \cdot g}{k^2}} \cdot \frac{T}{2} = \pi$$

or $$T = 2\pi \sqrt{\frac{k^2}{GM \cdot g}}$$

Also $$\omega = \frac{2\pi}{T} = \sqrt{\frac{GM \cdot g}{k^2}} \qquad ...(3.8)$$

Alternately, Eq. 3.7 may be compared with a standard equation of simple harmonic motion,

i.e. $$\frac{d^2\theta}{dt^2} + \omega^2\theta = 0 \text{ where } \omega = \sqrt{\frac{GM \cdot g}{k^2}} \text{ and then } T = \frac{2\pi}{\omega}$$

$$= 2\pi \sqrt{\frac{k^2}{GM \cdot g}}$$

From this equation time period of oscillation of a floating body can be calculated provided metacentric height and the radius of gyration are known.

It is to be noted that an increase in the metacentric height increases the stability of a ship, but reduces the time period of oscillation. A reduced value of time period is undesirable because it is uncomfortable for the passengers and can harm the ship structure. Thus a compromise in this respect has to be made.

Example 3.15 A ship floating in the sea water and weighing 28000 kN has its centre of buoyancy 1·4 m below its centre of gravity. The moment of inertia of the ship at the water line about fore and aft axis is 9200 m^4. If the radius of gyration of the ship is 3 m and the specific weight of the sea water is 10·2 kN/m^3, determine the time period of rolling of the ship.

Solution $I = 9200 \text{ m}^4$ $W = 28000 \text{ kN}$

$k = 3 \text{ m}$ $w = 10{\cdot}2 \text{ kN/m}^3$

We have,

$$T = 2\pi \sqrt{\frac{k^2}{GM \, . \, g}}$$

where $GM = \frac{I}{V} - BG$

$$V = \frac{\text{Weight of the ship}}{\text{Specific weight of sea water}} = \frac{28000}{10{\cdot}2} = 2745 \text{ m}^2$$

$$GM = \frac{9200}{2745} - 1{\cdot}4 = 3{\cdot}351 - 1{\cdot}4 = 1{\cdot}951 \text{ m}$$

and $$T = 2\pi \sqrt{\frac{3^2}{1{\cdot}951 \times 9{\cdot}81}}$$

$$= 4{\cdot}308 \text{ s.}$$

PROBLEMS

1. State and prove the principle of Archimedes.
2. What are buoyancy and centre of buoyancy ?
3. What is buoyant force ? How is it caused ?
4. What are the conditions of equilibrium of a freely floating body ?
5. Define the terms metacentre and metacentric height.
6. Deduce an expression for the metacentric height of a floating body.
7. For floating bodies derive the expression

$$BM = \frac{I}{V}$$

where BM = distance between the metacentre and centre of buoyancy
I = moment of inertia of liquid line area about longitudinal axis
V = volume of the liquid displaced by the body

8. State the conditions of equilibrium of a floating body and a submerged body.
9. How would you determine the metacentric height experimentally ?
10. What do you mean by rolling of a ship ? How is the period of rolling or oscillation of a ship determined ?
11. The length of a wooden cylinder is two times its diameter. If it floats in water with its axis vertical, find whether the equilibrium is stable. The specific gravity of the wood is 0·6. (Unstable)
12. A wooden log 1 m in diameter and 4 m long floats in water. The specific gravity of the log is 0·6. Calculate the depth of log in water. (578 mm)

13. It is required to float a wooden cylinder of 400 mm diameter and of specific gravity 0·6 in an oil of specific gravity 0·8. Determine the maximum length of the cylinder so that the cylinder may float vertically. (326·5 mm)

14. A rectangular pontoon weighing 700 kN is 10 m long, 7 m wide and 2·2 m deep. A boiler of 5 m diameter and weighing 650 kN is carried on the upper deck of the pontoon. The centre of gravity of each lies at its respective geometric centre and in the same vertical line. The whole unit floats in the sea water of specific gravity 1·025. Determine the metacentric height. (250 mm)

15. A right solid cone with apex angle 60° floats in a liquid with its apex downwards. The specific gravity of the cone is S with respect to liquid in which it floats. What range of S is compatible with the stable equilibrium of the cone ? (0·422 to 1)

16. A solid cylinder is 200 mm in diameter and 1 m long. The base of the cylinder is 25 mm thick and made of a material of specific gravity 8. The specific gravity of the remaining portion of the cylinder is 0·5. Can the cylinder float vertically in water ? If not, what will be the maximum permissible length for stable equilibrium. (Unstable, 907 mm)

17. A metallic sphere is completely immersed in a water reservoir and is chained to the bottom. The diameter of the sphere is 1·5 m and the chain has a tension of 7·84 kN. What will be the weight of the sphere when taken out of the reservoir into the air. (9·5 kN)

18. A body weighs 30 N and 15 N when weighed under submerged conditions in liquid of specific gravities 0·9 and 1·4 respectively. Find its weight in air. (57 N)

19. A solid cylinder of diameter D and length L has a relative density S. Obtain an expression for the maximum value of L/D for which the cylinder can float in water in stable equilibrium with its axis vertical. $\left(\frac{L}{D} < \frac{1}{\sqrt{8S\,(1-S)}}\right)$

20. A ship floating in the sea water and weighing 1×10^5 kN has its centre of buoyancy 2·4 m below its centre of gravity. The moment of inertia of the ship at the water line about fore and aft axis is 32000 m^4. Specific gravity of sea water is 1·025. If the time period of rolling of the ship is 20 seconds, find its radius of gyration. (9 m)

21. The weight of a ship 56 m long and 8 m broad is 12 MN. On moving a weight of 150 kN across the deck through a distance of 5 m, the ship is tilted through 3°. The second moment of the area of the water line section about its fore and aft axis is 75 percent of that of the circumscribing rectangle. If the centre of buoyancy of the ship is 1·6 m below the water line, determine the metacentric height and the position of centre of gravity of the ship. Mass density of the sea water is 1025 kg/m^3. (1·2 m, 1·3 m)

Chapter 4

FLUID KINEMATICS

4.1 INTRODUCTION

Fluid kinematics is the study of fluid motion without the consideration of any forces or energies which cause the motion. In the movement of the solids, it is sufficient to measure the velocity of the body as a whole. But fluids rarely follow a set pattern during the flow at different points of observation. However, in fluid flows the velocity describes the speed as well as the direction of the flow at a given point and instant. Thus velocity is a vector quantity since it has magnitude and direction. A vector is represented by an arrow, the length of which is proportional to the magnitude of the velocity and the orientation of the arrow indicates the direction of the flow.

Velocity, a vector quantity, may be resolved in any desired directions such as x, y and z in the rectilinear system. Similarly, resultant can be obtained by combining the velocity components at a given point.

Thus if u, v and w are the velocity components in the x-, y- and z-directions respectively, the velocity V is obtained vectorially,

$$\overline{V} = ui + vj + wk$$

where i, j and k are the unit vectors in the directions of coordinate axes.

4.2 LANGRANGIAN AND EULERIAN METHOD

As stated earlier, fluids differ from solids in that the relative positions of the particles vary from time to time. At any instant of time, each fluid particle has its own velocity and acceleration at a point which may differ with space and time. Thus it is necessary to observe the motion of fluid particles at different points in space and at successive instants of time to completely describe a fluid motion.

In **Langrangian** method, an individual particle is followed during its course of motion through space and its behaviour is observed. In the rectilinear Cartesian coordinate system, a, b and c are adopted as independent variables. If the initial position of a fluid particle is described by $x_0 = a$, $y_0 = b$, $z_0 = c$ at the given initial time $t_0 = t$, then the exact spatial position (x, y, z) of the fluid particle at different times in terms of its initial position is given by,

$$x = f_1\,(a, b, c, t)$$
$$y = f_2\,(a, b, c, t)$$
$$z = f_3\,(a, b, c, t) \qquad \text{...(4.1)}$$

From these equations, the instantaneous velocity components and acceleration components at any instant of time can be determined by taking derivatives with respect to time.

$$u = \frac{dx}{dt} \qquad a_x = \frac{du}{dt} = \frac{d^2x}{dt^2}$$
$$v = \frac{dy}{dt} \qquad a_y = \frac{dv}{dt} = \frac{d^2y}{dt^2}$$
$$w = \frac{dz}{dt} \qquad a_z = \frac{dw}{dt} = \frac{d^2z}{dt^2} \qquad \text{...(4.2)}$$

where u, v and w are the x, y and z components of velocity and a_x, a_y and a_z are the x, y and z components of acceleration.

In **Eulerian** method, an individual particle is not followed, instead the behaviour of the fluid particles at various points is studied. x, y, z and t are chosen as independent variables. Then, the velocity components of the fluid motion at various points (x, y, z) in the space at any given time t, are given by,

$$u = F_1\,(x, y, z, t)$$
$$v = F_2\,(x, y, z, t)$$
$$w = F_3\,(x, y, z, t) \qquad \text{...(4.3)}$$

It can be observed that the Langrangian method provides the complete description of paths of fluid particles. However, it is very difficult to solve the equations mathematically whereas in the Eulerian method, it may be done with much simplicity. Fortunately, in almost all the fluid mechanics problems, the general state of motion of various points in the flow field are of greater practical significance rather than the knowledge of the behaviour of each particle. Hence most of the problems of the fluid mechanics are analysed by the Eulerian method.

4.3 TYPES OF FLOW LINES

Streamlines

A **Streamline** is defined as a continuous line drawn through the fluid in such a way that its direction at every point indicates the direction of the velocity vector at the point at a particular instant. This implies that a tangent at any point on a streamline will be in the direction of flow at that instant. Since a fluid particle always travels tangential to a streamline, there can be no flow across a streamline.

Two streamlines for a flow pattern in the xy plane are shown in Fig. 4.1. At any point, the tangent to a streamline is in the direction of the velocity vector V

If at a point, the components of the differential displacement ds in time dt are (Fig. 4.2)

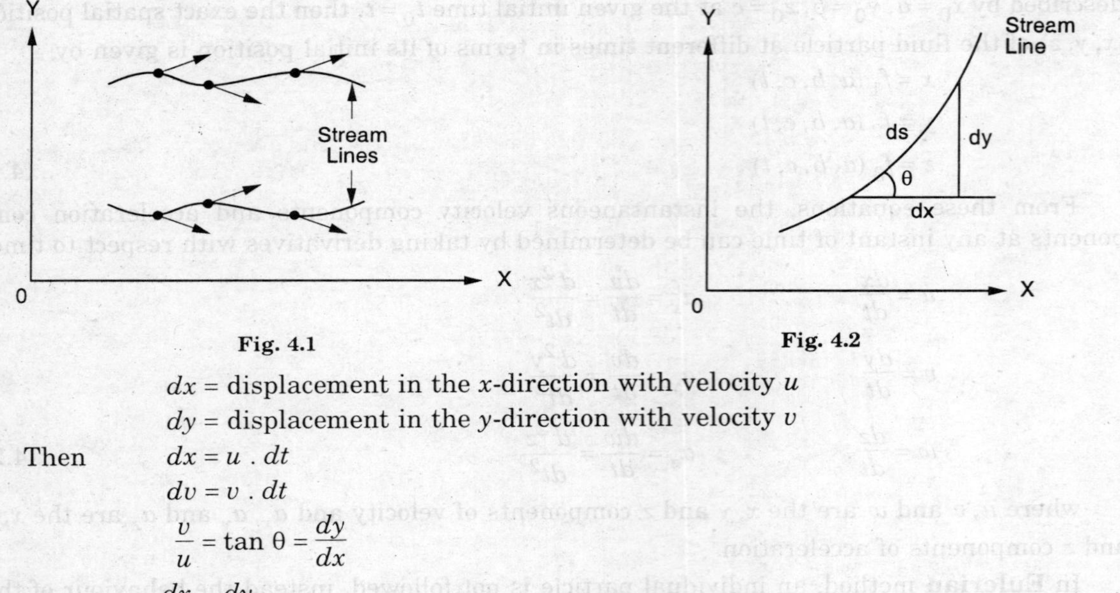

Fig. 4.1 Fig. 4.2

dx = displacement in the x-direction with velocity u

dy = displacement in the y-direction with velocity v

Then $dx = u \cdot dt$

$dv = v \cdot dt$

$$\frac{v}{u} = \tan\theta = \frac{dy}{dx}$$

or $$\frac{dx}{u} = \frac{dy}{v} \qquad ...(4.4)$$

which is the differential equation for streamlines in the xy plane. Similarly, in a 3-D flow, we have

$$\frac{dx}{u} = \frac{dy}{v} = \frac{dz}{w} \qquad ...(4.5)$$

It is possible to draw infinite number of streamlines in a flow field. However, for practical purposes a small number of streamlines may be sufficient. Usually, streamlines are drawn in such a way that the quantity of flow is same through any two adjacent streamlines. For a coverging passage, streamlines also coverage indicating that the flow is constant between any two adjacent streamlines.

A cylindrical passage or tube formed by a bundle of streamlines in a flow field is called a **stream tube** as shown in Fig. 4.3. The surface of the stream tube is known as the **stream surface.**

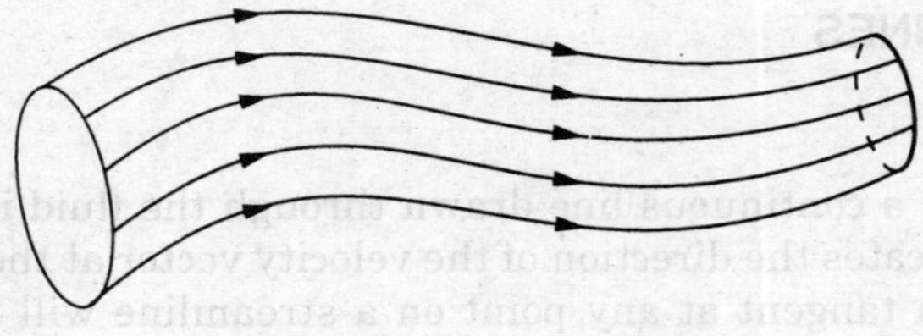

Fig. 4.3

As a stream tube is bounded by streamlines, no fluid can pass through the tube wall. Thus a streamtube behaves like a solid tube. In complicated flow problems, it is always convenient to divide the whole flow field into a large number of stream tubes to get a clear picture of the actual flow pattern. Such a technique is used in plotting flow nets (to be discussed later). In simpler flow problems, the solid boundaries may serve as the periphery of a stream tube, *e.g.* the flow of fluids in conduits.

Streaklines

A **streakline** is a line made by a series of fluid particles that have passed through a fixed point in the flow field *e.g.* if a dye is injected into a flowing field, the dye trails are called streaklines (Fig. 4.4a). The smoke emanating from a chimney also constitute streaklines (Fig. 4.4b).

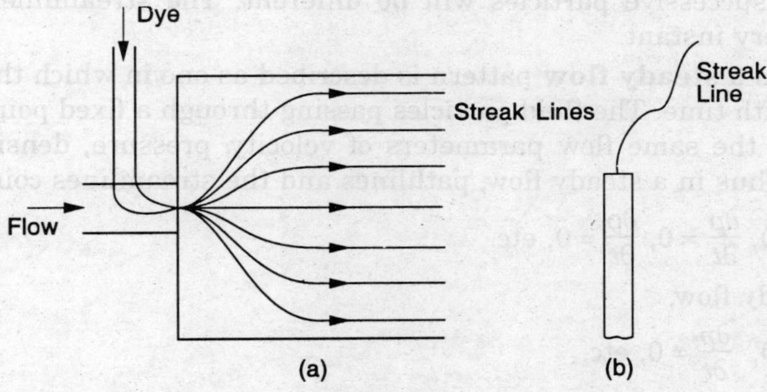

Fig. 4.4

Pathlines

A **pathline** is defined as the path traced out by a fluid particle in a flow field over a period of time. A path line can intersect itself as shown in Fig. 4.5.

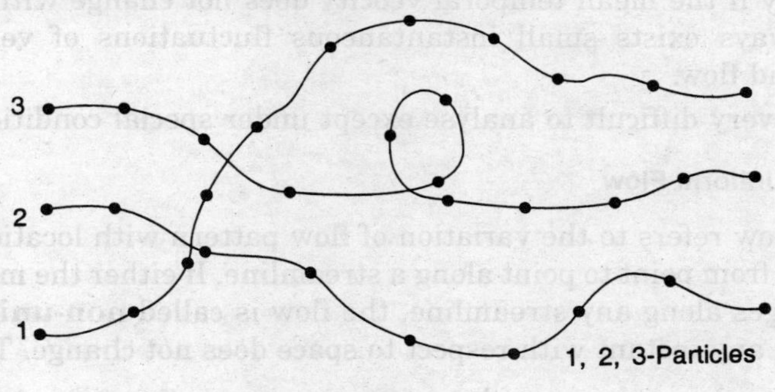

Fig. 4.5

4.4 TYPES OF FLOW

The flows are classified as :

1. Steady and unsteady flow
2. Uniform and non-uniform flow
3. Compressible and incompressible flow
4. Ideal and real flow
5. Rotational and irrotational flow

6. One-, two- and three-dimensional flow
7. Laminar and turbulent flow

1. Steady and Unsteady Flow

Unsteadiness of flow refers to the variation of the flow pattern with time *i.e.* in the **unsteady flow**, the velocity at any fixed point changes from instant to instant. For example, the flow pattern made by clouds of smoke that comes out of a chimney, will change continuously in form with time. This can be verified by taking two successive photographs of the same region. Pathlines traced by successive particles will be different. The streamlines will also have differing patterns every instant.

On the other hand a **steady flow** pattern is described as one in which there is no change in the flow pattern with time. The fluid particles passing through a fixed point have the same flow pattern or have the same flow parameters of velocity, pressure, density, temperature, surface tension etc. Thus in a steady flow, pathlines and the streamlines coincide and

$$\frac{\partial V}{\partial t} = 0, \frac{\partial p}{\partial t} = 0, \frac{\partial \rho}{\partial t} = 0, \text{ etc.}$$

and in an unsteady flow,

$$\frac{\partial V}{\partial t} \neq 0, \quad \frac{\partial p}{\partial t} \neq 0, \text{ etc.}$$

In experience, truly steady flow is rarely encountered. Usually, a seemingly steady flow of a fluid is found to fluctuate incessantly about a mean steady motion. However, over a certain finite interval of time, the mean of velocity fluctuations is zero. Such motions in which the temporal (of time) mean values of the velocities do not change with time may be considered steady, though in reality the velocities are fluctuating all the time. Thus a turbulent flow may be considered steady if the mean temporal veocity does not change with time in spite of the fact that there always exists small instantaneous fluctuations of velocity due to irratic behaviour of the fluid flow.

Usteady flow is very difficult to analyse except under special conditions.

2. Uniform and Non-Uniform Flow

Non-uniformity of flow refers to the variation of flow pattern with location *i.e.* the velocity at any instant changes from point to point along a streamline. If either the magnitude or direction of the velocity changes along any streamline, the flow is called **non-uniform.** In a **uniform flow,** the velocity at any instant with respect to space does not change. Truly uniform motion is seldom observed.

The streamlines in the vicinity of a solid surface always conform to the outline of boundary surface and the solid surface is considered to consist of streamlines. For example, flow around airfoils, inside a conduit etc. In the equation form, a uniform flow has $(\partial V/\partial s) = 0$ and in a non-uniform flow $(\partial V/\partial s) \neq 0$.

Thus, it can be observed that the variation of flow pattern with time is the unsteadiness of flow whereas the variation with location is the non-uniformity of flow. Both types of flows can coexist independent of each other.

Steady uniform flow : Flow through a long pipe of constant cross-sectional area at a constant rate (Fig. 4.6).

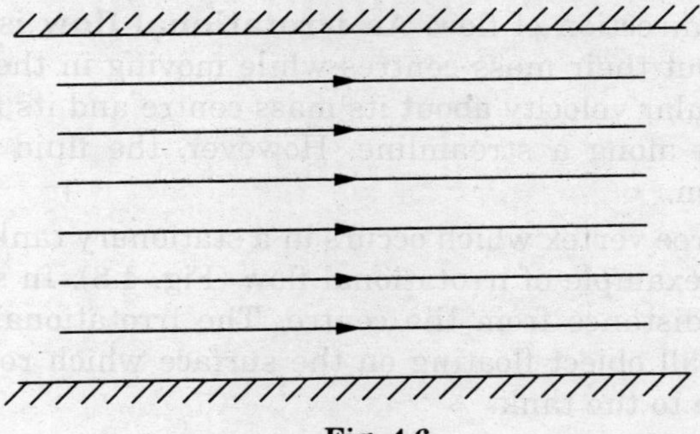

Fig. 4.6

Steady non-uniform flow : Flow through a tapering pipe of increasing or decreasing cross-section at constant rate (Fig. 4.7).

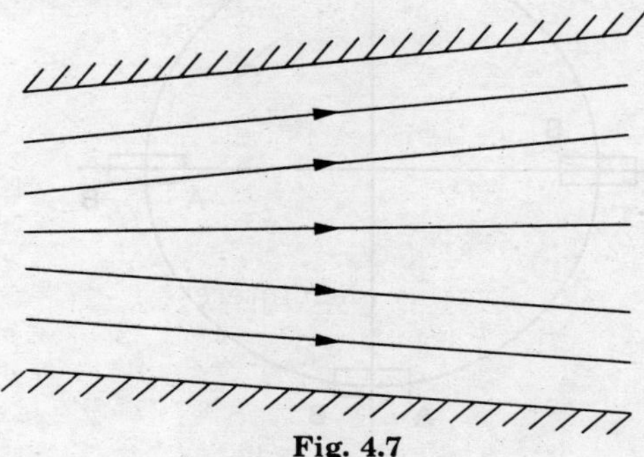

Fig. 4.7

Unsteady uniform flow : Varying flow rate through a pipe line of constant cross-sectional area.

Unsteady non-uniform flow : Varying flow rate through a pipe line of increasing or decreasing cross-sectional area.

3. Compressible and Incompressible Flow

Flow is said to be **incompressible** if the change in density is negligible with the variation in pressure and temperature. All liquid flows and gas flows at low velocities are considered incompressible flows. When the density change is considerable, the flow may be termed as **compressible.** Gas flow above 60-90 m/s may be considered as a compressible flow.

4. Ideal and Real Flow

An **ideal flow** is the flow of an inviscid (non-viscous) fluid. No shear stress exists between two adjacent layers or between the fluid and the boundary. The velocity gradient perpendicular to the direction of flow is zero.

On the other hand, the flow of real (viscous) fluids is called **real flow.** Shear stresses exists between two layers which retard the motion of one layer relative to that of the other.

5. Rotational and Irrotational Flow

A flow is called a **rotational flow** if the fluid elements also rotate about their mass centres apart from moving in the direction of flow. An **irrotational flow** is one in which the fluid elements do not rotate about their mass centres while moving in the direction of flow. Each fluid element has zero angular velocity about its mass centre and its net rotation is zero as it moves from point to point along a streamline. However, the fluid elements may undergo distortion during the motion.

A whirlpool motion or free vertex which occurs in a stationary tank while the liquid drains out from the bottom is an example of irrotational flow (Fig. 4.8). In such a flow, the velocity varies inversely with the distance from the centre. The irrotationality of the flow may be verified by observing a small object floating on the surface which rotates in a circular path but does not rotate relative to the tank.

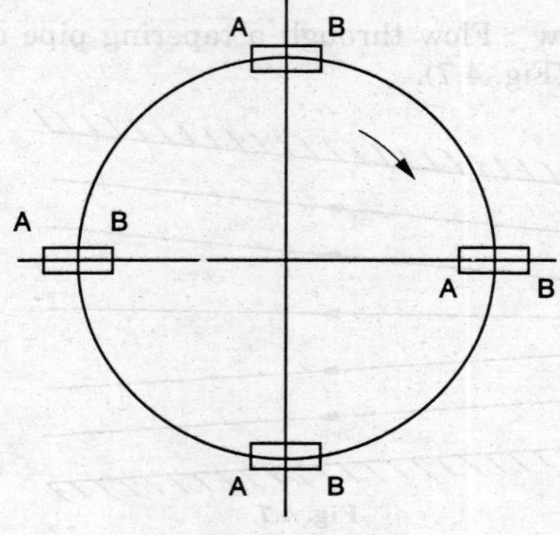

Fig. 4.8

A liquid in a rotating tank known as forced vortex is an example of rotating flow (Fig. 4.9). In this case, the velocity is directly proportional to the distance from the centre. The rotation of the fluid element may be verified by observing an object floating on the surface which rotates in the same manner as the tank itself.

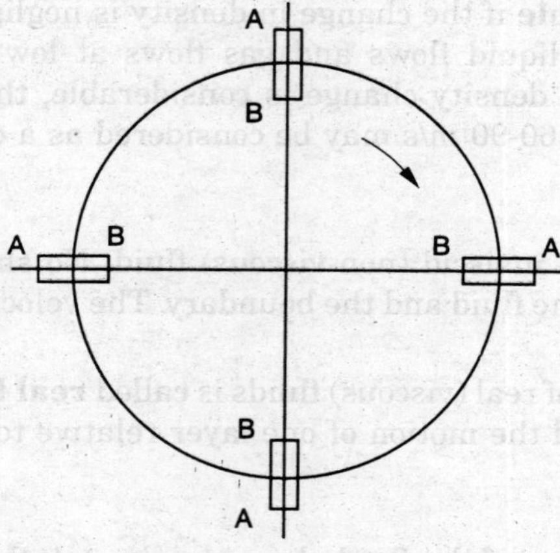

Fig. 4.9

Consider a flow through a pipe of constant cross-sectional area (Fig. 4.10). If the flow is of an ideal (non-viscous) fluid, the velocity of the fluid would be same throughout the cross-section of the pipe whereas if the fluid is the real one (viscous) the velocity is maximum at the centre and zero at the pipe surface. A fluid particle originally assumed to be a square will undergo no change in its shape in the first case while traveling through the pipe and the flow is said to be irrotational (Fig. 4.10a). In the second case the fluid particle undergoes a change in its shape due to the variation in velocity. One of its median (axis) lines has been rotated, the other remaining in its position (Fig 4.10b). Net effect is that the fluid particle has rotated about its mass centre and the flow is considered to be rotational.

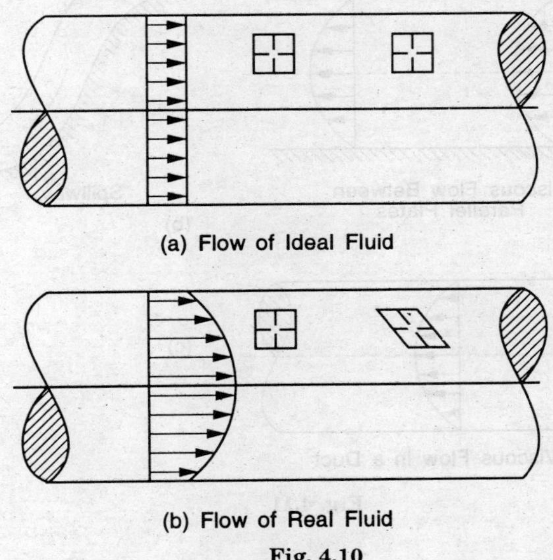

(a) Flow of Ideal Fluid

(b) Flow of Real Fluid

Fig. 4.10

6. One-, Two-, and Three-dimensional Flow

In a fluid flow, the flow and fluid parameters such as velocity, pressure, temperature, density and viscosity may vary in one, two or in all the three directions.

If these parameters vary only in one direction, the flow is said to be a function of only one of the coordinate directions and time and the flow is known as **one-dimensional flow.** In a steady flow as the flow parameters do not change with time, the flow is a function of one coordinate direction only. In such a flow, all fluid parameters are constant throughout a cross-section normal to the flow and the streamlines are straight and parallel. At a different cross-section these may differ, but again throughout that cross-section, these have to be constant. In other words, the parameters are a function of one coordinate direction. Flow of ideal fluid in a duct is an example of one-dimensional flow (Fig. 4.11a). Mathematically,

$$u = f(x, t)$$

A **two-dimensional flow** is one in which the flow parameters vary in two directions and thus is a function of the space coordinates and time. In a steady flow, the flow parameters are functions of two coordinate directions only. Examples of a two-dimensional flow are viscous flow between parallel plates, flow over a spillway etc. (Fig. 4.11b). Mathematically.

$$u = f_1(x, y, t)$$

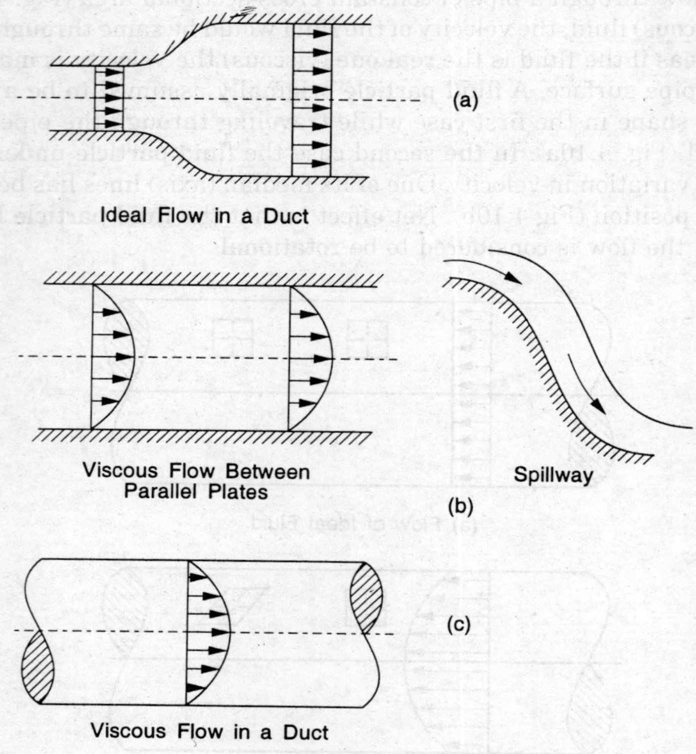

Fig. 4.11

$$v = f_2(x, y, t)$$

In a **three-dimensional flow,** flow parameters change in all the three coordinate directions and thus are functions of three space coordinates and time in an unsteady flow. In a steady flow, parameters are functions of three coordinate directions only. Thus gradients of the fluid or flow parameters exist in all the three directions. Flow in a stream, flow at inlet to a nozzle, and flow within fluid machines are examples of three-dimensional flows. Fig. 4.11c illustrates a three-dimensional flow. Mathematicaly,

$$u = f_1(x, y, z, t)$$
$$v = f_2(x, y, z, t)$$
$$w = f_3(x, y, z, t)$$

It is to be noted that the flow of only ideal fluids between finite boundaries is one-dimensional since the velocity distribution is considered uniform. In a real fluid flow in a duct, the velocity distribution across any cross-section is non-uniform due to viscous effects and hence the flow is three-dimensional. However, if average values across any cross-section are considered, it may be called one dimensional.

7. Laminar and Turbulent Flow

A **laminar flow** is one in which fluid layers flow alongside one another at the same or at slightly different velocities. If a non soluble dye is injected into a laminar flow, it appears as a single thread which does not disperse throughout the field.

In a **turbulent flow,** there is a mixing of the fluid and there are small velocity fluctuations, parallel and transverse to the mean velocity. The injected dye will quickly disperse throughout the fluid. In such flows, eddies or vortices of different sizes and shapes are present. Fig. 4.12a and b show the laminar and turbulent flows respectively.

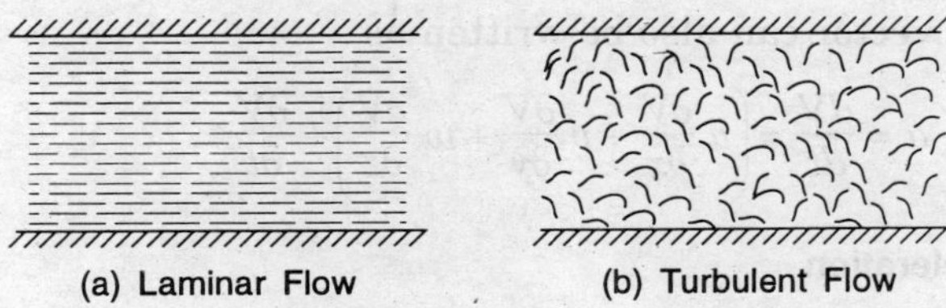

(a) Laminar Flow (b) Turbulent Flow

Fig. 4.12

If a cigarette is held in still air, the straight filaments of smoke rising for a few inches is an example of a laminar flow whereas the haphazard motion of the smoke above the straight filaments is an example of turbulent flow. In a laminar flow, usually, we have low velocities, small flow passages and highly viscous fluids whereas in a turbulent flow, high velocities, larger flow passages and low viscosity fluids are associated. In most of the flow problems, turbulent flow is much more common than laminar flow.

4.5 VELOCITY AND ACCELERATION

Let V be the resultant velocity at any point in a fluid flow with its components u, v and w being functions of space coordinates (in x-, y- and z-directions) and time (t). Mathematically,

$$u = f_1(x, y, z, t)$$

$$v = f_2(x, y, z, t)$$

$$w = f_3(x, y, z, t)$$

Then

$$V = ui + vj + wk \qquad \text{...(4.6)}$$

$$= \sqrt{u^2 + v^2 + w^2}$$

Let a denote the total acceleration with a_x, a_y and a_z being the respective components.

$$a_x = \frac{du}{dt} = \left(\frac{\partial u}{\partial x}\cdot\frac{\partial x}{\partial t} + \frac{\partial u}{\partial y}\cdot\frac{\partial y}{\partial t} + \frac{\partial u}{\partial z}\cdot\frac{\partial z}{\partial t}\right) + \frac{\partial u}{\partial t}$$

$$= \left(\frac{\partial u}{\partial x}\,.\,u + \frac{\partial u}{\partial y}\,.\,v + \frac{\partial u}{\partial z}\,.\,w\right) + \frac{\partial u}{\partial t}$$

$$= \left(u\frac{\partial u}{\partial x} + v\frac{\partial u}{\partial y} + w\frac{\partial u}{\partial z}\right) + \frac{\partial u}{\partial t}$$

Similarly,

$$a_y = \left(u\frac{\partial v}{\partial x} + v\frac{\partial v}{\partial y} + w\frac{\partial v}{\partial z}\right) + \frac{\partial v}{\partial t}$$

$$a_z = \left(u\frac{\partial w}{\partial x} + v\frac{\partial w}{\partial y} + w\frac{\partial w}{\partial z}\right) + \frac{\partial w}{\partial t} \qquad \text{...(4.7)}$$

Also total acceleration,

$$a = a_x\, i + a_y\, j + a_z\, k$$
$$= \sqrt{{a_x}^2 + {a_y}^2 + {a_z}^2} \quad \text{...(4.8)}$$

For steady flow, $\frac{\partial V}{\partial t} = 0$

or $\frac{\partial u}{\partial t} = 0, \frac{\partial v}{\partial t} = 0$ and $\frac{\partial w}{\partial t} = 0$

The acceleration vector can also be written as

$$a = \frac{dV}{dt} = \left(u\frac{\partial V}{\partial x} + v\frac{\partial V}{\partial y} + w\frac{\partial V}{\partial z} \right) + \frac{\partial V}{\partial t} \quad \text{...(4.9)}$$

Local (Temporal) Acceleration

The rate of change of velocity with respect to time is known as **local or temporal** acceleration.

$\frac{\partial u}{\partial t}, \frac{\partial v}{\partial t}$ and $\frac{\partial w}{\partial t}$ are the local accelerations.

Convective Acceleration

The rate of change of velocity with respect to position of the fluid particles in a fluid flow is known as **convective** acceleration. The terms in the parenthesis of the above equations are known as convective accelerations.

4.6 RATE OF FLOW (DISCHARGE)

The rate of flow or **discharge** is defined as the volume of a fluid flowing per second through a section of a pipe or a channel. It is denoted by Q (or q). Consider a pipe of cross-sectional area A, through which a fluid with an average velocity V is flowing. The discharge is given by

$$Q = \text{Area} \times \text{Average velocity}$$
$$= A \times V \quad \text{...(4.10)}$$

Units of discharge are m^3/s (also called **cumecs**). It is also expressed as litres per second (l/s).

1000 litres = 1 m^3/s

4.7 CONTINUITY EQUATION

An equation based on the principle of conservation of mass is known as **continuity equation.** For a fluid passing through a pipe, the mass of fluid passing across different sections is the same if no fluid is added or removed from the pipe.

Consider two cross-sections 1-1 and 2-2 of a pipe (Fig. 4.13)

Let V_1 = velocity at section 1-1

A_1 = area at section 1-1

ρ_1 = density at section 1-1

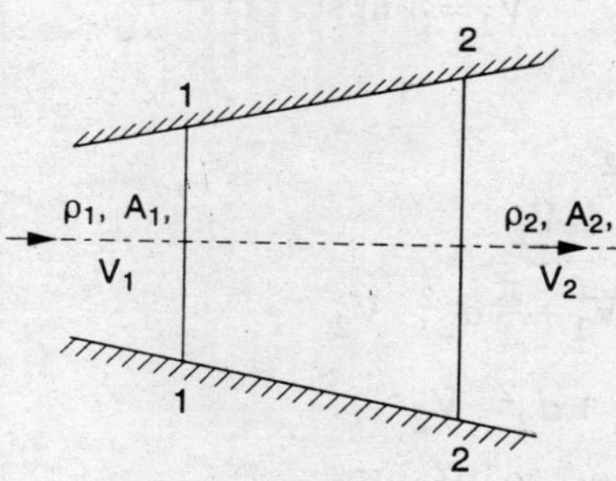

Fig. 4.13

V_2, A_2 and ρ_2 are the corresponding values at section 2-2. Mass flow rate at section 1-1

$$= \text{Density} \times \text{Volume flow rate}$$
$$= \text{Density} \times \text{Area} \times \text{Velocity}$$
$$= \rho_1 \,.\, A_1 \,.\, V_1$$

Mass flow rate at section 2-2 = $\rho_2 \,.\, A_2 \,.\, V_2$

According to law of conservation of mass, the mass flow rate at the two cross-sections must be equal.

i.e. $\rho_1 \,.\, A_1 \,.\, V_1 = \rho_2 \,.\, A_2 \,.\, V_2$ ...(4.11)

This is called **continuity equation** and is applicable to compressible as well as incompressible fluids. However, in case of incompressible fluids, $\rho_1 = \rho_2$,

Therefore, $A_1 \,.\, V_1 = A_2 \,.\, V_2$ ...(4.12)

Example 4.1 A pipe of enlarging cross-section has diameters 80 mm and 130 mm at sections 1-1 and 2-2 respectively. The velocity of water at section 1-1 is 8 m/s. Determine the velocity at section 2-2, the discharge and the mass flow rate.

Solution $d_1 = 0{\cdot}08$ m $\qquad V_1 = 8$ m/s

$d_2 = 0{\cdot}12$ m

$$A_1V_1 = A_2V_2$$

$$\frac{\pi}{4}d_1^2V_1 = \frac{\pi}{4}d_2^2V_2$$

$$V_2 = V_1\left(\frac{d_1}{d_2}\right)^2 = 8 \times \left(\frac{0{\cdot}08}{0{\cdot}12}\right)^2 = 3{\cdot}56 \text{ m/s}$$

$$Q = A_1V_1 = \frac{\pi}{4} \times (0{\cdot}08)^2 \times 8 = 0{\cdot}0402 \text{ m}^3\text{/s}$$

$$m = \rho \,.\, Q = 1000 \times 0{\cdot}0402 = 40{\cdot}2 \text{ kg/s}$$

Example 4.2 A main pipe 600 mm in diameter branches into two pipes of diameters 400 mm and 300 mm. The average velocity in the main pipe is 4 m/s. Determine the velocity in the 300 mm diameter pipe if the average velocity in the 400 mm diameter pipe is 3 m/s.

Solution $d = 0{\cdot}6$ m $\qquad V = 4$ m/s

$d_1 = 0{\cdot}4$ m $\qquad V_1 = 3$ m/s

$d_2 = 0{\cdot}3$ m

$$Q = Q_1 + Q_2$$

$$AV = A_1V_1 + A_2V_2$$

$$\frac{\pi}{4}d^2 \,.\, V = \frac{\pi}{4}d_1^2 \,.\, V_1 + \frac{\pi}{4}d_2^2 \,.\, V_2$$

$$d^2 \,.\, V = d_1^2 \,.\, V_1 + d_2^2 \,.\, V_2$$

$$(0{\cdot}6)^2 \times 4 = (0{\cdot}4)^2 \times 3 + (0{\cdot}3)^2 \times V_2$$

$$V_2 = \frac{1{\cdot}44 - 0{\cdot}48}{0{\cdot}09}$$

$$= 10{\cdot}67 \text{ m/s}$$

Example 4.3 A pipe coverages uniformly from 800 mm to 400 mm over a length of 1·5 m. The discharge through the pipe is 100 l/s. Determine the convective acceleration at the middle of the pipe.

If the discharge through the pipe increases uniformly from 100 l/s to 200 l/s in 40 seconds, determine the total acceleration at the middle of the pipe after 20 seconds.

Solution $u_1 = \dfrac{Q}{\frac{\pi}{4} \times d^2} = \dfrac{0{\cdot}1}{\frac{\pi}{4} \times (0{\cdot}8)^2} = 0{\cdot}199$ m/s

$$u_2 = \frac{0{\cdot}1}{\frac{\pi}{4} \times (0{\cdot}4)^2} = 0{\cdot}796 \text{ m/s}$$

As the cross-section decreases linearly, the increase in velocity is also linear. Therefore, velocity at any cross-section,

$$u = u_1 + (u_2 - u_1) \,.\, \frac{x}{l} = 0{\cdot}199 + (0{\cdot}796 - 0{\cdot}199)\, \frac{x}{1{\cdot}5} = 0{\cdot}199 + 0{\cdot}398\, x$$

$$\text{Convective acceleration} = u\, \frac{\partial u}{\partial x}$$

$$= (0{\cdot}199 + 0{\cdot}398x) \times 0{\cdot}398$$

At the middle of the pipe, $x = 0{\cdot}75$ m

$$\text{Convective acceleration} = (0{\cdot}199 + 0{\cdot}398 \times 0{\cdot}75) \times 0{\cdot}398$$

$$= 0{\cdot}198 \text{ m/s}$$

The discharge increases uniformly from 100 l/s to 200 l/s in 40 seconds. Thus discharge after 20 seconds,

$$Q = Q_1 + (Q_2 - Q_1)\, \frac{t}{T} = 0{\cdot}1 + (0{\cdot}2 - 0{\cdot}1) \times \frac{20}{40} = 0{\cdot}15 \text{ m}^3\text{/s}$$

$$u_1 = \frac{0{\cdot}15}{\frac{\pi}{4} \times (0{\cdot}8)^2} = 0{\cdot}2985 \text{ m/s}$$

$$u_2 = \frac{0.15}{\frac{\pi}{4} \times (0.4)^2} = 1.194 \text{ m/s}$$

$$u = u_1 (u_2 - u_1) \cdot \frac{x}{l} = 0.2985 + (1.194 - 0.2985) \frac{x}{1.5} = 0.2985 + 0.597\,x$$

$$\text{Convective acceleration} = u\,\frac{\partial u}{\partial x} = (0.2985 + 0.597x) \times 0.597$$

At the middle of the pipe, $x = 0.75$ m

$$\text{Convective acceleration} = (0.2985 + 0.597 \times 0.75) \times 0.597$$
$$= 0.446 \text{ m/s}$$

$$\text{Local acceleration} = \frac{\partial u}{\partial t}$$
$$= \frac{\text{Change in velocity in time } \partial t}{\text{time } \partial t}$$
$$= \frac{1}{40}\left(\frac{Q_1}{(\pi/4)\,d^{\,2}} - \frac{Q_2}{(\pi/4)\,d^{\,2}}\right)$$
$$= \frac{(Q_1 - Q_2)}{40 \times (\pi/4)\,d^{\,2}}$$
$$= \frac{(0.2 - 0.1)}{40 \times (\pi/4)\,0.6^2} \text{ (diameter of middle of pipe = 0.6 m)}$$
$$= 0.00884 \text{ m/s}^2$$

Total acceleration = Convective acceleration + Local acceleration
$$= 0.446 + 0.00884$$
$$= 0.45484 \text{ m/s}^2$$

Example 4.4 A three-dimensional flow is given by

$$V = (y^2 + z^2)\,i + (x^2 + z^2)\,j + (x^2 + y^2)\,k$$

Determine the components of acceleration at a point (2, 3, 4).

Solution
$u = y^2 + z^2$
$v = x^2 + z^2$
$w = x^2 + y^2$

$$a_x = \left(u\,\frac{\partial u}{\partial x} + v\,\frac{\partial u}{\partial y} + w\,\frac{\partial u}{\partial z}\right) + \frac{\partial u}{\partial t}$$
$$= 0 + 2y\,(x^2 + z^2) + 2z\,(x^2 + y^2) + 0$$
$$= 6 \times 20 + 8 \times 13$$
$$= 224$$

$$\text{Similarly, } a_y = \left(u\,\frac{\partial v}{\partial x} + v\,\frac{\partial v}{\partial y} + w\,\frac{\partial v}{\partial z}\right) + \frac{\partial v}{\partial t}$$
$$= 2x\,(y^2 + z^2) + 0 + 2z\,(x^2 + y^2)$$

$$= 4 \times 25 + 8 \times 13$$
$$= 204$$
$$a_z = \left(u \frac{\partial w}{\partial x} + v \frac{\partial w}{\partial y} + w \frac{\partial w}{\partial z} \right) + \frac{\partial w}{\partial t}$$
$$= 2x (y^2 + z^2) + 2y (x^2 + z^2) + 0$$
$$= 4 \times 25 + 6 \times 20$$
$$= 220$$

Example 4.5 The components of velocity of a flow field are given by

$u = yz + t,\ v = xz - t,\ w = xy$

Determine the velocity and acceleration at A (1, 3, 5) after 2 seconds.

Solution $V = \sqrt{u^2 + v^2 + w^2}$

$$= \sqrt{(yz + t)^2 + (xz - t)^2 + (xy)^2}$$
$$= \sqrt{17^2 + 3^2 + 3^2}$$
$$= 17{\cdot}52$$

$V = (yz + t)\, i + (xz - t)\, j + (xy)\, k$

Acceleration is given by

$$a = \frac{dV}{dt} = \left(u \frac{\partial V}{\partial x} + v \frac{\partial V}{\partial y} + w \frac{\partial V}{\partial z} \right) + \frac{\partial V}{\partial t} \qquad \text{(Eq. 4.9)}$$

$$a = (yz + t)(zj + yk) + (xz - t)(zi + xk) + xy\,(yi + xj) + (i - j)$$
$$= 17\,(5j + 3k) + 3\,(5i + k) + 3\,(3i + j) + (i - j)$$
$$= (85j + 51k + 15i + 3k + 9i + 3j) + (i - j)$$
$$= (24i + 88j + 54k) + (i - j)$$

The convective acceleration components are 24, 88 and 54

The local acceleration components are 1 and – 1 along x and y directions.

The total acceleration at A (1, 3, 5)

$$a = \sqrt{24^2 + 88^2 + 54^2} = 106$$

Example 4.6 In a two dimensional flow, the velocity components are given by

$$u = \frac{x}{x^2 + y^2} \quad \text{and} \quad v = \frac{y}{x^2 + y^2}$$

Determine the acceleration components.

Solution $a_x = u \dfrac{\partial u}{\partial x} + v \dfrac{\partial u}{\partial y}$

$$= \frac{x}{x^2 + y^2} \frac{\partial}{\partial x} \left(\frac{x}{x^2 + y^2} \right) + \frac{y}{x^2 + y^2} \frac{\partial}{\partial y} \left(\frac{x}{x^2 + y^2} \right)$$
$$= \frac{x}{x^2 + y^2} \times \frac{(x^2 + y^2) \times 1 - x\,(2x)}{(x^2 + y^2)^2} + \frac{y}{x^2 + y^2} \frac{\partial}{\partial y} [x\,(x^2 + y^2)^{-1}]$$
$$= \frac{x}{x^2 + y^2} \times \frac{y^2 - x^2}{(x^2 + y^2)^2} + \frac{y}{x^2 + y^2} \times \left(-\frac{2xy}{(x^2 + y^2)^2} \right)$$

$$= \frac{xy^2 - x^3}{(x^2 + y^2)^3} - \frac{2xy}{(x^2 + y^2)^3}$$

$$= \frac{-xy^2 - x^3}{(x^2 + y^2)^3}$$

$$= -\frac{x\,(y^2 + x^2)}{(x^2 + y^2)^3}$$

$$= -\frac{x}{(x^2 + y^2)^2}$$

$$a_y = u\,\frac{\partial v}{\partial x} + v\,\frac{\partial v}{\partial y}$$

$$= \frac{x}{x^2 + y^2}\,\frac{\partial}{\partial x}\left(\frac{y}{x^2 + y^2}\right) + \frac{y}{x^2 + y^2}\,\frac{\partial}{\partial y}\left(\frac{y}{x^2 + y^2}\right)$$

$$= \frac{x}{x^2 + y^2}\,\frac{\partial}{\partial x}\,[y(x^2 + y^2)^{-1}] + \frac{y}{x^2 + y^2} \times \frac{(x^2 + y^2) \times 1 - y\,(2y)}{(x^2 + y^2)^2}$$

$$= \frac{x}{x^2 + y^2} \times \left(-\frac{2xy}{(x^2 + y^2)^2}\right) + \frac{y}{x^2 + y^2} \times \frac{x^2 - y^2}{(x^2 + y^2)^2}$$

$$= -\frac{2x^2y}{(x^2 + y^2)^3} + \frac{y\,(x^2 - y^2)}{(x^2 + y^2)^3}$$

$$= \frac{-2x^2y + x^2y - y^3}{(x^2 + y^2)^3}$$

$$= \frac{y\,(y^2 + x^2)}{(x^2 + y^2)^3}$$

$$= -\frac{y}{(x^2 + y^2)^2}$$

4.8 CONTINUITY EQUATION IN 3-D (DIFFERENTIAL FORM)

Consider a three-dimensional element of dimensions δx, δy and δz, whose centre is at $P\,(x, y, z)$ as shown in Fig. 4.14.

In the x-direction,

Mass flow rate per unit time through the face normal to the x-axis at P

$$= \text{Density} \times \text{Volume flow rate}$$

$$= \text{Density} \times \text{Velocity} \times \text{Area}$$

$$= \rho \,.\, u\,(\delta y \,.\, \delta z)$$

Mass flow rate entering the left hand side face

$$= \rho \,.\, u \,.\, \delta y \,.\, \delta z - \left[\frac{\partial}{\partial x}\,(\rho u)\,\frac{\delta x}{2}\right] \delta y \,.\, \delta z$$

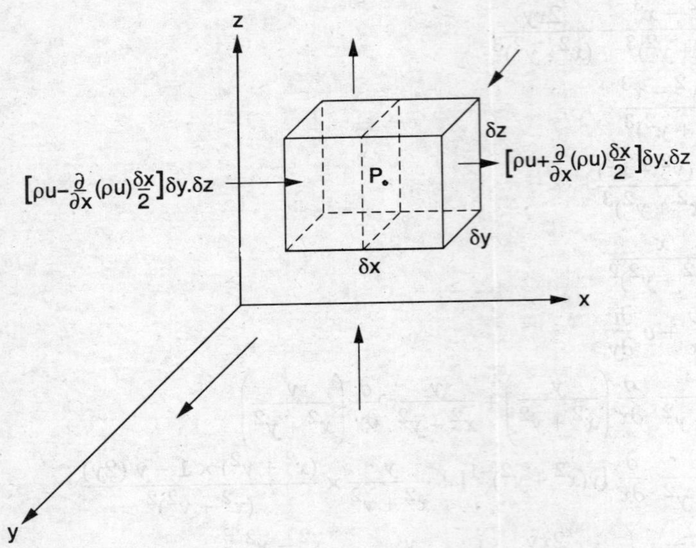

Fig. 4.14

ρ and u both are assumed to vary continuously throughout the flow.

Mass flow rate leaving the right hand side face

$$= \rho \,.\, u \,.\, \delta y \,.\, \delta z + \left[\frac{\partial}{\partial x}(\rho u)\frac{\partial x}{2}\right]\delta y \,.\, \delta z$$

The net gain in mass per unit time by the fluid element in x-direction

$$= -\frac{\partial}{\partial x}(\rho u)\,\delta x \,.\, \delta y \,.\, \delta z$$

Similarly, net gain in the y-direction

$$= -\frac{\partial}{\partial y}(\rho v)\,\delta x \,.\, \delta y \,.\, \delta z$$

and net gain in the z-direction

$$= -\frac{\partial}{\partial z}(\rho w)\,.\,\delta x \,.\, \delta y \,.\, \delta z$$

The total gain in mass per unit time within the element is the sum of net mass gains in the x-, y- and z-directions, *i.e.*

$$-\left[\frac{\partial(\rho u)}{\partial x} + \frac{\partial(\rho v)}{\partial y} + \frac{\partial(\rho w)}{\partial z}\right]\delta x \,.\, \delta y \,.\, \delta z$$

According to the principle of conservation of mass flow, total gain of mass within the element must be equal to the time rate of increase of mass.

Time rate of increase of mass

$$= \frac{\partial}{\partial t}(\rho \,.\, \delta x \,.\, \delta y \,.\, \delta z)$$

Therefore,

$$-\left[\frac{\partial(\rho u)}{\partial x}+\frac{\partial(\rho v)}{\partial y}+\frac{\partial(\rho w)}{\partial z}\right]\delta x\,.\,\delta y\,.\,\delta z=\frac{\partial}{\partial t}(\rho\,.\,\delta x\,.\,\delta y\,.\,\delta z)$$

or

$$\frac{\partial(\rho u)}{\partial x}+\frac{\partial(\rho v)}{\partial y}+\frac{\partial(\rho w)}{\partial z}+\frac{\partial\rho}{\partial t}=0 \qquad ...(4.13)$$

which is the continuity equation for a three-dimensional flow. This equation is valid for,

1. Viscous or inviscid flow as viscosity does not appear in the equation.
2. For compressible fluids. If the flow is incompressible, density variations becomes negligible and the continuity equation becomes,

$$\frac{\partial u}{\partial x}+\frac{\partial v}{\partial y}+\frac{\partial w}{\partial z}=0$$

3. For uniform and non-uniform flows because u, v and w are the local values for continuity at a point. Whenever expressions for u, v and w as functions of x, y, z and t are used to describe the motion of an incompressible flow in a three-dimensional flow, they must satisfy the continuity equation.

In a two-dimensional flow,

$$\frac{\partial(\rho u)}{\partial x}+\frac{\partial(\rho v)}{\partial y}=0 \qquad ...(4.14)$$

or for an incompressible flow

$$\frac{\partial u}{\partial x}+\frac{\partial v}{\partial y}=0 \qquad ...(4.15)$$

4. For unsteady fluids. If the fluid is steady, the term $\partial\rho/\partial t$ vanishes and then

$$\frac{\partial(\rho u)}{\partial x}+\frac{\partial(\rho v)}{\partial y}+\frac{\partial(\rho w)}{\partial z}=0 \qquad ...(4.16)$$

4.9 CONTINUITY EQUATION IN POLAR COORDINATES

Consider a fluid element of the following dimensions (Fig. 4.15) :

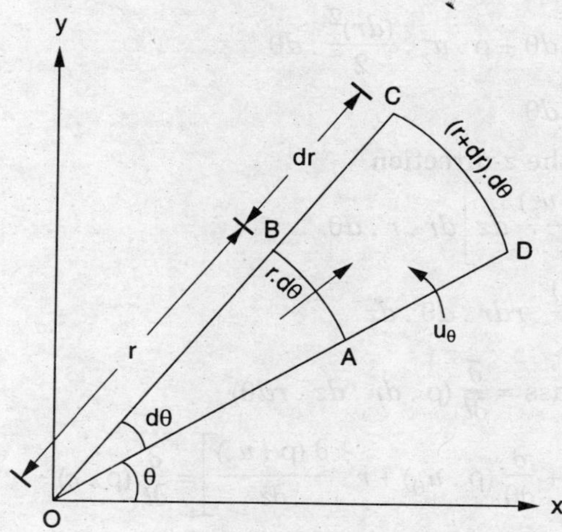

Fig. 4.15

$AB = rd\theta$, $\quad CD = (r + dr)\, d\theta$, $\quad AD = BC = dr$

Depth of the element perpendicular to the paper = dz

Mass flow rate entering at radius r in the radial direction

$$= \text{Density} \times \text{Volumetric flow rate}$$

$$= \text{Density} \times \text{Velocity} \times \text{Area}$$

$$= \rho \,.\, u_r \,.\, (rd\theta \,.\, dz)$$

Mass flow rate leaving at radius $(r + dr)$

$$= \left[\rho .\, u_r + \frac{\partial}{\partial r}(\rho \,.\, u_r)\, dr\right](r + dr) \,.\, d\theta \,.\, dz$$

$\therefore$ Net mass of fluid gained by the fluid element

$$= \rho \,.\, u_r \,.\, r \,.\, d\theta \,.\, dz$$

$$- \left(\rho u_r \,.\, r \,.\, d\theta\, dz + \rho u_r \,.\, dr \,.\, d\theta\, dz + \frac{\partial}{\partial r}(\rho u_r)\, dr \,.\, r \,.\, d\theta \,.\, dz + \text{smaller terms}\right)$$

$$= -\left(\rho u_r \,.\, dr \,.\, d\theta\, dz + \frac{\partial}{\partial r}(\rho u_r)\, dr \,.\, r \,.\, d\theta\, dz\right)$$

$$= -\frac{\partial}{\partial r}[(\rho \,.\, u_r) \,.\, r]\, dr \,.\, d\theta \,.\, dz$$

Mass flow rate entering in the angular direction = $\rho \,.\, u_\theta \,.\, dr \,.\, dz$

Mass flow rate leaving in the angular direction

$$= \left[\rho \,.\, u_\theta + \frac{\partial(\rho \,.\, u_\theta)}{\partial\theta}\, d\theta\right] .\, dr \,.\, dz$$

$$\text{Net mass gained} = -\frac{\partial(\rho u_\theta)}{\partial\theta}\, d\theta \,.\, dr \,.\, dz$$

Mass flow rate entering in the z-direction

$$= \rho \,.\, u_z \,.\, dr\left[\left(r + \frac{dr}{2}\right) d\theta\right]$$

$$= \rho \,.\, u_z \,.\, dr \,.\, r \,.\, d\theta + \rho \,.\, u_z \,.\, \frac{(dr)^2}{2} \,.\, d\theta$$

$$= \rho \,.\, u_z \,.\, dr \,.\, r \,.\, d\theta \qquad \text{...(Canceling } dr^2 \text{ terms)}$$

Mass flow rate leaving in the z-direction

$$= \left[\rho \,.\, u_z + \frac{\partial(\rho \,.\, u_z)}{\partial z}\, dz\right] dr \,.\, r \,.\, d\theta$$

$$\therefore \text{ Mass gained} = -\frac{\partial(\rho \,.\, u_z)}{\partial z} \,.\, rdr \,.\, d\theta \,.\, dz$$

$$\text{Time rate of increase of mass} = \frac{\partial}{\partial t}(\rho \,.\, dr \,.\, dz \,.\, rd\theta)$$

$$\therefore \qquad -\left[\frac{\partial}{\partial r}(\rho \,.\, u_r \,.\, r) + \frac{\partial}{\partial\theta}(\rho \,.\, u_\theta) + r \,.\, \frac{\partial(\rho \,.\, u_z)}{\partial z}\right] = \frac{\partial}{\partial t}(\rho \,.\, r)$$

Dividing throughout by r,

$$\frac{1}{r} \cdot \frac{\partial}{\partial r}(\rho \, . \, u_r \, . \, r) + \frac{1}{r} \cdot \frac{\partial}{\partial \theta}(\rho \, . \, u_\theta) + \frac{\partial (\rho \, . \, u_z)}{\partial z} + \frac{\partial \rho}{\partial t} = 0$$

For a steady flow,

$$\frac{1}{r} \cdot \frac{\partial}{\partial r}(\rho \, . \, u_r \, . \, r) + \frac{1}{r} \cdot \frac{\partial}{\partial \theta}(\rho \, . \, u_\theta) + \frac{\partial (\rho \, . \, u_z)}{dz} = 0 \quad ...(4.17)$$

For a two-dimensional flow,

$$\frac{\partial}{\partial r}(\rho \, . \, u_r \, . \, r) + \frac{\partial}{\partial \theta}(\rho \, . \, u_\theta) = 0 \quad ...(4.18)$$

For a two-dimensional incompressible flow,

$$\frac{\partial}{\partial r}(u_r \, . \, r) + \frac{\partial u_\theta}{\partial \theta} = 0 \quad ...(4.19)$$

4.10 STREAM FUNCTION

A **stream function** (ψ) is a mathematical device to describe a particular pattern of flow.

For a fluid flow streamlines can be drawn in the flow field to show the flow direction. Also as the flow cannot cross the streamlines, the flow between two streamlines remains constant. This means that the streamlines may converge or diverge at different cross-sections depending upon the velocity of flow and the area of cross-section, so that the discharge remains constant between two streamlines. For unit thickness of flow, the spacing between two streamlines will be inversely proportional to the velocity (for incompressible fluids). If a particular value (known as **stream function**) is assigned to a streamline, the difference between the values of two streamlines (or stream functions) will give the volumetric flow rate between the two.

Consider a point $P\,(xy)$ on a streamline in a two dimensional steady flow having unit thickness perpendicular to the xy plane (Fig. 4.16). Let ψ be the value assigned to the streamline passing through this point P. Let there be another streamline with a stream function value $\psi + d\psi$ in the vicinity of the first streamline. Assume a point Q on this

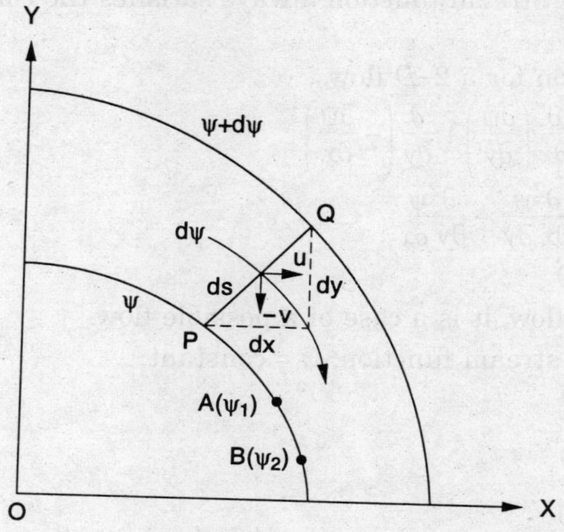

Fig. 4.16

streamline distance ds apart from P. The discharge through the spacing between the two streamlines will be $d\psi$, being the difference between the two stream functions. This is also the discharge passing through the line joining P and Q. If the elementary components of the distance between PQ be dx and dy, the variation in the value of stream function at point P may be written as

$$d\psi = \frac{d\psi}{\partial x} \, . \, dx + \frac{\partial \psi}{\partial y} \, . \, dy \qquad \text{...(4.20)}$$

which is the discharge through the section PQ.

Let ψ_1 and ψ_2 be the values of stream functions assigned to two points A and B located on the same streamline. Then the flow across AB will be

$$\psi_1 - \psi_2 = 0 \qquad \text{(the flow cannot cross } AB\text{)}$$

or

$$\psi_1 = \psi_2$$

which means that the two points on a streamline must have the same value of stream function. In general, we can say that a streamline has only one value of stream function whereas two different streamlines have two different values of stream functions.

The flow between two streamlines can also be evaluated in the following way :

Let u = velocity at P in the x direction

v = velocity of P in the y directioin

Flow rate $d\psi = - vdx + udy$

From (i) and (ii)

$$- v = \frac{\partial \psi}{\partial x} \text{ and } u = \frac{\partial \psi}{\partial y}$$

or

$$u = \frac{\partial \psi}{\partial y} \text{ and } v = - \frac{\partial \psi}{\partial x} \qquad \text{...(4.21)}$$

Thus a stream function is a scalar function of space and time. Its partial derivative with respect to x gives the velocity in the y-direction (negative) and with respect to y gives the velocity in the x-direction. Stream function always satisfies the continuity equation as shown below :

The continuity equation for a 2–D flow

$$\frac{\partial u}{\partial x} + \frac{\partial v}{\partial y} = \frac{\partial}{\partial x}\left(\frac{\partial \psi}{\partial y}\right) + \frac{\partial}{\partial y}\left(-\frac{\partial \psi}{\partial x}\right)$$

$$= \frac{\partial^2 \psi}{\partial x \, \partial y} - \frac{\partial^2 \psi}{\partial y \, \partial x}$$

$$= 0$$

Thus if ψ exists for a flow, it is a case of a possible flow.

For a line of constant stream function, ψ = constant.

or

$$d\psi = 0$$

$$\frac{\partial \psi}{\partial x} dx + \frac{\partial \psi}{\partial y} dy = 0$$

$$- v \, . \, dx + u \, . \, dy = 0$$

$$u \, . \, dy = v \, . \, dx$$

$$\frac{dy}{dx} = \frac{v}{u}$$

Slope of stream line $= \frac{v}{u}$ ...(4.22)

In polar co-ordinates, we can write

$$\partial\psi = \frac{\partial\psi}{\partial r} dr + \frac{\partial\psi}{\partial\theta} d\theta$$

The flow can be expressed as $d\psi = -\, u_\theta \,.\, dr + u_r \,.\, r d\theta$

On comparing, we have

$$u_r \,.\, r = \frac{\partial\psi}{\partial\theta} \quad \text{or} \quad u_r = \frac{1}{r}\frac{\partial\psi}{\partial\theta} \qquad (4.23)$$

and $$-\, u_\theta = \frac{\partial\psi}{\partial r} \quad \text{or} \quad u_\theta = -\frac{1}{r}\frac{\partial\psi}{\partial r} \qquad ...(4.23a)$$

4.11 VELOCITY POTENTIAL FUNCTION

In an open channel flow, such as a stream or river, the flow or the velocity depends upon the slope of the bed. If the flow is unidirectional and the slope is only in the direction of flow, then along a transverse section, the base has a constant elevation with respect to an assumed datum. At any other section the elevation will be different with respect to the same datum. The value may be assigned to these elevations. The value assigned to a particular cross-section is known its **velocity potential function.** The more the potential difference between two cross-sections, the more is the velocity of the flow.

In a three-dimensional flow, the flow has the components in all the three directions. A **velocity potential function** (ϕ) is defined as a scalar function of position and time so that in a steady flow, its negative derivative in any direction gives the velocity in that direction,

i.e. $$u = -\frac{\partial\phi}{\partial x},\ v = -\frac{\partial\phi}{\partial y},\ w = -\frac{\partial\phi}{\partial z} \qquad ...(4.24)$$

The negative sign signifies that the flow is always in the direction of decreasing ψ. Lines of constant potential function are called **equipotential lines**

$$\phi = \text{constant}$$

$$d\phi = 0$$

$$\frac{\partial\phi}{\partial x} dx + \frac{\partial\phi}{\partial y} dy = 0$$

$$-\, u \,.\, dx - v \,.\, dy = 0$$

$$-\,(u \,.\, dx + v \,.\, dy) = 0$$

$$u \,.\, dx + v \,.\, dy = 0$$

$$\frac{dy}{dx} = -\frac{u}{v}$$

or slope of equipotential line $= -\frac{u}{v}$ ...(4.25)

Now product of slopes of equipotential lines and streamlines

$$= -\frac{u}{v} \times \frac{v}{u} = -1$$

which shows that at any point in a flow, the streamlines and equipotential lines intersect each other orthogonally.

In polar co-ordinates,

$$u_r = \frac{\text{Decrease in } \phi \text{ in elementary distance } \partial r \text{ in } r\text{-direction}}{\text{Distance } \partial r}$$

$$= -\frac{\partial \phi}{\partial r} \qquad \text{...(4.26)}$$

$$u_\theta = \frac{\text{Decrease in } \phi \text{ in elementary distance } r \,.\, \partial\theta \text{ in } \theta\text{-direction}}{\text{Distance } r \,.\, \partial\theta}$$

$$= -\frac{1}{r} \cdot \frac{\partial \phi}{\partial \theta} \qquad \text{...(4.27)}$$

4.12 FLOW NETS

An orthogonal network (grid) obtained by drawing a series of streamlines and equipotential lines is called a **flow net.** It is customary to keep the change in their values same between adjacent equipotential lines and adjacent streamlines. The streamlines are drawn by dividing the space between the boundaries into a few equal number of flow channels. Then the equipotential lines are drawn orthogonal to streamlines intersecting the boundaries taking care that the flow fields are roughly squares in the region of uniform flow.

As the velocities at various points in the flow are inversely proportional to the spacing between the streamlines and equipotential lines, the network is useful to estimate the velocity at different points in the flow field when it is not convenient to use analytical methods.

Example 4.7 Determine the stream function at point P (2, 3) for a two dimensional flow described by

$$u = 5x^3 \text{ and } v = -15x^2y$$

Solution

$$\frac{d\psi}{\partial y} = u = 5x^3 \qquad \text{...(i)}$$

$$\frac{\partial \psi}{\partial x} = -v = 15x^2y \qquad \text{...(ii)}$$

Integrating (i), $\psi = 5x^3y + f(x) + C_1$ ...(iii)

Differentiating w.r.t. x, $\dfrac{\partial \psi}{\partial x} = 15x^2y + f'(x)$

Comparing it with (ii),

$$15x^2y + f'(x) = 15x^2y$$

or $\quad f'(x) = 0$

or $\quad f(x) = C_2$

From (iii), $\psi = 5x^3y + C_1 + C_2$

$$= 5x^3y + C$$

For $\psi = 0$ at the origin *i.e.* at $x = 0, y = 0$,

$$C = 0,$$

and $\psi = 5x^3y$

Alternately,

$$d\psi = \frac{\partial \psi}{\partial x}\, dx + \frac{\partial \psi}{\partial y}\, dy$$

$$= -\,v\,.\,dx + u\,.\,dy$$

$$= (15x^2y\,.\,dx + 5x^3\,.\,dy)$$

$$\psi = 5x^3y + C$$

For $\psi = 0$ at the origin, $C = 0$,

and $\psi = 5x^3$

Stream function at point P (2, 3),

$$\psi = 5\,.\,(2)^3\,.\,(3) = 120.$$

Example 4.8 A two-dimensional flow is described by

$$u = \frac{-\,y}{b^2} \quad \text{and} \quad v = \frac{x}{a^2}$$

Determine the equation of a streamline passing through $(a, 0)$.

Solution

$$d\psi = \frac{\partial \psi}{\partial x}\, dx + \frac{\partial \psi}{\partial y}\, dy$$

$$= -\,v\,.\,dx + u\,.\,dy$$

$$= \left(-\,\frac{x}{a^2}\,.\,dx - \frac{y}{b^2}\,.\,dy\right)$$

$$\psi = -\,\frac{x^2}{2a^2} - \frac{y^2}{2b^2} + C$$

For $\psi = 0$ at the origin, $C = 0$,

and $\psi = -\,\dfrac{1}{2}\left(\dfrac{x^2}{a^2} + \dfrac{y^2}{b^2}\right)$

As the streamline passes through $(a, 0)$,

$$\psi = -\,\frac{a^2}{2a^2} - 0 = -\,\frac{1}{2}$$

Thus equation of the required streamline is

$$-\,\frac{1}{2}\left(\frac{x^2}{a^2} + \frac{y^2}{b^2}\right) = -\,\frac{1}{2}$$

or

$$\left(\frac{x^2}{a^2} + \frac{y^2}{b^2}\right) = 1$$

Example 4.9 The velocity components in a two-dimensional flow are given as

$$u = \frac{y^3}{3} - x^2y + 2x$$

$$v = xy^2 - \frac{x^3}{3} - 2y$$

Obtain the expressions for the stream function and the velocity potential function.

Solution $d\psi = \frac{\partial \psi}{\partial x} dx + \frac{\partial \psi}{\partial y} dy$

$$= - v \, . \, dx + u \, . \, dy$$

$$= - \left(xy^2 - \frac{x^3}{3} - 2y \right) dx + \left(\frac{y^3}{3} - x^2y + 2x \right) dy$$

$$= \frac{x^3}{3} dx + \frac{y^3}{3} dy + 2\,(y \, . \, dx + x \, . \, dy) - (xy^2 \, . \, dx + x^2y \, . \, dy)$$

$$\psi = \frac{x^4}{12} + \frac{y^4}{12} + 2xy - \frac{x^2y^2}{2} + C$$

For $\psi = 0$ at the origin, $C = 0$,

and $\psi = \frac{x^4}{12} + \frac{y^4}{12} + 2xy - \frac{x^2y^2}{2}$

$$d\phi = \frac{\partial \phi}{\partial x} dx + \frac{\partial \phi}{\partial y} dy$$

$$= - u \, . \, dx - v \, . \, dy$$

$$= - \left(\frac{y^3}{3} - x^2y + 2x \right) dx - \left(xy^2 - \frac{x^3}{3} - 2y \right) dy$$

$$= - \frac{y^3}{3} \, . \, dx + x^2y \, . \, dx - 2x \, . \, dx - xy^2 \, . \, dy + \frac{x^3}{3} \, . \, dy + 2y \, . \, dy$$

$$= - \left(\frac{y^3}{3} \, . \, dx + xy^2 \, . \, dy \right) + \left(x^2y \, . \, dx + \frac{x^3}{3} \, . \, dy \right) - 2x \, . \, dx + 2y \, . \, dy$$

$$\phi = - \frac{xy^3}{3} + \frac{x^3y}{3} - x^2 + y^2 + C$$

For $\phi = 0$ at the origin, $C = 0$

$$\phi = - \frac{xy^3}{3} + \frac{x^3y}{3} - x^2 + y^2$$

Example 4.10 Determine the stream function for the following velocity field $u = 2y$ and $v = - 2x$ and plot a set of streamlines.

Solution $d\psi = \frac{\partial \psi}{\partial x} dx + \frac{\partial \psi}{\partial y} dy$

$$= - v \, . \, dx + u \, . \, dy$$

$$= 2x \, . \, dx + 2y \, . \, dy$$

$$\psi = x^2 + y^2 + C$$

For $\psi = 0$ at the origin, $C = 0$,

and $\psi = x^2 + y^2$

The form of equation for the streamlines indicate that streamlines are a set of concentric circles with centre (0, 0) and radius $\sqrt{x^2 + y^2}$. In the 1st quadrant (say at 1, 1) as u is positive ($u = 2y = 2$) and v is negative ($v = -2x = -2$). Therefore, streamlines have clockwise direction (Fig. 4.17).

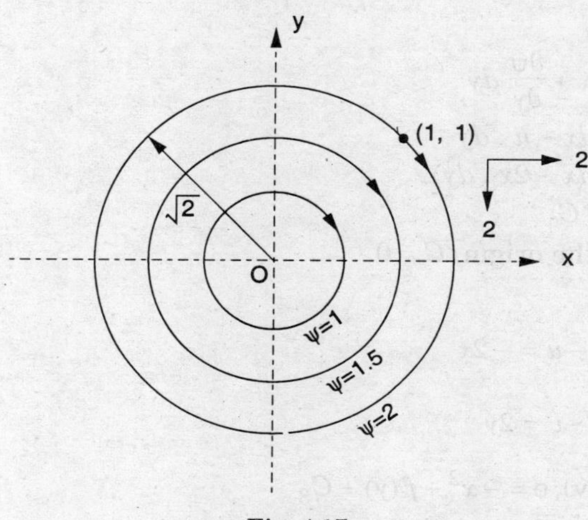

Fig. 4.17

Example 4.11 The velocity components of a flow are

$u = 2x$ and $v = -2y$

Is it a case of physically possible flow ? If so, find the expression for stream function. Also determine the velocity potential function. Plot the flow net. Show that the streamlines and velocity potential lines intersect orthogonally at the point (1, 1).

Solution The flow is physically possible if the continuity equation is satisfied.

$$\frac{\partial u}{\partial x} + \frac{\partial v}{\partial y} = 0$$

or $$2 - 2 = 0$$

Thus the flow is physically possible.

$$\frac{\partial \psi}{\partial y} = u = 2x \quad \text{...(i)}$$

$$\frac{\partial \psi}{\partial x} = -v = 2y \quad \text{...(ii)}$$

Integrating (i), $\psi = 2xy + f(x) + C_1$ ...(iii)

Differentiating w.r.t x, $\frac{\partial \psi}{\partial x} = 2y + f'(x)$

Comparing with (ii), $2y + f'(x) = 2y$

or $f'(x) = 0$

or $f(x) = C_2$

From (iii), $\psi = 2xy + C_1 + C_2$

$$= 2x^3y + C$$

For $\psi = 0$ at the origin *i.e.* at $x = 0, y = 0$,

$C = 0$,

and $\psi = 2xy$

Alternatively,

$$d\psi = \frac{\partial \psi}{\partial x}\, dx + \frac{\partial \psi}{\partial y}\, dy$$

$$= -v \,.\, dx + u \,.\, dy$$

$$= (2y \,.\, dx + 2x \,.\, dy)$$

$$\psi = 2xy + C$$

For $\psi = 0$ at the origin, $C = 0$,

and $\psi = 2xy$

$$\frac{\partial \phi}{\partial x} = -u = -2x \quad \text{...(iv)}$$

$$\frac{\partial \phi}{\partial x} = -v = 2y \quad \text{...(v)}$$

integrating (iv), $\phi = -x^2 + f(y) + C_3$ ...(vi)

Differentiating w.r.t. x, $\frac{\partial \phi}{\partial x} = f'(y)$

Comparing with (v), $f'(y) = 2y$

or $f(y) = y^2 + C_4$

From (vi), $\phi = -x^2 + y^2 + C_3 + C_4$

$$= y^2 - x^2 + C$$

For $\psi = 0$ at the origin *i.e.* at $x = 0, y = 0$,

$C = 0$,

and $\phi = y^2 - x^2$

Alternately,

$$d\phi = \frac{\partial \phi}{\partial x}\, dx + \frac{\partial \phi}{\partial y}\, dy$$

$$= -u \,.\, dx - v \,.\, dy$$

$$= -2x \,.\, dx + 2y \,.\, dy$$

$$\phi = -x^2 + y^2 + C$$

For $\phi = 0$ at the origin, $C = 0$,

and $\phi = y^2 - x^2$

The flow net is drawn by giving equal increments for ψ and ϕ values (say 10 units).

Thus for ψ values :

$2xy$	xy	$x-$	1	2	3	4	5	6	7
10	5	$y-$	5	2·5	1·67	1·25	1	0·83	0·71
20	10	$y-$	10	5·0	3·34	2·50	2	1·67	1·43
30	15	$y-$	15	7·5	5·00	3·75	3	2·50	2·14
40	20	$y-$	20	10·0	6·67	5·00	4	3·30	2·86
50	25	$y-$	25	12·5	8·33	6·25	5	4·17	3·57

For ϕ values :

$y^2 - x^2$	y^2	$x-$	1	2	3	4	5	6	7
10	$10 + x^2$	$y-$	3·31	3·74	4·36	5·09	5·92	6·78	7·68
20	$20 + x^2$	$y-$	4·58	4·90	5·39	6·00	6·71	7·48	8·31
30	$30 + x^2$	$y-$	5·57	5·83	6·24	6·78	7·42	8·12	8·89
40	$40 + x^2$	$y-$	6·40	6·63	7·00	7·48	8·06	8·72	9·43
50	$50 + x^2$	$y-$	7·14	7·35	7·68	8·12	8·66	9·27	9·95

Fig. 4.18 shows the required flow net.

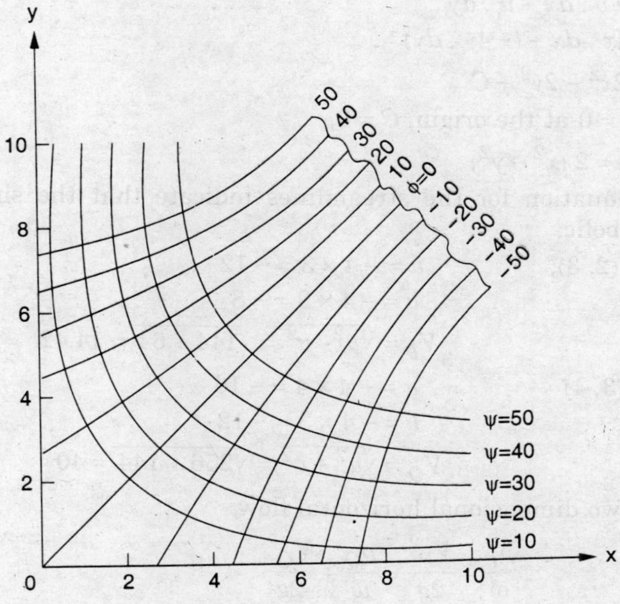

Fig. 4.18

Slope of streamlines at (1, 1),

$$\left(\frac{dy}{dx}\right)_\psi = \frac{v}{u} = -\frac{2y}{2x} = -1$$

Slope of velocity potential lines at (1, 1),

$$\left(\frac{dy}{dx}\right)_\phi = -\frac{u}{v} = \frac{2x}{2y} = 1$$

$$\left(\frac{dy}{dx}\right)_\psi \times \left(\frac{dy}{dx}\right)_\phi = -1 \times 1 = -1$$

which shows that the streamlines and potential lines intersect orthogonally.

Example 4.12 The velocity potential function for a two-dimensional flow field is given by $\phi = 4xy$. Find an expression for the stream function and define the shape of the flow field ?

Also find the velocities at points P (2, 3) and Q (3, 4). What is the pressure difference between P and Q and the discharge between these points ?

Solution

$$u = -\frac{\partial \phi}{\partial x} = -4y$$

$$v = -\frac{\partial \phi}{\partial x} = -4x$$

$$d\psi = \frac{\partial \psi}{\partial x}dx + \frac{\partial \psi}{\partial y}dy$$

$$= -v \,.\, dx + u \,.\, dy$$

$$= 4x \,.\, dx + (-4y \,.\, dy)$$

$$\psi = 2x^2 - 2y^2 + C$$

For $\psi = 0$ at the origin, $C = 0$,

and $\psi = 2\,(x^2 - y^2)$

The equation for the streamlines indicate that the shape of the flow field is hyperbolic.

At P (2, 3),

$$u = -4 \times 3 = -12$$

$$v = -4 \times 2 = -8$$

$$V_P = \sqrt{u^2 + v^2} = \sqrt{144 + 64} = 14{\cdot}42$$

At Q (3, 4)

$$u = -4 \times 4 = -16$$

$$v = -4 \times 3 = -12$$

$$V_Q = \sqrt{u^2 + v^2} = \sqrt{256 + 144} = 40$$

In a two dimensional horizontal flow,

$$\frac{p_P}{w} + \frac{V_P^{\,2}}{2g} = \frac{p_Q}{w} + \frac{V_Q^{\,2}}{2g}$$

$$\frac{p_P - p_Q}{w} = \frac{V_Q^{\,2} - V_P^{\,2}}{2g} = \frac{40^2 - 14{\cdot}42^2}{2 \times 9{\cdot}81} = 70{\cdot}95 \text{ units}$$

Discharge,

$$Q = \psi_{Q\,(3,\,4)} - \psi_{P\,(2,\,3)} = 2\,(3^2 - 4^2) - 2\,(2^2 - 3^2) = -\,14 + 10 = -\,4 \text{ or } 4 \text{ units}$$

4.13 TYPES OF MOTION

A moving fluid element can undergo the following types of motion :

(i) Linear (pure) translation
(ii) Linear deformation
(iii) Angular (shear) deformation
(iv) Rotation

Linear Translation

When a fluid element moves bodily from one position to another. New axes of the fluid element remain parallel to the original axes. The particle is neither rotated nor deformed (Fig. 4.19).

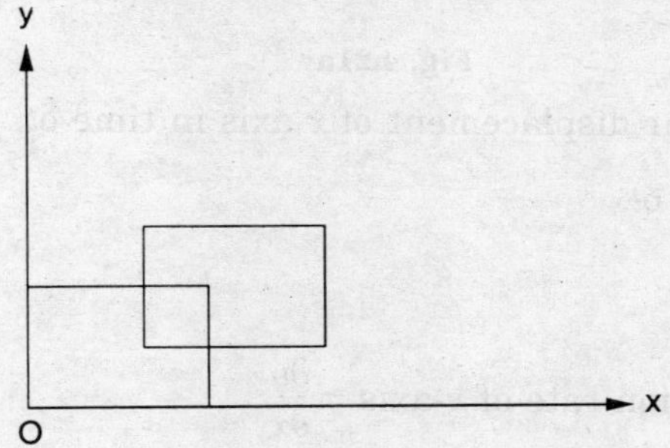

Fig. 4.19

Linear Deformation

When the fluid element is deformed while in motion. Axes of the fluid element remain parallel and do not rotate. However, their lengths are changed as shown in Fig. 4.20.

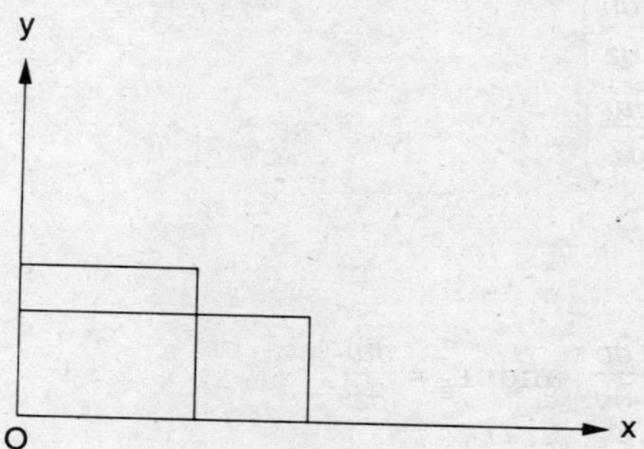

Fig. 4.20

Angular Deformation

When a fluid particle is deformed in such a way that the two axes rotate by the some amount in the opposite directions with respect to the original positions it is said to undergo angular deformation. It is equal to the average change in the angle between the two adjacent sides (Fig. 4.21a).

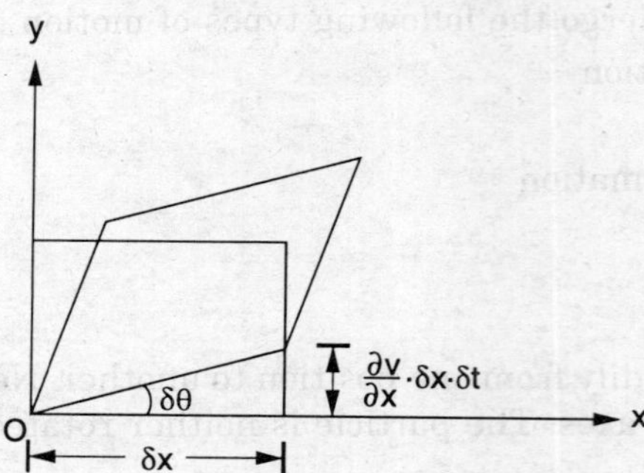

Fig. 4.21a

Let $\delta\theta$ be the angular displacement of x-axis in time δt.

Then $\delta\theta \,.\, \delta x = \dfrac{\partial v}{\partial x} \,.\, \delta x \,.\, \delta t$

$$\frac{\delta\theta}{\delta t} = \frac{\partial v}{\partial x} \qquad \text{...(i)}$$

or angular deformation rate of x-axis $= \dfrac{\partial v}{\partial x}$

Similarly, angular deformation rate of y-axis $= \dfrac{\partial u}{\partial y}$

Shear strain rate = Average angular deformation rate

$$\gamma_{xy} = \frac{1}{2}\left(\frac{\partial v}{\partial x} + \frac{\partial u}{\partial y}\right)$$

Similarly, $$\gamma_{yz} = \frac{1}{2}\left(\frac{\partial w}{\partial y} + \frac{\partial v}{\partial z}\right)$$

and $$\gamma_{zx} = \frac{1}{2}\left(\frac{\partial x}{\partial z} + \frac{\partial w}{\partial x}\right) \qquad \text{...(4.28)}$$

Dilatency (Linear Strain)

It is defined as

$$\varepsilon_x = \frac{\partial u}{\partial x}, \quad \varepsilon_y = \frac{\partial v}{\partial y} \quad \text{and} \quad \varepsilon_z = \frac{\partial w}{\partial z} \qquad \text{...(4.29)}$$

Rotation

When both the axes of a moving fluid element rotate in the same direction, it is said to undergo rotation (Fig. 4.21b). The average of the sum of angular velocities of the two sides is that net rotation of the fluid element.

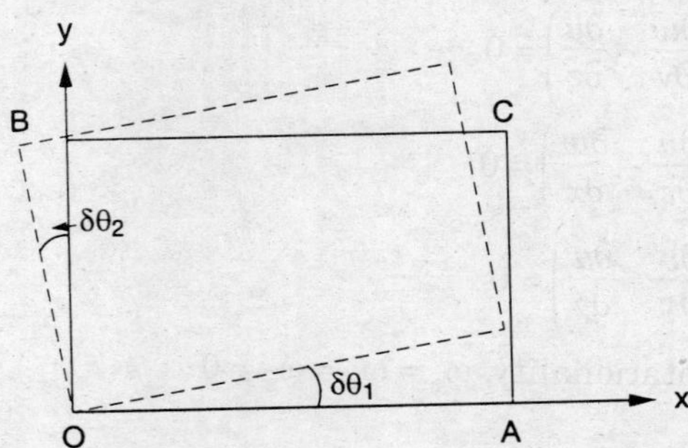

Fig. 4.21b

Angular velocity of side $OA = \lim_{\delta t \to 0} \dfrac{\delta\theta_1}{\delta t}$

or $$\omega_{OA} = \frac{\partial v}{\partial x} \quad \text{[as in (i)]}$$

Angular velocity of side $OB = \lim_{\delta t \to 0} \dfrac{\delta\theta_2}{\delta t}$

or $$\omega_{OB} = -\frac{\partial u}{\partial y} \quad (u \text{ negative on LHS of } y\text{-axis})$$

The average of the two angular velocities of the fluid elements is called **rotation** of the fluid element.

$$\omega_z = \frac{1}{2}(\omega_{OA} + \omega_{OB})$$
$$= \frac{1}{2}\left(\frac{\partial v}{\partial x} - \frac{\partial u}{\partial y}\right)$$

Similarly, $$\omega_x = \frac{1}{2}\left(\frac{\partial w}{\partial y} - \frac{\partial v}{\partial z}\right)$$
$$\omega_y = \frac{1}{2}\left(\frac{\partial u}{\partial z} - \frac{\partial w}{\partial x}\right) \quad \text{...(4.30)}$$

The resultant rotation

$$\omega = \sqrt{\omega_x^2 + \omega_y^2 + \omega_z^2} \quad \text{...(4.31)}$$

In vector equation

$$\omega = (\omega_x\, i + \omega_y\, j + \omega_z\, k) \quad \text{...(4.32)}$$

When $\dfrac{\partial v}{\partial x} = \dfrac{\partial u}{\partial y}$, there is no rotation of the fluid particle. Usually, a steady, irrotational flow is known as **potential flow**. Thus condition of irrotationality in a two-dimensional flow (in the xy plane) will be

$$\frac{\partial v}{\partial x} = \frac{\partial u}{\partial y} \text{ or } \omega_z = 0 \quad \text{...(4.33)}$$

i.e. no rotation about the z-axis.

In a three-dimensional flow, similarily, we can have

$$\omega_x = \frac{1}{2}\left(\frac{\partial w}{\partial y} - \frac{\partial v}{\partial z}\right) = 0$$

$$\omega_y = \frac{1}{2}\left(\frac{\partial u}{\partial z} - \frac{\partial w}{\partial x}\right) = 0$$

$$\omega_z = \frac{1}{2}\left(\frac{\partial v}{\partial x} - \frac{\partial u}{\partial y}\right) = 0$$

and condition of irrotationality, $\omega_x = \omega_y = \omega_z = 0$

$$\frac{\partial v}{\partial x} = \frac{\partial u}{\partial y}$$

$$\frac{\partial w}{\partial y} = \frac{\partial v}{\partial z}$$

$$\frac{\partial u}{\partial z} = \frac{\partial w}{\partial x} \qquad \text{...(4.34)}$$

Substituting the values of u, v and w, we get

$$\frac{\partial^2 \phi}{\partial y \,.\, \partial x} = \frac{\partial^2 \phi}{\partial x \,.\, \partial y}$$

$$\frac{\partial^2 \phi}{\partial z \,.\, \partial y} = \frac{\partial^2 \phi}{\partial y \,.\, \partial z}$$

$$\frac{\partial^2 \phi}{\partial x \,.\, \partial z} = \frac{\partial^2 \phi}{\partial z \,.\, \partial x}$$

As these relations are true, this means the velocity potential function satisfies the **conditions of irrotationality** of flow. In other words, if ϕ exists and the flow is possible, it must be irrotational.

Further, we have the continuity equation for a steady incompressible fluid flow,

$$\frac{\partial u}{\partial x} + \frac{\partial v}{\partial y} + \frac{\partial w}{\partial z} = 0$$

Substituting the values of u, v and w,

$$\frac{\partial}{\partial x}\left(-\frac{\partial \phi}{\partial x}\right) + \frac{\partial}{\partial y}\left(-\frac{\partial \phi}{\partial y}\right) = \frac{\partial}{\partial z}\left(-\frac{\partial \phi}{\partial z}\right) = 0$$

$$\frac{\partial^2 \phi}{\partial x^2} + \frac{\partial^2 \phi}{\partial y^2} + \frac{\partial^2 \phi}{\partial z^2} = 0 \qquad \text{...(4.35)}$$

This is known as **Laplace equation.** Thus if a velocity potential function (ϕ) satisfies the laplace equation, it must be a possible case of irrotational flow usually known as **potential flow.**

Example 4.13 In a two dimensional flow, the velocity components are given by

$$u = \frac{x}{x^2 + y^2} \text{ and } v = \frac{y}{x^2 + y^2}$$

Show that the flow is irrotational.

Solution The flow is irrotational if the rotation about z-axis is zero,

$$\omega_z = \frac{1}{2}\left(\frac{\partial v}{\partial x} - \frac{\partial u}{\partial y}\right) = 0$$

or $$\frac{\partial v}{\partial x} = \frac{\partial u}{\partial y}$$

or $$\frac{\partial}{\partial x}\left(\frac{y}{x^2 + y^2}\right) = \frac{\partial}{\partial y}\left(\frac{x}{x^2 + y^2}\right)$$

or $$\frac{\partial}{\partial x}[y\,(x^2 + y^2)^{-1}] = \frac{\partial}{\partial y}[x\,(x^2 + y^2)^{-1}]$$

or $$-\frac{2xy}{(x^2 + y^2)^2} = -\frac{2xy}{(x^2 + y^2)^2}$$

which s true. Thus the flow is irrotational.

Example 4.14 Two velocity components of a three-dimensional incompressible fluid flow are given by

$u = x^2 + y^2 + z^2$ and $v = -(xy + yz + zx)$

Determine the third component of velocity. Can stream function and velocity potential function be defined for such a flow ?

Solution The continuity equation is

$$\frac{\partial u}{\partial x} + \frac{\partial v}{\partial y} + \frac{\partial w}{\partial z} = 0$$

$$2x + (-x - z) + \frac{\partial w}{\partial z} = 0$$

$$\frac{\partial w}{\partial z} = z - x$$

$$w = \frac{z^2}{2} - xz + f(x, y) + C$$

where $f(x, y)$ can be a function of x or y *i.e.* independent of z and C a numerical constant.

As the continuity equation is satisfied, the stream function can be defined for such a flow. The velocity potential function is defined only if the flow is irrotational *i.e.* if

$$\frac{\partial u}{\partial y} - \frac{\partial v}{\partial x} = 0$$

$$2y - (-y - z) = 0$$

$$3y + z = 0$$

which is not true. There is no need to verify the other two equations. The flow is not irrotational. Hence velocity potential function cannot be defined for such a flow.

Example 4.15 A two-dimensional incompressible flow is represented by

$u = x - 4y$ and $v = -y - 4x$

Determine the expression for the stream function. Show that the velocity potential function exists for this flow. Find the expression for the same.

Solution $d\psi = \frac{\partial \psi}{\partial x} dx + \frac{\partial \psi}{\partial y} dy$

$= - v \,.\, dx + u \,.\, dy$

$= (y + 4x) \,.\, dx + (x - 4y) \,.\, dy$

$\psi = (y \,.\, dx + x \,.\, dy) + 4x \,.\, dx - 4y \,.\, dy$

$= xy + 2x^2 - 2y^2 + C$

For $\psi = 0$ at the origin, $C = 0$,

and $\psi = xy + 2\,(x^2 - y^2)$

The velocity potential function exists if the flow is irrotational.

$$\frac{\partial u}{\partial y} - \frac{\partial v}{\partial x} = 0$$

or $\quad - 4 - (- 4) = 0$

Thus the velocity potential function exists.

$d\phi = \frac{\partial \phi}{\partial x} dx + \frac{\partial \phi}{\partial y} dy$

$= - u \,.\, dx - v \,.\, dy$

$= - (x - 4y) \,.\, dx - (- y - 4x) \,.\, dy$

$= - x \,.\, dx + y \,.\, dy + 4\,(y \,.\, dx + x \,.\, dy)$

$\phi = - \frac{x^2}{2} + \frac{y^2}{2} + 4xy$

For $\phi = 0$ at the origin, $C = 0$,

and $\phi = \frac{1}{2}\,(y^2 - x^2) + 4xy$

4.14 CIRCULATION

In a flow field, the inherent tendency of a fluid to circulate in a closed path is called **circulation.** The meaning of this statement can be fully understood from the following examples :

Consider a linear flow field *i.e.* the flow is unidirectional. The streamlines are straight and parallel to each other. The width of the flow is perpendicular to the paper and is uniform. Now, if it is assumed that the flow within an area *abcd* is suddenly enclosed in a rectangular box as shown in Fig. 4.22, the flow within the box will have no circulatory motion. This is because

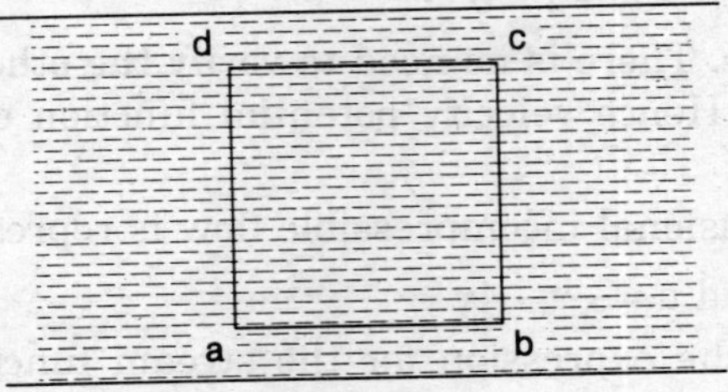

Fig. 4.22

the fluid near the wall ab strikes the wall bc with the same impact as the fluid near the wall dc at the wall bc. Thus the fluid strikes the wall bc throughout with a uniform impact.

Now consider the case of a circular (vortex) motion (Fig. 4.23). If it is a forced vortex, the velocity of flow on a streamline is directly proportional to its radial distance from the centre. As the mass of the elementary fluid along any streamline is proportional to its radial distance, it may be said that the momentum or the impact of the fluid along any streamline is proportional to r^2. This means that the impact of the fluid near the wall dc is more as compared to that of the fluid near the wall ab. This will tend to move the fluid at point c towards b within the box and thus tending to circulate the flow in a closed path.

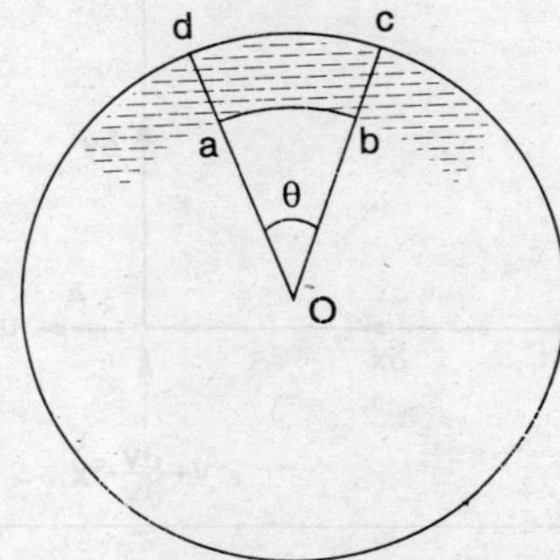

Fig. 4.23

In case of free vortex, the velocity is inversely proportional to the radial distance and the mass of the elementary fluid directly proportional to its radial distance. This means that the impact of the fluid striking the wall bc will be same throughout and the fluid has no tendency to circulate.

Thus circulation is the flow in a closed curve. Its magnitude is given by the line integral of the product of the velocity component tangent to an element of the closed path and the infinitesimal path length (length of the small element). Consider a closed curve and an elementary length ds on this curve as shown in Fig. 4.24. The velocity vector V makes an angle α with the tangent to the length ds.

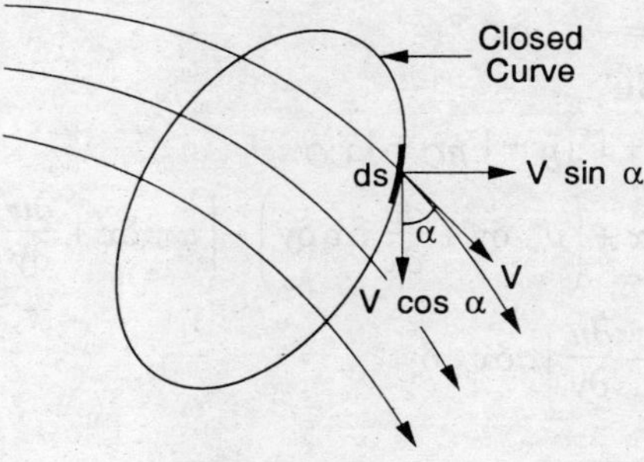

Fig. 4.24

Line integral for the element $ds = V \cos \alpha \,.\, ds$

and circulation,

$$\Gamma = \oint V \cos \alpha \, . \, ds$$

the line integral indicates that the summation is to be done all around the curve.

In a two-dimensional flow, consider a rectangular closed curve *OABC* of sides δ_x and δ_y as shown in Fig. 4.25

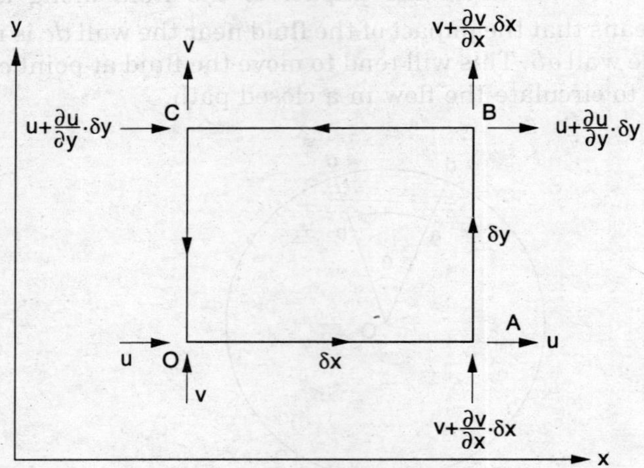

Fig. 4.25

Let $u = x$-component of velocity at O

$v = y$-component of velocity at O

then circulation along $OA = u \, . \, \delta x$

Circulation along $AB = \left(v + \frac{\partial v}{\partial x} \, . \, \delta x \right) \delta y$

Circulation along $BC = -\left(u + \frac{\partial u}{\partial y} \, . \, \delta y \right) \delta x$

Circulation along $CO = -\, v \, . \, \delta y$

Circulation along $OABC$,

$$\Gamma = \Gamma_{OA} + \Gamma_{AB} + \Gamma_{BC} + \Gamma_{CO}$$

$$= u \, . \, \delta x + \left(v \, . \, \delta y + \frac{\partial v}{\partial x} \delta x \, \delta y \right) - \left(u \, . \, \delta x + \frac{\partial u}{\partial y} \delta y \, \delta x \right) - v \, . \, \delta y$$

$$= \left(\frac{\partial v}{\partial x} - \frac{\partial u}{\partial y} \right) . \, \delta x \, . \, \delta y \qquad \text{...(4.36)}$$

Vorticity

It is defined as the circulation per unit area

$$\frac{\Gamma}{\partial x \, . \, \partial y} = \frac{\partial v}{\partial x} - \frac{\partial u}{\partial y}$$

$$= 2\omega_z$$

or $\qquad$ Vorticity = 2 × Rotation = $2\omega_z$ ...(4.37)

Thus the circulation per unit area is equal to twice the rotation about an axis perpendicular to the area and is called **vorticity.**

Again considering the forced vortex motion, in which $v = r \cdot \omega$ (Fig. 4.26)

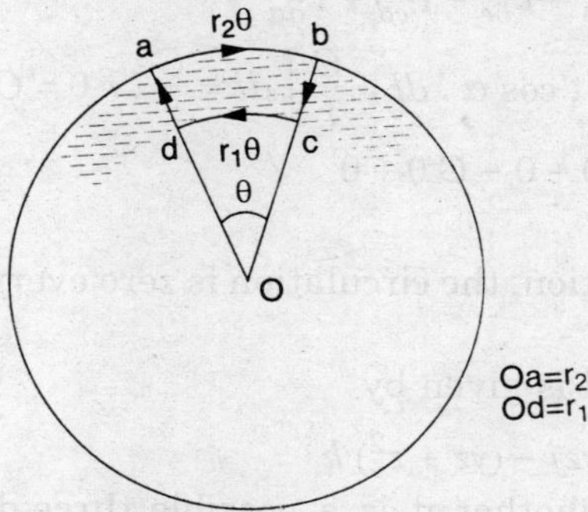

Fig. 4.26

$$\Gamma = \oint v \cdot \cos\alpha \cdot dl = \oint r\,\omega \cdot \cos\alpha \cdot dl$$

Here $\cos\alpha = 1$, ω constant. For a closed curved path along a stream line,

$$\oint dl = 2\pi r$$

$$\therefore \qquad \Gamma = r\,\omega \cdot 2\pi r = 2\pi r^2 \cdot \omega \qquad \text{...(4.38)}$$

i.e. circulation increases in direct proportion to the square of the radius of any circular path.

Circulation around a closed path, not including the centre may be calculated as below :

$$\Gamma_{abcd} = \Gamma_{ab} + \Gamma_{bc} + \Gamma_{cd} + \Gamma_{da}$$

$$\Gamma_{ab} = r_2\omega \cdot r_2\theta = r_2^{\,2}\omega\,\theta$$

$$\Gamma_{bc} = 0$$

$$\Gamma_{cd} = -\,r_1^{\,2}\,\omega\,\theta$$

$$\Gamma_{da} = 0$$

$$\Gamma_{abcd} = r_2^{\,2}\,\omega\,\theta - r_1^{\,2}\,\omega\,\theta$$

$$= \omega\,\theta\,(r_2^{\,2} - r_1^{\,2}) \qquad \text{...(4.39)}$$

In case of a free vortex, we have $v \cdot r$ = constant (= C). Thus circulation around *ay* streamlines along a concentric circle.

$$\Gamma = \int v \cos\alpha \cdot dl$$

$$= \frac{C}{r} \cos\alpha \int dl$$

$$= \frac{C}{r} \cdot 1 \cdot 2\pi r$$

$$= 2\pi C \qquad \text{...(4.40)}$$

i.e. the circulation is independent of the radius and depends upon the vortex constant C only. Circulation around a closed path excluding the vortex centre,

$$\Gamma_{abcd} = \Gamma_{ab} + \Gamma_{bc} + \Gamma_{cd} + \Gamma_{da}$$

$$\Gamma_{ab} = \int v \cdot \cos\alpha \cdot dl = \frac{C}{r}\int dl = \frac{C}{r} \cdot r\,\theta = C\,\theta$$

$$\Gamma = C\,\theta + 0 - C\,\theta + 0$$

$$= 0$$

Thus in a free vortex motion, the circulation is zero every where except at the centre, where the velocity is infinite.

Example 4.16 A flow field is given by

$$V = xyi + 2yzj - (yz + z^2)\,k$$

Find out whether it is a possible three-dimensional steady incompressible flow. Is the flow rotational ? If yes, find the angular velocity, vorticity, shear strain and dilatency at point P (3, 4, 5).

Solution The flow is physically possible if the continuity equation is satisfied.

$$\frac{\partial u}{\partial x} + \frac{\partial v}{\partial y} + \frac{\partial w}{\partial z} = 0$$

ie $\quad y + 2z - (y + 2z) = 0$

which is true. Hence the flow is a three-dimensional steady (because, the expression is not a function of t) incompressible.

The flow is irrotational if all the three conditions given by Eq. 4. are satisfied.

$$\frac{\partial v}{\partial x} = \frac{\partial u}{\partial y}$$

$$0 = x$$

which is not true. There is no need of checking the other two conditions. Thus the flow is rotational.

$$\omega = \frac{1}{2}(\omega_x\, i + \omega_y\, j + \omega_z\, k)$$

$$= \frac{1}{2}\left[\left(\frac{\partial w}{\partial y} - \frac{\partial v}{\partial z}\right)i + \left(\frac{\partial u}{\partial z} - \frac{\partial w}{\partial x}\right)j + \left(\frac{\partial v}{\partial x} - \frac{\partial u}{\partial y}\right)k\right]$$

$$= \frac{1}{2}\,[(-z - 2y)\,i + (0 - 0)\,j + (0 - x)\,k]$$

$$= -\frac{1}{2}\,[(4 + 6)\,i + 2k]$$

$$= -\frac{1}{2}\,(10i + 2k)$$

Vorticity = $2\omega = -(10i + 2k)$

Shear stain (Eq. 4.28),

$$\gamma_{xy} = \frac{1}{2}\left(\frac{\partial v}{\partial x} + \frac{\partial u}{\partial y}\right) = \frac{1}{2}(0 + x) = \frac{1}{2} \times 2 = 1$$

$$\gamma_{yz} = \frac{1}{2}\left(\frac{\partial w}{\partial y} + \frac{\partial v}{\partial z}\right) = \frac{1}{2}(-z + 2y) = \frac{1}{2}(-4 + 6) = 1$$

$$\gamma_{zx} = \frac{1}{2}\left(\frac{\partial u}{\partial z} + \frac{\partial w}{\partial x}\right) = \frac{1}{2}(0 + 0) = 0$$

Dilatency,

$$\varepsilon_x = \frac{\partial u}{\partial x} = y = 4$$

$$\varepsilon_y = \frac{\partial w}{\partial y} = 2z = 2 \times 5 = 10$$

$$\varepsilon_z = \frac{\partial w}{\partial z} = -(y + 2z) = -(4 + 2 \times 5) = -14$$

Example 4.17 A velocity field is prescribed by

$u = 8y - 4x$ and $v = 4y - 3x$

Determine the circulation around a closed curve defined by

$x_1 = 3$, $x_2 = 6$, $y_1 = 2$ and $y_2 = 5$

Solution $\Gamma_{ABCD} = \int_{ABCD} (u \,.\, dx + v \,.\, dy)$

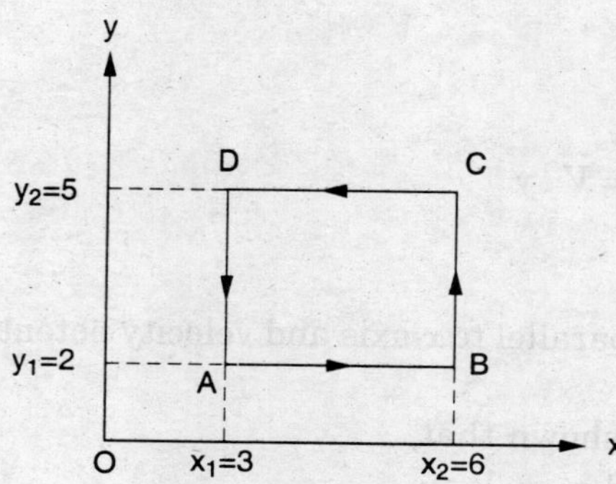

Fig. 4.27

$$= \int_{AB} (u \,.\, dx + v \,.\, dy) + \int_{BC} (u \,.\, dx + v \,.\, dy) + \int_{CD} (u \,.\, dx + v \,.\, dy) + \int_{DA} (u \,.\, dx + v \,.\, dy)$$

But along AB and CD, $v = 0$

Along BC and DA, $u = 0$

$$\Gamma_{ABCD} = \int_{AB} (u \,.\, dx + 0) + \int_{BC} (0 + v \,.\, dy) + \int_{CD} (u \,.\, dx + 0) + \int_{DA} (0 + v \,.\, dy)$$

$$= \int_{AB}^{6} (8y - 4x)\, dx + \int_{BC}^{5} (4y - 3x)\, dy + \int_{CD}^{3} (8y - 4x)\, dx + \int_{DA}^{2} (4y - 3x)\, dy$$

$$= [8yx - 2x^2] + [2y^2 - 3xy] + [8yx - 2x^2] + [2y^2 - 3xy]$$

$$= [8y_1 (x_2 - x_1) - 2 (x_2^2 - x_1^2)] + [2 (y_2^2 - y_1^2) - 3x_2 (y_2 - y_1)]$$

$$= [8y_2 (x_1 - x_2) - 2 (x_1^2 - x_2^2)] + [2(y_1^2 - y_2^2) - 3x_1 (y_1 - y_2)]$$

Now $x_1 = 3 \qquad y_1 = 2$

$x_2 = 6 \qquad y_2 = 5$

$$\Gamma_{ABCD} = (48 - 54) + (42 - 54) + (-120 + 54) + (-42 + 27)$$

$$= -6 - 12 - 66 - 15$$

$$= -99$$

$$\text{Circulation/unit area} = \frac{-99}{(x_2 - x_1)(y_2 - y_1)} = -\frac{99}{3 \times 3} - 11$$

4.15 BASIC FLOW PATTERNS

Some basic flow patterns are discussed below :

1. Rectilinear Flow

Horizointal flow

$$u = V$$

$$\frac{d\psi}{dy} = V$$

$$\psi = V \,.\, y + C$$

For $\psi = 0$ at the origin, $\psi = V \,.\, y$

Also

$$\phi = -V \,.\, x$$

Thus the streamlines are parallel to x-axis and velocity potential lines are parallel to y-axis (Fig. 4.28a).

Vertical flow : It can be shown that,

$$\psi = -V \,.\, x \text{ and } \phi = -V \,.\, y$$

The streamlines are parallel to y-axis and velocity potential lines parallel to x-axis.

Inclined flow : The velocity components in x and y directions are (Fig. 4.28b),

$$u = V \cos \alpha \quad \text{and} \quad v = V \sin \alpha$$

$$d\psi = \frac{\partial \psi}{\partial x} dx + \frac{\partial \psi}{\partial y} dy$$

$$= -v \,.\, dx + u \,.\, dy$$

$$\psi = uy - vx + C$$

For $\psi = 0$ at the origin

$$\psi = uy - vx$$

$$= V (y \cos \alpha - x \sin \alpha)$$

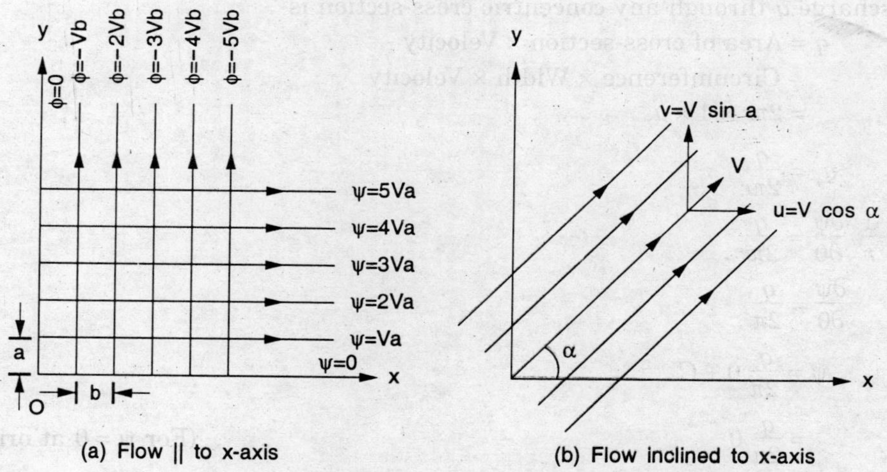

Fig. 4.28

$$\text{Similarly, } d\phi = \frac{\partial \phi}{\partial x}\,dx + \frac{\partial \phi}{\partial y}\,dy$$

$$= -u\,.\,dx - v\,.\,dy$$

$$\phi = -ux - vy + C$$

For $\phi = 0$ at the origin,

$$\phi = -ux - vy$$

$$= -V(x\cos\alpha + y\sin\alpha)$$

2. Source

A point in a plane from where fluid discharges radially out in all directions at a constant rate is known as a **source** (Fig. 4.29). The volume flow rate per unit width per unit time is known as the strength of the source (some authors define the source strength as $q/2\pi$).

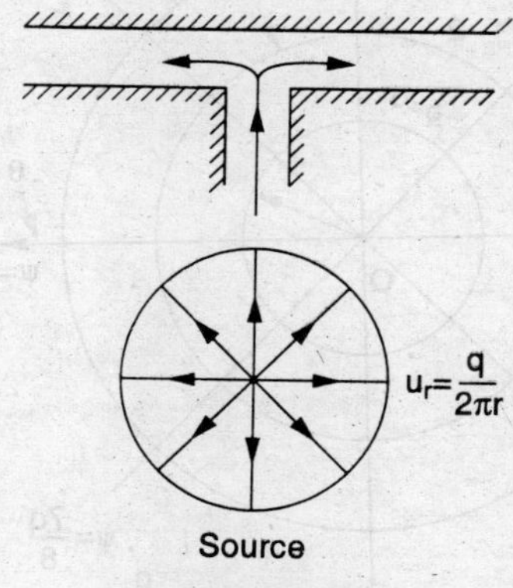

Fig. 4.29

The discharge q through any concentric cross-section is

$$q = \text{Area of cross-section} \times \text{Velocity}$$
$$= \text{Circumference} \times \text{Width} \times \text{Velocity}$$
$$= 2\pi r \times 1 \times u_r$$

or $$u_r = \frac{q}{2\pi r} \quad \text{...(i)}$$

$$\frac{1}{r}\frac{\partial\psi}{\partial\theta} = \frac{q}{2\pi r} \quad \text{(Eq. 4.23)}$$

$$\frac{\partial\psi}{\partial\theta} = \frac{q}{2\pi}$$

$$\psi = \frac{q}{2\pi}\theta + C$$

$$= \frac{q}{2\pi}\theta \quad \text{(For } \psi = 0 \text{ at origin, } C = 0\text{)}$$

When $\theta = 0$, $\psi = 0$

When $\theta = 2\pi$, $\psi = \frac{q}{2\pi} . 2\pi = q$

Also $$\psi = \frac{q}{2\pi}\theta = \frac{u_r . 2\pi r}{2\pi}\theta = r . u_r \theta = K . \theta \quad \text{[using (i)]}$$

where K is a constant. Thus expression for ψ shows that the streamlines are radial lines.
Also

$$\frac{\partial\phi}{\partial r} = -u_r = -\frac{q}{2\pi r} \quad \text{(using Eq. 4.26)}$$

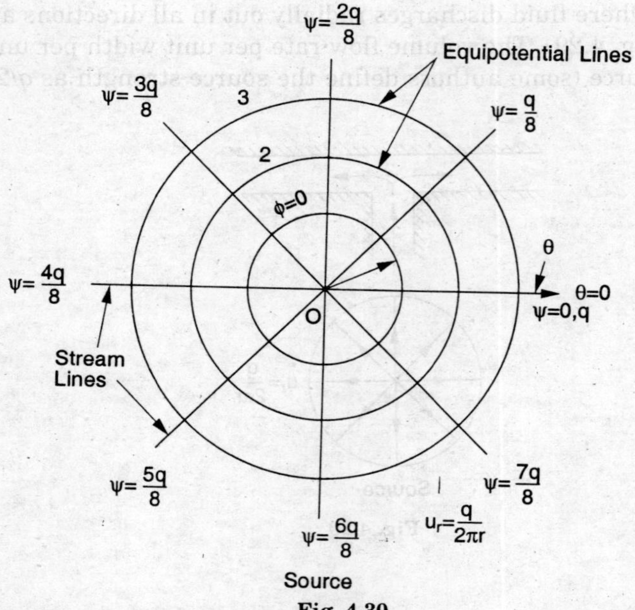

Fig. 4.30

$$\phi = -\frac{q}{2\pi}\log_e r + C$$

$$= -r \, . \, u_r \log_e r \qquad (\text{For } \phi = 0 \text{ at } r = 1)$$

$$= -K \log_e r$$

which indicates that the equipotential lines are concentric circles with O as centre. The streamlines and the equipotential lines intersect orthogonally (Fig. 4.30).

3. Sink

A **sink** is a source with the flow in the opposite direction *i.e.* in a sink the flow is radially inwards (Fig. 4.31). Thus sink is a source with negative strength. The pattern of streamlines and equipotential lines are the same. The equation for ψ and ϕ can be obtained by replacing q with $-q$.

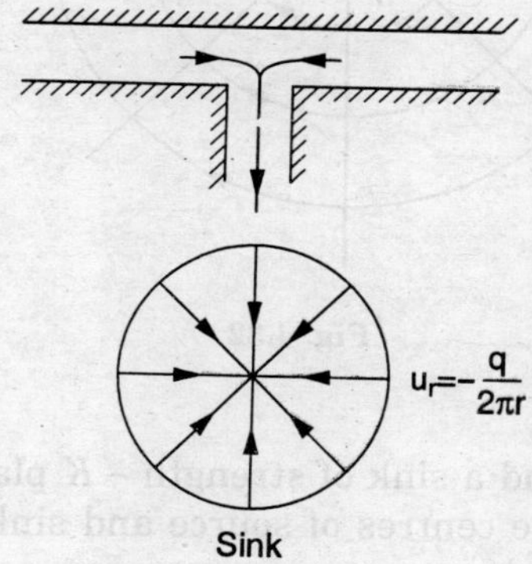

Fig. 4.31

4. Free Vortex Flow

For a free vortex flow

$$u_\theta r = \text{Constant} = C \qquad \text{(Eq. 5.38)}$$

Then $$u_\theta = \frac{C}{r}$$

or $$-\frac{\partial \psi}{\partial r} = \frac{C}{r} \qquad \text{(Eq. 4.23a)}$$

or $$\psi = -C \log_e r$$

Also $$u_\theta = \frac{C}{r}$$

$$-\frac{1}{r}\frac{\partial \phi}{\partial \theta} = \frac{C}{r} \qquad \text{(Eq. 4.27)}$$

or $$\frac{\partial \phi}{\partial \theta} = -C$$

or $$\phi = -C\,\theta$$

The pattern of streamlines and equipotential lines are similar to that for the source except that ψ and ϕ lines are interchanged as shown in Fig. 4.32.

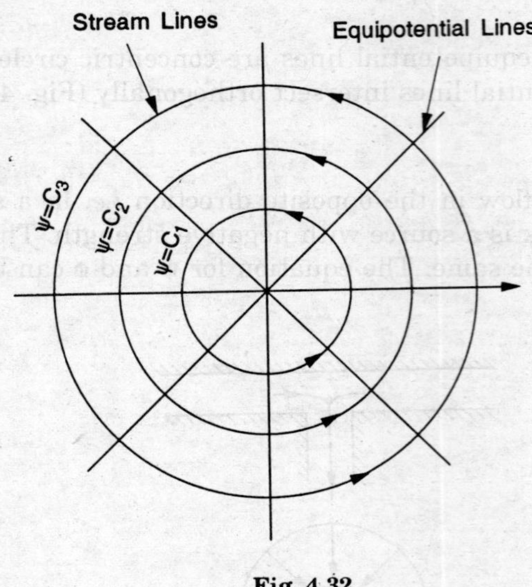

Fig. 4.32

5. Source and Sink Pair

Consider a source of strength K and a sink of strength $-K$ placed at a distance $2a$ apart. Let the midpoint of the line joining the centres of source and sink be the origin for x and y axes (Fig. 4.33). At any point $P(r, \theta)$

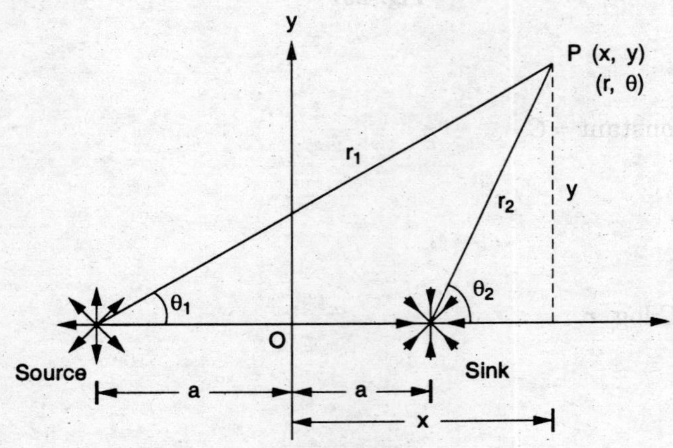

Fig. 4.33

$$\tan(\theta_2 - \theta_1) = \frac{\tan\theta_2 - \tan\theta_1}{1 - \tan\theta_1 \tan\theta_2}$$

$$= \frac{\dfrac{y}{x-a} - \dfrac{y}{x+a}}{1 + \dfrac{y}{x-a} \cdot \dfrac{y}{x+a}}$$

$$= \frac{\dfrac{yx + ay - yx + ya}{x^2 - a^2}}{\dfrac{x^2 - a^2 + y^2}{x^2 - a^2}}$$

$$= \frac{2ay}{x^2 - a^2 + y^2}$$

$$\theta_2 - \theta_1 = \tan^{-1} \frac{2ay}{x^2 - a^2 + y^2}$$

For the source, $\psi = K\,\theta_1$

For the sink, $\psi = -K\,\theta_2$

For a source and sink pair,

$$\psi = K\,\theta_1 - K\,\theta_2 = K\,(\theta_2 - \theta_1)$$

$$= -K \tan^{-1} \frac{2ay}{x^2 - a^2 + y^2}$$

$$-\frac{\psi}{K} = \tan^{-1} \frac{2ay}{x^2 - a^2 + y^2}$$

$$\tan\left(-\frac{\psi}{K}\right) = \tan^{-1} \frac{2ay}{x^2 - a^2 + y^2}$$

or $$x^2 - a^2 + y^2 = 2ay \tan\left(-\frac{\psi}{K}\right)$$

$$= 2ayp \qquad \left[\text{assuming } \tan\left(-\frac{\psi}{K}\right) = p\right]$$

$$x^2 + (y^2 - 2ayp) - a^2 = 0$$

$$x^2 + (y^2 - 2ayp + a^2p^2) - (a^2p^2 + a^2) = 0$$

$$x^2 + (y - ap)^2 = a^2\,(1 + p^2)^2$$

which is the equation of concentric circles with centre $(0, ap)$ *i.e.* centres on y-axis and radius $a\sqrt{1 + p^2}$ (Fig. 4.34).

$$\phi = K \log_e r_1 - K \log_e r_2 = K \log_e \frac{r_1}{r_2}$$

which is the equation of eccentric non-intersecting circles with centres on the x-axis (Fig. 4.34).

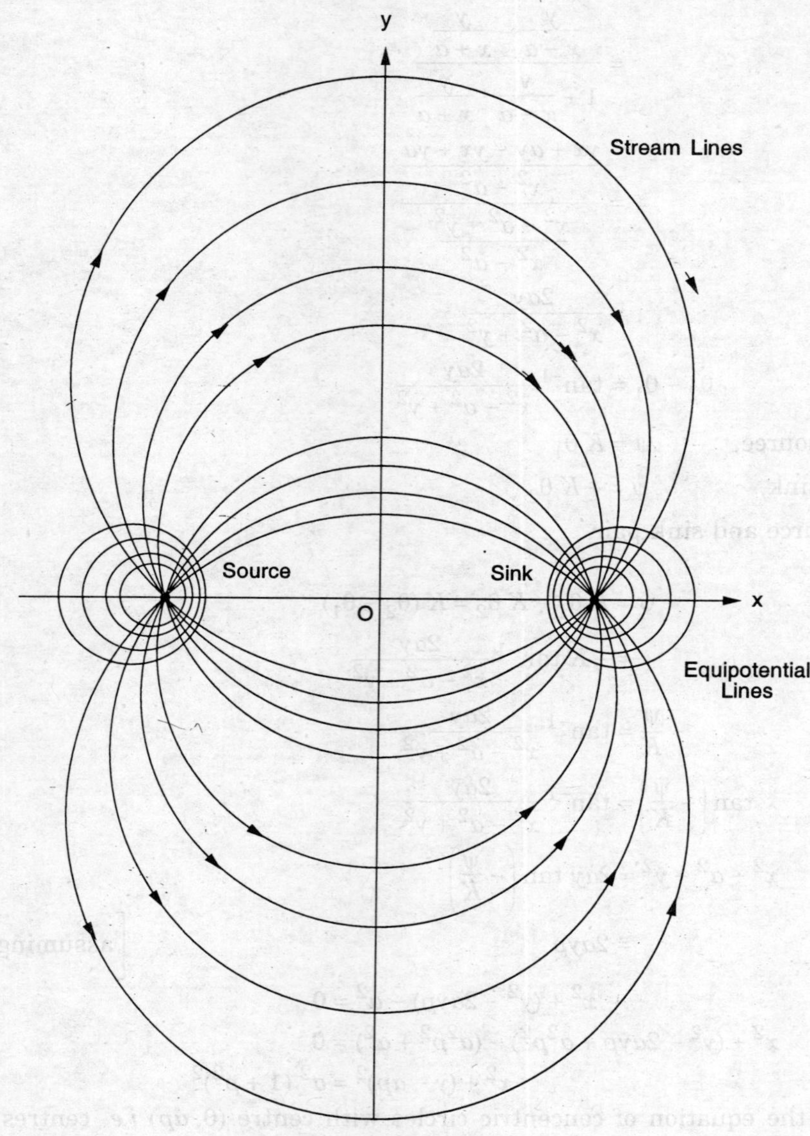

Fig. 4.34

6. Doublet

A **doublet** is defined as a special case of a source and a sink pair when the two approach each other and the product $\delta a \,.\, q$ (known as the strength of the doublet μ) remains constant.

$$\mu = \delta_a \,.\, q \text{ or } q = \frac{\mu}{\delta a} \qquad ...(i)$$

In a doublet source and sink approach each other.

Let $\theta_1 = \theta$

and $\theta_2 = \theta + \delta\,\theta$

Refer Fig. 4.35

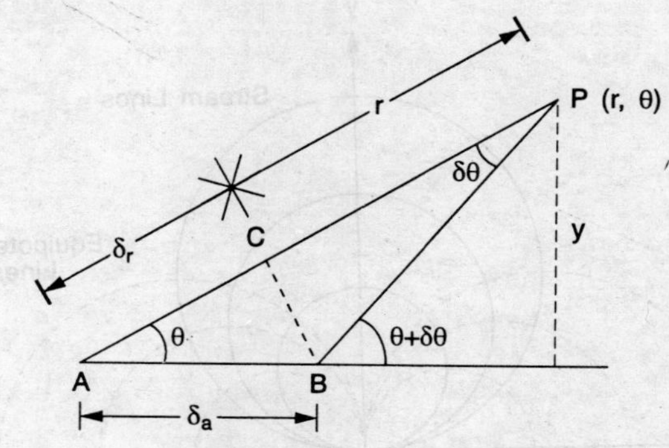

Fig. 4.35

$$\psi = K\,\theta_1 - K\,\theta_2 = -K\,[\theta - (\theta + \delta\theta)] = -K\,.\,\delta\theta$$

$$= -K\,.\,\frac{BC}{r} \qquad \left(\text{As } \delta\theta \text{ is small, } \delta\theta = \frac{BC}{r}\right)$$

$$= -K\,.\,\frac{AB\sin\theta}{r}$$

$$= -\frac{q}{2\pi}\cdot\frac{\delta_a\sin\theta}{r}$$

$$= -\frac{1}{2\pi}\left(\frac{\mu}{\delta_a}\right)\frac{\delta_a\sin\theta}{r} \qquad (\text{from } (i))$$

$$= -\frac{\mu}{2\pi}\cdot\frac{\sin\theta}{r}$$

$$= -\frac{\mu}{2\pi}\cdot\frac{y/r}{r} \qquad (\text{As } \delta_a \to 0,\ \sin\theta = y/r)$$

$$= -\frac{\mu}{2\pi}\cdot\frac{y}{r^2}$$

$$= -\frac{\mu}{2\pi}\cdot\frac{y}{x^2+y^2} \qquad (r^2 + x^2 + y^2)$$

$$x^2 + y^2 + \frac{\mu}{2\pi\psi}\,.\,y = 0$$

$$x^2 + \left[y^2 + \left(\frac{\mu}{2\pi\psi}\right)^2 + \frac{\mu}{2\pi\psi}\,y\right] - \left(\frac{\mu}{2\pi\psi}\right)^2 = 0$$

$$x^2 + \left(y + \frac{\mu}{4\pi\psi}\right)^2 = \left(\frac{\mu}{4\pi\psi}\right)^2$$

Which shows that the streamlines are circles with centre $\left(0, \frac{\mu}{4\pi\psi}\right)$ and radius $\frac{\mu}{4\pi\psi}$. Thus the streamlines are circles tangent to the x-axis with entres on y-axis (Fig. 4.36).

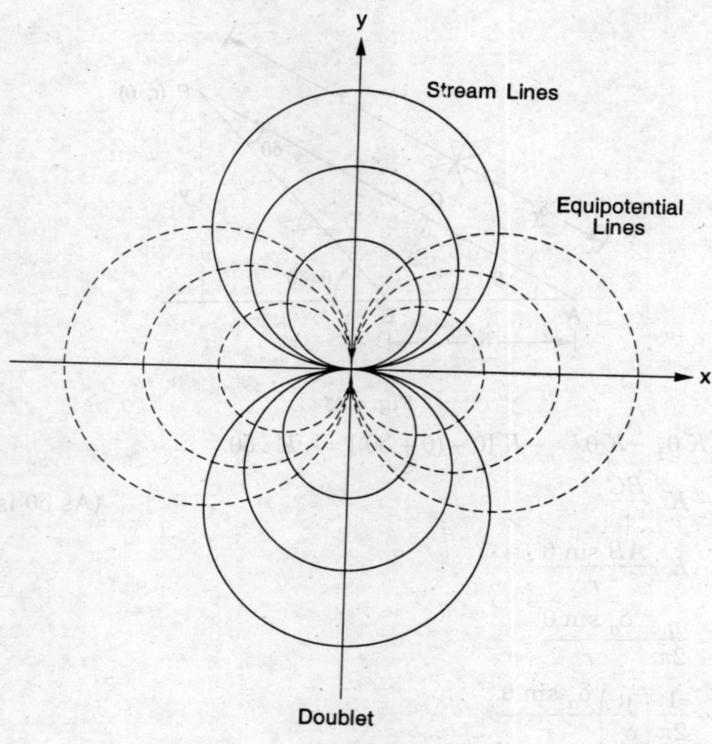

Fig. 4.36

$$\phi = K \log \frac{r_1}{r_2} = \frac{q}{2\pi} \log_e \left(\frac{r + \delta r}{r}\right)$$

$$= \frac{q}{2\pi} \log_e \left(1 + \frac{\delta r}{r}\right)$$

$$= \frac{q}{2\pi} \left[\frac{\delta r}{r} + \frac{1}{2}\left(\frac{\delta r}{r}\right)^2 + ...\right]$$

$$= \frac{q}{2\pi} \cdot \frac{\delta r}{r} + \text{smaller terms}$$

$$= \frac{q}{2\pi} \cdot \frac{\delta_a \cos\theta}{r} = \frac{\mu}{2\pi\delta_a} \cdot \frac{\delta_a \cos\theta}{r} = \frac{\mu}{2\pi} \cdot \frac{\cos\theta}{r} = \frac{\mu}{2\pi} \cdot \frac{x}{r^2}$$

$$= \frac{\mu}{2\pi} \cdot \frac{x}{x^2 + y^2}$$

$$x^2 + y^2 - \frac{\mu}{2\pi\phi} \cdot x = 0$$

$$y^2 + \left(x - \frac{\mu}{4\pi\phi}\right)^2 = \left(\frac{\mu}{4\pi\phi}\right)^2$$

This is the equation of a circle with centre at $\left(\frac{\mu}{4\pi\phi}, 0\right)$ and radius $\frac{\mu}{4\pi\phi}$ · Thus equipotential lines are a family of circles tangent to the y-axis and centres on x-axis (Fig. 4.35).

7. Doublet in a Uniform Flow

Stream function is given by

$$\psi = \psi_{\text{uniform flow}} + \psi_{\text{doublet}}$$

$$= V \cdot y - \frac{\mu}{2\pi r} \sin\theta$$

$$= V r \sin\theta - \frac{\mu}{2\pi r} \sin\theta$$

$$= \sin\theta \left(V \cdot r - \frac{\mu}{2\pi r}\right)$$

For the dividing streamline $\psi = 0$, *i.e.*

$$\psi = \sin\theta \left(V \cdot r - \frac{\mu}{2\pi r}\right) = 0$$

Either $\sin\theta = 0$ which is the x-axis.

or $$V \cdot r = \frac{\mu}{2\pi r}$$

or $$r^2 = \frac{\mu}{2\pi V} = \text{Constant}$$

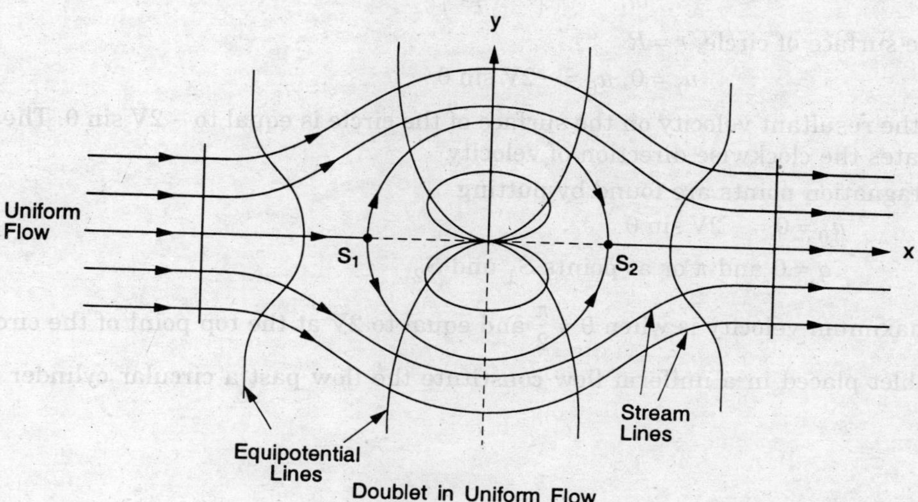

Doublet in Uniform Flow

Fig. 4.37

These observations show that the dividing streamline is a circle with centre at the doublet and the radius $R = \sqrt{\dfrac{\mu}{2\pi V}}$ · Flow field due to doublet lies completely within the circle. Uniform flow remains outside and does not mingle with the doublet flow (Fig. 4.37). Let R be the radius of the doublet circle.

$$R^2 = \frac{\mu}{2\pi V} \cdot r$$

or $$\frac{\mu}{2\pi} = R^2 V$$

Thus $$\psi = \sin\theta \left(V \cdot r - \frac{R^2 V}{r} \right) = V \left(r - \frac{R^2}{r} \right) \sin\theta$$

$$\phi = V \cdot x + \frac{\mu}{2\pi r} \cos\theta$$

$$= V \cdot r \cos\theta + \frac{\mu}{2\pi r} \cos\theta$$

$$= V \left(r + \frac{R^2}{r} \right) \cos\theta$$

Velocity components $u_r = \dfrac{1}{2} \dfrac{\partial \psi}{\partial \theta}$

$$= \left[V \left(1 - \frac{R^2}{r^2} \right) \right] \frac{d}{d\theta} (\sin\theta)$$

$$= V \left(1 - \frac{R^2}{r^2} \right) \cos\theta$$

$$u_\theta = -\frac{\partial \psi}{\partial r} = -V \left(1 + \frac{R^2}{r^2} \right) \sin\theta$$

On the surface of circle, $r = R$

$$u_r = 0,\ u_\theta = -2V \sin\theta$$

Thus the resultant velocity on the surface of the circle is equal to $-2V \sin\theta$. The negative sign indicates the clockwise direction of velocity.

The stagnation points are found by putting

$$u_\theta = 0 = -2V \sin\theta$$

i.e. $q = 0$ and π or at points S_1 and S_2

The maximum velocity is when $\theta = \dfrac{\pi}{2}$ and equal to 2V at the top point of the circle.

A doublet placed in a uniform flow constitute the flow past a circular cylinder of radius R.

PROBLEMS

1. What are the methods of describing a fluid flow ? How do the Langrangian and Eulerian methods differ ?
2. Define the terms : streamlines, streaklines and pathlines.
3. Distinguish between :
   (a) Steady and unsteady flow (b) Compressible and incompressible flow
   (c) Uniform and non-uniform flow (d) Laminar and turbulent flow
   (e) Rotational and irrotational flow
4. What do you understand by convective and local accelerations ?
5. What is equation of continuity ? Derive it for a three-dimensional flow in the differential form.
6. Define the terms stream function and velocity potential function.
7. Show that the streamlines and equipotential lines are perpendicular to each other.
8. If stream function exists for a flow, does it mean velocity potential function also exists ?
9. What is a flow net ? How is it drawn ?
10. What are the types of motion which a fluid element can undergo ?
11. What do you mean by circulation and vorticity ?
12. Discuss and sketch the following flow patterns :
    (a) Source and sink (b) Source and sink pair
    (c) Doublet (d) Doublet in a uniform flow
13. A pipe 300 mm in diameter branches into two pipes of diameters 150 mm and 200 mm. The average velocity of water in the 300 mm diameter pipe is 2·5 m/s. Determine the velocity of water in the 150 mm diameter pipe if the average velocity in the 200 mm diameter pipe is 2 m/s. (6·44 m/s)
14. A flow field is represented by
$$V = (3 + 2xy + 4t^2)\, i + (xy^2 + 3t)\, j$$
Determine the velocity and acceleration at point (1, 2) after 2 seconds.
(25·05 m/s, 186 m/s^2)
15. Two velocity components of a three-dimensional incompressible flow field are given as
$u = xy$ and $v = 2yz$
Determine the third component. $[-yz - z^2 + f(x, y, t)]$
16. Find out which of the velocity component sets given below satisfy the continuity equation :
    (a) $u = cx,\ v = -cy$ (b) $u = x + y,\ v = x - y$
    (c) $u = 2x^2 + yz,\ v = -2xy + 3y^3 + 3yz,\ w = -(3/2)z^2 - 2xy - 6yz$ (*a* and *b*)
17. A velocity field is given by
$$u = x^2y,\ v = 2yz - xy^2 \text{ and } w = x^2 - z^2$$
Is it a case of kinematically possible flow for an incompressible fluid ? (yes)

18. The velocity components in a fluid flow are given by

$$u = 2xy \text{ and } v = a^2 + x^2 - y^2$$

show that the flow is possible. Determine the stream function also.

($\psi = xy^2 - a^2x - x^3/3 + C$)

19. The velocity components for a two-dimensional flow are

$$u = -y^3 \text{ and } v = -3x$$

Obtain the equation of streamline passing through the point (2, 3).

($2y^3 - 9x^2 - 18 = 0$)

20. A stream function is given as $\psi = 3x^2y - y^3$. Find the velocity at point (2, 3).

(39 units)

21. A velocity field is given by

$$V = 2x^3 i - 6x^2 yj$$

Determine the equation to the streamlines. Also sketch the streamlines. ($yx^3 = C$)

22. State which of the following velocity potential functions are the possible cases of irrotational flow :

(a) $\phi = \left(r - \frac{2}{r}\right) \sin \theta$ (b) $\phi = \sin x$

(c) $\phi = 3x^2 - 3x + 3y^2 + 16t + 2zt$ (d) $\phi = x + y + z$ (d)

23. State the stream functions which are possible irrotational flow fields :

(a) $\psi = 2xy + y$ (b) $\psi = A(x^2 - y^2)$

(c) $\psi = Ax^2y^2$ (d) $\psi = 2Axy$

(e) $\psi = A\theta$ (f) $\psi = A \cos xy$ (a, b, d and e)

24. Find whether the following cases represent possible cases of irrotational flow :

(a) $u = x + y, v = x - y$ (b) $u = 2x, v = x^2 - y^2$

(c) $u = -x, v = 2y, w = 5 - z$ (a and c are irrotational)

25. In a two-dimensional flow, the velocity components are

$$u = ax \text{ and } v = by$$

Determine the equation of streamlines. ($X^b = Cy^a$)

26. In a steady fluid flow, the velocity components are given by

$$u = 2kx, v = 2ky, w = -4kz$$

Obtain the equation of a streamline passing through the point (1, 0, 1).

($y = 0$ and $z = 1/x^2$)

27. A velocity field is given by

$$u = x^2 + 2xy \text{ and } v = -(y^2 + 2xy)$$

Find the circulation around a closed curve given by $x = 1, x = 3$ and $y = 1, y = 4$.

(– 86)

Chapter 5

FLUID DYNAMICS

5.1 INTRODUCTION

Fluid dynamics is the study of fluid motion with the consideration of forces or energies which cause the motion. In solid mechanics, solution to problems is obtained by applying the basic laws to a fixed quantity of matter. However, in fluid mechanics, analysis is made by considering a fixed region of finite boundaries rather than a fixed mass. The fixed region is known as the **control volume** and the bounded surface as the **control surface.** The size of the control volume may be chosen according to the convenience which may be infinitesimally small or large.

5.2 EULER'S EQUATION OF MOTION

Euler's equation of motion is obtained by considering the motion of a fluid element along a stream line (Fig. 5.1). Let dA be the cross-sectional area, ds the length and θ the angle of the fluid element with the vertical. Various forces acting on the fluid element are

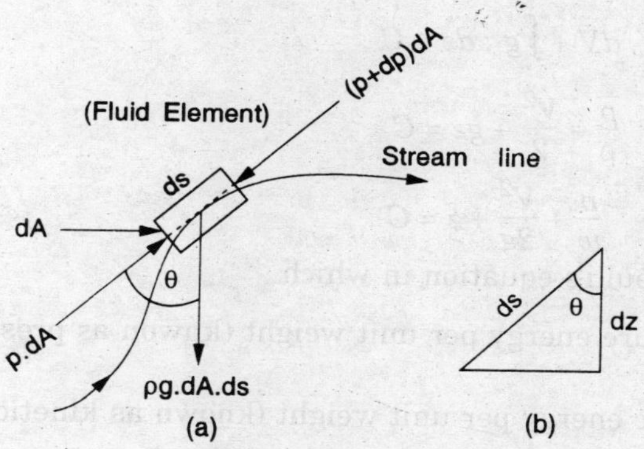

Fig. 5.1

1. Pressure force $p \cdot dA$ in the direction of flow
2. Pressure force $(p + dp) \cdot dA$ in the opposite direction of flow and
3. Weight of the fluid element $(\rho g \cdot dA \cdot ds)$ acting vertically downward

From Newton's second law of motion,

$$\text{Force} = \text{Mass} \times \text{Acceleration}$$

$$p \cdot dA - (p + dp) \cdot dA - \rho g \cdot dA \cdot ds \cos\theta = \rho dA \cdot ds \times a$$

or

$$-dp - \rho g \cdot ds \cos\theta = \rho \cdot ds \times a \qquad \text{...(i)}$$

Now, acceleration $a = \dfrac{dV}{dt}$ where V is the velocity.

In general, V is a function of position and time, *i.e.*

$$V = \frac{\partial V}{\partial t} \cdot dt + \frac{\partial V}{\partial s} \cdot ds$$

$$a = \frac{\partial V}{\partial t} = \frac{\partial V}{\partial t}\frac{dt}{dt} + \frac{\partial V}{\partial s}\frac{\partial s}{\partial t}$$

$$= \frac{\partial V}{\partial t} + V\frac{\partial V}{\partial s} \qquad \left(V = \frac{\partial s}{\partial t}\right)$$

In a steady flow, $\dfrac{\partial V}{\partial t} = 0$

$$a = V\frac{dV}{ds} \qquad \textbf{(Partial derivative becomes total derivative)}$$

(i) becomes,

$$-dp - \rho g \cdot ds \cdot \frac{dz}{ds} = \rho \cdot ds \cdot V\frac{dV}{ds} \qquad \left(\cos\theta = \frac{dz}{ds}\right)$$

$$-dp - \rho g \cdot dz = \rho V \cdot dV$$

$$\frac{dp}{\rho} + V \cdot dV + g \cdot dz = 0 \qquad \text{...(5.1)}$$

This is the **Euler's equation** of motion.

5.3 BERNOULLI'S EQUATION

The integration of Euler's equation leads to the derivation of Bernoulli's equation.

$$\frac{1}{\rho}\int dp + \int V \cdot dV + \int g \cdot dz = C$$

$$\frac{p}{\rho} + \frac{V^2}{2} + gz = C$$

$$\frac{p}{w} + \frac{V^2}{2g} + z = C \qquad \text{...(5.2)}$$

This is known as Bernoulli's equation in which

$\dfrac{p}{w}$ = pressure energy per unit weight (knwon as pressure head)

$\dfrac{V^2}{2g}$ = kinetic energy per unit weight (known as kinetic head)

z = potential energy per unit weight (known as potential head)

If the equation is applied to two sections of fluid flow,

$$\frac{p_1}{w}+\frac{V_1^2}{2g}+z_1=\frac{p_2}{w}+\frac{V_2^2}{2g}+z_2 \quad ...(5.3)$$

The equation is applicable to an ideal fluid (frictionless flow of the fluid with zero viscosity), steady, incompressible and irrotational flow.

Bernoulli's equation may be modified to apply to the real fluids by considering the losses during the flow between two sections *i.e.*

$$\frac{p_1}{w}+\frac{V_1^2}{2g}+z_1=\frac{p_2}{w}+\frac{V_2^2}{2g}+z_2+h_L \quad ...(5.4)$$

where h_L = head loss between sections 1 and 2.

5.4 ENERGY EQUATION

Energy equation is derived by considering the various energies of the fluid in a control volume (Fig. 5.2). A fluid flowing through a control volume consists of following energies per unit weight at section '1' of the passage :

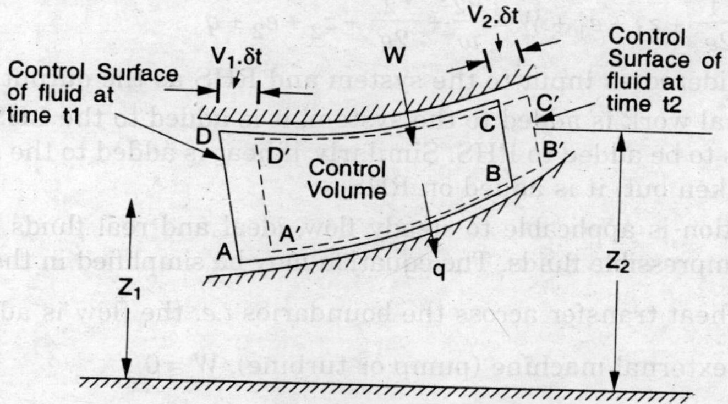

Fig. 5.2

1. Internal energy, e_1
2. Kinetic energy,

$$\frac{1}{2}mV_1^2=\frac{1}{2}\frac{W}{g}.V_1^2$$

Kinetic energy per unit weight $=\dfrac{V_1^2}{2g}$

3. Potential energy, z_1
4. Energy due to flow weork

Let A_1 = area of the control volume at any cross-section

V_1 = velocity of flow

p_1 = pressure of the fluid

w = specific weight of the fluid

Then flow work or the work done in moving the fluid during time dt,

$$W = \text{Force} \times \text{Distance moved}$$
$$= \text{Force} \times \text{Velocity} \times \text{Time}$$
$$= (p_1 . A_1) . (V_1 . t)$$

Weight of the fluid moved during the period

$$= w \times \text{Volume of fluid moved}$$
$$= w \times (\text{Area} \times \text{Velocity} \times \text{Time})$$
$$= w \times A_1 \times V_1 \times \delta t$$

$$\text{Flow work per unit weight} = \frac{p_1 A_1 V_1 \delta t}{w . A_1 V_1 \delta t} = \frac{p_1}{w}$$

Similarly, the quantities can be defined at section '2'.

Further, during the flow between two chosen sections, mechanical work W and heat q may also transfer into or out of the system. Assuming that the work is added to the system whereas the heat is dissipated to the surroundings, the general equation for an incompressible steady fluid flow between the two sections can be written as (energies are expressed per unit weight)

$$\frac{p_1}{w} + \frac{V_1^2}{2g} + z_1 + e_1 + W = \frac{p_2}{w} + \frac{V_2^2}{2g} + z_2 + e_2 + q \qquad ...(5.5)$$

LHS may be considered as input to the system and RHS as the output.

Thus, if mechanical work is added to the system, it is added to the LHS and if it is taken out of the system, it is to be added to RHS. Similarly, if heat is added to the system, it is added on LHS and if it is taken out, it is added on RHS.

This energy equation is applicable to steady flow, ideal and real fluids, liquids and gases, compressible and incompressible fluids. The equation may be simplified in the following way :

- If there is no heat transfer across the boundaries *i.e.* the flow is adiabatic, $q = 0$.
- If there is no external machine (pump or turbine), $W = 0$.
- If there are no friction losses (*i.e.* fluid is ideal), no heat transfer and no external work there is no change in the internal energy of the fluid *i.e.* $e = 0$.

Then the energy equation reduces to

$$\frac{p_1}{w} + \frac{V_1^2}{2g} + z_1 = \frac{p_2}{w} + \frac{V_2^2}{2g} + z_2$$

$$\frac{p}{w} + \frac{V^2}{2g} + z = C$$

which is the Bernoulli's equation as defined in section 5.3.

Example 5.1 Water flows through a pipe with diameters 300 mm and 450 mm at its two ends where the corresponding pressures are 120 kN/m^2 and 400 kN/m^2 respectively. If the rate of flow through the pipe is 50 l/s, find the difference in the datum head.

Solution Let 1 and 2 represent the upper end and the lower end respectively.

$$Q = 50 \text{ l/s} = 0{\cdot}05 \text{ m}^3\text{/s}$$

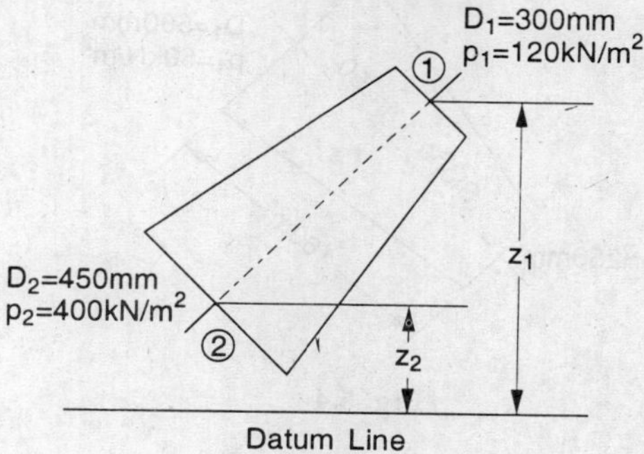

Fig. 5.3

$$p_1 = 120 \text{ kN/m}^2$$

$$p_2 = 400 \text{ kN/m}^2$$

$$V_1 = \frac{Q}{A_1} = \frac{0{\cdot}05}{\frac{\pi}{4}(0{\cdot}3)^2} = 0{\cdot}707 \text{ m/s}^2$$

$$V_2 = \frac{Q}{A_2} = \frac{0{\cdot}05}{\frac{\pi}{4}(0{\cdot}45)^2} = 0{\cdot}314 \text{ m/s}^2$$

$$\frac{p_1}{w} + \frac{V_1^2}{2g} + z_1 = \frac{p_2}{w} + \frac{V_2^2}{2g} + z_2$$

$$\frac{120}{9{\cdot}81} + \frac{(0{\cdot}707)^2}{2 \times 9{\cdot}81} + z_1 = \frac{400}{9{\cdot}81} + \frac{0{\cdot}314^2}{2 \times 9{\cdot}81} + z_2$$

$$z_1 - z_2 = \frac{1}{9{\cdot}81}\left[(400 - 120) + \frac{(0{\cdot}314)^2 - (0{\cdot}707)^2}{2}\right]$$

$$= \frac{1}{9{\cdot}81}(280 - 0{\cdot}2)$$

$$= 28{\cdot}52 \text{ m}$$

Example 5.2 Oil of specific gravity 0·9 flows through a taper pipe having a slope of 1 in 50 upwards. The rate of oil flowing through the pipe is 80 l/s. The pipe is 160 m long and has diameters 500 mm at the upper end and 250 mm at the lower end. What will be the pressure at the lower end if the pressure at the upper end is 50 kN/m^2.

Solution Let 1 and 2 represent the upper end and the lower end respectively.

$$Q = 80 \text{ l/s} = 0{\cdot}08 \text{ m}^3/\text{s}$$

$$p_1 = 50 \text{ kN/m}^2$$

$$p_2 = ?$$

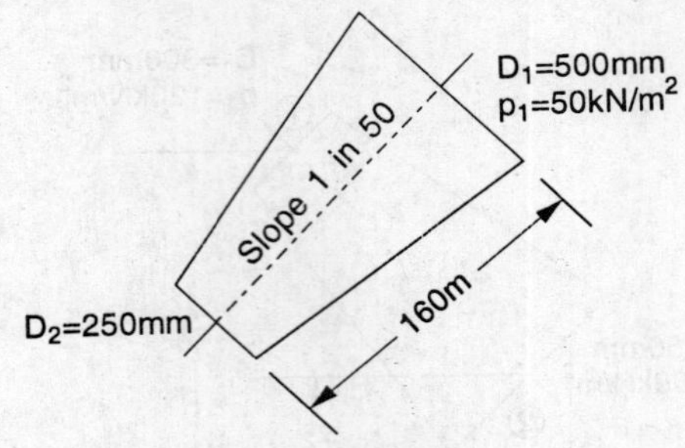

Fig. 5.4

$$V_1 = \frac{Q}{A_1} = \frac{0{\cdot}08}{\frac{\pi}{4}(0{\cdot}5)^2} = 0{\cdot}407 \text{ m/s}^2$$

$$V_2 = \frac{Q}{A_2} = \frac{0{\cdot}08}{\frac{\pi}{4}(0{\cdot}25)^2} = 1{\cdot}63 \text{ m/s}^2$$

$$z_1 = \frac{160}{50} = 3{\cdot}2 \text{ m} \quad \text{and } z_2 = 0$$

$$\frac{p_2}{w} + \frac{V_1^2}{2g} + z_1 = \frac{p_2}{w} + \frac{V_2^2}{2g} + z_2$$

$$\frac{50}{0{\cdot}9 \times 9{\cdot}81} + \frac{(0{\cdot}407)^2}{2 \times 9{\cdot}81} + 3{\cdot}2 = \frac{p_2}{0{\cdot}9 \times 9{\cdot}81} + \frac{(1{\cdot}63)^2}{2 \times 9{\cdot}81} + 0$$

$$\frac{p_2}{0{\cdot}9} = \frac{50}{0{\cdot}9} + \frac{1}{2}(0{\cdot}407^2 - 1{\cdot}63^2) + 31{\cdot}392$$

$$p_2 = 0{\cdot}9\,(55{\cdot}56 - 1{\cdot}197 + 31{\cdot}392)$$

$$= 77{\cdot}18 \text{ kN/m}^2$$

Example 5.3 The flow through a conical vertical tube of length 2·4 m is in the downward direction. The velocity of flow at the inlet end is 8 m/s and at the outlet end 3 m/s. The pressure head at the inlet end is 3 m of liquid and the loss of head in the tube is $0{\cdot}3\,(V_1 - V_2)^2/2g$, where V_1 and V_2 are the velocities at the inlet and outlet ends of the tube respectively. Find the pressure head at the lower end.

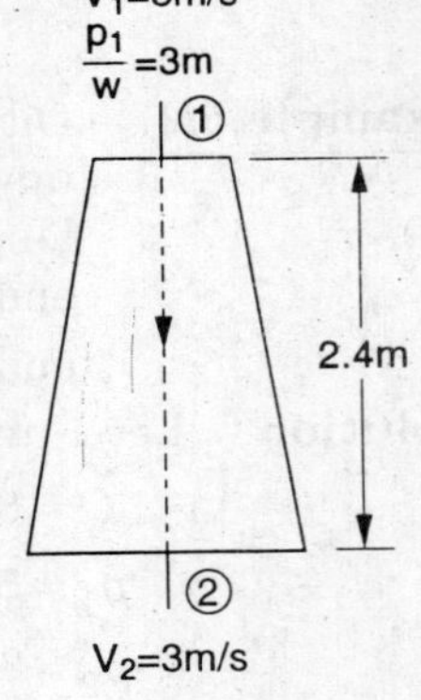

Fig. 5.5

Solution Let 1 represents the upper (inlet) end and 2 the lower (outlet) end of the tube.

$$\frac{p_1}{w} = 3 \text{ m of liquid}$$

$$V_1 = 8 \text{ m/s}, \quad V_2 = 3 \text{ m/s}$$

$$z_1 = 2{\cdot}4 \text{ m and } z_2 = 0$$

$$\frac{p_1}{w} + \frac{{V_1}^2}{2g} + z_1 = \frac{p_2}{w} + \frac{{V_2}^2}{2g} + z_2 + h_L$$

$$3 + \frac{8^2}{2 \times 9{\cdot}81} + 2{\cdot}4 = \frac{p_2}{w} + \frac{3^2}{2 \times 9{\cdot}81} + 0 + \frac{0{\cdot}3\,(8-3)^2}{2 \times 9{\cdot}81}$$

$$\frac{p_2}{w} = 5{\cdot}4 + \frac{1}{2 \times 9{\cdot}81}(8^2 - 3^2 - 0{\cdot}3 \times 2{\cdot}5)$$

$$= 5{\cdot}4 + 2{\cdot}42$$

$$= 7{\cdot}82 \text{ m of liquid}$$

Example 5.4 The diameter of a pipe line carrying an oil of specific gravity 0·85 changes from 240 mm at the lower end to 480 mm at the upper end. The level of the upper end is 3 m higher than the lower level. The pressures at the upper and lower levels are 100 kN/m^2 and 160 kN/m^2 respectively. Determine the loss of head and the direction of flow if the discharge through the pipe is 250 l/s.

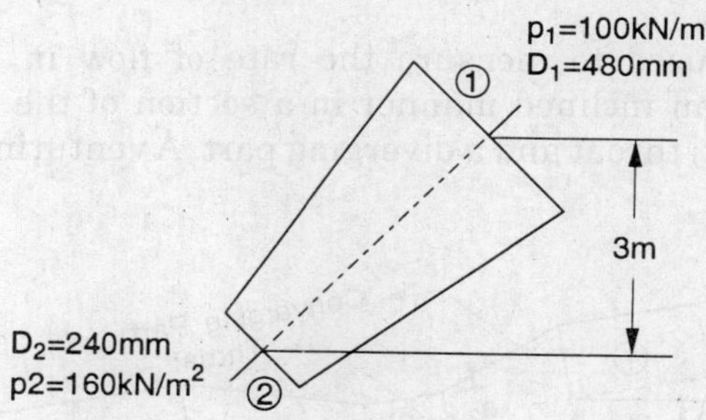

Fig. 5.6

Solution $p_1 = 100 \text{ kN/m}^2$

$$V_1 = \frac{Q}{A_1} = \frac{0{\cdot}25}{\frac{\pi}{4}(0{\cdot}48)^2} = 1{\cdot}382 \text{ m/s}^2$$

$$z_1 = 3 \text{ m}$$

Total energy (head) at the upper end

$$\frac{p_1}{w} + \frac{{V_1}^2}{2g} + z_1 = \frac{100}{0{\cdot}85 \times 9{\cdot}81} + \frac{(1{\cdot}382)^2}{2 \times 9{\cdot}81} + 3$$

$$= 11{\cdot}993 + 0{\cdot}097 + 3$$

$$= 15{\cdot}09 \text{ m}$$

$$p_2 = 160 \text{ kN/m}^2$$

$$V_2 = \frac{Q}{A_2} = \frac{0.25}{\frac{\pi}{4}(0.24)^2} = 5.526 \text{ m/s}^2$$

$$z_2 = 0$$

Total energy (head) at the lower end

$$\frac{p_2}{w} + \frac{V_2^2}{2g} + z_2 = \frac{160}{0.85 \times 9.81} + \frac{(5.526)^2}{2 \times 9.81} + 0$$

$$= 19.188 + 1.557$$

$$= 20.745 \text{ m}$$

As the total head at the lower end is more than that at the upper end, the flow is from the lower end to the upper end.

Loss of head $= 20.745 - 15.09$

$= 5.655$ m

5.5 PRACTICAL APPLICATIONS OF BERNOULLI'S EQUATION

Bernoulli's equation is applicable to all problems of incompressible flow involving different kinds of energies. Its application to some measuring devices is illustrated below :

Venturimeter

A venturimeter is a device used to measure the rate of flow in a pipe. It may be fixed horizontally, vertically or in an inclined manner in a section of the pipe. It consists of three parts : a short converging part, throat and a diverging part. A venturimeter fitted to an inclined pipe is shown in Fig. 5.7.

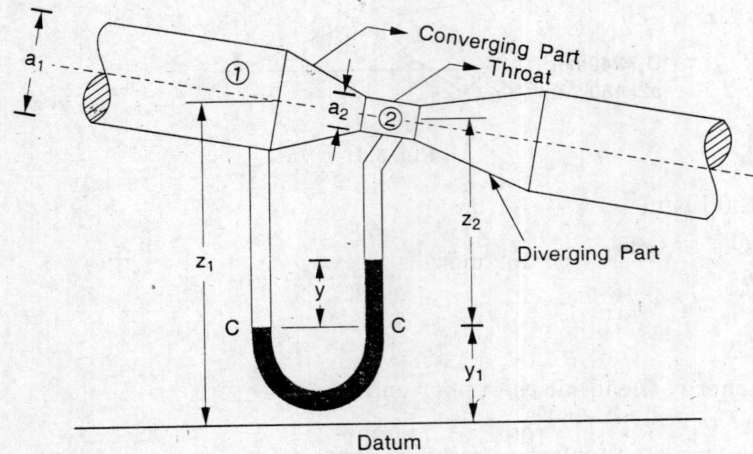

Fig. 5.7

Let at section 1 (main pipe),

a_1 = area

p_1 = pressure

V_1 = velocity of fluid

z_1 = datum head

and a_2, p_2, V_2 and z_2 are the corresponding values at section 2 (throat).

Applying Bernoulli's equation at sections 1 and 2,

$$\frac{p_1}{w} + \frac{V_1^2}{2g} + z_1 = \frac{p_2}{w} + \frac{V_2^2}{2g} + z_2$$

(w is the specific weight of the liquid in the pipe)

or $$\frac{1}{2g}(V_2^2 - V_1^2) = \left(\frac{p_1}{w} + z_1\right) - \left(\frac{p_2}{w} + z_2\right)$$

or $$\frac{V_2^2}{2g}\left(1 - \frac{V_1^2}{V_2^2}\right) = h \qquad ...(i)$$

h is the difference in piezometric head between 1 and 2 *i.e.* if two piezometric tubes are inserted at sections 1 and 2, the difference in their levels from the datum line will be h and is known as **venturi head.**

From continuity equation,

$$a_1 V_1 = a_2 V_2$$

$$V_1 = \frac{a_2 V_2}{a_1}$$

(i) Becomes $$\frac{1}{2g}\left(V_2^2 - \frac{a_2^2 V_2^2}{a_1^2}\right) = h$$

$$\frac{V_2^2}{2g}\left(\frac{a_1^2 - a_2^2}{a_1^2}\right) = h$$

$$V_2 = \frac{a_1}{\sqrt{a_1^2 - a_2^2}}\sqrt{2gh} \qquad ...(5.6)$$

Flow rate (discharge) through the pipe,

$$Q = a_2 V_2 = \frac{a_1 a_2}{\sqrt{a_1^2 - a_2^2}}\sqrt{2gh} \qquad ...(5.7)$$

Equation 5.7 gives the discharge under ideal conditions (theoretical discharge). Actual discharge is less than the theoretical discharge due to non-uniformity of velocity distribution in the pipe as well as friction and other losses *i.e.*

$$Q = C_d \cdot \frac{a_1 a_2}{\sqrt{a_1^2 - a_2^2}}\sqrt{2gh} \qquad ...(5.7a)$$

The expression can be put in terms of diameters also,

$$Q = C_d \cdot \frac{\left(\frac{\pi}{4} d_1^2\right)\left(\frac{\pi}{4} d_2^2\right)}{\left(\frac{\pi}{4} d_1^2\right)\sqrt{1 - \left(\frac{d_2^2}{d_1^2}\right)^2}} \sqrt{2gh}$$

$$= C_d \cdot \frac{\frac{\pi}{4} d_2^2}{\sqrt{1 - \left(\frac{d_2}{d_1}\right)^4}} \sqrt{2gh} \quad \text{...(5.8)}$$

where C_d is known as the coefficient of discharge, the value of which lies between 0·96 and 0·98 or it is found from the following relation :

$$C_d = \sqrt{\frac{h - h_f}{h}} = \sqrt{1 - \frac{h_f}{h}} \quad \text{...(5.9)}$$

where h_f is the friction head loss between the inlet and the throat.

In case some heavy liquid is used in the manometer connected to the venturimeter, the difference in the venturi head h is found by equating the heads at the horizontal level C–C in the two limbs of the manometer. Assuming that the heavier liquid is mercury,

$$\frac{p_1}{w} + (z_1 - y_1) = \frac{p_2}{w} + (z_2 - y_1 - y) + y \cdot \frac{w_m}{w}$$

$$\frac{p_1}{w} + z_1 = \frac{p_2}{w} + (z_2 - y) + y \cdot \frac{w_m}{w}$$

$$\left(\frac{p_1}{w} + z_1\right) - \left(\frac{p_2}{w} + z_2\right) = y \cdot \frac{w_m}{w} - y$$

or
$$h = y\left(\frac{S_m}{S} - 1\right) \quad \text{...(5.10)}$$

where h = reading of the manometer

S_m = specific gravity of mercury

and S = specific gravity of the liquid in the pipe.

In case the manometer contains a lighter liquid of specific gravity S_L (*i.e.* manometer inverted), h is given by

or
$$h = y\left(1 - \frac{S_L}{S}\right) \quad \text{...(5.11)}$$

If the manometer is horizontal, $z_1 = z_2$ and h is still given by the same relation.

Orifice Meter

An orifice meter is used to measure flow in a pipe line. A circular plate with a sharp edged circular orifice at the centre is fitted in the pipe. A vena-contracta (contracted section of the jet) is formed at a small distance from the plate as shown in Fig. 5.8. An orifice meter occupies

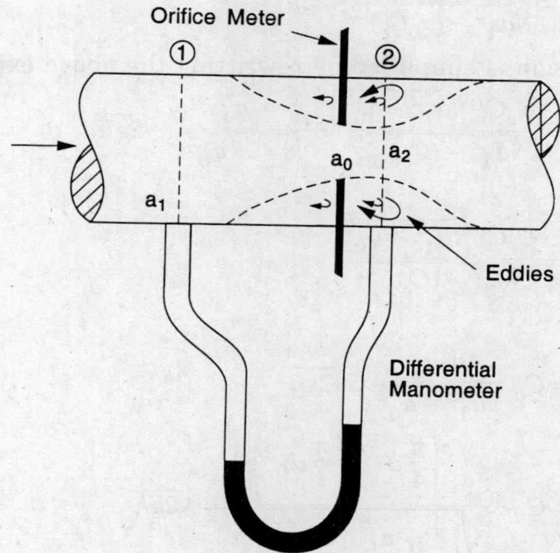

Fig. 5.8

less space as compared to a venturimeter and is simpler and cheaper. Orifice diameter lies between $0{\cdot}4d$ to $0{\cdot}8d$. Usual value is $0{\cdot}5d$.

Applying Bernoulli's equation at sections 1 and 2,

$$\frac{p_1}{w} + \frac{V_1^2}{2g} + z_1 = \frac{p_2}{w} + \frac{V_2^2}{2g} + z_2$$

Solving as in case of venturimeter,

$$V_2 = \frac{a_1}{\sqrt{a_1^2 - a_2^2}} \sqrt{2gh}$$

This is the theoretical velocity at the vena-cointracta. The actual velocity is less than this due to friction losses *i.e.*

$$V_{2a} = C_v \, . \, \frac{a_1}{\sqrt{a_1^2 - a_2^2}} \sqrt{2gh}$$

The discharge,

$$Q = a_2 V_{2a} = C_v \frac{a_1 a_2}{\sqrt{a_1^2 - a_2^2}} \sqrt{2gh}$$

Let

$$C_c = \frac{a_2}{a_0} \quad \text{or} \quad a_2 = a_0 \, . \, C_c$$

where a_0 is the area of the orifice meter and C_c, the coefficient of contraction.

Thus

$$Q = \frac{C_c\, C_v\, a_0\, a_1}{\sqrt{a_1^{\,2} - (a_0 C_c)^2}} \sqrt{2gh}$$

In practice, the expression is simplified by rewritting the above expression as

$$Q = \frac{C_c C_v \sqrt{a_1^{\,2} - a_0^{\,2}}}{\sqrt{a_1^{\,2} - (C_c a_0)^2}} \cdot \frac{a_0 a_1}{\sqrt{a_1^{\,2} - a_0^{\,2}}} \cdot \sqrt{2gh}$$

and defining

$$C_d = \frac{C_c C_v \sqrt{a_1^{\,2} - a_0^{\,2}}}{\sqrt{a_1^{\,2} - (C_c a_0)^2}}$$

so that

$$Q = C_d \frac{a_0 a_1}{\sqrt{a_1^{\,2} - a_0^{\,2}}} \cdot \sqrt{2gh} \qquad \text{...(5.12)}$$

$$= C_d \cdot \frac{\left(\frac{\pi}{4} d_1^{\,2}\right)\left(\frac{\pi}{4} d_0^{\,2}\right)}{\sqrt{\left(\frac{\pi}{4} d_1^{\,2}\right)^2 - \left(\frac{\pi}{4} d_0^{\,2}\right)^2}} \sqrt{2gh}$$

$$= C_d \cdot \frac{\frac{\pi}{4} d_2^{\,2}}{\sqrt{1 - \left(\frac{d_0}{d_1}\right)^4}} \sqrt{2gh} \qquad \text{...(5.13)}$$

C_d for an orifice meter is much smaller than that for a venturimeter (approximate value 0·67)

Pitot Tube

A **pitot tube** is used to measure the velocity of flow at any point in a channel or pipe. In its simplest form which is useful to measure the velocity of flow in channels, a pitot tube consists of a glass tube bent at right angle as shown in Fig. 5.9. The lower limb known as the body is placed in the direction of flow whereas the other end known as stem is open to atmosphere. The liquid rises up in the tube through some height (say h).

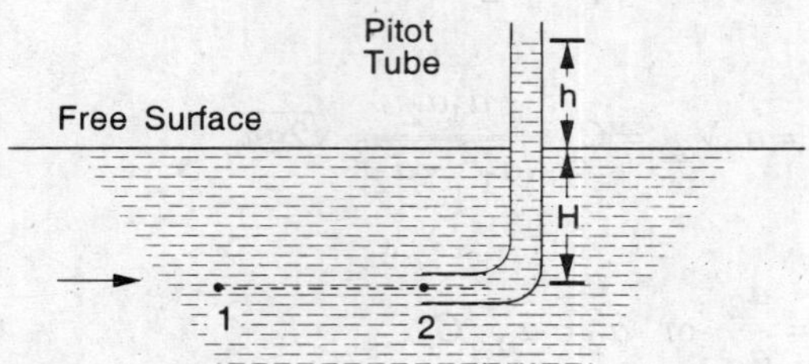

Fig. 5.9

The pressure in the free flow far away from the tube is called the static pressure and the pressure at the tip of the tube where the fluid is brought to a state of zero velocity is called stagnation pressure and the point as **stagnation point.** At the stagnation point the velocity head converts into the pressure head and thus the rise of liquid in the stem from the tip of the pitot tube is the sum of the static head and the velocity head.

Consider two points 1 and 2 at the same level in a free channel such that point 1 is far away from the tube and 2 at the stagnation point.

Applying Bernoulli's equation at points 1 and 2,

$$\frac{p_1}{w} + \frac{V_1^2}{2g} + Z_1 = \frac{p_2}{w} + \frac{V_2^2}{2g} + Z_2$$

But $$z_1 = z_2 \text{ and } V_2 = 0$$

$$\frac{p_1}{w} + \frac{V_1^2}{2g} = \frac{p_2}{w} \qquad \text{...(i)}$$

Now $$\frac{p_1}{w} = H \text{ and } \frac{p_2}{w} = H + h$$

(i) becomes $$H + \frac{V_1^2}{2g} = H + h$$

$$\frac{V_1^2}{2g} = h$$

$$V_1 = \sqrt{2gh}$$

This is the theoretical velocity. The actual velocity is given by

$$V_1 = C_v \sqrt{2gh} \qquad \text{...(5.14)}$$

where C_v is the coefficient of pitot tube (≈ 0·95). It can be observed that the liquid level in the vertical limb rises above the free surface by an amount equal to the velocity head.

When a pitot tube is used to determine the velocity of flow at a point in a pipe line, the static pressure has to be measured in addition to the stagnation pressure. As the velocity in a direction perpendicular to the flow is zero, the static pressure can be measured by making a static pressure opening in the pipe wall if the static pressure head difference between the level of the tip of the tube and the upper wall of the pipe is negligible. In case of high pressure in the pipes, pressure tappings of the pitot tube can be connected to the two limbs of a *U*-tube differential manometer which gives the difference between the stagnation and the static pressures (Fig. 5.10a and b).

A convenient way to use a pitot static tube is by having two circular concentric tubes one inside the other as shown in Fig. 5.10c. The inner tube which is open-ended senses the fluid stagnation pressure. A number of holes are drilled in the surface of the outer tube in the horizontal portion. As the holes are perpendicular to the direction of the flow, the outer tube senses the static pressure at that level. The leads from the two limbs can be connected to a differential manometer as usual. The following relation is used to obtain the velocity head h.

$$h = y\left(\frac{S_m}{S} - 1\right)$$

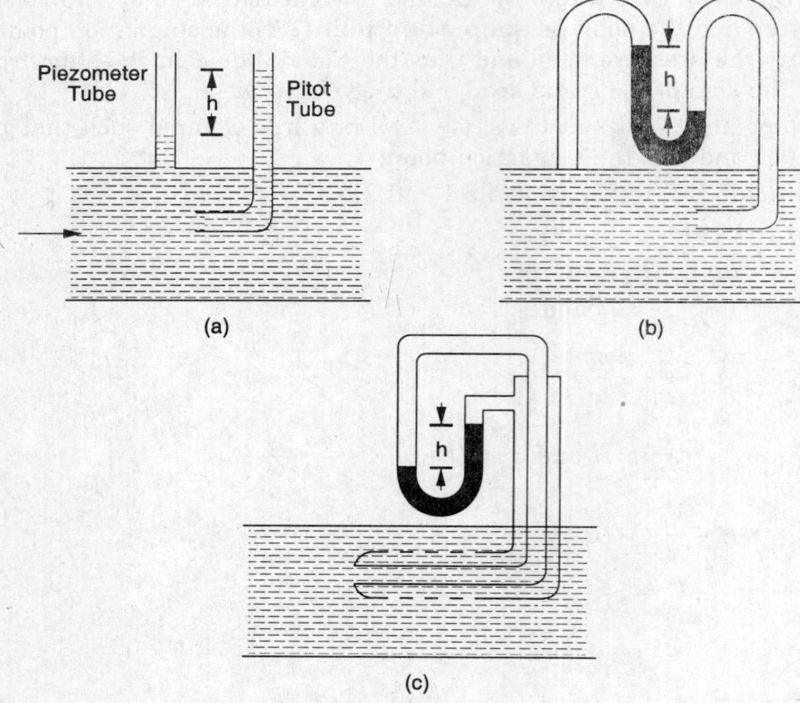

Fig. 5.10

Example 5.5 An oil of specific gravity 0·85 flows through a venturimeter with inlet and throat diameters 160 mm and 80 mm respectively. The oil-mercury differential manometer indicates a reading of 200 mm. Determine the flow of oil through the venturimeter. Take coefficient of discharge of the venturimeter as 0·985.

Solution $h = y\left(\frac{S_m}{S} - 1\right) = 0{\cdot}2 \times \left(\frac{13{\cdot}6}{0{\cdot}85} - 1\right) = 3$ m of oil

$$Q = C_d \cdot \frac{\frac{\pi}{4} d_2^{\,2}}{\sqrt{1 - \left(\frac{d_2}{d_1}\right)^2}} \sqrt{2gh}$$

$$= 0{\cdot}985 \times \frac{\frac{\pi}{4}(0{\cdot}08)^2}{\sqrt{1 - \left(\frac{0{\cdot}08}{0{\cdot}16}\right)^4}} \sqrt{2 \times 9{\cdot}81 \times 3}$$

$$= 0{\cdot}985 \times 0{\cdot}00519 \times 7{\cdot}672$$

$$= 0{\cdot}03922 \text{ m}^3\text{/s}$$

$$= 39{\cdot}22 \text{ l/s}$$

Example 5.6 A horizontal venturimeter used to measure the flow of water has inlet and throat diameters as 400 mm and 150 mm respectively. The pressure intensity at the inlet is 150 kN/m^2 and the vacuum head at the throat is 300 mm of mercury. Assuming that the head lost between the inlet and the throat is 3·6 percent, determine the value of the coefficient of discharge of the venturimeter and the rate of flow.

Solution $h_f = 0{\cdot}036\, h$ or $\dfrac{h_f}{h} = 0{\cdot}036$

$$C_d = \sqrt{1 - \frac{h_f}{h}} = \sqrt{1 - 0{\cdot}036} = 0{\cdot}982$$

h = Difference in the pressure head in the two limbs

$$= \frac{p_1}{w} - \left(h_2 \cdot \frac{w_m}{w}\right) = \frac{p_1}{w} - (h_2 \cdot S_m) = \frac{150}{9{\cdot}81} - (-0{\cdot}3 \times 13{\cdot}6) = 19{\cdot}37 \text{ m}$$

$$Q = C_d \cdot \frac{\frac{\pi}{4} d_2^2}{\sqrt{1 - \left(\frac{d_2}{d_1}\right)^4}} \sqrt{2gh}$$

$$= 0{\cdot}982 \times \frac{\frac{\pi}{4}(0{\cdot}12)^2}{\sqrt{1 - \left(\frac{0{\cdot}12}{0{\cdot}4}\right)^4}} \sqrt{2 \times 9{\cdot}81 \times 19{\cdot}37}$$

$$= 0{\cdot}982 \times 0{\cdot}0114 \times 19{\cdot}495$$

$$= 0{\cdot}218 \text{ m}^3/\text{s}$$

Example 5.7 A vertical pipe line carries oil of specific gravity 0·85. A 400 mm × 160 mm venturimeter is provided in the pipe line and the difference in the elevation of the throat section and the entrance section of this venturimeter is 500 mm. The gauge deflection of the *U*-tube mercury manometer connected to the venturimeter is 300 mm. Assuming the coefficient of the meter as 0·98 determine the discharge of the oil through the pipe and the pressure difference between the entrance and the throat sections.

Solution $h = y\left(\dfrac{S_m}{S} - 1\right) = 0{\cdot}3 \times \left(\dfrac{13{\cdot}6}{0{\cdot}85} - 1\right) = 4{\cdot}5 \text{ m}$

$$Q = C_d \cdot \frac{\frac{\pi}{4} d_2^2}{\sqrt{1 - \left(\frac{d_2}{d_1}\right)^4}} \sqrt{2gh}$$

$$= 0{\cdot}98 \times \frac{\frac{\pi}{4}(0{\cdot}16)^2}{\sqrt{1 - \left(\frac{0{\cdot}16}{0{\cdot}4}\right)^4}} \sqrt{2 \times 9{\cdot}81 \times 4{\cdot}5}$$

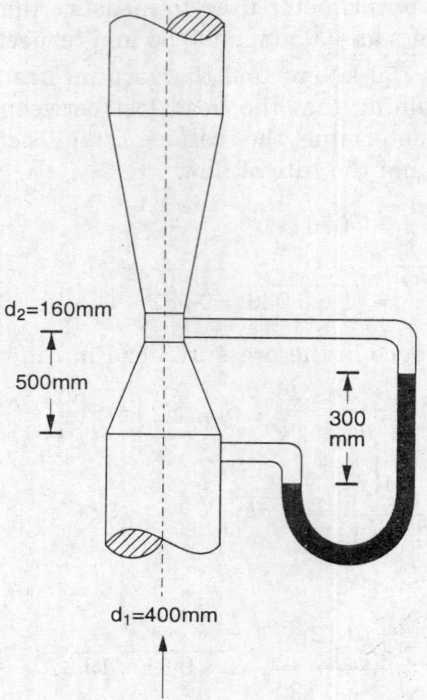

Fig. 5.11

$$= 0{\cdot}98 \times 0{\cdot}02037 \times 9{\cdot}396$$

$$= 0{\cdot}1876 \text{ m}^3/\text{s}$$

$$h = \left(\frac{p_1}{w} + z_1\right) - \left(\frac{p_2}{w} + z_2\right) = 4{\cdot}5$$

$$\frac{p_1 - p_2}{w} = (z_2 - z_1) + 4{\cdot}5$$

$$= 0{\cdot}5 + 4{\cdot}5$$

$$= 5 \text{ m}$$

$$p_1 - p_2 = 5 \times 9{\cdot}81 = 49{\cdot}05 \text{ kN/m}^2$$

Example 5.8 An inclined pipe line 600 mm in diameter is inclined with the horizontal at an angle of 40° and carries oil of specific gravity 0·85. A venturimeter connected to the pipe has throat diameter as 300 mm and the distance between the mouth and the throat 800 mm. Reading of the manometer which has mercury as the liquid is 60 mm. Assuming the coefficient of the meter as 0·98 find the discharge through the pipe.

Solution The effect of inclination of the venturimeter is incorporated in the reading of the manometer.

$$h = y\left(\frac{S_m}{S} - 1\right) = 0{\cdot}06 \times \left(\frac{13{\cdot}6}{0{\cdot}85} - 1\right) = 0{\cdot}9 \text{ m}$$

$$Q = C_d \cdot \frac{\frac{\pi}{4} {d_2}^2}{\sqrt{1 - \left(\frac{d_2}{d_1}\right)^4}} \sqrt{2gh}$$

$$= 0{\cdot}98 \times \frac{\frac{\pi}{4}(0{\cdot}3)^2}{\sqrt{1 - \left(\frac{0{\cdot}3}{0{\cdot}6}\right)^4}} \sqrt{2 \times 9{\cdot}81 \times 0{\cdot}9}$$

$$= 0{\cdot}98 \times 0{\cdot}073 \times 4{\cdot}202$$

$$= 0{\cdot}3 \text{ m}^3/\text{s}$$

Example 5.9 The following data relate to an orifice meter :

Diameter of the pipe = 300 mm

Diameter of the orifice = 150 mm

Reading of the differential manometer = 300 mm of mercury

Specific gravity of the oil flowing = 0·9

Coefficient of discharge of the meter = 0·6

Calculate the discharge of the oil through the pipe.

Solution

$$h = y\left(\frac{S_m}{S} - 1\right) = 0{\cdot}3 \times \left(\frac{13{\cdot}6}{0{\cdot}9} - 1\right) = 4{\cdot}23 \text{ m}$$

$$Q = C_d \cdot \frac{\frac{\pi}{4} {d_2}^2}{\sqrt{1 - \left(\frac{d_2}{d_1}\right)^4}} \sqrt{2gh}$$

$$= 0{\cdot}6 \times \frac{\frac{\pi}{4}(0{\cdot}15)^2}{\sqrt{1 - \left(\frac{0{\cdot}15}{0{\cdot}3}\right)^4}} \sqrt{2 \times 9{\cdot}81 \times 4{\cdot}23}$$

$$= 0{\cdot}6 \times 0{\cdot}01825 \times 9{\cdot}11$$

$$= 0{\cdot}0998 \text{ m}^3/\text{s or } 99{\cdot}8 \text{ l/s}$$

Example 5.10 A pitot tube is used to measure the velocity of flow of an oil through a pipe. The difference of mercury level in a differential *U*-tube manometer connected to the two tappings of a pitot tube is 120 mm. Assuming the coefficient of the pitot tube to be 0·98 and specific gravity of oil as 0·85, determine the velocity of flow.

Solution

$$h = y\left(\frac{S_m}{S} - 1\right) = 0{\cdot}12 \times \left(\frac{13{\cdot}6}{0{\cdot}85} - 1\right) = 1{\cdot}8 \text{ m}$$

$$V = C_v \sqrt{2gh}$$

$$= 0{\cdot}98 \times \sqrt{2 \times 9{\cdot}81 \times 1{\cdot}8}$$
$$= 5{\cdot}82 \text{ m/s}$$

Example 5.11 A pitot tube inserted in a pipe of 400 mm diameter pipe records a stagnation pressure of 10 kN/m^2 at the centre of the pipe. The static vacuum pressure in the pipe is 100 mm of mercury. Determine the rate of flow of water through the pipe assuming that the mean velocity of flow is 0·8 times the velocity at the centre. Take coefficient of the pitot tube as 0·98.

Solution h = Stagnation pressure head – Static pressure head

$$= \frac{p}{w} - h_m \cdot \frac{w_m}{w} = \frac{p}{w} - h_m \cdot S_m$$

$$= \left(\frac{10}{9{\cdot}81} - (-0{\cdot}1 \times 13{\cdot}6)\right) \text{ m of water}$$

$$= 2{\cdot}38 \text{ m of water}$$

Velocity at the centre

$$= C_v \cdot \sqrt{2gh} = 0{\cdot}98 \times \sqrt{2 \times 9{\cdot}81 \times 2{\cdot}38} = 6{\cdot}697 \text{ m/s}$$

Mean velocity = $0{\cdot}8 \times 6{\cdot}697 = 5{\cdot}357$ m/s

Rate of flow of water = Velocity × Area of the pipe

$$= 5{\cdot}357 \times \frac{\pi}{4} (0{\cdot}4)^2$$

$$= 0{\cdot}673 \text{ m}^3/\text{s}$$

5.6 KINETIC ENERGY AND MOMENTUM CORRECTION FACTORS (CORIOLI'S COEFFICIENTS)

In the derivation of Bernoulli's equation, it is assumed that the velocity across a single stream tube is uniform. However, as there is appreciable variation in the velocity distribution due to viscosity and boundary resistance, kinetic energy correction factor (α) and momentum correction factor (β) need to be applied to find the correct amount of kinetic energy or momentum at a given cross-section.

Kinetic Energy Correction Factor (α)

It is the ratio of the kinetic energies of the fluid per second based on actual velocity to that based on the average velocity. Refer Fig. 5.12,

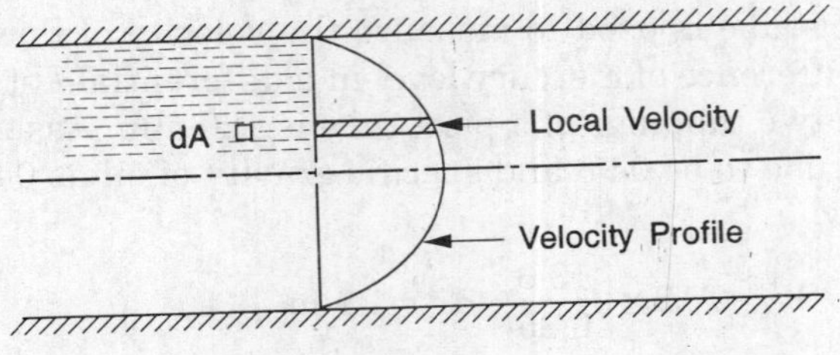

Fig. 5.12

Let u = actual velocity at a point (local velocity)

$\overline{u}$ = average velocity at the section under consideration

dA = elementary area at the point

A = area of the whole cross-section

$$\beta = \frac{\text{Kinetic energy per second based on actual velocity}}{\text{Kinetic energy per second based on average velocity}}$$

$$= \frac{\int \frac{1}{2} \times \text{Elementary mass flow rate} \times u^2}{\int \frac{1}{2} \times \text{Total mass flow rate} \times \overline{u}^2}$$

$$= \frac{\int \frac{1}{2} (\rho \,.\, dA \,.\, u)\, u^2}{\frac{1}{2} (\rho \,.\, A \,.\, \overline{u})\, \overline{u}^2}$$

$$= \frac{\rho \int u^3 \,.\, dA}{\frac{\rho}{2} A \,.\, \overline{u}^3}$$

$$= \frac{1}{A \,.\, \overline{u}^3} \int u^3 \,.\, dA \qquad \text{...(5.15)}$$

Momentum Correction Factor (β)

It is defined as the ratio of momentums of the flow per second based on actual velocity to that based on the average velocity.

$$\beta = \frac{\text{Momentum per second based on actual velocity}}{\text{Momentum per second based on average velocity}}$$

$$= \frac{\int \text{Elementary mass flow rate} \times u}{\int \text{Total mass flow rate} \times \overline{u}}$$

$$= \frac{\int (\rho \,.\, dA \,.\, u)\, u}{(\rho \,.\, A \,.\, \overline{u})\, \overline{u}}$$

$$= \frac{\int u^2 \,.\, dA}{A \,.\, \overline{u}^2}$$

$$= \frac{1}{A \,.\, \overline{u}^2} \int u^2 \,.\, dA \qquad \text{...(5.16)}$$

Example 5.12 Find the kinetic energy and the momentum correction factors for the laminar flow through a pipe given by

$$u = -\frac{1}{4\mu} \cdot \frac{\delta p}{\delta x} (R^2 - r^2)$$

Solutin $u = -\frac{1}{4\mu} \cdot \frac{\delta p}{\delta x} (R^2 - r^2)$

The velocity is maximum at the axis where $r = 0$

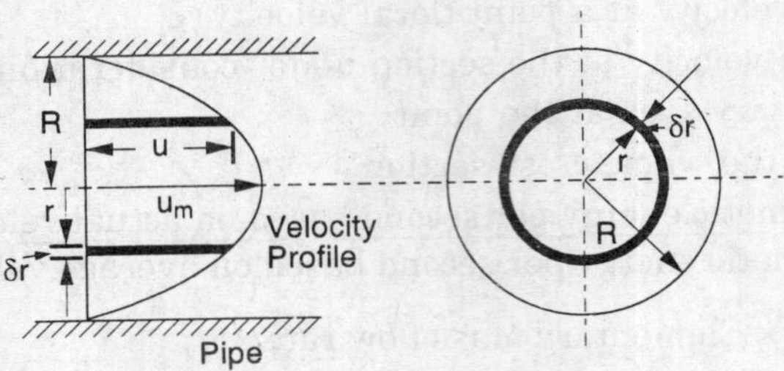

Fig. 5.13

$$u_m = -\frac{1}{4\mu} \cdot \frac{\delta p}{\delta x} R^2$$

$$\frac{u}{u_m} = \frac{R^2 - r^2}{R^2}$$

$$u = u_m \left[1 - \left(\frac{r}{R} \right)^2 \right]$$

Consider an elementary area da in the form of a ring at a radius r and of thickness dr.

$$\bar{u} = \frac{Q}{A} = \frac{\int_0^R u \,.\, da}{A}$$

$$= \frac{1}{\pi R^2} \int_0^R u_m \left[1 - \left(\frac{r}{R} \right)^2 \right] \times 2\pi r \,.\, dr$$

$$= \frac{1}{\pi R^2} \int_0^R \frac{u_m}{R^2} [R^2 - r^2] \times 2\pi r \,.\, dr$$

$$= \frac{2u_m}{R^4} \int_0^R [R^2 r - r^3] \,.\, dr$$

$$= \frac{2u_m}{R^4} \left(\frac{R^2 r^2}{2} - \frac{r^2}{4} \right)_0^R$$

$$= \frac{2u_m}{R^4} \left(\frac{R^4}{2} - \frac{R^4}{4} \right)$$

$$= \frac{u_m}{2}$$

Kinetic energy correction factor

$$\alpha = \frac{1}{A\bar{u}^3} \int_0^R u^3 \,.\, dA$$

$$= \frac{1}{\pi R^2 \bar{u}^3} \int_0^R u_m{}^3 \left[1 - \left(\frac{r}{R}\right)^2\right]^3 . 2\pi r . dr$$

$$= \frac{2u_m{}^3}{R^2 \bar{u}^3} \int_0^R \left[1 - \left(\frac{r^2}{R^2}\right)^3 - 3 \times 1 \times \frac{r^2}{R^2}\left(1 - \frac{r^2}{R^2}\right)\right] . r . dr$$

$$= \frac{2}{R^2}\left(\frac{u_m}{\bar{u}}\right)^3 \int_0^R \left[1 - \frac{3r^2}{R^2} + \frac{3r^4}{R^4} - \frac{r^6}{R^6}\right] r . dr$$

$$= \frac{2}{R^2}(2)^3 \int_0^R \left[r - \frac{3r^3}{R^2} + \frac{3r^5}{R^4} - \frac{r^7}{R^6}\right] dr$$

$$= \frac{16}{R^2}\left[\frac{r^2}{2} - \frac{3r^4}{4R^2} + \frac{r^6}{2R^4} - \frac{r^8}{8R^6}\right]_0^R$$

$$= \frac{16}{R^2} \frac{R^2}{8}$$

$$= 2$$

Momentum correction factor

$$\beta = \frac{1}{A \bar{u}^2} \int_0^R u^2 . dA$$

$$= \frac{1}{\pi R^2 \bar{u}^2} \int_0^R u_m{}^2 \left[1 - \left(\frac{r}{R}\right)^2\right]^2 . 2\pi r . dr$$

$$= \frac{2u_m{}^2}{R^2 \bar{u}^2} \int_0^R \left[1 - 2 \times 1 \times \frac{r^2}{R^2} + \frac{r^4}{R^4}\right] . r . dr$$

$$= \frac{2}{R^2}\left(\frac{u_m}{\bar{u}}\right)^2 \int_0^R \left[1 - \frac{2r^3}{R^2} + \frac{r^5}{R^4}\right] . dr$$

$$= \frac{8}{R^2}\left[\frac{r^2}{2} - \frac{r^4}{2R^2} + \frac{r^6}{6R^4}\right]_0^R$$

$$= \frac{8}{R^2} \frac{R^2}{6}$$

$$= \frac{4}{3} = 1{\cdot}33$$

Example 5.13 Turbulent flow through a pipe is given by

$$\frac{u}{u_m} = \left(\frac{y}{R}\right)^{\frac{1}{7}}$$

Determine the kinetic energy and momentum correction factors.

Solution $\dfrac{u}{u_m} = \left(\dfrac{y}{R}\right)^{\frac{1}{7}}$

$$\overline{u} = \frac{Q}{A} = \frac{\int_0^R u \,.\, da}{A}$$

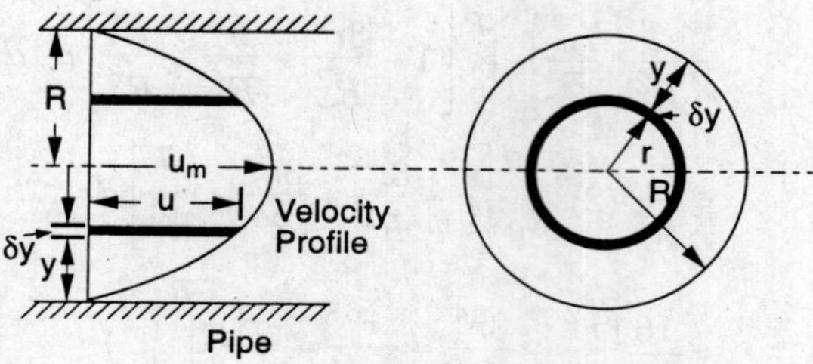

Fig. 5.14

$$= \frac{1}{\pi R^2} \int_0^R u_m \left(\frac{y}{R}\right)^{\frac{1}{7}} \times 2\pi\,(R - y) \,.\, dy$$

$$= \frac{2u_m}{R^2 \,.\, R^{1/7}} \int_0^R y^{1/7} \,.\, (R - y) \,.\, dy$$

$$= -\frac{2u_m}{R^{15/7}} \int_0^R (Ry^{1/7} - y^{8/7}) \,.\, dy$$

$$= \frac{2u_m}{R^{15/7}} \left(\frac{R \,.\, R^{8/7}}{8/7} - \frac{R^{15/7}}{15/7}\right)_0$$

$$= \frac{2u_m}{R^{15/7}} \left(\frac{7R^{15/7}}{8} - \frac{7R^{15/7}}{15/7}\right)$$

$$= \frac{2u_m}{R^{15/7}} \cdot \frac{49R^{15/7}}{120}$$

$$= \frac{49u_m}{120}$$

Kinetic energy correction factor

$$\alpha = \frac{1}{A\,\overline{u}^3} \int_0^R u^3 \,.\, dA$$

$$= \frac{1}{\pi R^2\, \overline{u}^3} \int_0^R u_m{}^3 \left(\frac{y}{R}\right)^{\frac{3}{7}} . \, 2\pi\,(R - y)\, dy$$

$$= \frac{2u_m{}^3}{R^2\, \overline{u}^3\, R^{3/7}} \int_0^R (Ry^{3/7} - y^{10/7})\, dy$$

$$= \frac{2}{R^{17/7}}\left(\frac{u_m}{\overline{u}}\right)^3\left[\frac{Ry^{10/7}}{10/7} - \frac{y^{17/7}}{17/7}\right]_0$$

$$= \frac{2}{R^{17/7}}\left(\frac{60}{49}\right)^3\left(\frac{7R^{17/7}}{10} - \frac{7R^{17/7}}{17}\right)$$

$$= 1{\cdot}06$$

Momentum correction factor

$$\beta = \frac{1}{A\,\overline{u}^2}\int_0^R u^2 \,.\, dA$$

$$= \frac{1}{\pi R^2\,\overline{u}^2}\int_0^R u_m{}^2\left(\frac{y}{R}\right)^{\frac{2}{7}} .\, 2\pi\,(R - y)\, dy$$

$$= \frac{2u_m{}^2}{R^2\,\overline{u}^2\,R^{2/7}}\int_0^R (Ry^{2/7} - y^{9/7})\, dy$$

$$= \frac{2}{R^{16/7}}\left(\frac{u_m}{\overline{u}}\right)^2\left[\frac{Ry^{9/7}}{9/7} - \frac{y^{16/7}}{16/7}\right]_0$$

$$= \frac{2}{R^{16/7}}\left(\frac{60}{49}\right)^2\left(\frac{7R^{16/7}}{9} - \frac{7R^{16/7}}{16}\right)$$

$$= 1{\cdot}02$$

5.7 VELOCITY OF FLOW IN A PIPE

Velocity of flow at any point in a pipe can be found by using

- a pitot tube with a vertical piezometric tube. Piezometric tube gives the static pressure head in the tube whereas the pitot tube gives the static as well as the velocity head. Thus the difference of the two depicts the velocity head.
- Pitot tube and vertical piezometric tube connected by a differential U-tube manometer.

  h is given by the relation

  $$h = y\left(\frac{S_h}{S} - 1\right)$$

- a composite tube consisting of two circular concentric tubes, one inside the other (Fig. 5.10c). The outer tube has some side holes. The pressure in this tube indicates the static head since the velocity in the perpendicular direction to the flow is zero. The inner tube gives the stagnation pressure head (static pressure head and velocity head). If the two tubes are connected to a manometer containing a heavy liquid, then h is given by

  $$h = y\left(\frac{S_h}{S} - 1\right)$$

5.8 THE MOMENTUM EQUATION

Momentum equation is based on Newton's second law of motion *i.e.* the force applied is equal to the rate of change of momentum and acts in the direction of force. Momentum is the product of mass and velocity of the fluid.

Let M = mass of the fluid

V = velocity of the fluid

Then momentum = MV

Rate of change of momentum = $\frac{d}{dt}(MV)$

According to Newton's second law,

Force, $$F = \frac{d}{dt}(MV) \qquad ...(5.17)$$

This is known as the **momentum equation.**

Also $$F \, . \, dt = d\,(MV) \qquad ...(5.18)$$

This is known as the **impulse-momentum equation.** The LHS of this equation represents the impulse of the force and the RHS gives the change in momentum and hence, the equation is known as impulse-momentum equation.

The principle of momentum equation may be applied to determine forces on pipe bends, reducers, moving vanes, jet propulsion etc. Consider a steady flow of fluid through a diverging stream tube entirely in xy-plane. Flow is assumed to be uniform and normal to the inlet and outlet areas (Fig. 5.15).

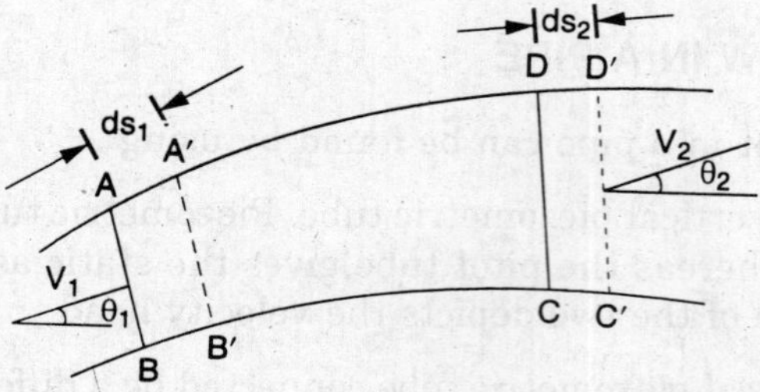

Fig. 5.15

Let V_1 = average velocity at entrance

V_2 = average velocity at exit

ρ_1 = density of fluid at the entrance

ρ_2 = density of fluid at exit

A_1 = cross-sectional area at entrance

A_2 = cross-sectional area at exit

Let the mass of fluid at the region $ABCD$ shifts to $A'B'C'D'$ in time t under the effect of external forces. As the area $A'B'CD$ is common to both the regions, $ABCD$ and $A'B'C'D'$, it does not undergo any momentum change. Therefore, the momentum change is only due to shifting of mass in the region $ABB'A'$ to the region $DCC'D'$.

From the principle of mass conservation,

Fluid mass within the region $ABCD$ = Fluid mass within the region $A'B'C'D'$

or Fluid mass in $ABB'A'$ + Fluid mass in $A'B'CD$

= Fluid mass in $A'B'CD$ + Fluid mass in $DCC'D'$

or Fluid mass in $ABB'A'$ = Fluid mass in $DCC'D'$

$$\rho_1 A_1 \cdot ds_1 = \rho_2 A_2 \cdot ds_2$$

Momentum of fluid mass in $ABB'A'$ = Mass × Average velocity

$$= (\rho_1 A_1 \cdot ds_1)\, V_1$$

$$= (\rho_1 A_1 V_1 \cdot dt)\, V_1$$

Momentum of fluid mass in $DCC'D' = (\rho_2 A_2 V_2 \cdot dt)\, V_2$

But $\rho_1 = \rho_2 = \rho$ for steady incompressible flow

and $A_1 V_1 = A_2 V_2 = Q$ from continuity equation

$\therefore$ change in momentum $= \rho Q\,(V_2 - V_1) \cdot dt$

By impulse momentum principle,

Impulse force = Change in momentum

$$F \cdot dt = \rho Q\,(V_2 - V_1)\, dt$$

$$F = \rho Q\,(V_2 - V_1)$$

$$= \frac{wQ}{g}(V_2 - V_1)$$

$$= m\,(V_2 - V_1) \qquad \text{...(5.19)}$$

The quantity $\frac{wQ}{g}$ or m is the mass flow rate and is called the momentum flux.

Component of the force in the x and y directions are

$$F_x = m\,(V_2 \cos\theta_2 - V_1 \cos\theta_1)$$

$$F_y = m\,(V_2 \sin\theta_2 - V_1 \sin\theta_1) \qquad \text{...(5.20)}$$

These components represent the forces exerted by the pipe bend on the fluid. The forces exerted by the fluid on the bend are equal and opposite as action and reactions are equal and opposite. Thus the force components exerted by the fluid on the bend are

$$F_x = m\,(V_1 \cos\theta_1 - V_2 \cos\theta_2)$$

$$F_y = m\,(V_1 \sin\theta_1 - V_2 \sin\theta_2) \qquad \text{...(5.21)}$$

Remember that these dynamic forces must be supplemented by the static pressure forces acting on the inlet and outlet sections.

Let p_1 and p_2 be the static pressures at the inlet and the outlet.

$$F_x = m\,(V_1 \cos\theta_1 - V_2 \cos\theta_2) + p_1 A_1 \cos\theta_1 - p_2 A_2 \cos\theta_2$$

$$F_y = m\,(V_1 \sin\theta_1 - V_2 \sin\theta_2) + p_1 A_1 \sin\theta_1 - p_2 A_2 \sin\theta_2 \qquad \text{...(5.22)}$$

The resultant force on the bend,

$$F_r = \sqrt{F_x^{\,2} + F_y^{\,2}} \qquad \text{...(5.23)}$$

and the direction of the resultant force is given by

$$\tan\theta = \frac{F_y}{F_x} \qquad \text{...(5.24)}$$

Example 5.14 A right angled bend in a horizontal plane discharge water into the atmosphere. The pressure at the inlet is 200 kN/m^2 where the cross-sectional area is 0·008 m^2. At the discharge end the velocity is 8 m/s and the area 0·003 m^2. Determine the resultant force exerted on the bend.

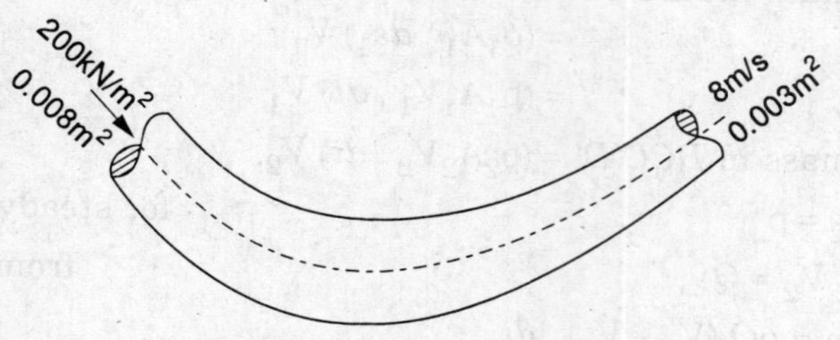

Fig. 5.16

Solution $Q = A_2V_2 = 0{\cdot}003 \times 8 = 0{\cdot}024 \text{ m}^3/\text{s}$

$$F_x = \frac{wQ}{g}(V_1\cos\theta_1 - V_2\cos\theta_2) + p_1A_1\cos\theta_1 - p_2A_2\cos\theta_2$$

$\theta_1 = 0°$, $\theta_2 = 90°$ and $p_2 = 0$

$$F_x = \frac{wQ}{g}V_1 + p_1A_1$$

We have

$$V_1 = \frac{Q}{A_1} = \frac{0{\cdot}024}{0{\cdot}008} = 3 \text{ m/s}$$

$$F_x = \frac{9{\cdot}81 \times 0{\cdot}024}{9{\cdot}81} \times 3 + 200 \times 0{\cdot}008 = 1{\cdot}672 \text{ kN}$$

$$F_y = \frac{wQ}{g}(V_1\sin\theta_1 - V_2\sin\theta_2) + p_1A_1\sin\theta_1 - p_2A_2\sin\theta_2$$

$$\therefore\ F_y = \frac{wQ}{g}(-V_2)$$

$$= \frac{9{\cdot}81 \times 0{\cdot}024}{9{\cdot}81}(-8)$$

$$= -0{\cdot}192 \text{ kN}$$

Resultant, $R = \sqrt{(1{\cdot}672)^2 + (0{\cdot}192)^2}$

$$= 1{\cdot}683 \text{ kN}$$

$$\theta = \tan^{-1}\left(\frac{0{\cdot}192}{1{\cdot}672}\right) = 6{\cdot}55°$$

Example 5.15 A nozzle is fitted to a pipe of diameter 100 mm. The quantity of water flowing through the pipe is 0·04 m^3/s. If the diameter of the nozzle is 40 mm, determine the force exerted by the nozzle on the water.

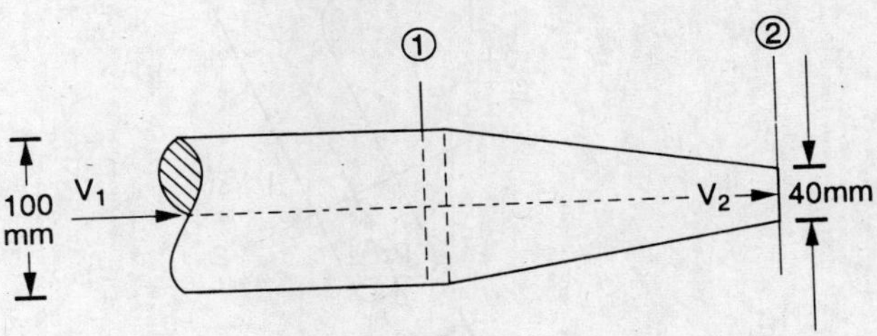

Fig. 5.17

Solution $A_1 = \frac{\pi}{4}(0{\cdot}1)^2 = 0{\cdot}007854 \text{ m}^2$

$$A_2 = \frac{\pi}{4}(0{\cdot}04)^2 = 0{\cdot}001256 \text{ m}^2$$

$$V_1 = \frac{Q}{A_1} = \frac{0{\cdot}04}{0{\cdot}007854} = 5{\cdot}09 \text{ m/s}$$

$$V_2 = \frac{Q}{A_2} = \frac{0{\cdot}04}{0{\cdot}001256} = 31{\cdot}83 \text{ m/s}$$

$$\frac{p_1}{w} + \frac{V_1^2}{2g} + z_1 = \frac{p_2}{w} + \frac{V_2^2}{2g} + z_2$$

$$\frac{p_1}{9{\cdot}81} + \frac{5{\cdot}09^2}{2 \times 9{\cdot}81} + 0 = 0 + \frac{31{\cdot}82^2}{2 \times 9{\cdot}81} + 0$$

$$p_1 = \frac{31{\cdot}82^2 - 5{\cdot}09^2}{2}$$

$$p_1 = 493{\cdot}27 \text{ kN/m}^2$$

$$F_x = \frac{wQ}{g}(V_2 \cos\theta_2 - V_1 \cos\theta_1) + p_2A_2 \cos\theta_2 - p_1A_1 \cos\theta_1$$

$$F_x = \frac{wQ}{g}(V_2 - V_1) + p_2A_2$$

$$F_x = \frac{9{\cdot}81 \times 0{\cdot}04}{9{\cdot}81}(31{\cdot}82 - 5{\cdot}09) + 0 + 493{\cdot}27 \times 0{\cdot}00785$$

$$= 1{\cdot}069 + 3{\cdot}872$$

$$= 4{\cdot}941 \text{ kN}$$

Example 5.16 A pipe bend of 400 mm at the inlet and 200 mm at the outlet turns the flow of water through 120° in a vertical plane. The flow through the bend is 200 litres per second and the pressure at the inlet is 100 kN/m^2. The axis at the

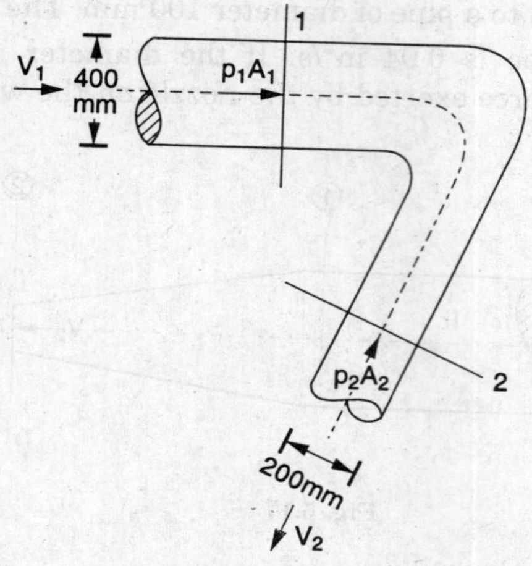

Fig. 5.18

inlet is horizontal and the exit is 1·2 m below the entrance section. If the volume of the bend is 0·18 m^3, determine the force exterted on the bend.

Solution $A_1 = \frac{\pi}{4}(0{\cdot}4)^2 = 0{\cdot}1257 \text{ m}^2$

$$A_2 = \frac{\pi}{4}(0{\cdot}2)^2 = 0{\cdot}0314 \text{ m}^2$$

$$V_1 = \frac{Q}{A_1} = \frac{0{\cdot}2}{0{\cdot}1257} = 1{\cdot}591 \text{ m/s}$$

$$V_2 = \frac{Q}{A_2} = \frac{0{\cdot}2}{0{\cdot}0314} = 6{\cdot}364 \text{ m/s}$$

$$\frac{p_1}{w} + \frac{V_1^2}{2g} + z_1 = \frac{p_2}{w} + \frac{V_2^2}{2g} + z_2$$

$$\frac{100}{9{\cdot}81} + \frac{1{\cdot}591^2}{2 \times 9{\cdot}81} + 1{\cdot}2 = \frac{p_2}{9{\cdot}81} + \frac{6{\cdot}364^2}{2 \times 9{\cdot}81} + 0$$

$$200 + 1{\cdot}591^2 + 1{\cdot}2 \times 19{\cdot}62 = 2p_2 + 6{\cdot}364$$

$$p_2 = 92{\cdot}79 \text{ kN/m}^2$$

$$F_x = \frac{wQ}{g}(V_1 \cos\theta_1 - V_2\cos\theta_2) + p_1A_1\cos\theta_1 - p_2A_2\cos\theta_2$$

$$= \frac{9{\cdot}81 \times 0{\cdot}2}{9{\cdot}81}(1{\cdot}59 - 6{\cdot}36 \cos 120°) + 100 \times 0{\cdot}1257$$

$$- 92{\cdot}81 \times 0{\cdot}0314 \cos 120°$$

$$= 0{\cdot}954 + 12{\cdot}57 + 1{\cdot}456$$

$$= 14{\cdot}98 \text{ kN}$$

$$F_y = \frac{wQ}{g}(V_1 \sin\theta_1 - V_2\sin\theta_2) + p_1A_1\sin\theta_1 - p_2A_2\sin\theta_2$$

– Weight of water contained in the bend

$$= \frac{wQ}{g} (- V_2 \sin \theta_2)_1 - p_2 A_2 \sin \theta_2 - w \times \text{Volume of the bend}$$

$$= \frac{9{\cdot}81 \times 0{\cdot}2}{9{\cdot}81} (- 6{\cdot}36 \sin 120°) - 92{\cdot}81 \times 0{\cdot}0314 \sin 120° - 9{\cdot}81 \times 0{\cdot}18$$

$$= - 1{\cdot}10 - 2{\cdot}524 - 1{\cdot}766$$

$$= - 5{\cdot}39 \text{ kN}$$

$$R = \sqrt{14{\cdot}98^2 + 5{\cdot}39^2}$$

$$= 15{\cdot}92 \text{ kN}$$

$$\tan \theta = \frac{5{\cdot}39}{14{\cdot}98} = 0{\cdot}36$$

$$\theta = 19{\cdot}79°$$

5.9 MOMENT OF MOMENTUM EQUATION

Moment of momentum equation is based on the principle of moment of momentum according to which the resulting torque acting on a rotating fluid is equal to the rate of change of moment of momentum.

Let Q = rate of flow of the fluid

V_1 = velocity of flow at section 1

V_2 = velocity of flow at section 2

r_1 = radius of curvature at section 1

r_2 = radius of curvature at section 2

ρ = density of the fluid

Momentum of the fluid at section 1 = Mass × Velocity

$= (\rho Q \,.\, V_1)$ per second

Moment of momentum per second at section 1

$= \rho Q \,.\, V_1 \,.\, r_1$

Similarly,

Moment of momentum per second at section 2

$= \rho Q \,.\, V_2 \,.\, r_2$

Rate of change of momentum $= \rho Q \, (V_2 \,.\, r_2 - V_1 \,.\, r_1)$

Applying the moment of momentum principle,

Resultant torque = Rate of change of momentum

$$T = \rho Q \, (V_2 \,.\, r_2 - V_1 \,.\, r_1) \quad \text{...(5.25)}$$

This is known as **moment of momentum** equation. It is used to find the torque exerted by water on sprinklers and to analyse the flow problems in turbines and centrifugal pumps.

Example 5.17 An unsymmetrical lawn sprinkler has two nozzles each of diameter 10 mm. The nozzles are at distances of 300 mm and 400 mm from the centre as shown

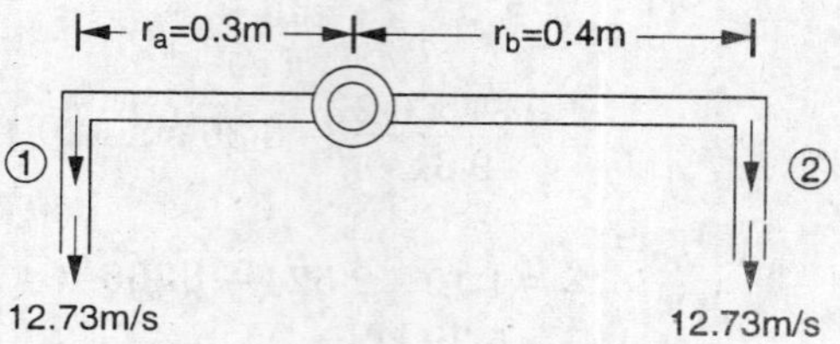

Fig. 5.19

in Fig. 5.19. Determine the speed of rotation of the sprinkler for a flow of 2 litres per second.

Solution $Q = 0{\cdot}002 \text{ m}^3/\text{s}$

$d = 0{\cdot}01$ m

Velocity of water through each nozzle

$$V = \frac{Q/2}{A} = \frac{Q/2}{(\pi/4)\, d^2} = \frac{0{\cdot}001}{(\pi/4)\,(0{\cdot}01)^2} = 12{\cdot}73 \text{ m/s}$$

Let ω be the angular velocity of the sprinkler and assume any direction of rotation of the sprinkler. Say it rotates clockwise.

Absolute velocity, $V_1 = V + \omega\, r_a = 12{\cdot}73 - 0{\cdot}3\, \omega$

Absolute velocity, $V_2 = V + \omega\, r_b = 12{\cdot}73 + 0{\cdot}4\, \omega$

As no external torque acts on the sprinkler, the resultant torque is zero.

$$T = \rho Q\, (V_2 \,.\, r_2 - V_1 \,.\, r_1) = 0$$

$$V_2 \,.\, r_2 - V_1 - r_1 = 0$$

$$0{\cdot}3\,(12{\cdot}73 - 0{\cdot}3\,\omega) = 0{\cdot}4\,(12{\cdot}73 + 0{\cdot}4\,\omega)$$

$$12{\cdot}73 - 0{\cdot}3\,\omega = 1{\cdot}333\,(12{\cdot}73 + 0{\cdot}4\,\omega)$$

$$= 16{\cdot}97 + 0{\cdot}533\,\omega$$

$$0{\cdot}833\,\omega = -\,4{\cdot}24$$

$$\omega = -\,5{\cdot}09 \text{ rad/s}$$

ω comes out to be negative. This indicates that the sprinkler rotates in the counter-clockwise direction.

$$\frac{2\pi N}{60} = 5{\cdot}09$$

$$N = 48{\cdot}6 \text{ rpm}$$

Example 5.18 A lawn sprinkler shown in Fig. 5.20 has 10 mm diameter nozzles at the end of rotating arms. The water is discharged into the atmosphere with a velocity of 12 m/s. Find the torque required to hold the rotating arm stationary. What is the speed of rotation of the arm when free to rotate.

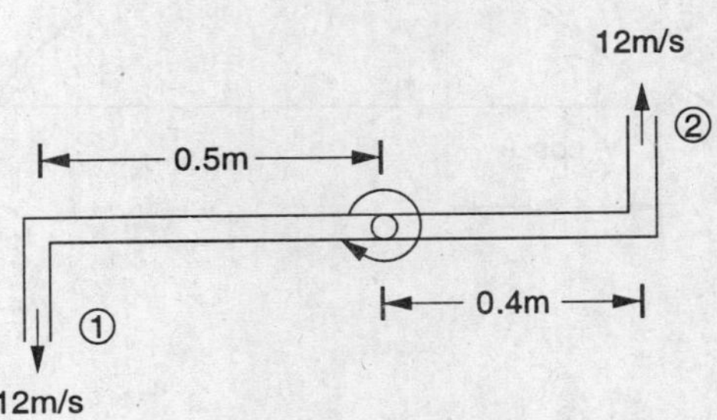

Fig. 5.20

Solution Total torque exerted by the water on sprinkler

$$= \rho Q\,(V_1 \cdot r_1 + V_2 \cdot r_2)$$

$$= \rho \,.\, A \,.\, V\,(r_2 + r_1)$$

$$= \rho \,.\, \frac{\pi}{4} d^2 \,.\, V^2\,(r_1 + r_2)$$

$$= 1000 \times \frac{\pi}{4} \times (0{\cdot}01)^2 \times 12^2 \times (0{\cdot}5 + 0{\cdot}4)$$

$$= 10{\cdot}18 \text{ N . m}$$

The sprinkler rotates in the counter-clockwise direction.

Absolute velocity, $V_1 = V - \omega\, r_a = 12 - 0{\cdot}5\,\omega$

Absolute velocity, $V_2 = V - \omega\, r_b = 12 - 0{\cdot}4\,\omega$

As no external torque acts on the sprinkler, the resultant torque is zero.

$$T = \rho Q\,(V_1 \cdot r_1 + V_2 \cdot r_2) = 0$$

$$V_1 \cdot r_1 + V_2 \cdot r_2 = 0$$

$$0{\cdot}5\,(12 - 0{\cdot}5\,\omega) + 0{\cdot}4\,(12 - 0{\cdot}4\,\omega) = 0$$

$$6 - 0{\cdot}25\,\omega + 4{\cdot}8 - 0{\cdot}16\,\omega = 0$$

$$0{\cdot}41\,\omega = 10{\cdot}8$$

$$\omega = 26{\cdot}34 \text{ rad/s}$$

$$\frac{2\pi N}{60} = 26{\cdot}34$$

$$N = 251{\cdot}5 \text{ rpm}$$

5.9 FREE LIQUID JET

A jet of liquid coming out from the nozzle in atmosphere is known as free liquid jet.

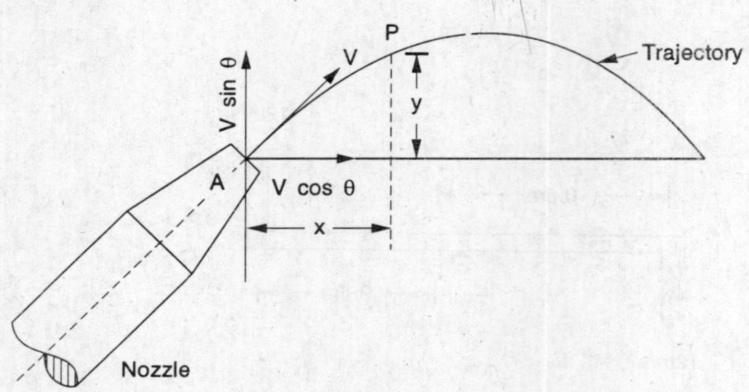

Fig. 5.21

Let a jet come out of the nozzle at A with velocity V making an angle θ with the horizontal direction (Fig. 5.21).

Let t = Time taken by a liquid particle in moving from A to P

Then horizontal component travelled by the liquid particle,

$$x = V \cos\theta \times t \quad \text{...(i)}$$

Verticle component travelled by the liquid particle,

$$y = V \sin\theta \times t - \frac{1}{2} g t^2$$

It may be noted that the horizontal component remains constant whereas the vertical component is affected by the gravity.

$$y = V \sin\theta \times \frac{x}{V\cos\theta} - \frac{1}{2} g \left(\frac{x}{V\cos\theta}\right)^2 \quad \left[t = \frac{x}{V\cos\theta} \text{ from (i)}\right]$$

$$= x \tan\theta - \frac{gx^2}{2V^2}\sec^2\theta \quad \text{...(5.26)}$$

This is the equation of a parabola. Therefore, the path travelled by the free jet in atmosphere is parabolic.

Maximum Height Attained by the Jet

Using the relation,

$$v_2^2 - v_1^2 = -2gH$$

(negative sign is used as g acts downwards and H in the upward direction).

v_1 and v_2 are the vertical components at the points 1 and 2.

At the highest point (2), $v_2 = 0$

$$-(V\sin\theta)^2 = -2gH$$

$$-V^2\sin^2\theta = -2gH$$

$$H = \frac{V^2 \sin^2\theta}{2g} \quad \text{...(5.27)}$$

Time of Flight

It is the time taken by the fluid particle reaching to point A from B. Let it be T seconds.

Using the relation,

$$y = V \sin\theta \times t - \frac{1}{2} g t^2$$

As the particle reaches at B, $y = 0$ and $t = T$

$$0 = V \sin\theta \times T - \frac{1}{2} g T^2$$

$$0 = V \sin\theta - \frac{1}{2} g T$$

$$T = \frac{2V \sin\theta}{g} \quad ...(5.28)$$

$$\text{Time to reach the highest point} = \frac{T}{2} = \frac{V \sin\theta}{g} \quad ...(5.29)$$

Horizontal Range of the Jet

Let the horizontal range be x

then x = Horizontal velocity component × Time taken

$$= V \cos\theta \times T$$

$$= V \cos\theta \times \frac{2V \sin\theta}{g}$$

$$= \frac{V^2 \times 2 \sin\theta \cos\theta}{g}$$

$$= \frac{V^2 \sin 2\theta}{g} \quad ...(5.30)$$

The range is maximum, if $\sin 2\theta = 1$

or $2\theta = 90^\circ$

or $\theta = 45^\circ$

$$\text{Then the maximum range} = \frac{V^2}{g} \quad ...(5.31)$$

Example 5.19 A jet of water from a nozzle is to be projected to the top of a wall 6 m high. The water comes out of the nozzle with a velocity of 15 m/s. The nozzle is at a distance of 12 m from the wall. Determine the angle of projection of the nozzle with the horizontal.

Solution $y = x \tan\theta - \frac{g x^2}{2V^2} \sec^2\theta$

$$6 = 12 \tan\theta - \frac{9{\cdot}81 \times 12^2}{2 \times 15^2} \sec^2\theta$$

$$\sec^2\theta - 3{\cdot}82 \tan\theta + 1{\cdot}91 = 0$$

$$(1 + \tan^2\theta) - 3{\cdot}82 \tan\theta + 1{\cdot}91 = 0$$

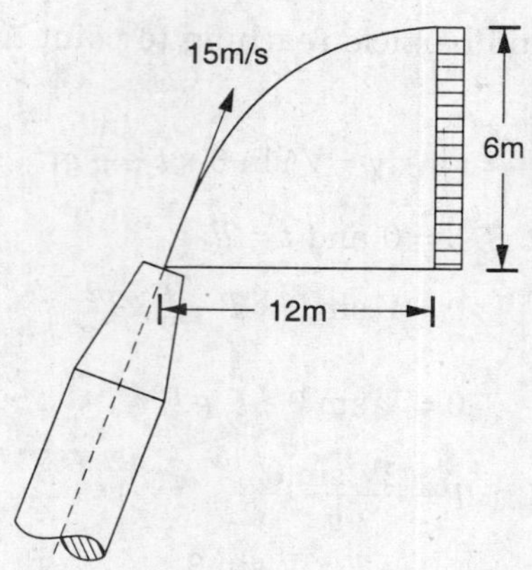

Fig. 5.22

$$\tan^2 \theta - 3{\cdot}82 \tan \theta + 2{\cdot}91 = 0$$

$$\tan \theta = \frac{3{\cdot}82 \pm \sqrt{(-3{\cdot}82)^2 - 4 \times 2{\cdot}91}}{2}$$

$$= \frac{3{\cdot}82 + 1{\cdot}72}{2} \text{ or } \frac{3{\cdot}82 - 1{\cdot}72}{2}$$

$$= 2{\cdot}77 \text{ or } 1{\cdot}05$$

$$\theta = 70{\cdot}1^\circ \text{ or } 46{\cdot}4^\circ$$

Example 5.20 Five nozzles each 20 mm in diameter are situated at a distance of 1·5 m from the ground level and are inclined at an angle of 45° to the horizontal. The jets of water strike the ground at a horizontal distance of 5 m. Determine the pressure head at the nozzles and the total discharge. The velocity coefficient of each nozzle is 0·98.

Solution $y = x \tan \theta - \frac{gx^2}{2V^2} \sec^2 \theta$

$$-1{\cdot}5 = 5 \tan 45^\circ - \frac{g \times 5^2}{2V^2} \sec^2 45^\circ$$

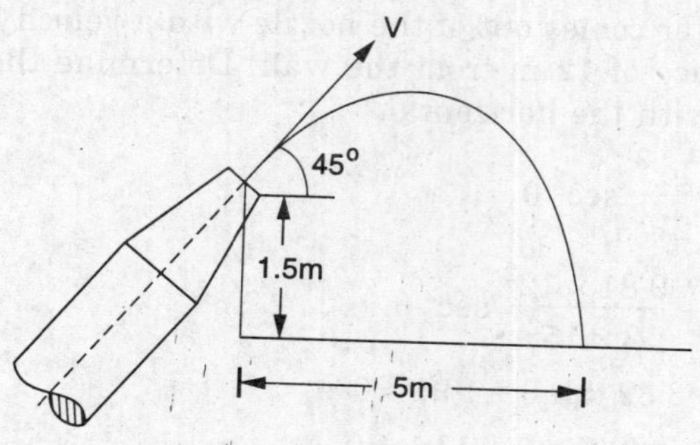

Fig. 5.23

$$-3V^2 = 10V^2 - 490{\cdot}5$$
$$13V^2 = 490{\cdot}5$$
$$V^2 = 37{\cdot}73$$
$$V = 6{\cdot}143 \text{ m/s}$$

Let h be the pressure head.

Then $$V = C_v\sqrt{2gh}$$

$$\therefore \quad C_v\sqrt{2gh} = 6{\cdot}143$$
$$0{\cdot}98\sqrt{2\times 9{\cdot}81\times h} = 6{\cdot}143$$
$$h = 1{\cdot}415 \text{ m}$$

Total discharge = Area × Velocity × Number of nozzles

$$= \frac{\pi}{4}\times(0{\cdot}02)^2\times 6{\cdot}143\times 5$$
$$= 0{\cdot}00965 \text{ m}^3/\text{s}$$
$$= 0{\cdot}65 \text{ l/s}$$

5.11 VORTEX MOTION

Vortex or cylindrical vortex motion is of two types, forced vortex and free vortex. A vortex formed with the aid of some external torque is called a **forced vortex** whereas a **free vortex** is obtained as a result of some previous vortex motion imparted to the fluid due to some internal action of the fluid. Fluid motion inside the runner of a hydraulic turbine or inside the impeller of a centrifugal pump are examples of forced vortex. Free vortex is formed in the casing of a pump after the fluid leaves the impeller or in the casing of a turbine before the fluid enters the runner.

Consider an elementary prism of fluid rotating at a uniform velocity about a vertical axis as shown in Fig. 5.24.

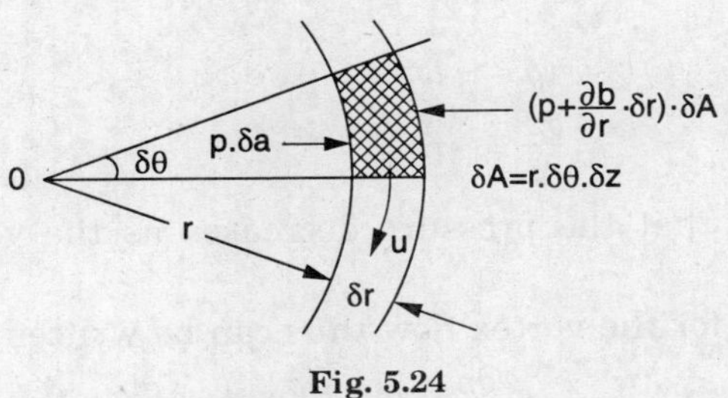

Fig. 5.24

Let p = radial pressure at radius r

u = tangential velocity of the elementary prism

A = cross-sectional area of the prism

z = depth of the prism

The equilibrium of the fluid prism may be considered in the tangential, radial and vertical directions as follows :

(i) In the tangential direction :.

The velocity is constant and thus there is no change in the pressure energy, *i.e.*

$$\frac{\partial p}{\partial \theta} = 0$$

(ii) In the radial direction :

Pressure force in the outward direction $= p \,.\, \delta A$

Pressure force in the inward direction $= \left(p + \frac{\partial p}{\partial r} \,.\, \delta r\right) . \,\delta A$

Centrifugal force in the outward direction = Mass of prism $\times \frac{u^2}{r}$

$$= \text{Volume} \times \text{Density} \times \frac{u^2}{r}$$

$$= (\delta A \,.\, \delta r) \,.\, \rho \,.\, \frac{u^2}{r}$$

Equating the forces in the radial direction,

$$\left(p + \frac{\partial p}{\partial r} \,.\, \delta r\right) . \,\delta A - p \,.\, \delta A = (\delta A \,.\, \delta r) \,.\, \rho \,.\, \frac{u^2}{r}$$

$$\frac{\partial p}{\partial r} = \rho \,.\, \frac{u^2}{r}$$

(iii) In the vertical direction :

Applying Bernoulli's equation to two points on the upper and lower faces of the prism

$$\frac{p_1}{w} + \frac{V_1^2}{2g} + z_1 = \frac{p_2}{w} + \frac{V_2^2}{2g} + z_2$$

There is zero velocity in the vertical direction. So pressure variation in the vertical direction is simply due to the change in datum of the two surfaces.

$$\frac{p_1 - p_2}{w} = z_2 - z_1$$

$$p = -\,wz = -\,\rho \,.\, g \,.\, z$$

$$\frac{\partial p}{\partial z} = -\,\rho \,.\, g$$

Negative sign signifies that the pressure decreases as the value of z increases in the vertical upward direction.

The equation of motion for the vortex flow thus can be written as

$$dp = \frac{\partial p}{\partial \theta} \,.\, d\theta + \frac{\partial p}{\partial r} \,.\, dr + \frac{\partial p}{\partial z} \,.\, dz$$

$$= 0 + \rho \,.\, \frac{u^2}{r} \,.\, dr - \rho \,.\, g \,.\, dz$$

$$\frac{dp}{\rho} = . \,\frac{u^2}{r} \,.\, dr - .\, g \,.\, dz \qquad ...(5.32)$$

Forced Cylindrical Vortex

When a fluid in a cylindrical container is rotated by applying an external torque, the fluid rotates as a solid mass and the motion is termed as forced cylindrical vortex or simply **forced vortex.** The whole of the fluid rotates with the same angular speed. The fluid particles which are away from the axis move with a higher tangential velocity than the particles which are near to the axis of rotation. No shear stresses exist in the fluid.

Such type of flow is also obtained if the vessel containing the fluid is rotated. As the vessel gains in speed, the fluid inside starts rotating and soon the cylinder and the fluid begin to rotate as one mass and attains the same angular velocity.

The tangential velocity is given by

$$u = \omega . r$$

and the equation of forced vortex becomes

$$\frac{dp}{\rho} = . \frac{u^2}{r} . dr - . g . dz$$

$$= \omega^2 r^2 . dr - g . dz$$

Energy difference between any two points is,

$$\frac{p_2 - p_1}{\rho} = \frac{1}{2}\omega^2 (r_2^2 - r_1^2) - g (z_2 - z_1) \qquad ...(5.33)$$

In a horizontal plane, $z_1 = z_2$, then

$$\frac{p_2 - p_1}{\rho} = \frac{1}{2}\omega^2 (r_2^2 - r_1^2)$$

$$= \frac{u_2^2 - u_1^2}{2} \qquad ...(5.34)$$

u_1 and u_2 are the tangential velocities of the fluid motion at the two points. The expression shows that on any horizontal plane, the fluid pressure increases with the radial distance from the axis of the vortex motion.

If $p_1 = 0$ where $r_1 = 0$,

$$\frac{p_2}{\rho} = \frac{1}{2}\omega^2 r_2^2$$

or simply, it can be written

$$\frac{p}{\rho} = \frac{1}{2}\omega^2 r^2 \qquad ...(5.35)$$

which is the equation of a paraboloid and it shows that the surface of constant pressure energy is of the shape of paraboloid (Fig. 5.25).

Also at the surface of constant pressure energy, $p_1 = p_2$, therefore, from Eq. 5.33,

$$z_1 - z_1 = \frac{1}{2g}\omega^2 (r_2^2 - r_1^2)$$

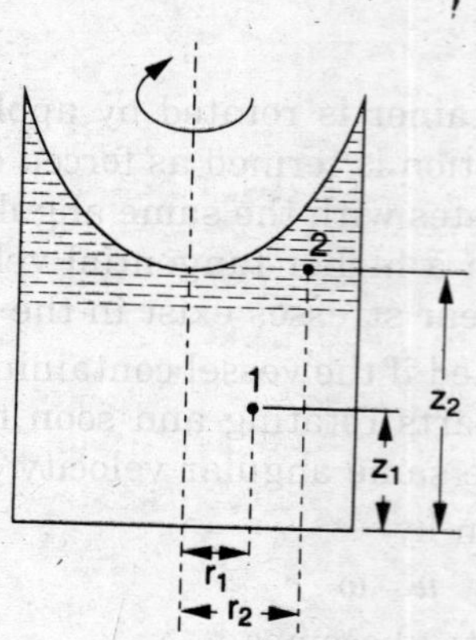

Fig. 5.25

At the vertex of the paraboloid, $r_1 = 0$ and $z_1 = 0$

$$z_2 = \frac{\omega^2 r_2^2}{2g}$$

or
$$z = \frac{\omega^2 r^2}{2g} \qquad \text{...(5.36)}$$

where r is the radial distance of a point on the surface of the paraboloid and z its height above the vertex of the paraboloid.

Thus in a forced vortex, the fluid rises above the vertex as given by Eq. 5.36.

If $p_1 = p_z$ at $r_1 = 0$ *i.e.* if the pressure at a point on the axis of rotation h_z m below the vertex (origin) of the paraboloid is p_z,

$$\frac{p_2 - p_z}{\rho} = \frac{1}{2}\,\omega^2 r^2$$

$$\frac{p}{\rho g} = \frac{p_z}{\rho g} + \frac{\omega^2 r^2}{2g}$$

or
$$h = h_z + z \qquad \text{...(5.37)}$$

which shows that the pressure head at any point in the cylinder is equal to the sum of the heights of two water columns : one above the horizontal plane passing through the origin of the paraboloid and the other below that plane above the required point.

The volume of a paraboloid of revolution is half of that of the cylinder enclosing it. Thus volume of the liquid above a horizontal plane through the vertex is equal to the volume of the paraboloid.

$$\text{Volume of paraboloid} = \frac{1}{2}\, a\,.\, z = \frac{1}{2}\,.\,\pi r^2\,.\,\frac{\omega^2 r^2}{2g}$$

Free Cylindrical Vortex

In case of a free cylindrical vortex motion is established naturally without the aid of any external torque due to some internal action or due to the previous motion given to it. Some initial disturbance in the fluid may determine the direction of rotation. There is no energy transfer across the streamlines.

When water flows through a hole at the bottom of a cylindrical vessel, the remaining water in the vessel moves in a spiral form. Now, if the hole is plugged, the water executes a simple rotation in horizontal planes. This is called cylindrical vortex flow. Other examples are

- Whirlpool in a river
- Flow of liuid in the casing of a centrifugal pump after leaving the impeller
- Flow of liquid in the casing of a reaction turbine before entering the runner
- Flow of liquid around a pipe bend.

Now, angular momentum = Momentum × Radial distance

= Mass × Tangential velocity × Radial distance

$= m \cdot u \cdot r$

Rate of change of momentum

$$= \frac{\partial}{\partial t}(m \cdot u \cdot r)$$

$$= m\frac{\partial}{\partial t}(u \cdot r) \qquad (m \text{ is constant})$$

Since no torque is needed to form a free cylindrical vortex, the rate of change of angular momentum is zero.

Therefore,

$$m\frac{\partial}{\partial t}(u \cdot r) = 0$$

$$\frac{\partial}{\partial t}(u \cdot r) = 0$$

On integration, $u \cdot r = C$

or

$$u = \frac{C}{r} \qquad \text{...(5.38)}$$

This value of the tangential velocity can be inserted in the general equation of the vortex flow to get an expression for the pressure variation in a free cylindrical vortex.

$$\frac{dp}{\rho} = \frac{u^2}{r} \cdot dr - g \cdot dz$$

$$= \frac{C^2}{r^3} \cdot dr - g \cdot dz$$

$$= C^2 (r^{-3}) \cdot dr - g \cdot dz$$

The pressure difference between two points,

$$\frac{p_2 - p_1}{\rho} = \frac{C^2}{-2}(r_2^{-2} - r_1^{-2}) + g(z_1 - z_2)$$

$$= g(z_1 - z_2) - \frac{C^2}{2}\left(\frac{1}{r_2^2} - \frac{1}{r_1^2}\right)$$

$$= gz_1 - gz_2 - \frac{u^2}{2} + \frac{u_1^{\,2}}{2} \qquad \left(\because \frac{C}{r} = u\right)$$

$$\frac{p_1}{\rho} + \frac{u_1^{\,2}}{2} + gz_1 = \frac{p_2}{\rho} + \frac{u_2^{\,2}}{2} + gz_2 \qquad ...(5.39)$$

which is Bernoulli's equation for fluid flows. This should have been true as there is no energy transfer in the flow.

In a horizontal plane, $z_1 = z_2$ and so

$$\frac{p_2 - p_1}{\rho} = \frac{u_1^{\,2} - u_2^{\,2}}{2}$$

$$= \frac{C^2}{2}\left(\frac{1}{r_1^{\,2}} - \frac{1}{r_2^{\,2}}\right) \qquad ...(5.40)$$

The equation gives the difference of pressure energy between the two streamlines considered. The fluid pressure increases with radial distance from the axis of the vortex motion. Also, it is an equation of a parabola where radius represents the base (Fig. 5.26). The upper surface of the fluid is of parabolic shape if it is free.

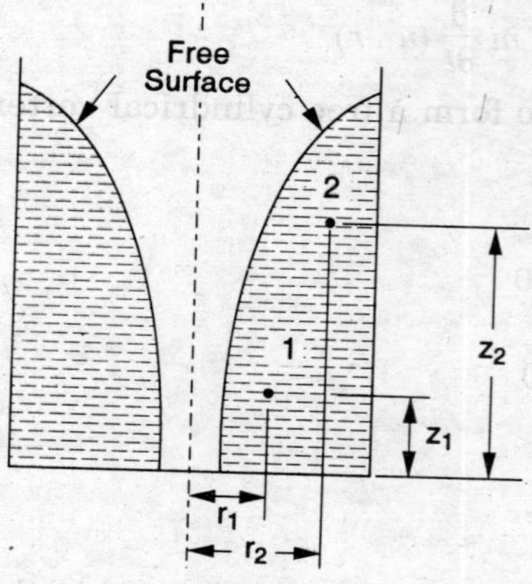

Fig. 5.26

As ur = Constant, the velocity varies inversely with the radius *i.e.* u is infinity where r is zero and u is zero where r is infinity. The point of infinite velocity is called singular point.

As the pressure reduces towards centre, a radial flow is established towards the centre where it is drained away, creating a suction behind it which induces more liquid to flow towards the centre. Thus a combination of circumferential and radial motion is established which is known as free spiral vortex. Therefore, a free cylindrical vortex cannot exist since it tries to convert itself into free spiral vortex.

Example 5.22 Show that in a case of a vortex, the rise of liquid level on the sides is equal to the fall of liquid at the axis of rotation.

Solution Let the rise of liquid on the sides $= x$

and the fall of liquid at the axis $= y$

We are to prove that $x = y$ (Fig. 5.27)

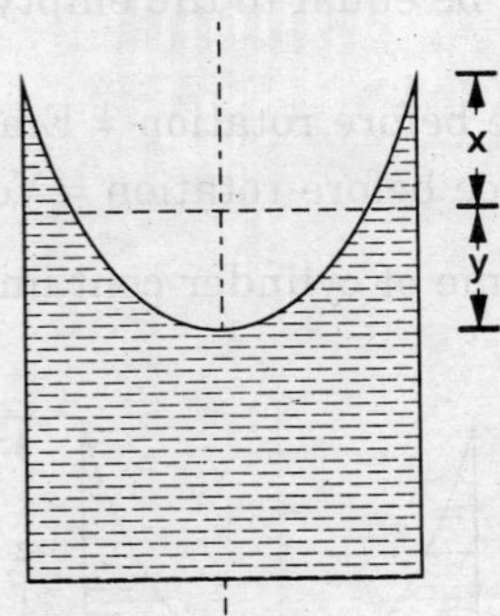

Fig. 5.27

Assume that the height of the vessel is upto the height of the liquid level after rotation. Let a be the area of cross section of the vessel.

Empty space before rotation = Empty space after rotation

= Volume of paraboloid

$$a \cdot x = \frac{1}{2} \cdot a\,(x + y)$$

$$x = \frac{x + y}{2}$$

$$2x = x + y$$

$$x = y$$

Thus the rise of the liquid level on sides is equal to the fall of the liquid level at the axis of rotation.

Example 5.23 An open cylindrical vessel of 160 mm internal diameter and 1 m high contains water upto a height of 750 mm from the bottom. Find

(i) Height of the paraboloid formed at the free surface of water if the vessel is rotated about its vertical axis at 250 rpm.

(ii) Maximum speed at which the vessel is rotated about its vertical axis so that no water spills out.

(iii) Speed at which the axial depth of water becomes zero.

Solution $H = 1$ m $\quad d = 0{\cdot}16$ m

$h = 0{\cdot}75$ m $\quad r = 0{\cdot}08$ m

(i) $N = 250$ rpm

$$\omega = \frac{2\pi N}{60} = \frac{2\pi \times 250}{60} = 26{\cdot}18 \text{ rpm}$$

Height of paraboloid,

$$z = \frac{\omega^2 r^2}{2g} = \frac{26{\cdot}18 \times 0{\cdot}08^2}{2 \times 9{\cdot}81} = 0{\cdot}224 \text{ m or } 224 \text{ mm}$$

(ii) When no water spills out of the vessel, the volume of water remains same before and after rotation of the veseel. Also, the volume of empty space of the vessel before rottion will be equal to the empty space after rotation *i.e.* volume of the paraboloid.

Empty space before rotation = Empty space after rotation

i.e. Volume of empty space before rotation = Volume of paraboloid

$$\pi r^2 (H - h) = \frac{1}{2} \times \text{Volume of cylinder containing the paraboloid}$$

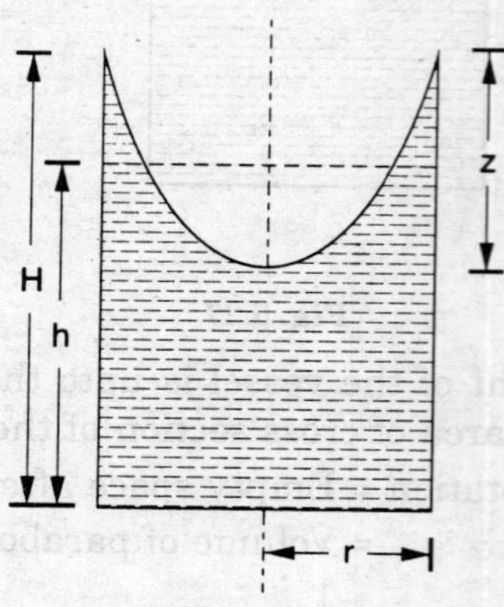

Fig. 5.28

$$= \frac{1}{2} \times \pi r^2 . z$$

$$H - h = \frac{1}{2} . \frac{\omega^2 r^2}{2g}$$

$$(1 - 0{\cdot}75) = \frac{1}{4} . \frac{\omega^2 \times (0{\cdot}08)^2}{9{\cdot}81}$$

$$0{\cdot}25 = \frac{\omega^2 \times (0{\cdot}08)^2}{4 \times 9{\cdot}81}$$

$$\omega^2 = 1532{\cdot}8$$

$$\omega = 39{\cdot}15 \text{ rad/s}$$

$$\frac{2\pi N}{60} = 39{\cdot}15$$

$$N = 373{\cdot}9 \text{ rpm}$$

Alternate Method

The rise of liquid at the sides of the vessel is equal to the fall of the liquid at the centre. So, height of the paraboloid is given by

z = Rise of liquid at the ends + Fall of liquid of the centre

$= 0{\cdot}25 + 0{\cdot}25$

$= 0{\cdot}5$ m

$$z = \frac{\omega^2 r^2}{2g}$$

$$0{\cdot}5 = \frac{\omega^2 \times 0{\cdot}08^2}{2 \times 9{\cdot}81}$$

$\omega = 39{\cdot}15$ rad/s as before

(iii) As depth of water at the axis of the paraboloid is to be zero, therefore,

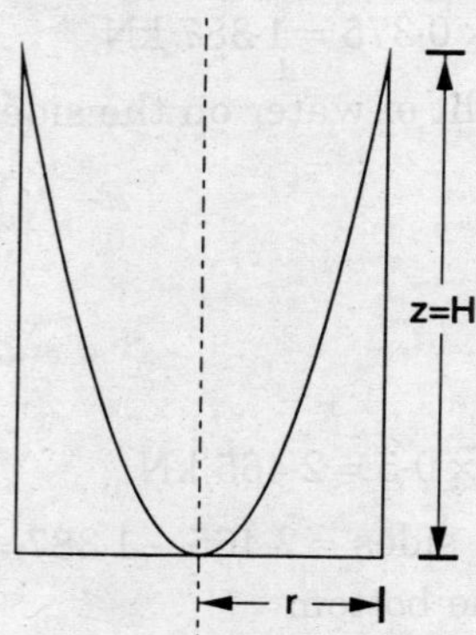

Fig. 5.29

$$z = H = 1 \text{ m}$$

$$z = \frac{\omega^2 r^2}{2g}$$

$$1 = \frac{\omega^2 \times 0{\cdot}08^2}{2 \times 9{\cdot}81}$$

$$\omega^2 = 3065{\cdot}6$$

$$\omega = 55{\cdot}37 \text{ rad/s}$$

$$\frac{2\pi N}{60} = 55{\cdot}37$$

$$N = 528{\cdot}7 \text{ rpm}$$

Example 5.24 For Example 5.23 case (iii), determine the change in total pressure force on the sides and the bottom of the vessel.

Solution (i) Pressure force on sides :

Let F_1 and F_2 be the total pressure forces before and after rotation of the vessel respectively.

Then $F_1 = w \, . \, a \, . \, \bar{h}$

where a = Surface area of the sides of the vessel in contact with water

$= 2\pi \, r \times$ Height of water in vessel

$= 2\pi \times 0{\cdot}08 \times 0{\cdot}75$

$= 0{\cdot}377 \text{ m}^2$

$\bar{h}$ = Height of CG of the above area

$$= \frac{1}{2} \times \text{Height of water}$$

$$= \frac{1}{2} \times 0{\cdot}75$$

$$= 0{\cdot}375 \text{ m}$$

$$F_1 = 9{\cdot}81 \times 0{\cdot}377 \times 0{\cdot}375 = 1{\cdot}387 \text{ kN}$$

After rotation, the height of water on the sides of the vessel becomes equal to the height of the vessel.

$$a = 2\pi \times 0{\cdot}08 \times 1$$

$$= 0{\cdot}05056 \text{ m}^2$$

$$\bar{h} = \frac{1}{2} \times 1 = 0{\cdot}5 \text{ m}$$

$$F_1 = 9{\cdot}81 \times 0{\cdot}5056 \times 0{\cdot}5 = 2{\cdot}465 \text{ kN}$$

Increase in pressure on sides = 2·465 – 1·387 = 1·078 kN

(ii) Pressure force on the bottom

$$F_1 = w \,.\, a \,.\, \bar{h} = 9{\cdot}81 \times \frac{\pi}{4} \times 0{\cdot}16^2 \times 0{\cdot}75 = 0{\cdot}147 \text{ kN} = 147 \text{ N}$$

Pressure on the bottom after rotation varies as a function of the radial distance. Consider an elementary ring of the bottom surface of the vessel of width dr at a radial distance r from the axis of the vessel (Fig. 5.30).

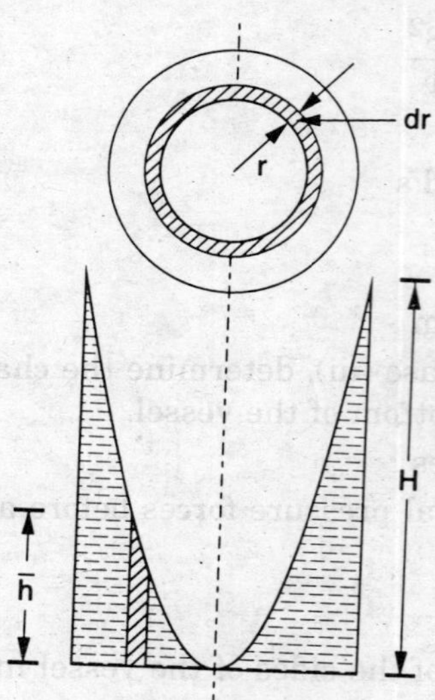

Fig. 5.30

Force on the ring,

$$dF_2 = w \,.\, a \,.\, \bar{h}$$

$$= w \,.\, (2\pi r \,.\, dr) \frac{\omega^2 r^2}{2g}$$

$$= \frac{w}{g} \pi \,.\, \omega^2 r^3 \,.\, dr$$

$$= \pi\rho \,.\, \omega^2 r^3 \, dr$$

Total pressure force on the bottom

$$F_2 = \int_0^r \pi \,\rho \,.\, \omega^2 \,.\, r^3 \,.\, dr = \pi \,\rho \,.\, \omega^2 \int_0^r r^3 \,.\, dr = \pi \,\rho \,.\, \omega^2 \,.\, \frac{r^4}{4}$$

$$= \pi \times 1000 \times (55{\cdot}37)^2 \times \frac{0{\cdot}08^4}{4} = 98{\cdot}6 \text{ N}$$

Decrease in pressure on the bottom surface = 147·2 – 98·6 = 48·6 N

Example 5.25 An open cylindrical vessel 160 mm of internal diameter and 400 mm high is completely filled with water. The water is rotated at (a) 250 rpm (b) 500 rpm and then brought to rest. Find the height of water in the vessel after rest in two cases.

Solution $d = 0{\cdot}16$ m $\qquad H = 0{\cdot}4$ m

$r = 0{\cdot}08$ m

$a = \frac{\pi}{4} \times (0{\cdot}16)^2 = 0{\cdot}02 \text{ m}^2$

(a) $N = 250$ rpm

$$\omega = \frac{2\pi N}{60} = \frac{2\pi \times 250}{60} = 26{\cdot}18 \text{ rad/s}$$

$$z = \frac{\omega^2 r^2}{2g} = \frac{26{\cdot}18^2 \times 0{\cdot}08^2}{2 \times 9{\cdot}81} = 0{\cdot}224 \text{ m}$$

Owing to rotation, water equal to the volume of the paraboloid is spilled out.

Volume of water in the vessel = Volume of vessel – Volume of paraboloid

$$= a \,.\, H - \frac{1}{2} a \,.\, z$$

$$= a\left(H - \frac{1}{2} z\right)$$

Height of water after the vessel comes to rest

$$= \frac{\text{Volume}}{\text{Area}} = \frac{a\left(H - \frac{1}{2} z\right)}{a} = H - \frac{1}{2} z = 0{\cdot}4 - \frac{0{\cdot}224}{2} = 0{\cdot}288 \text{ m}$$

(b) $N = 500$ rpm

$$\omega = \frac{2\pi \times 500}{60} = 53{\cdot}36 \text{ rad/s}$$

$$z = \frac{\omega^2 r^2}{2g} = \frac{53{\cdot}36^2 \times 0{\cdot}08^2}{2 \times 9{\cdot}81} = 0{\cdot}894 \text{ m}$$

As the height of the paraboloid is more than the height of the vessel, the shape of the free surface is as shown in Fig. 5.31.

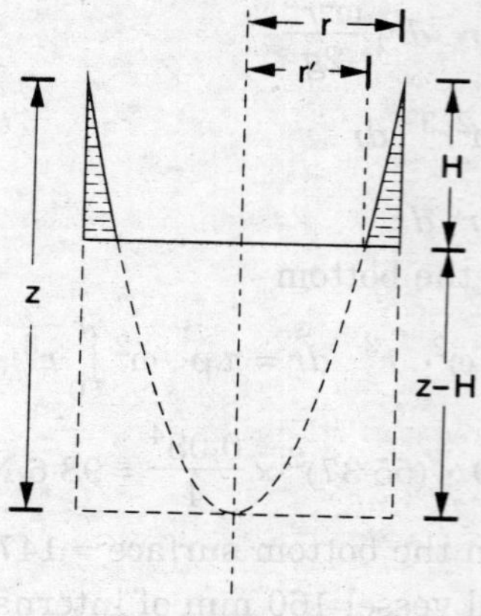

Fig. 5.31

Volume of water spilled out = Total volume of paraboloid – Volume of paraboloid below the bottom surface of vessel

$$= \frac{1}{2} \cdot a \cdot z - \frac{1}{2} a' (z - H)$$

where area a' corresponds to radius r' (Fig. 5.31) and can be found out as below :

Height of paraboloid below the vessel $= \frac{\omega^2 r'^2}{2g}$

$$z - H = \frac{\omega^2 r'^2}{2g}$$

$$0{\cdot}894 - 0{\cdot}4 = \frac{52{\cdot}36^2 \times r'^2}{2 \times 9{\cdot}81}$$

$$r' = 0{\cdot}059 \text{ m}$$

$$a' = \pi\,(0{\cdot}059)^2 = 0{\cdot}011 \text{ m}^2$$

Volume of water spilled out

$$= \frac{1}{2} \times 0{\cdot}02 \times 0{\cdot}894 - \frac{1}{2} \times 0{\cdot}011 \times (0{\cdot}894 - 0{\cdot}4)$$

$$= 0{\cdot}00623 \text{ m}^3$$

and volume of water in the vessel

$$= a \cdot H - 0{\cdot}00623$$

$$= 0{\cdot}02 \times 0{\cdot}04 - 0{\cdot}00623$$

$$= 0{\cdot}00178 \text{ m}^3$$

$$\therefore \text{ Height of water} = \frac{0{\cdot}00178}{a} = \frac{0{\cdot}00178}{0{\cdot}02} = 0{\cdot}089 \text{ m}$$

Example 5.26 A closed cylindrical vessel 160 mm of internal diameter and 1 m high contains water upto a height of 750 mm from the bottom. Find

(a) Height of paraboloid if the vessel is rotated at 450 rpm about its axis.

(b) Speed at which the water just touches the top lid of the vessel.

(c) Speed when the axial depth of water becomes zero.

Solution $d = 0{\cdot}16$ m $\quad H = 1$ m

$r = 0{\cdot}08$ m $\quad h = 0{\cdot}75$ m

$$a = \frac{\pi}{4} \times (0{\cdot}16)^2 = 0{\cdot}02 \text{ m}^2$$

(a) $N = 450$ rpm

$$\omega = \frac{2\pi N}{60} = \frac{2\pi \times 450}{60} = 47{\cdot}12 \text{ rad/s}$$

Let r' = Radius of the paraboloid at the top lid of vessel

Volume of air before rotation = Volume of air after rotation

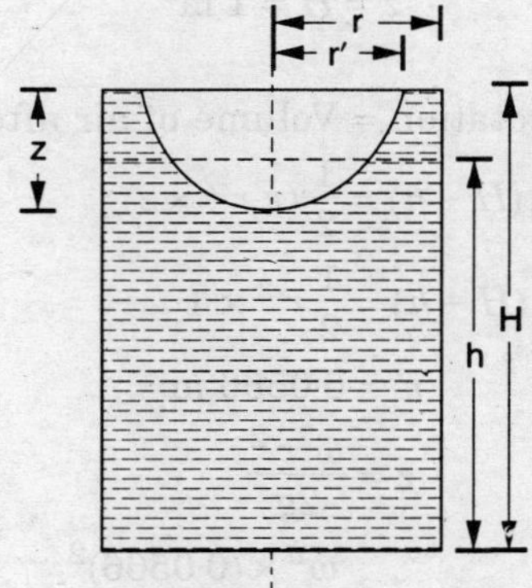

Fig. 5.32

$$a\,(H - h) = \frac{1}{2}\,(\pi\, r'^2 \times z)$$

$$= \frac{1}{2}\,\pi\, r'^2 \times \frac{\omega^2 r'^2}{2g}$$

$$= \frac{\pi}{4} \cdot \frac{\omega^2}{g} \cdot r'^4$$

or $$0{\cdot}02\,(1 - 0{\cdot}75) = \frac{\pi}{4} \cdot \frac{47{\cdot}12^2}{9{\cdot}81} \cdot r'^4$$

$$r' = 0{\cdot}0728 \text{ m}$$

$$z = \frac{\omega^2 r'^2}{2g} = \frac{47{\cdot}12^2 \times 0{\cdot}0728^2}{2 \times 9{\cdot}81} = 0{\cdot}6 \text{ m or } 600 \text{ mm}$$

(b) When the water just touches the top lid :

Volume of air before rotation = Volume of air after rotation

$$a\,(H - h) = \frac{1}{2}\,(\pi\, r^2 \times z)$$

$$\pi\, r^2\,(H - h) = \frac{1}{2}\,\pi\, r^2 \times \frac{\omega^2 r^2}{2g}$$

$$H - h = \frac{1}{4} \cdot \frac{\omega^2 r^2}{g}$$

$$1 - 0.75 = \frac{1}{4} \cdot \frac{\omega^2 \times 0.08^2}{9.81}$$

$$\omega^2 = 1532.8$$

$$\omega = 39.15$$

$$\frac{2\pi N}{60} = 39.15$$

$$N = 373.9 \text{ rpm}$$

(c) When the depth of water at the axis of rotation is zero,

$$z = H = 1 \text{ m}$$

Volume of air before rotation = Volume of air after rotation

$$\pi r^2 (H - h) = \frac{1}{2} (\pi r'^2 \times z)$$

$$r^2 (H - h) = \frac{1}{2} r'^2 \times 1$$

$$r' = 0.0566 \text{ m}$$

Since
$$z = \frac{\omega^2 r'^2}{2g}$$

$$1 = \frac{\omega^2 \times (0.0566)^2}{2 \times 9.81}$$

$$\omega^2 = 6131.2$$

$$\omega = 78.3$$

$$\frac{2\pi N}{60} = 78.3$$

$$N = 747.7 \text{ rpm}$$

Example 5.27 Prove that in case of a closed cylindrical vessel, the speed at which the water touches the top lid is $(2h/H)$ times the speed at which the air touches the base, where H is the height of the vessel and h is the height of the air column in the vessel.

Solution Let the subscript t denotes the quantities when the water touches the top lid and subscript b when the water touches the base.

z_t = Height of paraboloid when water touches the top lid = $2h$

z_b = Height of paraboloid when air touches the base = H

$$z_t = \frac{\omega_t^2 r_t^2}{2g} \text{ and } z_b = \frac{\omega_b^2 r_b^2}{2g}$$

$$\frac{z_t}{z_b} = \frac{\omega_t}{\omega_b} \cdot \left(\frac{r_t}{r_b}\right)^2 = \frac{N_t}{N_b} \cdot \left(\frac{r_t}{r_b}\right)^2$$

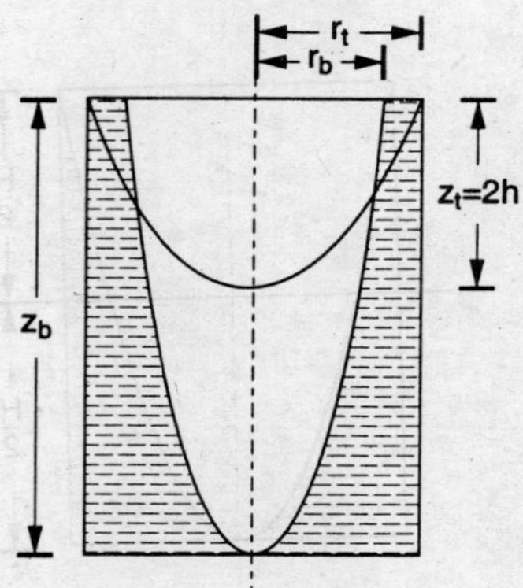

Fig. 5.33

$$\frac{N_t}{N_b} = \frac{r_b}{r_t}\sqrt{\frac{z_t}{z_b}} \qquad \text{...(i)}$$

But volume of air when water touches the top lid

= Volume of air when the air touches the base

$$\frac{1}{2}a_t^{\,2}\,.\,z_t^{\,2} = \frac{1}{2}a_b^{\,2}\,.\,z_b^{\,2}$$

$$\frac{1}{2}\pi\, r_t^{\,2}\,.\,z_t^{\,2} = \frac{1}{2}\pi\, r_b^{\,2}\,.\,z_b^{\,2}$$

$$\left(\frac{r_b}{r_t}\right)^2 = \frac{z_t}{z_b}$$

From (i),

$$\frac{N_t}{N_b} = \sqrt{\frac{z_t}{z_b}}\,.\,\sqrt{\frac{z_t}{z_b}} = \frac{z_t}{z_b} = \frac{2h}{H}$$

Example 5.28 Prove that in a closed cylindrical vessel, half filled with water, the water touches the top lid of the vessel at the same speed at which the air touches the bottom surface of the vessel.

Solution The height of the water level in the vessel is $H/2$. We know that the rise of water level on the sides of the vessel is equal to the fall of liquid at the axis of rotation. Therefore, the rise of $H/2$ height on the sides of the vessel will also result in fall $H/2$ height at the axis. Thus the water touches the top lid at the same speed when the air touches the base of the vessel.

Also from the previous example

$$\therefore \text{Ratio} \qquad \frac{2h}{H} = \frac{2}{H}\cdot\frac{H}{2} = 1 \qquad \left(\because h = \frac{1}{2}H\right)$$

which means the two speeds are equal.

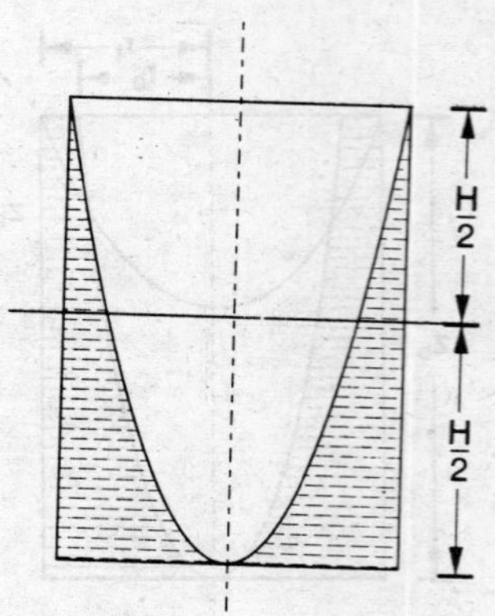

Fig. 5.34

Example 5.29 A closed cylindrical vessel 160 mm of internal diameter is completely filled with water and rotated at 1200 rpm about its vertical axis. Determine the difference in pressure for a horizontal plane between

(a) its circumference and the centre

(b) its circumference and at a radial distance of 40 mm.

Solution $d_2 = 0{\cdot}16$ m $\qquad N = 1200$ rpm

$r_2 = 0{\cdot}08$ m

$$\omega = \frac{2\pi \times 1200}{60} = 125{\cdot}7 \text{ rad/s}$$

(a) In a horizontal plane,

$$\frac{p_2 - p_1}{w} = \frac{\omega_2 (r_2^2 - r_1^2)}{2g}$$

In this case $\qquad r_1 = 0$

$$\frac{p_2 - p_1}{9{\cdot}81} = \frac{\omega_2 r_2^2}{2g}$$

$$= \frac{125{\cdot}7^2 \times 0{\cdot}08^2}{2 \times 9{\cdot}81}$$

$$p_2 - p_1 = 50{\cdot}56 \text{ kN/m}^2$$

(b) In this case $r_1 = 0{\cdot}04$

$$\frac{p_2 - p_1}{9{\cdot}81} = \frac{125{\cdot}7^2 \times (0{\cdot}08^2 - 0{\cdot}04^2)}{2 \times 9{\cdot}81}$$

$$p_2 - p_1 = 37{\cdot}92 \text{ kN/m}^2$$

5.12 RELATIVE EQUILIBRIUM OF FLUIDS

In fluid statics, shear stresses are absent and thus variation of pressure is simple to compute.

When a body of confined liquid as a whole is subjected to a uniformly accelerated motion such that no layer moves relative to an adjacent layer, the shear stresses are zero throughout the fluid. The fluid is said to be in relative equilibrium.

In such cases, variation of pressure can be determined by writing the equation of motion for an appropriate free body.

Uniform Linear Acceleration

Let a liquid in an open vessel be subjected to a uniform linear acceleration as shown in Fig. 5.35. After some time, when the liquid is adjusted to the acceleration, it acts as a solid. The distance between any two fluid particles remains fixed and no shear stresses involve.

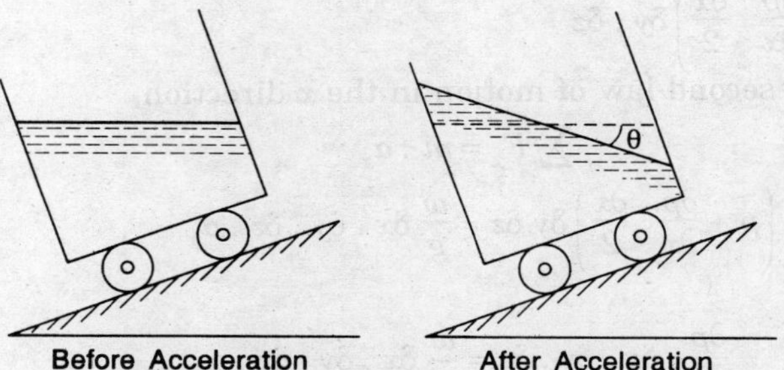

Fig. 5.35

Assuming that the acceleration has components in the x (horizontal) and z (vertical) directions. There is no component of the acceleration in the y direction. Consider an element of fluid in the form of a small rectangular prism having edges δx, δy and δz as shown in Fig. 5.36. Pressure at the centre of the element is p.

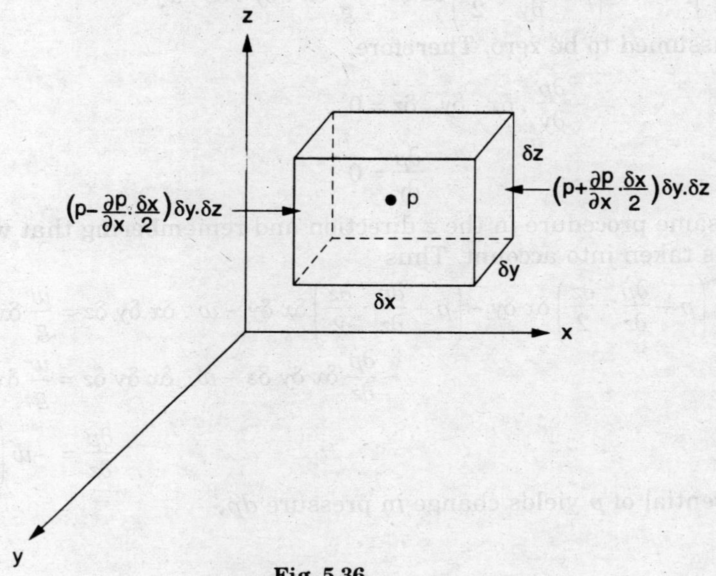

Fig. 5.36

Then pressure intensity on the left hand face of element

$$= p - \frac{\partial p}{\partial x} \cdot \frac{\delta x}{2}$$

Pressure force on the left hand face of element

$$= \left(p - \frac{\partial p}{\partial x} \cdot \frac{\delta x}{2} \right) \delta y \, . \, \delta z$$

Pressure intensity on the right hand face of element

$$= p + \frac{\partial p}{\partial x} \cdot \frac{\delta x}{2}$$

Pressure force on the right hand face of element

$$= \left(p + \frac{\partial p}{\partial x} \cdot \frac{\delta x}{2} \right) \delta y \, . \, \delta z$$

Applying Newton's second law of motion in the x direction,

$$\sum F_x = m \, . \, a_x$$

$$\left(p - \frac{\partial p}{\partial x} \cdot \frac{\partial x}{2} \right) \delta y \, \delta z - \left(p + \frac{\partial p}{\partial x} \cdot \frac{\partial x}{2} \right) \delta y \, \delta z = \frac{w}{g} \delta x \, . \, \delta y \, . \, \delta z \, . \, a_x$$

$$- \frac{\partial p}{\partial x} \, . \, \delta x \, . \, \delta y \, . \, \delta z = \frac{w}{g} \delta x \, . \, \delta y \, . \, \delta z \, . \, a_x$$

$$\frac{\partial p}{\partial x} = - w \frac{a_x}{g}$$

Similarly, in the y direction

$$\sum F_y = m \, . \, a_y$$

$$\left(p - \frac{\partial p}{\partial y} \cdot \frac{\partial y}{2} \right) \delta x \, \delta z - \left(p + \frac{\partial p}{\partial y} \cdot \frac{\partial y}{2} \right) \delta x \, \delta z = \frac{w}{g} \delta x \, . \, \delta y \, . \, \delta z \, . \, a_y$$

But a_y is assumed to be zero. Therefore,

$$- \frac{\partial p}{\partial y} \, . \, \delta x \, . \, \delta y \, . \, \delta z = 0$$

$$\frac{\partial p}{\partial y} = 0$$

Apply the same procedure in the z direction and remembering that weight of the element $w \, . \, \delta x \, . \, \delta y \, . \, \delta z$ is taken into account. Thus

$$\left(p - \frac{\partial p}{\partial z} \cdot \frac{\partial z}{2} \right) \delta x \, \delta y - \left(p + \frac{\partial p}{\partial z} \cdot \frac{\partial z}{2} \right) \delta x \, \delta y - w \, . \, \delta x \, \delta y \, \delta z = \frac{w}{g} \delta x \, \delta y \, \delta z \, . \, a_z$$

$$- \frac{\partial p}{\partial z} \delta x \, \delta y \, \delta z - w \, . \, \delta x \, \delta y \, \delta z = \frac{w}{g} \delta x \, \delta y \, \delta z \, . \, a_z$$

$$\frac{\partial p}{\partial z} = - w \left(1 + \frac{a_z}{g} \right)$$

Total differential of p yields change in pressure dp,

$$dp = \frac{dp}{dx} \cdot dx + \frac{dp}{dy} \cdot dy + \frac{dp}{dz} \cdot dz$$

$$= -w \frac{a_x}{g} \cdot dx + 0 - w\left(1 + \frac{a_z}{g}\right) dz$$

$$= -w \frac{a_x}{g} \cdot dx - w\left(1 + \frac{a_z}{g}\right) dz \qquad ...(5.41)$$

This is the equation for pressure variation when the acceleration has x and z components or when the acceleration is along some inclined plane (upwards).

To find the slope of the free liquid surface, $\frac{\partial z}{\partial x}$ can be evaluated from Eq. 5.41.

($dp = 0$, because pressure is constant on the free surface)

$$\frac{\partial z}{\partial x} = \tan \theta = -\frac{a_x}{g + a_z} \qquad ...(5.42)$$

Since the expression comes out to be negative, it means the free surface is sloping downwards in the direction of acceleration (Fig. 5.35).

The lines of constant pressure are parallel to the free surface.

Uniform Vertical Acceleration

When an open vessel filled with liquid is moved vertically upward with a uniform acceleration a_z and it has no component in the x direction, the pressure at a point in the fluid mass is given by Eq. 5.41, putting $a_x = 0$,

$$dp = -w\left(1 + \frac{a_z}{g}\right) dz \qquad \text{(z taken upwards)} \; ...(5.43)$$

If p is taken zero at datum ($z = 0$),

$$p = -\rho g\left(1 + \frac{a_z}{g}\right) z$$

$$= -\rho\,(g + a_z)\, z \qquad ...(5.44)$$

Negative sign indicates that pressure decreases in the upward direction. The slope of the constant pressure lines is given by equation by putting $a_x = 0$

$$\frac{\partial z}{\partial x} = 0$$

i.e. the free surface will have zero slope or it is horizontal. The lines of constant pressure thus are horizontal in this case.

Uniform Horizontal Acceleration

Consider an open vessel moving in the x direction with a uniform acceleration a_x, the pressure at a point in the fluid mass is given by Eq. 5.41, putting $a_z = 0$,

or pressure head at a point $= \frac{a_x}{g} \cdot dx + w \cdot dz \qquad ...(5.45)$

Slope of constant pressure lines from equation

$$\frac{\partial z}{\partial x} = \tan \theta = \frac{a_x}{g} \qquad ...(5.46)$$

Example 5.21 An open rectangular tank 10 m × 6 m in cross-section and 4 m deep contains water upto a height of 3 m from the bottom. It is accelerated at 3 m/s^2
(a) horizontally (b) vertically upwards (c) vertically downwards
Determine in each case the slope of the free surface and total pressure on the bottom and sides of the tank.

Solution (a) Horizontal acceleration (Fig. 5.37)

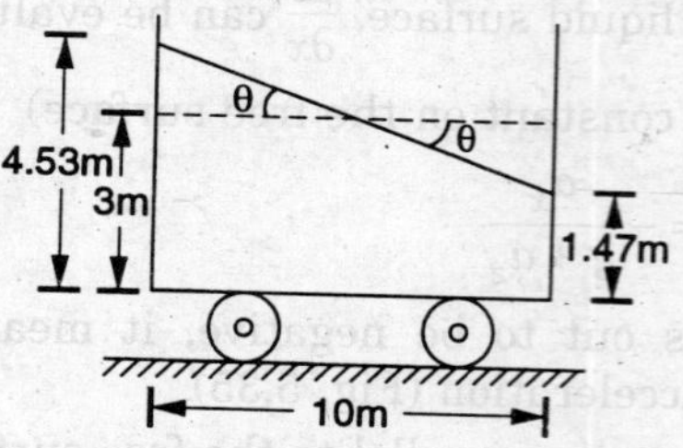

Fig. 5.37

$$\tan \theta = \frac{a_x}{g} = \frac{3}{9.81} = 0.306$$

$$h_l = (3 + 5 \tan \theta) = 3 + 5 \times 0.306 = 4.53 \text{ m}$$

$$h_r = (3 - 5 \tan \theta) = 3 - 5 \times 0.306 = 1.47 \text{ m}$$

$$P_l = wA\bar{x} = 9.81 \times (4.53 \times 6) \times \frac{4.53}{2} = 603.9 \text{ kN}$$

$$P_r = wA\bar{x} = 9.81 \times (1.47 \times 6) \times \frac{1.47}{2} = 63.6 \text{ kN}$$

The force on the long sides remains the same as before the acceleration since the area of contact of water remains the same and addition of the pressure force on the left half is equal to the reduction of pressure force on the right side.
Thus pressure force on each long side

$$= wA\bar{x} = 9.81 \times (10 \times 3) \times \frac{3}{2} = 441.5 \text{ kN}$$

The total force on the bottom again remains the same due to the reason that the addition of force on the left half is equal to the reduction of the force on the right half.
Thus pressure force on the bottom

$$= wA\bar{x} = 9.81 \times (10 \times 6) \times 3 = 1765.8 \text{ kN}$$

(b) Vertical upward acceleration

$$P_l = P_r = wA\bar{x} = \rho g A\bar{x}$$

$$= 10^3 \times (9.81 + 3) \times (6 \times 3) \times \frac{3}{2} = 345.9 \times 10^3 \text{ N} = 345.9 \text{ kN}$$

$$P_{\text{long side}} = \rho gA\bar{x} = 10^3 \times (9{\cdot}81 + 3) \, . \, (10 \times 3) \times \frac{3}{2} = 576{\cdot}5 \times 10^3 \text{ N} = 576{\cdot}5 \text{ kN}$$

$$P_{\text{bottom}} = \rho gA\bar{x} = 10^3 \times (9{\cdot}81 + 3) \, . \, (10 \times 6) \times 3 = 2305{\cdot}8 \times 10^3 \text{ N} = 2305{\cdot}8 \text{ kN}$$

(c) Vertical downward acceleration

$$P_l = P_r = wA\bar{x} = \rho gA\bar{x}$$

$$= 10^3 \times (9{\cdot}81 - 3) \times (6 \times 3) \times \frac{3}{2} = 183{\cdot}9 \times 10^3 \text{ N} = 183{\cdot}9 \text{ kN}$$

$$P_{\text{long side}} = \rho gA\bar{x} = 10^3 \times (9{\cdot}81 - 3)\,(10 \times 3) \times \frac{3}{2} = 306{\cdot}5 \times 10^3 \text{ N} = 306{\cdot}5 \text{ kN}$$

$$P_{\text{bottom}} = \rho gA\bar{x} = 10^3 \times (9{\cdot}81 - 3)\,(10 \times 6) \times 3 = 1225{\cdot}8 \times 10^3 \text{ N} = 1225{\cdot}8 \text{ kN}$$

Example 5.22 **A rectangular tank containing water is accelerated at 4 m/s² upwards on inclined plane 30° to the horizontal. Determine the slope of free liquid surface.**

Solution Refer Fig. 5.38,

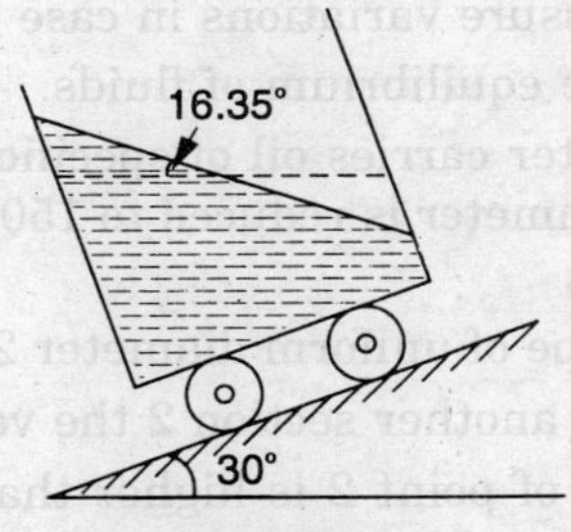

Fig. 5.38

Horizontal and vertical components of the acceleration are

$$a_x = 4 \cos 30° = 3{\cdot}464 \text{ m/s}$$

$$a_z = 4 \sin 30° = 2 \text{ m/s}^2$$

$$\tan \theta = \frac{a_x}{g + a_z}$$

$$= \frac{3{\cdot}464}{9{\cdot}81 + 2}$$

$$= 0{\cdot}2933$$

$$\theta = 16{\cdot}35°$$

PROBLEMS

1. What is meant by potential head, velocity head and datum head of a fluid.
2. Obtain the Euler's equation of motion. How is Bernoulli's equation derived from it ?
3. State and prove Bernoulli's equation. Also mention the assumptions made.

4. What are practical applications of Bernoulli's equation ?
5. What are the limitations of Bernoulli's equation ?
6. What is a venturimeter ? Deduce an expression for the rate of flow through a venturimeter.
7. Define an orifice meter. Why is coefficient of discharge of an orifice meter is smaller than that of a venturimeter ?
8. What is a pitot tube ? How can you determine the velocity of flow of a liquid with its help ?
9. What are kinetic energy and momentum correction factors.
10. State and derive the momentum equation.
11. What is moment of momentum equation ? Where is it applicable ?
12. What do you mean by free jet of liquid ? Derive the expression for the path travelled by a free jet out of a nozzle.
13. What is vortex motion ? Derive a general expression for the same.
14. Define a forced and a free vortex. Give examples.
15. Derive equations for pressure variations in case of a forced and free vortex motion.
16. Explain the term relative equilibrium of fluids.
17. A pipe 200 mm in diameter carries oil of specific gravity 0·83 at a velocity of 2 m/s. At another section the diameter is reduced to 150 mm. Determine the velocity at this section and the flow rate. (3·56 m/s, 62·9 l/s)
18. Water flows through a pipe of uniform diameter 250 mm. At section 1 of the pipe the pressure is 50 kN/m^2. At another section 2 the velocity is 1·25 m/s and the pressure 20 kN/m^2. The elevation of point 2 is higher than that of 1 by 2 m. Determine the direction of flow and the head loss between the two sections. (From 1 to 2, 1·058 m)
19. A tapering pipe has diameters 400 mm and 200 mm at sections A and B respectively. The water flows through the pipe at the rate of 50 litres per second. The level of section A is 5 m higher than that of section B. What is the intensity of pressure at section B if at section A it is 500 kN/m^2. (547·86 kN/m^2)
20. The inlet and throat diameters of a horizontal venturimeter are 300 mm and 150 mm respectively. The reading of the differential gauge connected to the inlet and the throat shows a reading of 185 mm of mercury. Assuming coefficient of discharge to be 0·975, find the discharge. (122·8 l/s)
21. A venturimeter with 300 mm main diameter and 150 mm throat diameter is provided in a vertical pipeline carrying oil of specific gravity 0·9. The flow of oil is upwards. The throat section is 300 mm above the entrance section of the venturimeter. The differential U-tube mercury manometer shows a gauge deflection of 250 mm. Determine the discharge and the pressure difference between the entrance and throat sections. The coefficient of meter is 0·98. (148·8 l/s, 3·69 m of oil)
22. A pipeline 300 mm in diameter carries oil of specific gravity 0·85 and is fitted with an orifice plate to measure the discharge. The diameter of the orifice is 150 mm. A U-tube manometer connected to the orifice meter shows a deflection of 200 mm. Find the discharge assuming the coefficient of the meter as 0·86. (116·6 l/s)

23. A lawn sprinkler shown in Fig. 5.39 has 10 mm diameter nozzles at the ends of rotating arm. The water is discharged at a speed of 8 m/s. Determine the speed of rotation of the arm when free to rotate. Also find the torque required to hold the rotating arm stationary. (507 rpm, 14·79 N . m)

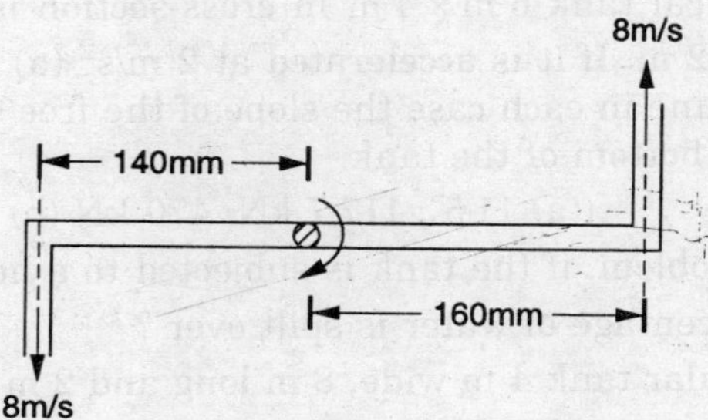

Fig. 5.39

24. It is desired to have an orifice in the side of a tank at such a height from the base so that the jet attains the maximum horizontal distance from the tank at the level of its base. If the height of liquid in the tank is maintained at 2·4 m, determine the height of free surface from the centre of the orifice. (1·2 m)

25. A nozzle with a diameter of 40 mm is placed at a height of 1·2 m above the ground level and is inclined upwards at an angle of 60° to the horizontal. The jet of water strikes the ground at a horizontal distance of 5 m. Determine the flow rate. (8·85 l/s)

26. An open cylindrical vessel 300 mm in diameter and 1·5 m high contains water upto a height of 1 m. Find
    (a) the height of paraboloid formed when the vessel is rotated at 300 rpm
    (b) the maximum speed at which the water is on the point of spilling out
    (c) the speed at which the height of water at the axis is zero.
    (1·132 m, 282 rpm, 345·4 rpm)

27. An open cylindrical vessel 240 mm in diameter and 500 mm high is completely filled with water. The vessel is rotated at 300 rpm and after some time brought to rest. Determine the depth of water in the vessel. Also determine the volume of water in the vessel if the vessel is rotated at 450 rpm. (138 mm, 0·0035 m^3)

28. The diameter of a cylindrical vessel is 240 mm and the height 1·2 m and it contains water upto a height of 700 mm. If it is rotated at 200 rpm, find the difference in total pressure force on the sides and the bottom before and after rotation. (3·513 kN, zero)

29. A closed cylindrical vessel 120 mm internal diameter and 750 mm high is filled with 2/3rd of water. Determine
    (a) the height of paraboloid if the vessel is rotated at 240 rpm about its axis
    (b) the speed at which the top of the cup formed by the air just touches the top lid of the vessel
    (c) the speed at which the bottom of the cup just touches the base of the vessel.
    (240 mm, 498·5 rpm, 748 rpm)

30. A closed cylindrical vessel 120 mm in diameter completely filled with water is rotated at 800 rpm. Find the difference in pressure for a horizontal plane between its circumference and the centre and between its circumference and at a radius of 40 mm. (12·63 kN, 7·019 kN)

31. An open cylindrical tank 6 m × 4 m in cross-section is 3 m deep. It contains water upto a height of 2 m. If it is accelerated at 2 m/s^2 (a) horizontally and (b) vertically upwards, determine in each case the slope of the free surface and the total pressure on the sides and bottom of the tank.
((a) 11·5°, 117·7 kN, 470 kN (b) zero slope, 141·7 kN, 568 kN)

32. For the above problem, if the tank is subjected to a horizontal acceleration of 392·4 mm/s^2, what percentage of water is spilt over ? (10 percent)

33. A closed rectangular tank 4 m wide, 8 m long and 2 m deep is completely filled with an oil of specific gravity 0·9. If the tank is moved with an acceleration of 2·4 m/s^2 in the horizontal direction, find the pressure difference between the front and rear top corners of the tank. (19·2 kN/m^2)

Chapter 6

ORIFICES AND MOUTHPIECES

6.1 INTRODUCTION

An **orifice** is an opening in the wall or bottom of a tank and serves to discharge the fluid contained in the tank. The top edge of the opening lies below the surface of the fluid in the tank.

A **mouthpiece** is a short length of pipe or tube fixed to an orifice. The pipe extension is usually 2 to 3 times the orifice diameter. Orifices as well as mouthpieces are also used for measuring the discharge from a tank or reservoir.

6.2 CLASSIFICATION OF ORIFICES

The classification of orifices is made as follows :

According to shape

An orifice may be **circular, triangular, rectangular** or **square** depending upon its cross-sectional area.

According to size

An orifice may be **small** or **large** depending upon the size of the orifice as compared to the head of the fluid on the upstream side. If the head of the fluid is quite high as compared to the dimensions of the orifice (approximately $H > 5d$) and the variation in the velocity is negligible from top to the bottom edge of the orifice, it is known as a **small orifice**. On the other hand, if the head is small the velocity of flow through the orifice varies from top to the bottom edge, it is a **large orifice**.

According to shape of Upstream Edge

An orifice may be **sharp-edged** or **bell-mouthed**.

According to discharge conditions

An orifice may be free discharging or **submerged (drowned)**. A free discharging orifice discharges into the atmosphere whereas a submerged orifice discharges into another tank of fluid. Further, a submerged orifice may be fully submerged or partially submerged.

6.3 FLOW THROUGH AN ORIFICE

Figure 6.1 shows a tank fitted with a small sharp-edged circular orifice with its sharp edge on the side of the tank and discharging free into the atmosphere. Let H be the head of the fluid above the centre line of the orifice. As the fluid flows through the orifice, it forms a jet of liquid and attains a parallel form at a distance $d/2$ from the plane of the orifice. This point where the stream lines first become parallel is known as **vena-contracta**. The area of cross-section of the jet at this point is always less than that of the orifice. Beyond this section, the jet diverges and is attracted in the downward direction due to the effect of gravity.

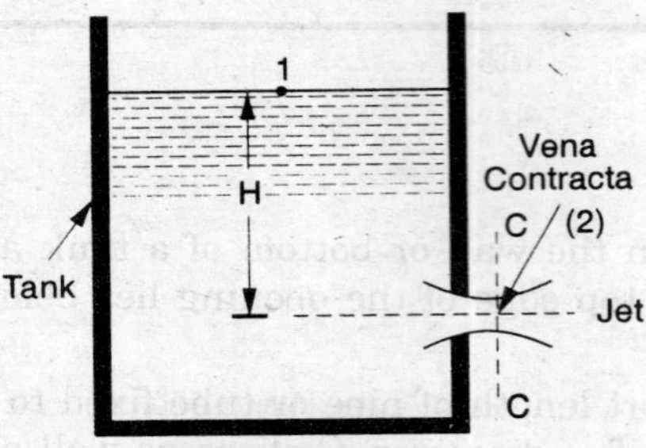

Fig. 6.1

Considering points 1 and 2 as shown in Fig. 6.1. Point 1 is at the free surface of fluid in the tank and point 2 at the vena contracta. As the cross-section of the tank is very large, the fluid at 1 is assumed to be stationary. Applying Bernoulli's equation at points 1 and 2.

$$\frac{p_1}{w} + \frac{V_1^2}{2g} + z_1 = \frac{p_2}{w} + \frac{V_2^2}{2g} + z_2$$

$$0 + 0 + H = 0 + \frac{V_2^2}{2g} + 0$$

or

$$V_2 = \sqrt{2gH} \quad \text{...(6.1)}$$

This is the theoretical velocity and shows that the velocity of flow through an orifice is equal to the velocity of free fall from the surface of the tank. This is known as **Torricelli's theorem**.

In real fluid flows, the actual velocity of flow through an orifice is always less than the theoretical velocity due to viscous effects of the fluids.

6.4 HYDRAULIC COEFFICIENTS

The hydraulic coefficients are :

1. Coefficient of velocity

2. Coefficient of contraction
3. Coefficient of discharge

1. Coefficient of Velocity

It is the ratio of the actual velocity of the jet at vena-contracta to the theoretical velocity. It is denoted by C_v. Mathematically,

$$C_v = \frac{\text{Actual velocity of jet at vena-contracta}}{\text{Theoretical velocity}}$$

$$= \frac{V}{\sqrt{2gH}} \quad ...(6.2)$$

The value of C_v lies between 0·95 to 0·99 depending upon the size and shape of the orifice and on the head of the fluid under which the flow takes place. Usually, C_v is taken as 0·98 for sharp edged orifices.

2. Coefficient of Contraction

It is the ratio of the area of the jet at the vena-contracta to the area of the orifice. It is denoted by C_c.

$$C_c = \frac{a_c}{a} \quad ...(6.3)$$

where a = area of the orifice

a_c = area of the jet at vena-contracta

$\boldsymbol{a_c}$ is also referred as the actual area and $\boldsymbol{a}$ as the theoretical area of the jet.

The value of C_v varies from 0·61 to 0·69 depending on the shape and size of the orifice and the head of the liquid. The average value is taken as 0·64.

3. Coefficient of Discharge

It is defined as the ratio of the actual discharge through an orifice to the theoretical discharge. It is denoted by C_d.

$$C_d = \frac{Q}{Q_{th}}$$

$$= \frac{\text{Actual velocity} \times \text{Actual area}}{\text{Theoretical velocity} \times \text{Theoretical area}}$$

$$= \frac{\text{Actual velocity}}{\text{Theoretical velocity}} \times \frac{\text{Actual area}}{\text{Theoretical area}}$$

$$= C_v \times C_c \quad ...(6.4)$$

The value of C_d varies from 0·62 to 0·65 and depends on the shape and size of the orifice and the head.

Example 6.1 An orifice is discharging water under a head of 12 m. The diameter of the orifice is 60 mm and the coefficients of discharge and velocity are 0·62 and 0·98 respectively. Determine the actual discharge and the actual velocity at the vena-contracta.

Solution

$$H = 12 \text{ m}$$

$$d = 0{\cdot}06 \text{ m}$$

$$C_d = 0{\cdot}62$$

$$C_v = 0{\cdot}98$$

$$\text{Actual discharge} = C_d \times \text{Theoretical discharge}$$

$$= C_d \times \text{Area} \times \text{Theoretical Velocity}$$

$$= C_d \times \frac{\pi}{4} \times d^2 \times \sqrt{2gH}$$

$$= 0{\cdot}62 \times \frac{\pi}{4} \times (0{\cdot}06)^2 \times \sqrt{2 \times 9{\cdot}81 \times 12}$$

$$= 0{\cdot}62 \times 0{\cdot}002827 \times 15{\cdot}344$$

$$= 0{\cdot}029 \text{ m}^3/\text{s}$$

$$\text{Actual velocity} = C_v \times \text{Theoretical velocity}$$

$$= C_v \times \sqrt{2gH}$$

$$= 0{\cdot}98 \times \sqrt{2 \times 9{\cdot}81 \times 12}$$

$$= 15{\cdot}037 \text{ m/s}$$

6.5 EXPERIMENTAL DETERMINATION OF HYDRAULIC COEFFICIENTS

The hydraulic coefficients are found as given below :

1. Coefficient of Discharge

The water in the tank is made to flow through the orifice under a constant head H and is collected in another measuring tank (Fig. 6.2). The rise of water level in the measuring tank in a specific time t is noted down.

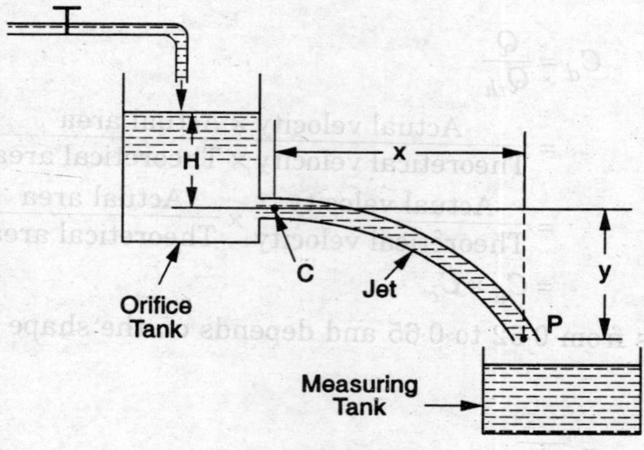

Fig. 6.2

Then the actual discharge,

$$Q = \frac{\text{Area of the measuring tank} \times \text{Rise of water level}}{\text{Time}}$$

and $$Q_{th} = \text{area of orifice} \times \sqrt{2gH} = a\sqrt{2gH}$$

$$C_d = \frac{Q}{Q_{th}} = \frac{Q}{a\sqrt{2gH}} \quad ...(6.5)$$

2. Coefficient of Velocity

Let a liquid particle take time t to travel along the jet from vena-contracta C to position P.

Let x = horizontal distance travelled by the particle in time t

y = vertical distance between points C and P

V = actual velocity of jet at vena-contracta

$$x = V \, . \, t \quad \text{or} \quad t^2 = \frac{x^2}{V^2} \quad ...(i)$$

$$y = \frac{1}{2}gt^2 \quad \text{or} \quad t^2 = \frac{2y}{g} \quad ...(ii)$$

From (i) and (ii),

$$\frac{x^2}{V^2} = \frac{2y}{g}$$

$$V^2 = \frac{gx^2}{2y}$$

$$V = \sqrt{\frac{gx^2}{2y}}$$

$$C_v = \frac{V}{V_{th}} = \sqrt{\frac{gx^2}{2y}} \frac{1}{\sqrt{2gH}} = \sqrt{\frac{gx^2}{2y \, . \, 2gH}} = \frac{x}{\sqrt{4yH}} \quad ...(6.6)$$

3. Coefficient of Contraction

We know that

$$C_d = C_v \, . \, C_c$$

$$C_c = \frac{C_d}{C_v} \quad ...(6.7)$$

Thus C_c is found from the known values of C_d and C_v

Example 6.2 A vertical sharp-edged orifice 100 mm in diameter discharges water at the rate of 60 l/s under a constant head of 8 m. A point on the jet, measured from the vena-contracta of the jet has co-ordinates 3·8 m horizontal and 0·48 m vertical. Determine the coefficients of velocity, discharge and contraction.

Solution Refer Fig. 6.2,

$x = 3{\cdot}8$ m $\qquad d = 0{\cdot}1$ m

$Q = 0{\cdot}06 \text{ m}^3/\text{s}$ $\quad H = 8 \text{ m}$

$y = 0{\cdot}48 \text{ m}$

Coefficient of velocity

$$C_v = \frac{x}{\sqrt{4yH}} = \frac{3{\cdot}8}{\sqrt{4 \times 0{\cdot}48 \times 8}} = 0{\cdot}97$$

Coefficient of discharge

$$C_d = \frac{\text{Actual discharge}}{\text{Theoretical discharge}} = \frac{0{\cdot}06}{\frac{\pi}{4} \times 0{\cdot}1^2 \times \sqrt{2 \times 9{\cdot}81 \times 8}}$$

Coefficient of contraction

$$C_c = \frac{C_d}{C_v} = \frac{0{\cdot}61}{0{\cdot}97} = 0{\cdot}629$$

Example 6.3 A sharp-edged circular orifice fixed to a water tank discharges water which strikes a wall 1·2 m away and 0·32 m vertically below the centre line of contracted section of the jet. The diameter of the orifice is 6 mm and the height of water in the tank is 1·2 m above the orifice. The actual discharge through the orifice is measured to be 16 litres in 188 seconds. Find the orifice coefficient and the power loss at the orifice.

Solution (Fig. 6.2),

The water strikes the wall at point P.

$x = 1{\cdot}2 \text{ m}$ $\quad y = 0{\cdot}32 \text{ m}$

$H = 1{\cdot}2 \text{ m}$ $\quad d = 0{\cdot}006 \text{ m}$

$$\text{Actual discharge} = \frac{0{\cdot}016}{188} = 8{\cdot}51 \times 10^{-5} \text{ m}^3/\text{s}$$

$$C_v = \frac{x}{\sqrt{4yH}} = \frac{12}{\sqrt{4 \times 0{\cdot}32 \times 1{\cdot}2}} = 0{\cdot}968$$

$$C_d = \frac{\text{Actual discharge}}{\text{Theoretical discharge}} = \frac{8{\cdot}51 \times 10^{-5}}{\frac{\pi}{4} \times 0{\cdot}006^2 \times \sqrt{2 \times 9{\cdot}81 \times 1{\cdot}2}} = 0{\cdot}62$$

$$C_c = \frac{C_d}{C_v} = \frac{0{\cdot}62}{0{\cdot}968} = 0{\cdot}64$$

$$h_f = H - \frac{V^2}{2g} = H\left(1 - \frac{V^2}{2gH}\right) = H(1 - C_v^2) = 1{\cdot}2(1 - 0{\cdot}968^2) = 0{\cdot}0756 \text{ m}$$

$$\text{Power loss} = wQh_f = 9{\cdot}81 \times 1000 \times 8{\cdot}51 \times 10^{-5} \times 0{\cdot}0756 = 0{\cdot}0631 \text{ W}$$

Example 6.4 A tank has two identical orifices in one of its vertical sides located at 2 m and 4 m below the water surface. Find the point at which the two jets intersect assuming a coefficient of velocity 0·96 for both the orifices.

Solution Let the point of intersection be at P (Fig. 6.3).

x = horizontal coordinate of point P

y_1 = vertical coordinate of point P from orifice 1

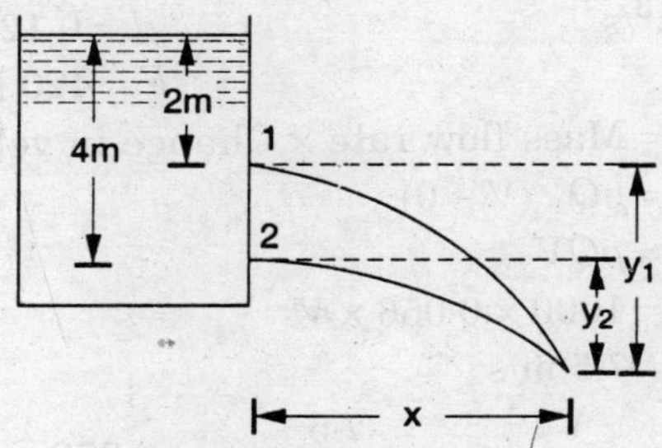

Fig. 6.3

y_2 = vertical coordinate of point P from orifice 2

$H_1 = 2$ m $\qquad C_v = 0{\cdot}96$

$H_2 = 4$ m

Since coefficient of velocity is same for the two orifices.

$$C_v = \frac{x}{\sqrt{4y_1 H_1}} = \frac{x}{\sqrt{4y_2 H_2}} \qquad \text{...(i)}$$

$$y_1 H_1 = y_2 H_2$$

$$y_1 = \frac{H_2}{H_1} \cdot y_2 = \frac{4}{2} X y_2 = 2y_2$$

From the geometry of the tank,

$$y_1 = y_2 + (4 - 2) = y_2 + 2 = \frac{y_1}{2} + 2$$

which gives,

$$y_1 = 4\text{m}$$

$$y_2 = \frac{y_1}{2} = 2\text{m}$$

It may be deduced that $H_1 = y_2$ and $H_2 = y_1$.

From (*i*),

$$0{\cdot}96 = \frac{x}{\sqrt{4 \times 4 \times 2}}$$

$$x = 5{\cdot}43\text{m}$$

Example 6.5 A 120 mm diameter orifice discharges 56 litres of water per second under a constant head of 3 m. A flat plate held normal to the jet downstream in the path of the jet requires a force of 420 N to resist the impart of the jet. Find the hydraulic coefficients.

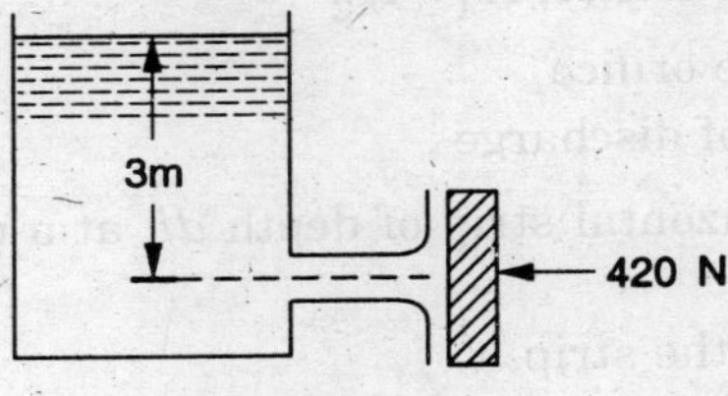

Fig. 6.4

Solution $Q = 0{\cdot}056 \text{ m}^3/\text{s}$ $d = 0{\cdot}12 \text{ m}$

$H = 3 \text{ m}$ $F = 420 \text{ N}$

$$F = \text{Mass flow rate} \times \text{Change in velocity}$$
$$= pQ\,.\,(V - 0)$$
$$= pQV$$
$$420 = 1000 \times 0{\cdot}056 \times V$$
$$V = 7{\cdot}5 \text{ m/s}$$
$$C_v = \frac{V}{\sqrt{2gH}} = \frac{7{\cdot}5}{\sqrt{2 \times 9{\cdot}81 \times 3}} = 0{\cdot}978$$
$$C_d = \frac{Q}{\frac{\pi}{4} \times d^2 \times \sqrt{2gH}} = \frac{0{\cdot}056}{\frac{\pi}{4} \times 0{\cdot}12^2 \times \sqrt{2 \times 9{\cdot}81 \times 3}} = 0{\cdot}645$$
$$C_c = \frac{C_d}{C_v} = \frac{0{\cdot}645}{0{\cdot}978} = 0{\cdot}66$$

6.6 DISCHARGE THROUGH LARGE RECTANGULAR ORIFICES

In case of large orifice ($H < 5$ times height of orifice), the velocity of the jet cannot be considered constant throughout the height of cross-section of the jet.

Let a large rectangular orifice be on one side of a tank discharging freely into atmosphere under a constant head as shown in Fig. 6.5.

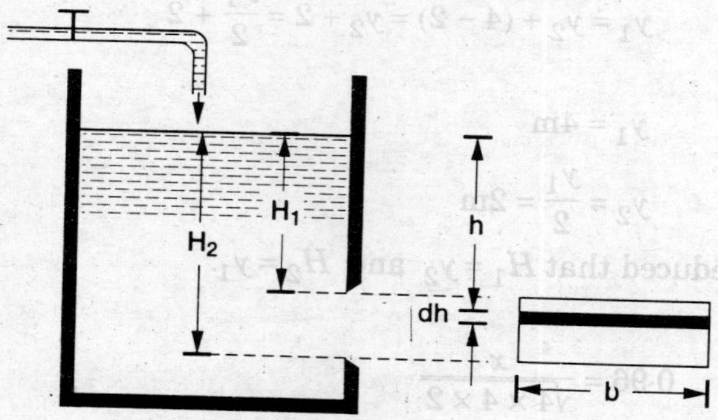

Fig. 6.5

Let H_1 = height of liquid above the upper edge of the orifice

H_2 = height of liquid above the lower edge of the orifice

d = depth of the orifice, $H_1 - H_2$

b = width of the orifice

C_d = coefficient of discharge

Consider an elementary horizontal strip of depth dh at a depth h below the free surface of the liquid.

Then the discharge through the strip,

$$dQ = C_d \times \text{Area of strip} \times \text{Velocity}$$
$$= C_d \times b \,.\, dh \times \sqrt{2gh}$$

Thus total discharge

$$Q = \int_{H_1}^{H_2} C_d \,.\, b \,.\, dh \sqrt{2gh}$$
$$= C_d \,.\, b \,.\, \sqrt{2gh} \int_{H_1}^{H_2} h^{1/2} \,.\, dh$$
$$= C_d \,.\, b \,.\, \sqrt{2g} \left[\frac{h^{3/2}}{3/2} \right]_{H_1}^{H_2}$$
$$= \frac{2}{3} C \,.\, b \,.\, \sqrt{2g}\ [H_2^{3/2} - H_1^{3/2}] \qquad ...(6.8)$$

Example 6.6 **Water is discharged from a tank by a rectangular orifice, 500 mm wide and 600 mm deep. The top edge of the orifice is 300 mm below the water surface in the tank. Taking the value of coefficient of discharge as 0·62, determine the discharge through the orifice and the percentage error if the orifice is treated as small orifice.**

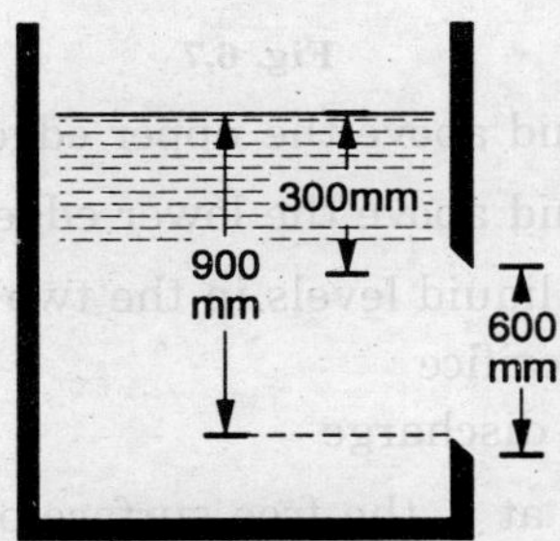

Fig. 6.6

Solution $H_1 = 0{\cdot}3$ m $\qquad b = 0{\cdot}5$ m

$H_2 = 0{\cdot}3 + 0{\cdot}6 = 0{\cdot}9$ m $\qquad C_d = 0{\cdot}62$

$$Q = \frac{2}{3} C_d \,.\, b \,.\, \sqrt{2g}\ [H_2{}^{3/2} - 1H_1{}^{3/2}]$$
$$= \frac{2}{3} \times 0{\cdot}62 \times 0{\cdot}5 \times \sqrt{2 \times 9{\cdot}81}\ [0{\cdot}9^{3/2} - 0{\cdot}3^{3/2}]$$
$$= 0{\cdot}6313 \text{ m}^3\text{/s}$$

If the orifice is treated as a small one,

$$Q = C_d \times \text{Area} \times \sqrt{2g \times \text{Mean head}}$$
$$= 0{\cdot}62 \times (0{\cdot}6 \times 0{\cdot}5) \times \sqrt{2 \times 9{\cdot}81 \times (0{\cdot}3 + 0{\cdot}3)}$$
$$= 0{\cdot}6382 \text{ m}^3\text{/s}$$

$$\text{Percentage error} = \frac{0.6382 - 0.6313}{0.6313} \times 100 = 1.09$$

6.7 DISCHARGE THROUGH SUBMERGED ORIFICES

The orifice may be fully submerged or partially submerged at the discharge end.

Fully Submerged Orifice

When the whole of the outlet side of the orifice is submerged under the liquid of the same kind, it is known as **fully submerged orifice** (Fig. 6.7).

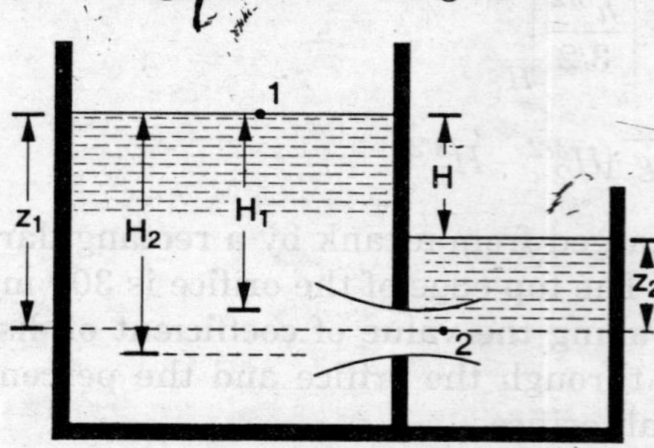

Fig. 6.7

Let H_1 = height of liquid above the upper edge of the orifice

H_2 = height of liquid above the lower edge of the orifice

H = difference in liquid levels in the two tanks

b = width of the orifice

C_d = coefficient of discharge

Applying Bernoulli's equation at 1, the free surface of the liquid in the first tank and 2, the vena-contracta (assuming the level at the vena-contracta as datum),

$$\frac{p_1}{w} + \frac{V_1^2}{2g} + z_1 = \frac{p_2}{w} + \frac{V_2^2}{2g} + z_2 \quad \text{...(i)}$$

where $\frac{p_1}{w} = 0, \frac{V_1^2}{2g} = 0$

z_1 = height of liquid level above the centre line of orifice

$$= H_1 + \frac{H_2 - H_1}{2}$$

$$= \frac{H_1 + H_2}{2}$$

$\frac{p_2}{w}$ = pressure head on the downstream side

= height of liquid level above the centre line of orifice

$= z_1 - H$

$$= \frac{H_1 + H_2}{2} - H$$

(i) becomes,

$$0 + 0 + \frac{H_1 + H_2}{2} = \left(\frac{H_1 + H_2}{2} - H\right) + \frac{V_2^2}{2g} + 0$$

$$\frac{V_2^2}{2g} = H$$

$$V_2 = \sqrt{2gH}$$

Discharge through the orifice = $C_d \times$ Area of orifice $\times$ Velocity

$$= C_d \cdot b\ (H_2 - H_1) \cdot \sqrt{2gH} \quad \text{...(6.9)}$$

Partially Submerged Orifice

When the outlet side of the orifice is **partially submerged,** the upper portion behaves as an orifice discharging free whereas the lower portion behaves as a submerged orifice. Only large orifices can behave like partially submerged orifices (Fig. 6.8).

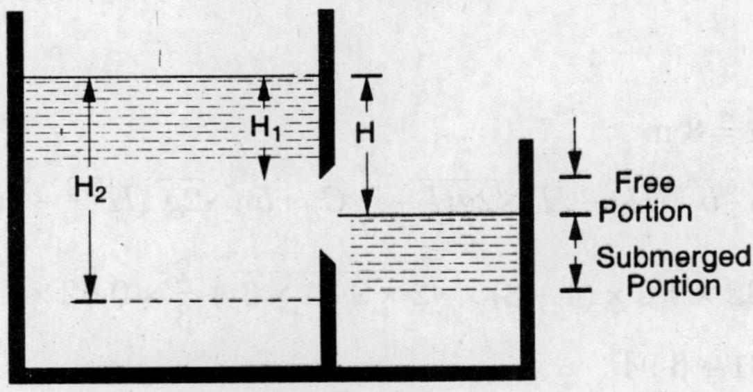

Fig. 6.8

Discharge through submerged portion,

$$Q_1 = C_d \cdot b \cdot (H_2 - H)\sqrt{2gH}$$

Discharge through free portion,

$$Q_2 = \frac{2}{3} C_d \cdot b \cdot \sqrt{2g}\ (H^{3/2} - H_1^{3/2})$$

Total discharge, $Q = Q_1 + Q_2$

$$Q = C_d \cdot b \cdot (H_2 - H)\sqrt{2gH} + \frac{2}{3} C_d \cdot b \cdot \sqrt{2g}\ (H^{3/2} - H_1^{3/2}) \quad \text{...(6.10)}$$

For convenience, the two coefficients of discharge are assumed to be the same.

Example 6.7 Determine the discharge through a totally submerged orifice 1 m wide and 0·8 m deep when the difference of water levels on the two sides of the orifice is 2 m. Coefficient of discharge is 0·62.

Solution Refer Fig. 6.7,

$H = 2$ m $\qquad b = 1$ m

$$H_2 - H_1 = 0{\cdot}8 \text{ m} \qquad C_d = 0{\cdot}62$$

$$Q = C_d \cdot b\,(H_2 - H_1)\sqrt{2gH}$$

$$= 0{\cdot}62 \times 1 \times 0{\cdot}8 \times \sqrt{2 \times 9{\cdot}81 \times 2}$$

$$= 3{\cdot}107 \text{ m}^3/\text{s}$$

Example 6.8 A large tank consists of a rectangular orifice 1·6 m wide and 2 m deep fitted on one of its sides. The water level on one side of the orifice is 2 m above the top edge of the orifice while on the other side it is 1 m below the top edge. Find the discharge through the orifice if $C_d = 0{\cdot}62$.

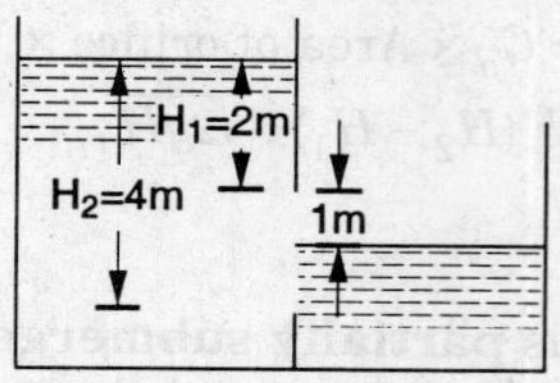

Fig. 6.9

Solution $H_1 = 2$ m

$H_2 = 4$ m

$H = 2 + 1 = 3$ m

$$Q = C_d \cdot b \cdot (H_2 - H)\sqrt{2gH} + \frac{2}{3} C_d \cdot b \cdot \sqrt{2g}\,(H^{3/2} - H_1^{\,3/2})$$

$$= 0{\cdot}62 \times 1{\cdot}6 \times (4 - 3) \times \sqrt{2 \times 9{\cdot}81 \times 3} + \frac{2}{3} \times 0{\cdot}62 \times 1{\cdot}6 \times \sqrt{2 \times 9{\cdot}81}\,(3^{3/2} - 2^{3/2})$$

$$= 7{\cdot}61 + 6{\cdot}94$$

$$= 14{\cdot}55 \text{ m}^3/\text{s}$$

6.8 TIME OF EMPTYING A TANK THROUGH AN ORIFICE

The tank may be

1. Rectangular, Square or Cylindrical
2. Hemispherical
3. Circular Horizontal

1. Rectangular, Square or Cylindrical Tank

Consider a tank of uniform cross-sectional area containing some liquid. The tank is fitted with an orifice at the bottom as shown in Fig 6.10.

Let H_1 = initial height of liquid

H_2 = final height of liquid

A = cross-sectional area of tank

a = cross-sectional area of orifice

T = Time to lower the level from H_1 to H_2

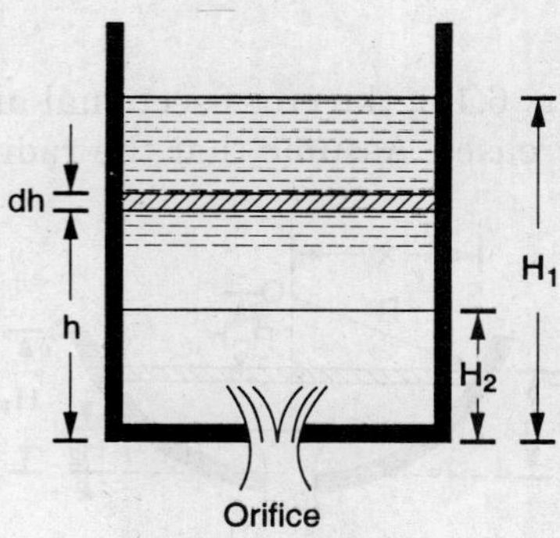

Fig. 6.10

Assume that at some instant, the height of liquid in the tank is h and the level lowers by a height dh in a small interval of time dt.

Volume passing through the orifice per unit time = $C_d \times$ Area $\times$ Velocity

$$= C_d \,.\, a \sqrt{2gh}$$

Volume passing through the orifice in time dt $= C_d \,.\, a \sqrt{2gh} \,.\, dt$

Also volume passing through the orifice = Change in the empty space in the tank

= Area of liquid surface $\times \, dh$

$$C_d \,.\, a \sqrt{2gh} \,.\, dt = -A \,.\, dh$$

(*-ve* sign indicates that with the increase in time, head on the orifice decreases).

$$dt = \frac{-A \,.\, dh}{C_d \,.\, a \sqrt{2gh}}$$

Therefore, time to lower the liquid head from H_1 to H_2,

$$T = \int_{H_1}^{H_2} \frac{-A}{C_d \,.\, a \sqrt{gh}} \,.\, dh$$

$$= \frac{-A}{C_d \,.\, a \sqrt{2g}} \int_{H_1}^{H_2} h^{-1/2} \,.\, dh$$

$$= \frac{-A}{C_d \,.\, a \sqrt{2g}} \left[\frac{h^{1/2}}{1/2} \right]_{H_1}^{H_2}$$

$$= \frac{-2A}{C_d \,.\, a \sqrt{2g}} (\sqrt{H_2} - \sqrt{H_1})$$

$$= \frac{2A(\sqrt{H_1} - \sqrt{H_2})}{C_d \,.\, a \sqrt{2g}} \qquad ...(6.11)$$

For emptying the tank completely, $H_2 = 0$

$$T = \frac{2A\sqrt{H_1}}{C_d \,.\, a \sqrt{2g}} \qquad ...(6.12)$$

2. Hemispherical Tank

In case of hemispherical tank (Fig. 6.11), the cross-sectional area is not constant. As the level of the liquid lowers, the area decreases. Assume that the radius of the tank is R.

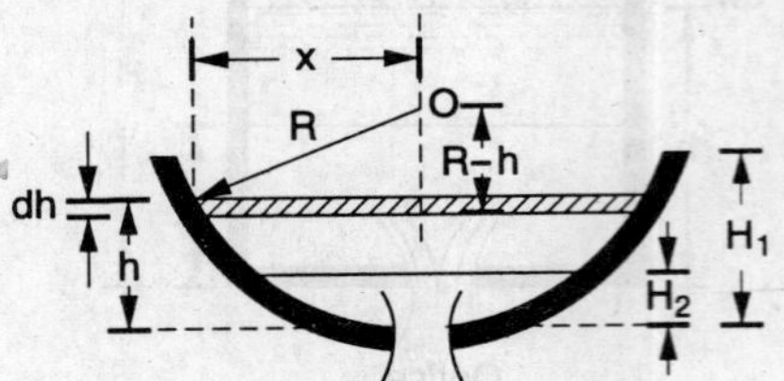

Fig. 6.11

Let the level of the liquid reduce from H_1 to H_2 in a small period of time dt.

Then change in the empty space in time dt

$$= \text{Area of liquid surface} \times dh$$

$$= \pi x^2 \,.\, dh$$

$$= \pi\,[R^2 - (R-h)^2]\,.\,dh$$

$$= \pi\,[R^2 - (R^2 + h^2 - 2Rh)]\,.\,dh$$

$$= \pi\,[2Rh - h^2]\,.\,dh$$

As in the previous case,

Volume passing through the orifice = Change in the empty space in the tank

$$C_d \,.\, a\sqrt{2gh}\,.\,dt = \pi\,(2Rh - h^2)\,.\,dh$$

$$dt = \frac{-\pi\,(2Rh - h^2)}{C_d \,.\, a\sqrt{2gh}}\,.\,dh$$

$$T = \int_{H_1}^{H_2} \frac{-\pi\,(2Rh - h^2)}{C_d \,.\, a\sqrt{2gh}}\,.\,dh$$

$$= \frac{-\pi}{C_d \,.\, a\sqrt{2g}} \int_{H_1}^{H_2} (2Rh^{1/2} - h^{3/2})\,.\,dh$$

$$= \frac{-\pi}{C_d \,.\, a\sqrt{2g}} \left[2R\frac{h^{3/2}}{3/2} - \frac{h^{5/2}}{5/2}\right]_{H_1}^{H_2}$$

$$= \frac{-\pi}{C_d \,.\, a\sqrt{2g}} \left[\frac{4}{3}Rh^{3/2} - \frac{2}{5}h^{5/2}\right]_{H_1}^{H_2} \qquad \text{...(6.13)}$$

$$= \frac{-\pi}{C_d \,.\, a\sqrt{2g}} \left[\frac{4}{3}R\,(H_2^{3/2} - H_1^{3/2}) - \frac{2}{5}\,(H_2^{5/2} - H_1^{5/2})\right]$$

To empty the tank completely, $H_2 = 0$

$$T = \frac{\pi}{C_d \,.\, a\sqrt{2g}} \left[\frac{4}{3}RH_1^{3/2} - \frac{2}{5}H_1^{5/2}\right] \qquad \text{...(6.14)}$$

3. Circular Horizontal Tank

Let R and L be the radius and length of the tank respectively (Fig. 6.12). As before, increase in the empty space in time dt

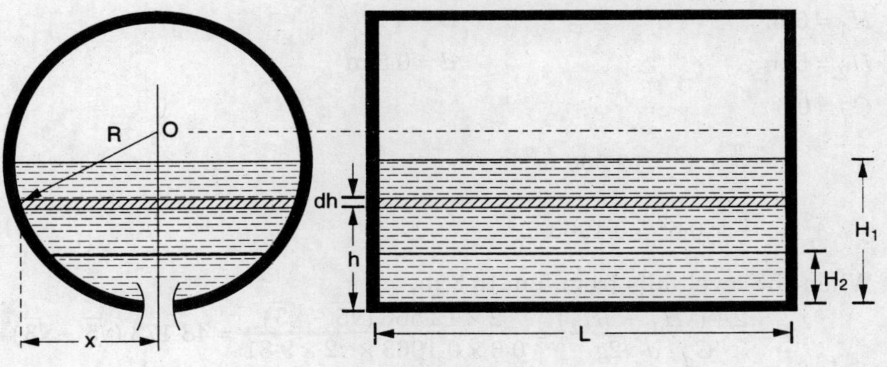

Fig. 6.12

$$= \text{Area of liquid surface} \times dh$$
$$= (2x \,.\, L) \,.\, dh$$
$$= 2\sqrt{R^2 - (R-h)^2} \,.\, L \,.\, dh$$
$$= 2\sqrt{R^2 - R^2 - h^2 + 2Rh} \,.\, L \,.\, dh$$
$$= 2L\sqrt{2Rh - h^2} \,.\, dh$$

Volume passing through the orifice = Change in the empty space in the tank

$$C_d \,.\, a\sqrt{2gh} \,.\, dt = -\,2L\sqrt{2Rh - h^2} \,.\, dh$$

$$dt = \frac{-\,2L\sqrt{2Rh - h^2}}{C_d \,.\, a\sqrt{2gh}} \,.\, dh$$

$$T = \int_{H_1}^{H_2} \frac{-\,2L\sqrt{2Rh - h^2}}{C_d \,.\, a\sqrt{2gh}} \,.\, dh$$

$$= \frac{-\,2L}{C_d \,.\, a\sqrt{2g}} \int_{H_1}^{H_2} \sqrt{\frac{2Rh - h^2}{h}} \,.\, dh$$

$$= \frac{-\,2L}{C_d \,.\, a\sqrt{2g}} \int_{H_1}^{H^2} (2R - h)^{1/2} \,.\, dh$$

$$= \frac{-\,2L}{C_d \,.\, a\sqrt{2g}} \left[\frac{(2R-h)^{3/2}}{3/2} \times (-1)\right]_{H_1}^{H_2}$$

$$= \frac{4L}{3C_d \,.\, a\sqrt{2g}} \left[(2R - H_2)^{3/2} - (2R - H_1)^{3/2}\right] \qquad \text{...(6.15)}$$

To empty the tank completely, $H_2 = 0$

$$T = \frac{4L}{3C_d \,.\, a\sqrt{2g}} \left[(2R)^{3/2} - (2R - H_1)^{3/2}\right] \qquad \text{...(6.16)}$$

Example 6.9 A circular tank 4 m in diameter has water upto a height of 5 m. An orifice of diameter 500 mm is provided at the bottom of the tank. Find the time taken by water level to fall from 5 m to 3 m and also the time to empty the tank completely. The coefficient of discharge is 0·6.

Solution $H_1 = 3$ m $\qquad D = 4$ m

$H_2 = 5$ m $\qquad d = 0·5$ m

$C_d = 0·6$

$$A = \frac{\pi}{4} \times 4^2 = 12·566 \text{ m}^2$$

$$a = \frac{\pi}{4} \times 0·5^2 = 0·1963 \text{ m}^2$$

When the level falls from 5 m to 3 m,

$$T = \frac{2A(\sqrt{H_1} - \sqrt{H_2})}{C_d \cdot a\sqrt{2g}} = \frac{2 \times 12·566(\sqrt{5} - \sqrt{3})}{0·6 \times 0·1963 \times \sqrt{2 \times 9·81}} = 48·173(\sqrt{5} - \sqrt{3}) = 24·28 \text{ s}$$

To empty the tank completely,

$$T = \frac{2A\sqrt{H_1}}{C_d \cdot a\sqrt{2g}} = \frac{2 \times 12·566 \times \sqrt{5}}{0·6 \times 0·1963 \times \sqrt{2 \times 9·81}} = 48·173 \times \sqrt{5} = 107·72 \text{ s}$$

Example 6.10 A hemispherical tank 2·4 m in radius contains water upto a height of 2 m. It is provided with an orifice of diameter 48 mm at its bottom. Determine the time required to fall the water level from 2 m to 1·4 m as well as to completely empty the tank. Take the value of C_d as 0·62.

Solution $H_1 = 2$ m

$H_2 = 1·4$ m

$R = 2·4$ m

$d = 0·048$ m

$C_d = 0·62$

$$a = \frac{\pi}{4} \times 0·048^2 = 1·8096 \times 10^{-3} \text{ m}^2$$

$$T = \frac{\pi}{C_d \cdot a\sqrt{2g}}\left[\frac{4}{3}R(H_1^{3/2} - H_2^{3/2}) - \frac{2}{5}(H_1^{5/2} - H_2^{5/2})\right]$$

$$= \frac{\pi}{0·62 \times 1·8096 \times 10^{-3} \times \sqrt{2 \times 9·81}}\left[\frac{4}{3} \times 2·4(2^{3/2} - 1·4^{3/2}) - \frac{2}{5}(2^{5/2} - 1·4^{5/2})\right]$$

$$= 632·08(3·7502 - 1·3351)$$

$$= 1526·5 \text{ s or 25 min. } 26·5 \text{ s}$$

To empty the tank completely,

$$T = \frac{\pi}{C_d \cdot a\sqrt{2g}}\left[\frac{4}{3}RH_1^{3/2} - \frac{2}{5}H_1^{5/2}\right]$$

$$= \frac{\pi}{0·82 \times 1·8096 \times 10^{-3} \times \sqrt{2 \times 9·81}}\left[\frac{4}{3} \times 2·4 \times 2^{3/2} - \frac{2}{5} \times 2^{5/2}\right]$$

$= 632{\cdot}08 \times 6{\cdot}7882$

$= 4290{\cdot}7$ s or $1h$ 11 min 30·7 s

Example 6.11 A water tank has an upper cylindrical portion 1·5 m in radius and 4 m high and a lower hemispherical portion. The tank is provided with an orifice of 120 mm in diameter at its bottom. Determine the time required to empty the tank completely if initially it is full of water. $C_d = 0{\cdot}62$ for the orifice.

Solution $A = \pi \times 1{\cdot}5^2 = 7{\cdot}0686 \text{ m}^2$

$$a = \frac{\pi}{4} \times 0{\cdot}12^2 = 0{\cdot}0113 \text{ m}^2$$

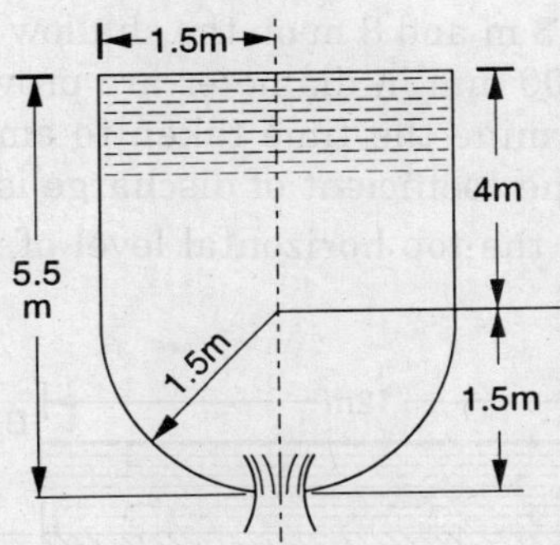

Fig. 6.13

The tank has two portions, cylindrical and hemispherical.

Let T_1 = time required to empty the upper cylindrical portion

T_2 = time required to empty the bottom hemispherical portion

$$T_1 = \frac{2A\,(\sqrt{H_1} - \sqrt{H_2})}{C_d \cdot a\,\sqrt{2g}} = \frac{2 \times 7{\cdot}0686\,(\sqrt{5{\cdot}5} - \sqrt{1{\cdot}5})}{0{\cdot}62 \times 0{\cdot}0113 \times \sqrt{2 \times 9{\cdot}81}} = 510{\cdot}4 \text{ s}$$

$$T_2 = \frac{\pi}{C_d \cdot a\,\sqrt{2g}}\left[\frac{4}{3}RH_1^{\,3/2} - \frac{2}{5}H_1^{\,5/2}\right]$$

$$= \frac{\pi}{0{\cdot}62 \times 0{\cdot}0113 \times \sqrt{2 \times 9{\cdot}81}}\left[\frac{4}{3} \times 1{\cdot}5 \times 1{\cdot}5^{3/2} - \frac{2}{5} \times 1{\cdot}5^{5/2}\right]$$

$$= 101{\cdot}22 \times (3{\cdot}674 - 1{\cdot}102)$$

$$= 260{\cdot}3 \text{ s}$$

$$T = T_1 + T_2 = 510{\cdot}4 + 260{\cdot}3 = 770{\cdot}73 \text{ or } 12 \text{ min } 50{\cdot}7 \text{ s}$$

Example 6.12 An orifice of 120 mm in diameter is provided at the bottom of a horizontal boiler drum 5 m long and 2·5 m in diameter full of water. Determine the time required to empty the drum completely. The coefficient of discharge is 0·6.

Solution $R = 2{\cdot}5/2 = 1{\cdot}25$ m $\quad L = 5$ m

$d = 0{\cdot}12$ m $\quad C_d = 0{\cdot}6$

$H_2 = 2{\cdot}5$ m

$H_2 = 0$

$$a = \frac{\pi}{4} \times 0{\cdot}12^2 = 0{\cdot}0113 \text{ m}^2$$

$$T = \frac{4L}{3C_d \cdot a\sqrt{2g}} [(2R)^{3/2} - (2R - H_1)^{3/2}]$$

$$= \frac{4 \times 5}{3 \times 0{\cdot}6 \times 0{\cdot}0113 \times \sqrt{2 \times 9{\cdot}81}} [(2 \times 1{\cdot}25)^{3/2} - (2 \times 1{\cdot}25 - 2{\cdot}5)^{3/2}]$$

$$= 221{\cdot}96 \times 3{\cdot}599$$

$$= 798{\cdot}9 \text{ s or } 13 \text{ min } 18{\cdot}9 \text{ s}$$

Example 6.13 A swimming pool 40 m long and 12 m wide has sloping bottom widthwise, the depths being 3 m and 8 m on the shallow and deep ends respectively. Two outlets each of 500 mm in diameter are provided at each of the shallow and deep ends. Determine the time taken to empty the pool if both the outlets are kept open. The coefficient of discharge is 0·6 for each opening.

Solution Let CD be parallel to the top horizontal level of water through the upper orifice (Fig. 6.14a).

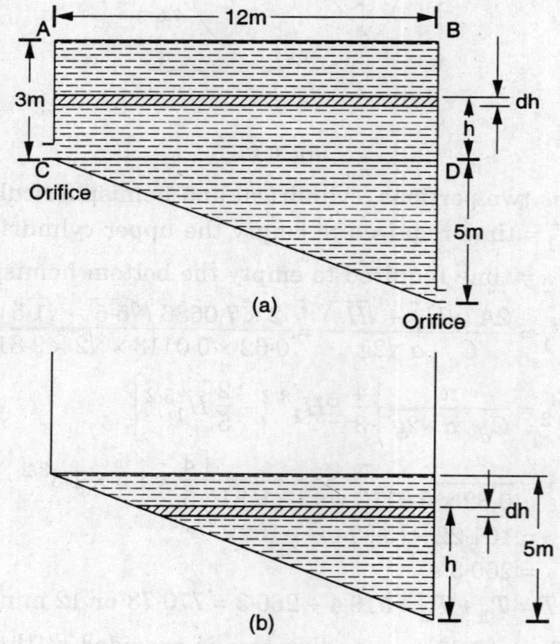

Fig. 6.14

Let L_1 = Time taken to lower the level from AB to CD

T_2 = Time taken to lower the level from CD to zero

$$A = 40 \times 12 = 480 \text{ m}^2$$

$$a = \frac{\pi}{4} \times 0{\cdot}5^2 = 0{\cdot}1963 \text{ m}^2$$

At an instant when the water level is h m above CD, let the water level fall by dh in a small interval of time dt. Then

Volume of water leaving the pool in time dt = Flow through the two orifices in time dt

$$-A\,.\,dh = C_d\,.\,a\,\sqrt{2g\,(h+5)}\,.\,dt + C_d\,.\,a\,\sqrt{2gh}\,.\,dt$$

$$dt = \frac{-A\,.\,dh}{C_d\,.\,a\,\sqrt{2g}\,.\,(\sqrt{h+5}+\sqrt{h})}$$

$$T_1 = \frac{-A}{C_d\,.\,a\,\sqrt{2g}}\int_3^0 \frac{dh}{(\sqrt{h+5}+\sqrt{h})}$$

$$= \frac{-A}{C_d\,.\,a\,\sqrt{2g}}\int_3^0 \frac{1}{(\sqrt{h+5}+\sqrt{h})}\cdot\frac{(\sqrt{h+5}-\sqrt{h})}{(\sqrt{h+5}-\sqrt{h})}\,dh$$

$$= \frac{-A}{C_d\,.\,a\,\sqrt{2g}}\int_3^0 \frac{(\sqrt{h+5}-\sqrt{h})}{5}\,dh$$

$$= \frac{-A}{C_d\,.\,a\,\sqrt{2g}}\times\frac{1}{5}\left[\frac{(h+5)^{3/2}}{3/2}-\frac{h^{3/2}}{3/2}\right]_3^0$$

$$= \frac{-A}{C_d\,.\,a\,\sqrt{2g}}\times\frac{2}{15}\,[(h+5)^{3/2}-h^{3/2}]_3^0$$

$$= \frac{480}{0{\cdot}6\times0{\cdot}1963\times\sqrt{2\times9{\cdot}81}}\times\frac{2}{15}\left[(h+5)^{3/2}-h^{3/2}\right]_0^3$$

$$= 122{\cdot}66\,[\{(3+5)^{3/2}-3^{3/2}\}-5^{3/2}]$$

$$= 122{\cdot}66\times6{\cdot}25$$

$$= 766{\cdot}7 \text{ s}$$

Let h be the height of water level above the bottom orifice at any instant when the water level has become lower than CD. Let the water level fall by dh during a small period of time dt (Fig. 6.14b).

Width of water surface $= 12\times\frac{h}{5}$

Area of water surface $= 12\times\frac{h}{12}\times40 = 96\,h$

Volume of water leaving the pool in time dt = Flow through the lower orifice in time dt

$$-A\,.\,dh = C_d\,.\,a\,\sqrt{2gh}\,.\,dt$$

$$dt = \frac{-A\,.\,dh}{C_d\,.\,a\,\sqrt{2gh}} = \frac{-96\,h\,.\,dh}{C_d\,.\,a\,\sqrt{2gh}} = \frac{-96}{C_d\,.\,a\,\sqrt{2g}}\,h^{1/2}\,.\,dh$$

$$T_2 = \frac{-96}{C_d\,.\,a\,\sqrt{2g}}\int_5^0 h^{1/2}\,.\,dh = \frac{-96}{C_d\,.\,a\,\sqrt{2g}}\left(\frac{h^{3/2}}{3/2}\right)_5^0$$

$$= \frac{-96}{0{\cdot}6\times0{\cdot}1963\times\sqrt{2\times9{\cdot}81}}\left(\frac{2}{3}\times h^{3/2}\right)_5^0$$

$$= -183{\cdot}99\left(\frac{2}{3}\times(0-5)^{3/2}\right)$$

$$= 1371{\cdot}4 \text{ s}$$

Total time = 766·7 + 1371·4 = 2138·1 s

Example 6.14 An orifice of 200 mm in diameter is fitted to the bottom of a cylindrical tank 4 m in diameter and 8 m high. A constant discharge of 100 litres of water per second is supplied into the tank and at the same time water is discharged through the orifice. Find the time taken to lower the water level in the tank from 6 m to 3 m above the bottom of the tank. The coefficient of discharge is 0·7. The top of the tank is open to the atmosphere.

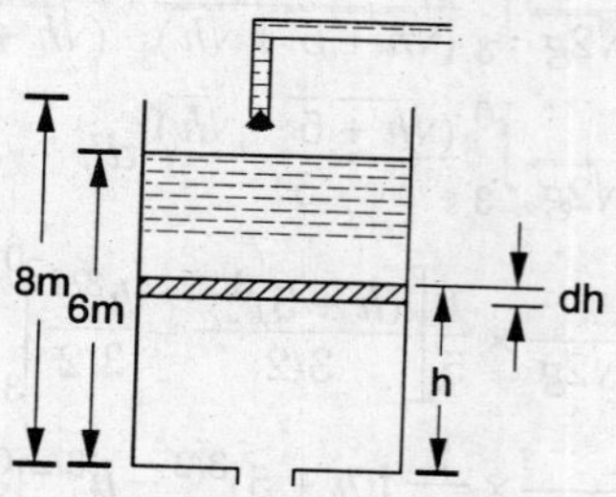

Fig. 6.15

Solution $H_1 = 6$ m $\quad d = 0{\cdot}2$ m

$H_2 = 3$ m $\quad D = 4$ m

$q = 0{\cdot}1$ m³/s $\quad C_d = 0{\cdot}7$

$$A = \frac{\pi}{4}\times 4^2 = 12{\cdot}566 \text{ m}^2$$

$$a = \frac{\pi}{4}\times 0{\cdot}2^2 = 0{\cdot}0314 \text{ m}^2$$

Volume of water leaving the tank in time dt = Flow through the orifice in time dt – flow into the tank in time dt

$$-A\,.\,dh = (Q-q)\,dt$$

$$= (C_d\,.\,a\,.\,\sqrt{2gh}-q)\,dt$$

$$= (p\,.\,\sqrt{h}-q)\,dt \qquad \text{...(Assuming } p = C_d\,.\,a\,.\,\sqrt{2g}\text{)}$$

$$dt = \frac{-A\,.\,dh}{p\,.\sqrt{h}-q}$$

Let $p\,.\,\sqrt{h} - q = y$

$$p\,.\,\frac{1}{2}h^{-1/2}\,dh = dy \qquad (q \text{ is constant})$$

$$dh = \frac{2\sqrt{h}}{p}\,dy$$

$$dt = \frac{-A}{y}\,.\,\frac{2\sqrt{h}}{p}\,dy$$

$$= \frac{-2A}{p^2} \cdot \frac{p\sqrt{h}}{y} dy$$

$$= \frac{-2A}{p^2} \cdot \frac{y+q}{y} dy$$

$$= \frac{-2A}{p^2} \cdot \left(1 + \frac{q}{y}\right) dy$$

$$T = \frac{-2A}{p^2} \int_{H_1}^{H_2} \left(1 + \frac{q}{y}\right) dy$$

$$= \frac{-2A}{p^2} \left(y + q \log_e y\right)_{H_1}^{H_2}$$

$$= \frac{2A}{p^2} \left(y + q \log_e y\right)_{H_2}^{H_1}$$

$$= \frac{2A}{p^2} \left[p \cdot \sqrt{h} - q + q \log_e (p \cdot \sqrt{h} - q)\right]_{H_2}^{H_1}$$

$$= \frac{2A}{p^2} \left[p\,(\sqrt{H_1} - \sqrt{H_2}) + q \log_e \frac{p\sqrt{H_2} - q}{p\sqrt{H_2} - q}\right]$$

$$p = C_d\, a \cdot \sqrt{2g} = 0{\cdot}7 \times 0{\cdot}0314 \times \sqrt{2 \times 9{\cdot}81} = 0{\cdot}0974$$

$$T = \frac{2A}{p^2} \left[p\,(\sqrt{H_1} - \sqrt{H_2}) + q \log_e \frac{q\sqrt{H_1} - q}{p\sqrt{H_2} - q}\right]$$

$$= \frac{2 \times 12{\cdot}566}{0{\cdot}0974^2} \left[0{\cdot}0974\,(\sqrt{6} - \sqrt{3}) + 0{\cdot}1 \log_e \frac{0{\cdot}0974\sqrt{6} - 0{\cdot}1}{0{\cdot}0974\sqrt{3} - 0{\cdot}1}\right]$$

$$= 2649 \left(0{\cdot}0699 + 0{\cdot}1 \log_e \frac{0{\cdot}13858}{0{\cdot}0687}\right)$$

$$= 371 \text{ s ro } 6 \text{ m } 11 \text{ s}$$

6.9 CLASSIFICATION OF MOUTHPIECES

The mouthpieces are classified

1. According to Shape

Cylindrical, convergent or convergent-divergent

2. According to Position

Internal or external

3. According to Nature of Discharge

Running free or running full. This classification is limited to internal mouthpieces. A mouthpiece is running free if the jet of liquid after contraction does not touch the sides. On the other hand in a mouthpiece running full, the liquid touches the walls of the mouthpiece completely after the contraction.

6.10 FLOW THROUGH AN EXTERNAL MOUTHPIECE

Fig. 6.16 shows a tank fitted with an external small tube (two or three times its diameter in length) known as orifice. The jet of liquid entering the mouthpiece contracts to form a vena-contracta at a section C–C. The jet again expands beyond this section to fill the mouthpiece completely.

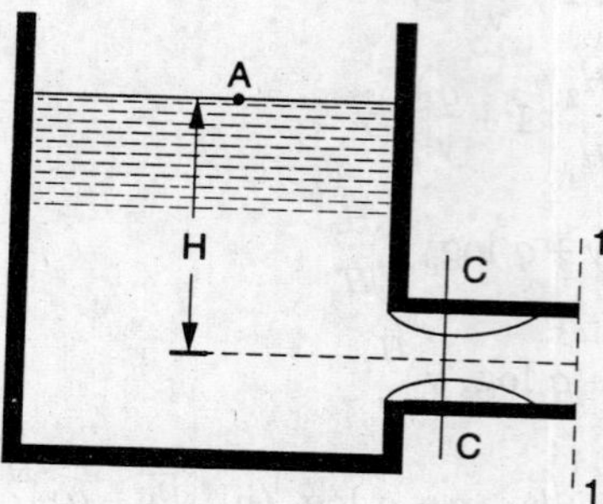

Fig. 6.16

Let H = height of liquid above the centre line of mouthpiece

a = area of mouthpiece at outlet

a_c = area of mouthpiece at vana-contracta

V_1 = velocity of liquid at the outlet

V_c = velocity of liquid at the vena-contracta

C_c = coefficient of contraction

The loss of head is due to sudden enlargement from C–C to 1 – 1. Thus

$$h_L = \frac{(V_c - V_1)^2}{2g}$$

where V_c is found by applying continuity equation to sections 1 – 1 and $C - C$, *i.e.*

$$a_c \cdot V_c = a_1 \cdot V_1$$

$$V_c = \frac{a_1 \cdot V_1}{a_c} = \frac{V_1}{a_c/a_1} = \frac{V_1}{C_c}$$

Assuming $C_c = 0{\cdot}62$,

$$V_c = \frac{V_1}{0{\cdot}62}$$

$$h_L = \frac{\left(\dfrac{V_1}{0{\cdot}62} - V_1\right)^2}{2g} = \frac{V_1^2}{2g}\left(\frac{1}{0{\cdot}62} - 1\right)^2 = \frac{0{\cdot}375\, V_1^2}{2g}$$

Applying Bernoulli's equation to points A and 1 – 1 (centre line of mouthpiece),

$$\frac{p_A}{w} + \frac{V_A^2}{2g} + z_A = \frac{p_1}{w} + \frac{V_1^2}{2g} + z_1 + h_L$$

$$0 + 0 + H = 0 + \frac{V_1^{\,2}}{2g} + 0 + \frac{0{\cdot}375\, V_1^{\,2}}{2g}$$

$$H = 1{\cdot}375\, \frac{V_1^{\,2}}{2g}$$

$$V_1 = \sqrt{\frac{2gH}{1{\cdot}375}}$$

$$= 0{\cdot}855\,\sqrt{2gH} \qquad \text{...(6.17)}$$

Theoretical velocity of liquid at the outlet, $V = \sqrt{2gH}$

$$C_v = \frac{\text{Actual velocity}}{\text{Theoretical velocity}} = \frac{0{\cdot}855\,\sqrt{2gH}}{\sqrt{2gH}} = 0{\cdot}855 \qquad \text{...(6.18)}$$

For a mouthpiece $C_c = 1$, as the area of the jet of liquid at the outlet is equal to the area of the mouthpiece at the outlet.

$$C_d = C_v \times C_c = 0{\cdot}855 \times 1 = 0{\cdot}855 \qquad \text{...(6.19)}$$

It may be observed that the value of C_d for a mouthpiece is higher than that for an orifice. Therefore, the discharge through a mouthpiece is more as compared to an orifice.

In case of orifices, the actual velocity at the vana-contracta is less than the theoretical velocity as there is some head loss due to friction of the orifice at the entrance. In the same way, there is some loss of head at the entrance of the mouthpiece which further decreases the coefficient of velocity and thus coefficient of discharge (upto 0·82). However, in case of mouthpieces for all practical purposes, this loss at the entrance is neglected and only the loss due to divergence of the jet is considered.

Example 6.15 An external cylindrical mouthpiece of diameter 100 mm discharges water under a constant head of 5 m. The coefficient of contraction for vena-contracta is 0·62 and coefficient of discharge. 0·85. The atmospheric pressure head is 10·3 m of water. Determine the discharge through the mouthpiece and the absolute pressure head of water at the vena-contracta.

Solution $H = 5$ m $\qquad C_d = 0{\cdot}62$

$H_a = 10{\cdot}3$ m $\qquad C_d = 0{\cdot}85$

$d = 0{\cdot}1$ m

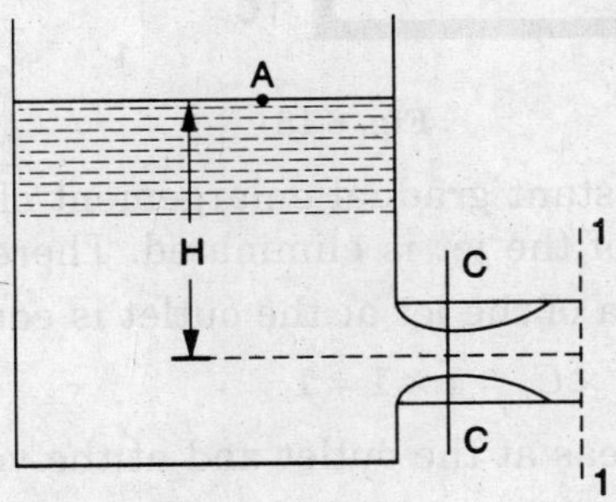

Fig. 6.17

$$Q = C_d \cdot a\sqrt{2gH} = 0{\cdot}85 \times \frac{\pi}{4} \times 0{\cdot}1^2 \times \sqrt{2 \times 9{\cdot}81 \times 5} = 0{\cdot}0661 \text{ m}^3\text{/s}$$

Apply Bernoulli's equation at A and C–C (Considering absolute pressures),

$$\frac{p_A}{w} + \frac{V_A^2}{2g} + z_A = \frac{p_c}{w} + \frac{V_c^2}{2g} + z_c$$

$$H_a + 0 + H = H_c + \frac{V_c^2}{2g} + 0$$

$$H_a + H = H_c + \left(\frac{V_1}{0{\cdot}62}\right)^2 \times \frac{1}{2g}$$

$$H_a + H = H_c + 2{\cdot}601 \frac{V_1^2}{2g}$$

$$H_a + H = H_c + 2{\cdot}601 \times \frac{H}{1{\cdot}375} \qquad \left(\because H = 1{\cdot}375 \frac{V_1^2}{2g}\right)$$

$$10{\cdot}3 + 5 = H_c + 2{\cdot}601 \times \frac{5}{1{\cdot}375}$$

$$H_c = 15{\cdot}3 - 9{\cdot}458$$

$$= 5{\cdot}84 \text{ m absolute}$$

6.11 FLOW THROUGH A CONVERGENT-DIVERGENT MOUTHPIECE

If a mouthpiece first converges upto vena-contracta and then diverges gradually, it is known as a convergent-divergent mouthpiece (Fig. 6.18). In this type of mouthpiece, there is no sudden

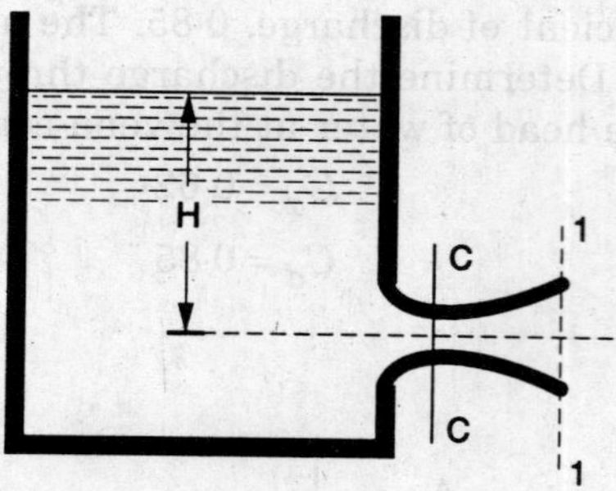

Fig. 6.18

enlargement of the jet due to constant gradual enlargement of the section. Thus the energy loss due to sudden enlargement of the jet is eliminated. Therefore, C_v is taken as 1 in this case. Since C_c is also 1 as the area of the jet at the outlet is equal to that of the mouthpiece,

$$C_d = C_v \times C_c = 1 \times 1 = 1$$

Ratio of the cross-sectional areas at the outlet and at the vena-contracta can be found as given below :

Let $\qquad H_1$ = head of liquid above the mouthpiece

H_a = atmospheric pressure head

H_c = absolute pressure head at the vena-contracta

Applying Bernoulli's equation at the free liquid surface (A) and the outlet (1 – 1), taking the centre line of the mouthpiece as the datum and writing in terms of absolute pressures,

$$\frac{p_A}{w} + \frac{V_A^2}{2g} + z_A = \frac{p_1}{w} + \frac{V_1^2}{2g} + z_1$$

$$H_a + 0 + H = H_a + \frac{V_1^2}{2g} + 0$$

$$V_1 = \sqrt{2gH}$$

Now apply Bernoulli's equation to the vena-contracta and the outlet

$$\frac{p_c}{w} + \frac{V_c^2}{2g} + z_c = \frac{p_1}{w} + \frac{V_1^2}{2g} + z_1$$

$$H_c + \frac{V_c^2}{2g} + 0 = H_a + \frac{(\sqrt{2gH})^2}{2g} + 0$$

$$\frac{V_c^2}{2g} = H_a + H - H_c$$

$$V_c = \sqrt{2g\,(H_a + H - H_c)} \qquad \text{...(6.20)}$$

Applying continuity equation to section 1 – 1 and $C – C$,

$$a \,.\, V_1 = a \,.\, C_c$$

$$\frac{a}{a_c} = \frac{V_c}{V_1} = \frac{\sqrt{2g\,(H_a + H - H_c)}}{\sqrt{2gH}} = \sqrt{1 + \frac{H_a - H_c}{H}} \qquad \text{...(6.21)}$$

Example 6.16 Water is discharged under a constant head of 3·6 m through a convergent-divergent mouthpiece with a throat diameter 48 mm. Find the maximum outlet diameter to avoid the separation of water flow if the maximum permissible vacuum pressure is 8 m. Also find the discharge.

Solution $H = 3{\cdot}6$ m $\qquad d = 0{\cdot}048$ m

$H_a - H_c = 8$ m (permissible vacuum pressure)

$$a_c = \frac{\pi}{4} \times 0{\cdot}048^2 = 0{\cdot}00181 \text{ m}^2$$

$$\frac{a}{a_c} = \sqrt{1 + \frac{H_a - H_c}{H}}$$

$$\left(\frac{d_1}{d_c}\right)^2 = \sqrt{1 + \frac{8}{3{\cdot}6}} = 1{\cdot}795$$

$$\frac{d_1}{d_c} = 1{\cdot}34$$

$$d_1 = 1{\cdot}34 \times 0{\cdot}048 = 0{\cdot}0643 \text{ m} = 64{\cdot}3 \text{ mm}$$

$$Q = a_c \times \sqrt{2gH} = 0{\cdot}00181 \times \sqrt{2 \times 9{\cdot}81 \times 3{\cdot}6} = 0{\cdot}0152 \text{ m}^3/\text{s}$$

Example 6.17 A convergent-divergent mouthpiece fitted to the side of a tank discharges 5 litres of water per second under a constant head of 1·8 m. Assuming the head loss in the divergent portion equivalent to 8 percent of the kinetic head at the outlet and the separation pressure head of water 2·5 m, find the throat and the exit diameters. The atmospheric pressure is 10·3 m of water.

Solution $Q = 0{\cdot}05 \text{ m}^3/\text{s}$ $\qquad H_c = 2{\cdot}5$ m (absolute)

$H = 1{\cdot}8$ m $\qquad H_a = 10{\cdot}3$ m (absolute)

$$h_L = 0{\cdot}08 \frac{V_1^2}{2g}$$

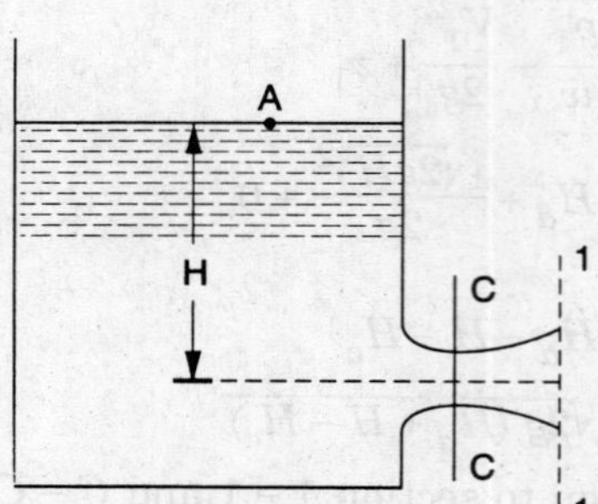

Fig. 6.19

Apply Bernoulli's equation to the free water surface (*A*) and the throat section (*C – C*),

$$\frac{p_A}{w} + \frac{V_A^2}{2g} + z_A = \frac{p_c}{w} + \frac{V_c^2}{2g} + z_c$$

$$H_a + 0 + H = H_c + \frac{V_c^2}{2g} + 0$$

$$10{\cdot}3 + 1{\cdot}8 = 2{\cdot}5 + \frac{V_c^2}{2g}$$

$$\frac{V_c^2}{2g} = 9{\cdot}6$$

$$V_c = \sqrt{2 \times 9{\cdot}81 \times 9{\cdot}6} = 13{\cdot}724 \text{ m/s}$$

$$Q = a_c V_c = \frac{\pi}{4} \times d_c^2 \times V_c$$

$$0{\cdot}005 = \frac{\pi}{4} \times d_c^2 \times 13{\cdot}724$$

$$d_c = 0{\cdot}0215 \text{ m or } 215 \text{ mm}$$

Apply Bernoulli's equation to the free water surface (*A*) and the outlet of the mouthpiece (1 – 1),

$$\frac{p_A}{w} + \frac{V_A^2}{2g} + Z_A = \frac{p_1}{w} + \frac{V_1^2}{2g} + Z_1 + h_L$$

$$H_a + 0 + H = H_a + \frac{V_1^2}{2g} + 0 + 0{\cdot}08 \times \frac{V_1^2}{2g}$$

$$1{\cdot}08\,\frac{V_1^2}{2g} = H$$

$$V_1 = \sqrt{\frac{2gH}{1{\cdot}08}} = \sqrt{\frac{2 \times 9{\cdot}81 \times 1{\cdot}8}{1{\cdot}08}} = 5{\cdot}718 \text{ m/s}$$

$$Q = a_1 V_1 = \frac{\pi}{4} \times d_1^2 \times V_c$$

$$0{\cdot}005 = \frac{\pi}{4} \times d_1^2 \times 5{\cdot}718$$

$$d_1 = 0{\cdot}03337 \text{ m or } 33{\cdot}37 \text{ mm}$$

6.12 DISCHARGE THROUGH AN INTERNAL MOUTHPIECE

If a short cylindrical tube is fitted to an orifice of a tank in such a way that it projects inwards, it is known as an **internal** or **Borda's** or **Re-entrant** mouthpiece. The internal mouthpiece is said to be

- Running free, when the length of the tube is equal to the diameter and the jet of liquid comes out without touching the sides of the tube.
- Running full, when the length of the tube is about three times its diameter and the jet comes out with its diameter equal to the diameter of the mouthpiece at the outlet.

Mouthpiece Running Free

Refer Fig. 6.20,

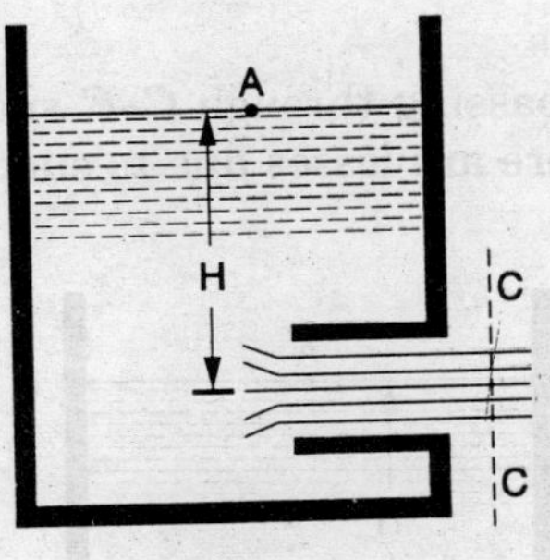

Fig. 6.20

Let H = height of liquid above centre line of mouthpiece

a = area of the mouthpiece

a_c = area of the contracted jet

V_c = velocity of the contracted jet

According to momentum equation,
The force applied on the mouthpiece

$$= \text{Mass flow rate} \times \text{Change in velocity}$$

$$\text{Pressure} \times \text{Area} = \frac{w}{g} \cdot Q \times \Delta V$$

$$wH \cdot a = \frac{w \cdot a_c V_c}{g} \times (V_c - 0)$$

$$= \frac{w \cdot a_c V_c^2}{g} \qquad \text{...(i)}$$

Applying Bernoulli's equation to the free surface (A) of the liquid and section C–C,

$$\frac{p_A}{w} + \frac{V_A^2}{2g} + z_A = \frac{p_c}{w} + \frac{V_c^2}{2g} + z_c$$

$$0 + 0 + H = 0 + \frac{V_c^2}{2g} + 0$$

$$V_c = \sqrt{2gH}$$

From (i),

$$wHa = \frac{w \cdot a_c}{g} 2gH$$

$$\frac{a_c}{a} = 0{\cdot}5$$

$$C_c = 0{\cdot}5$$

As there is no loss of head, $C_v = 1$

$$C_d = C_v \times C_c = 1 \times 0{\cdot}5 = 0{\cdot}5$$

and the discharge,

$$Q = C_d \times a \sqrt{2gH} \qquad \text{...(6.22)}$$

$$= 0{\cdot}5\, a \sqrt{2gH}$$

Mouthpiece Running Full

In this case, the jet of liquid after passing through C–C suddenly enlarges to fill the whole area of the mouthpiece and thus there are losses due to sudden enlargement (Fig. 6.21),

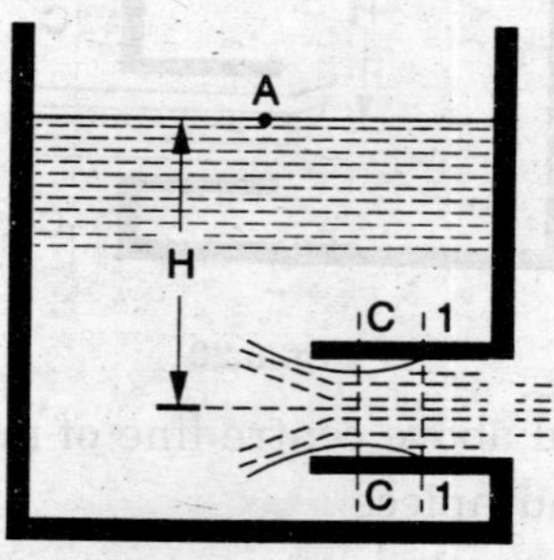

Fig. 6.21

$$h_L = \frac{(V_c - V_1)^2}{2g}$$

But from continuity equation,

$$a_c \cdot V_c = a \cdot V_1$$

$$V_c = \frac{a}{a_c} \cdot V_1 = 2V_1$$

$$\left(\text{From the 1st case when the mouthpiece runs free, } C_c = \frac{a}{a_c} = 2\right)$$

$$h_L = \frac{(2V_1 - V_1)^2}{2g} = \frac{V_1^2}{2g}$$

Applying Bernoulli's equation to free surface of liquid (A) in the tank and 1 – 1,

$$\frac{p_A}{w} + \frac{V_A^2}{2g} + z_A = \frac{p_1}{w} + \frac{V_1^2}{2g} + z_1 + h_L$$

$$0 + 0 + H = 0 + \frac{V_1^2}{2g} + 0 + \frac{V_1^2}{2g}$$

$$2V_1^2 = 2gH$$

$$V_1 = \sqrt{gH}$$

Here V is the actual velocity since losses are considered.

Theoretical velocity $= V = \sqrt{2gH}$ (*i.e.* neglecting the losses)

$$C_v = \frac{V_1}{V_{th}} = \frac{\sqrt{gH}}{\sqrt{2gH}} = \frac{1}{\sqrt{2}} = 0.707$$

$C_c = 1$ as the area at the outlet is equal to area of the mouthpiece.

$C_d = C_c \times C_v = 1 \times 0.707 = 0.707$

Discharge, $Q = C_d \times a \times \sqrt{2gh}$

$$= 0.707 \times a \times \sqrt{2gH} \quad \text{...(6.23)}$$

Example 6.18 An internal mouthpiece 120 mm in diameter discharges water under a constant head of 8 m. Calculate the discharge through the mouthpiece when it is running (i) free (ii) full.

Solution $H = 8$ m $\quad d = 0.12$ m

$$a = \frac{\pi}{4} \times 0.12^2 = 0.0113 \text{ m}^2$$

When the mouthpiece is running free,

$$Q = 0.5 \times a \times \sqrt{2gH} = 0.5 \times 0.0113 \times \sqrt{2 \times 9.81 \times 8} = 0.0708 \text{ m}^3/\text{s}$$

when the mouthpiece is running full,

$$Q = 0.707 \times a \times \sqrt{2gH} = 0.707 \times 0.0113 \times \sqrt{2 \times 9.81 \times 8} = 0.1001 \text{ m}^3/\text{s}$$

PROBLEMS

1. Define an orifice. How is it different from a mouthpiece ?
2. How are the orifices classified ?
3. What are hydraulic coefficients ? Explain them briefly.
4. How do you determine the hydraulic coefficients experimentally ?
5. Define vena-contracta.
6. Derive an expression for the discharge through a large rectangular orifice.
7. What is meant by fully submerged and partially submerged orifices ? Deduce expressions for the discharge through each of them.
8. How do you classify mouthpieces ?
9. Why is the coefficient of discharge of an orifice as low as 0·6 to 0·62 ? How can this be increased ?
10. Why is a tank fitted with an external mouthpiece emptied quicker than by simply through a sharp-edged orifice ?
11. What is a convergent-divergent mouthpiece ? What is its function ?
12. Distinguish between an internal mouthpiece running free and running full.
13. The discharge of oil through a 25 mm diameter orifice under a head of 5·5 m strikes a wall 1·5 m away and 120 mm vertically below the centre line of the orifice. The rate of flow of the oil is 0·003 m^3/s. Determine the coefficient of velocity, coefficient of contraction and the coefficient of discharge. (0·923, 0·637, 0·588)
14. The head of water over an orifice of diameter 80 mm is 10 m. The rise of level in the collecting tank 2·5 m × 1 m in dimensions is 1·5 m in 85 seconds. Determine the coefficient of discharge. (0·627)
15. A closed tank is partially filled with water upto a height of 9 m. At the bottom of the tank an orifice 15 mm in diameter is provided. Air is pumped into the tank. Find the pressure required for a flow of 1·5 litres per second through the orifice. The coefficient of discharge is 0·62. (85·64 kN/m^2)
16. A tank has two identical orifices in one of its vertical sides each with the same value of C_v. The orifice 1 is at depth h_1 below the liquid surface in the tank and the orifice 2 at depth h_2. Show that the point of intersection of the two jets is h_2 below orifice 1 and h_1 below orifice 2. Also determine the distance of the point of intersection from the plane of the orifices. ($2 C_v \sqrt{H_1 H_2}$)
17. What is the discharge through a rectangular orifice 4 m wide and 2 m deep fitted to a water tank. The water level in the tank is 6 m above the upper edge of the orifice. Assume coefficient of discharge to be 0·62. (58·08 m^3/s)
18. A tank is fitted with an external cylindrical mouthpiece on one of its sides. Derive an expression for the minimum pressure developed in the mouthpiece assuming a coefficient of contraction to be 0.62. (0·89 H negative)
19. A conical tank is held with its axis vertical and vertex downwards. The apex angle of the tank is 60° and the bottom of the tank is provided with a 120 mm diameter

plate with an orifice of 100 mm diameter. The tank contains water upto a height of 2 m above the orifice. Determine the time taken to lower the water level by 1 m and the velocity at the beginning and at the end of the process.

(79·96 s, 2·83 mm/s, 7·27 mm/s)

20. A rectangular tank 4 m in height is full of water. A small orifice with its centre at a depth of h m below the free surface of liquid is provided in one of its vertical sides. Determine the value of h to obtain the maximum horizontal distance of the point of strike of the jet one the ground. Also calculate the maximum horizontal distance if the value of coefficient of velocity is 0·98. (2 m, 4·8 m)
21. A circular tank 1 m in diameter has an orifice of 40 mm diameter at its bottom. If the initial height of water in the tank is 4 m, what will be the height after 90 seconds. (3·27 m)
22. Determine the percentage increase in the discharge if a short external tube of the same diameter is added to a circular orifice. Also find the limiting condition of head under which the orifice with external tube will run full. Assume that separation occurs at 2·5 m of water for a barometric pressure of 10·33 m of water and that for a sharp-edged circular orifice coefficient of contraction is 0·62 and coefficient of velocity 0·98. (40·8%, 10·58 m)
23. A properly shaped converging external mouthpiece of 75 mm diameter discharges water under a head of 5 m. Find the discharge. (43·76 l/s)
24. Water from a reservoir is released into a canal through a rectangular sharp-edged opening 1 m wide and 1·5 m high. The water levels in the canal and the reservoir are 1 m and 3 m respectively above the bottom edge of the opening. Find the discharge. (5·7 m^3/s)
25. A Borda's mouthpiece 100 mm in diameter flows full under a head of 3 m. Determine the discharge, the diameter at the vena-contracta and the pressure at the vena-contracta. (0·0452 m^3/s, 71 mm, 30 kN/m^2 vacuum)
26. A convergent-divergent mouthpiece fitted in the vertical side of a tank discharges water into the atmosphere at the rate of 0·36 m^3/s. The head over the centre of the mouthpiece is 2 m. Determine the throat and the exit diameters assuming that the loss in the divergent portion of the mouthpiece is one-fifth of the velocity head at the exit. (23·6 mm, 36·2 mm)
27. A convergent-divergent mouthpiece fitted in the vertical side of a tank has throat and exit diameters as 40 mm and 80 mm respectively. For a maximum vacuum pressure of 7·5 m of water, determine the maximum head of water for steady flow. (500 mm)

Chapter 7

NOTCHES AND WEIRS

7.1 INTRODUCTION

A **notch** is an opening in the side of a tank or vessel in such a way that the liquid surface is below the top edge of the opening. Usually it is made of metallic plate and used for measuring the rate of flow of a liquid from a tank or a small channel.

A **weir** is a regular concrete or masonary structure in an open channel or stream over which the liquid flow takes place. It is used to measure the flow of water in rivers or streams. The conditions of flow are practically the same as for a notch.

The difference between a notch and a weir is that a notch is of small size whereas a weir is of larger size. Also, a notch is generally made of a metallic plate and a weir of masonary or concrete.

The sheet of water flowing through a notch or weir is known as **nappe** or **vein**. The bottom edge of a notch or the top edge of a weir over which the water flows is called **sill** or **crest.**

7.2 CLASSIFICATION OF NOTCHES AND WEIRS

The notches are classified into

1. Rectangular notch
2. Triangular notch
3. Trapezoidal notch
4. Stepped notch

The weirs are classified according to

1. Shape :
   – Rectangular weir
   – Triangular weir
   – Trapezoidal (Cippoletti) weir
2. Shape of crest :

– Narrow crested weir
– Broad crested weir
– Sharp crested weir
– Ogee shaped weir

3. Nature of discharge
– Ordinary weir
– Submerged or drowned weir

7.3 DISCHARGE OVER A RECTANGULAR NOTCH OR WEIR

Consider a rectangular notch or weir provided in a channel carrying water (Fig. 7.1).

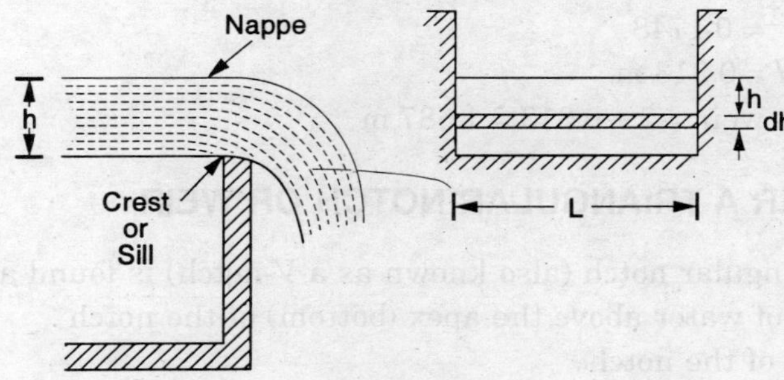

Fig. 7.1

Let H = height of water over the crest
L = length of notch or weir
C_d = coefficient of discharge

Consider an elementary horizontal strip of water of thickness dh and length L at a depth h from the free surface of water.

The discharge through the strip,

$$dQ = C_d \times \text{Area of strip} \times \text{Theoretical velocity}$$

$$= C_d \times L \;\; dh \times \sqrt{2gh}$$

Total discharge,

$$Q = \int_0^H C_d \,.\, L \sqrt{2gh} \,.\, dh$$

$$= C_d \,.\, L \sqrt{2g} \int_0^H h^{1/2} \,.\, dh$$

$$= C_d \,.\, L \sqrt{2g} \left[\frac{h^{3/2}}{3/2} \right]_0^H$$

$$= \frac{2}{3} C_d \,.\, L \sqrt{2g}\, H^{3/2} \qquad \text{...(7.1)}$$

Example 7.1 A rectangular weir of length 5 m is built across a rectangular channel. The maximum depth of water on the upstream side of the weir is 2 m and the discharge is 1600 l/s. Assuming the value of C_d to be 0·62, determine the height of the weir.

Solution $L = 5$ m

$Q = 1·6$ m^3/s

$C_d = 0·62$

$$Q = \frac{2}{3} C_d \,.\, L\sqrt{2g}\, H^{3/2}$$

$$1·6 = \frac{2}{3} \times 0·62 \times 5 \times \sqrt{2 \times 9·81} \times H^{3/2} = 9·155\, H^{3/2}$$

$$H^{3/2} = 0·1748$$

$$H = 0·313 \text{ m}$$

Height of the weir = 2 – 0·313 = 1·687 m

7.4 DISCHARGE OVER A TRIANGULAR NOTCH OR WEIR

The discharge over a triangular notch (also known as a *V*-notch) is found as follows :

Let H = head of water above the apex (bottom) of the notch

θ = angle of the notch

C_d = coefficient of discharge

Consider a horizontal strip of water of thickness dh at a depth h from the free water surface (Fig. 7.2).

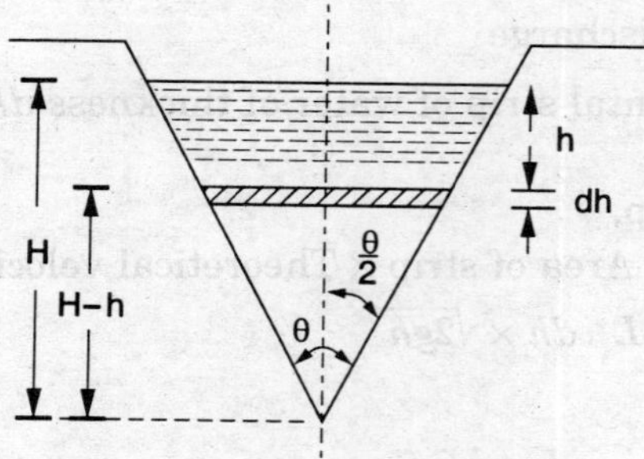

Fig. 7.2

$$dQ = C_d \times \text{Area of strip} \times \text{Theoretical velocity}$$

$$= C_d \times (\text{Width of strip} \times \text{Thickness of strip}) \times \text{Theoretical velocity}$$

$$= C_d \times 2\left[(H - h)\tan\frac{\theta}{2}\right] dh \times \sqrt{2gh}$$

$$= C_d \times 2\,(H - h)\tan\frac{\theta}{2} \,.\, dh \times \sqrt{2gh}$$

Total discharge,

$$Q = \int_0^H C_d \times 2\,(H - h)\tan\frac{\theta}{2}\,.\,dh \times \sqrt{2gh}$$

$$= 2C_d\sqrt{2g}\tan\frac{\theta}{2}\int_0^H (H - h)\sqrt{h}\,.\,dh$$

$$= 2C_d\sqrt{2g}\tan\frac{\theta}{2}\int_0^H (Hh^{1/2} - h^{3/2})\,.\,dh$$

$$= 2C_d\sqrt{2g}\tan\frac{\theta}{2}\left[H\,\frac{h^{3/2}}{3/2} - \frac{h^{5/2}}{5/2}\right]_0^H$$

$$= 2C_d\sqrt{2g}\tan\frac{\theta}{2}\left[\frac{2}{3}H\,.\,H^{3/2} - \frac{2}{5}H^{5/2}\right]$$

$$= 2C_d\sqrt{2g}\tan\frac{\theta}{2}\left[\frac{4}{15}H^{5/2}\right]$$

$$= \frac{8}{15}C_d\sqrt{2g}\tan\frac{\theta}{2}\,.\,H^{5/2} \qquad ...(7.2)$$

If $\theta = 90°$, *i.e.* for a right angled V-notch,
assuming $C_d = 0{\cdot}6$

$$Q = \frac{8}{15} \times 0{\cdot}6 \times \sqrt{2 \times 9{\cdot}81} \times 1 \times H^{3/2}$$

$$= 1{\cdot}417\ \mathrm{H}^{5/2} \qquad ...(7.3)$$

A triangular notch is preferred over a rectangular notch due to the following reasons :

1. The expression for discharge for a right angled notch is very simple.
2. For a given triangular notch, only one reading of head H is required to compute the discharge.
3. For low discharges, a triangular notch gives more accurate results.
4. Ventilation of a triangular notch is not required.

Example 7.2 Water flows over a 1·2 m wide and 180 mm deep rectangular weir and afterwards passes through a triangular right-angled weir. Determine the depth over the triangular weir taking C_d for rectangular and triangular weirs as 0·62 and 0·6 respectively.

Solution $L = 1{\cdot}2$ m

$H = 0{\cdot}18$ m

The discharge over the rectangular weir is given by

$$Q = \frac{2}{3}C_d\,.\,L\sqrt{2g}\,H^{3/2}$$

$$= \frac{2}{3} \times 0{\cdot}62 \times 1{\cdot}2 \times \sqrt{2 \times 9{\cdot}81} \times 0{\cdot}18^{3/2}$$

$$= 0{\cdot}1678\ \mathrm{m^3/s}$$

This is the discharge that passes over the triangular notch. Thus

$$Q = \frac{8}{15} \times 0{\cdot}6 \times \sqrt{2 \times 9{\cdot}81} \times 1 \times H^{5/2} = 1{\cdot}417\ H^{5/2}$$

$$0{\cdot}1678 = 1{\cdot}417\ H^{5/2}$$

$$H^{5/2} = 0{\cdot}1184$$

$$H = 0{\cdot}426 \text{ m}$$

Example 7.3 The discharge through a rectangular channel 2 m wide is 0·3 m^3/s. A right angled *V*-notch is used to measure this discharge. Determine the position of the apex of the notch from the bed of the channel if the maximum depth of water does not exceed 1·2 m. Take $C_d = 0{\cdot}6$

Solution $Q = 0{\cdot}3$ m^3/s

$C_d = 0{\cdot}6$

$$Q = \frac{8}{15} \times 0{\cdot}6 \times \sqrt{2 \times 9{\cdot}81} \times 1 \times H^{5/2} = 1{\cdot}417\ H^{5/2}$$

$$0{\cdot}3 = 1{\cdot}417\ H^{5/2}$$

$$H^{5/2} = 0{\cdot}2117$$

$$H = 0{\cdot}537 \text{ m}$$

Position of the apex of notch from the bed = 1·2 – 0·537 = 0·663 m

(The width of the channel is superfluous data).

7.5 DISCHARGE OVER A TRAPEZOIDAL NOTCH OR WEIR

A **trapezoidal notch** is a combination of a rectangular and a triangular notch or weir (Fig. 7.3). Therefore, the total discharge is the sum of the two discharges.

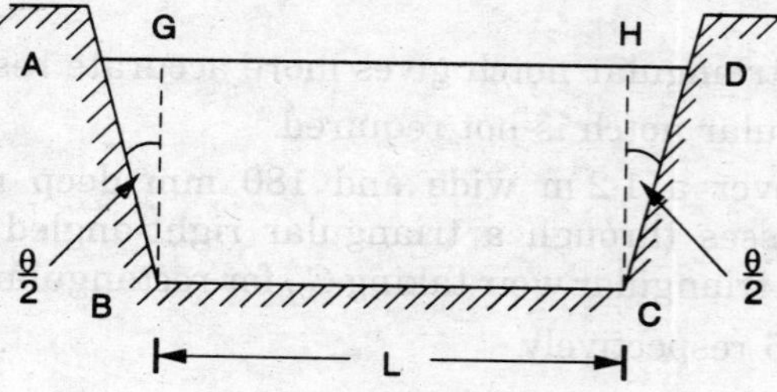

Fig. 7.3

Let L = length of the rectangular portion (crest) of the notch

H = height of water over the notch

C_{dr} = coefficient of discharge for rectangular notch

C_{dt} = coefficient of discharge for triangular notch

The discharge through rectangular portion *BCGH*,

$$Q_r = \frac{2}{3} C_d L \sqrt{2g}\ H^{3/2}$$

The discharge through the two triangular portions *ABG* and *DCH*, each with angle θ/2 is equal to the discharge through a single triangular notch of angle θ.

$$Q_t = \frac{8}{15} C_d \sqrt{2g} \tan\frac{\theta}{2} H^{5/2}$$

Total discharge through trapezoidal notch,

$$Q = Q_r + Q_t$$

$$= \frac{2}{3} C_{dr} L \sqrt{2g} H^{3/2} + \frac{8}{15} C_{dt} \sqrt{2g} \tan\frac{\theta}{2} H^{5/2} \quad ...(7.4)$$

Example 7.4 A trapezoidal notch 1·5 m wide at the top and 600 mm wide at the bottom is 500 mm high. The head of water on the notch is 400 mm. Determine the discharge through the notch assuming the value of coefficient of discharge for the rectangular portion as 0·62 and for triangular portion 0·6.

Solution $H = 0{\cdot}4$ m $\qquad C_{dr} = 0{\cdot}62$

$L = 0{\cdot}6$ m $\qquad C_{dt} = 0{\cdot}6$

Refer Fig. 7.4,

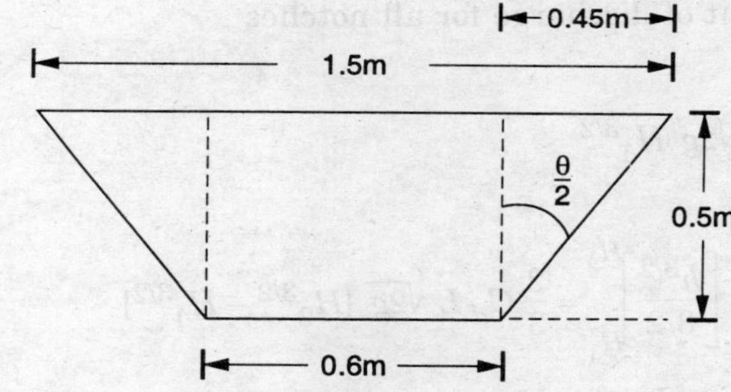

Fig. 7.4

$$\tan\frac{\theta}{2} = \frac{(1{\cdot}5 - 0{\cdot}6)/2}{0{\cdot}5} = 0{\cdot}9$$

$$Q = \frac{2}{3} C_{dr} L \sqrt{2g} H^{3/2} + \frac{8}{15} C_{dt} \sqrt{2g} \tan\frac{\theta}{2} H^{5/2}$$

$$= \frac{2}{3} \times 0{\cdot}62 \times 0{\cdot}6 \times \sqrt{2 \times 9{\cdot}81} \times 0{\cdot}4^{3/2} + \frac{8}{15} \times 0{\cdot}6 \times \sqrt{2 \times 9{\cdot}81} \times 0{\cdot}9 \times 0{\cdot}4^{5/2}$$

$$= 0{\cdot}2779 + 0{\cdot}1291$$

$$= 0{\cdot}407 \text{ m}^3\text{/s}$$

7.6 DISCHARGE OVER A STEPPED NOTCH

A **stepped notch** consists of a number of rectangular notches. The total discharge through it is equal to the sum of the discharges through the different rectangular notches.

Consider a stepped notch as shown in Fig. 7.4.

Let H_1 = height of water above the sill of notch

L_1 = length of notch 1

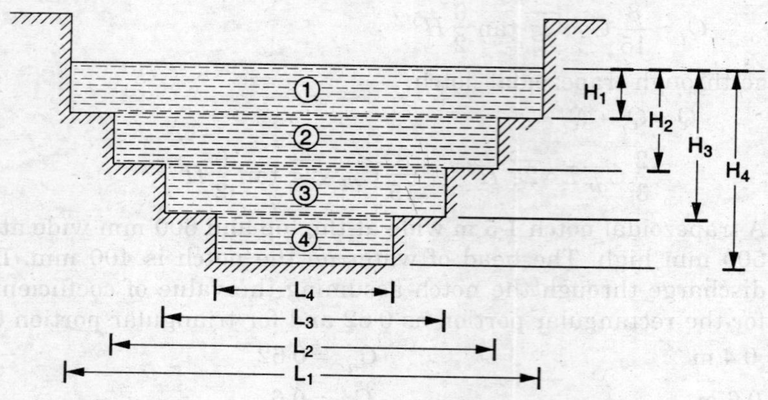

Fig. 7.5

Similarly, let H_2 and L_2 be the height and length of the notch 2 and so on.

C_d = coefficient of discharge for all notches

Discharge over notch 1,

$$Q_1 = \frac{2}{3} C_d L_1 \sqrt{2g} \, H_1^{3/2}$$

Discharge over notch 2,

$$Q_2 = C_d L \sqrt{2g} \left[\frac{h^{3/2}}{3/2} \right]_{H_1}^{H_2} = \frac{2}{3} C_d L \sqrt{2g} \, [H_2^{3/2} - H_1^{3/2}]$$

Similarly,

Discharge over notch 3,

$$Q_3 = \frac{2}{3} C_d L \sqrt{2g} \, [H_3^{3/2} - H_2^{3/2}]$$

Discharge over notch 4,

$$Q_4 = \frac{2}{3} C_d L \sqrt{2g} \, [H_4^{3/2} - H_3^{3/2}]$$

Total discharge,

$$Q = Q_1 + Q_2 + Q_3 + Q_4$$

$$= \frac{2}{3} C_d L_1 \sqrt{2g} \, H_1^{3/2} + \frac{2}{3} C_d L \sqrt{2g} \, [H_2^{3/2} - H_1^{3/2}]$$

$$+ \frac{2}{3} C_d L \sqrt{2g} \, [H_3^{3/2} - H_2^{3/2}] + \frac{2}{3} C_d L \sqrt{2g} \, [H_4^{3/2} - H_3^{3/2}] \quad ...(7.5)$$

Example 7.5 Fig. 7.6 shows a stepped rectangular notch. Find the discharge over the notch taking a coefficient of discharge as 0·62 for all the portions.

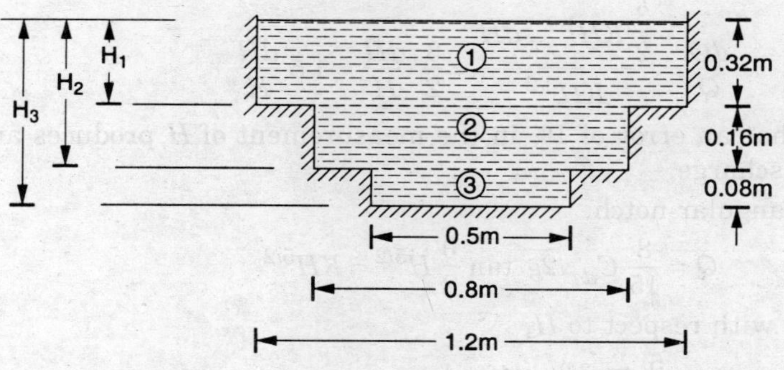

Fig. 7.6

Solution $Q = Q_1 + Q_2 + Q_3$

$$Q = \frac{2}{3} C_d L_1 \sqrt{2g} \,.\, H_1^{3/2} + \frac{2}{3} C_d L_2 \sqrt{2g}\, [H_2^{3/2} - H_1^{3/2}] + \frac{2}{3} C_d L_3 \sqrt{2g}\, [H_3^{3/2} - H_2^{3/2}]$$

$$= \frac{2}{3} \times 0{\cdot}62 \times 1{\cdot}2 \sqrt{2 \times 9{\cdot}81} \times 0{\cdot}32^{3/2} + \frac{2}{3} \times 0{\cdot}62 \times 0{\cdot}8 \times \sqrt{2 \times 9{\cdot}81}\, [0{\cdot}48^{3/2} - 0{\cdot}32^{3/2}]$$

$$+ \frac{2}{3} \times 0{\cdot}62 \times 0{\cdot}5 \sqrt{2 \times 9{\cdot}81}\, [0{\cdot}56^{3/2} - 0{\cdot}48^{3/2}]$$

$$= \frac{2}{3} \times 0{\cdot}62 \times 4{\cdot}43 \times [(1{\cdot}2 \times 0{\cdot}32^{3/2}) + 0{\cdot}8\,(0{\cdot}48^{3/2} - 0{\cdot}32^{3/2}) + 0{\cdot}5\,(0{\cdot}56^{3/2} - 0{\cdot}48^{3/2})]$$

$$= 1{\cdot}831 \times [0{\cdot}2172 + 0{\cdot}1212 + 0{\cdot}0433]$$

$$= 1{\cdot}831 \times 0{\cdot}3817$$

$$= 0{\cdot}6989 \text{ m}^3\text{/s}$$

7.7 EFFECT ON DISCHARGE DUE TO ERROR IN MEASUREMENT OF HEAD

From the expressions of discharge over a rectangular and triangular notches, it is observed that for a rectangular notch it is proportional to $H^{3/2}$ and for a triangular notch, it is $H^{5/2}$. This indicates that to measure the discharge accurately, it is essential to measure the value of H accurately. The following cases will indicate the extent of error in the measurement of discharge with the error in the measurement of head.

For a rectangular notch,

$$Q = \frac{2}{3} C_d L_1 \sqrt{2g}\, H^{3/2} = KH^{3/2} \quad ...(i)$$

where $K = \frac{2}{3} C_d L \sqrt{2g}$ = Constant for a given notch

Differentiating with respect to H, $dQ = \frac{3}{2} KH^{1/2} \,.\, dH$ ...(ii)

Dividing (ii) by (i),

$$\frac{dQ}{Q} = \frac{\frac{3}{2} KH^{1/2} \cdot dH}{KH^{3/2}} = \frac{3}{2} \frac{dH}{H} = 1{\cdot}5 \frac{dH}{H}$$

which shows that an error of 1% in the measurement of H produces an error of 1·5% in measurement of discharge.

In case of a triangular notch,

$$Q = \frac{8}{15} C_d \sqrt{2g} \tan \frac{\theta}{2} H^{5/2} = KH^{5/2}$$

Differentiating with respect to H,

$$\frac{dQ}{Q} = \frac{\frac{5}{2} KH^{3/2} \cdot dH}{KH^{5/2}} = \frac{5}{2} \frac{dH}{H} = 2{\cdot}5 \frac{dH}{H}$$

Thus an error of 1% in the head gives an error of 2·5% in the discharge.

7.8 TIME REQUIRED TO EMPTY A RESERVOIR

Time required to empty a reservoir with a rectangular or triangular notch or weir can be found as follows :

1. Rectangular Notch or Weir

(Refer Fig. 7.7),

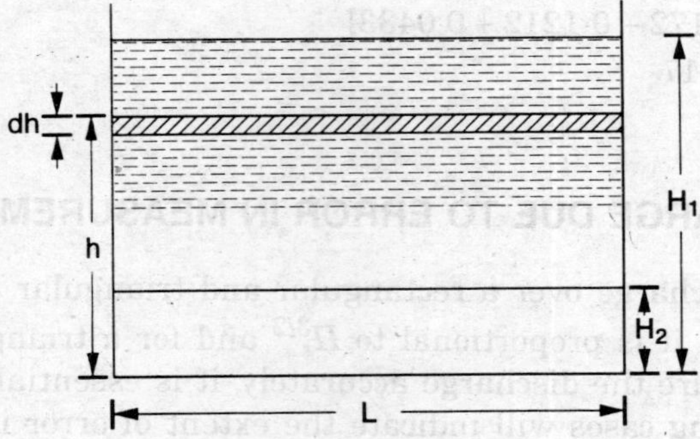

Fig. 7.7

Let L = length of crest of notch or weir

A = cross-sectional area of tank

H_1 = initial height of liquid above the crest

H_2 = final height of liquid above the crest

C_d = coefficient of discharge

T = time taken to lower the height from H_1 to H_2

Let h be the height of liquid surface above the crest at any instant and let it fall through a height dh in a small interval of time dT.

Decrease in the volume of liquid in the tank = Discharge × Time

$$-A \,.\, dh = Q \times dT \qquad ...(7.6)$$

(*-ve* sign indicates that h decreases with increase in T)

$$-A \,.\, dh = C_d \,.\, L \sqrt{2g} \,.\, h^{3/2} \,.\, dT$$

$$dT = \frac{-A \,.\, dh}{\frac{2}{3} C_d \,.\, L \sqrt{2g} \,.\, h^{3/2}}$$

Total time,

$$T = \int_{H_1}^{H_2} \frac{-A \,.\, dh}{\frac{2}{3} C_d \,.\, L \sqrt{2g} \,.\, h^{3/2}}$$

$$= \frac{-A}{\frac{2}{3} C_d \,.\, L \sqrt{2g}} \int_{H_1}^{H_2} h^{-3/2} \,.\, dh$$

$$= \frac{-3A}{2\, C_d \,.\, L \sqrt{2g}} \left[\frac{h^{-1/2}}{1/2} \right]_{H_1}^{H_2}$$

$$= \frac{-3A}{2\, C_d \,.\, L \sqrt{2g}} (-2) \left(\frac{1}{\sqrt{h}} \right)_{H_1}^{H_2}$$

$$= \frac{3A}{C_d \,.\, L \sqrt{2g}} \left(\frac{1}{\sqrt{H_2}} - \frac{1}{\sqrt{H_1}} \right) \qquad ...(7.7)$$

2. Triangular Notch or Weir

(Refer Fig. 7.8),

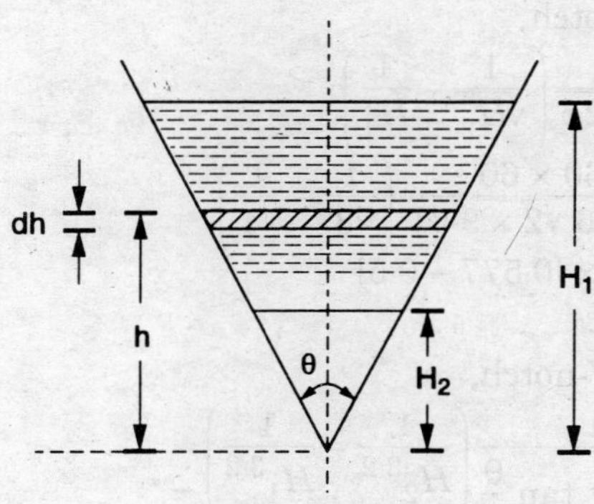

Fig. 7.8

Let H_1 = initial height of liquid above the apex of notch

H_2 = final height of liquid above the apex of notch

θ = angle of notch

A = cross-sectional area of the tank

As before,

$$-A \cdot dh = Q \cdot dT$$

$$-A \cdot dh = \frac{8}{15} C_d \sqrt{2g} \cdot \tan \frac{\theta}{2} h^{5/2} \cdot dT$$

$$dT = \frac{-A \cdot dh}{\frac{8}{15} C_d \sqrt{2g} \cdot \tan \frac{\theta}{2} h^{5/2}}$$

Total time,

$$T = \int_{H_1}^{H_2} \frac{-A \cdot dh}{\frac{8}{15} C_d \sqrt{2g} \cdot \tan \frac{\theta}{2} h^{5/2}}$$

$$= \frac{-A}{\frac{8}{15} C_d \sqrt{2g} \tan \frac{\theta}{2}} \int_{H_1}^{H_2} h^{-5/2} \cdot dh$$

$$= \frac{-15A}{8 C_d \sqrt{2g} \tan \frac{\theta}{2}} \left[\frac{h^{-3/2}}{-3/2} \right]_{H_1}^{H_2}$$

$$= \frac{5A}{8 C_d \sqrt{2g} \tan \frac{\theta}{2}} \left(\frac{1}{h^{3/2}} \right)_{H_1}^{H_2}$$

$$= \frac{5A}{4 C_d \sqrt{2g} \tan \frac{\theta}{2}} \left(\frac{1}{H_2^{3/2}} - \frac{1}{H_1^{3/2}} \right) \quad \text{...(7.8)}$$

Example 7.6 Calculate the time required to lower the water level from 4 m to 3 m in a reservoir of dimensions 60 m × 60 m by (i) a rectangular notch of length 1·6 m and (ii) a right-angled *V*-notch. $C_d = 0·62$ in each case.

Solution For a rectangular notch,

$$T = \frac{3A}{C_d \cdot L \sqrt{2g}} \left(\frac{1}{\sqrt{H_2}} - \frac{1}{H_1} \right)$$

$$= \frac{3 \times 60 \times 60}{0·62 \times 1·6 \sqrt{2 \times 9·81}} \left(\frac{1}{\sqrt{3}} - \frac{1}{\sqrt{4}} \right)$$

$$= 2457·58 \times [0·577 - 0·5]$$

$$= 189·23 \text{ s}$$

For a right-angled *V*-notch,

$$T = \frac{5A}{4 C_d \sqrt{2g} \tan \frac{\theta}{2}} \left(\frac{1}{H_2^{3/2}} - \frac{1}{H_1^{3/2}} \right)$$

$$= \frac{5 \times 60 \times 60}{4 \times 0·62 \times \sqrt{2 \times 9·81} \tan \frac{90°}{2}} \left(\frac{1}{3^{3/2}} - \frac{1}{4^{3/2}} \right)$$

$$= 1638·4 \times [0·1924 - 0·125]$$

$$= 110·5 \text{ s}$$

7.9 VELOCITY OF APPROACH

Velocity of approach is defined as the velocity with which the water approaches the weir or notch before flowing over it.

Let V_a = velocity of approach

The head equivalent to V_a,

$$H_a = \frac{V_a^{\ 2}}{2g} \qquad ...(7.9)$$

This is the additional head that acts on the water flowing over the notch. In that case the initial height of water over the notch becomes $(H + H_a)$ and the final h.

Thus the more accurate relation for a rectangular notch or weir will be

$$Q = \frac{2}{3} C_d L \sqrt{2g}\ [h^{3/2}]_{H_a}^{H+H_1}$$

$$= \frac{2}{3} C_d L \sqrt{2g}\ [(H + H_a)^{3/2} - H_a^{\ 3/2}] \qquad ...(7.10)$$

Similarly, the expressions for discharge for other types of notches and weirs can be found.

V_a can be determined by finding the discharge over the notch or weir neglecting the velocity of approach. If Q is the discharge over the notch or weir and A is the cross-sectional area of the channel on the upstream side of the notch of weir, then

$$V_a = \frac{Q}{A} \qquad ...(7.11)$$

Example 7.7 A rectangular weir of crest length 600 mm is used to measure the flow of water in a rectangular channel 600 mm wide and 500 mm deep. The water level in the weir is 60 mm above the crest of weir. Find the discharge in the channel taking into account the velocity of approach also. $C_d = 0{\cdot}62$.

Solution For weir, $L = 0{\cdot}6$ m

$H = 0{\cdot}06$ m

$$Q = \frac{2}{3} C_d L \sqrt{2g}\ H^{3/2} = \frac{2}{3} \times 0{\cdot}62 \times 0{\cdot}6\ \sqrt{2 \times 9{\cdot}81} \times 0{\cdot}06^{3/2} = 0{\cdot}01615 \text{ m}^3/\text{s}$$

Velocity of approach,

$$V_a = \frac{Q}{A} = \frac{0{\cdot}01615}{0{\cdot}6 \times 0{\cdot}5} = 0{\cdot}0538 \text{ m/s} \qquad (A = \text{Area of channel} = 0{\cdot}6 \times 0{\cdot}5 \text{ m}^2)$$

Head due to velocity of approach

$$h_a = \frac{V_a^2}{2g} = \frac{0{\cdot}0538^2}{2 \times 9{\cdot}81} = 0{\cdot}0001476 \text{ m}$$

$$Q = \frac{2}{3} C_d L \sqrt{2g}\ [(H + H_a)^{3/2} - H_a^{\ 3/2}]$$

$$= \frac{2}{3} \times 0{\cdot}62 \times 0{\cdot}6\ \sqrt{2 \times 9{\cdot}81}\ [(0{\cdot}06 + 0{\cdot}0001476)^{3/2} - 0{\cdot}0001476^{3/2}]$$

$$= 1{\cdot}09864 \times [0{\cdot}01475 - 0{\cdot}00000179]$$

$$= 1{\cdot}09864 \times 0{\cdot}014748$$

$= 0.0162 \text{ m}^3/\text{s}$

7.10 SUPPRESSED WEIRS

For a rectangular weir,

$$Q = \frac{2}{3} C_d L \sqrt{2g} H^{3/2} \quad \text{without velocity of approach}$$

$$Q = \frac{2}{3} C_d L \sqrt{2g} [(H + H_a)^{3/2} - H_a^{3/2}] \quad \text{with velocity of approach}$$

These relations are applicable to suppressed weirs *i.e.* the weirs in which the crest length is equal to the width of the channel. However, if the weir is not suppressed *i.e.* the width is shorter than the width of the channel, the effect of the end contraction is considered.

Francis Formula

On the basis of his experiments, Francis established that each end contraction reduces the effective length of the crest of weir by $0.1 \times H$ times where H is the head over the weir (Fig. 7.9). Thus the actual discharge decreases.

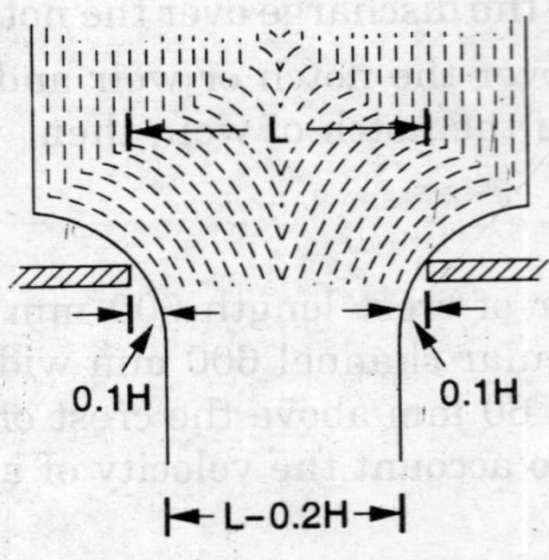

Fig. 7.9

In a rectangular weir, as there are two end contractions, the effective length

$$= L - 0.1 \times 2 \times H = L - 0.2\,H$$

$$Q = \frac{2}{3} C_d (L - 0.1\,nH) \sqrt{2g} H^{3/2}$$

$$= \frac{2}{3} C_d (L - 0.2\,H) \sqrt{2g} H^{3/2} \quad ...(7.12)$$

Assuming $C_d = 0.623$ and $g = 9.81 \text{ m/s}^2$

$$Q = \frac{2}{3} \times 0.623 \times \sqrt{2 \times 9.81} \times (L - 0.2\,H) \times H^{3/2}$$

$$= 1.84 (L - 0.2\,H) H^{1/2} \quad ...(7.13)$$

If end contractions are suppressed,

$$Q = 1.84\,L\,H^{3/2} \quad ...(7.14)$$

If velocity of approach is considered,

$$Q = 1{\cdot}84\, L\, [(H + H_a)^{3/2} - H_a^{\ 3/2}] \quad ...(7.15)$$

Bazin Formula

Bazin proposed the following empirical relation for the discharge over a rectangular weir on the basis of his experiments :

$$Q = m \,.\, L \,.\, \sqrt{2g} \,.\, H^{3/2} \quad ...(7.16)$$

where $m = 0{\cdot}405 + \dfrac{0{\cdot}003}{H}$

H is the height of water over the weir.

If velocity of approach is considered,

$$Q = m\, L \,.\, \sqrt{2g} \,.\, (H + H_a)^{3/2} \quad ...(7.17)$$

where $m = 0{\cdot}405 + \dfrac{0{\cdot}003}{H + H_a}$

Example 7.8 The head of water over a rectangular weir is 360 mm. The length of the crest of the weir with end contractions suppressed is 1·6 m. Determine the discharge using the (i) Francis's formula and (ii) Bazin's formula.

Solution $L = 1{\cdot}6$ m

$H = 0{\cdot}36$ m

Discharge by Francis's formula.

$$Q = 1{\cdot}84\, LH^{3/2} = 1{\cdot}84 \times 1{\cdot}6 \times 0{\cdot}36^{3/2} = 0{\cdot}636 \text{ m}^3/\text{s}$$

Discharge by Bazin's formula,

$$Q = m \,.\, L \,.\, \sqrt{2g} \,.\, H^{3/2}$$

where $m = 0{\cdot}405 + \dfrac{0{\cdot}003}{H} = 0{\cdot}405 + \dfrac{0{\cdot}003}{0{\cdot}36} = 0{\cdot}4133$

$$Q = 0{\cdot}4133 \times 1{\cdot}6 \times \sqrt{2 \times 9{\cdot}81} \times 0{\cdot}36^{3/2} = 0{\cdot}633 \text{ m}^3/\text{s}$$

Example 7.9 A 50 m long weir is divided into 10 equal bays by vertical posts each 500 mm wide. Determine the discharge over the weir by using Francis's formula if the head over the crest is 1·5 m and velocity of approach is 3 m/s.

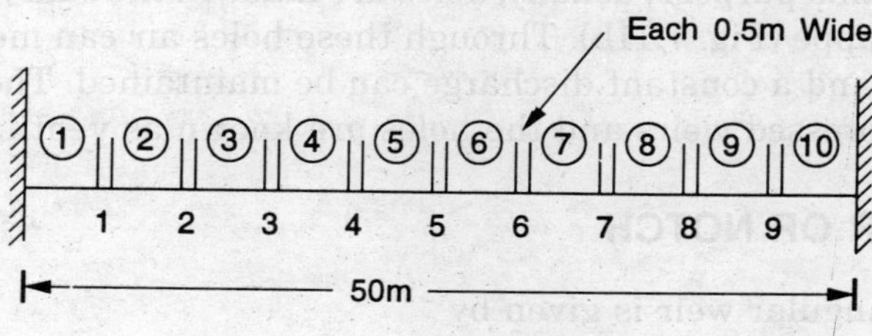

Fig. 7.10

Solution $L = 50 - (10 - 1) \times 0{\cdot}5 = 45{\cdot}5$ m

$n = 2 \times 10 = 20$

$$H_a = \frac{V_a^2}{2g} = \frac{3^2}{2 \times 9{\cdot}81} = 0{\cdot}4587$$

$$Q = 1{\cdot}84\ [L - 0{\cdot}1\ n\ (H + H_a)]\ [(H + H_a)^{3/2} - H_a^{3/2}]$$

$$= 1{\cdot}84\ [45{\cdot}5 - 0{\cdot}1 \times 20 \times (1{\cdot}5 + 0{\cdot}4587)]\ [(1{\cdot}5 + 0{\cdot}4587)^{3/2} - 0{\cdot}4587^{3/2}]$$

$$= 1{\cdot}84 \times 41{\cdot}583 \times 2{\cdot}431$$

$$= 186\ \text{m}^3/\text{s}$$

7.11 VENTILATION OF SUPPRESSED WEIR

In suppressed weirs, the flowing sheet of water (nappe) touches the side walls of the channel. The air trapped under the nappe is gradually sucked away by the nappe creating a sub-atmospheric pressure in the region. As a result, the nappe clings to the weir wall (Fig. 7.11a). Owing to the negative pressure under the nappe, the discharge is increased from the actual value.

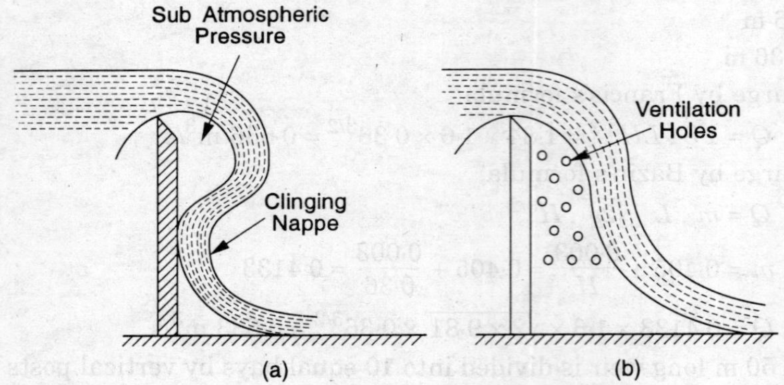

Fig. 7.11

To have the accurate discharge over the weir, it is necessary that the pressure under the nappe is atmospheric. For this purpose, usually, holes are made in the side walls of the channel just below the springing nappe (Fig. 7.11b). Through these holes air can move in the region of sub-atmospheric pressure and a constant discharge can be maintained. The method is known as **ventilation** of the suppressed weirs and the holes are known as ventilation holes.

7.12 CIPPOLETTI WEIR OR NOTCH

The discharge over a rectangular weir is given by

$$Q = \frac{2}{3} C_d\ L\ \sqrt{2g}\ H^{3/2} \qquad \text{with end contractions suppressed}$$

$$Q = \frac{3}{3} C_d\ (L - 0{\cdot}2\ H)\ \sqrt{2g}\ H^{3/2} \qquad \text{with no end contractions suppressed}$$

(Eq. 7.12)

A Cippoletti weir is a trapezoidal weir and is designed in such a way that the factor of end contractions on a rectangular weir is incorporated in the design itself.

$$Q = \frac{2}{3} C_d (L - 0{\cdot}2\,H) \sqrt{2g}\, H^{3/2}$$

$$= \frac{2}{3} C_d L\sqrt{2g}\, H^{3/2} - \frac{2}{15} C_d \sqrt{2g}\, H^{5/2} \qquad ...(7.18)$$

Thus the discharge decreases by $\frac{2}{15} C_d \sqrt{2g}\, H^{5/2}$ due to end contractions. This decrease is compensated by making the slope of the sides of the trapezoidal weir in such a way that the discharge through two triangular portions is equal to this decrease.

Let θ/2 be the angle of the slope. The discharge through the *V*-notch with angle θ

$$Q = \frac{8}{15} C_d \sqrt{2g} \tan\frac{\theta}{2} H^{5/2}$$

$$\frac{8}{15} C_d \sqrt{2g} \tan\frac{\theta}{2} H^{5/2} = \frac{2}{15} C_d \sqrt{2g}\, H^{5/2}$$

$$\tan\frac{\theta}{2} = \frac{1}{4}$$

$$\theta = 14^\circ\ 2'$$

The relation shows that the slope of the sides should be 1 horizontal to 4 vertical (Fig. 7.12). Then the discharge will be given by

$$Q = \frac{2}{3} C_d L \sqrt{2g}\, H^{3/2}$$

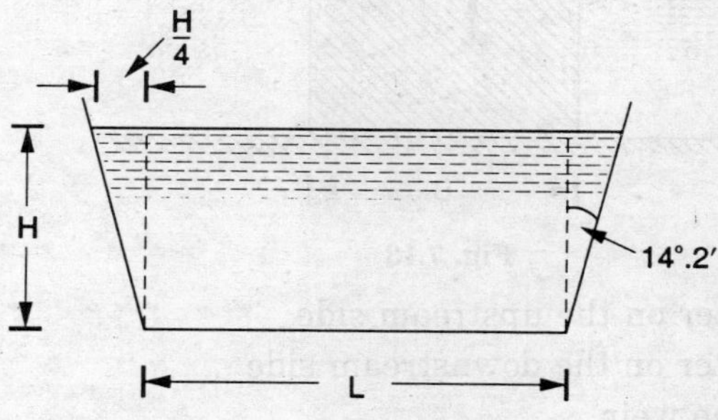

Fig. 7.12

Example 7.10 Determine the discharge over a Cippoletti weir of crest length 600 mm when the head over the weir is 400 mm. The coefficient of discharge is 0·6.

If the channel is 800 mm wide and 600 deep and the velocity of approach is also considered, what will be the discharge ?

Solution $L = 0{\cdot}6$ m

$H = 0{\cdot}4$ m

$C_d = 0{\cdot}6$

$$Q = \frac{2}{3} C_d L \sqrt{2g}\, H^{3/2} = \frac{2}{3} \times 0{\cdot}6 \times 0{\cdot}6 \sqrt{2 \times 9{\cdot}81} \times 0{\cdot}4^{3/2} = 0{\cdot}269 \text{ m}^3/\text{s}$$

$$V_a = \frac{\text{Discharge}}{\text{Area of channel}} = \frac{Q}{A} = \frac{0{\cdot}269}{0{\cdot}8 \times 0{\cdot}6} = 0{\cdot}56 \text{ m/s}$$

Head due to velocity of approach,

$$H_a = \frac{V_a^{\,2}}{2g} = \frac{0{\cdot}56^2}{2 \times 9{\cdot}81} = 0{\cdot}016 \text{ m}$$

$$Q = \frac{2}{3} C_d L \sqrt{2g}\, [(H + H_a)^{3/2} - H_a^{\,3/2}]$$

$$= \frac{2}{3} \times 0{\cdot}6 \times 0{\cdot}6 \times \sqrt{2 \times 9{\cdot}81}\, [(0{\cdot}4 + 0{\cdot}016)^{3/2} - 0{\cdot}016^{3/2}]$$

$$= 1{\cdot}0632 \times 0{\cdot}2663$$

$$= 0{\cdot}283 \text{ m}^3/\text{s}$$

7.13 DISCHARGE OVER A BROAD CRESTED WEIR

A weir having a broad crest or sill is known as a **broad crested weir.** Figure 7.13 represents a broad crested weir. Crest width is such that $B > H/2$.

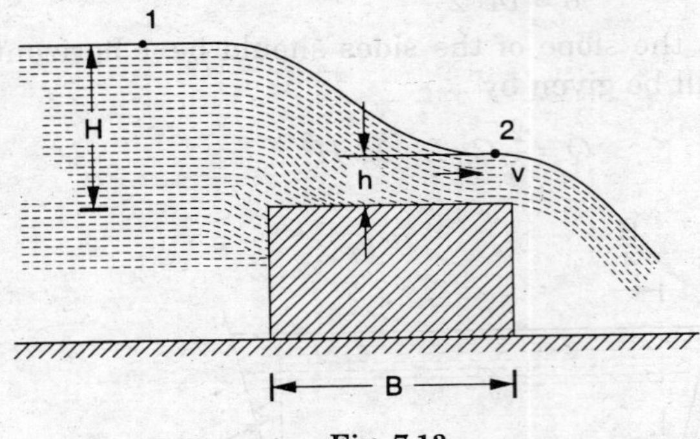

Fig. 7.13

Let H = head of water on the upstream side

h = head of water on the downstream side

L = length of the weir

C_d = Coefficient of discharge

Applying Bernoulli's equation at 1 (where water is assumed to be still) and 2 (where velocity is v),

$$0 + 0 + H = 0 + \frac{v^2}{2g} + h$$

$$\frac{v^2}{2g} = H - h$$

$$v = \sqrt{2g\,(H - h)}$$

Discharge = C_d × Area of flow × Velocity

$$= C_d \times L \times h \times \sqrt{2g\,(H-h)}$$

$$= C_d \,.\, L \sqrt{2g\,(Hh^2 - h^3)}$$

The discharge is maximum if $(Hh^2 - h^3)$ is maximum.

$$\frac{d}{dh}\,(Hh^2 - h^3) = 0$$

$$2Hh - 3h^2 = 0$$

$$2H - 3h = 0$$

$$h = \frac{2}{3}H$$

Maximum discharge $= C_d\, L \sqrt{2g\, H\left(\frac{2}{3}H\right)^2 - \left(\frac{2}{3}H\right)^3}$

$$= C_d\, L \sqrt{2g} \sqrt{\frac{4}{27}H^3}$$

$$= \frac{2}{3\sqrt{3}}\, C_d\, L \sqrt{2 \times 9{\cdot}81}\, H^{3/2}$$

$$= 1{\cdot}705\, C_d\, LH^{3/2} \qquad \text{...(7.19)}$$

Example 7.11 A 40 m long broad crested weir has 400 mm of water above its crest. Determine the maximum discharge over the weir neglecting velocity of approach. $C_d = 0{\cdot}6$.

Solution $L = 40$ m

$H = 0{\cdot}4$ m

$C_d = 0{\cdot}6$

Maximum discharge $= 1{\cdot}705\, C_d\, LH^{3/2} = 1{\cdot}705 \times 0{\cdot}6 \times 40 \times 0{\cdot}4^{3/2} = 10{\cdot}35$ m^3/s

7.14 OGEE WEIR

In an **Ogee weir**, the crest of the weir rises upto a height of $0{\cdot}115\,H$ and then falls according to

$$y = \frac{x^{1{\cdot}85}}{2H^{0{\cdot}85}} \qquad \text{...(7.20)}$$

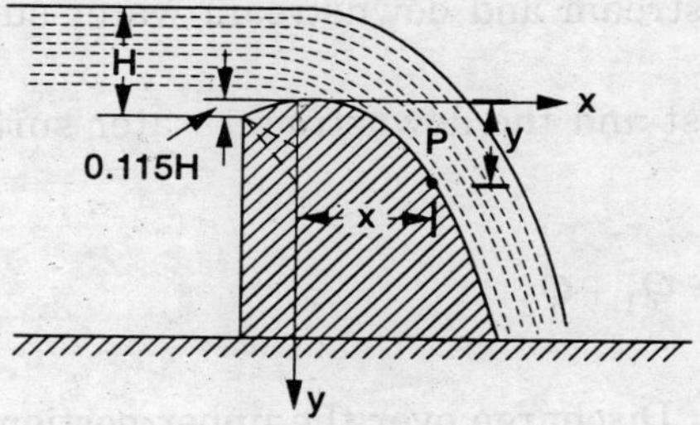

Fig. 7.14

Proper ventilation is required to have a constant discharge over a suppressed weir. However, if it is not possible to do so, the weir profile can be made conforming to the shape of the nappe on the downstream. Such a weir is called an **Ogee weir** or a **spillway** (Fig. 7.14). The discharge over an Ogee weir is the same as for a rectangular weir,

$$Q = \frac{2}{3} C_d \,.\, L \sqrt{2g}\, H^{3/2} \qquad ...(7.21)$$

Example 7.12 Determine the discharge over an Ogee weir 4 m long and having a head of 500 mm of water. Take $C_d = 0{\cdot}62$.

Solution $L = 4$ m

$H = 0{\cdot}5$ m

$C_d = 0{\cdot}62$

$$Q = \frac{2}{3} C_d \,.\, L \sqrt{2g}\, H^{3/2} = \frac{2}{3} \times 0{\cdot}62 \times 4 \sqrt{2 \times 9{\cdot}81} \times 0{\cdot}5^{3/2} = 2{\cdot}59 \text{ m}^3/\text{s}$$

7.15 SUBMERGED WEIR

A **submerged weir** is also known as a **drowned weir.** In this type of weir, the water level on the drowned side of the weir is above the top surface of the weir (Fig. 7.15).

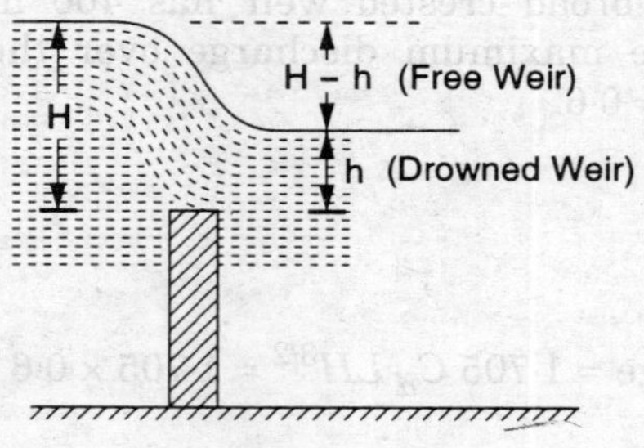

Fig. 7.15

The total discharge is determined by splitting up the height of water into two parts.

Let H = height of water on the upstream side

h = height of water on the downstream side

L = length of the weir

Then

The portion between the upstream and downstream water surface is considered as free weir and

The portion between the crest and the downstream water surface is treated as drowned weir.

Then total discharge,

$$Q = Q_1 + Q_2$$

where

$$Q_1 = \text{Discharge over the upper portion}$$

$$= \frac{2}{3} C_{d1} L \sqrt{2g} (H - h)^{3/2}$$

and $\quad Q_2$ = Discharge through the drowned portion

$$= C_d \times \text{Area of flow} \times \text{Velocity of flow}$$

$$= \frac{2}{3} C_{d2} L \sqrt{2g} (H - h)^{3/2}$$

$$Q = \frac{2}{3} C_{d1} L \sqrt{2g} (H - h)^{3/2} + C_{d2} L \cdot h \sqrt{2g (H - h)} \qquad ...(7.22)$$

Example 7.13 The heights of water on the upstream and downstream sides in a submerged weir 3 m long are 160 mm and 80 mm respectively. Determine the discharge over the weir assuming the discharge coefficients for the free and the drowned portions to be 0·6 and 0·8 respectively.

Solution $Q = \frac{2}{3} C_d \cdot L \sqrt{2g} (H - h)^{3/2} + C_d \cdot L \cdot h \sqrt{2g (H - h)}$

$$= \frac{2}{3} \times 0{\cdot}6 \times 3 \sqrt{2 \times 9{\cdot}81} (0{\cdot}16 - 0{\cdot}08)^{3/2} + 0{\cdot}8 \times 3 \times 0{\cdot}08 \times \sqrt{2 \times 9{\cdot}81 (0{\cdot}16 - 0{\cdot}08)}$$

$$= 0{\cdot}1203 + 0{\cdot}068$$

$$= 0{\cdot}1883 \text{ m}^3/\text{s}$$

PROBLEMS

1. Define a notch and a weir. What is the difference between the two ?
2. How are the notches classified ?
3. Deduce an expression for the discharge over a
   (a) Rectangular notch
   (b) Triangular notch.
4. Why is a triangular notch preferred over a rectangular notch ?
5. Establish the formula for the discharge over a trapezoidal notch or weir.
6. Define a stepped notch. How do you determine the discharge over such a notch ?
7. What is the effect on the discharge due to an error in the measurement of head in case of (i) a rectangular notch and (ii) a triangular notch ?
8. Derive an expression for the time required to empty a tank with a rectangular notch. In what way the expression is changed for a triangular notch ?
9. Define velocity of approach. Write the expression for discharge over a rectangular weir with velocity of approach.
10. What do you understand by suppressed weirs ? What is the effect of end contractions ?
11. What do you mean by ventilation of suppressed weirs ?
12. Define a Cippoletti weir. What is the discharge over such a weir ?
13. Define the following :
    (a) A broad-crested weir

(b) Ogee weir

(c) Submerged weir.

14. The discharge through a rectangular weir is 0·35 m^3/s. The head of water is half the length of the weir. Assuming the value of coefficient of discharge as 0·6, determine the length of the weir. (800 mm)

15. During an experiment on a right-angled notch, 100 litres of water is collected in 120 seconds. The head on the sill is 50 mm. Find the coefficient of discharge of the notch. (0·631)

16. Water flows over a 1 m wide and 150 mm deep rectangular weir and afterwards passes through a triangular right-angled weir. Assuming the values of coefficient of discharge for rectangular and triangular weirs as 0·62 and 0·59 respectively, determine the depth over the triangular weir. (0·357 m)

17. Find the discharge through a trapezoidal notch which is 1·2 m wide at the top and 0·45 m at the bottom and is 0·3 m in height. The head of water on the notch is 225 mm. Assume coefficient of discharge to be 0·6. (0·1276 m^3/s)

18. While measuring the discharge over a right-angled *V*-notch, an error of 1·5 mm is made. If 600 litres of water is measured in 10 seconds, find the percentage error in the discharge. The coefficient of discharge for the notch is 0·6. (1·33 percent)

19. A right-angled *V*-notch is fitted in the side of a rectangular tank 1·5 m wide and 3 m long. If the water flows freely over the notch and there is no inflow into the tank, determine the time taken to reduce the head in the tank from 300 mm to 75 mm. The coefficient of discharge may be assumed as 0·62. (1 m 27 s)

20. Water flows through a rectangular channel 1 m wide and 500 mm deep and then over a Cipolletti weir of crest length 600 mm. Determine the discharge over the weir if the water level in the channel is 225 mm above the weir crest. Assume the value of coefficient of discharge as 0·6 and make correction for the velocity of approach. (0·1153 m^3/s)

21. Determine the maximum discharge over a 4 m long broad crested weir if the head of water on the upstream side is 800 mm and the coefficient of discharge 0·6. (2·928 m^3/s)

22. Find the discharge over an Ogee weir 8·5 m in length and having 800 mm head of water. Assume the value of C_d as 0·62. (11·14 m^3/s)

23. The crest of a submerged weir is 1 m above the bed of the channel and it spans over the entire width of the channel which is 10 m. The heights of water on the upstream and downstream sides of the weir are 1·8 m and 1·2 m respectively. Assuming the coefficient of discharge for the drowned portion as 0·8 and for the free portion as 0·6, determine the discharge. Neglect the velocity of approach. (13·72 m^3/s)

Chapter 8

DIMENSIONAL ANALYSIS

8.1 INTRODUCTION

All physical phenomena are controlled by many interrelated physical quantities. Normally, it is convenient to transform them into dimensionless quantities which are also interrelated. Number of dimensionless quantities so formed is always lesser than the total number of physical quantities. Owing to the reduction in the number of quantities of the variables upon which a physical phenomenon depends, it becomes easier to correlate the experimental data. It is more convenient to say that the friction factor for an incompressible fluid flow in a smooth pipe is a function of Reynolds number, rather than to say that the pressure drop per unit length depends upon density, viscosity, pipe diameter and velocity of the fluid.

Dimensional analysis is a mathematical technique and it suggests which variables or parameters affecting a physical phenomenon are to be grouped together to obtain dimensionless quantities. It provides a functional relationship between them. For example, in a hydraulic press the mechanical advantage is given by the ratio of ram area to plunger area (a dimensionless number) and is independent of the overall size of the press as long as the ratio of the two areas is the same. Moreover, it is also independent of the fluid properties. The concept of dimensional analysis along with an understanding of the type of flow, help in generalisation of the experimental data. The generalisation is helpful because.

- One is able to describe a phenomenon in its entirety and is not constrained to discuss merely the specialised experiment that is performed.
- One is able to perform fewer, though highly selective experiments, to go into the depth of a problem. Thus time and money are saved.
- Results of investigation are easily communicable to others in a more compact and meaningful way.

However, dimensionless quantities obtained by dimensional analysis indicate only how to organise a set of experiments and in what way the resulting experimental data are to be plotted. It does not provide any clue regarding the variation of the one dimensionless quantity with another, which is possible only by experimentation or in some cases by analytical means.

8.2 DIMENSIONS

A phenomenon involves a number of physical quantities. Some of the quantities may be independent of each other whereas the others may be expressed in terms of the independent quantities.

The quantities that are independent of each other are called **primary** or **fundamental quantities** and the quantities which can be expressed in terms of the primary quantities are called **secondary** or **derived quantities.** Primary quantities are **mass, length** and **time** and are denoted by M, L and T respectively. These cannot be expressed in simpler terms. All other quantities such as velocity, force, power etc. are expressible in terms of primary quantities and thus are called derived of secondary quantities. The expression of a secondary quantity in terms of primary quantity is called the **dimension** of the physical quantity.

There are two systems of dimensioning a physical quantity : the **mass-length-time** system *i.e.* the M-L-T system and the **force-length-time** system *i.e.* the F-L-T system. In the former system, M or mass is taken as a fundamental quantity whereas in the latter, F or force is considered as a fundamental quantity. However, the M-L-T system is more commonly used and the same will be used in our discussion. Various physical quantities with their dimensions are given in the Table 8.1.

Table 8.1

	Quantity	Symbol	Dimension
1.	Fundamental Quantities		
	Mass	M	M
	Length	L	L
	Time	T	T
2.	Derived Quantities		
	(i) Geometric quantities		
	Area	A	L^2
	Volume	V	L^3
	(ii) Kinematic quantities		
	Velocity	V	LT^{-1}
	Angular velocity	ω	T^{-1}
	Acceleration	a	LT^{-2}
	Angular acceleration	α	T^{-2}
	Gravity	g	LT^{-2}
	Discharge	Q	L^3T^{-1}
	Kinematic viscosity	ν	L^2T^{-1}

(iii) Dynamic quantities

Density	ρ	ML^{-3}
Force	F	MLT^{-2}
Weight	W	MLT^{-2}
Specific weight	w	$ML^{-2}T^{-2}$
Dynamic viscosity	μ	$ML^{-1}T^{-1}$
Pressure intensity	p	$ML^{-1}T^{-2}$
Surface tension	σ	MT^{-2}
Modulus of elasticity	E, K	$ML^{-1}T^{-2}$
Work	W	ML^2T^{-2}
Energy	E	ML^2T^{-2}
Power	P	ML^2T^{-3}
Torque	T	ML^2T^{-3}
Momentum	M	MLT^{-1}

The method of finding the dimensions of various quantities is explained below :

1. KInematic quantites

(a) Acceleration $= \dfrac{\text{Velocity}}{\text{Time}} = \dfrac{L}{T} \cdot \dfrac{1}{T} = LT^{-2}$

(b) Angular acceleration $= \dfrac{\text{Angle (in radians)}}{\text{Time}^2} = \dfrac{1}{T^2} = T^{-2}$

(c) Discharge = Area × Velocity $= L^2 \times \dfrac{L}{T} = L^3T^{-1}$

2. Dynamic quantities

(a) Density $= \dfrac{\text{Mass}}{\text{Volume}} = \dfrac{M}{L^3} = ML^{-3}$

(b) Force = Mass × Acceleration $= MLT^{-2}$

(c) Weight = Force (due to gravity) $= MLT^{-2}$

(d) Specific weight $= \dfrac{\text{Weight}}{\text{Volume}} = \dfrac{MLT^{-2}}{L^3} = ML^{-2}T^{-2}$

(e) Dynamic viscosity $= \dfrac{\text{Shear stress}}{\partial u/\partial y} = \dfrac{\text{Force/Area}}{\partial u/\partial y} = \dfrac{MLT^{-2}/L^2}{LT^{-1}/L} = ML^{-1}T^{-1}$

(f) Pressure intensity $= \dfrac{\text{Force}}{\text{Area}} = \dfrac{MLT^{-2}}{L^2} = ML^{-1}T^{-2}$

(g) Surface tension $= \dfrac{\text{Surface tension force}}{\text{Length}} = \dfrac{MLT^{-2}}{L} = MT^{-2}$

(h) Modulus of elasticity $= \dfrac{\text{Stress}}{\text{Strain}} = \dfrac{\text{Force}}{\text{Area}} = \dfrac{MLT^{-2}}{L^2} = ML^{-1}T^{-2}$

(Strain has no unit)

(i) Work = Force × Distance $= MLT^{-2} \times L = ML^2T^{-2}$

(j) Energy = Work $= ML^2T^{-2}$

(k) Torque = Force × Radius $= MLT^{-2} \times L = ML^2T^{-2}$

(l) Power $= \dfrac{\text{Work}}{\text{Time}} = \dfrac{ML^2T^{-2}}{T} = ML^2T^{-3}$

(m) Momentum = Mass × Velocity $= MLT^{-1}$

(n) Kinematic viscosity $= \dfrac{\text{Dynamic viscosity}}{\text{Density}} = \dfrac{ML^{-1}T^{-1}}{ML^{-3}} = L^2T^{-1}$

8.3 DIMENSIONAL HOMOGENEITY

Dimensional homogeneity implies that in any equation describing a physical phenomenon, the dimensions of each term on the two sides of an equation are same. A dimensionally homogeneous equation is independent of the system of units adopted. Thus, a dimensionally homogeneous equation can be used in all systems of units without changing it in any way.

The equation for velocity of a fluid through a nozzle is,

$$v = \sqrt{2gH}$$

Substituting the dimensions of quantities on both sides,

$$\left[\frac{L}{T}\right] = \left[\sqrt{\frac{L}{T^2} \times L}\right]$$

$$[LT^{-1}] = \left[\sqrt{\frac{L^2}{T^2}}\right]$$

$$= [LT^{-1}]$$

As the dimensions on both sides are the same, it is a dimensionally homogeneous equation.

The equation of time period T of a simple pendulum of length l is,

$$[T] = \left[2\pi\sqrt{\frac{l}{g}}\right]$$

$$= \left[\sqrt{\frac{L}{LT^{-2}}}\right]$$

$$= [\sqrt{T^2}]$$

$$= [T]$$

Since the dimensions of both sides are same, it is again a dimensionally homogeneous equation. 2π being a constant has a dimension of unity. Rectangular brackets are used to indicate that the dimensions of the quantities are being considered. Some other examples of dimensionally homogeneous equations are

- $Q = A \cdot V$ Discharge equation
- $\Delta p = \dfrac{32\, \mu V l}{d^2}$ Laminar flow in pipes
- $h_f = \dfrac{4\, f l V^2}{2gd}$ Darcy-Weisbach equation
- $Q = \dfrac{2}{3} C_d \sqrt{2g} \cdot L H^{3/2}$ Flow over rectangular weir

A dimensionally homogeneous equation can always be reduced to a non-dimensional form. For example, the above equation may be transformed into non-dimensional forms as given below :

- $\dfrac{Q}{AV} = 1$
- $\dfrac{\Delta p \cdot d}{\mu V} = 32 \dfrac{l}{d}$
- $\dfrac{h_f}{V^2/g} = 2f \dfrac{l}{d}$
- $\dfrac{Q}{\sqrt{g} \cdot L H^{3/2}} = \dfrac{2\sqrt{2}}{3} C_d$

It may also be observed that some of the equations are transformed into non-dimensional forms in more than one way. The second equation can also be written as

$$\frac{(\Delta p)\, d^2}{\mu V l} = 32$$

which is a dimensionless form. In general, the number of dimensionless quantities that can be obtained is equal to the number of variables involved minus the number of fundamental quantities involved in a phenomenon.

There can be some equations which are not dimensionally homogeneous. They are usually the equations that give empirical relations such as

$$N_S = \frac{N\sqrt{P}}{H^{5/4}}$$ Specific speed of turbine

$$Q = 1{\cdot}84\,(L - 0{\cdot}1\, nH)\, H^{3/2}$$ Francis formula

$$V = \frac{1}{N} R^{2/3} \sqrt{i}$$ Manning equation

It can be verified that the dimensions on the two sides of the equations are not the same. Thus, they are dimensionally non-homogeneous equations. These equations are valid only in a particular system of units.

8.4 DIMENSIONAL ANALYSIS METHODS

There are two methods of dimensional analysis :

1. Rayleigh method

2. Buckingham π-method

1. Rayleigh Method

The following procedure is adopted :

- Select independent variables likely to affect the value of dependent variable.
- Write the functional equation in the following form,
  $X = f(X_1, X_2, X_3, ...)$
  where X is the dependent variable and $X_1, X_2, X_3, ...$ are the independent variables.
- Write the above equation in the form
  $X = K \,.\, X_1^a, X_2^b, X_3^c, ...$
  where K is a dimensionless constant and is determined either from physical characteristics of the problem or from experimental values.
- The values of $a, b, c, ...$ etc. are found by assuming the equation to be dimensionally homogeneous *i.e.* by comparing the powers of the primary dimensions on both sides.
- If the number of exponents to be evaluated is more than equations formed, express the values of some exponents in terms of the others.

Thus, the expression for the dependent variable is obtained. However, this method is useful only where independent variables are not more than three or four. To obtain an expression for the dependent variable for higher number of independent variables, it is a difficult method.

2. Buckingham π-Method

As stated above, Releigh method is not helpful when the number of independent variables is more than three or four. This difficulty is eliminated in Buckingham π-method which is equally useful irrespective of the number of independent variables.

According to this method, if a physical phenomenon involves n quantities or variables, both independent and dependent and m fundamental dimensions, the n quantities can be arranged into $(n - m)$ independent dimensionless terms or groups. Each term is called a π-term. Let $X_1, X_2, X_3, ..., X_n$ be the variables involved in a physical phenomenon. The variables may be denoting velocity, pressure, viscosity etc. The functional relationship between them may be expressed as

$$f'(X_1, X_2, X_3, ..., X_n) = 0$$

According to π-theorem, the quantities X_1, X_2, X_3 etc. may be arranged into $(n - m)$ dimensionless independent groups. The functional relationship of the π-terms can be written as,

$$f(\pi_1, \pi_2, \pi_3, ..., \pi_{n-m}) = 0$$

Each of the π-terms is dimensionless, independent of the system and contains $(m + 1)$ variables. The proof of this theorem is beyond the scope of this book.

Method of Determining π-Terms

Following procedure may be adopted :

- Select m number of quantities out of the n quantities of variables such that they have different dimensions and contain m fundamental dimensions among them, not necessarily in all the quantities. Generally, the dependent variable is not chosen for the purpose.
- Use this selected m number of quantities as repeating variables to determine π-terms *i.e.* each π-term is found by associating the repeating variables with one of the other quantities.

  Let X_1, X_2 and X_3 be the repeating variables and have three fundamental dimensions of M, L and T in them collectively. Then the π-terms can be written as,

$$\pi_1 = X_1^{a_1} . X_2^{b_1} . X_3^{c_1} . X_4$$

$$\pi_2 = X_1^{a_2} . X_2^{b_2} . X_3^{c_2} . X_5$$

$$\pi_3 = X_1^{a_3} . X_2^{b_3} . X_3^{c_3} . X_6$$

....

....

$$\pi_{n-m} = X_1^{a_{n-m}} . X_2^{b_{n-m}} . X_3^{c_{n-m}} . X_n$$

- Find the exponents of each π-term on the assumption that each term is dimensionless, *i.e.* by putting dimensions of X quantities in the above equations. On the left-hand side, the exponents of M, L and T are zero.
- For each π-term, three equations with three unknowns are obtained, which are solved simultaneously to determine the unknowns. For a phenomenon involving only two fundamental dimensions, only two repeating variables are needed. Then each π-term would provide two simultaneous equations to be solved to determine the unknowns.

Sometimes, a dimensionless group can be obtained by judgement only. An angle as a variable is a π-term. Similarly, the ratio of two quantities having same dimensions gives a π-term.

Selection of Repeating Variables

The following points should be kept in mind while selecting the repeating variables :

- They must contain all the fundamental dimensions involved in the phenomenon collectively. They may not conttain all the dimensions in each of them.
- Any two of the repeating variables should not have the same dimensions *i.e.* they should not make a dimensionless group among themselves.
- Usually, a dependent variable is not taken as a repeating variable.
- In fluid flow problems, preferably choose one variable out of geometric characteristics, *i.e.* L, D, H etc. One out of kinematic or flow characteristics (related to time) *i.e.* velocity, acceleration etc. and one variable out of dynamic characteristics (related to force or mass) or fluid properties *i.e.* ρ, μ etc.
- While choosing a repeating variable out of fluid properties, preferably choose one with

more influence on the phenomenon. In a viscous flow through pipes, choose viscosity μ as a repeating variable out of ρ and μ but in a turbulent flow μ is not so important and thus ρ should be selected as a repeating variable.

Alteration in π-Terms

To derive a given particular result or equation for a phenomenon, following operations may be performed :

- Any π-term may be multiplied by another π-term to get a single π-term.
- Any π-term may be multiplied by any power of the term. For example π_1 may be replaced by any one of the following :

$$\pi_1^{2}, \pi_1^{-2}, \pi_1^{1/2}, \pi_1^{-1/2} \text{ etc.}$$

- Any π-term may be replaced by a new π-term obtained by multiplying or dividing the previous π-term by a numerical constant. For example π_1 may be replaced by $3\pi_1$ or $\pi_1/5$ etc.

It is to be noted that the maximum number of dimensionless groups n_m that can be obtained for any phenomenon with n variables and m fundamental dimensions is equal to

$$\frac{\lfloor n}{\lfloor m+1 \cdot \lfloor n-m-1}$$

If $n = 6$ and $m = 3$, then the maximum number of dimensionless groups is 15. But some of the dimensionless groups will not be independent. According to π-theorem, there are minimum of $(n - m)$ independent dimensionless groups in dimensional analysis. Assume that the pressure difference δp in a pipe depends upon four variables, diameter D, length L, velocity V and viscosity μ. Here $n = 5$ and $m = 3$ and thus the maximum number of dimensionless groups is 5. These groups are

(i) $\dfrac{\delta pL}{\mu V}$ (ii) $\dfrac{\delta pD}{\mu V}$ (iii) $\dfrac{\delta pD^2}{\mu VL}$ (iv) $\dfrac{\delta pL^2}{\mu VD}$ (v) $\dfrac{D}{L}$

Out of these five dimensionless groups, the second is a combination of the first and the last terms whereas the third is a combination of the second and the last terms. Thus these are not independent dimensionless groups. The others are the possible independent dimensionless groups. By using dimensional analysis, two $(n - m = 2)$ of these dimensionless groups may be obtained. These are either the first and the last groups or the second and the last depending upon the choice of repeating variables.

8.5 CHOICE OF VARIABLES

To investigate a physical phenomenon, the first requirement is to select the variables that affect the phenomenon. Usually, this is not a simple task. In any problem on the fluid flow, some of the variables are easily known whereas the others may be doubtful. For example in a centrifugal pump, the rate of discharge Q is a function of density ρ, viscosity μ and pressure p of the fluid and diameter D and speed N of the impeller. The result obtained by dimensional analysis is

$$\frac{Q}{ND^3} = f\left(\frac{p}{\rho N^2 D^2}, \frac{\mu}{\rho N D^2}\right)$$

Now if somehow the viscosity μ of the fluid is not included in the list of variables, the result obtained will be

$$\frac{Q}{ND^3} = f\left(\frac{p}{\rho N^2 D^2}\right)$$

i.e. the last dimensionless group is missing. When the curve of Q/ND^3 *vs* . $p/\rho N^2 D^2$ is plotted, it does not come in an orderly way. Points are found scattered on the graph meaninglessly (Fig. 8.1a). This happens as the parameter $\mu/\rho N D^2$ is also varied unknowingly along with $p/\rho N^2 D^2$ that leads to vague results. Such results indicate to the possibility of some more variables to exist. If this happens, a careful study of the phenomenon is again undertaken to find out the missing variables. Such parameters are called **omitted variables.**

Sometimes, it is also possible that some more variable, such as compressibility K of the fluid is also included in the list of variables which, in fact, is not required. In that case the result obtained by dimensional analysis will be

$$\frac{Q}{ND^3} = f\left(\frac{p}{\rho N^2 D^2}, \frac{\mu}{\rho N D^2}, \frac{K}{\rho N^2 D^2}\right)$$

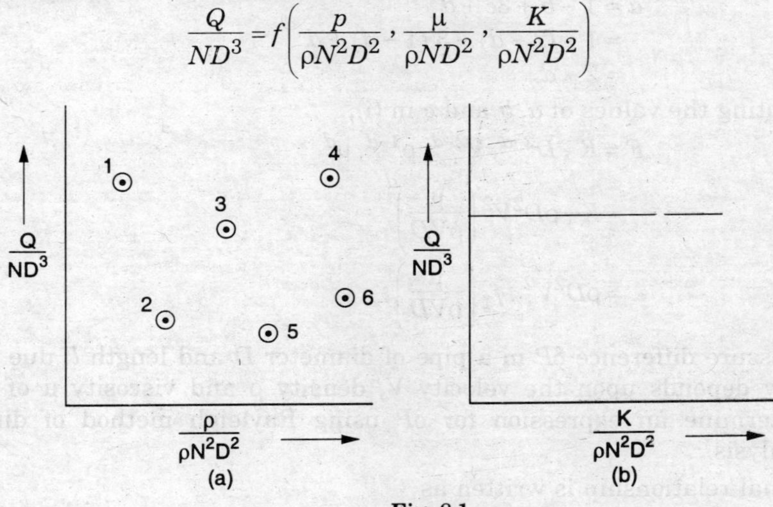

Fig. 8.1

i.e. the addition of another dimensionless group to the original result. When the graphs are plotted on the basis of a number of experiments, the plot of Q/ND^3 *vs* . $K/\rho N^2 D^2$, (keeping the other two dimensionless groups constant), is a straight line parallel to the axis of $K/\rho N^2 D^2$ (Fig. 8.1b). This shows that the compressibility K does not influence the phenomenon. Therefore, K may be omitted from the list of variables. Such variables are called **superfluous variables.**

Thus, it is concluded that there is no clear-cut method to find the variables which influence a phenomenon. One has to depend upon guesswork along with an insight into the phenomenon based on logic and experience.

Example 8.1 Find an expression for the relationship of drag force F on a smooth sphere of diameter D, moving with a uniform velocity V in a fluid of density ρ and dynamic viscosity μ. Use Rayleigh method.

Solution Using Rayleigh method, the functional relationship may be written as

$$F = f(D, V, \rho, \mu) = K \,.\, D^a, V^b, \rho^c, \mu^d \qquad \text{...(i)}$$

where K is a dimensionless constant. Substituting the dimensions on both sides,

$$MLT^{-2} = K \,.\, (L)^a \,.\, (LT^{-1})^b \,.\, (ML^{-3})^c \,.\, (ML^{-1}T^{-1})^d$$

Equating the powers of M, L and T,

Power of M, $1 = c + d$

Power of L, $1 = a + b - 3c - d$

Power of T, $-2 = -b - d$

Now, there are four unknowns and three equations. Thus, the exponents of three unknowns are to be expressed in terms of the fourth. In aerodynamics, the dynamic viscosity is very important variable. Therefore, expressing the other unknowns in terms of d,

$$c = 1 - d$$

$$b = 2 - d$$

$$\begin{aligned} a &= 1 - b + 3c + d \\ &= 1 - (2 - d) + 3(1 - d) + d \\ &= 2 - d \end{aligned}$$

Substituting the values of a, b and c in (i),

$$\begin{aligned} F &= K \,.\, D^{2-d}, V^{2-d}, \rho^{1-d}, \mu^d \\ &= K \,.\, \rho D^2 V^2 \left(\frac{\mu}{\rho V D}\right)^d \\ &= \rho D^2 V^2 \,.\, f_1\left(\frac{\mu}{\rho V D}\right) \end{aligned}$$

Example 8.2 Pressure difference δP in a pipe of diameter D and length l, due to viscous flow depends upon the velocity V, density ρ and viscosity μ of the fluid. Determine an expression for δP using Rayleigh method of dimensional analysis.

Solution Functional relationship is written as

$$\delta P = f(D, l, V, \rho, \mu) = KD^a \,.\, l^b \,.\, V^c \,.\, \rho^d \,.\, \mu^e \qquad \text{...(i)}$$

K is a dimensionless constant. Substituting the dimensions on both sides,

$$ML^{-1}T^{-2} = K(L)^a \,.\, (L)^b (LT^{-1})^c \,.\, (ML^{-3})^d \,.\, (ML^{-1}T^{-1})^e$$

Equate the powers of M, L and T,

Power of M, $1 = d + e$

Power of L, $-1 = a + b + c - 3d - e$

Power of T, $-2 = -c - e$

There are five unknowns and three equations. Express a, c and e in terms of b and d,

$$e = 1 - d$$

$$c = 2 - 1 + d = 1 + d$$

$$a = -1 - b - c + 3d + e = -1 - b - (1 + d) + 3d + (1 - d) = -1 - b + d$$

Substitute these values of a, c and e in (i)

$$\delta P = KD^{-1-b+d} \,.\, l^b \,.\, V^{1+d} \,.\, e^d \,.\, \mu^{1-d}$$

$$= K \,.\, \frac{\mu V}{D} \cdot \left(\frac{l}{D}\right)^b \cdot \left(\frac{\rho V D}{\mu}\right)$$

$$= \frac{\mu V}{D} \cdot f_1\left(\frac{l}{D}, \frac{\rho V D}{\mu}\right)$$

Experience shows that the pressure difference δP is proportional to the length l of the pipe,

$$\delta P = \frac{\mu V}{D} \cdot \frac{l}{D} f_2\left(\frac{\rho V D}{\mu}\right) = \frac{\mu V l}{D^2} \cdot f_2\,(R_e)$$

Also,

$$h_f = \frac{\delta P}{w} = \frac{\mu V l}{D^2} \cdot \frac{1}{w} f_2\,(R_e) = \frac{\mu V l}{w D^2} f_2\,(R_e)$$

Example 8.3 A partially submerged ship moves in water. The resistance force R depends upon length of the ship l, speed V, mass density ρ, acceleration due to gravity g and viscosity of water μ. Show that R is given by

$$R = \rho l^2 V^2 f\left(\frac{\mu}{\rho V l}, \frac{gl}{V^2}\right)$$

Solution Using Rayleigh method, functional relationship is written as

$$R = f'\,(l, V, g, \rho, \mu) = K \,.\, l^a \,.\, V^b \,.\, g^c \,.\, \rho^d \,.\, \mu^e \qquad \text{...(i)}$$

where K is a dimensionless constant. Substituting the dimensions on both sides.

$$MLT^{-2} = K(L)^a \,.\, (LT^{-1})^b \,.\, (LT^{-2})^c \,.\, (ML^{-3})^d \,.\, (ML^{-1}T^{-1})^e$$

Equating the powers of M, L and T

Power of M, $1 = d + e$

Power of L, $1 = a + b + c - 3d - e$

Power of T, $-2 = -b - 2c - e$

There are five unknowns and three equations. Expressing a, b and d in terms of c and e,

$$d = 1 - e$$

$$b = 2 - 2c - e$$

$$a = 1 - b - c + 3d + e = 1 - (2 - 2c - e) - c + 3\,(1 - e) + e = 2 + c - e$$

Substituting these values in (i),

$$R = K \,.\, l^{2+c-e} \,.\, V^{2-2c-e} \,.\, g^c \,.\, \rho^{1-e} \,.\, \mu^e$$

$$= K \,.\, \rho l^2 V^2 \left(\frac{\mu}{\rho V l}\right)^c \cdot \left(\frac{gl}{V^2}\right)^e$$

$$= \rho l^2 V^2 f\left(\frac{\mu}{\rho V l}, \frac{gl}{V^2}\right)$$

Example 8.4 Determine the form of equation for the discharge Q through a sharp edged triangular notch that depends upon the notch head H, acceleration due to gravity g, central angle α and velocity of approach V. Use Buckingham π-method.

Solution Functional relationship may be written as

$$Q = f'(H, g, V, \alpha)$$

or $$f''(Q, H, g, V, \alpha) = 0$$

Number of variables, $n = 5$

Number of fundamental dimensions, $m = 2$

Number of repeating variables $= n - m = 5 - 2 = 3$

Number of π-terms, $= 3$

Functional relationship of π-terms may be written as

$$f(\pi_1, \pi_2, \pi_3) = 0 \quad \text{...(i)}$$

Let g and H be the repeating variables, then

$$\pi_1 = g^{a_1} . H^{b_1} . Q$$

$$\pi_2 = g^{a_2} . H^{b_2} . V$$

$$\pi_3 = \alpha$$

As α is dimensionless itself, the term π_3 is directly written as equal to α. However, it may be written in the usual way $\pi_3 = g^{a_3} . H^{b_3} . \alpha$. In the end it will be found that the superscripts a_3 and b_3 come out to be zero.

Inserting the fundamental dimensions,

$$\pi_1 = g^{a_1} . H^{b_1} . Q$$

$$L^0T^0 = (LT^{-2})^{a_1} . (L)^{b_1} . (L^3T^{-1})$$

Equating the powers of L and T,

$$0 = a_1 + b_1 + 3$$

$$0 = -2a_1 - 1$$

Solving,

$$a_1 = -\frac{1}{2} \quad \text{and} \quad b_1 = -\frac{5}{2}$$

$$\pi_1 = \frac{Q}{\sqrt{g}H^{5/2}}$$

Similarly,

$$\pi_2 = g^{a_2} . H^{b_2} . V$$

$$L^0T^0 = (LT^{-2})^{a_2} . (L)^{b_2} . (LT^{-1})$$

Equating the powers of L and T,

Power of L, $0 = a_2 + b_2 + 1$

Power of T, $0 = -2a_2 - 1$

Solving,

$$a_2 = -\frac{1}{2} \quad \text{and} \quad b_1 = -\frac{1}{2}$$

$$\therefore \quad \pi_2 = \frac{V}{\sqrt{gH}}$$

$\therefore$ (i) becomes,

$$f\left(\frac{Q}{\sqrt{g}\,H^{5/2}}, \frac{V}{\sqrt{gH}}, \alpha\right) = 0$$

$$\frac{Q}{\sqrt{g}\,H^{5/2}} = f_1\left(\frac{V}{\sqrt{gH}}, \alpha\right)$$

$$Q = \sqrt{g}\,H^{5/2} \,.\, f_1\left(\frac{V}{\sqrt{gH}}, \alpha\right)$$

$$= CH^{5/2}$$

where $\quad C = \sqrt{g} \,.\, f_1\left(\dfrac{V}{\sqrt{gH}}, \alpha\right)$

which is the required from of equation.

Suppose we had taken H and V as the repeating variables instead of g and H. Then

$$\pi_1 = H^{a_1} \,.\, V^{b_1} \,.\, Q$$

$$\pi_2 = H^{a_2} \,.\, V^{b_2} \,.\, V$$

$$\pi_3 = \alpha$$

Inserting the fundamental dimensions,

$$\pi_1 = H^{a_1} \,.\, V^{b_1} \,.\, Q$$

$$L^0T^0 = (L)^{a_1} \,.\, (LT^{-1})^{b_1} \,.\, (L^3T^{-1})$$

Equating the powers of L and T,

Power of L, $\quad 0 = a_1 + b_1 + 3$

Power of T, $\quad 0 = -b_1 - 1$

solving,

$$a_1 = -2 \quad \text{and} \quad b_1 = -1$$

$$\pi_1 = \frac{Q}{H^2V}$$

Similarly,

$$\pi_2 = H^{a_2} \,.\, V^{b_2} \,.\, g$$

$$L^0T^0 = (L)^{a_2} \,.\, (LT^{-1})^{b_2} \,.\, (LT^{-2})$$

Equating the powers of L and T,

$$0 = a_2 + b_2 + 1$$

$$0 = -b_2 - 2$$

solving,

$$a_2 = 1 \quad \text{and} \quad b_2 = -2$$

and $$\pi_2 = \frac{gH}{V^2}$$

and (i) becomes,

$$f\left(\frac{Q}{H^2V}, \frac{gH}{V^2}, \alpha\right) = 0$$

$$f_2\left(\frac{Q}{H^2V}, \frac{V^2}{\sqrt{gH}}, \alpha\right) = 0$$

$$\frac{Q}{H^2V} = f_3\left(\frac{V^2}{\sqrt{gH}}, \alpha\right)$$

$$Q = VH^2 \cdot f_3\left(\frac{V}{\sqrt{gH}}, \alpha\right)$$

It is observed from the two forms of Q that the unknown function f_3 has the same parameters as f_1, but it cannot be the same function. The latter form of Q is not very useful, as the velocity of approach is usually, neglected in case of V-notches and thus the expression becomes zero. Therefore, it is recommended that a term of minor importance should not be selected as a repeating variable.

Example 8.5 A fluid of mass density ρ and viscosity μ flows with velocity V through a pipe of diameter D, length l and roughness k of the surface. Using dimensional analysis show that the head loss due to friction h_f is given by

$$h_f = \frac{\Delta p}{w} = \frac{4flV^2}{2gD} \text{ where } f \text{ is a coefficient of friction.}$$

Solution Functional relationship can be written as

$$\Delta p = f'(D, l, V, k, \rho, \mu)$$

or $f''(\Delta p, D, l, V, k, \rho, \mu) = 0$

Number of variables $= 7$

Number of fundamental dimensions, $m = 3$

Number of repeating variables $= m = 3$

Number of π-terms, $(n - m) = 7 - 3 = 4$

Functional relationship may be expressed as

$$f_1(\pi_1, \pi_2, \pi_3, \pi_4) = 0 \quad \text{...(i)}$$

Let D, V and ρ be the three repeating variables.

Therefore,

$$\pi_1 = D^{a_1} \,.\, V^{b_1} \,.\, \rho^{c_1} \,.\, \Delta p$$

$$\pi_2 = D^{a_2} \,.\, V^{b_2} \,.\, \rho^{c_2} \,.\, l$$

$$\pi_3 = D^{a_3} . V^{b_3} . \rho^{c_3} . k$$

$$\pi_4 = D^{a_4} . V^{b_4} . \rho^{c_4} . \mu$$

Inserting the dimensions,

$$\pi_1 = D^{a_1} . V^{b_1} . \rho^{c_1} . \Delta p$$

$$M^0L^0T^0 = (L)^{a_1} . (LT^{-1})^{b_1} . (ML^{-3})^{c_1} . (ML^{-1}T^{-2})$$

Equating the powers of M, L and T,

Power of M, $0 = c_1 + 1$

Power of L, $0 = a_1 + b_1 - 3c_1 - 1$

Power of T, $0 = -b_1 - 2$

Solving these equations,

$$a_1 = 0, b_1 = -2 \quad \text{and} \quad c_1 = -1$$

$$\pi_1 = D^0 . V^{-2} . \rho^{-1} . \Delta p = \frac{\Delta p}{\rho V^2}$$

Similarly,

$$\pi_2 = D^{a_2} . V^{b_2} . \rho^{c_2} . l$$

$$M^0L^0T^0 = (L)^{a_2} . (LT^{-1})^{b_1} . (ML^{-3})^{c_2} . L$$

Equating the powers of M, L and T,

Power of M, $0 = c_2$

Power of L, $0 = a_2 + b_2 - 3c_2 + 1$

Power of T, $0 = -b_2$

Solving these equations,

$$a_2 = 0, b_2 = -2 \quad \text{and} \quad c_2 = -1$$

$$\therefore \quad \pi_2 = D^{-1} . V^0 . \rho^0 . l = \frac{l}{D}$$

$$\pi_3 = D^{a_3} . V^{b_3} . \rho^{c_3} . k$$

$$M^0L^0T^0 = (L)^{a_3} . (LT^{-1})^{b_3} . (ML^{-3})^{c_3} . L$$

Equating the powers of M, L and T,

Power of M, $0 = c_3$

Power of L, $0 = a_3 + b_3 - 3c_3 + 1$

Power of T, $0 = -b_3$

Solving these equations,

$$a_3 = 0, b_3 = -2 \quad \text{and} \quad c_3 = -1$$

$$\therefore \quad \pi_3 = \frac{k}{D}$$

$$\pi_4 = D^{a_4} . V^{b_4} . \rho^{c_4} . \mu$$

$$M^0L^0T^0 = (L)^{a_4} . (LT^{-1})^{b_4} . (ML^{-3})^{c_4} . ML^{-1}T^1$$

Equating the powers of L and T,

Power of M, $0 = c_4 + 1$

Power of L, $0 = a_4 + b_4 - 3c_4 - 1$

Power of T, $0 = -b_4 - 1$

Solving these equations,

$$a_4 = -1, b_4 = -1 \quad \text{and} \quad c_4 = -1$$

$$\pi_4 = \frac{\mu}{\rho VD}$$

Therefore, (i) becomes,

$$f_1\left(\frac{\Delta p}{\rho V^2}, \frac{l}{D}, \frac{k}{D}, \frac{\mu}{\rho VD}\right) = 0$$

$$f_2\left(\frac{\Delta p}{\rho V^2}, \frac{l}{D}, \frac{k}{D}, \frac{\rho VD}{\mu}\right) = 0 \qquad \text{(Reversing } \pi_4\text{)}$$

$$\frac{\Delta p}{\rho V^2} = f_3\left(\frac{l}{D}, \frac{k}{D}, \frac{\rho VD}{\mu}\right)$$

Experiments show that the pressure coefficient $\Delta p/\rho V^2$ is a function of l/D.

$$\frac{\Delta p}{\rho V^2} = \frac{l}{D} f_4\left(\frac{k}{D}, \frac{\rho VD}{\mu}\right)$$

$$\Delta p = \frac{\rho l V^2}{D} f_4\left(\frac{k}{D}, \frac{\rho VD}{\mu}\right) = \frac{\rho l V^2}{D} f_4\left(\frac{k}{D}, R_e\right)$$

$$h_f = \frac{\Delta p}{w} = \frac{\Delta p}{\rho g} = \frac{\rho l V^2}{\rho g D} f_4\left(\frac{k}{D}, R_e\right)$$

$$= \frac{4}{2} \cdot \frac{lV^2}{gD} \frac{1}{2} f_2\left(\frac{k}{D}, R_e\right)$$

$$= \frac{4flV^2}{2gD}$$

where f is a coefficient of friction and is given by

$$f = \frac{1}{2} f_4\left(\frac{k}{D}, R_e\right)$$

a function of Reynolds number and roughness factor.

Example 8.6 The discharge Q, through a rotating system depends upon head H, speed N, power P, density ρ, viscosity μ and diameter D. Show that

$$\frac{Q}{ND^3} = f\left(\frac{gH}{N^2D^2}, \frac{P}{\rho gHND^3}, \frac{\mu}{\rho ND^2}\right)$$

Solution Functional relationship can be written as

$$Q = f'(H, P, N, D, \rho, \mu, g)$$

or

$$f''(Q, H, P, N, D, \rho, \mu, g) = 0$$

Number of variables, $n = 8$
Number of fundamental dimensions, $m = 3$
Number of repeating variables $= m = 3$
Number of π-terms $= n - m = 5$.
Assume D, ρ and N to be the repeating variables.

$$\pi_1 = D^{a_1} \rho^{b_1} N^{c_1} . Q$$

$$M^0L^0T^0 = (L)^{a_1} . (ML^{-3})^{b_1} . (T^{-1})^{c_1} L^3T^{-1}.$$

Equating the powers of M, L and T,

For M, $0 = b_1$

For L, $0 = a_1 - 3b_1 + 3$

For T, $0 = - c_1 - 1$

On solving $a_1 = - 3, b_1 = 0$ and $c_1 = - 1$

$$\therefore \quad \pi_1 = \frac{Q}{ND^3}$$

Similarly, $\pi_2 = D^{a_2} \rho^{b_2} N^{c_2} . H$

$$M^0L^0T^0 = (L)^{a_2} (ML^{-3})^{b_2} . (T^{-1})^{c_2} . L$$

$$0 = b_2$$

$$0 = a_2 - 3b_2 + 1$$

$$0 = - c_2$$

Thus $a_2 = - 1, b_2 = 0$ and $c_2 = 0$

$$\therefore \quad \pi_2 = \frac{H}{D}$$

(The term π_2 can also be written directly by observation)

$$\pi_3 = D^{a_3} . \rho^{b_3} . N^{c_3} . P$$

$$M^0L^0T^0 = (L)^{a_3} . (ML^{-3})^{b_3} . (T^{-1})^{c_3} . ML^2T^{-3}$$

$$0 = b_3 + 1$$

$$0 = a_3 - 3b_3 + 2$$

$$0 = - c_3 - 3$$

On solving $a_3 = - 5, b_3 = - 1$ and $c_3 = - 3$

$$\pi_3 = \frac{P}{\rho N^3 D^5}$$

$$\pi_4 = D^{a_4} \rho^{b_4} N^{c_4} . \mu$$

$$M^0L^0T^0 = (L)^{a_4} . (ML^{-3})^{b_4} (T^{-1})^{c_4} . ML^{-1}T^{-1}$$

$$0 = b_4 + 1$$

$$0 = a_4 - 3b_4 - 1$$

$$0 = -c_4 - 1$$

On solving $a_4 = -2,\ b_4 = -1$ and $c_4 = -1$

$$\pi_4 = \frac{\mu}{\rho N D^2}$$

$$\pi_5 = D^{a_5} . \rho^{b_5} . N^{c_5} . g$$

$$M^0 L^0 T^0 = (L)^{a_5} . (ML^{-3})^{b_5} (T^{-1})^{c_5} (LT^{-2})$$

$$0 = b_5$$

$$0 = a_5 - 3b_5 + 1$$

$$0 = -c_5 - 2$$

or $a_5 = -1,\ b_5 = 0$ and $c_5 = -2.$

$$\therefore \quad \pi_5 = \frac{g}{DN^2}$$

Thus functional relationship is written as,

$$f_1\left(\frac{Q}{ND^3}, \frac{H}{D}, \frac{P}{\rho N^3 D^5}, \frac{\mu}{\rho N D^2}, \frac{g}{DN^2}\right) = 0$$

Combining the second and the last terms,

$$f_2\left(\frac{Q}{ND^3}, \frac{gH}{N^2 D^2}, \frac{P}{\rho N^3 D^5}, \frac{\mu}{\rho N D^2}\right) = 0$$

Dividing the third term by the second term,

$$f_3\left(\frac{Q}{ND^3}, \frac{gH}{N^2 D^2}, \frac{P}{\rho g H N D^3}, \frac{\mu}{\rho N D^2}\right) = 0$$

or

$$\frac{Q}{ND^3} = f\left(\frac{gH}{N^2 D^2}, \frac{P}{\rho g H N D^3}, \frac{\mu}{\rho N D^2}\right) = 0$$

which is the required expression.

- $\mu/\rho ND^2$ is a form of Reynolds number. It can be written in the form $\rho ND^2/\mu$ or $\rho VD/\mu$, because $V \propto u \propto ND$.
- g has not been mentioned as a variable in the given statement; however it has been included because it occurs in the result. Had g not been taken, the result would have been,

$$f_4\left(\frac{Q}{ND^3}, \frac{H}{D}, \frac{P}{\rho N^3 D^5}, \frac{\mu}{\rho N D^2}\right) = 0$$

- It is to be noted, that gH also could have been taken as a single variable indicating the energy per unit mass.
- Also g and H could have been replaced by pressure p, because $p/\rho = gH$.

  The result would have been,

$$f_5\left(\frac{Q}{ND^3}, \frac{p}{\rho N^2 D^2}, \frac{P}{\rho N^3 D^5}, \frac{\mu}{\rho N D^2}\right) = 0$$

But $$\frac{p}{\rho N^2 D^2} = \frac{\rho g H}{\rho n^2 D^2} = \frac{gH}{N^2 D^2}$$

i.e. the same result.

Example 8.7 The diameter and the width of the runner of a turbine are D and B respectively and it rotates at speed N. The working head is H. If density and the viscosity of the working fluid be ρ and μ respectively, show that the power developed is given by,

$$P = \rho N^3 D^5 f\left(\frac{B}{D}, \frac{ND}{\sqrt{gH}}, \text{Re}\right)$$

Solution Functional relationship is,

$$P = f'(H, N, D, \rho, \mu, g, B)$$

or $$f''(P, H, N, D, \rho, \mu, g, B).$$

The problem is similar to the previous example. Q is missing here, therefore a π-term containing Q can be deleted from the solution of the previous example. Width B has been added here, for which a π-term has to be added. The π-term can be directly written as B forms a dimensionless number with D.

$$f_1\left(\frac{B}{D}, \frac{gH}{N^2D^2}, \frac{\mu}{\rho N D^2}, \frac{P}{\rho N^3 D^5}\right) = 0$$

$$f_2\left(\frac{B}{D}, \frac{N^2D^2}{gH}, \frac{\rho N D^2}{\mu}, \frac{P}{\rho N^3 D^5}\right) = 0$$

(Reversing the 2nd and the 3rd π-terms)

$$f_3\left(\frac{B}{D}, \frac{ND}{\sqrt{gH}}, Re, \frac{P}{\rho N^3 D^5}\right) = 0$$

(Taking square-root of the 2nd π-term)

$$\frac{P}{\rho N^3 D^5} = f\left(\frac{B}{D}, \frac{ND}{\sqrt{gH}}, Re\right)$$

$$P = \rho N^3 D^5 f\left(\frac{B}{D}, \frac{ND}{\sqrt{gH}}, Re\right)$$

Example 8.8 (a) Determine an expression for the specific speed of a turbine.

(b) Deduce an expression for the specific speed of a centrifugal pump.

Solution (a) We know from the definition of the specific speed of a turbine that it involves three variables, power P, speed N and head H. To obtain a relationship among these three variables, we need the following π-terms out of the four π-terms obtained in example 8.6.

i.e. $$f\left(\frac{gH}{N^2D^2}, \frac{P}{\rho N^3 D^5}\right) = 0$$

The result is obtained if the variables are chosen as P, H, N, D, ρ and g. In case more variables are added, more π-terms are obtained which can be deleted later on.

Combining the two π-terms to eliminate D,

$$\frac{(\text{2nd term})^{1/2}}{(\text{1st term})^{5/4}} = \frac{p^{1/2}}{(\rho N^3 D^5)^{1/2}} \cdot \frac{(N^2D^2)^{5/4}}{(gH)^{5/4}}$$

$$= \frac{N\sqrt{P}}{(gH)^{5/4} \,.\, \rho^{1/2}}$$

Thus for complete similarity between the model (hypothetical turbine) and the prototype, this dimensionless number must be same in the two,

i.e.
$$\left(\frac{N\sqrt{P}}{(gH)^{5/4} \,.\, \rho^{1/2}}\right)_m = \left(\frac{N\sqrt{P}}{(gH)^{5/4} \,.\, \rho^{1/2}}\right)_p$$

Assuming g and ρ to be the same for the model and the prototype, we can write,

$$\left(\frac{N\sqrt{P}}{H^{5/4}}\right)_m = \left(\frac{N\sqrt{P}}{H^{5/4}}\right)_p$$

Now, specific speed of a turbine is defined as the speed of a geometrically similar turbine (model) which produces unit power when working under unit head.

Thus $N = N_s$, when $P = 1$ and $H = 1$ for the model.

i.e.
$$N_s = \frac{N\sqrt{P}}{H^{5/4}}$$

(b) On the same lines, expression for the specific speed of a pump can also be deduced. Essential π-terms needed are,

$$f\left(\frac{Q}{ND^3} \cdot \frac{gH}{N^2D^2}\right) = 0$$

Combining the terms to eliminate D,

$$\frac{(\text{1st term})^{1/2}}{(\text{2nd term})^{3/4}} = \frac{Q^{1/2}}{(ND^3)^{1/2}} \cdot \left(\frac{N^2D^2}{gH}\right)^{3/4}$$

$$= \frac{N\sqrt{Q}}{(gH)^{5/4}}$$

This should be same for the specific pump (model) and the prototype.

ie
$$\left(\frac{N\sqrt{Q}}{(gH)^{3/4}}\right)_m = \left(\frac{N\sqrt{Q}}{(gH)^{3/4}}\right)_p$$

Specific speed of a centrifugal pump is defined as the speed of a geometrically similar pump that delivers unit discharge and produces unit head.

Thus when $Q = 1$, $H = 1$, $N = N_s$ for the model.

Also assuming g to be the same for the two.

$$N_s = \frac{N\sqrt{Q}}{H^{3/4}}$$

Example 8.9 A fluid flow situation depends upon velocity V, density ρ, pressure drop Δp, gravity g, length l, diameter D, viscosity μ, surface tension σ and bulk modulus of elasticity K. Find a set of π-parameters with the help of dimensional analysis.

Solution Functional relationship can be written as,

$$f'(V, \rho\, \Delta p, g, l, \mu, \sigma, K, D) = 0$$

Number of variables $n = 9$

Number of fundamental dimensions, $m = 3$

Number of repeating variables $= m = 3$

Number of π-terms $= n - m$

$= 6$

Let l, V and ρ be the repeating variables.

Then
$$\pi_1 = l^{a_1} V^{b_1} \rho^{c_1} \Delta p$$
$$\pi_2 = l^{a_2} V^{b_2} \rho^{c_2} g$$
$$\pi_3 = l^{a_3} V^{b_3} \rho^{c_3} \mu$$
$$\pi_4 = l^{a_4} V^{b_4} \rho^{c_4} \sigma$$
$$\pi_5 = l^{a_5} V^{b_5} \rho^{c_5} K$$
$$\pi_6 = \frac{D}{l}$$ (Writing directly)

Inserting the dimensions and equating the powers of M, L and T,

$$\pi_1 = l^{a_1} V^{b_1} . \rho^{c_1} . \Delta p$$
$$M^0L^0T^0 = (L)^{a_1} . (LT^{-1})^{b_1} . (ML^{-3})^{c_1} . ML^{-1}T^2$$

For M, $c_1 + 1 = 0$

For L, $a_1 + b_1 - 3c_1 - 1 = 0$

For T, $-b_1 - 2 = 0$

On solving $a_1 = 0$, $b_1 = -2$ and $c_1 = -1$.

$$\therefore \quad \pi_1 = \frac{\Delta p}{\rho V^2}$$
$$\pi_2 = l^{a_2} . V^{b_2} . \rho^{c_2} . g$$
$$M^0L^0T^0 = (L)^{a_2} . (LT^{-1})^{b_2} . (ML^{-3})^{c_2} . LT^{-2}$$
$$c_2 = 0$$
$$a_2 + b_2 - 3c_2 + 1 = 0$$
$$-b_2 - 2 = 0$$

On solving $a_2 = 1$, $b_2 = -2$ and $c_2 = 0$

$$\therefore \quad \pi_2 = \frac{gl}{V^2}$$
$$\pi_3 = l^{a_3} V^{b_3} \rho^{c_3} \mu$$
$$M^0L^0T^0 = (L)^{a_3} . (LT^{-1})^{b_3} . (ML^{-3})^{c_3} . ML^{-1}T^{-1}$$
$$c_3 + 1 = 0$$
$$a_3 + b_3 - 3c_3 - 1 = 0$$

$$-b_3 - 1 = 0$$

On solving $a_3 = -1, b_3 = -1, c_3 = -1$

$$\pi_3 = \frac{\mu}{\rho V l}$$

$$\pi_4 = l^{a_4} V^{b_4} \rho^{c_4} \sigma$$

$$M^0L^0T^0 = (L)^{a_4} . (LT^{-1})^{b_4} (ML^{-3})^{c_4} . MT^{-2}$$

$$c_4 + 1 = 0$$

$$a_4 + b_4 - 3c_4 = 0$$

$$-b_4 - 2 = 0$$

$$a_4 = -1, b_4 = -2, c_4 = -1$$

$$\pi_4 = \frac{\sigma}{\rho l V^2}$$

$$\pi_5 = l^{a_5} . V^{b_5} . \rho^{c_5} . k$$

$$M^0L^0T^0 = (L)^{a_5} . (LT^{-1})^{b_5} . (ML^{-3})^{c_5} . ML^{-1}T^{-2}$$

$$c_5 + 1 = 0$$

$$a_5 + b_5 - 3c_5 - 1 = 0$$

$$-b_5 - 2 = 0$$

$$a_5 = 0, b_5 = -2, c_5 = -1$$

$$\pi_5 = \frac{K}{\rho V^2}$$

Hence $f\left(\frac{\Delta p}{\rho V^2}, \frac{gl}{V^2}, \frac{\mu}{\rho V l}, \frac{\sigma}{\rho l V^2}, \frac{k}{\rho V^2}, \frac{D}{l}\right) = 0$

or $f_1\left(\frac{V}{\sqrt{(\Delta p/\rho)}}, \frac{V}{\sqrt{lg}}, \frac{\rho V l}{\mu}, \frac{V}{\sqrt{\sigma/gl}}, \frac{V}{\sqrt{K/\rho}}, \frac{D}{l}\right) = 0$

(Reversing and square-root of 1st, 2nd, 4th and 5th terms and reversing only of 3rd terms)

or $f_1\left(\text{Eu, Fr, Re, We, M}, \frac{D}{l}\right) = 0$

where Eu is the Euler number
Fr is the Froude number
We is the Weber number
Re is the Reynolds number
M is the Mach number.

These are dimensionless numbers and are defined in section 8.8.

The functional relationship of π-terms can also be written as,

$$\Delta p = \rho V^2 f_2 \left(\text{Fr, Re, We, M}, \frac{D}{l} \right)$$

Thus Euler number is nothing but a pressure coefficient and in most of the cases, the pressure coefficient or the pressure difference is associated with other forces, *i.e.* it is a function of other dimensionless numbers and is a dependable variable. In brief, pressure coefficient is a function of,

Re and k — in case of pipes
Re and M — if compressible fluids are used
Re, We and M — for small size orifices with compressible fluids
Re and M — for large size orifices with compressible fluids
Re — for large size orifices with incompressible fluids
and M — for large pressure drops with low viscosity fluids.

8.6 DATA PRESENTATION

The dimensional analysis is useful in reducing the number of variables by forming independent dimensionless groups. If the number of dimensionless groups so obtained is two, the relationship between them is shown by a curve as shown in Fig. 8.2a. If they are three, a number of curves are plotted between any two groups keeping the third constant at different values (Fig. 8.2b). If the number of dimensionless groups is four, say π_1, π_2, π_3 and π_4, the relationship among them is plotted as given below :

- Keep π_4 constant at some particular value, say $p_4 = C_1$. Draw a number of curves π_1 vs. π_2 for different values of π_3 as shown in Fig. 8.2c.
- Keep π_4 constant at some other particular value, say $\pi_4 = C_2$. Repeat curves π_1 *vs.* π_2 for the same values of π_3 (Fig. 8.2d).
- Keep π_4 constant at some other particular values, say $\pi_4 = C_3, C_4$ ... and draw the curves π_1 vs. π_2 as before (Fig. 8.2e and f).

8.7 DYNAMIC FORCES

The following are known as **dynamic forces** that influence a physical phenomenon :

Inertia Force

It is the force resisting the acceleration of a fluid particle.

$$F_i = \text{Mass} \times \text{Acceleration}$$

$$= (\text{Density} \times \text{Volume}) \times \frac{\text{Velocity}}{\text{Time}}$$

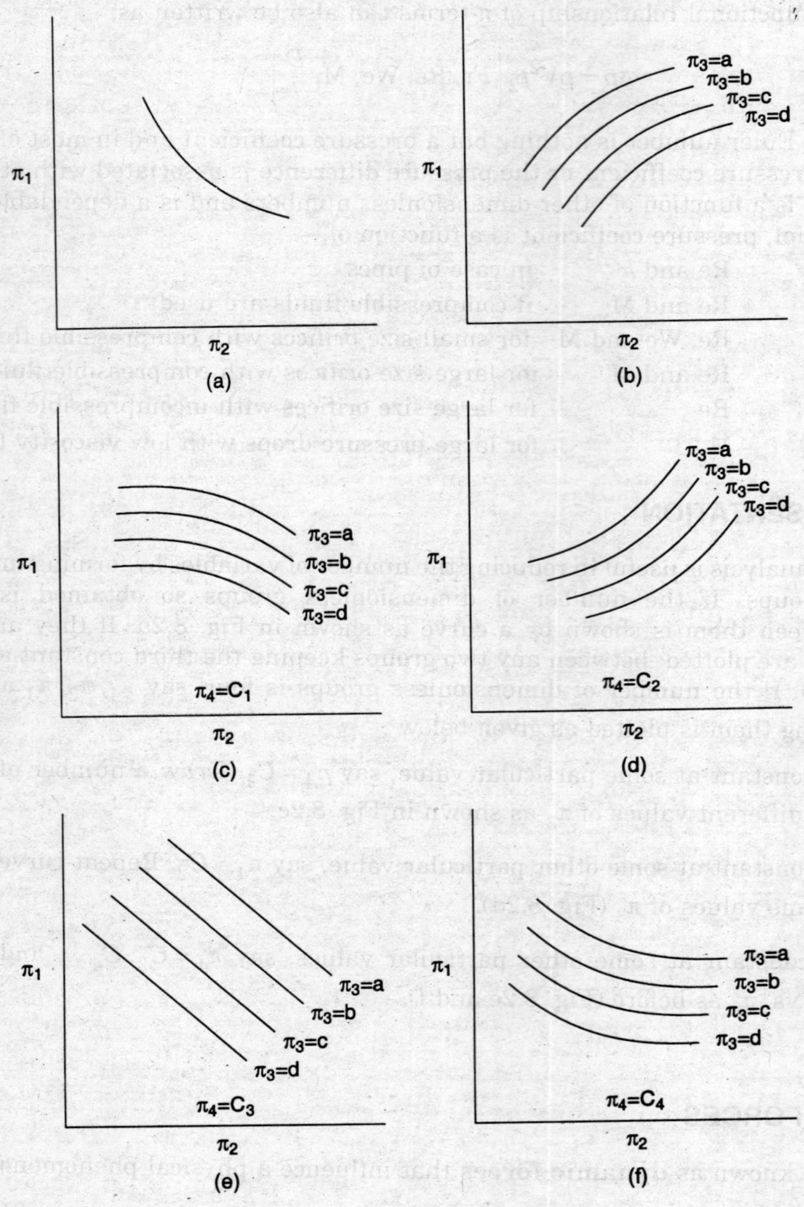

Fig. 8.2

$= \text{Density} \times \text{Discharge} \times \text{Velocity}$

$= \text{Density} \times (\text{Area} \times \text{Velocity}) \times \text{Velocity}$

$= \rho \,.\, L^2 \,.\, V^2$

Viscous Force

It is due to viscosity of the fluid.

$$F_v = \text{Shear stress} \times \text{Area}$$

$$= \left(\mu \cdot \frac{du}{dy}\right) \times A = \left(\mu \cdot \frac{V}{L}\right) \times L^2 = \mu VL$$

Pressure Force

$$F_p = \text{Pressure} \times \text{Area} = p \cdot A = p \cdot L^2$$

Gravity Force

$$F_g = \text{Mass} \times g = (\text{Density} \times \text{Volume}) \times g = \rho \cdot L^3 \cdot g$$

Surface Tension Force

It is due to surface tension effects.

$$F_s = \text{Surface tension per unit length} \times \text{Length} = \sigma \cdot L$$

Elastic Force

As the density varies due to pressure, there is elastic compression and the corresponding force is known as elastic force.

$$F_e = \text{Elastic stress} \times \text{Area} = K \cdot L^2$$

8.8 DIMENSIONLESS NUMBERS

In a fluid flow phenomenon, all the six forces mentioned above are rarely present at a time. Some of the forces may have more influence on a phenomenon than the other forces. These dominating forces govern the flow. However, inertia force is always present in a flow, as it is associated with the motion of the fluid particles. The viscous force predominates of all the other forces in a laminar pipe flow, the gravitational force in the flow through a nozzle and the elastic force in the supersonic flow.

The ratio of inertia force to each of the other forces provides a dimensionless number as discussed below :

1. Reynolds Number

It is the ratio of the inertia force to the viscous force of flowing fluid.

$$\text{Re} = \frac{\text{Inertia force}}{\text{Viscous force}} = \frac{\rho \cdot L^2 \cdot V^2}{\mu \cdot V \cdot L} = \frac{\rho \cdot L \cdot V}{\mu} = \frac{L \cdot V}{\mu/\rho} = \frac{LV}{\nu}$$

In pipe flows, L is replaced by diameter of the pipe and thus $\text{Re} = \dfrac{VD}{\nu}$.

2. Froude Number

It is defined as the square root of the ratio of inertia force to the gravity force.

$$\text{Fr} = \sqrt{\frac{\text{Inertia force}}{\text{Gravity force}}} = \sqrt{\frac{\rho \cdot L^2 \cdot V^2}{\rho \cdot L^3 \cdot g}} = \sqrt{\frac{V^2}{Lg}} = \frac{V}{\sqrt{L \cdot g}}$$

3. Euler Number

It is defined as the square root of the ratio of inertia force to the pressure force.

$$\text{Eu} = \sqrt{\frac{\text{Inertia force}}{\text{Pressure force}}} = \sqrt{\frac{\rho \,.\, L^2 \,.\, V^2}{p \,.\, L^2}} = \frac{V}{\sqrt{p/\rho}}$$

Twice of the ratio of the pressure force to the inertia force is called the **pressure coefficient.**

$$C_P = 2 \times \frac{\text{Pressure force}}{\text{Inertia force}} = 2 \times \frac{pL^2}{\rho L^2 V^2} = \frac{p}{\rho V^2/2}$$

4. Weber Number

It is defined as the square root of the ratio of the inertia force to the surface tension force.

$$\text{We} = \sqrt{\frac{\text{Inertia force}}{\text{Surface tensin force}}} = \sqrt{\frac{\rho \,.\, L^2 \,.\, V^2}{\sigma \,.\, L}} = \sqrt{\frac{\rho \,.\, L \,.\, V^2}{\sigma}} = \frac{V}{\sqrt{\sigma/\rho \,.\, L}}$$

5. Mach Number

It is the square root of the ratio of the inertia force of a flowing fluid to the elastic force.

$$\text{M} = \sqrt{\frac{\text{Inertia force}}{\text{Elastic force}}} = \sqrt{\frac{\rho \,.\, L^2 \,.\, V^2}{K \,.\, L^2}} = \sqrt{\frac{\rho \,.\, V^2}{K}} = \frac{V}{\sqrt{K/\rho}}$$

All the numbers except Reynolds number are the square roots of the ratios between the inertia force and each of the other forces respectively. Also, note that a higher value of a dimensionless number signifies the dominance of the inertia force. Thus, a higher value of the Reynolds number indicates a large inertia force compared to the viscous force and the flow to be non-viscous or turbulent. Similarly, a low value of the Reynolds number indicates a small inertia force compared to the viscous force and the flow is known as viscous or laminar.

8.9 MODEL TESTING

A lot of time, money and energy go into the design, construction and erection of hydraulic structures and machines. To minimise the chances of failure, it is desired that the tests be performed on small size models of the structures or the machines. In rare cases, the models can be of larger size also as the model of a watch.

A structure of machine of the actual size is called a **prototype** and a small replica of the prototype is known as its **model** (Fig. 8.3). Model testing is economical and easy as the design and the size of the model can conveniently be changed a number of times. The results obtained from the performance of the model, the performance of the prototype can be predicted. However, this is possible if similarity between the prototype and its model is completely established.

9.10 SIMILITUDE

Similitude is defined as the complete similarity between the prototype and its model. It is attained if the following three type of similarities are established :

1. Geometric similarity

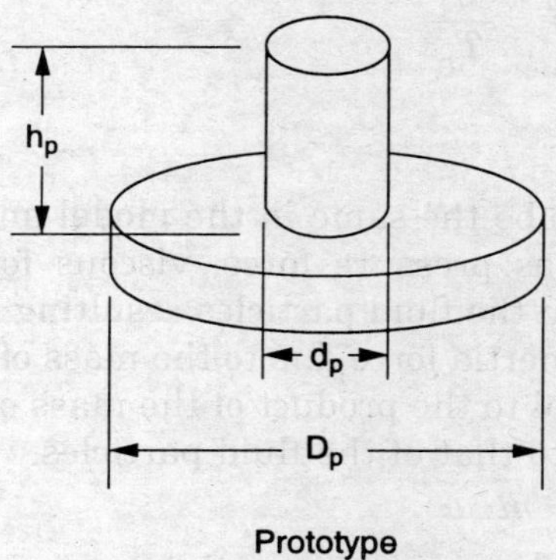

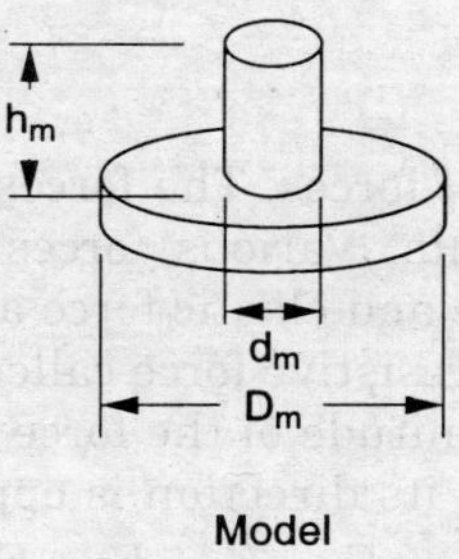

Fig. 8.3

2. Kinematic similarity
3. Dynamic similarity

Geometric Similarity

A model and its prototype are geometrically similar if the ratios of the corresponding length dimensions are equal. Then

Length scale ratio, $$L_r = \frac{L_m}{L_p}$$ (*L* may be length, height, width etc.)

Area scale ratio $$A_r = \frac{L_m^{\ 2}}{L_p^{\ 2}} = L_r^{\ 2}$$

Volume scale ratio $$V_r = \frac{L_m^{\ 3}}{L_p^{\ 3}} = L_r^{\ 3}$$

Kinematic Similarity

It implies the similarity of motion or the ratios of velocity, acceleration, discharge etc. are same in the model as well as in the prototype at all homologous points (geometrically corresponding points). It also indicates that the path of the moving points is same at the homologous points. The scale ratios are

Time scale ratio, $$T_r = \frac{T_m}{T_p}$$

Velocity scale ratio, $$V_r = \frac{L_m/T_m}{L_p/T_p} = \frac{L_m}{L_p} \cdot \frac{T_p}{T_m} = \frac{L_r}{T_r}$$

Acceleration scale ratio, $$a_r = \frac{L_m/T_m^{\ 2}}{L_p/T_p^{\ 2}} = \frac{L_r}{T_r^{\ 2}}$$

Discharge scale ratio, $$Q_r = \frac{L_m^{\ 3}/T_m}{L_p^{\ 3}/T_p} = \frac{L_r^{\ 3}}{T_r}$$

Dynamic Similarity

It is the similarity of forces. The forces are to be the same in the model and the prototype at all homologous points. Various forces such as pressure force, viscous force, gravity force, surface tension force and elastic force act upon the fluid particles resulting in the acceleration of the fluid mass. A resistive force called the inertia force due to the mass of the fluid particles also exists. The magnitude of the force is equal to the product of the mass of the particles and the acceleration and its direction is opposite to that of the fluid particles.

$$F_1 = F_p + F_v + F_g + F_s + F_e = m\ .\ a$$

where m is the mass of the liquid particles.

For dynamic similarity,

$$\frac{(F_i)_m}{(F_i)_p} = \frac{(F_p + F_v + F_g + F_s + F_e)_m}{(F_p + F_v + F_g + F_s + F_e)_p} = \frac{(\Sigma F)_m}{(\Sigma F)_p} = \frac{(m\ .\ a)_m}{(m\ .\ a)_p}$$

For complete similarity between the model and the prototype, the ratios of the individual component forces must also be equal to the ratio of the inertia forces.

$$\frac{(F_i)_m}{(F_i)_p} = \frac{(F_p)_m}{(F_p)_p} = \frac{(F_s)_m}{(F_s)_p} = \frac{(F_g)_m}{(F_g)_p} \ \ldots \text{ etc.}$$

This can also be written as

$$\left(\frac{F_i}{F_p}\right)_m = \left(\frac{F_i}{F_p}\right)_p$$

$$\left(\frac{F_i}{F_g}\right)_m = \left(\frac{F_i}{F_g}\right)_p$$

$$\left(\frac{F_i}{F_v}\right)_m = \left(\frac{F_i}{F_v}\right)_p \quad \text{etc.}$$

These ratios are the dimensionless numbers as defined earlier. It is to be noted that dynamic similarity is possible only if there are geometric and kinematic similarities. In the absence of geometric similarity, the ratios of the velocities or the accelerations etc. at the homologous points in the model and the prototype are changed and thus there remains no kinematic similarity. The equilibrium of forces is also upset.

8.11 SIMILITUDE LAWS

Various laws are defined depending upon the importance of each dimensionless number as discussed below :

Reynolds Model Law

This law is applicable where the viscous force is important. Usually, the flow is incompressible and the Mach number is less than three. There is no free surface so that the effects of gravity, surface tension, pressure etc. are not predominant. Examples are :

- Incompressible fluid flow in a pipe
- Flow around deeply submerged bodies like submarines
- Flow in an open channel as long as there is no fluid activity at the surface, like waves and hydraulic jump
- Aeroplane motion
- Structures immersed in a fluid flow.

Reynolds model law establishes similarity of flows in the model and the prototype in such situations and we have

$$(\text{Re})_m = (\text{Re})_p$$

$$\left(\frac{\rho LV}{\mu}\right)_m = \left(\frac{\rho LV}{\mu}\right)_p$$

Froude Model Law

The gravitational force is important in cases where gravity is the motivating force and where there is activity at the free surface of water, such as wave formation. Examples are :

- Flow through a nozzle or orifice
- Flow over weirs and notches
- Open-channel flow when waves are formed
- Motion of a ship through water, waves are set up
- Flow over the spillway of a dam.

$$(\text{Fr})_m = (\text{Fr})_p$$

$$\left(\frac{V}{\sqrt{Lg}}\right)_m = \left(\frac{V}{\sqrt{Lg}}\right)_p$$

$$\left(\frac{V}{\sqrt{L}}\right)_m = \left(\frac{V}{\sqrt{L}}\right)_p \quad \text{if } g \text{ is same for the two}$$

For similarity of flow, Froude model law is applicable.

Euler Model Law

This law is applicable when the pressure force predominates. Other forces such as gravity, surface tension, viscous etc. are insignificant or absent. Example are :

- Water hammer in pipes *i.e.* pressure rise due to sudden closure of valves
- Flow in an enclosed pipe when turbulence is high
- Cavitation in hydraulic machines

For similarity of flow, Euler model law is applicable.

$$(\text{Eu})_m = (\text{Eu})_p$$

$$\left(\frac{V}{\sqrt{p/\rho}}\right)_m = \left(\frac{V}{\sqrt{p/\rho}}\right)_p$$

Weber Model Law

When passages are narrow, surface tension force is important. For example

- Flow of thin sheet of liquid over a smooth surface
- Fluid rise in capillary tubes
- Atomisation of fluid through nozzles
- Blood flow in vanes.

For similarity of flow,

$$(F_s)_m = (F_s)_p$$

$$\left(\frac{V}{\sqrt{\sigma/\rho L}}\right)_m = \left(\frac{V}{\sqrt{\sigma/\rho L}}\right)_p$$

This law is rarely met in practice as small passages are not usually met in practice.

Mach Model Law

When high velocity flows are there and density variations are important, force due to elastic compression becomes dominant. For example :

- Under water testing of torpedoes
- Unsteady flows such as water hammer problems
- Aeroplanes flying with a velocity higher than the velocity of sound.

For similarity of flow,

$$\left(\frac{V}{\sqrt{K/p}}\right)_m = \left(\frac{V}{\sqrt{K/\rho}}\right)_p$$

Example 8.10 (a) With Reynolds number as the criterion of similarity, obtain the scale ratios for velocity, discharge, time, acceleration and force.

(b) For Froudes model law, determine the scale ratios for velocity discharge, time, acceleration, force, pressure, work and power.

Solution (a) For Reynolds model law,

$$\left(\frac{\rho VL}{\mu}\right)_m = \left(\frac{\rho VL}{\mu}\right)_p$$

$$\therefore \quad \frac{V_m}{V_p} = \frac{\mu_m}{\mu_p} \cdot \frac{\rho_p}{\rho_m} \cdot \frac{L_p}{L_m}$$

$$V_r = \frac{\mu_r}{L_r \cdot \rho_r}$$

where $V_r = \dfrac{V_m}{V_p}$, $\mu_r = \dfrac{\mu_m}{\mu_p}$, $L_r = \dfrac{L_m}{L_p}$ and $\rho_r = \dfrac{\rho_m}{\rho_p}$

$$Q_r = \frac{Q_m}{Q_p} = \frac{(A \times V)_m}{(A \times V)_p}$$

$$= \frac{{L_m}^2}{{L_p}^2} \cdot \frac{V_m}{V_p}$$

$$= {L_r}^2 \,.\, \frac{\mu_r}{L_r \,.\, \rho_r} = \frac{L_r \,.\, \mu_r}{\rho_r}$$

$$T_r = \frac{T_m}{T_p} = \frac{(L/V)_m}{(L/V)_p} \qquad \left(\begin{array}{l}\because V = L/T \\ \therefore T = L/V\end{array}\right)$$

$$= \frac{L_m}{L_p} \cdot \frac{V_p}{V_m}$$

$$= L_r \,.\, \frac{L_r \,.\, \rho_r}{\mu_r} = \frac{{L_r}^2 \, \rho_r}{\mu_r}$$

$$a_r = \frac{a_m}{a_p} = \frac{(V/T)_m}{(V/T)_p}$$

$$= \frac{V_m}{V_p} \cdot \frac{T_p}{T_m}$$

$$= \frac{\mu_r}{L_r \,.\, \rho_r} \cdot \frac{\mu_r}{{L_r}^2 \,.\, \rho_r} = \frac{{\mu_r}^2}{{\rho_r}^2 \, {L_r}^3}$$

$$F_r = \frac{F_m}{F_p} = \frac{(m \times a)_m}{(m \times a)_p}$$

$$= \frac{(\rho L^3)_m}{(\rho L^3)_p} \cdot \frac{a_m}{a_p}$$

$$= \rho_r \,.\, {L_r}^3 \,.\, \frac{{\mu_r}^2}{{\rho_r}^2 \,.\, {L_r}^3} = \frac{{\mu_r}^2}{\rho_r}$$

(b) For Froude model law,

$$\left(\frac{V}{\sqrt{Lg}}\right)_m = \left(\frac{V}{\sqrt{Lg}}\right)_p$$

or $$\frac{V_m}{V_p} = \sqrt{\frac{L_m}{L_p}} = {L_r}^{0.5}$$

$$Q_r = \frac{Q_m}{Q_p} = \frac{A_m}{A_p} \cdot \frac{V_m}{V_p}$$

$$= {L_r}^2 \,.\, {L_r}^{0.5} = {L_r}^{2.5}$$

$$T_r = \frac{T_m}{T_p} = \frac{(L/V)_m}{(L/V)_p}$$

$$= L_r \cdot \frac{1}{L_r^{0.5}} = L_r^{0.5}$$

$$a_r = \frac{a_m}{a_p} = \frac{(V/T)_m}{(V/T)_p}$$

$$= \frac{V_m}{V_p} \cdot \frac{T_p}{T_m} = L_r^{0.5} \cdot \frac{1}{L_r^{0.5}} = 1$$

$$F_r = \frac{F_m}{F_p} = \frac{(m \times a)_m}{(m \times a)_p}$$

$$= \frac{(\rho L^3)_m}{(\rho L^3)_p} \cdot a_r$$

$$= L_r^3 \times 1 = L_r^3$$

$$p_r = \frac{p_m}{p_p} = \frac{(F/A)_m}{(F/A)_p}$$

$$= \frac{F_r}{A_r} = \frac{L_r^3}{L_r^3} = L_r$$

$$W_r = \frac{W_m}{W_p} = \frac{(F \times L)_m}{(F \times L)_p}$$

$$= L_r^3 \cdot L_r = L_r^4$$

$$P_r = \frac{P_m}{P_p} = \frac{(W/T)_m}{(W/T)_p}$$

$$= \frac{W_r}{T_r} = \frac{L_r^4}{L_r^{0.5}} = L_r^{3.5}$$

Example 8.11 A 1 : 20 scale model of a submarine is to be tested in a wind tunnel. Speed of the prototype is to be 12 m/s. What will be the speed of air in the wind tunnel ? Kinematic viscosities of the sea water and the air are 120 m^2/s and 160 m^2/s respectively and mass densities of the sea water and the air are 1030 kg/m^3 and 1·25 kg/m^3 respectively.

Also find the ratio of drag of the model and the prototype.

Solution Submarine is to move far below the surface of water, *i.e.* it is totally submerged. Therefore, similarity between the model and the prototype will be ensured if Reynolds number is the same for the two.

$$\frac{L_m}{L_p} = \frac{1}{20} \qquad \frac{\nu_m}{\nu_p} = \frac{160}{120}$$

$V_p = 12$ m/s $\qquad \dfrac{\rho_m}{\rho_p} = \dfrac{1.25}{1030}$

Applying Reynolds model law,

$$\left(\frac{\rho VL}{\mu}\right)_m = \left(\frac{\rho VL}{\mu}\right)_p$$

$$\left(\frac{VL}{\nu}\right)_m = \left(\frac{VL}{\nu}\right)_p$$

or
$$V_m = V_p \cdot \frac{\nu_m}{\nu_p} \cdot \frac{L_p}{L_m}$$

$$= 12 \times \frac{160}{120} \times 20$$

$$= \mathbf{320\ m/s}$$

Ratio of the drag forces is given by,

$$(F_d)_r = \frac{(F_d)_m}{(F_d)_p} = \frac{\mu_r^2}{\rho_r} \qquad \text{(Example 8.10)}$$

$$= \frac{(\nu\rho)_m^2}{(\nu\rho)_p^2} \cdot \frac{\rho_p}{\rho_m} \qquad \left(\nu = \frac{\mu}{\rho} \text{ or } \mu = \nu\rho\right)$$

$$= \left(\frac{\nu_m}{\nu_p}\right)^2 \cdot \frac{\rho_m}{\rho_p}$$

$$= \left(\frac{160}{120}\right)^2 \cdot \frac{1.25}{1030}$$

$$= \mathbf{0.00216}$$

The ratio of drag forces can also be found out,

$$F = m \times a$$

$$= \rho L^3 \frac{V}{t} = \rho \frac{L}{t} \cdot L^2 V$$

$$= \rho L^2 V^2$$

ie
$$\frac{F_m}{F_p} = \frac{\rho_m}{\rho_p} \cdot \left(\frac{L_m}{L_p}\right)^2 \cdot \left(\frac{V_m}{V_p}\right)^2$$

$$= \left(\frac{1.25}{1030}\right) \cdot \left(\frac{1}{20}\right)^2 \cdot \left(\frac{320}{12}\right)^2$$

$$= \mathbf{0.00216}$$

Example 8.12 A spillway model has a ratio of 1 : 40. Width of the model is 400 mm. Prototype is 20 m high and expected head on it is 2 m. Find,

(a) Height on the model

(b) Head of the model

(c) Expected flow per m-width of phototype if flow over the model at a particular head is 15 litres per second.

(d) Negative pressure in the prototype if in the model it is 100 mm.

Solution $L_r = \frac{L_m}{L_p} = \frac{1}{40}$ Height $H_p = 20$ m

Head $h_p = 2$ m $Q_m = 15$ l/s

Negative pressure head,

$h'_m = 0{\cdot}1$ m $B_m = 0{\cdot}4$ m

(a) $$\frac{H_m}{H_p} = L_r = \frac{1}{40}$$

or $$H_m = \frac{20}{40} = \mathbf{0{\cdot}5\ m}$$

(b) $$\frac{h_m}{h_p} = L_r = \frac{1}{40}$$

$$h_m = \frac{2}{40} = \mathbf{0{\cdot}05\ m}$$

(c) $$\frac{B_m}{B_p} = L_r = \frac{1}{40}$$

$$\frac{0{\cdot}4}{B_p} = \frac{1}{40}$$

$$B_p = 40 \times 0{\cdot}4 = 16 \text{ m}$$

For flow over spillway, Froude model law is applicable.

Thus $$\frac{Q_m}{Q_p} = L_r^{2{\cdot}5}$$ (Example 8.10)

or $$\frac{15}{Q_p} = \left(\frac{1}{40}\right)^{2{\cdot}5}$$

$$Q_p = 151789 \text{ lit/s}$$

Discharge/unit width $= \frac{Q_p}{B_p}$

$$= \frac{151789}{16} = \mathbf{9487\ l/s}$$

(d) $$\frac{h'_m}{h'_p} = L_r = \frac{1}{40}$$

$$\frac{0{\cdot}1}{h'_p} = \frac{1}{40}$$

$$h_p = 40 \times 0{\cdot}1$$

$$= \mathbf{4\ m\ negative.}$$

PROBLEMS

1. What do you mean by dimensional analysis ? In what way does it help in generalising a problem ?
2. Define the fundamental and derived quantities.
3. What is meant by dimensional homogeneity ? Give examples.
4. What are the methods adopted in dimensional analysis ? Discuss.
5. What are the guidelines in selecting the repeating variables ?
6. How can the π-terms be altered without changing their basic character ?
7. In what way does the dimensional analysis help in presenting experimental data ?
8. What are omitted and superfluous variables ? Give examples.
9. What are dynamic forces ? What are dimensionless numbers ?
10. What is meant by similitude ? Describe its types.
11. What is the importance of model testing ?
12. For a complete similarity between the model and the prototype it is enough to ensure dynamic similarity. Comment.
13. What are similarity laws ? What is their importance in model testing ? Elaborate giving examples in each case.
14. Show that the frictional torque T of a disc of diameter D, rotating at speed N in a fluid of density ρ and viscosity μ is given by

$$T = D^5 N^5 \rho \,.\, f\left(\frac{\mu}{D^2 N \rho}\right)$$

Use Rayleigh method of dimensional analysis.

15. The efficiency η of a fan depends on the discharge Q, dynamic viscosity μ and density ρ of the fluid and diameter D and the angular velocity ω of the rotor. Derive an expression for η using Reyleigh method of dimensional analysis.

$$\left[\eta = f\left(\frac{\mu}{\rho\omega D^2}, \frac{Q}{\omega D^3}\right)\right]$$

16. The thrust of a ship propeller F is dependent upon its diameter D, the rotational speed N, the fluid viscosity μ, the fluid density ρ, the speed of the propeller through the fluid V, called the advance velocity and the acceleration due to gravity g. Determine an expression for F using the method of dimensional analysis.

$$\left[F = \rho D^2 V^2 \,.\, f\left(\frac{V}{ND}, \frac{V\rho D}{\mu}, \frac{gD}{V^2}\right)\right]$$

17. The discharge Q consumed by an oil ring depends upon the internal diameter D, rotational speed N, density of the oil ρ, viscosity μ, specific weight w and surface tension σ. Show that the discharge is given by

$$Q = ND^3 \,.\, f\left(\frac{\mu}{\rho N D^2}, \frac{w}{\rho N^2 D}, \frac{\sigma}{\rho N^2 D^3}\right)$$

18. The resisting force F of a supersonic plane depends upon the velocity of the aircraft V, its length L, air viscosity μ, air density ρ and the bulk modulus of air K. Show that the resisting force is a function of Reynolds number and Mach number.
19. The drag force F of a vessel floating in water is given by speed V, length L, the dynamic viscosity μ and the density ρ of water and acceleration due to gravity g. Using Buckingham π-method of dimensional analysis, deduce an expression for F.
$$\left[F = \rho L^2 V^2 \cdot f\left(\frac{Lg}{V^2}, \frac{\mu}{\rho VL}\right)\right]$$
20. The discharge and the velocity of flow over the model of a spillway of scale 1 : 49 are $2{\cdot}5\ m^3/s$ and $1{\cdot}75$ m/s respectively. Find the discharge and the velocity for the prototype. (42017 m^3/s, 12·25 m/s)

Chapter 9

VISCOUS FLOW

9.1 INTRODUCTION

Viscosity is a property of fluids that resists the motion of one layer of fluid relative to the other. All real fluids are viscous in nature. Tangential or shear stresses are produced in the moving fluids due to viscosity. It has been accepted that there are two kinds of fluid motion, **laminar** and **turbulent.** In a laminar flow, the fluid particles move in unmixing layers and follow a smooth continuous path. There is no transverse displacement of the fluid particles and they remain in their respective layers in an orderly sequence. In turbulent flow, the fluid motion is irregular. Streamlines are intertwined and the pathlines are irratic curves. There are cross components of velocity in the lateral direction. There is intermingling of fluid particles. The flow of blood in human beings and the ground water movement are laminar flows whereas the flow in municipal water supply pipes is turbulent.

9.2 REYNOLDS EXPERIMENTS

In 1883, Osborne Reynolds, an English scientist performed experiments on an apparatus shown in Fig. 9.1. The apparatus consisted of a fairly long glass tube (approximately 1·5 m

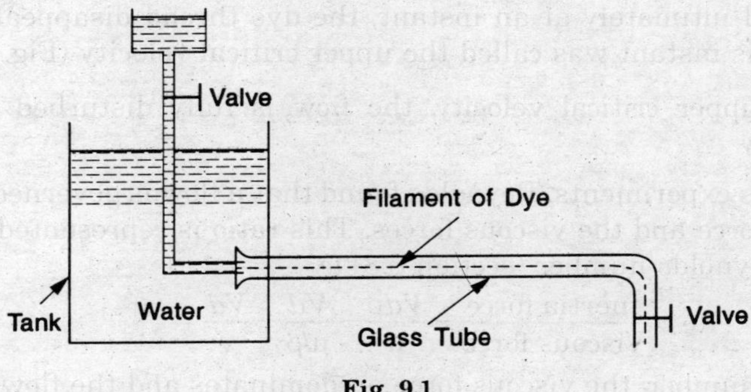

Fig. 9.1

long and 50 mm in diameter) fitted to a constant head tank filled with water. The glass tube had a bell mouth entrance. The flow of water through the glass tube was controlled by a regulating valve. A small tank containing a dye (of same specific weight as water) supplied a jet of dye at the centre of the glass tube. Before carrying out the experiments, the water in the tank was made to stand for several hours so that it became completely at rest. When the water was made to flow from the tank through the glass tube into the atmosphere, the liquid dye was also introduced into the flow. It was observed that

- When the velocity of flow was low, a fine thread of the dye was carried by the flowing water. The dye thread moved so steadily that it hardly appeared to be in motion. This was termed as laminar flow (Fig. 9.2a).

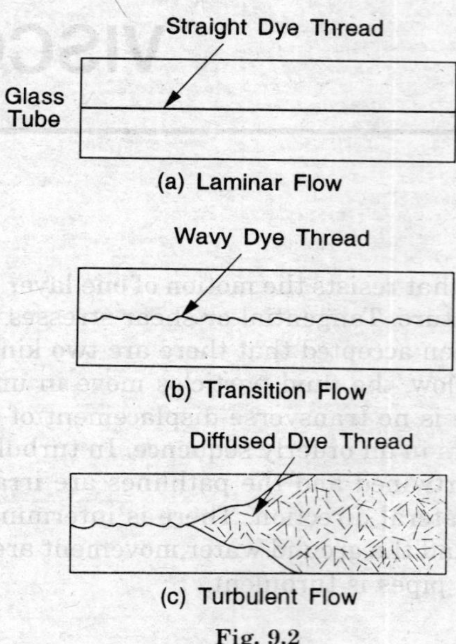

Fig. 9.2

- As the velocity was increased, a stage came when the dye thread started becoming irregular and wavy. This velocity was called lower critical velocity (Fig. 9.2b).
- On further increasing the velocity, the irregularity and the waviness of the dye thread increased and ultimately at an instant, the dye thread disappeared or diffused. The velocity at this instant was called the upper critical velocity (Fig. 9.2c).
- Beyond the upper critical velocity, the flow is fully disturbed and was known as turbulent flow.

On the basis of his experiments, Reynolds found that a flow is governed by the relationship between the inertia force and the viscous forces. This ratio is represented by a dimensionless number known as Reynolds number (section 8.8) *ie*

$$\text{Re} = \frac{\text{Inertia force}}{\text{Viscous force}} = \frac{Vd\rho}{\mu} = \frac{Vd}{\mu/\rho} = \frac{Vd}{\nu} \qquad ...(9.1)$$

At low Reynolds number, the viscous force predominates and the flow is laminar whereas at higher Reynolds number, the inertia force is much higher as compared to viscous force and the flow is turbulent. The values of lower critical Reynolds number are

2320 for circular pipes
1000 for parallel walls
50 for open channels

However, in case of ordinary pipes with usual degree of roughness, the lower critical Reynolds number is taken as 2000 and the flow is categorised as under :

For Re < 2000, the laminar flow
For Re > 2000 and < 4000, the transition flow
For Re > 4000, the turbulent flow

Reynolds also established on the basis of his experiments that the loss of pressure head in a pipe is proportional to velocity for the laminar flow and proportional to V^n for the turbulent flow (n varies from 1·75 to 2·0).

9.3 FLOW OF VISCOUS FLUID IN CIRCULAR PIPES

Consider a cylindrical free body of fluid having radius r and length dx in a horizontal circular pipe of radius R (Fig. 9.3). Since the fluid motion is steady and uniform, the particles of fluid are not accelerated.

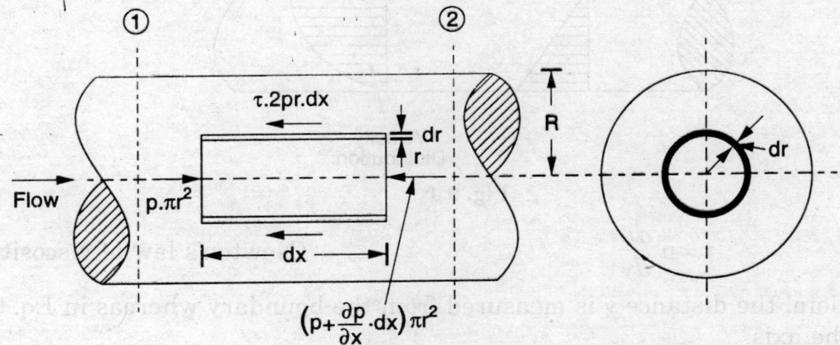

Fig. 9.3

If p is the intensity of pressure on the left end of the fluid element, the intensity of pressure on the right end will be $\left(p + \frac{\partial p}{\partial x} \,.\, dx\right)$.

The various forces acting on the fluid element are :

1. The pressure force, $p \,.\, \pi r^2$ on the left end
2. The pressure force, $\left(p + \frac{\partial p}{\partial x} \,.\, dx\right) . \,\pi r^2$ on the right end
3. The shear force, $\tau\,(2\pi r \,.\, dx)$ towards left on the surface of the fluid element (τ is the shear stress).

As the flow is steady, the sum of the forces in the direction of flow must be zero, *i.e.*

$$p\pi r^2 - \left(p + \frac{\partial p}{\partial x}\,dx\right)\pi r^2 - 2\pi r\,.\,dx\,.\,\tau = 0$$

or
$$-\frac{\partial p}{\partial x}\,dx\,\pi r^2 = 2\pi r\,.\,dx\,.\,\tau$$

or
$$\tau = -\frac{\partial p}{\partial x}\frac{r}{2} \qquad ...(9.2)$$

At the pipe wall,

$$\tau_0 = -\frac{\partial p}{\partial x}\frac{R}{2} \qquad ...(9.2a)$$

This shows that the flow occurs only if pressure gradient exists in the direction of flow. The negative sign indicates that the pressure decreases in the direction of flow. The equation also indicates that the shear stress varies linearly across the section (Fig. 9.4) having zero value at the centre.

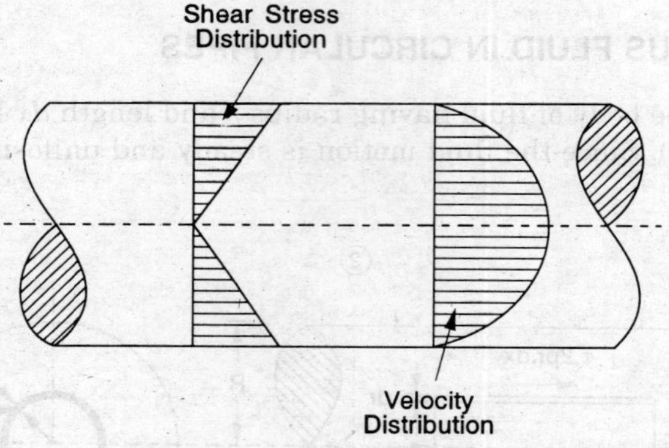

Fig. 9.4

Also $\tau = \mu \frac{du}{dy}$ (Newton's law of viscosity) ...(9.3)

In this equatioin, the distance y is measured from the boundary whereas in Eq. (9.2), r is measured from the axis.

Since $y = (R - r)$ or $dy = -dr$, the above equation may be written as

$$\tau = -\mu \frac{du}{dr} \qquad ...(9.4)$$

From Eq. (9.2) and (9.4)

$$\mu \frac{du}{dr} = \frac{\partial p}{\partial x}\frac{r}{2}$$

$$du = \frac{1}{2\mu}\left(\frac{\partial p}{\partial x}\right) r\,.\,dr$$

For steady uniform flow, $\frac{\partial p}{\partial x}$ must have a constant value. Integrating the above equation with respect to r.

$$\mu = \frac{1}{4\mu}\left(\frac{\partial p}{\partial x}\right) r^2 + C$$

C is a constant of integration and its value is obtained from the boundary conditions, *i.e.* at $r = R$, $u = 0$.

$$O = \frac{1}{4\mu}\left(\frac{\partial p}{\partial x}\right) R^2 + C$$

$$C = -\frac{1}{4\mu}\left(\frac{\partial p}{\partial x}\right) R^2$$

Then,

$$\mu = \frac{1}{4\mu}\left(\frac{\partial p}{\partial x}\right) r^2 - \frac{1}{4\mu}\left(\frac{\partial p}{\partial x}\right) R^2$$

$$= -\frac{1}{4\mu}\frac{\partial p}{\partial x}(R^2 - r^2) \qquad ...(9.5)$$

This equation shows that the velocity distribution is parabolic (Fig. 9.4). The surface of the velocity distribution is a paraboloid of revolution.

$$u_{max} = -\frac{1}{4\mu}\frac{\partial p}{\partial x} R^2 \text{ at the axis} \qquad ...(9.6)$$

and u in terms of u_{max},

$$u = -\frac{1}{4\mu}\left(\frac{\partial p}{\partial x}\right) R^2 \left[1 - \left(\frac{r}{R}\right)^2\right]$$

$$= u_{max}\left[1 - \left(\frac{r}{R^2}\right)\right] \qquad ...(9.7)$$

The average or the mean value of the velocity can be obtained by dividing the discharge through the pipe by the cross-sectional area.

Consider an elementary ring of thickness *dr* at a radius *r*.

Then $dQ = u \times 2\pi r \, . \, dr$

$$= -\frac{1}{4\mu}\frac{\partial p}{\partial x}(R^2 - r^2)\, 2\pi r \, dr$$

$$\therefore \quad Q = \int_0^R dQ$$

$$= -\int_0^R \frac{1}{4\mu}\frac{\partial p}{\partial x}(R^2 - r^2)\, 2\pi r \, . \, dr$$

$$= -\frac{\pi}{2\mu}\frac{\partial p}{\partial x}\int_0^R (R^2 r - r^3)\, . \, dr$$

$$= -\frac{\pi}{2\mu}\frac{\partial p}{\partial x}\left[\frac{R^2 r^2}{2} - \frac{r^4}{4}\right]_0^R$$

$$= -\frac{\pi}{2\mu}\frac{\partial p}{\partial x}\frac{R^4}{4}$$

$$= -\frac{\pi}{8\mu}\frac{\partial p}{\partial x} R^4 \qquad ...(9.8)$$

Average velocity,

$$\bar{u} = \frac{Q}{a} = \frac{\left(-\frac{\pi}{8\mu}\frac{\partial p}{\partial x}R^4\right)}{\pi R^2} = -\frac{1}{8\mu}\frac{\partial p}{\partial x}R^2 \quad \text{...(9.9)}$$

Ratio of average velocity to maximum velocity,

$$\frac{\bar{u}}{u_{\max}} = \frac{-\frac{1}{8\mu}\frac{\partial p}{\partial x}R^2}{-\frac{1}{4\mu}\frac{\partial p}{\partial x}R^2} = \frac{1}{2} = 0{\cdot}5 \quad \text{...(9.10)}$$

Integrating Eq. 9.9 with respect to x to determine the pressure difference between two sections 1 and 2 (Fig. 9.3),

$$\int_1^2 dp = -\frac{8\mu\bar{u}}{R^2}\int_1^2 dx$$

$$p_2 - p_1 = -\frac{8\mu\bar{u}}{R^2}(x_2 - x_1)$$

$$p_1 - p_2 = \frac{32\mu\bar{u}l}{D^2} \quad \text{...(9.11)}$$

Where D is the diameter of pipe and l is the length. Eq. 9.11 is known as **Hagen-Poiseuille Equation.**

The above equation can also be written as

$$\frac{p_1 - p_2}{w} = \frac{32\mu\bar{u}l}{wD^2}$$

$$h_f = \frac{32\mu\bar{u}l}{wD^2} \quad \text{...(9.12)}$$

where h_f is the friction head loss between two sections.

Example 9.1 A crude oil of viscosity 0·92 poise and relative density 0·88 flows through a horizontal pipe. The length of the pipe is 6 m and the diameter 80 mm. Determine the pressure difference between the two ends of the pipe if 900 N of the oil is collected in a tank in 20 seconds.

Solution $\mu = 0{\cdot}92$ poise $= 0{\cdot}092$ Ns/m^2 $W = 900/20 = 45$ N/s

$D = 0{\cdot}08$ m $l = 6$ m

$$\text{Weight of oil collected/s} = \frac{900}{20} = 45 \text{ N/s}$$

$$\text{Volume of oil collected/s, } Q = \frac{45}{\text{Specific weight}}$$

$$= \frac{45}{0{\cdot}88 \times 1000 \times 9{\cdot}81}$$

$$= 0{\cdot}005213 \text{ m}^3/\text{s}$$

$$V = \bar{u} = \frac{Q}{A} = \frac{0{\cdot}005213}{\frac{\pi}{4}\times(0{\cdot}08)^2} = 1{\cdot}037 \text{ m/s}$$

$$R_e = \frac{\rho VD}{\mu} = \frac{0{\cdot}88 \times 1000 \times 1{\cdot}037 \times 0{\cdot}08}{0{\cdot}092} = 793{\cdot}5$$

As R_e is less than 2000, the flow is laminar.

$$p_1 - p_2 = \frac{32\mu\bar{u}l}{D^2} = \frac{32 \times 0{\cdot}092 \times 1{\cdot}037 \times 6}{0{\cdot}08^2} = 2862 \text{ N/m}^2$$

Example 9.2 The maximum velocity of flow of a laminar flow in a pipe of diameter 240 mm is 2·8 m/s. Determine the mean velocity and the radius at which it occurs. Also find the velocity at 60 mm from the wall of the pipe.

Solution $D = 0{\cdot}24$ m $\qquad u_{max} = 2{\cdot}8$ m/s

$R = 0{\cdot}24/2 = 0{\cdot}12$ m

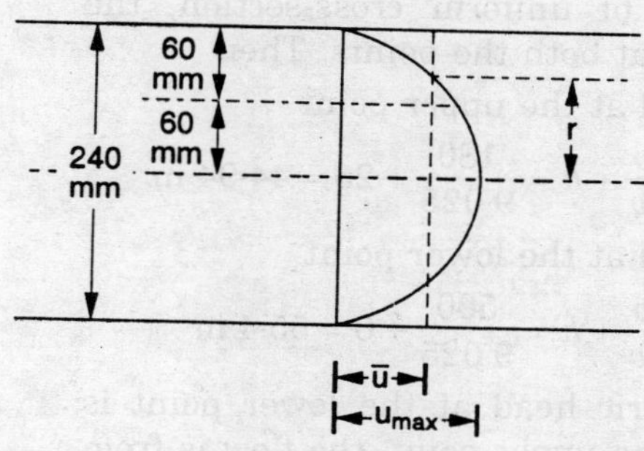

Fig. 9.5

Mean velocity,

$$\bar{u} = \frac{1}{2} \times u_{max} = \frac{1}{2} \times 2{\cdot}8 = 1{\cdot}4 \text{ m/s}$$

$$u = u_{max}\left[1 - \left(\frac{r}{R}\right)^2\right]$$

$$1{\cdot}4 = 2{\cdot}8\left[1 - \left(\frac{r}{0{\cdot}12}\right)^2\right]$$

$$1 - \left(\frac{r}{0{\cdot}12}\right)^2 = 0{\cdot}5$$

$$\left(\frac{r}{0{\cdot}12}\right)^2 = 0{\cdot}5$$

$$r = 0{\cdot}0849 \text{ m or } 84{\cdot}9 \text{ mm}$$

Velocity at 60 mm from the wall,

$$r = R - 0{\cdot}06 = 0{\cdot}12 - 0{\cdot}06 = 0{\cdot}06 \text{ m}$$

and $$u = 2{\cdot}8\left[1 - \left(\frac{0{\cdot}06}{0{\cdot}12}\right)^2\right] = 2{\cdot}8 \times 0{\cdot}75 = 2{\cdot}1 \text{ m/s}$$

Example 9.3 An oil of relative density 0·92 and viscosity 1·6 poise flows through a 24 mm diameter vertical pipe. Two pressure gauges fixed at 25 m apart read 500 kN/m^2 and 180 N/m^2, the lower value being of the gauge at the higher level. Determine the direction and the rate of flow through the pipe.

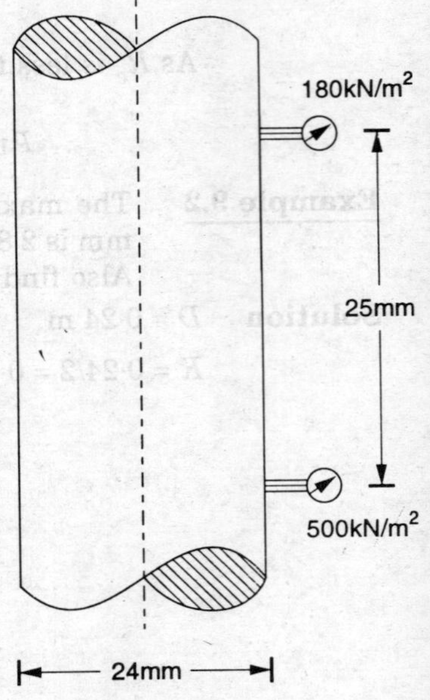

Fig. 9.6

Solution $w = 0{\cdot}92 \times 9{\cdot}81 \times 1000 = 9025$ N/m^3

or $9{\cdot}025$ kN/m^3

$\mu = 0{\cdot}16$ N.s/m

$\rho = 0{\cdot}92 \times 1000 = 920$ kg/m^3

As the pipe is of uniform cross-section, the velocity is same at both the points. Then

Piezometric head at the upper point

$$= \frac{p}{w} + h = \frac{180}{9{\cdot}025} + 25 = 44{\cdot}94 \text{ m}$$

Piezometric head at the lower point

$$= \frac{p}{w} + h = \frac{500}{9{\cdot}025} + 0 = 55{\cdot}4 \text{ m}$$

As the piezometric head at the lower point is higher than at the upper point, the flow is from lower to the upper point.

$$h_f = \frac{32\mu\bar{u}l}{wD^2}$$

$$55{\cdot}4 - 44{\cdot}94 = \frac{32 \times 0{\cdot}16 \times \bar{u} \times 25}{9025 \times (0{\cdot}024)^2}$$

$$\bar{u} = 0{\cdot}4248 \text{ m/s}$$

$$R_e = \frac{\rho VD}{\mu} = \frac{920 \times 0{\cdot}4248 \times 0{\cdot}024}{0{\cdot}16} = 58{\cdot}62$$

which is less than 2000. Hence the flow is laminar and the above expression is valid.

$$Q = \text{Average velocity} \times \text{Area}$$

$$= 0{\cdot}4248 \times \frac{\pi}{4} \times (0{\cdot}024)^2$$

$$= 1{\cdot}922 \times 10^{-4} \text{ m}^3\text{/s}$$

Example 9.4 An oil of specific gravity 0·8 and viscosity 8 poise flows through a horizontal pipe of 400 mm diameter. If the pressure drop in a length of 80 m is 1600 kN/m^2, find

(i) the oil flow per second
(ii) the centre line velocity
(iii) the total frictional drag over 80 m length
(iv) the power required to continue the flow
(v) the velocity gradient at the wall of the pipe
(vi) the velocity and the shear stress at 6 mm from the pipe wall

Solution $p_1 - p_2 = 1600 \times 10^3 \text{ N/m}^2$ $\quad S = 0{\cdot}8$

$\mu = 8 \text{ poise} = 0{\cdot}8 \text{ N.s/m}^2$

$l = 80 \text{ m}$

$D = 0{\cdot}4 \text{ m}$

$$p_1 - p_2 = \frac{32\mu\overline{u}l}{D^2}$$

$$1600 \times 10^3 = \frac{32 \times 0{\cdot}8 \times \overline{u} \times 80}{(0{\cdot}04)^2}$$

$$\overline{u} = 1{\cdot}25 \text{ m/s}$$

$$\text{Re} = \frac{\rho VD}{\mu} = \frac{(0{\cdot}8 \times 1000) \times 1{\cdot}25 \times 0{\cdot}4}{0{\cdot}8} = 500$$

Re is less than 2000, thus the flow is laminar.

(i) $Q = A \times \overline{u} = \frac{\pi}{4} \times (0{\cdot}04)^2 \times 1{\cdot}25 = 0{\cdot}00157 \text{ m}^3\text{/s}$

(ii) $u_{max} = 2\overline{u} = 2 \times 1{\cdot}25 = 2{\cdot}5 \text{ m/s}$

(iii) Wall shear stress, $\tau_0 = \frac{\partial p}{\partial x} \cdot \frac{R}{2} = \frac{1600}{80} \times \frac{0{\cdot}04/2}{2} = 200 \text{ N/m}^2$

Friction drag, F_D = Shear stress × Wall area

$$= \tau_0 \times \pi DL$$
$$= 200 \times \pi \times 0{\cdot}04 \times 80$$
$$= 2010{\cdot}6 \text{ N or } 2{\cdot}0106 \text{ kN}$$

(iv) Power required to continue the flow,

$$P = \text{Friction drag} \times \text{Average velocity}$$
$$= F_D \times u$$
$$= 2{\cdot}0106 \times 1{\cdot}25$$
$$= 2{\cdot}513 \text{ W}$$

(v) Velocity gradient at the wall of the pipe,

$$\tau_0 = \mu\left(\frac{\partial u}{\partial y}\right)_{y=0}$$

$$\left(\frac{\partial u}{\partial y}\right)_{y=0} = \frac{\tau_0}{\mu} = \frac{200}{0.8} = 250$$

(vi) At 6 mm from the pipe wall,

$$r = R - 0.006 = 0.02 - 0.006 = 0.014 \text{ m}$$

$$\mu = -\frac{1}{4\mu}\frac{\partial p}{\partial x}(R^2 - r^2) = \frac{1}{4 \times 0.8} \times \frac{1600 \times 10^3}{80}(0.02^2 - 0.014^2) = 1.2 \text{ m/s}$$

As the shear stress varies linearly across the section,

$$\frac{\tau}{r} = \frac{\tau_0}{R}$$

$$\frac{\tau}{0.014} = \frac{200}{0.02}$$

$$\tau = 140 \text{ kN/m}^2$$

Example 9.5 A fluid of specific gravity 1·1 and viscosity 9 poise flows through a circular pipe of diameter 125 mm. The maximum shear stress at the wall of the pipe is 200 N/m^2. Determine the pressure gradient, the average velocity and the Reynolds number of the flow.

Solution

$$\mu = 9 \text{ poise} = 0.9 \text{ N.s/m}^2$$

$$D = 0.125 \text{ m}$$

$$\rho = S \,.\, \rho_w = 1.1 \times 1000 = 1100 \text{ kg/m}^3$$

$$\tau_0 = -\frac{\partial p}{\partial x} \cdot \frac{R}{2}$$

$$200 = -\frac{\partial p}{\partial x} \times \frac{(0.125/2)}{2}$$

$$\frac{\partial p}{\partial x} = -\,6400 \text{ N/m}^2$$ (*-ve* sign indicates that p decreases with increase in x)

$$\bar{u} = -\frac{1}{8\mu}\frac{\partial p}{\partial x}R^2 = -\frac{1}{8 \times 0.9} \times (-\,6400) \times \left(\frac{0.125}{2}\right)^2 = 3.472 \text{ m/s}$$

$$\text{Re} = \frac{\rho VD}{\mu} = \frac{1.1 \times 1000 \times 3.472 \times 0.125}{0.9} = 530.5$$

Example 9.6 A pipe 50 mm in diameter and 400 m long slopes upwards at the rate of 1 in 40. An oil of specific gravity 0·84 and viscosity 0·8 N.s/m^2 is pumped at the rate of 4 litres per second through this pipe. Determine

(i) the pressure difference between the two ends

(ii) power of the pump assuming an overall efficienty of 70 percent

(iii) the centre line velocity

(iv) the velocity gradient at the pipe wall.

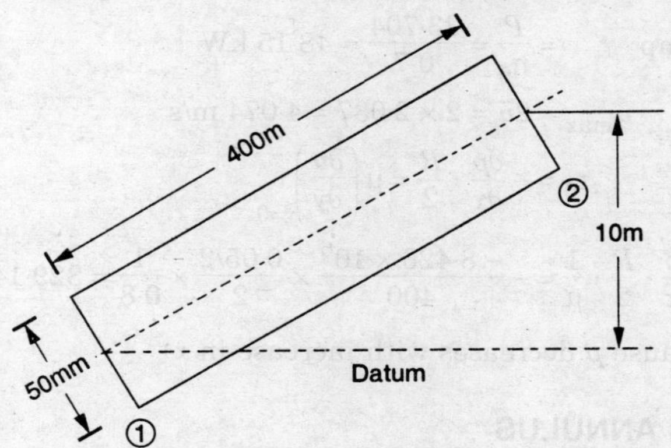

Fig. 9.7

Solution $Q = 0{\cdot}004\ \text{m}^3/\text{s}$ $\quad L = 400$ m

$\mu = 0{\cdot}8\ \text{N.s/m}^2$ $\quad D = 0{\cdot}05$ m

$w = 0{\cdot}84 \times 9{\cdot}81 \times 1000 = 8240{\cdot}4\ \text{N/m}^3$

$$\bar{u} = \frac{0{\cdot}004}{(\pi/4) \times (0{\cdot}05)^2} = 2{\cdot}037\ \text{m/s}$$

$$\text{Re} = \frac{\rho VD}{\mu} = \frac{(0{\cdot}84 \times 1000) \times 2{\cdot}037 \times 0{\cdot}05}{0{\cdot}8} = 107$$

As Re < 2000, the flow is laminar.

Applying Bernoulli's equation between the two ends,

$$\frac{p_1}{w} + \frac{V_1^2}{2g} + z_1 = \frac{p_2}{w} + \frac{V_2^2}{2g} + z_2 + h_f$$

$$\frac{p_1}{w} + \frac{V_1^2}{2g} + 0 = \frac{p_2}{w} + \frac{V_1^2}{2g} + \frac{1}{40} \times 400 + h_f \qquad (V_1 = V_2 \text{ and } z_1 = 0)$$

$$\frac{p_1 - p_2}{w} = 10 + \frac{32\mu\bar{u}L}{wD^2}$$

$$\frac{p_2 - p_2}{8240{\cdot}4} = 10 + \frac{32 \times 0{\cdot}8 \times 2{\cdot}037 \times 400}{8240{\cdot}4 \times (0{\cdot}05)^2}$$

$$p_1 - p_2 = 10 \times 8240{\cdot}4 + \frac{32 \times 0{\cdot}8 \times 2{\cdot}037 \times 400}{(0{\cdot}05)^2}$$

$$= 82404 + 8{\cdot}3436 \times 10^6$$

$$= 8{\cdot}426 \times 10^6\ \text{N/m}^2 \text{ or } 8{\cdot}426\ \text{MN/m}^2$$

$$P = wQh = wQ\,\frac{p_1 - p_2}{w} = Q\,(p_1 - p_2)$$

$$= 0{\cdot}004 \times 8{\cdot}426 \times 10^6 = 33704\ \text{W} = 33{\cdot}704\ \text{kW}$$

Power of the pump $= \frac{P}{\eta_0} = \frac{33{\cdot}704}{0{\cdot}7} = 48{\cdot}15$ kW

$$u_{max} = 2\bar{u} = 2 \times 2{\cdot}037 = 4{\cdot}074 \text{ m/s}$$

$$\tau_0 = -\frac{\partial p}{\partial x} \cdot \frac{R}{2} = \mu\left(\frac{\partial u}{\partial y}\right)_{y=0}$$

$$\left(\frac{\partial u}{\partial y}\right)_{y=0} = -\frac{\partial p}{\partial x} \cdot \frac{R}{2} \cdot \frac{1}{\mu} = -\frac{-8{\cdot}426 \times 10^6}{400} \times \frac{0{\cdot}05/2}{2} \times \frac{1}{0{\cdot}8} = 329{\cdot}14 \text{ s}^{-1}$$

($\partial p/\partial x$ is -ve, because p decreases with increase in x)

9.4 FLOW THROUGH AN ANNULUS

Consider an annulus of an outer radius R and the inner radius r (Fig. 9.8). Let the flow through it be steady laminar of an incompressible fluid. Consider a fluid element of the shape of a small concentric cylinder at a radial distance r and of thickness dr. The various forces acting on the fluid element are :

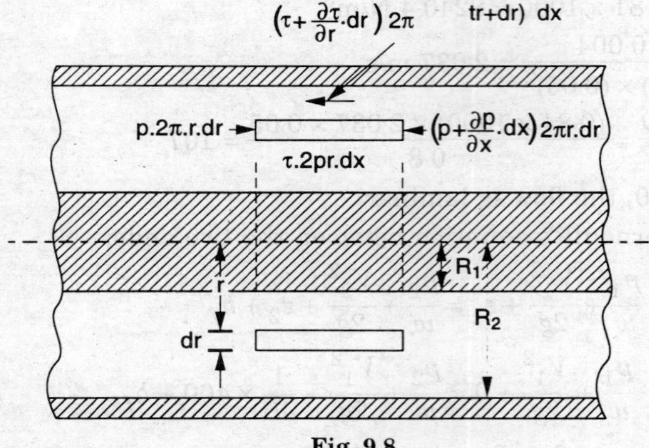

Fig. 9.8

Net pressure force on the left end $= p \,.\, 2\pi r \,.\, dr$

Net pressure force on the right end $= \left(p + \frac{\partial p}{\partial x} \,.\, dr\right) 2\pi r \,.\, dr$

Shear force towards right $= \tau \,.\, 2\pi r \,.\, dx$ (on inner curved surface)

Shear force towards left $= \left(\tau + \frac{\partial \tau}{\partial r} \,.\, dr\right) 2\pi\, (r + dr) \,.\, dx$ (on outer curved surface)

Since the flow is steady and uniform, the total force has to be zero.

$$\left(p + \frac{\partial p}{\partial x} \,.\, dx\right) 2\pi r \,.\, dr - p \,.\, 2\pi r \,.\, dr + \left(\tau + \frac{\partial \tau}{\partial r} \,.\, dr\right) 2\pi\, (r + dr) \,.\, dx - \tau \,.\, 2\pi r \,.\, dx = 0$$

$$\left(\frac{\partial p}{\partial x} \,.\, \partial x\right) 2\pi r \,.\, dr + 2\pi\tau \,.\, dr \,.\, dx + 2\pi \frac{\partial \tau}{\partial r} \,.\, dr \,.\, r \,.\, dx + \text{small terms (negligible)} = 0$$

Dividing by $2\pi \,.\, dx \,.\, dr$

$$r\frac{\partial p}{\partial x} + \tau + r\frac{\partial \tau}{\partial r} = 0$$

$$r\frac{\partial p}{\partial x} + r\frac{\partial \tau}{\partial r} + \tau = 0$$

$$r\frac{\partial p}{\partial x} + \frac{d\,(r\,.\,\tau)}{dr} = 0 \qquad \left(\frac{d\,(r\,.\,\tau)}{dr} = r\frac{\partial \tau}{\partial r} + \tau\right)$$

But $$\tau = -\,\mu\,\frac{du}{dr} \qquad \text{(Eq. 9.4)}$$

$\therefore$ $$r\frac{\partial p}{\partial x} + \frac{d}{dr}\left(-r\,.\,\mu\,\frac{du}{dr}\right) = 0$$

$$\frac{d}{dr}\left(\mu r\frac{du}{dr}\right) = r\frac{\partial p}{\partial x}$$

$$\frac{d}{dr}\left(r\frac{du}{dr}\right) = \frac{r}{\mu}\,\frac{\partial p}{\partial x} \qquad (\mu \text{ is constant})$$

Integrating with respect to r,

$$r\frac{du}{dr} = \frac{r^2}{2\mu}\,\frac{\partial p}{\partial x} + C_1$$

$$\frac{du}{dr} = \frac{r}{2\mu}\,\frac{\partial p}{\partial x} + \frac{C_1}{r} \qquad \ldots\text{(i)}$$

Integrating again with respect to r,

$$\mu = \frac{r^2}{4\mu}\,\frac{\partial p}{\partial x} + C_1 \log r + C_2 \qquad \ldots\text{(ii)}$$

Now the boundary conditions are

$$u = 0 \quad \text{at} \quad r = R_1 = R_2$$

$$0 = \frac{{R_1}^2}{4\mu}\,\frac{\partial p}{\partial x} + C_1 \log R_1 + C_2 \qquad \ldots\text{(iii)}$$

$$0 = \frac{{R_2}^2}{4\mu}\,\frac{\partial p}{\partial x} + C_1 \log R_2 + C_2 \qquad \ldots\text{(iv)}$$

Equating (iii) from (iv),

$$\frac{{R_1}^2}{4\mu}\,\frac{\partial p}{\partial x} + C_1 \log R_1 = \frac{{R_2}^2}{4\mu}\,\frac{\partial p}{\partial x} + C_1 \log R_2$$

$$\frac{{R_1}^2 - {R_2}^2}{4\mu}\,\frac{\partial p}{\partial x} = -\,C_1\,(\log R_1 - \log R_2)$$

$$\frac{{R_1}^2 - {R_2}^2}{4\mu}\,\frac{\partial p}{\partial x} = -\,C_1 \log \frac{R_1}{R_2}$$

$$C_1 = -\,\frac{{R_1}^2 - {R_2}^2}{4\mu\,.\,\log \dfrac{R_1}{R_2}}\,.\,\frac{\partial p}{\partial x} \qquad \ldots\text{(v)}$$

Velocity u is maximum when

$$\frac{du}{dr} = 0$$

Thus from (i),

$$0 = \frac{r}{2\mu}\frac{\partial p}{\partial x} + \frac{C_1}{r}$$

$$C_1 = -\frac{r^2}{2\mu}\frac{\partial p}{\partial x} \quad \text{...(vi)}$$

Comparing (v) and (vi),

$$\frac{r^2}{2\mu}\frac{\partial p}{\partial x} = \frac{R_1^2 - R_2^2}{4\mu \log \frac{R_1}{R_2}}\frac{\partial p}{\partial x}$$

$$r = \sqrt{\frac{R_1^2 - R_2^2}{2 \log \frac{R_1}{R_2}}} \quad \text{...(9.13)}$$

The maximum velocity occurs at a radius given by Eq. 9.13.

Inserting the value of C_1 from (v) into (iii),

$$0 = \frac{R_1^2}{4\mu}\frac{\partial p}{\partial x} - \frac{R_1^2 - R_2^2}{4\mu \,.\, \log \frac{R_1}{R_2}}\frac{\partial p}{\partial x}\log R_1 + C_2$$

$$C_2 = -\frac{1}{4\mu}\frac{\partial p}{\partial x}\left[R_1^2 - \frac{R_1^2 - R_2^2}{\log \frac{R_1}{R_2}}\log R_1\right] \quad \text{...(vii)}$$

Inserting the values of C_1 and C_2 from (v) and (vii) into (ii),

$$u = \frac{r_2}{4\mu}\frac{\partial p}{\partial x} - \frac{R_1^2 - R_2^2}{4\mu \,.\, \log \frac{R_1}{R_2}}\frac{\partial p}{\partial x}\log r - \frac{1}{4\mu}\frac{\partial p}{\partial x}\left[R_1^2 - \frac{R_1^2 - R_2^2}{\log \frac{R_1}{R_2}}\log R_1\right]$$

$$= -\frac{1}{4\mu}\frac{\partial p}{\partial x}\left[R_1^2 - r^2 - \frac{R_1^2 - R_2^2}{\log \frac{R_1}{R_2}}(\log R_1 - \log r)\right]$$

$$= -\frac{1}{4\mu}\frac{\partial p}{\partial x}\left[R_1^2 - r^2 - \frac{R_1^2 - R_2^2}{\log \frac{R_1}{R_2}}\log \frac{R_1}{r}\right] \quad \text{...(9.14)}$$

The discharge through the annulus,

$$Q = \int_{R_1}^{R_2} (2\pi r \,.\, dr) \,.\, u$$

$$= -\int_{R_1}^{R_2} 2\pi r \,.\, \frac{1}{4\mu} \frac{\partial p}{\partial x} \left[{R_1}^2 - r^2 - \frac{({R_1}^2 - {R_2}^2)^2}{\log \dfrac{R_1}{R_2}} \cdot \log \frac{R_1}{r} \right] dr$$

$$= -\frac{\pi}{8\mu} \cdot \frac{\partial p}{\partial x} \left[{R_1}^4 - r^4 - \frac{({R_1}^2 - {R_2}^2)^2}{\log \dfrac{R_1}{R_2}} \right] \qquad \text{...(9.15)}$$

The average velocity,

$$\overline{u} = \frac{Q}{\pi ({R_1}^2 - {R_2}^2)} = \frac{1}{8\mu} \frac{\partial p}{\partial x} \left[{R_1}^2 + r^2 - \frac{{R_1}^2 - {R_2}^2}{\log \dfrac{R_1}{R_2}} \right] \qquad \text{...(9.16)}$$

The shear stress,

$$\tau = -\mu \,.\, \frac{\partial u}{\partial r}$$

$$= \mu \,.\, \frac{\partial}{\partial r} \left[\frac{1}{4\mu} \frac{\partial p}{\partial x} \left\{ {R_1}^2 - r^2 - \frac{{R_1}^2 - {R_2}^2}{\log \dfrac{R_1}{R_2}} \log \frac{R_1}{r} \right\} \right]$$

$$= \frac{1}{4} \frac{\partial p}{\partial x} \left[-2r - \frac{{R_1}^2 - {R_2}^2}{\log \dfrac{R_1}{R_2}} \left(-\frac{1}{r} \right) \right]$$

$$= -\frac{1}{4} \frac{\partial p}{\partial x} \left[2r - \frac{1}{r} \cdot \frac{{R_1}^2 - {R_2}^2}{\log \dfrac{R_1}{R_2}} \right] \qquad \text{...(9.17)}$$

Example 9.7 A circular pipe of radius R_1 has a fixed co-axial cylindrical solid axial core of radius R_2. Through the annular passage, an incompressible viscous fluid under a pressure gradient $\left(-\dfrac{\partial p}{\partial x} \right)$ flows. Assuming the flow to be laminar under steady state condition, determine the radius at which shear stress in the stream is zero.

Solution $\tau = \dfrac{1}{4} \dfrac{\partial p}{\partial x} \left[2r - \dfrac{1}{r} \cdot \dfrac{{R_1}^2 - {R_2}^2}{\log \dfrac{R_1}{R_2}} \right]$

The shear stress is to be zero, therefore,

$$\frac{1}{4}\frac{\partial p}{\partial x}\left[2r - \frac{1}{r}\cdot\frac{R_1^2 - R_2^2}{\log \frac{R_1}{R_2}}\right] = 0$$

or
$$2r - \frac{1}{r}\cdot\frac{R_1^2 - R_2^2}{\log \frac{R_1}{R_2}} = 0$$

or
$$2r^2 - \frac{R_1^2 - R_2^2}{\log \frac{R_1}{R_2}} = 0$$

or
$$r^2 = \frac{R_1^2 - R_2^2}{2 \log \frac{R_1}{R_2}}$$

or
$$r = \sqrt{\frac{R_1^2 - R_2^2}{2 \log \frac{R_1}{R_2}}}$$

9.5 FLOW OF VISCOUS FLUID BETWEEN TWO PARALLEL PLATES

Consider two parallel fixed plates located at a distance h apart as shown in Fig. 9.9. A fluid element of length dx, thickness dy and unit width is considered as a free body. In this case, the shear stress on the upper and lower surfaces of the fluid element will be different.

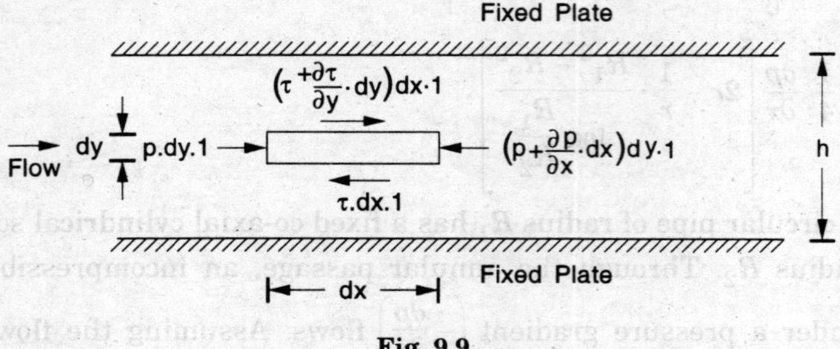

Fig. 9.9

The various forces acting on the fluid element are :

1. The pressure force, $p \,.\, dy \,.\, 1$ on the left end
2. The pressure force, $\left(p + \frac{\partial p}{\partial x} \,.\, dx\right) . \,dy \,.\, 1$ on the right end
3. The shear force, $\tau \,.\, dx \,.\, 1$ on the lower surface
4. The shear force, $\left(\tau + \frac{\partial \tau}{\partial y} \,.\, dy\right) . \,dx \,.\, 1$ on the upper surface.

The fluid above the upper surface of the fluid element moves faster relative to it which means that the surface of the fluid element moves backwards relative to the surface of the fluid above it and thus the shear force acts in the forward direction. As the flow is uniform and steady, there is no acceleration and the resultant force in the direction of flow is zero.

Thus

$$p \, . \, dy - \left(p + \frac{\partial p}{\partial x} dx\right) dy - \tau \, . \, dx + \left(\tau + \frac{\partial \tau}{\partial y} dy\right) dx = 0$$

$$-\frac{\partial p}{\partial x} + \frac{\partial \tau}{\partial y} = 0$$

$$\frac{\partial p}{\partial x} = \frac{\partial \tau}{\partial y}$$

$$= \frac{\partial}{\partial y}\left(\mu \, . \, \frac{du}{dy}\right)$$

$$= \mu \, . \, \frac{d^2u}{dy^2}$$

$$\frac{d^2u}{dy^2} = \frac{1}{\mu}\frac{\partial p}{\partial x}$$

Integrating the above equation with respect to y,

$$\frac{du}{dy} = \frac{1}{\mu}\frac{\partial p}{\partial x} y + C_1$$

Integrating again,

$$u = \frac{1}{\mu}\frac{\partial p}{\partial x}\frac{y^2}{2} + C_1 y + C_2 \qquad \text{...(9.18)}$$

where C_1 and C_2 are the constants of integration, the values of which are obtained from two boundary conditions.

In case both the plates are fixed,

$u = 0$ at $y = 0$

and $u = 0$ at $y = h$

$$0 = 0 + C_1 \times 0 + C_2$$

or $C_2 = 0$

and $$0 = \frac{1}{\mu}\frac{\partial p}{\partial x}\frac{h^2}{2} + C_1 h + C_2$$

$$C_1 = -\frac{1}{2\mu}\frac{\partial p}{\partial x} h$$

Substituting the values of C_1 and C_2 in Eq. 9.18,

$$u = \frac{1}{\mu}\frac{\partial p}{\partial x}\frac{y^2}{2} - \frac{1}{2\mu}\frac{\partial p}{\partial x} h \, . \, y$$

$$= -\frac{1}{2\mu}\frac{\partial p}{\partial x}(h \, . \, y - y^2) \qquad \text{...(9.19)}$$

This is the equation of a parabola indicating that the velocity distribution across the section of parallel plates is parabolic (Fig. 9.10).

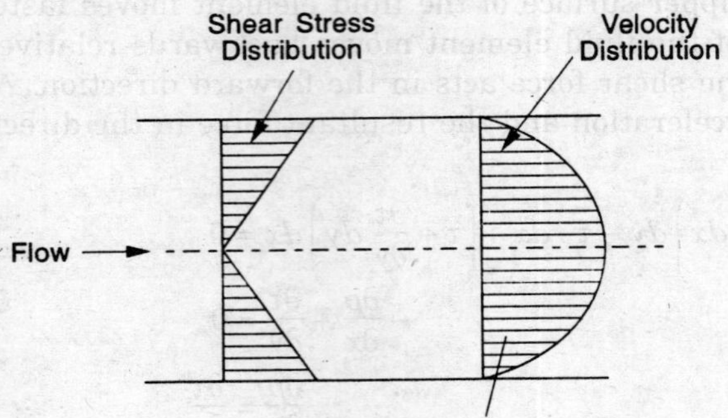

Fig. 9.10

The velocity is maximum when $\frac{d}{dy}(hy - y^2) = 0$

or $h - 2y = 0$

or $y = h/2$

$$u_{max} = -\frac{1}{2\mu}\frac{\partial p}{\partial x}\left[h\frac{h}{2} - \left(\frac{h}{2}\right)^2\right]$$

$$= -\frac{1}{2\mu}\frac{\partial p}{\partial x}h^2 \qquad ...(9.20)$$

To find the average velocity, discharge has to be found out. Discharge through the strip

$$dQ = \text{Velocity} \times \text{Area}$$

$$= -\frac{1}{2\mu}\frac{\partial p}{\partial x}(h \cdot y - y^2) \times (dy \cdot 1)$$

$$Q = \int_0^h -\frac{1}{2\mu}\frac{\partial p}{\partial x}(h \cdot y - y^2)\,dy$$

$$= -\frac{1}{2\mu}\frac{\partial p}{\partial x}\int_0^h (h \cdot y - y^2)\,dy$$

$$= -\frac{1}{2\mu}\frac{\partial p}{\partial x}\left[h \cdot \frac{y^2}{2} - \frac{y^3}{3}\right]$$

$$= -\frac{1}{2\mu}\frac{\partial p}{\partial x}\left[\frac{h^3}{2} - \frac{h^3}{3}\right]$$

$$= -\frac{1}{12\mu}\frac{\partial p}{\partial x}h^3 \qquad ...(9.21)$$

$$\bar{u} = \frac{Q}{a} = \frac{-\frac{1}{12\mu}\frac{\partial p}{\partial x}h^3}{h \times 1} = -\frac{1}{12\mu}\frac{\partial p}{\partial x}h^2$$

$$\frac{\partial p}{\partial x} = -\frac{12\mu\bar{u}}{h^2} \qquad ...(9.22)$$

Integrating this equation with respect to x between two section 1 and 2,

$$\int_1^2 dp = -\int_1^2 \frac{12\mu\bar{u}}{h^2}\, dx$$

$$p_2 - p_1 = -\frac{12\mu\bar{u}l}{h^2} \qquad (x_2 - x_1 = l)$$

$$p_1 - p_2 = \frac{12\mu\bar{u}l}{h^2} \qquad \ldots(9.23)$$

Also the ratio,

$$\frac{\bar{u}}{u_{max}} = \frac{-\dfrac{1}{12\mu}\dfrac{\partial p}{\partial x}h^2}{-\dfrac{1}{8\mu}\dfrac{\partial p}{\partial x}h^2} = \frac{2}{3} \qquad \ldots(9.24)$$

Shear stress distribution can be obtained by using Newton's law of viscosity.

$$\tau = \mu\frac{\partial u}{\partial y}$$

$$= \mu\frac{\partial}{\partial y}\left[-\frac{1}{2\mu}\frac{\partial p}{\partial x}(hy - y^2)\right]$$

$$= -\frac{1}{2}\frac{\partial p}{\partial x}\cdot\frac{\partial}{\partial y}(hy - y^2)$$

$$= -\frac{1}{2}\frac{\partial p}{\partial x}(h - 2y) \qquad \ldots(9.25)$$

i.e. shear stress varies linearly with y (Fig. 9.10). It is maximum where $y = 0$.

$$\tau_{max} = \tau_0 = -\frac{1}{2}\frac{\partial p}{\partial p}h \qquad \text{at the plate surface} \qquad \ldots(9.26)$$

Bottom Plate Fixed, Upper Plate Moving

This type of flow is also known as **Couette flow.** For this type of flow, the boundary conditions are

$$u = 0 \qquad \text{at } y = 0$$

and

$$u = V \qquad \text{at } y = h$$

(assuming that the upper plate moves with velocity V)

Substituting these values in Eq. 9.18,

$$C_2 = 0$$

$$V = \frac{1}{\mu}\frac{\partial p}{\partial x}\frac{h^2}{2} + C_1 h$$

$$C_1 = \frac{V}{h} - \frac{1}{2\mu}\frac{\partial p}{\partial x}h$$

Thus Eq. 9.18 can be written as

$$u = \frac{1}{\mu}\frac{\partial p}{\partial x}\frac{y^2}{2} + \left(\frac{V}{h} - \frac{1}{2\mu}\frac{\partial p}{\partial x}h\right)y$$

$$u = \frac{V}{h}y - \frac{1}{2\mu}\frac{\partial p}{\partial x}(hy - y^2) \qquad \ldots(9.27)$$

which shows that the velocity distribution in Couette flow depends upon the velocity of the moving plate V as well as the pressure gradient $\frac{\partial p}{\partial x}$ which can be positive or negative (Fig. 9.11).

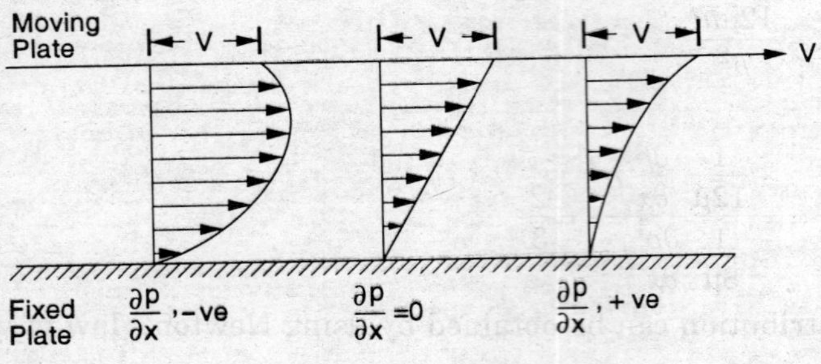

Fig. 9.11

If $\frac{\partial p}{\partial x} = 0,\ u = \frac{Vy}{h}$

and the flow is known as simple shear flow or plain Couette flow.

The discharge is given by,

$$Q = \int_0^h \left[\frac{V}{h} y - \frac{1}{2\mu} \frac{\partial p}{\partial x} (hy - y^2)\right] dy$$

$$= \left[\frac{V}{h}\frac{y^2}{2} - \frac{1}{2\mu}\frac{\partial p}{\partial x}\left(h\frac{y^2}{2} - \frac{y^3}{3}\right)\right]$$

$$= \frac{V}{h}\frac{h^2}{2} - \frac{1}{2\mu}\frac{\partial p}{\partial x}\left(\frac{h^3}{2} - \frac{h^3}{3}\right)$$

$$= V\frac{h}{2} - \frac{h^3}{12\mu}\frac{\partial p}{\partial x} \qquad ...(9.28)$$

The shear stress distribution is given by

$$\tau = \mu \frac{\partial}{\partial y}\left[\frac{V}{h} y - \frac{1}{2\mu}\frac{\partial p}{\partial x}(hy - y^2)\right]$$

$$= \mu\left[\frac{V}{h} - \frac{1}{2\mu}\frac{\partial p}{\partial x}(h - 2y)\right]$$

$$= \mu\frac{V}{h} - \frac{1}{2}\frac{\partial p}{\partial x}(h - 2y) \qquad ...(9.29)$$

Example 9.8 A 100 mm thick wall has a horizontal crack 30 mm wide and 2 mm deep. Determine the rate of leakage of water through the crack if the difference of pressure between the two ends of the crack is 0·4 kN/m^2. The viscosity of water is 0·01 poise.

Solution $\mu = 0{\cdot}001$ N.s/m^2 $\qquad l = 0{\cdot}1$ m

$h = 0{\cdot}002$ m $\qquad$ width $= 0{\cdot}03$ m

$$p_1 - p_2 = 0{\cdot}4 \times 1000 = 400 \text{ N/m}^2$$

$$p_1 - p_2 = \frac{12\mu \bar{u} l}{h^2}$$

$$400 = \frac{12 \times 0{\cdot}001 \times \bar{u} \times 0{\cdot}1}{0{\cdot}002^2}$$

$$\bar{u} = 1{\cdot}333 \text{ m/s}$$

$$Q = \bar{u} \times \text{area} = 1{\cdot}333 \times 0{\cdot}03 \times 0{\cdot}002 = 8 \times 10^{-5} \text{ m}^3\text{/s}$$

Example 9.9 The maximum velocity of oil flow between two parallel plates 80 mm apart is 1·2 m/s. Assuming the viscosity of oil to be 22 poise, determine

(i) the discharge per metre width
(ii) the pressure gradient
(iii) the pressure difference between two points 24 m apart
(iv) the shear stress at the plates
(v) the velocity gradient at the plates
(vi) the velocity at 18 mm from the plate.

Solution $\mu = 22$ poise $= 2{\cdot}2$ N.s/m^2

$h = 0{\cdot}8$ m

(i) $$\bar{u} = \frac{2}{3} u_{max} = \frac{2}{3} \times 1{\cdot}2 = 0{\cdot}8 \text{ m/s}$$

$$Q = \bar{u} \times h \times 1 = 0{\cdot}8 \times 0{\cdot}08 \times 1 = 0{\cdot}064 \text{ m}^3\text{/s}$$

(ii) $$\frac{\partial p}{\partial x} = -\frac{12\mu\bar{u}}{h^2} = \frac{12 \times 2{\cdot}2 \times 0{\cdot}8}{0{\cdot}08^2} = 3300$$

(iii) From (ii)

$$\frac{p_1 - p_2}{x_2 - x_1} = \frac{12\mu\bar{u}}{h^2} = 3300$$

$$\therefore \quad p_1 - p_2 = 3300 \times 24 = 79200 \text{ N/m}^2$$

(iv) $$\tau_0 = -\frac{1}{2}\frac{\partial p}{\partial x} h = -\frac{1}{2} \times (-3300) \times 0{\cdot}08 = 132 \text{ N/m}^2$$

(v) $$\tau_0 = \mu \left(\frac{\partial u}{\partial y}\right)_{y=0}$$

$$\left(\frac{\partial u}{\partial y}\right)_{y=0} = \frac{\tau_0}{\mu} = \frac{132}{2{\cdot}2} = 60 \text{ s}^{-1}$$

(vi) $$u = -\frac{1}{2\mu}\left(\frac{\partial p}{\partial x}\right)(hy - y^2)$$

$$= -\frac{1}{2 \times 2{\cdot}2} \times (-3300) \times (0{\cdot}08 \times 0{\cdot}018 - 0{\cdot}018^2) = 0{\cdot}837 \text{ m/s}$$

Example 9.10 The space between two horizontal plates 12 mm apart is filled with a liquid of viscosity 0·9 poise. The lower plate is stationary and the upper plate moves at a velocity of 1·2 m/s. The pressure difference between two sections 40 m apart is 48 kN/m^2. Determine the velocity distribution, the discharge per unit width and the shear stress on the upper plate.

Solution $\mu = 0{\cdot}8$ poise $= 0{\cdot}08$ N.s/m^2

$V = 1{\cdot}2$ m/2

$h = 0{\cdot}012$ m

$$\frac{\partial p}{\partial x} = -\frac{p_2 - p_1}{x_2 - x_1} = -\frac{48000}{40} = -1200 \text{ N/m}^2\text{/m}$$

$$u = \frac{V}{h} y - \frac{1}{2\mu}\frac{\partial p}{\partial x}(hy - y^2)$$

$$= \frac{V}{h} y + \frac{1}{2\mu}\left(-\frac{\partial p}{\partial x}\right)(hy - y^2)$$

$$= \frac{1{\cdot}2}{0{\cdot}012} y + \frac{1}{2 \times 0{\cdot}08} \times (-1200)\, y\,(0{\cdot}012 - y)$$

$$= y\,(100 + 90 - 7500y)$$

$$= y\,(190 - 7500y)$$

$$Q = V\frac{h}{2} - \frac{h^3}{12\mu}\left(\frac{\partial p}{\partial x}\right) = 1{\cdot}2 \times \frac{0{\cdot}012}{2} - \frac{0{\cdot}012^3}{12 \times 0{\cdot}08} \times (-1200)$$

$$= 0{\cdot}0072 + 0{\cdot}00216$$

$$= 0{\cdot}00935 \text{ m}^3\text{/s}$$

$$\tau_0 = \mu\frac{V}{h} - \frac{1}{2}\frac{\partial p}{\partial x}(h - 2y) = 0{\cdot}08 \times \frac{1{\cdot}2}{0{\cdot}012} - \frac{1}{2}(-1200) \times (0{\cdot}012 - 2 \times 0{\cdot}012)$$

$$= 8 - 7{\cdot}2 = 0{\cdot}8 \text{ N/m}^2$$

($y = h$, at the upper plate)

9.6 POWER ABSORBED IN BEARINGS

In a bearing, a very thin layer of lubricating oil is maintained between its stationary surface and that of the revolving shaft. If a highly viscous oil is used for lubrication of bearings, a great resistance and hence a great power loss will occur. On the other hand, a light oil may not be able to maintain a required oil film between the staionary and rotating surfaces. Thus an oil of proper viscosity should be used for the lubrication of bearings. Power absorbed or power loss due to viscous resistance in different types of bearing is as follows :

Journal Bearings

Consider a journal bearing as shown in Fig. 9.12 in which a shaft of diameter D rotates at N rpm. Let L be the length of the bearing and t, be the thickness of the oil film.

As the thickness of oil film is very small, a linear velocity distribution along the thickness can be assumed.

$$\frac{du}{dy} = \frac{V - 0}{t} = \frac{V}{t} = \frac{\pi DN}{60t}$$

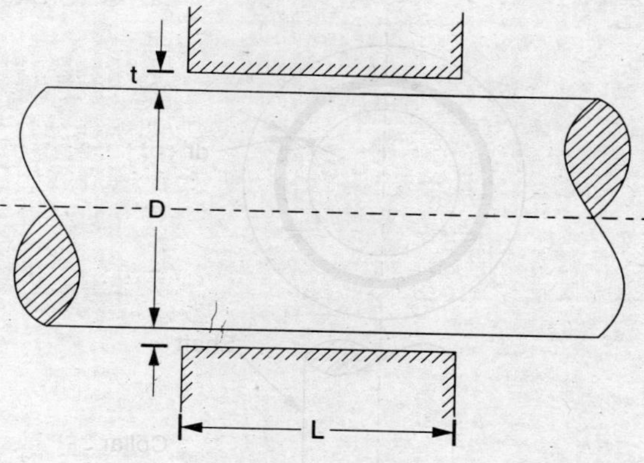

Fig. 9.12

Now power absorbed,

$$P = T \times \omega = \frac{2\pi NT}{60}$$

where

$$T = F \times \frac{D}{2}$$

$$= (\tau \times \text{Surface area}) \times \frac{D}{2}$$

$$= \left(\mu \frac{du}{dy} \,.\, \pi DL\right) \times \frac{D}{2}$$

$$= \mu \frac{\pi DN}{60t} \,.\, \pi DL \,.\, \frac{D}{2}$$

$$= \frac{\mu\pi^2 D^3 NL}{120t} \qquad \ldots(9.30)$$

Collar Bearings

A collar bearing takes up the axial thrust of a shaft. The face of the collar is separated from the bearing surface by an oil film of uniform thickness (Fig. 9.13).

Let R_1 = internal radius of the collar

R_2 = external radius of the collar

N = speed of the shaft in rpm

t = thickness of the oil film

Consider an elementary ring of thickness dr at a radial distance r from the axis of the shaft. As the thickness of the oil film is small, a linear velocity distribution along the thickness can be assumed.

$$\frac{du}{dy} = \frac{V - 0}{t} = \frac{V}{t} = \frac{2\pi rN}{60t}$$

(V is the tangential velocity of the shaft at radius r).

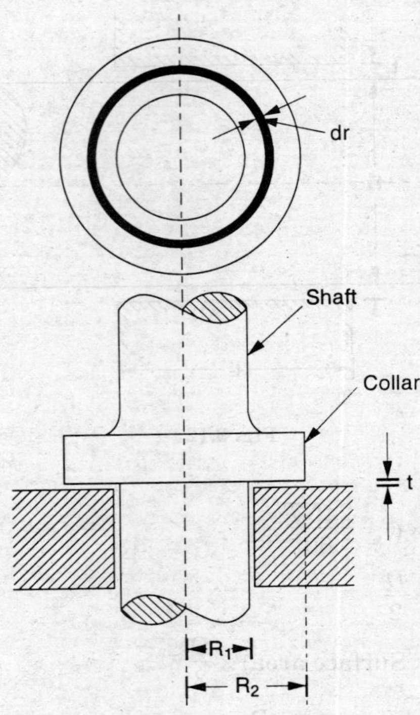

Fig. 9.13

Shear force on the elementary ring,

$$F = \text{Shear stress} \times \text{Area of the ring},$$

$$= \tau \,.\, 2\pi r \,.\, dr = \mu \frac{du}{dy} \,.\, 2\pi r \,.\, dr = \mu \frac{2\pi N}{60t} \,.\, 2\pi r^2 \,.\, dr = \frac{\mu}{15t} \,.\, \pi^2 N r^2 \,.\, dr$$

Shear torque,

$$\delta T = \left(\frac{\mu}{15t} \,.\, \pi^2 N r^2 \,.\, dr \right) . \, r$$

Total torque

$$T = \int_{R_1}^{R_2} \frac{\mu}{15t} \,.\, \pi^2 N r^3 \,.\, dr$$

$$= \frac{\mu}{15t} \,.\, \pi^2 N \int_{R_1}^{R_2} r^3 \,.\, dr$$

$$= \frac{\mu}{15t} \,.\, \pi^2 N \left(\frac{r^4}{4} \right)_{R_1}^{R_2}$$

$$= \frac{\mu}{60t} \,.\, \pi^2 N \,(R_2{}^4 - R_1{}^4) \qquad ...(9.31)$$

Power absorbed,

$$P = T \times \omega = \frac{2\pi N T}{60}$$

Foot Step Bearing

Figure 9.14 shows a foot step bearing in which one end of the shaft rests in the bearing. A thin film of oil separates the two. The bearing withstands the vertical load (axial thrust) of the shaft.

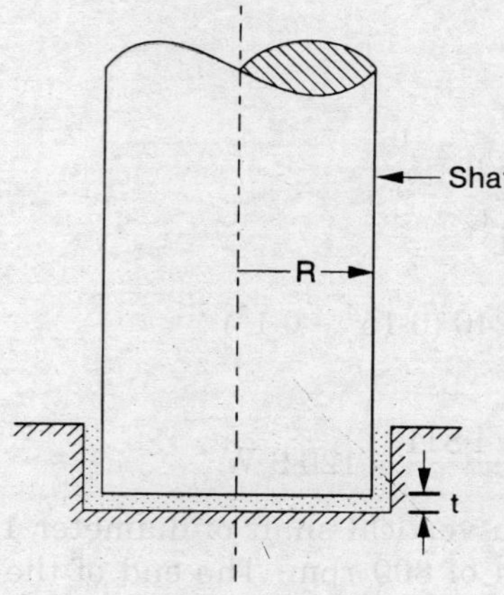

Fig. 9.14

A little consideration shows that the function of a foot step bearing is the same as that of a collar bearing. Ths, the same relations are valid in this case also with the modification that $R_1 = 0$ and $R_2 = R$.

$$T = \frac{\mu}{60t} \cdot \pi^2 NR^4 \qquad \text{...(9.32)}$$

$$P = \frac{2\pi NT}{60}$$

Example 9.11 A shaft 80 mm in diameter rotates in a 160 mm long bearing at a speed 50 rpm. The two surfaces are uniformly separated by a clearance of 0·5 mm. The clearance has a lubricating oil of dynamic viscosity 0·04 poise. What is the power absorbed in the bearing if velocity distribution in the oil is assumed to be linear ?

Solution $\mu = 0{\cdot}04$ poise $= 0{\cdot}004$ N.s/m^2

$D = 0{\cdot}08$ m

$N = 50$ rpm

$L = 0{\cdot}16$ m

$t = 0{\cdot}5$ mm $= 0{\cdot}0005$ m

$$T = \frac{\mu\pi^2 D^3 NL}{120t} = \frac{0{\cdot}004 \times \pi^2 \times 0{\cdot}08^3 \times 50 \times 0{\cdot}16}{120 \times 0{\cdot}0005} = 0{\cdot}002695 \text{ Nm}$$

$$P = \frac{2\pi NT}{60} = \frac{2\pi \times 50 \times 0{\cdot}002695}{60} = 0{\cdot}0141 \text{ W}$$

Example 9.12 A collar bearing used to take the thrust of a shaft has external and internal diameters 300 mm and 200 mm respectively. An oil film of thickness 0·3 mm and of dynamic viscosity 0·9 poise is maintained between the collar surface and the bearing. Determine the power lost in overcoming the viscous resistance of the oil if the shaft rotates at 240 rpm.

Solution $R_1 = 0{\cdot}2/2 = 0{\cdot}1$ m

$R_2 = 0{\cdot}3/2 = 0{\cdot}15$ m

$N = 240$ rpm

$t = 0{\cdot}3$ mm $= 0{\cdot}0003$ m

$\mu = 0{\cdot}9$ poise $= 0{\cdot}09$ N.s/m^2

$$T = \frac{\mu}{60t} \cdot \pi^2 N (R_2{}^4 - R_1{}^4)$$

$$= \frac{0{\cdot}09}{60 \times 0{\cdot}0003} \cdot \pi^2 \times 240\ (0{\cdot}15^4 - 0{\cdot}1^4)$$

$$= 4{\cdot}811 \text{ N.m}$$

$$P = \frac{2\pi NT}{60} = \frac{2\pi \times 240 \times 4{\cdot}811}{60} = 120{\cdot}9 \text{ W}$$

Example 9.13 The lower end of a vertical shaft of diameter 120 mm rotates in a foot step bearing at a speed of 800 rpm. The end of the shaft and the surface of the bearing are both flat and are separated by an oil film of thickness 0·4 mm. Find the power absorbed in the bearing if the viscosity of the oil is 1·5 poise.

Solution $D = 0{\cdot}12$ m

$N = 800$ rpm

$t = 0{\cdot}4$ mm $= 0{\cdot}0004$ m

$\mu = 1{\cdot}5$ poise $= 0{\cdot}15$ N.s/m^2

$$T = \frac{\mu}{60t} \cdot \pi^2 N R^4$$

$$= \frac{0{\cdot}15}{60 \times 0{\cdot}0004} \cdot \pi^2 \times 800 \times \left(\frac{0{\cdot}12}{2}\right)^4$$

$$= 0{\cdot}6396 \text{ N.m}$$

$$P = \frac{2\pi NT}{60} = \frac{2\pi \times 800 \times 0{\cdot}6396}{60} = 53{\cdot}58 \text{ W}$$

9.7 MOVEMENT OF PISTON IN DASHPOT

A dashpot is used to dampen the vibrations of machines. In its simplest form, it contains a piston moving in a cylinder containing oil of viscosity μ (Fig. 9.15). As the piston moves due to vibrations, oil from one side of the piston transfers to the other through a small clearance provided for the purpose between the piston and the cylinder walls. Depending upon the viscosity of the oil, the resistance to the movement of the piston and thus to the vibration of the machines is provided.

Let L = length of the piston

D = diameter of the piston

V = velocity of the piston

$\bar{u}$ = average velocity of the oil in the clearance

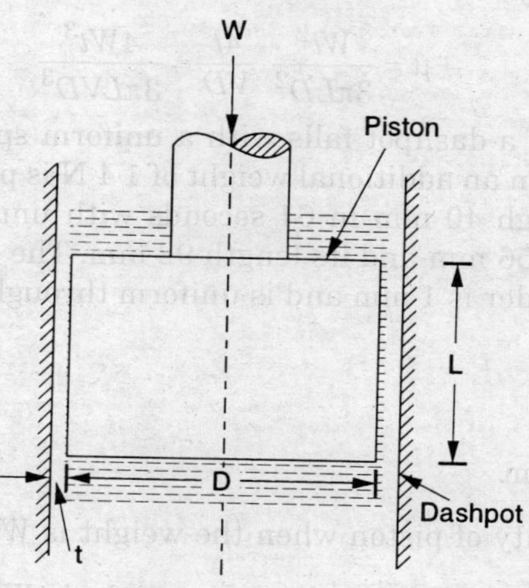

Fig. 9.15

t = clearance between the piston and the dashpot

μ = viscosity of the oil

W = weight of the piston

The flow of oil through the clearance of the dashpot is similar to that of flow between two parallel plates. Thus the difference of pressure between the two ends of the piston is given by the relation

$$\Delta p = \frac{12\mu \bar{u} L}{t^2} \quad \ldots(i)$$

This difference of pressure intensities on the two ends must be equal to the pressure exerted due to the weight of the piston which is equal to

$$\frac{W}{\frac{\pi}{4} D^2} = \frac{4W}{\pi D^2} \quad \ldots(ii)$$

Equating (i) and (ii),

$$\frac{12\mu \bar{u} L}{t^2} = \frac{4W}{\pi D^2}$$

$$\mu = \frac{Wt^2}{3\pi \bar{u} L D^2} \quad \ldots(9.33)$$

u, the average velocity of oil in the clearance may be replaced in terms of V, the velocity of the piston by using the continuity equation *i.e.*

Rate of flow in the clearance = Rate of flow in the dashpot

Average velocity × Clearance area = Velocity of piston × Area of piston

$$\bar{u} \times \pi D t = V \times \frac{\pi}{4} D^2$$

$$\bar{u} = \frac{VD}{4t}$$

The Eq. 9.33 is modified to

$$\mu = \frac{Wt^2}{3\pi LD^2} \frac{4t}{VD} = \frac{4Wt^3}{3\pi LVD^3} \quad \text{...(9.34)}$$

Example 9.14 The piston of a dashpot falls with a uniform speed and covers 40 mm in 80 seconds. When an additional weight of 1·4 N is placed on the top of the piston, it falls through 40 mm in 64 seconds with uniform speed. The diameter of the piston is 56 mm and its length 92 mm. The clearance between the piston and the cylinder is 1 mm and is uniform throughout. Determine the viscosity of the oil.

Solution $D = 0{\cdot}056$ m

$L = 0{\cdot}092$ m

$t = 1$ mm $= 0{\cdot}001$ m

V = velocity of piston when the weight is $W = \frac{40}{80} = 0{\cdot}5$ mm/s $= 0{\cdot}0005$ m/s

V' = velocity of piston when the weight is $W + 1{\cdot}4$

$$\mu = \frac{4Wt^3}{3\pi LVD^3} = \frac{4\,(W + 1{\cdot}4)\,t^3}{3\pi LV'D^3}$$

$$\frac{W}{W + 1{\cdot}4} = \frac{V}{V'} = \frac{40/80}{40/64} = 0{\cdot}8$$

$$W = 0{\cdot}8\,W + 1{\cdot}12$$

$$W = 5{\cdot}6 \text{ N}$$

$$\mu = \frac{4 \times 5{\cdot}6 \times (0{\cdot}001)^3}{3\pi \times 0{\cdot}92 \times 0{\cdot}0005 \times (0{\cdot}056)^3} = 0{\cdot}2942 \text{ N.s/m}^2$$

9.8 VISCOMETERS

A viscometer is an apparatus used to determine the viscosity of liquid. They are of the following types :

1. Capillary Tube Viscometer
2. Rotating Cylinder Viscometer
3. Falling Sphere Viscometer
4. Efflux Viscometer

1. Capillary Tube Viscometer

Fig. 9.16 shows the arrangement of a capillary type viscometer. In this method, the liquid is filled in a constant head tank and is allowed to pass through a capillary tube. The liquid is collected in a measuring tank for a given period of time and the rate of flow of the liquid is determined. The pressure head is measured at a far away point from the tank.

As the pressure head at the outlet is atmospheric, h is the difference of pressure head for the length L of the capillary tube.

Let d = diameter of the tube

Q = volume flow rate of the liquid

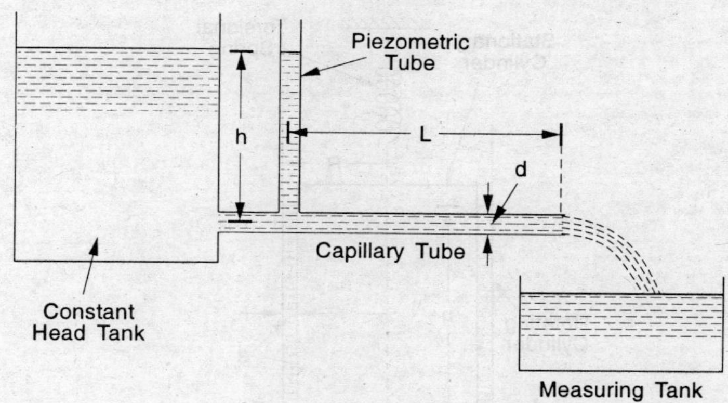

Fig. 9.16

L = length of the tube
w = specific weight of the liquid
μ = coefficient of viscosity of the liquid

According to Hagen Poiseville's equation

$$h = \frac{32\mu L}{wd^2}\,\overline{u} \qquad \text{(Eq. 9.12)}$$

$$= \frac{32\mu L}{wd^2}\;\frac{Q}{\frac{\pi}{4}d^2}$$

$$= \frac{128\mu QL}{\pi wd^2}$$

$$\mu = \frac{\pi whd^2}{128QL} \qquad \ldots(9.35)$$

2. Rotating Cylinder Viscometer

In this method, two concentric cylinders of radii r and R are used. The annular space between the cylinders (a on the side and t at the bottom) is filled with the liquid (Fig. 9.17). The outer cylinder is rotated at a fixed speed N and the inner cylinder is held stationary. The inner cylinder is suspended by a torsion spring which measures the torque T required to hold the inner cylinder stationary.

The total viscous torque = Viscous torque due to liquid in the annular space + Viscous torque due to liquid at the bottom

or $$T = T_a + T_b$$

Now $$T_a = \text{Viscous drag} \times \text{radius}$$

$$= (\text{Shear stress} \times \text{Area}) \times \text{Radius}$$

$$= \left(\mu\,\frac{du}{dy}\right)(2\pi r\,.\,h)\,r$$

As the annular space is quite small, the velocity gradient $\frac{du}{dy}$ may be simplified as

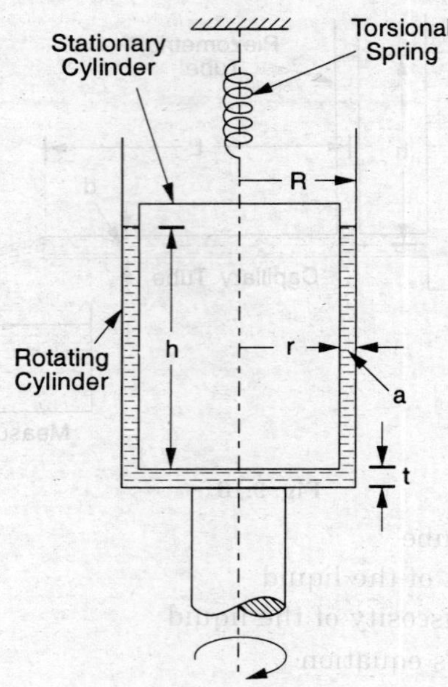

Fig. 9.17

$$\frac{du}{dy} = \frac{\text{Tangential velocity}}{\text{Space thickness}} = \frac{u}{a} = \frac{\pi DN}{60a} = \frac{2\pi RN}{60a}$$

$$T_a = \mu \frac{2\pi RN}{60a} 2\pi rhr = \frac{\mu\pi^2 Nr^2 Rh}{15a}$$

$$T_b = \frac{\mu\pi^2 Nr^4}{60t} \qquad \text{(Eq. 9.32)}$$

$$T = \frac{\mu\pi^2 Nr^2 Rh}{15a} + \frac{\mu\pi^2 Nr^4}{60t}$$

$$= \frac{\mu\pi^2 N}{15}\left[\frac{r^2 Rh}{a} + \frac{r^4}{4t}\right]$$

$$\mu = \frac{T}{\frac{\pi^2 N}{15}\left[\frac{r^2 Rh}{a} + \frac{r^4}{4t}\right]} \qquad \text{...(9.36)}$$

where T is the torque measured by the torsional spring.

3. Falling Sphere Viscometer

This method makes use of the Stoke's law. Fig. 9.18 shows a falling sphere viscometer with a tall vertical transparent cylinderical tank filled with liquid whose viscosity is to be found. The cylindrical tank is maintained at a constant temperature by enclosing it inside a constant temperature bath tub. A steel ball or sphere is made to fall into the test liquid. The sphere accelerates under the gravitational force until it attains the maximum (terminal) velocity V.

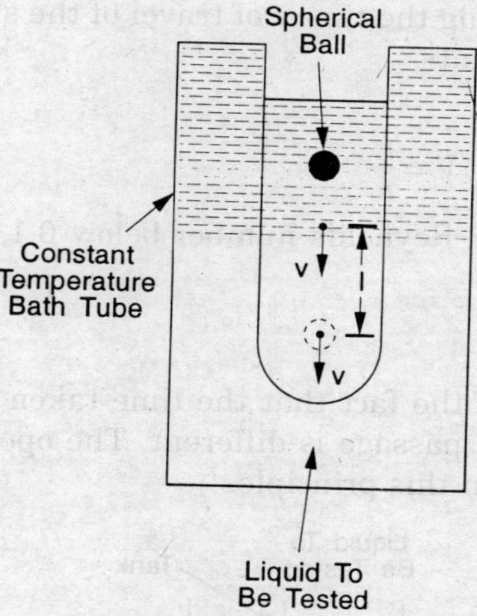

Fig. 9.18

The buoyant and the viscous drag forces balance the gravity force and onwards the sphere falls with a uniform velocity V.

Let ρ_L = density of the liquid

ρ_S = density of the sphere

d = diameter of the spherical ball

V = velocity of the falling ball

Now, weight of sphere, W = Volume of sphere × Density × g

$$= \frac{\pi}{6} d^3 \cdot \rho_S \cdot g$$

Buoyant force, F_B = Weight of the liquid displaced by the sphere

= Mass of liquid displaced × g

= Volume of liquid displaced × Density of liquid × g

$$= \frac{\pi}{6} d^3 \times \rho_L \times g$$

Drag force,

$$F_D = 3\pi \mu V \cdot d \qquad \text{(drag of sphere)}$$

Thus for the equilibrium,

$$F_D + F_B = W$$

$$3\pi \mu V d + \frac{\pi}{6} d^3 \rho_L \cdot g = \frac{\pi}{6} d^3 \rho_S \cdot g$$

$$3\pi \mu V d + \frac{\pi}{6} d^3 g (\rho_S - \rho_L)$$

$$\mu = \frac{g d^2}{18V} (\rho_S - \rho_L) \qquad \ldots(9.37)$$

Velocity V is known by noting the time t of travel of the sphere through a known vertical distance l between two fixed points.

Then

$$V = \frac{l}{t}$$

Stoke's law is applicable for Reynolds number below 0·1, where the wall does not affect the terminal velocity of fall.

4. Efflux Viscometers

These viscometers make use of the fact that the time taken by a fixed quantity of different liquids to pass through a small passage is different. The operation of Saybolt, Redwood and Engler viscometers are based on this principle.

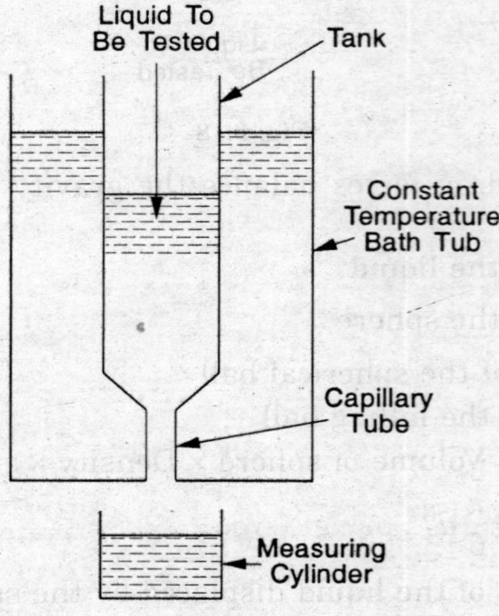

Fig. 9.19

Fig. 9.19 shows a Saybolt viscometer. It consists of a tank having a short capillary tube at its bottom. The liquid whose viscosity is to be found, is filled in this tank. The tank is surrounded by a constant temperature bath that keeps the temperature of the liquid constant during the testing. The time taken by 60000 mm^3 of liquid to flow out of the reservoir is noted down. The initial height of the liquid in the tank is fixed at a predetermined level. The kinematic viscosity of the liquid is found from the time measured using either an empirical formula or a calibration chart.

Example 9.15 The specific gravity of oil used in a capillary tube viscometer having a tube diameter of 48 mm is 0·85. The difference of pressure head between two points 1·6 m apart is found to be 0·4 m of water. The weight of the oil collected in a measuring tank is 500 N in 100 seconds. What is the viscosity of the oil ?

Solution $h = 0{\cdot}4$ m

$d = 0{\cdot}048$ m

$L = 1{\cdot}6$ m

$$w = 0{\cdot}85 \times 9{\cdot}81 \times 1000 = 8338{\cdot}5 \text{ N/m}^3$$

$$Q = \frac{\text{Weight of oil collected/s}}{\text{Weight density}} = \frac{500/100}{8338{\cdot}5} = 0{\cdot}0005996 \text{ m}^3\text{/s}$$

$$\mu = \frac{\pi whd^2}{128QL} = \frac{\pi \times 8338{\cdot}5 \times 0{\cdot}4 \times (0{\cdot}048)^4}{128 \times 0{\cdot}0005996 \times 1{\cdot}6} = 0{\cdot}453 \text{ N.s/m}^2$$

Example 9.16 **The radii of the cylinders in a rotating cylinder viscometer are 36 mm and 34 mm. The outer cylinder is rotated steadily at 240 rpm. The annular space between the two cylinders is filled with a certain liquid upto a depth of 100 mm and the torque produced on the inner cylinder is $1{\cdot}2 \times 10^{-4}$ N.m. Calculate the viscosity of the liquid. The bottom clearance is 2 mm.**

Solution

$$T = \frac{\mu\pi^2 N}{15}\left[\frac{r^2 Rh}{a} + \frac{r^4}{4t}\right]$$

$$1{\cdot}2 \times 10^{-4} = \frac{\mu\pi^2 \times 240}{15}\left[\frac{0{\cdot}034^2 \times 0{\cdot}036 \times 0{\cdot}1}{0{\cdot}002} + \frac{0{\cdot}034^4}{4 \times 0{\cdot}002}\right]$$

$$= \mu \times 157{\cdot}91\,(0{\cdot}0020808 + 0{\cdot}0000167)$$

$$= \mu \times 157{\cdot}91 \times 0{\cdot}0020975$$

$$\mu = 3{\cdot}623 \times 10^{-4} \text{ N.s/m}^2$$

Example 9.17 **In a falling sphere viscometer, a lubricating oil of density 800 kg/m³ is placed in a 100 mm inside diameter tube. A steel ball of diameter 8 mm and of density 9000 kg/m³ travels a distance of 1 m in 20 seconds. Find the viscosity of the oil.**

Solution $d = 0{\cdot}008$ m

$V = 1/20 = 0{\cdot}05$ m/s

$\rho_S = 9000$ kg/m³

$\rho_L = 800$ kg/m³

$$\mu = \frac{gd^2}{18V}(\rho_S - \rho_L) = \frac{9{\cdot}81 \times (0{\cdot}008)^2}{18 \times (1/20)}(9000 - 800) = 5{\cdot}72 \text{ N.s/m}^2$$

Reynolds number should be checked if it is less than 0·1

$$\text{Re} = \frac{\rho Vd}{\mu} = \frac{800 \times 0{\cdot}05 \times 0{\cdot}008}{5{\cdot}72} = 0{\cdot}056$$

PROBLEMS

1. How do you differentiate between viscous and turbulent flows ?
2. Describe the experimental set-up of Osborne Reynolds. What were his observations ?
3. What is Reynolds number ? What are upper and lower critical Reynolds numbers ?
4. Show that for a steady laminar flow in a circular pipe
   (a) the velocity distribution across the section is parabolic and
   (b) the average velocity is half the maximum local velocity.

5. What is Hagen-Poiseuille equation ? What assumptions are made while deriving this equation ?
6. Derive expressions for discharge, average velocity of flow and shear stress distribution for a viscous flow through an annulus.
7. Show that for a steady laminar flow through two fixed parallel plates
   (a) the velocity distribution is parabolic
   (b) the average velocity is two-third of the maximum velocity.
8. What is a Couette flow ?
9. Deduce expressions for the power absorbed in overcoming viscous resistance in case of
   (a) a journal bearing
   (b) a collar bearing.
10. What is a dashpot ? Find an expression to determine the coefficient of viscosity in a dashpot ?
11. What is a viscometer ? Describe any two types of viscometers.
12. An oil of specific gravity 0·9 and viscosity 1·43 poise flows through a pipe 3 m long and 25 mm in diameter. The velocity of flow is 1/10th of the critical velocity for which Reynolds number is 2500. Determine the velocity of flow through the pipe, the head required to maintain the flow and the power required to overcome the viscous resistance. (1·589 m/s, 3·59 m of oil, 24·73 W)
13. A 500 m long and 50 mm diameter pipe slopes upwards at 1 in 50. An oil of specific gravity 0·9 and viscosity 8 poise is pumped at the rate of 0·3 m^3/s. Is the flow laminar ? If so, determine
   (a) the pressure difference to maintain the flow
   (b) the power required assuming an overall efficiency of 60 percent of the pump ?
   (c) the centreline velocity
   (d) the velocity gradient at the pipe wall.

   (The flow is laminar, 12088kN/m^2, 60·5 kW, 5·1 m/s, 377·5 s^{-1})
14. An oil of specific gravity 0·9 and viscosity 0·9 poise flows through a horizontal circular pipe 12 m in length and 120 mm in diameter. If 1·57 kN of oil is collected in a tank in 50 seconds, find the difference of pressure at the two ends of the pipe.

   (753·6 N/m^2)
15. A 30 mm diameter pipe is used to pump a lubricating oil of viscosity 0·1 N.s/m^2 and specific gravity 0·9. The pressure drop per metre length of the pipe is 20 kN/m^2. Determine the mass flow rate and the shear stress at the pipe wall.

   (3·578 kg/s, 150 N/m^2)
16. In a pipe of 200 mm diameter, the maximum velocity for a laminar flow is found to be 1·5 m/s. Determine the average velocity and the radius at which it occurs. Also find the velocity at 40 mm from the pipe wall. (0·75 m/s, 70·7 mm, 0·96 m/s)
17. In a laminar flow through a 60 mm diameter circular pipe, a pitot tube held at a radial distance of 20 mm from the axis indicates a velocity of 0·6 m/s. Determine the maximum velocity, the mean velocity and the discharge.

   (0·675 m/s, 0·3375 m/s, 0·964 l/s)

18. For a laminar flow in a circular pipe, find the distance from the boundary where the local velocity equals the average velocity. (0·293 R)

19. An oil of viscosity 30 poise flows between two parallel plates kept 60 mm apart. The maximum velocity is found to be 1 m/s. Determine

    (a) the discharge per metre width

    (b) the pressure gradient

    (c) the shear stress and the velocity gradient at the plates

    (d) the velocity at 10 mm from the plates.

    (0·04 m^3/s, 6666·6, 200 N/m^2, 66·67 s^{-1}, 0·5556 m/s)

20. The wall of a water tank 600 mm thick has a horizontal crack 3 m wide and 0·15 mm deep. The crack is 3 m below the water surface. Determine the rate of leakage, the maximum velocity and the velocity gradient at the bounday. The viscosity of water is 0·01 poise. (4.14×10^{-5} m^3/s, 0·138 m/s, 3680 s^{-1})

21. A shaft 120 mm in diameter rotates at 80 rpm in a 300 mm long bearing. The two surfaces are uniformly separated by a clearance of 0·6 mm. The clearance has a lubricating oil of dynamic viscosity 0·05 poise. Calculate the power absorbed in the bearing. Assume linear velocity distribution in the oil. (0·357 W)

22. A collar bearing used to take the thrust of a shaft has external and internal diameters as 200 mm and 150 mm respectively. An oil film of 0·2 mm thickness and 0·85 poise viscosity is maintained between the collar surface and the bearing. If the shaft rotates at 200 rpm, determine the power lost in overcoming the friction. (20·02 W)

23. An arrangement to damp out the vibrations consists of a piston moving in a cylinder having oil. The piston falls with a uniform speed and covers 60 mm in 90 seconds. When an additional weight of 1·6 N is placed on the top of the piston, it falls through 60 mm in 75 seconds. The diameter of the piston is 50 mm and the length 72 mm. The clearance between the cylinder and the piston is 0·8 mm and is uniform throughout. Determine the viscosity of the oil. (0·2898 N.s/m^2)

34. The viscosity of an oil of specific gravity 0·9 is measured by using a capillary tube of 50 mm diameter. The difference of pressure head between two points 1 m apart is 0·2 mm of water. What is the viscosity of oil if the weight of oil collected in a measuring tank is 300 N in 75 seconds. (0·598 N.s/m^2)

25. In a viscous liquid, a sphere of 1 mm diameter falls through 268 mm in 80 seconds. The specific gravities of the sphere and the liquid are 7 and 0·96 respectively. Determine the dynamic viscosity of the liquid. (0·9826 N.s/m^2)

Chapter 10

TURBULENT FLOW

10.1 INTRODUCTION

In a pipe flow, when the Reynolds number is greater than 4000, the flow is termed as turbulent. This type of flow is irregular and chaotic with violent mixing of the fluid. Consequently, mass, momentum and heat transfer are increased. The velocity profiles become more flatter as compared to velocity profile for the laminar flow and continue to do so with increase in the Reynolds number. The ratio of the mean velocity to the maximum velocity increases. Experiments show that whereas the ratio of mean to maximum velocity is approximately 0·5 for laminar flow, it is 0·8 to 0·84 for turbulent flow (Fig. 10.1). In a turbulent flow, the slope of the velocity profile at the wall is quite high. As a result, the shear stress and the skin friction drag are very high. For the same rate of discharge, viscosity and pipe size, there are much greater frictional losses in a turbulent flow than in a laminar flow due to the formation of eddies, mixing and curving of pathlines.

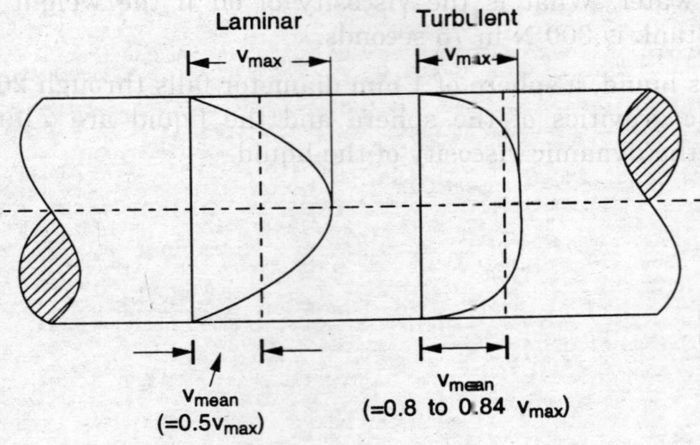

Fig. 10.1

10.2 LOSS OF HEAD IN PIPES

Consider a uniform horizontal pipe having a steady flow of water. Consider a control volume between sections 1–1 and 2–2 (Fig. 10.2).

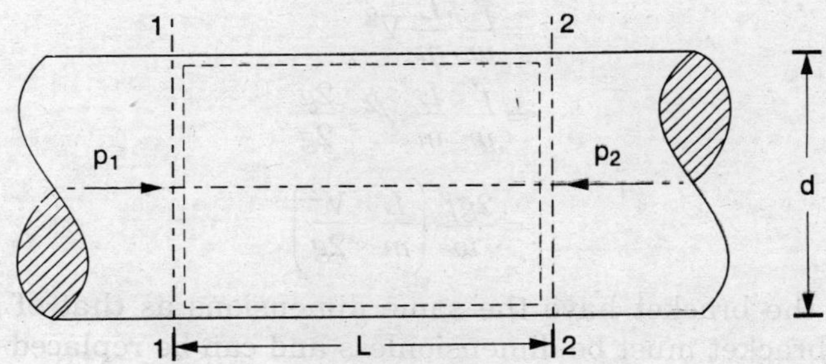

Fig. 10.2

Let L = length of the pipe between 1–1 and 2–2

d = diameter of pipe

A = area of cross-section of the pipe

p_1 = pressure intensity at section 1–1

p_2 = pressure intensity at section 2–2

f = frictional resistance per unit wetted area

V = average flow velocity

P = wetted perimeter

Owing to the frictional resistance to the flow, the pressure continuously decreases along the length of the pipe. Thus it may be written,

$$p_1 A = p_2 A + \text{Frictional resistance}$$

$$(p_1 - p_2) \,.\, A = \text{Frictional resistance}$$

The experiments have shown that the frictional resistance is proportional to

- pipe length
- wetted perimeter
- V^n where V is the average flow velocity and n is an index which depends upon the material and the nature of the pipe surface. Its value varies between 1·5 and 2·0. For commercial pipes, its value is taken as 2 for turbulent flows.

$$(p_1 - p_2)\, A \propto PLV^2 = f'\, PLV^2$$

where f' is the frictional resistance per unit wetted area.

$$\frac{p_1 - p_2}{w} = \frac{f'\, PLV^2}{wA}$$

$$h_f = \frac{f'}{w}\left(\frac{P}{A}\right) LV^2$$

A/P is known as the **hydraulic mean depth** or **hydraulic radius** and is denoted by m and has units of length. Therefore,

$$h_f = \frac{f'}{w}\left(\frac{1}{m}\right)LV^2 \qquad ...(10.1)$$

$$= \frac{f'}{w}\frac{L}{m}V^2$$

$$= \frac{f'}{w}\frac{L}{m}V^2 \cdot \frac{2g}{2g}$$

$$= \frac{2gf'}{w}\left(\frac{L}{m}\cdot\frac{V^2}{2g}\right)$$

the terms inside the bracket have the same dimensions as that of h and therefore the quantity outside the bracket must be dimensionless and can be replaced by another constant (say f)

$$h_f = f\frac{L}{m}\frac{V^2}{2g} \qquad ...(10.1a)$$

For a circular pipe,

$$m = \frac{A}{P} = \frac{\frac{\pi}{4}d^2}{\pi d} = \frac{d}{4}$$

$$h_f = f\frac{4L}{d}\frac{V^2}{2g} = \frac{4fLV^2}{2gd} \qquad ...(10.2)$$

The factor f is known as **Darcy coefficient of friction.**
($4f$ is usually referred as the **friction factor**).

The above equation is known as **Darcy-Weisbach equation** and is valid for all types of flows if a proper value of f is chosen.

The coefficient of friction f can be excpressed in terms of shear stress as follows :

$(p_1 - p_2)A$ = Friction (shear) resistance force due to shear stress on the pipe walls

= Shear stress × Surface area

$= \tau_0 \times \pi DL$

$$p_1 - p_2 = \tau_0 \frac{\pi dL}{\frac{\pi}{4}d^2} = \frac{4\tau_0 L}{d}$$

$$\frac{p_1 - p_2}{w} = \frac{4\tau_0 L}{wd}$$

$$\frac{4fLV^2}{2gd} = \frac{4\tau_0 L}{\rho gd}$$

$$f = \frac{2\tau_0}{\rho V^2} \qquad ...(10.3)$$

$$\tau_0 = \frac{f\rho V^2}{2} \qquad ...(10.3a)$$

10.3 SHEAR STRESS IN TURBULENT FLOW

In turbulent flows, the velocity fluctuations cause momentum transfer between adjacent layers resulting in additional shear stresses of high magnitudes. A number of semi-empirical theories have been developed to determine the magnitude of the turbulent shear. Some of these theories have been discussed below :

1. Boussinesq's Theory

In 1877, J. Boussinesq developed an expression for the shear stress similar to that for viscous shear stress.

$$\tau_t = \eta \frac{d\bar{u}}{dy} \qquad \text{...(10.4)}$$

where τ_t = Shear stress due to turbulence

$\bar{u}$ = Time average (temporal mean) velocity at a distance y from the boundary

η = Eddy viscosity

(Eddy kinematic viscosity is expressed in the same way as kinematic viscosity, *i.e.* eddy kinematic viscosity $\varepsilon = \eta/\rho$)

Thus the total shear stress

$$\tau = \tau_0 + \tau_t = \mu \frac{d\bar{u}}{dy} + \eta \frac{d\bar{u}}{dy} \qquad \text{...(10.5)}$$

For laminar flows $\eta = 0$. For turbulent flows, the value of η can vary and can be several thousand times that of μ. Since it is not easily possible to predict the value of η, the equation has limited use.

2. Reynolds Theory

In 1886, Reynolds developed an expression for the turbulent shear stress between two layers of a fluid at a small distance apart.

$$\tau_t = \rho u' v' \qquad \text{...(10.6)}$$

where u' = Fluctuating component of velocity in the x-direction

v' = Fluctuating component of velocity in the y-direction

As u' and v' are fluctuating, τ_t also fluctuates. To find the average turbulent shear stress, the time average is taken, *i.e.*

$$\bar{\tau}_t = \bar{\rho}\,\bar{u}'\bar{u}' = \rho \bar{u}'\bar{v}' \qquad \text{...(10.7)}$$

The shear stress so obtained is known as Reynolds stress.

3. Prandtl Mixing Length Theory

In 1925, Prandtl used the Reynolds expression to find the shear stress and forwarded a mixing length hypothesis to determine the value of $\bar{u}'\bar{v}'$. According to Prandtl, the mixing length l is that transverse distance between two layers through which the fluid particles from one layer reach the other and mix with the fluid particles of that layer in such a way that their momentum along x-axis remains same. It was also assumed that

$$\bar{u}' = l \frac{du}{dy}$$

and v' is of the same order as u'

$$\therefore \qquad \overline{v'} = \overline{u'} = l\,\frac{du}{dy}$$

Thus $$\overline{u'}\overline{v'} = l^2\left(\frac{du}{dy}\right)^2$$

and from Eq. 10.7,

$$\bar{\tau} = \rho l^2\left(\frac{du}{dy}\right)^2 \qquad \text{...(10.8)}$$

The total shear stress

$$\bar{\tau} = \mu\,\frac{d\bar{u}}{dy} + \rho l^2\left(\frac{du}{dy}\right)^2 \qquad \text{...(10.9)}$$

However, the viscous shear part is negligibly small except near the boundary and therefore the shear can be assumed due to turbulence alone.

10.4 HYDRAULICALLY SMOOTH AND ROUGH BOUNDARIES

In general, if the average height of the irregularities (k) projecting from the surface of a boundary is large, the boundary is said to be rough and if this value is small, the boundary is known as smooth. However, for proper distinction between the rough and the smooth boundaries, the flow characteristics are also considered.

For turbulent flow in pipes, now it is established that there is a laminar sublayer (of height δ') in the immediate neighbourhood of the boundary where the viscous shear forces predominates. Depending upon the value of δ' and k (Fig. 10.3), the boundary is known as

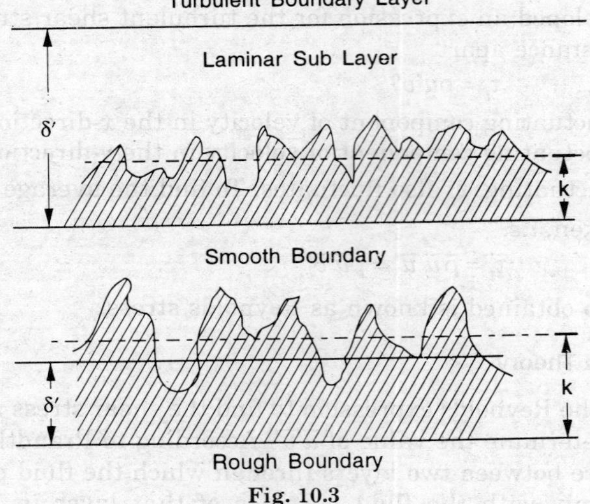

Fig. 10.3

- **smooth,** if the value of k is much less than δ'. Though the eddies present in the turbulent flow try to penetrate the laminar sublayer, they are unable to reach the surface irregularities due to higher thickness of the laminar sublayer and hence the boundary behaves like a smooth boundary.

- **rough,** if the value of k is much more than δ'. In this case the irregularities of the surface are above the laminar sublayer and the eddies of the turbulent zone easily come in contact with the irregularities of the surface and a lot of energy is lost.

According to Nikuradse's experiments, the boundary is

– smooth, if $\frac{k}{\delta'} < 0{\cdot}25$

– in transitioin, if $\frac{k}{\delta'} > 0{\cdot}25 < 6{\cdot}0$

– rough, if $\frac{k}{\delta'} > 6{\cdot}0$

In terms of roughness Reynolds number $\frac{u_* k}{\nu}$

if $\frac{u_* k}{\nu} < 4$, the boundary is smooth

if $\frac{u_* k}{\nu} > 4$ and < 100, the boundary is in transition

if $\frac{u_* k}{\nu} > 100$, the boundary is rough.

10.5 VELOCITY DISTRIBUTION IN PIPES

For the solution of Eq. 10.7,

$$\bar{\tau} = \rho l^2 \left(\frac{du}{dy}\right)^2$$

Prandtl assumed a constant shear stress in the vicinity of pipe wall. He also assumed the mixing length l to be a linear function of the distance y from the pipe wall.

i.e. $l \propto y$ or $l = ky$

where k is known as Karman constant, the value of which is suggested as 0·4.

$$\tau = \rho\,(ky)^2 \left(\frac{du}{dy}\right)^2 = \rho\,.\,k^2 y^2 \left(\frac{du}{dy}\right)^2$$

$$\frac{du}{dy} = \sqrt{\frac{\tau_0}{\rho\,.\,k^2 y^2}} = \frac{1}{ky}\sqrt{\frac{\tau_0}{\rho}} \qquad \ldots(10.10)$$

In the above equation, the quantity $\sqrt{\frac{\tau_0}{\rho}}$ has the dimensions

$$\sqrt{\frac{\tau_0}{\rho}} = \sqrt{\frac{ML^{-1}T^{-2}}{M/L^{-3}}} = \frac{L}{T}$$

i.e. that of velocity and is known as the **shear (friction) velocity** and is denoted by u_*.

$$u_* = \sqrt{\frac{\tau_0}{\rho}} \qquad \ldots(10.11)$$

Equation (10.10) becomes,

$$\frac{du}{dy} = \frac{u_*}{ky}$$

u_* is constant for a given case of turbulent flow.

$$du = \frac{u_*}{k}(y^{-1})\,dy$$

$$u = \frac{u_*}{k}\log y + C \qquad \text{...(10.12)}$$

The expression shows that in a turbulent flow, the velocity distribution is logarithmic in nature. The constant C can be determined as follows :

(i) At $y = R$ (radius of pipe), $u = u_{max}$

$$u = \frac{u_*}{k}\log_e R + C$$

$$C = u_{max} - \frac{u_*}{k}\log_e R$$

Thus equation 10.12 becomes,

$$u = \frac{u_*}{k}\log_e R + u_{max} - \frac{u_*}{k}\log_e R$$

$$= u_{max} + \frac{u_*}{k}\log_e\left(\frac{y}{R}\right)$$

$$= u_{max} + \frac{u_*}{0{\cdot}4}\log_e\left(\frac{y}{R}\right)$$

$$= u_{max} + 2{\cdot}5\, u_*\log_e\left(\frac{y}{R}\right) \qquad \text{...(10.13)}$$

This is known as **Prantdl's** universal velocity distribution equation and is applicable to smooth as well as rough pipe boundaries.

Eq. (10.13) is also written as

$$\frac{u_{max} - u}{u_*} = -\,2{\cdot}5\log_e\left(\frac{y}{R}\right)$$

$$= 2{\cdot}5\log_e\left(\frac{R}{y}\right)$$

$$= 2{\cdot}5 \times 2{\cdot}3\log_{10}\left(\frac{R}{y}\right)$$

$$= 5{\cdot}75\log_{10}\left(\frac{R}{y}\right) \qquad \text{...(10.14)}$$

The difference $(u_{max} - u)$ is known as the **velocity defect.**

(ii) For the velocity distribution given by Eq. (10.12) *i.e.*,

$$u = \frac{u_*}{k} \log_e y + C$$

At $y = 0$, u is $-\infty$ which means that at a certain distance y' from the boundary wall, the velocity is zero.

or
$$0 = \frac{u_*}{k} \log_e y' + C$$

$$C_1 = -\frac{u_*}{k} \log_e y'$$

$$u = \frac{u_*}{k} \log_e y - \frac{u_*}{k} \log_e y'$$

$$= \frac{u_*}{k} \log_e \frac{y}{y'}$$

$$\frac{u}{u_*} = 2{\cdot}5 \log_e \frac{y}{y'}$$

$$= 5{\cdot}75 \log_{10} \frac{y}{y'} \quad \text{(for } k = 0.4\text{)} \qquad \ldots(10.15)$$

(a) Smooth Boundary

In case of a smooth boundary there exists a laminar sublayer adjacent to the wall though the main flow is turbulent. In this sublayer the velocity distribution is parabolic and not logarithmic.

Let δ' be the thickness of this sublayer

Assuming the velocity distribution in the sublayer to be linear, the shear stress in the vicinity of the pipe wall

$$\tau_0 = \mu \frac{du}{dy} = \mu \frac{u}{y}$$

$$\frac{\tau_0}{\rho} = \frac{\mu}{\rho} \frac{u}{y}$$

$$u_*^2 = \nu \frac{u}{y}$$

$$\frac{u}{u_*} = \frac{u_* y}{\nu} \qquad \ldots(10.16)$$

At the sublayer boundary, $y = \delta'$ and $u = u'$

$$\frac{u'}{u_*} = \frac{u_* \delta'}{\nu} \qquad \ldots(10.17)$$

Nikuradse's experiments of turbulent flow in smooth pipes found the value of

$$\frac{u_* \delta'}{\nu} = 11{\cdot}6 \left[= \frac{u'}{u_*} \right]$$

or $$\delta' = \frac{11.6\nu}{u_*} \qquad \text{...(10.18)}$$

and the distance y' (where the velocity is zero) is equal to

$$y' = \frac{\delta'}{107} = \frac{11.6\nu}{107u_*} = 0.108\frac{\nu}{u_*}$$

Thus from Eq. 10.15,

$$\frac{u}{u_*} = 5.75\log_{10}\left(\frac{y}{0.108\frac{\nu}{u_*}}\right)$$

$$= 5.75\log_{10}\left(\frac{u_*y}{\nu} \times 9.259\right)$$

$$= 5.75\log_{10}\frac{u_*y}{\nu} + 5.75\log_{10}9.259$$

$$= 5.75\log_{10}\frac{u_*y}{\nu} + 5.5 \qquad \text{...(10.19)}$$

At the centre line of the pipe,

$$\frac{u}{u_*} = 5.75\log_{10}\frac{u_*R}{\nu} + 5.5 \qquad \text{...(10.19a)}$$

(b) Rough Boundary

In pipes with rough boundaries, the laminar sublayer is very small in thickness. The irregularities at the surface are above this sublayer, thus destroying the sublayer completely. The experiments conducted by Nikuradse and others showed that y' can be assumed to be proportional to the height of protrussions k. For pipes coated with uniform sand, y' is found to be $k/30$.

Inserting this value of y' in Eq. 10.15,

$$\frac{u}{u_*} = 5.75\log_{10}\left(\frac{y}{k/30}\right)$$

$$= 5.75\log_{10}\left(\frac{y}{k} \times 30\right)$$

$$= 5.75\log_{10}\frac{y}{k} + 5.75\log_{10}30$$

$$= 5.75\log_{10}\frac{y}{k} + 8.5 \qquad \text{...(10.20)}$$

At the centre line of the pipe,

$$\frac{u}{u_*} = 5.75\log_{10}\frac{R}{k} + 8.5 \qquad \text{...(10.20a)}$$

Example 10.1 In a badly corroded pipe of diameter 100 mm, the velocity of flow increases by 15 percent as a pitot tube is moved from a point 8 mm to a point 12 mm from the wall. Determine the height of roughness elements.

Solution The velocity distribution near the rough boundaries is given by

$$\frac{u}{u_*} = 5.75 \log_{10} \frac{y}{k} + 8.5 \quad \text{(Eq. 10.20)}$$

Let u = velocity at a distance of 8 mm from the wall and

$1.15u$ = velocity at a distance of 12 mm from the wall

Then

$$\frac{u}{u_*} = 5.75 \log_{10} \frac{8}{k} + 8.5$$

and

$$\frac{1.15u}{u_*} = 5.75 \log_{10} \frac{12}{k} + 8.5$$

Dividing (i) by (ii),

$$\frac{1}{1.15} = \frac{5.75 \log_{10} \frac{8}{k} + 8.5}{5.75 \log_{10} \frac{12}{k} + 8.5}$$

$$5.75 \log_{10} \frac{12}{k} + 8.5 = 1.15 \left[5.75 \log_{10} \frac{8}{k} + 8.5 \right]$$

$$5.75 \log_{10} \frac{12}{k} + 8.5 = 6.613 \log_{10} \frac{8}{k} + 9.775$$

$$5.75 \log_{10} 12 - 5.75 \log_{10} k + 8.5 = 6.613 \log_{10} 8 - 6.613 \log_{10} k + 9.775$$

$$6.205 - 5.75 \log_{10} k + 8.5 = 5.972 - 6.613 \log_{10} k + 9.775$$

$$0.863 \log_{10} k = 1.042$$

$$\log_{10} k = 1.2074$$

$$k = 16.12 \text{ mm}$$

Example 10.2 Water flows through a pipe of diameter 120 mm. The velocities at the pipe axis and 40 mm from the pipe axis are 4 m/s and 3 m/s respectively. Determine the wall shear stress.

Solution $D = 0.12$ m $\qquad u_{max} = 4$ m/s

$R = 0.06$ m $\qquad u = 3$ m/s

$y = R - r = 0.06 - 0.04 = 0.02$ m

$$\frac{u_{max} - u}{u_*} = 5.75 \log_{10} \left(\frac{R}{y} \right)$$

$$\frac{4 - 3}{u_*} = 5.75 \log_{10} \left(\frac{0.06}{0.02} \right)$$

$$u_* = 0.3646$$

$$\sqrt{\frac{\tau_0}{\rho}} = 0{\cdot}3646$$

$$\frac{\tau_0}{1000} = 0{\cdot}1329$$

$$\tau_0 = 132{\cdot}9 \text{ N/m}^2$$

Example 10.3 The average height of surface protrusions of a pipeline carrying water is 0·12 mm. The shear stress developed is 5 N/m^2. If the kinematic viscosity of water is 0·01 stoke, determine whether the pipe surface behaves as smooth, rough or in transition.

Solution $\nu = 0{\cdot}01 \times 10^{-4}$ m^2/s

$\tau_0 = 5$ N/m^2

$$u_* = \sqrt{\frac{\tau_0}{\rho}} = \sqrt{\frac{5}{1000}} = 0{\cdot}0707 \text{ m/s}$$

Roughness Reynolds number is

$$\frac{u_* \,.\, k}{\nu} = \frac{0{\cdot}0707 \times (0{\cdot}12 \times 10^{-3})}{0{\cdot}01 \times 10^{-4}} = 14{\cdot}14$$

As it lies between 4 and 100, the surface of the pipe acts as in **transition.**

Example 10.4 The discharge of water through a smooth pipe of 100 mm diameter and 1200 m long is 10 l/s. The coefficient of friction is given by the relation $f = \frac{0{\cdot}079}{(\text{Re})^{1/4}}$ where Re is the Reynolds number. The kinematic viscosity of the water is 0·015 stokes. Determine

(i) loss of head
(ii) wall shear stress
(iii) centre-line velocity
(iv) velocity and shear stress at 25 mm the wall of the pipe
(v) thickness of the laminar sublayer.

Solution $\nu = 0{\cdot}15$ stoke $= 0{\cdot}015 \times 10^{-4}$ m^2/s

$Q = 0{\cdot}01$ m^3/s

$d = 0{\cdot}1$ m $\qquad L = 1200$ m

Mean velocity,

$$V = \frac{Q}{A} \; \frac{0{\cdot}01}{(\pi/4)\,(0{\cdot}1)^2} = 1{\cdot}273 \text{ m/s}$$

Reynolds number,

$$\text{Re} = \frac{Vd}{\nu} = \frac{1{\cdot}273 \times 0{\cdot}1}{0{\cdot}015 \times 10^{-4}} = 84883$$

$$f = \frac{0{\cdot}079}{(\text{Re})^{1/4}} = \frac{0{\cdot}079}{84883} = 0{\cdot}004628$$

(i) $$h_f = \frac{4fLV^2}{2gd} = \frac{4 \times 0.004628 \times 1200 \times 1.273^2}{2 \times 9.81 \times 0.1} = 18.36 \text{ m}$$

(ii) $$\tau_0 = \frac{fV^2\rho}{2} = \frac{0.004628 \times 1.273^2 \times 1000}{2} = 3.75 \text{ N/m}^2$$

(iii) $$\frac{u}{u_*} = 5.75 \log_{10} \frac{u_* y}{\nu} + 5.5$$

$$u_* = \sqrt{\frac{\tau_0}{\rho}} = \sqrt{\frac{3.75}{1000}} = 0.0612$$

The velocity is maximum at the centre line *i.e.* at $y = d/2 = 0.05$ m

$$\frac{u_{max}}{0.0611} = 5.75 \log_{10} \left(\frac{0.0611 \times 0.05}{0.015 \times 10^{-4}} \right) + 5.5 = 24.532$$

$$u_{max} = 1.501$$

(iv) $$\tau = -\frac{\partial p}{\partial x} \times \frac{r}{2} \qquad \text{(Eq. 9.2)}$$

But $$\tau_0 = -\frac{\partial p}{\partial x} \times \frac{R}{2}$$

$\therefore$ $$\frac{\tau}{\tau_0} = \frac{r}{R}$$

or $$\tau = 3.75 \times \frac{0.025}{0.05} = 1.875 \text{ N/m}^2$$

$$\frac{u}{u_*} = 5.75 \log_{10} \frac{u_* y}{\nu} + 5.5 \qquad \text{(Eq. 10.19)}$$

$$\frac{u}{0.0612} = 5.75 \log_{10} \frac{0.0612 \times 0.025}{0.015 \times 10^{-4}} + 5.5$$

$$u = 1.3915 \text{ m/s}$$

(v) $$\delta' = \frac{11.6\nu}{u_*} \qquad \text{(Eq. 10.18)}$$

$$= \frac{11.6 \times 0.015 \times 10^{-4}\nu}{0.0612}$$

$$= 0.000284 \text{ m} \quad \text{or} \quad 0.284 \text{ mm}$$

10.6 VELOCITY DISTRIBUTION IN TERMS OF AVERAGE VELOCITY

Consider an annular ring of area $dA = 2\pi r \, . \, dr$ at a distance y from the pipe wall (Fig. 10.4). Then the average velocity is given by

$$\overline{u} = \frac{Q}{A} = \frac{1}{A} \int_0^R \text{Area of ring} \times \text{Velocity}$$

$$= \frac{1}{\pi R^2} \int_0^R 2\pi r dr \, . \, u$$

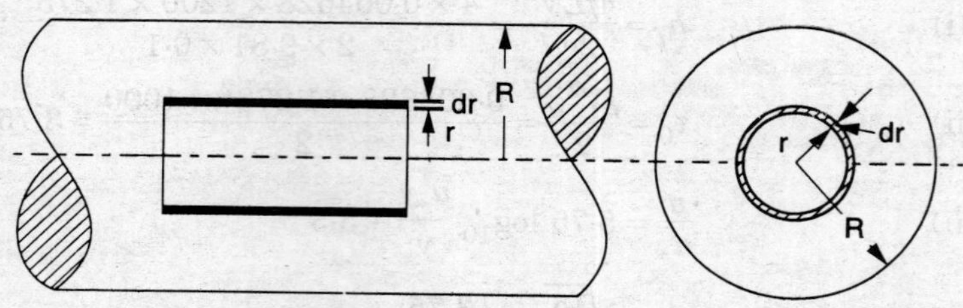

Fig. 10.4

Smooth Pipes

For smooth pipes

$$u = \left[5.75 \log_{10} \frac{u_* y}{\nu} + 5.5 \right] u_* \qquad \text{(Eq. 10.19)}$$

$$= \left[5.75 \log_{10} \frac{u_* (R-r)}{\nu} + 5.5 \right] u_*$$

$$\overline{u} = \frac{1}{\pi R^2} \int_0^R \left[5.75 \log_{10} \frac{u_* (R-r)}{\nu} + 5.5 \right] u_* \times 2\pi r dr$$

Integrating and simplifying the above equation,

$$\frac{\overline{u}}{u_*} = 5.75 \log_{10} \frac{u_* R}{\nu} + 1.75 \qquad \text{...(10.21)}$$

Also

$$\frac{u}{u_*} - \frac{\overline{u}}{u_*} = \left[5.75 \log_{10} \frac{u_* (R-r)}{\nu} + 5.5 \right] - \left[5.75 \log_{10} \frac{u_* R}{\nu} + 1.75 \right]$$

$$\frac{u - \overline{u}}{u_*} = 5.75 \log_{10} \left[\frac{u_* (R-r)}{\nu} - \log_{10} \frac{u_* R}{\nu} \right] + 3.75$$

$$= 5.75 \log_{10} \left[\frac{u_* (R-r)}{\nu} \cdot \frac{\nu}{u_* R} \right] + 3.75$$

$$= 5.75 \log_{10} \frac{(R-r)}{R} + 3.75$$

$$= 5.75 \log_{10} \frac{y}{R} + 3.75 \qquad \text{...(10.22)}$$

At the centre line of the pipe,

$$\frac{u - \overline{u}}{u_*} = 5.75 \log_{10} \frac{R}{R} + 3.75 = 3.75 \qquad \text{...(10.22a)}$$

Rough Pipes

Ror rough pipes

$$u = \left[5{\cdot}75 \log_{10} \frac{y}{k} + 8{\cdot}5\right] u_*$$

$$\overline{u} = \frac{1}{\pi R^2} \int_0^R \left[5{\cdot}75 \log_{10} \frac{(R-r)}{k} + 8{\cdot}5\right] u_* \times 2\pi r dr$$

Integrating and simplifying,

$$\frac{\overline{u}}{u_*} = 5{\cdot}75 \log_{10} \frac{R}{k} + 4{\cdot}75 \qquad \text{...(10.23)}$$

and $$\frac{u}{u_*} - \frac{\overline{u}}{u_*} = \left[5{\cdot}75 \log_{10} \frac{y}{k} + 8{\cdot}5\right] - \left[5{\cdot}75 \log_{10} \frac{R}{k} + 4{\cdot}75\right]$$

$$\frac{u - \overline{u}}{u_*} = 5{\cdot}75 \log_{10} \left[\frac{y}{k} - \frac{R}{k}\right] + 3{\cdot}75$$

$$= 5{\cdot}75 \log_{10} \left[\frac{y/k}{R/k}\right] + 3{\cdot}75$$

$$= 5{\cdot}75 \log_{10} \frac{y}{R} + 3{\cdot}75 \qquad \text{...(10.24)}$$

At the centre line of the pipe,

$$\frac{u - \overline{u}}{u_*} = 5{\cdot}75 \log_{10} \frac{R}{R} + 3{\cdot}75 = 3{\cdot}75 \qquad \text{...(10.24a)}$$

Eqs. 10.22 and 10.24 are the same indicating that the velocity distribution in the two types of pipes is the same.

Example 10.5 For turbulent flow in a pipe 400 mm in diameter, find the distance from the pipe wall at which the local velocity is equal to the average velocity.

Solution $R = D/2 = 0{\cdot}4/2 = 0{\cdot}2$ m

$$\frac{u - \overline{u}}{u_*} = 5{\cdot}75 \log_{10} \frac{y}{R} + 3{\cdot}75 \qquad \text{(Eq. 10.22)}$$

At the required point,

Local velocity = Average velocity

or $u = \overline{u}$

$$5{\cdot}75 \log_{10} \frac{y}{R} + 3{\cdot}75 = 0$$

$$\log_{10} \frac{y}{R} = -\,0{\cdot}652$$

$$\frac{y}{R} = 0{\cdot}2228$$

$$y = 0{\cdot}2228 \times 0{\cdot}2 = 0{\cdot}446 \text{ m or } 44{\cdot}6 \text{ mm}$$

10.7 VELOCITY DISTRIBUTION BY POWER LAW

The velocity distribution in pipes for turbulent flow given in the previous sections is logarithmic in nature and hence not convenient to use. Nikuradse presented the following distribution after carrying out experiments on smooth pipes at different Reynolds numbers :

$$\frac{u}{u_{max}} = \left(\frac{y}{R}\right)^{1/n} \quad \text{...(10.25)}$$

where exponent $1/n$ is a function of Reynolds number and decreases with increase in the value of Re.

$$\frac{1}{n} = \frac{1}{6} \quad \text{for} \quad \text{Re} = 4000$$

$$= \frac{1}{7} \quad \text{for} \quad \text{Re} = 110{,}000$$

$$= \frac{1}{10} \quad \text{for} \quad \text{Re} = 2{,}000{,}000$$

Thus for Reynolds number 110,000, $\frac{1}{n} = \frac{1}{7}$ and the velocity distribution law becomes

$$\frac{u}{u_{max}} = \left(\frac{y}{R}\right)^{1/7} \quad \text{...(10.26)}$$

This is known as 1/7th power law of velocity distribution for smooth pipes,

10.8 FRICTION COEFFICIENTS FOR SMOOTH AND ROUGH PIPES

It can be shown by dimensioinal analysis (Example 8.5) that the friction coefficient f is a function of Reynolds number and k/D ratio where k is the average height of pipe wall roughness protrusions, *i.e.*

$$f = \phi\left[\text{Re} \, . \, \frac{k}{D}\right]$$

This is the general equation which is applicable to laminar as well as turbulent flow in pipes.

1. For Laminar Flow

Friction head loss in laminar flow

$$h_f = \frac{32\mu\bar{u}l}{wD^2} = \frac{32\mu\bar{u}l}{\rho g D^2} \quad \text{(Eq. 9.12)}$$

Assuming that Darcy equation 10.2 is also applicable to laminar flow and remembering that average flow velocity $u = V$.

$$h_f = \frac{4flV^2}{2gD} = \frac{32\mu Vl}{\rho g D^2}$$

$$f = \frac{16}{\rho VD/\mu} = \frac{16}{\text{Re}} \quad \text{...(10.27)}$$

Where Re is the Reynolds number.

2. For Turbulent Flow

Friction coefficient for smooth and rough pipes is found as follows :

For Smooth Pipes

In case of smooth pipes, the value of laminar sublayer is large as compared to average height of the surface roughness k. Therefore, the coefficient of friction f is a factor of Reynolds number only.

- For Re, 4000 to 1×10^5, f is found by an empirical relation given by Blausis.

$$f = \frac{0.0791}{\text{Re}^{1/4}} \qquad \text{...(10.28)}$$

- For Re > 1×10^5, the value of f is obtained from equation

$$\frac{\bar{u}}{u_*} = 5.75 \log_{10} \frac{u_* R}{\nu} + 1.75 \qquad \text{(Eq. 10.21)}$$

But

$$u_* = \sqrt{\frac{\tau_0}{\rho}} \qquad \text{(Eq. 10.11)}$$

$$= \sqrt{\frac{2\tau_0}{\rho} \cdot \frac{2V^2}{2V^2}}$$

$$= \sqrt{\frac{2\tau_0}{\rho V^2} \cdot \frac{V^2}{2}}$$

$$= \sqrt{\frac{fV^2}{2}} \qquad \text{(using Eq. 10.3)}$$

$$= V\sqrt{\frac{4f}{8}} \qquad \text{...(10.29)}$$

As V is the average velocity, replacing V by $\bar{u}$,

$$u_* = \bar{u}\sqrt{\frac{4f}{8}}$$

Inserting this value of u_* in Eq. (10.21),

$$\frac{\bar{u}}{\bar{u}\sqrt{\frac{4f}{8}}} = 5.75 \log_{10} \left(\frac{\bar{u}}{\nu}\sqrt{\frac{4f}{8}}\right) R + 1.75$$

$$\frac{\sqrt{8}}{4f} = 5.75 \log_{10} \left(\frac{\bar{u}}{\nu}\,\frac{D}{2}\sqrt{\frac{4f}{8}}\right) + 1.75$$

$$\frac{1}{\sqrt{4f}} = \frac{1}{\sqrt{8}}\left[5.75 \log_{10} \frac{\bar{u}D}{\nu}\sqrt{4f} - 5.75 \log_{10} 2\sqrt{8} + 1.75\right]$$

$$= 2.03 \log_{10} \text{Re}\sqrt{4f} - 2.03 \log_{10} 2\sqrt{8} + 0.618$$

$$= 2.03 \log_{10} \text{Re}\sqrt{4f} - 1.528 + 0.618$$

$$= 2.03 \log_{10} \text{Re}\sqrt{4f} - 0.91 \qquad \text{...(10.30)}$$

This equation is valid upto Re = 4×10^6.

Karman-Prantdl resistance equation for turbulent flow in smooth pipes is,

$$f = \frac{1}{\sqrt{4f}} = 2.0 \log_{10} R_e \sqrt{4f} - 0.8 \qquad \text{...(10.31)}$$

Nikuradse's experimental data showed that this equation is valid in the range Re = 5×10^4 to 4×10^7.

Nikuradse's equation for Re between 4×10^3 to 3.2×10^6,

$$f = 0.0008 + \frac{0.05525}{\text{Re}^{0.237}} \qquad \text{...(10.32)}$$

The general graphical representation between Re and f is shown in Fig. 10.5.

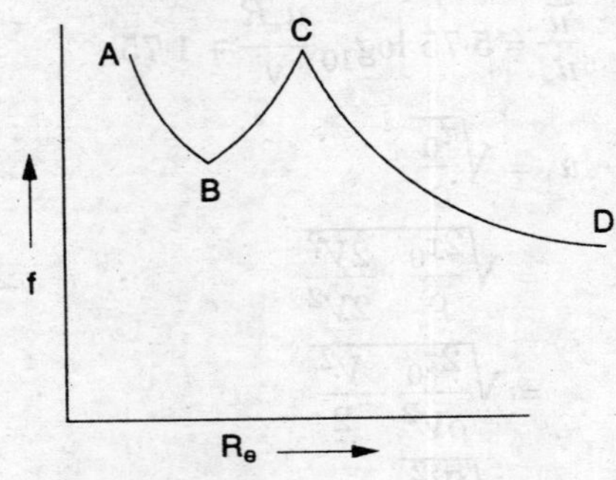

Fig. 10.5

AB = laminar flow

BC = transition flow

CD = turbulent flow

For Rough Pipes

$$\frac{\bar{u}}{u_*} = 5.75 \log_{10}\left(\frac{R}{k}\right) + 4.75 \qquad \text{...(10.33)}$$

$$\frac{\bar{u}}{\bar{u}\sqrt{\frac{4f}{8}}} = 5.75 \log_{10}\left(\frac{R}{k}\right) + 4.75$$

$$\frac{1}{\sqrt{4f}} = \frac{1}{\sqrt{8}}\left[5.75 \log_{10}\left(\frac{R}{k}\right) + 4.75\right]$$

$$= 2.03 \log_{10}\left(\frac{R}{k}\right) + 1.68 \qquad \text{...(10.34)}$$

However, Nikuradse recommended the following relation based on his experimental data,

$$\frac{1}{\sqrt{4f}} = 2{\cdot}0 \log_{10}\left(\frac{R}{k}\right) + 1{\cdot}74 \qquad \text{...(10.35)}$$

The relation shows that the coefficient of friction in case of rough pipes depends only upon the relative roughness and is independent of Reynolds number.

For Commercial Pipes

Colebrook and White presented the following empirical relation to find the friction factor for commercial pipes both for smooth as well as rough pipes

$$\frac{1}{\sqrt{4f}} - 2{\cdot}0 \log_{10} \text{Re} \sqrt{\frac{R}{k}} = 1{\cdot}74 - 2 \log_{10}\left(1 + 18{\cdot}7 \frac{\sqrt{R/k}}{\text{Re}\sqrt{4f}}\right) \qquad \text{...(10.36)}$$

Example 10.6 The centre line velocity in a pipe of diameter 400 mm is 2·6 m/s whereas at a point 120 mm from the centre it is 3·2 m/s. If the flow in the pipe is turbulent, determine the discharge through the pipe, coefficient of friction and the height of roughness projections.

Solution $d = 0{\cdot}4$ m

$R = 0{\cdot}2$ m $\qquad u_{max} = 2{\cdot}6$ m/s

$y = 0{\cdot}2 - 1{\cdot}2 = 0{\cdot}08$ m $\qquad u = 2{\cdot}2$ m/s

$$\frac{u\text{max} - u}{u_*} = 5{\cdot}75 \log_{10}\left(\frac{R}{y}\right) \qquad \text{(Eq. 10.14)}$$

$$\frac{2{\cdot}6 - 2{\cdot}2}{u_*} = 5{\cdot}75 \log_{10}\left(\frac{0{\cdot}2}{0{\cdot}08}\right) = 2{\cdot}288$$

$$u_* = 0{\cdot}1748 \text{ m/s}$$

$$\frac{u_{max} - \bar{u}}{u_*} = 3{\cdot}75 \qquad \text{(Eq. 10.24)}$$

$$\frac{26 - \bar{u}}{0{\cdot}1748} = 3{\cdot}75$$

$$2{\cdot}6 - \bar{u} = 0{\cdot}656$$

$$\bar{u} = 1{\cdot}945 \text{ m/s}$$

$$Q = \frac{\pi}{4}(0{\cdot}4)^2 \times 1{\cdot}945 = 0{\cdot}244 \text{ m}^3\text{/s}$$

$$u_* = V\sqrt{\frac{4f}{8}} \qquad \text{(Eq. 10.29)}$$

$$0{\cdot}1748 = 1{\cdot}945\sqrt{\frac{4f}{8}}$$

$$f = 0{\cdot}0162$$

$$\frac{1}{\sqrt{4f}} = 2{\cdot}0 \log_{10}\left(\frac{R}{k}\right) + 1{\cdot}74 \qquad \text{(Eq. 10.35)}$$

$$\frac{1}{\sqrt{4 \times 0.0162}} = 2.0 \log_{10}\left(\frac{0.02}{k}\right) + 1.74$$

$$3.928 = 2.0 \log_{10}\left(\frac{0.2}{k}\right) + 1.74$$

$$\log_{10}\left(\frac{0.2}{k}\right) = 1.094$$

$$\frac{0.2}{k} = 12.42$$

$$k = 0.0161 \text{ m}$$

Example 10.7 The flow of water through a rough pipe of diameter 400 mm and length 5000 m is 0·4 m³/s. The average height of roughness is 0·45 mm. Determine the power required to maintain the flow.

Solution $D = 0.4$ m $Q = 0.4 \text{ m}^3/\text{s}$

$R = 0.2$ m $L = 5000$ m

$k = 0.45 \times 10^{-3}$ m

$P = wQh_f$

To find h_f, it is required to find f and V first.

$$\frac{1}{\sqrt{4f}} = 2.0 \log_{10}\left(\frac{R}{k}\right) + 1.74 \qquad \text{(Eq. 10.35)}$$

$$= 2.0 \log_{10}\left(\frac{0.2}{0.45 \times 10^{-3}}\right) + 1.74$$

$$= 7.036$$

$$f = 0.00505$$

$$V = \frac{0.4}{(\pi/4) \times (0.4)^2} = 3.183 \text{ m/s}$$

$$h_f = \frac{4fLV^2}{2gD} = \frac{4 \times 0.00505 \times 5000 \times 3.183^2}{2 \times 9.81 \times 0.4} = 130.4 \text{ m}$$

$$P = wQh_f = 9.81 \times 0.4 \times 130.4 = 511.67 \text{ kW}$$

Example 10.8 A pipe 600 mm in diameter and 1200 mm long is used to discharge water at the rate of 0·06 m³/s. If the kinematic viscosity of water is 0·018 stokes, determine

(i) head loss due to friction

(ii) wall shear stress

(iii) centre line velocity

(iv) laminar sublayer thickness

Solution $L = 1200$ m $Q = 0.06 \text{ m}^3/\text{s}$

$D = 0.6$ m $\nu = 0.018 \text{ stokes} = 0.018 \times 10^{-4} \text{ m}^2/\text{s}$

$R = 0.3$ m

To know the type of flow, Reynolds number is found.

$$V = \frac{0{\cdot}06}{\frac{\pi}{4} \times (0{\cdot}6)^2} = 0{\cdot}2122 \text{ m/s}$$

$$\text{Re} = \frac{V \,.\, D}{\nu} = \frac{0{\cdot}2122 \times 0{\cdot}6}{0{\cdot}018 \times 10^{-4}} = 70733$$

As R is more than 4000, the flow is turbulent and f is given by

$$f = \frac{0{\cdot}0791}{(70733)^{1/4}} = 0{\cdot}00485 \qquad \text{(Eq. 10.28)}$$

$$h_f = \frac{4fLV^2}{2gD}$$

$$= \frac{4 \times 0{\cdot}00485 \times 1200 \times (0{\cdot}2122)^2}{2 \times 9{\cdot}81 \times 0{\cdot}6}$$

$$= 0{\cdot}089 \text{ m}$$

$$\tau_0 = \frac{f\rho V^2}{2} \qquad \text{(Eq. 10.3a)}$$

$$= \frac{0{\cdot}00485 \times 1000 \times (0{\cdot}2122)^2}{2}$$

$$= 0{\cdot}109 \text{ N/m}^2$$

$$\frac{u_{max}}{u_*} = 5{\cdot}75 \log_{10} \frac{u_* R}{\nu} + 5{\cdot}5 \qquad \text{(Eq. 10.19a)}$$

where $$u_* = \sqrt{\frac{\tau_0}{\rho}} \qquad \text{(Eq. 10.11)}$$

$$= \sqrt{\frac{0{\cdot}109}{1000}}$$

$$= 0{\cdot}01045$$

$$\frac{u_{max}}{0{\cdot}01045} = 5{\cdot}75 \log_{10} \frac{0{\cdot}01045 \times 0{\cdot}3}{0{\cdot}018 \times 10^{-4}} + 5{\cdot}5$$

$$= 24{\cdot}14$$

$$u_{max} = 0{\cdot}252 \text{ m/s}$$

$$\delta' = \frac{11{\cdot}6 \times \nu}{u_*} \qquad \text{(Eq. 10.18)}$$

$$= \frac{11{\cdot}6 \times 0{\cdot}018 \times 10^{-4}}{0{\cdot}01045}$$

$$= 0{\cdot}002 \text{ m} = 2 \text{ mm}$$

Example 10.9 Water flows through a rough pipe of 600 mm diameter and 1000 m length at the rate of 0·6 m^3/s. If the average height of roughness is 0·018 mm, determine

(i) coefficient of friction

(ii) wall shear stress

(iii) centre line velocity

(iv) velocity at a distance of 250 mm from the pipe wall.

Solution $D = 0{\cdot}6$ m $\quad L = 1000$ m

$R = 0{\cdot}3$ m $\quad k = 0{\cdot}018 \times 10^{-3}$ mm

$Q = 0{\cdot}6$ m^2/s

$$\frac{1}{\sqrt{4f}} = 2{\cdot}0 \log_{10}\left(\frac{R}{k}\right) + 1{\cdot}74 \qquad \text{(Eq. 10.35)}$$

$$\frac{1}{\sqrt{4f}} = 2{\cdot}0 \log_{10}\left[\frac{0{\cdot}3}{0{\cdot}018 \times 10^{-3}}\right] + 1{\cdot}74 = 10{\cdot}184$$

$$\sqrt{4f} = 0{\cdot}0982$$

$$f = 0{\cdot}00241$$

$$\tau_0 = \frac{f\rho V^2}{2} \qquad \text{(Eq. 10.3a)}$$

where $$V = \frac{0{\cdot}6}{(\pi/4)\,(0{\cdot}6)^2} = 2{\cdot}122 \text{ m/s}$$

$$\tau_0 = \frac{0{\cdot}00241 \times 1000 \times (2{\cdot}122)^2}{2} = 5{\cdot}426 \text{ N/m}^2$$

$$\frac{u_{max}}{u_*} = 5{\cdot}75 \log_{10}\frac{R}{k} + 8{\cdot}5 \qquad \text{(Eq. 10.20a)}$$

$$u_* = \sqrt{\frac{\tau_0}{\rho}} \qquad \text{(Eq. 10.11)}$$

$$= \sqrt{\frac{5{\cdot}426}{1000}}$$

$$= 0{\cdot}0737 \text{ m/s}$$

$$\frac{u_{max}}{0{\cdot}0737} = 5{\cdot}75 \log_{10}\frac{0{\cdot}3}{0{\cdot}018 \times 10^{-3}} + 8{\cdot}5 = 32{\cdot}77$$

$$u_{max} = 2{\cdot}414 \text{ m/s}$$

$$\frac{u}{u_*} = 5{\cdot}75 \log_{10}\frac{y}{k} + 8{\cdot}5 \qquad \text{(Eq. 10.20)}$$

$$\frac{u_{max}}{0{\cdot}0737} = 5{\cdot}75 \log_{10}\frac{0{\cdot}25}{0{\cdot}018 \times 10^{-3}} + 8{\cdot}5 = 32{\cdot}32$$

$$u = 2{\cdot}382 \text{ m/s}$$

PROBLEMS

1. What are characteristics of a turbulent flow ?
2. Deduce an expression for the head loss due to friction in a pipe.
3. Derive an equation to find the coefficient of friction in terms of shear stress.
4. What is the friction factor ? How does it differ from coefficient of friction ?
5. Define eddy viscosity as given by Boussinesq. Describe momentum exchange method to find it.
6. What is mixing length concept of Prandtl ? How does it vary with the distance from the wall in a turbulent flow ?
7. On what basis are smooth pipes different from rough pipes ?
8. Deduce Prandtl's universal velocity distribution equation. Show how is it applicable to smooth as well as rough pipe boundaries ?
9. Write a short note on Nikuradse's experiments and their significance.
10. Deduce expressions for velocity distribution in a smooth as well as a rough pipe in terms of average velocity.
11. Show that frictional coefficient in case of a laminar flow depends only upon Reynolds number.
12. Show that difference of local velocity u and the average velocity $\overline{u}$ for turbulent flow through rough or smooth pipes is the same and is given by

$$\frac{u - \overline{u}}{u_*} = 5{\cdot}75 \log_{10} \frac{y}{R} + 3{\cdot}75$$

13. For turbulent flow in a pipe of radius R, find the distance from the pipe wall where the local velocity is equal to (i) the average velocity (ii) half the mean velocity if the shear velocity is 1/20th of the mean velocity. (0·2228 R, 0·00406 R)
14. For turbulent flow in a pipe of radius 180 mm, the velocity at the axis is 7 m/s and at a distance 120 mm from the pipe wall it is 6 m/s. Determine the shear friction velocity. (0·36 m/s)
15. The velocity at the centreline of a 300 mm diameter pipe is 6 m/s and that at 50 mm from the wall 5·25 m/s. Find the mean velocity, the coefficient of friction and the discharge. (272 mm/s, 0·006, 352 l/s)
16. The velocity at the axis of 600 mm pipe carrying oil is 4·5 m/s and this value is reduced to 4·2 m/s at a radial distance 100 mm from the axis. Find the discharge in the pipe. (0·958 m^3/s)
17. In a rough 100 mm diameter pipe, a pitot tube is moved from a point 15 mm from the wall to a point 30 mm from the wall and the velocity of flow is found to increase by 10 percent. Calculate the relative roughness R/k and the coefficient of friction. (113·5, 0·0073)
18. A 100 mm diameter rough pipe gives 0·0065 and 0·0075 as coefficients of friction after five and ten years of its installation respectively. What will be the value of coefficient of friction after 15 years of service ? (0·0083)

19. The local velocities of water at the axis and at the mid radius of a 250 mm diameter pipe are 2·31 m/s and 2·09 m/s respectively. Determine the discharge and the pipe roughness k. (0·09 m^3/s, 2·58 mm)
20. The diameters at the inlet and the outlet of a horizontal conical pipe are 200 mm and 150 mm respectively. The length of the pipe is 2 m and the rate of flow is 0·4 m^3/s. Assuming coefficient of friction to be 0·01, determine the loss of head due to pipe friction. (71·4 mm)
21. What power is required per kilometre length to maintain a flow of 0·6 m^3/s of water in a 600 mm diameter rough pipe where k is 3 mm ? (68·28 kW)

Chapter 11

PIPE FLOW

11.1 INTRODUCTION

A pipe is a closed conduit, generally, circular in cross-section, used to carry a liquid under pressure. The liquid is under pressure only if it fills the cross-section of the pipe completely and there is no free surface. In case the pipe is not running full such as sewer pipes, culverts etc., the atmospheric pressure prevails above the liquid level and pipe behaves as an open channel. In this chapter, the pipes running full and under pressure will only be considered.

11.2 LOSS OF ENERGY (HEAD) IN PIPES

As the water or any other liquid flows in a pipe, the head continuously goes on reducing due to various kinds of losses. The major loss is due to the friction of the surface of the pipe. There are other losses in a pipe flow such as due to sudden enlargement or contraction of area, losses at the entrance and exit of the pipe, loss due to any obstruction in the path of flow etc.

11.3 MAJOR ENERGY LOSSES

Major energy losses are due to friction and are given by

- Darcy Weisback formula
- Chezy's formula

Darcy Weisback Formula

This has been derived in Section 10.2 and is given by Eq. (10.2)

$$h_f = \frac{4\,fLV^2}{2gd}$$

where f = Coefficient of friction and is a function of Reynolds number

$$= \frac{16}{\text{Re}} \text{ for Re} < 2000 \text{ (laminar flow)}$$

$$= \frac{0{\cdot}0791}{\text{Re}^{1/4}} \text{ for Re between 4000 and } 1 \times 10^5$$

($4f$ is usually known as friction factor)

Chezy's Formula

Equation 10.1 is.

$$h_f = \frac{f' L}{w\, m} V^2$$

or

$$V = \sqrt{\frac{w}{f'}} \sqrt{m \frac{h_f}{L}}$$

$$= C \sqrt{m.i} \qquad \text{...(11.1)}$$

where C is a constant known as Chezy's constant

and $i = \dfrac{h_f}{L}$ (*i.e.* loss of head per unit length)

m = hydraulic mean depth or hydraulic radius.

This equation is known as Chezy's formula. This formula can be used when the mean velocity of flow through the pipe and the Chezy's constant C are known.

For a circular pipe,

$$m = \frac{A}{P} = \frac{\frac{\pi}{4} d^2}{\pi d} = \frac{d}{4}$$

Example 11.1 The velocity of water in a pipe of diameter 250 mm and length 100 m is 3.5 m/s. Determine the head lost due to friction using

(i) Darcy-Weisbach formula and

(ii) Chezy's formula taking the value of $C = 58$.

The kinematic viscosity of water is 0.011 stoke.

Solution $d = 0{\cdot}25$ m $\qquad L = 100$ m

$V = 3{\cdot}5$ m/s $\qquad \upsilon = 0{\cdot}011 \times 10^{-4}$ m²/s

(i) According to Darcy-Weisbach formula

$$h_f = \frac{4\,fLV^2}{2gD}$$

where

$$f = \frac{0{\cdot}0791}{(\text{Re})^{1/4}} = \frac{0{\cdot}0791}{\left(\dfrac{VD}{\nu}\right)^{1/4}} = \frac{0{\cdot}0791}{\left(\dfrac{3{\cdot}5 \times 0{\cdot}25}{0{\cdot}011 \times 10^{-4}}\right)^{1/4}} = 0{\cdot}00265$$

$$h_f = \frac{4 \times 0{\cdot}00265 \times 100 \times (3{\cdot}5)^2}{2 \times 9{\cdot}81 \times 0{\cdot}25}$$

$$= 2{\cdot}646 \text{ m}$$

(ii) By Chezy's formula, $C = 58$

$$V = C\sqrt{m.i}$$

where $$m = \frac{A}{P} = \frac{(\pi/4)\times d^2}{\pi d} = \frac{d}{4} = \frac{0.25}{4} = 0.0625$$

$$3.5 = 58 \times \sqrt{0.0625 \times i}$$

$$i = 0.0583$$

$$\frac{h_f}{L} = 0.0583$$

$$\frac{h_f}{100} = 0.0583$$

$$h_f = 5.83 \text{ m}$$

Example 11.2 An oil of specific gravity 0·75 flows through a pipe of 400 mm diameter and 1200 m length at the rate of 400 litres per second. Determine the head lost due to friction and the power required to maintain the flow. Assume the kinematic viscosity of oil as 0·28 stoke.

Solution $d = 0.4$ m $\quad Q = 0.4\ \text{m}^3.\text{s}$

$L = 1200$ m $\quad \upsilon = 0.28 \times 10^{-4}\ \text{m}^2/\text{s}$

$$V = \frac{Q}{A} = \frac{0.4}{(\pi/4)\times 0.4^2} = 3.183 \text{ m/s}$$

$$\text{Re} = \frac{Vd}{\nu} = \frac{3.183 \times 0.4}{0.28 \times 10^{-4}} = 45473$$

Using Weisbach formula,

$$f = \frac{0.0791}{(R_e)^{1/4}} = \frac{0.0791}{(45473)^{1/4}} = 0.00542$$

$$h_f = \frac{4fLV^2}{2gd} = \frac{4 \times 0.00542 \times 1200 \times (3.183)^2}{2 \times 9.81 \times 0.4} = 33.586 \text{ m}$$

$$P = wQh_f = 0.75 \times 9.81 \times 0.4 \times 33.586 = 98.84 \text{ kW}$$

11.4 MINOR ENERGY (HEAD) LOSSES

The minor energy or head losses are

1. Loss of head due to sudden enlargment
2. Loss of head due to sudden contraction
3. Loss of head at the entrance of a pipe
4. Loss of head at the exit of a pipe
5. Loss of head due to bend in a pipe
6. Loss of head due to obstruction in a pipe
7. Loss of head due to various pipe fittings

In long pipes, the minor losses are usually small as compared to the major losses (friction losses) and may be neglected. However, in short pipes, these losses are comparable with the major losses and hence are not ignored.

The minor losses are discussed below.

1. Sudden Enlargement

Consider a sudden enlargement in a pipe through which a liquid is flowing (Fig. 11.1). Owing to sudden increase in the diameter of the pipe, the liquid flowing is not able to follow the abrupt change of boundary and separates from it. Eddies are formed in the enlargement area which cause the loss of energy or head. Consider two sections 1-1 and 2-2, just before and after the enlargement respectively.

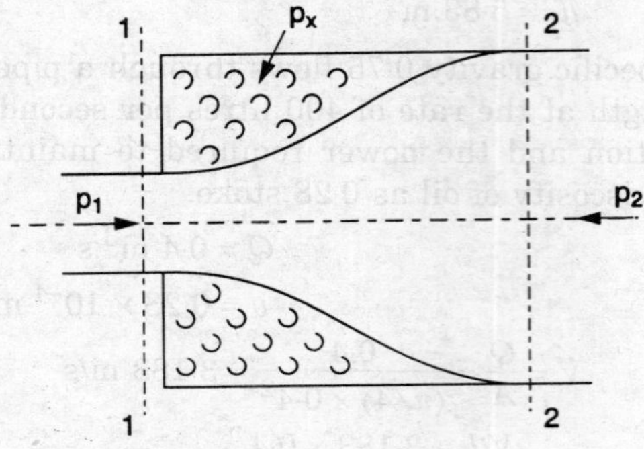

Fig. 11.1

Let p_1 = intensity of pressure at section 1-1

A_1 = area of pipe at section 1-1

V_1 = velocity of flow at section 1-1

p_2 , A_2 , V_2 are the corresponding values at section 2-2

h_e = loss of head due to sudden enlargement

Apply Bernoulli's equation to section 1-1 and 2-2,

$$\frac{p_1}{w} + \frac{V_1^2}{2g} + z_1 = \frac{p_2}{w} + \frac{V_2^2}{2g} + z_2 + h_e$$

$$h_e = \left(\frac{p_1 - p_2}{w}\right) + \left(\frac{V_1^2 - V_2^2}{2g}\right) \quad \text{...(i)}$$

The term $\left(\frac{p_1 - p_2}{w}\right)$ is evaluated by considering the control volume between sections 1-1 and 2-2 and applying the momentum equation *i.e.* the force applied in the direction of flow is equal to the rate of change of momentum.

i.e. F = Mass flow rate × change in velocity

= Density × Volume flow rate × Change in velocity

or $p_1A_1 - p_2A_2 + p_x (A_2 - A_1) = \rho Q (V_2 - V_1)$

where p_x is the pressure in the area where eddies are formed. This pressure is taken equal to p_1 since the liquid in the corners may be assumed to be at rest and the pressure constant throughout this section. Experimently also, this pressure has been found to be approximately equal to p_1.

Thus the above equation becomes,

$$p_1 A_1 - p_2 A_2 + p_1 (A_2 - A_1) = \rho\,(A_2 V_2)\,(V_2 - V_1)$$

$$p_1 A_1 - p_2 A_2 + p_1 A_2 - p_1 A_1 = \rho\, A_2 V_2\,(V_2 - V_1)$$

$$(p_1 - p_2) A_2 = \frac{w}{g} \,.\, A_2 V_2\,(V_2 - V_1)$$

$$\frac{p_1 - p_2}{w} = \frac{V_2\,(V_2 - V_1)}{g}$$

(i) becomes,

$$h_e = \frac{V_2\,(V_2 - V_1)}{g} + \frac{{V_1}^2 - {V_2}^2}{2g}$$

$$= \frac{2{V_2}^2 - 2V_1 V_2 + {V_1}^2 - {V_2}^2}{2g}$$

$$= \frac{{V_1}^2 + {V_2}^2 - 2V_1 V_2}{2g}$$

$$= \frac{(V_1 - V_2)^2}{2g} \quad \text{...(11.2)}$$

2. Sudden Contraction

In case of sudden contraction, the streamlines converge to a minimum cross-section and then expand to fill the full area of the pipe (Fig. 11.2). The minimum cross-section is the vena-contracta. During contraction of the flow, the loss of head is extremely small and is usually neglected. The main loss of head is due to the sudden enlargement beyond the vena-contracta.

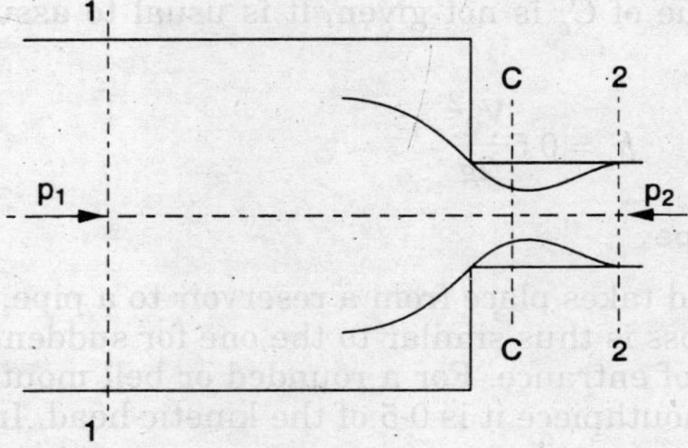

Fig. 11.2

Let V_c = velocity of flow at section C-C

V_2 = velocity of flow at section 2-2

A_c = area of flow at section C-C

A_2 = area of flow at section 2-2

h_c = loss of head due to sudden contraction

Thus h_c is in fact loss of head due to sudden enlargement from section C-C to 2-2.

$$h_c = \frac{(V_c - V_2)^2}{2g}$$

$$= \frac{1}{2g}\left[V_2\left(\frac{V_c}{V_2} - 1\right)\right]^2$$

$$= \frac{V_2^2}{2g}\left(\frac{V_c}{V_2} - 1\right)^2$$

$$= \frac{V_2^2}{2g}\left(\frac{A_2}{A_c} - 1\right)^2 \qquad (A_c V_c = A_2 V_2)$$

$$= \frac{V_2^2}{2g}\left(\frac{A_2}{C_c A_2} - 1\right)^2 \qquad (A_c = C_c A_2)$$

$$= \frac{V_2^2}{2g}\left(\frac{1}{C_c} - 1\right)^2 \qquad \ldots(11.3)$$

$$= \frac{k \cdot V_2^2}{2g} \qquad \ldots(11.3a)$$

where $k = \frac{1}{C_c} - 1$

For $C_c = 0{\cdot}62$, $k = 0{\cdot}375$

and $$h_c = 0{\cdot}375 \frac{V_2^2}{2g} \qquad \ldots(11.4)$$

However, if the value of C_c is not given, it is usual to assume the value of k as 0.5 and head loss,

$$h_c = 0{\cdot}5 \frac{V_2^2}{2g} \qquad \ldots(11.5)$$

3. Entrance (Inlet) of a Pipe

When the flow of a liquid takes place from a reservoir to a pipe, the streamlines first contract and then expand. The loss is thus similar to the one for sudden contraction. This loss of head depends upon the type of entrance. For a rounded or bell mouthed entrance, it is very small and for a sharp edged mouthpiece it is 0·5 of the kinetic head. In general, this loss is taken as

$$h_i = 0{\cdot}5 \frac{V^2}{2g} \qquad \ldots(11.6)$$

where V is the velocity of liquid in the pipe.

4. Exit of Pipe

The kinetic energy of the liquid is dissipated in the form of a free jet or in the tank of a reservoir depending upon whether the pipe discharges the liquid into the atmosphere or into a tank. In general, this loss is taken to be

$$h_{exit} = \frac{V^2}{2g} \qquad \text{...(11.7)}$$

where V is the velocity of liquid at the outlet of the pipe.

5. Bend

The loss of head due to bend in a pipe is written as

$$h_{bend} = \frac{kV^2}{2g} \qquad \text{...(11.8)}$$

where k is the coefficient of bend and depends upon the angle and the radius of curvature of the bend and the diameter of the pipe.

6. Obstruction in a Pipe

Let there be an obstruction in the flow as shown in Fig. 11.3. As the liquid flows past the obstruction (section 1-1), a vena-contracta is formed after which the flow again becomes uniform in the main pipe. The loss of head is there due to the enlargement of area of flow beyond vena-contracta. Thus

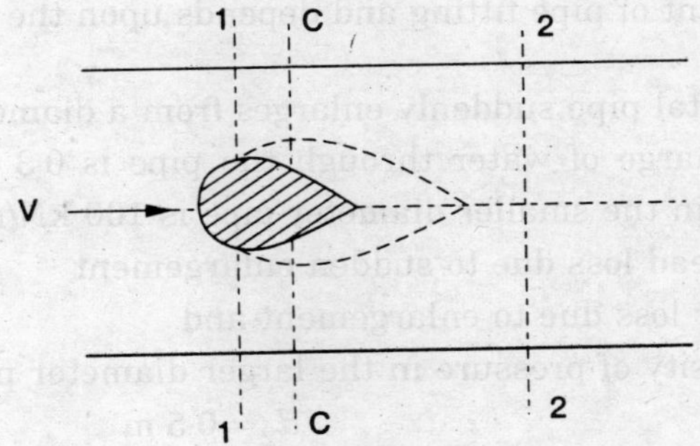

Fig. 11.3

$$h_{obs} = \frac{(V_c - V)^2}{2g} \qquad \text{...(i)}$$

Let V = velocity of flow in the main pipe (section 2-2)

V_c = velocity of flow at vena-contracta (section C-C)

A = area of cross-section of the main pipe (section 2-2)

a = maximum area of obstruction (section 1-1)

$(A - a)$ = area of pipe at section 1-1

A_c = area at the vena-contracta (section C-C)

C_c = coeffcient of contraction

$$\frac{\text{Area at the vena-contracta}}{\text{Area at section 1-1}} = \frac{A_c}{A-a}$$

$$A_c = C_c\,(A-a) \qquad \text{...(ii)}$$

Thus from (i)

$$h_{obs} = \frac{V}{2g}\left(\frac{V_c}{V}-1\right)^2$$

$$= \frac{V^2}{2g}\left(\frac{A}{A_c}-1\right)^2 \qquad (A_cV_c = AV)$$

$$= \frac{V^2}{2g}\left(\frac{A}{C_c\,(A-a)}-1\right)^2 \qquad \text{[From (ii)]} \qquad \text{...(11.9)}$$

7. Pipe Fittings

The loss of head due to various pipe fittings such as valves and couplings etc. is represented by

$$h_{fit} = \frac{kV^2}{2g} \qquad \text{...(11.10)}$$

where k is the coefficient of pipe fitting and depends upon the type of fitting and V is the velocity of flow.

Example 11.3 A horizontal pipe suddenly enlarges from a diameter of 250 mm to 500 mm. The discharge of water through the pipe is 0·3 m^3/s and the intensity of pressure in the smaller diameter pipe is 100 kN/m^3. Determine

(i) the head loss due to sudden enlargement

(ii) power loss due to enlargement and

(iii) intensity of pressure in the larger diameter pipe.

Solution $d_1 = 0{\cdot}25$ m $\qquad d_2 = 0{\cdot}5$ m

$Q = 0{\cdot}3$ m^3/s $\qquad p_1 = 100$ kN/m^2

$$V_1 = \frac{Q}{A_1} = \frac{0{\cdot}3}{(\pi/4)\times(0{\cdot}25)^2} = 6{\cdot}111 \text{ m/s}$$

$$V_2 = \frac{Q}{A_2} = \frac{0{\cdot}3}{(\pi/4)\times(0{\cdot}5)^2} = 1{\cdot}528 \text{ m/s}$$

$$h_e = \frac{(V_1-V_2)^2}{2g} = \frac{(6{\cdot}111-1{\cdot}528)^2}{2\times 9{\cdot}81} = 1{\cdot}07 \text{ m}$$

$$p = wQh_e = 9{\cdot}81\times 0{\cdot}3\times 1{\cdot}071 = 3{\cdot}151 \text{ kW}$$

Applying Bernoulli's equation before and after sudden enlargement,

$$\frac{p_1}{w}+\frac{V_1^2}{2g}+z_1 = \frac{p_2}{w}+\frac{V_2^2}{2g}+z_2+h_e$$

$$\frac{100}{9.81}+\frac{6.111^2}{2\times 9.81}+0=\frac{p_2}{9.81}+\frac{1.528^2}{2\times 9.81}+0+1.071$$

$$100\times 2+6.11^2=2p_2+1.528^2+1.071\times 2\times 9.81$$

$$p_2=107\ \text{kN/m}^2$$

Example 11.4 The diameter of a horizontal pipe suddenly contracts from 600 mm to 300 mm as a result of which the pressure changes from 100 kN/m^2 to 80 kN/m^2. Assuming the coefficient of contraction to be 0.62, determine the flow rate of water.

Solution $d_1 = 0.6$ m $\qquad d_2 = 0.3$ m

$p_1 = 100$ kN/m^2 $\qquad p_2 = 80$ kN/m^2

$C_c = 0.62$

$$h_c=\frac{V_2^2}{2g}\left(\frac{1}{C_c}-1\right)^2=\frac{V_2^2}{2g}\left(\frac{1}{0.62}-1\right)^2=0.375\frac{V_2^2}{2g}$$

Applying Bernoulli's equation before and after contraction,

$$A_1V_1=A_2V_2$$

$$V_1=\left(\frac{A_2}{V_2}\right)V_2=\left(\frac{D_2}{D_1}\right)^2 V_2=\left(\frac{0.3}{0.6}\right)^2 V_2=0.25V_2$$

$$\frac{p_1}{w}+\frac{V_1^2}{2g}+z_1=\frac{80}{w}+\frac{V_2^2}{2g}+z_2+h_e$$

$$\frac{100}{9.81}+\frac{(0.25V_2)^2}{2\times 9.81}+0=\frac{80}{9.81}+\frac{V_2^2}{2\times 9.81}+0+0.375\frac{V_2^2}{2\times 9.81}$$

$$100\times 2+0.0625V_2^2=80\times 2+1.375V_2^2$$

$$1.3125V_2^2=40$$

$$V_2=5.52\ \text{m/s}$$

$$Q=A_2V_2=\frac{\pi}{4}(0.3)^2\times 5.52=0.39\ \text{m}^3/\text{s}$$

11.5 HYDRAULIC GRADIENT AND TOTAL ENERGY LINES

Hydraulic gradient line (HGL) is a line joining the piezometric heads $\left(\frac{p}{w}+z\right)$ at various points in a flow with respect to some reference line.

Total energy line (TEL) or energy gradient line (EGL) with respect to an arbitrary datum is a line joining the total heads $\left(\frac{p}{w}+\frac{V^2}{2g}+z\right)$ at different points of the flow.

Remember that

- Total energy line is always above the hydraulic gradient line and the vertical intercept between the two is equal to the velocity head $\left(\frac{V^2}{2g}\right)$.
- Total energy line always drops in the direction of flow due to various head losses.
- Hydraulic gradient line may fall or rise depending upon the pressure changes.

Example 11.5 A horizontal pipe connected to a water tank at one end discharges freely into the atmosphere at the other end. The pipe is 50 m long and its diameter is 200 mm for the first 30 m of its length which suddenly enlarges to 400 mm for the rest of the length. The height of water in the tank is 10 m above the axis of the pipe. Determine the discharge through the pipe considering all the losses. The coefficient of friction is 0.01.

Draw the hydraulic gradient and energy gradient lines also.

Solution $L_1 = 30$ m $\quad L_2 = 20$ m

$d_1 = 0{\cdot}2$ m $\quad d_2 = 0{\cdot}4$ m

$H = 10$ m $\quad f = 0{\cdot}01$

$$A_1 V_1 = A_2 V_2$$

$$V_1 = \left(\frac{A_2}{A_1}\right) V_2 = \left(\frac{d_2}{d_1}\right)^2 V_2 = \left(\frac{0{\cdot}4}{0{\cdot}2}\right)^2 V_2 = 4V_2$$

Applying Bernoulli's equation to the free water level in the tank and at the outlet of pipe (Fig. 11.4a),

$$\frac{p_1}{w} + \frac{V_1^2}{2g} + z_1 = \frac{p_2}{w} + \frac{V_2^2}{2g} + z_2 + \text{losses}$$

$$0 + 0 + 10 = 0 + \frac{V_2^2}{2g} + 0 + h_i + h_{f1} + h_e + h_{f2}$$

where $$h_i = \frac{0{\cdot}5V_1^2}{2g} = \frac{0{\cdot}5 \times (4V_2)^2}{2g} = 8\frac{V_2^2}{2g}$$

$$h_{f1} = \frac{4\,fL_1V_1^2}{2gd_1} = \frac{4 \times 0{\cdot}01 \times 30 \times (4V_2)^2}{2g \times 0{\cdot}2} = 96\frac{V_2^2}{2g}$$

$$h_e = \frac{(V_1 - V_2)^2}{2g} = \frac{(4V_2 - V_2)^2}{2g} = 9\frac{V_2^2}{2g}$$

$$h_{f2} = \frac{4\,fL_2V_2^2}{2gd_2} = \frac{4 \times 0{\cdot}01 \times 20 \times V_2^2}{2g \times 0{\cdot}4} = 2\frac{V_2^2}{2g}$$

$$10 = \frac{V_2^2}{2g}(1 + 8 + 96 + 9 + 2)$$

$$\frac{V_2^{\,2}}{2 \times 9{\cdot}81} = \frac{10}{116}$$

$$V_2 = 1{\cdot}3 \text{ m/s}$$

$$Q = A_2 V_2 = \frac{\pi}{4} \times 0{\cdot}4^2 \times 1{\cdot}3 = 0{\cdot}163 \text{ m}^3\text{/s}$$

The various head losses are

$$h_{fi} = 8\,\frac{V_2^2}{2g} = 8 \times \frac{1{\cdot}3^2}{2 \times 9{\cdot}81} = 8 \times 0{\cdot}086 = 0{\cdot}689 \text{ m}$$

$$h_{f1} = 96\,\frac{V_2^2}{2g} = 96 \times 0{\cdot}086 = 8{\cdot}27 \text{ m}$$

$$h_e = 9\,\frac{V_2^2}{2g} = 9 \times 0{\cdot}086 = 0{\cdot}775 \text{ m}$$

$$h_{f2} = 2\,\frac{V_2^2}{2g} = 2 \times 0{\cdot}086 = 0{\cdot}172 \text{ m}$$

The energy gradient line is drawn as follows (Fig. 11.4b) :

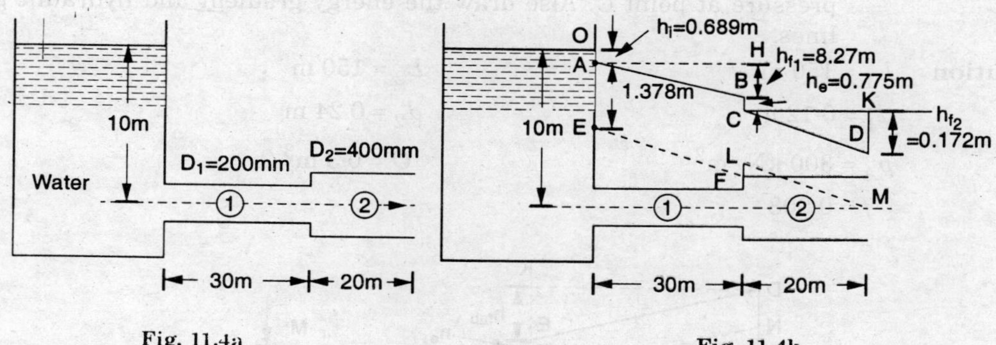

Fig. 11.4a

Fig. 11.4b

- Take OA equal to the head loss at the inlet *i.e.* 0.689 m.
- Draw a horizontal line *AB* through *A* equal to the length of the first pipe. Locate a point *B* vertically below *H* taking *BH* equal to the loss of head due to friction in the first pipe *i.e.* 8·27 m.
- Join *AB*.
- Take *BC* equal to the head loss due to sudden enlargment *i.e.* 0·775 m.
- Draw a horizontal line *CK* through *C* equal to the length of the second pipe. Locate a point *D* vertically below *K* taking *DK* equal to the loss of head due to friction in the second pipe *i.e.* 0·172 m.
- Join *CD*.

- The line *OABCD* is the required energy gradient line.

The hydraulic gradient line is drawn as under :

- Take *AE* equal to the velocity head *i.e.*

$$AE = \frac{V_1^2}{2g} = \frac{(4V_2)^2}{2g} = 16 \times 0{\cdot}086 = 1{\cdot}378 \text{ m}$$

- Draw *EF* parallel to *AB*.
- At the mid point *M* of the exit of pipe, the hydraulic gradient is zero (there is no pressure or potential head). Draw a line *ML* parallel to *DC*.
- Join *LE*.
- The line *EFLM* represents the required hydraulic gradient line.

Example 11.6 A pipeline *ABC* is 300 m long and is laid on an upward slope of 1 in 50. The portion *AB* is 150 m long and 120 mm in diameter. At *B* the diameter suddenly enlarges to 240 mm and remains so for the rest of the length *BC* of the pipe. At the lower end, 30 litres of water per second is pumped into the pipe under a pressure of 300 kN/m² which is discharged at the upper end into a closed tank. Assuming a coefficient of friction of 0.008, determine the pressure at point *C*. Also draw the energy gradient and hydraulic gradient lines.

Solution $L_1 = 150$ m $\quad L_2 = 150$ m

$d_1 = 0{\cdot}12$ m $\quad d_2 = 0{\cdot}24$ m

$p_A = 300$ kN/m² $\quad Q = 0{\cdot}3$ m³/s

$f = 0{\cdot}008$

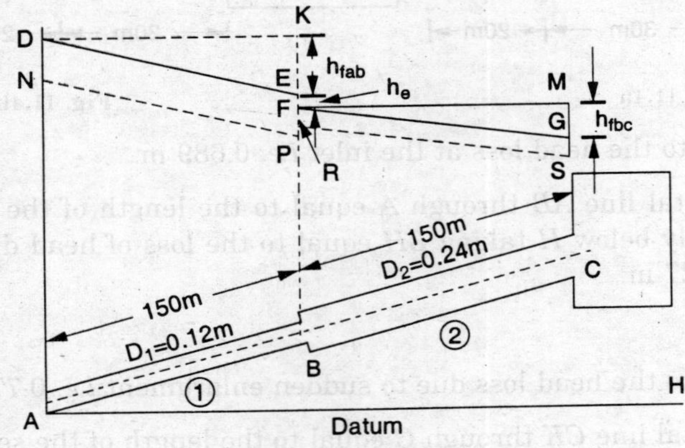

Fig. 11.5

$$V_1 = \frac{Q}{A_1} = \frac{0{\cdot}3}{(\pi/4) \times (0{\cdot}12)^2} = 26{\cdot}53 \text{ m/s}$$

$$V_2 = \frac{Q}{A_2} = \frac{0 \cdot 3}{(\pi/4) \times (0 \cdot 24)^2} = 0 \cdot 663 \text{ m/s}$$

Applying Bernoulli's equation to the inlet of pipe (*A*) and inlet of the tank (*C*) (Fig. 11.5),

$$\frac{p_A}{w} + \frac{V_A^{\ 2}}{2g} + z_A = \frac{p_C}{w} + \frac{V_C^{\ 2}}{2g} + z_C + h_{fAB} + h_e = h_{fBC}$$

where
$$h_{fAB} = \frac{4\,fL_1V_1^{\ 2}}{2gd_1} = \frac{4 \times 0 \cdot 008 \times 150 \times 2 \cdot 65^2}{2 \times 9 \cdot 81 \times 0 \cdot 12} = 14 \cdot 32 \text{ m}$$

$$h_e = \frac{(V_1 - V_2)^2}{2g} = \frac{(2 \cdot 65 - 0 \cdot 663)^2}{2 \times 9 \cdot 81} - 0 \cdot 201 \text{ m}$$

$$h_{fBC} = \frac{4\,fL_2V_2^{\ 2}}{2gd_2} = \frac{4 \times 0 \cdot 008 \times 150 \times 0 \cdot 663^2}{2 \times 9 \cdot 81 \times 0 \cdot 24} = 0 \cdot 448 \text{ m}$$

$$\frac{300}{9 \cdot 81} + \frac{2 \cdot 65^2}{2 \times 9 \cdot 81} + 0 = \frac{p_C}{9 \cdot 81} + \frac{0 \cdot 663^2}{2 \times 9 \cdot 81} + \frac{300}{50} + 14 \cdot 32 + 0 \cdot 201 + 0 \cdot 448$$

$$30 \cdot 58 + 0 \cdot 358 = \frac{p_C}{9 \cdot 81} + 0 \cdot 022 + 20 \cdot 97$$

$$\frac{p_C}{9 \cdot 81} = 9 \cdot 95$$

$$p_C = 97 \cdot 6 \text{ kN/m}^2$$

Assuming the datum line to pass through *A*, the Energy gradient line is drawn as follows :

The total energy at *A*

$$\frac{p_A}{w} + \frac{V_A^2}{2g} + Z_A = \frac{300}{9 \cdot 81} = \frac{2 \cdot 65^2}{2 \times 9 \cdot 81} + 0 = 30 \cdot 94 \text{ m}$$

- Draw a horizontal line *AH* through *A*.
- Draw the centre line of the pipe with a rising slope of 1 in 50. Thus at *B*, the rise is 3 m and at *C*, it is 6m.
- Take a point *D* vertically above *A* at a height of total energy at *A i.e.* 30·94 m.
- Draw a horizontal line through *D* which meets a vertical line through *B* at *K*.
- Take a point *E* on the vertical line through *B* and 14·32 m below *K* (equal to the friction head loss in pipe *AB*).
- Locate a point *F* below *E* at a distance of 0·201 m *i.e.* equal to the loss due to sudden enlargement.
- Draw a horizontal line through *F* meeting a vertical line through *C* at *M*.
- Locate a point *G* below *M* at a distance of 0·448 m *i.e.* equal to the frictional head loss in pipe *BC*.

- *DEFG* represents the required energy gradient line.

  To draw the hydraulic gradient line,

- First draw a line *NP* below and parallel to *DE* at a distance equal to velocity head in pipe *AB* *i.e.* 0·358 m.
- Draw a line *RS* below and parallel to *FG* at a distance equal to velocity head in pipe *BC* *i.e.* 0·022m.
- *NPRS* is the required hydraulic gradient line.

11.6 PIPES IN SERIES

When pipes of different lengths and diameters are connected end to end to form a pipe line, they are known as pipes in series or compound pipes (Fig. 11.6).

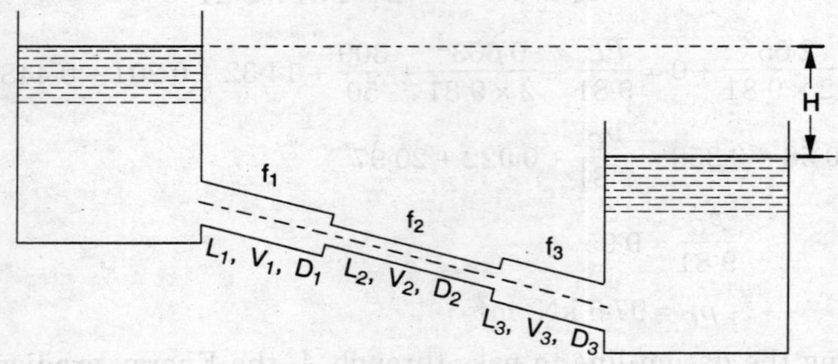

Fig. 11.6

Let L_1, L_2, L_3 = lengths of pipes 1, 2 and 3 respectively

V_1, V_2, V_3 = velocities of flow through pipes 1, 2, and 3 respectively

d_1, d_2, d_3 = diameters of pipes 1, 2, and 3 respectively

f_1, f_2, f_3 = coefficients of friction for pipes 1, 2, and 3 respectively

H = difference of water level in the two tanks

By equation of continuity,

$$Q = A_1 \cdot V_1 = A_2 \cdot V_2 = A_3 \cdot V_3$$

Now,

Difference of liquid surface levels = Sum of all the head losses in the pipes

H = Head loss at entrance to pipe 1
+ Friction head loss in pipe 1
+ Head loss due to contraction of area
+ Friction head loss in pipe 2
+ Head loss due to enlargement of area
+ Friction head loss in pipe 3
+ Head loss at the exit

$$H = \frac{0.5V_1^2}{2g} + \frac{4f_1L_1V_1^2}{2gd_1} + \frac{0.5V_2^2}{2g} + \frac{4f_2L_2V_2^2}{2gd_2} + \frac{(V_2 - V_3)^2}{2g} + \frac{4f_3L_3V_3^2}{2gd_3} + \frac{V_3^2}{2g} \quad ...(11.11)$$

If minor losses are neglected,

$$H = \frac{4f_1L_1V_1^2}{2gd_1} + \frac{4f_2L_2V_2^2}{2gd_2} + \frac{4f_3L_3V_3^2}{2gd_3} \quad ...(11.12)$$

If coefficient of friction is same for all pipes, $f_1 = f_2 = f_3 = f$.

Then

$$H = \frac{4f}{2g}\left(\frac{L_1V_1^2}{d_1} + \frac{L_2V_2^2}{d_2} + \frac{L_3V_3^2}{d_3}\right) \quad ...(11.12a)$$

Example 11.7 Three pipes connected in series have their diameters 400 mm, 250 mm and 500 mm, lengths 400 m, 250 m and 300 m and coefficients of friction as 0·006, 0·0065 and 0·0055 respectively. The pipes join two water tanks having a difference of 20 m in their levels. Determine the discharge of water considering all the losses. Also determine the discharge if minor losses are neglected.

Solution $d_1 = 0.4$ m $\quad L_1 = 400$ m $\quad f_1 = 0.006$

$d_2 = 0.25$ m $\quad L_2 = 250$ m $\quad f_2 = 0.0065$

$d_3 = 0.5$ m $\quad L_3 = 300$ m $\quad f_3 = 0.0055$

$H = 20$ m

From continuity equation (Fig. 11.6),

$$A_1V_1 = A_2V_2 = A_3V_3$$

$$V_2 = \frac{A_1}{A_2}V_1 = \left(\frac{d_1}{d_2}\right)^2 V_1 = \left(\frac{0.4}{0.25}\right)^2 V_1 = 2.56V_1$$

$$V_3 = \frac{A_1}{A_3}V_1 = \left(\frac{d_1}{d_3}\right)^2 V_1 = \left(\frac{0.4}{0.5}\right)^2 V_1 = 0.64V_1$$

$$H = h_i + h_{f1} + h_c + h_{f2} + h_e + h_{f3} + h_{exit}$$

where

$$h_i = \frac{0.5V_1^2}{2g}$$

$$h_{f1} = \frac{4fL_1V_1^2}{2gd_1} = \frac{4 \times 0.006 \times 400 \times V_1^2}{2g \times 0.4} = 24\frac{V_1^2}{2g}$$

$$h_c = \frac{0.5V_2^2}{2g} = \frac{0.5\,(2.56V_1)^2}{2g} = 3.28\frac{V_1^2}{2g}$$

$$h_{f2} = \frac{4fL_2V_2^2}{2gd_2} = \frac{4 \times 0.0065 \times 250 \times (2.56V_1)^2}{2g \times 0.25} = 170.4\,\frac{V_1^2}{2g}$$

$$h_e = \frac{(V_2 - V_3)^2}{2g} = \frac{(2.56V_1 - 0.64V_1)^2}{2g} = 3.69\frac{V_1^2}{2g}$$

$$h_{f3} = \frac{4fL_3V_3^2}{2gd_3} = \frac{4 \times 0{\cdot}0055 \times 300 \times (0{\cdot}64V_1)^2}{2g \times 0{\cdot}5} = 5{\cdot}41\,\frac{V_1^2}{2g}$$

$$h_{exit} = \frac{V_3^2}{2g} = \frac{(0{\cdot}64V_1)^2}{2g} = 0{\cdot}41\,\frac{V_1^2}{2g}$$

$$20 = \frac{V_1^2}{2g}\,(0{\cdot}5 + 24 + 3{\cdot}28 + 170{\cdot}4 + 3{\cdot}69 + 5{\cdot}41 + 0{\cdot}41)$$

$$\frac{V_1^2}{2 \times 9{\cdot}81} = \frac{20}{207{\cdot}69}$$

$$V_1 = 1{\cdot}375 \text{ m/s}$$

$$Q = A_1V_1 = \frac{\pi}{4}(0{\cdot}4)^2 \times 1{\cdot}375 = 0{\cdot}1726 \text{ m}^3/\text{s}$$

When minor losses are neglected,

$$H = h_{f1} + h_{f2} + h_{f3}$$

$$20 = \frac{V_1^2}{2g}\,(24 + 170{\cdot}4 + 5{\cdot}41)$$

$$\frac{V_1^2}{2g} = \frac{20}{199{\cdot}8}$$

$$V_1 = 1{\cdot}4 \text{ m/s}$$

$$Q = A_1V_1 = \frac{\pi}{4}(0{\cdot}4)^2 \times 1{\cdot}4 = 0{\cdot}176 \text{ m}^3/\text{s}$$

11.7 EQUIVALENT PIPE

An equivalent pipe is defined as a pipe of uniform diameter having loss of head and flow equal to the loss of head and flow of a compound pipe made up of several pipes of different diameters and lengths. The uniform diameter of the equivalent pipe is known as the equivalent diameter of the compound pipe.

Let d_1, d_2, d_3 = diameters of pipes 1, 2, and 3 respectively

L_1, L_2, L_3 = lengths of pipes 1, 2 and 3 respectively

V_1, V_2, V_3 = velocity of flow through pipes 1, 2, and 3 respectively

D = diameter of the equivalent pipe

L = length of the equivalent pipe

V = velocity of flow through the equivalent pipe

H = total head loss

f = friction coefficient for all the pipes

Q = discharge through the pipe

Neglecting minor losses, we can write,

Head loss in the equivalent pipe = Head loss in the compound pipe

$$\frac{4fLV^2}{2gd} = \frac{4fL_1V_1^2}{2gd_1} + \frac{4fL_2V_2^2}{2gd_2} + \frac{4fL_3V_3^2}{2gd_3}$$

$$\frac{LV^2}{d} = \frac{L_1V_1^2}{d_1} + \frac{L_2V_2^2}{d_2} + \frac{L_3V_3^2}{d_3}$$

$$\frac{L}{d}\left(\frac{Q}{A}\right)^2 = \frac{L_1}{d_1}\left(\frac{Q}{A_1}\right)^2 + \frac{L_2}{d_2}\left(\frac{Q}{A_2}\right)^2 + \frac{L_3}{d_3}\left(\frac{Q}{A_3}\right)^2$$

$$(\because Q = AV = A_1V_1 = A_2V_2 = A_3V_3)$$

$$\frac{L}{d}\left(\frac{Q}{(\pi/4)d^2}\right)^2 = \frac{L_1}{d_1}\left(\frac{Q}{(\pi/4)d_1^2}\right)^2 + \frac{L_2}{d_2}\left(\frac{Q}{(\pi/4)d_2^2}\right)^2 + \frac{L_3}{d_3}\left(\frac{Q}{(\pi/4)d_3^2}\right)^2$$

$$\frac{L}{d^3} = \frac{L_1}{d_1^3} + \frac{L_2}{d_2^3} + \frac{L_3}{d_3^3}$$

This is known as **Dupit's equation**. From this equation either the diameter of the equivalent pipe can be calculated if the length is known or vice-versa.

11.8 PIPES IN PARALLEL

When a main pipe line is divided into two or more parallel pipes which again join together downstream to form the main pipe, then the branch pipes are known as pipes connected in parallel (Fig. 11.7).

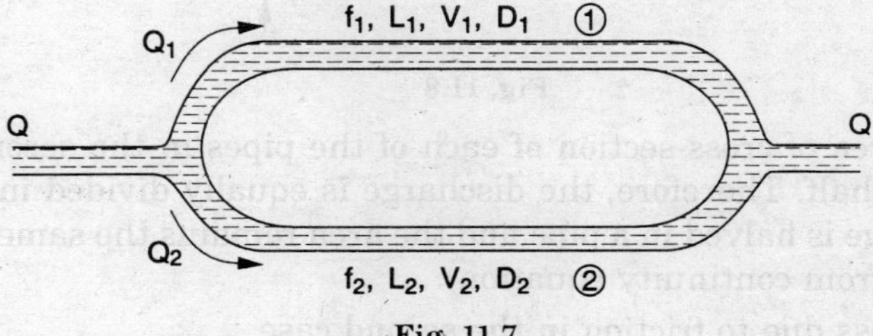

Fig. 11.7

When two pipes are joined in parallel, the discharge in the main pipe is equal to the sum of discharges through the parallel pipes,

$$Q = Q_1 + Q_2$$

As the pressure heads at the ends of the parallel pipes can not be different since they are to merge into a single pipe, the head loss in each branch pipes must be same.

Loss of head in pipe 1 = Loss of head in pipe 2

$$h_f = \frac{4f_1L_1V_1^2}{2gd_1} = \frac{4f_2L_2V_2^2}{2gd_2} \qquad ...(11.14)$$

If $f_1 = f_2 = f$, then,

$$\frac{L_1V_1^2}{d_1} = \frac{L_2V_2^2}{d_2} \qquad ...(11.14a)$$

Example 11.8 A pipe line discharges water from a tank into the atmosphere. Prove that if a pipe line of the same diameter and with the same coefficient of friction is added to a second half of its length, the discharge is increased by 26·5 percent, if minor losses are neglected.

If the diameter of a pipe line is 300 mm, length 2 km, head in the tank 5m, coefficient of friction 0·01, compute the increase in discharge if a pipe line of the same diameter and with the same coefficient of friction is added to the first in the second half of the length.

Solution When the pipe is single, let the velocity in the pipe be V,

The head loss due to friction in the first case

$$h_f = \frac{4fLV^2}{2gd} \qquad \text{...(i)}$$

When the second pipe is added to the second half of the first pipe (Fig. 11.8), let the velocity in the first half be V' and discharge Q'.

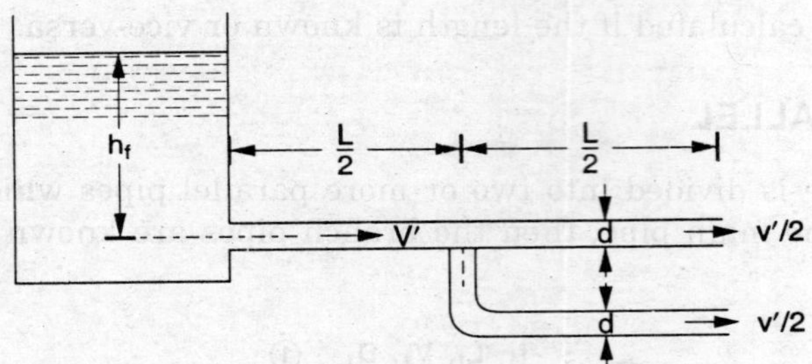

Fig. 11.8

Now, the area of cross-section of each of the pipes in the second half is same as in the first half. Therefore, the discharge is equally divided in the two pipes. As the discharge is halved in a pipe and the area remains the same, the velocity must be halved (from continuity equation).

The head loss due to friction in the second case

= Head loss in the first half + head loss in either of the two pipes

$$h_f = \frac{4f\,(L/2)V'^2}{2gd} + \frac{4f\,(L/2)\,(V'/2)^2}{2gd}$$

$$= \frac{4fLV'^2}{2gd}\left(\frac{1}{2} + \frac{1}{8}\right)$$

$$= \frac{5}{8}\left(\frac{4fLV'^2}{2gd}\right)$$

As the head of water in the two cases remains the same,

$$h_{f1} = h_{f2}$$

$$\frac{4fLV^2}{2gd} = \frac{5}{8}\left(\frac{4fLV'^2}{2gd}\right)$$

$$V^2 = \frac{5}{8} V'^2$$

$$\frac{V'}{V} = \sqrt{\frac{8}{5}} = 1{\cdot}265$$

Increase in velocity in the first half

$$\frac{V' - V}{V} = \frac{1{\cdot}265 - 1}{1} = 0{\cdot}265$$

As the velocity in the first half is increased by 0·265 times, thus the discharge is also increased by 0·265 times or by 26·5 percent.

From the given data, velocity in the first case is found from the relation (i),

$$5 = \frac{4 \times 0{\cdot}01 \times 2000 \times V^2}{2 \times 9{\cdot}81 \times 0{\cdot}3}$$

$$= 13{\cdot}592 V^2$$

$$V = 0{\cdot}6065 \text{ m/s}$$

$$Q = AV = \frac{\pi}{4} \times 0{\cdot}3^2 \times 0{\cdot}6065$$

$$= 0{\cdot}0429 \text{ m}^3/\text{s}$$

Increase in discharge $= 0.0429 \times 0.265 = 0.0114 \text{ m}^3/\text{s}$

Example 11.9 Two sharp ended pipes join two reservoirs which have a difference of level of 12 m. The pipes are 60 mm and 120 mm in diameters and each 120 m long. The friction factor for each pipe is 0·3. Determine the rate of flow of each pipe.

If the two pies are to be substituted by a single pipe of the same length having same friction factor and which gives the same discharge, determine its diameter.

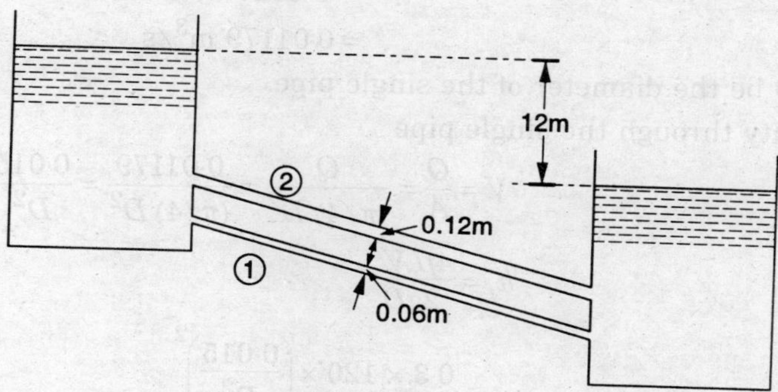

Fig. 11.9

Solution **First case**

$d_1 = 0{\cdot}06$ m $\qquad L = 120$ m

$d_2 = 0{\cdot}12$ m $\qquad 4f = 0{\cdot}3$

$H = 12$ m

Loss of head in either of the two pipes = 12 m

Considering the pipe of 60 mm diameter,

$$h_f = \frac{4fLV_1^{\,2}}{2gd_1}$$

$$12 = \frac{0{\cdot}3 \times 120 \times V_1^{\,2}}{2 \times 9{\cdot}81 \times 0{\cdot}06}$$

$$= 30{\cdot}58V_1^{\,2}$$

$$V_1 = 0{\cdot}626 \text{ m/s}$$

$$Q_1 = \frac{\pi}{4} \times (0{\cdot}06)^2 \times 0{\cdot}626 = 0{\cdot}00177 \text{ m}^3/\text{s}$$

Discharge through the pipe of 120 mm diameter,

$$h_f = \frac{4fLV_2^2}{2gd_2}$$

$$12 = \frac{0{\cdot}3 \times 120 \times V_2^{\,2}}{2 \times 9{\cdot}81 \times 0{\cdot}12}$$

$$= 15{\cdot}29V_2^2$$

$$V_2 = 0{\cdot}886 \text{ m/s}$$

$$Q_2 = \frac{\pi}{4} \times (0{\cdot}012)^2 \times 0{\cdot}886 = 0{\cdot}01002 \text{ m}^3/\text{s}$$

Second case

Discharge through the single pipe $= Q_1 + Q_2$

$$= 0{\cdot}00177 + 0{\cdot}01002$$

$$= 0{\cdot}01179 \text{ m}^3/\text{s}$$

Let D be the diameter of the single pipe.

Velocity through the single pipe

$$V = \frac{Q}{A} = \frac{Q}{(\pi/4)\,D^2} = \frac{0{\cdot}01179}{(\pi/4)\,D^2} = \frac{0{\cdot}015}{D^2}$$

$$h_f = \frac{4fLV^2}{2gD}$$

$$12 = \frac{0{\cdot}3 \times 120 \times \left(\dfrac{0{\cdot}015}{D^2}\right)^2}{2 \times 9{\cdot}81 \times D}$$

$$= \frac{0{\cdot}0004129}{D^5}$$

$$D^5 = \frac{0{\cdot}0004129}{D^5}$$

$$D = 0{\cdot}128 \text{ m}$$

Example 11.10 Two reservoirs are connected by a pipe of 300 mm diameter and 3 km long. The difference of levels between the two reservoirs is 50 m. The pipe is tapped at a distance of 1 km from the beginning of pipe and the water is drawn at a rate of 200 l/s. Determine the rate of flow of water into the lower reservoir assuming coefficient of friction as 0.005 for the pipe.

Solution $L_1 = 1000$ m $\qquad h_f = 50$ m

$L_2 = 2000$ m $\qquad Q = 0{\cdot}2\ \text{m}^3/\text{s}$

$f = 0{\cdot}005$

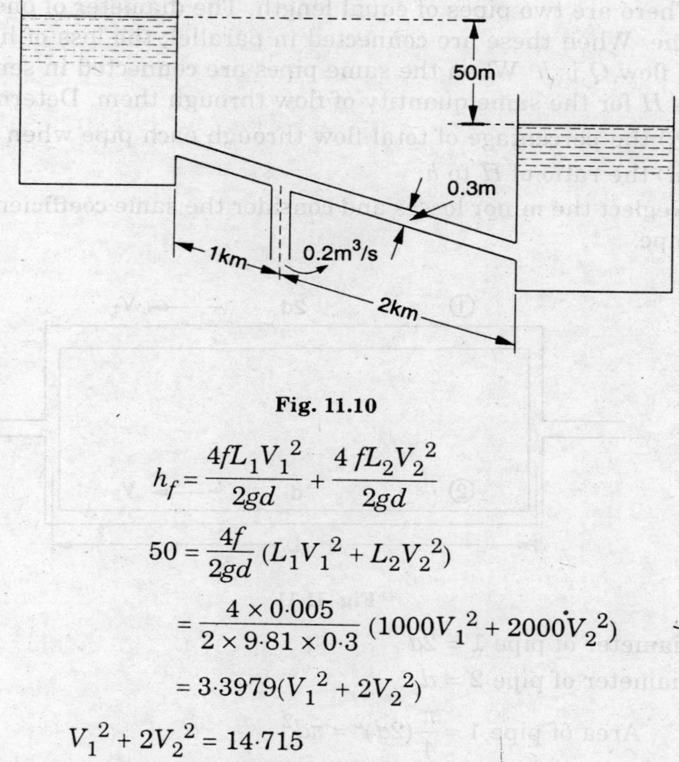

Fig. 11.10

$$h_f = \frac{4fL_1V_1^{\,2}}{2gd} + \frac{4\,fL_2V_2^{\,2}}{2gd}$$

$$50 = \frac{4f}{2gd}\,(L_1V_1^{\,2} + L_2V_2^{\,2})$$

$$= \frac{4 \times 0{\cdot}005}{2 \times 9{\cdot}81 \times 0{\cdot}3}\,(1000V_1^{\,2} + 2000V_2^{\,2})$$

$$= 3{\cdot}3979(V_1^{\,2} + 2V_2^{\,2})$$

$$V_1^{\,2} + 2V_2^{\,2} = 14{\cdot}715$$

Let the discharge through the pipe after the tapping be Q.

Then the discharge through the pipe before the tapping= $Q + 0{\cdot}2$

Thus

$$\left(\frac{Q + 0{\cdot}2}{(\pi/4)(0{\cdot}3)^2}\right)^2 + 2\left(\frac{Q}{(\pi/4)\,(0{\cdot}3)^2}\right)^2 = 14{\cdot}715$$

$$\frac{(Q + 0{\cdot}2)^2 + 2Q^2}{[(\pi/4)\,(0{\cdot}3)^2]^2} = 14{\cdot}715$$

$$\frac{(Q + 0{\cdot}2)^2 + 2Q^2}{0{\cdot}005} = 14{\cdot}715$$

$$(Q + 0{\cdot}2)^2 + 2Q^2 = 0{\cdot}07352$$
$$Q^2 + 0{\cdot}04 + 0{\cdot}4Q + 2Q^2 = 0{\cdot}07352$$
$$3Q^2 + 0{\cdot}4Q - 0{\cdot}03352 = 0$$
$$Q = \frac{-0{\cdot}4 \pm \sqrt{0{\cdot}16 + 0{\cdot}403}}{6}$$
$$= \frac{-0{\cdot}4 \pm 0{\cdot}75}{6}$$
$$= 0{\cdot}05832 \text{ m}^3/\text{s} \quad \text{(neglecting negative value)}$$

Example 11.11 There are two pipes of equal length. The diameter of one is half of the other one. When these are connected in parallel, the loss of head of each pipe for a flow Q is h. When the same pipes are connected in series, the loss of head is H for the same quantity of flow through them. Determine

(i) the percentage of total flow through each pipe when placed in parallel.

(ii) the ratio of H to h.

Neglect the minor losses and consider the same coefficient of friction for each pipe.

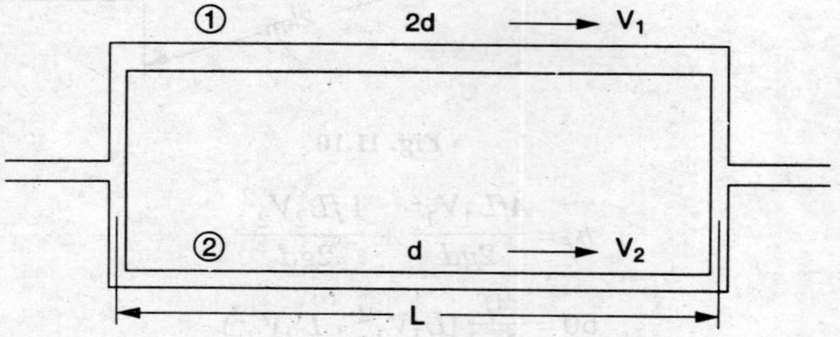

Fig. 11.11

Solution Let diameter of pipe 1 = $2d$

diameter of pipe 2 = d

$$\text{Area of pipe 1} = \frac{\pi}{4}(2d)^2 = \pi d^2$$

$$\text{Area of pipe 2} = \frac{\pi}{4}d^2$$

When the pipes are in parallel,

Let Q_1 be the flow through pipe 1 and Q_2 through pipe 2.

$$\text{Velocity in pipe 1}, V_1 = \frac{Q_1}{A_1} = \frac{Q_1}{\pi d^2}$$

$$\text{Velocity in pipe 2}, V_2 = \frac{Q_2}{A_2} = \frac{4Q_2}{\pi d^2}$$

$$h = \frac{4\,fLV_1^2}{2g\,(2d)} = \frac{4\,fLV_2^2}{2gd}$$

$$V_1^2 = 2V_2^2$$

$$\left(\frac{Q_1}{\pi d^2}\right)^2 = 2\left(\frac{4Q_2}{\pi d^2}\right)^2$$

$$\left(\frac{Q_1}{Q_2}\right)^2 = 32$$

$$\frac{Q_1}{Q_2} = 5{\cdot}657$$

$$Q_1 = 5{\cdot}657 Q_2$$

$$Q_1 + Q_2 = Q$$

$$5{\cdot}657\ Q_2 + Q_2 = Q$$

$$Q_2 = 0{\cdot}15Q$$

$$Q_1 = Q - 0{\cdot}15Q = 0{\cdot}85Q$$

Thus flow in pipe 1 = 85% of total flow

Flow in pipe 1 = 15% of total flow

When the pipes are in series,

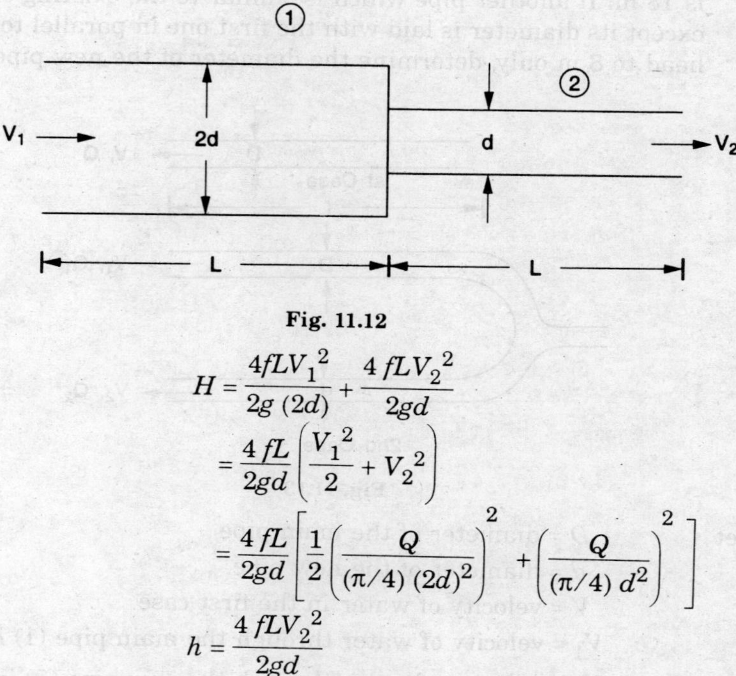

Fig. 11.12

$$H = \frac{4fLV_1^2}{2g\,(2d)} + \frac{4\,fLV_2^2}{2gd}$$

$$= \frac{4\,fL}{2gd}\left(\frac{V_1^2}{2} + V_2^2\right)$$

$$= \frac{4\,fL}{2gd}\left[\frac{1}{2}\left(\frac{Q}{(\pi/4)\,(2d)^2}\right)^2 + \left(\frac{Q}{(\pi/4)\,d^2}\right)^2\right]$$

$$h = \frac{4\,fLV_2^2}{2gd}$$

$$= \frac{4fL}{2gd}\left(\frac{0.15Q}{(\pi/4)\,d^2}\right)^2$$

$$\frac{H}{h} = \frac{\dfrac{4fL}{2gd}\left[\dfrac{1}{2}\left(\dfrac{Q}{(\pi/4)\,(2d)^2}\right)^2 + \left(\dfrac{Q}{(\pi/4)\,d^2}\right)^2\right]}{\dfrac{4fL}{2gd}\left(\dfrac{0.15Q}{(\pi/4)\,d^2}\right)^2}$$

$$= \frac{\dfrac{4fL}{2gd}\left(\dfrac{Q}{(\pi/4)\,d^2}\right)^2\left[\dfrac{1}{2}\left(\dfrac{1}{2^2}\right)^2 + 1\right]}{\dfrac{4fL}{2gd}\left(\dfrac{Q}{(\pi/4)\,d^2}\right)^2 (0.15)^2}$$

$$= \frac{\dfrac{1}{32} + 1}{0.15^2}$$

$$= \frac{33}{32 \times 0.0225}$$

$$= 45.83$$

Example 11.12 Water is forced through a 500 mm diameter main and the friction head loss is 18 m. If another pipe which is similar to the existing one in every respect except its diameter is laid with the first one in parallel to reduce the friction head to 8 m only, determine the diameter of the new pipe.

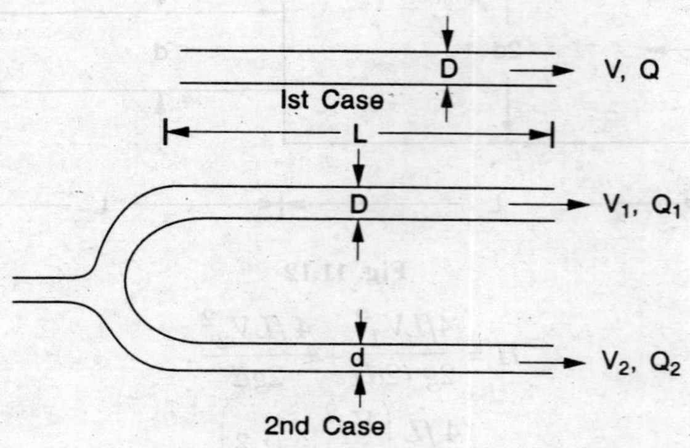

Fig. 11.13

Solution Let

D = diameter of the main pipe

d = diameter of the new pipe.

V = velocity of water in the first case.

V_1 = velocity of water through the main pipe (1) in the second case

V_2 = velocity of water through the new pipe (2) in the second case

Q = flow in the first case

Q_1 = flow through the main pipe (1) in the second case

Q_2 = flow through the new pipe (2) in the second case

h_f = friction head loss in the first case

h_f' = friction head loss in either pipe in the second case

Friction head loss in the second case,

$$h_f' = \frac{4\,fLV_1^2}{2gD} = \frac{4\,fLV_2^2}{2gd}$$

$$\frac{V_1}{V_2} = \sqrt{\frac{D}{d}} \quad \ldots\text{(i)}$$

Friction head loss in the first case,

$$h_f = \frac{4\,fLV^2}{2gD} = 18$$

Friction head loss in the second case (considering the main pipe),

$$h_f' = \frac{4\,fLV_1^2}{2gD} = 8$$

$$\frac{h_f}{h_f'} = \left(\frac{V}{V_1}\right)^2 = \frac{18}{8} = 2{\cdot}25$$

$$\frac{V}{V_1} = 1{\cdot}5$$

$$\frac{\dfrac{Q}{(\pi/4)\,D^2}}{\dfrac{Q_1}{(\pi/4)\,D^2}} = 1{\cdot}5$$

$$\frac{Q}{Q_1} = 1{\cdot}5$$

$$Q_1 = 0{\cdot}6667Q$$

$$Q_2 = Q - 0{\cdot}667Q = 0{\cdot}3333Q$$

$$\frac{Q_1}{Q_2} = \frac{0{\cdot}6667}{0{\cdot}3333} = 2$$

$$\frac{\dfrac{\pi}{4}\,D^2{\cdot}V_1}{\dfrac{\pi}{4}\,d^2{\cdot}V_2} = 2$$

$$\frac{D^2}{d^2}{\cdot}\sqrt{\frac{D}{d}} = 2 \qquad \left[\text{from (i) } \frac{V_1}{V_2} = \sqrt{\frac{D}{d}}\right]$$

$$\left(\frac{D}{d}\right)^{5/2} = 2$$

$$\frac{D}{d} = 1{\cdot}3195$$

$$d = \frac{500}{1{\cdot}32} = 378{\cdot}9 \text{ m}$$

Example 11.13 Three reservoirs are connected by pipes as shown in Fig. 11.14. Assuming a coefficient of friction for each pipe as 0·008, find the discharge in each pipe.

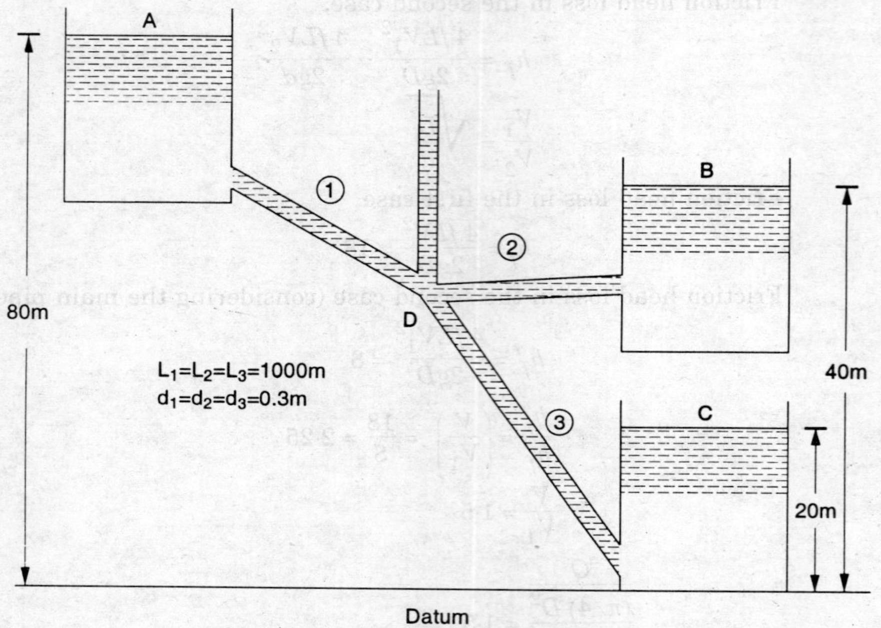

Fig. 11.14

Solution The flow in pipe 2 depends upon the piezometric level at section *D*. If it is more than level in reservoir *B* *i.e.* more than 40 m, the flow is from *D* to *B*, otherwise from *B* to *D*.

Assuming that the piezometric head at *D* is 40 m *i.e.* there is no flow either into or from reservoir *B*,

Head loss in pipe 1 = 80 – 40 = 40m

$$h_{f1} = \frac{4\, fL_1 V_1^2}{2gd_1}$$

$$40 = \frac{4 \times 0{\cdot}008 \times 1000 \times V_1^2}{2 \times 9{\cdot}81 \times 0{\cdot}3}$$

$$= 5{\cdot}437 V_1^2$$

$$V_1^2 = 7{\cdot}3575$$

$$V_1 = 2{\cdot}713 \text{ m/s}$$

$$Q_1 = A_1V_1 = \frac{\pi}{4} \times 0{\cdot}3^2 \times 2{\cdot}713 = 0{\cdot}1917 \text{ m}^3/\text{s}$$

Head loss in pipe 3 = 40 – 20 = 20 m

$$h_{f3} = \frac{4\,fL_3V_3^{\,2}}{2gd_3}$$

$$20 = \frac{4 \times 0{\cdot}008 \times 1000 \times V_3^{\,2}}{2 \times 9{\cdot}81 \times 0{\cdot}3}$$

$$= 5{\cdot}437V_3^2$$

$$V_3^2 = 3{\cdot}678$$

$$V_3 = 1{\cdot}918 \text{ m/s}$$

$$Q_3 = A_3V_3 = \frac{\pi}{4} \times 0{\cdot}3^2 \times 1{\cdot}918 = 0{\cdot}1356 \text{ m}^3/\text{s}$$

Now, $Q_1 > Q_2$, which means some discharge must go towards B. This is possible only if the level at D is higher than that at B or > 40 m.

Let h_D = pressure head at D

Applying Bernoulli's equation to levels at A and D,

$$\frac{p_A}{w} + z_A = \frac{p_D}{w} + z_D + h_{f1}$$

$$0 + 80 = h_D + z_D + h_{f1} \qquad \ldots\text{(i)}$$

Applying Bernoulli's equation to levels at D and B

$$\frac{p_D}{w} + z_D = \frac{p_B}{w} + z_B + h_{f2}$$

$$h_D + z_D = 0 + 40 + h_{f2}$$

$$h_D + z_D = 40 + h_{f2} \qquad \ldots\text{(ii)}$$

From (i) and (ii),

$$80 - h_{f1} = 40 + h_{f2}$$

$$h_{f1} + h_{f2} = 40 \qquad \ldots\text{(iii)}$$

$$h_{f1} = \frac{4\,fL_1V_1^{\,2}}{2gd_1} = \frac{4 \times 0{\cdot}008 \times 1000 \times V_1^2}{2 \times 9{\cdot}81 \times 0{\cdot}3} = 5{\cdot}437V_1^2$$

$$h_{f2} = \frac{4fL_2V_2^{\,2}}{2gd_2} = \frac{4 \times 0{\cdot}008 \times 1000 \times V_2^{\,2}}{2 \times 9{\cdot}81 \times 0{\cdot}3} = 5{\cdot}437V_2^2$$

Thus

$$5{\cdot}437V_1^{\,2} + 5{\cdot}437V_2^{\,2} = 40$$

$$5{\cdot}437(V_1^{\,2} + V_2^{\,2}) = 40$$

$$V_1^{\,2} + V_2^{\,2} = 7{\cdot}357$$

$$V_2 = \sqrt{7{\cdot}357 - V_1^{\,2}}$$

Applying Bernoulli's equation at D and C,

$$h_D + Z_D = 0 + 20 + h_{f3}$$
$$= 20 + h_{f3} \quad \text{...(iv)}$$

From (i) and (iv),

$$80 - h_{f1} = 20 + h_{f3}$$
$$h_{f1} + h_{f3} = 60$$
$$5{\cdot}437V_1^2 + 5{\cdot}437V_3^2 = 60$$
$$5{\cdot}437(V_1^2 + V_3^2) = 60$$
$$V_1^2 + V_3^2 = 11{\cdot}035$$
$$V_3 = \sqrt{11{\cdot}035 - V_1^2}$$

From continuity equation,

$$Q_1 = Q_2 + Q_3$$
$$A_1V_1 = A_2V_2 + A_3V_3$$

But $\quad A_1 = A_2 + A_3$

$\therefore \quad V_1 = V_2 + V_3$

$$V_1 = \sqrt{7{\cdot}357 - V_1^2} + \sqrt{11{\cdot}035 - V_1^2}$$

By trial and error,

$$V_1 = 2{\cdot}64 \text{ m/s}$$
$$V_2 = \sqrt{7{\cdot}357 - 2{\cdot}64^2} = 0{\cdot}622 \text{ m/s}$$
$$V_3 = \sqrt{11{\cdot}035 - 2{\cdot}64^2} = 2{\cdot}016 \text{ m/s}$$
$$Q_1 = \frac{\pi}{4} \times 0{\cdot}3^2 \times 2{\cdot}64 = 0{\cdot}187 \text{ m}^3/\text{s}$$
$$Q_2 = \frac{\pi}{4} \times 0{\cdot}3^2 \times 0{\cdot}622 = 0{\cdot}044 \text{ m}^3/\text{s}$$
$$Q_3 = \frac{\pi}{4} \times 0{\cdot}3^2 \times 2{\cdot}016 = 0{\cdot}143 \text{ m}^3/\text{s}$$

11.9 SYPHON

A syphon is a long bent pipe used to transfer a liquid from one reservoir at a higher level to another reservoir at a lower lever when the two reservoirs are separated by a high level ground or a hill between the two (Fig. 11.15).

At the highest point B (known as summit), the pressure is less than the pressure at point A since the point B is above the point A. As the pressure at A is atmospheric, the pressure at B is below the atmospheric pressure. Theoretically, the pressure at B can be reduced upto 10·3 m of water *i.e.* zero atmospheric pressure head. However, at -7·6 m (or 10·3-7·6 = 2·7 m absolute) of water head, the dissolved air and gases in the water come out from the water and collect at the summit thus obstructing the flow. Also the height of the inlet leg (rising portion

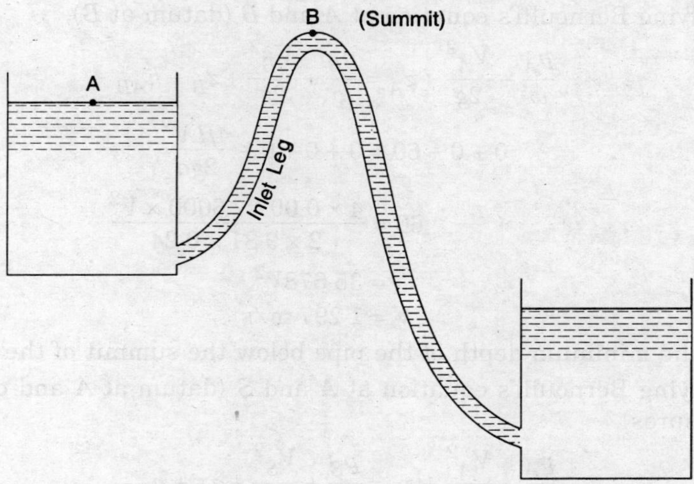

Fig. 11.15

of the syphon) is made the shortest possible to reduce the friction losses in the pipe which results in additional reduction of the pressure head.

Example 11.14 The surface level of two water reservoirs differ by 60 m. They are connected by a 240 mm diameter and 6000 m long pipe. At a distance of 600 m from the upper reservoir, the pipe crosses a ridge, the summit of which is 8 m above the water level in reservoir. If the absolute pressure head at the summit of syphon is not to fall below 2·8 m of water (absolute), determine the minimum depth of the pipe below the summit of the ridge. Also find the discharge through the pipe.

The coefficient of friction is 0·007 for the pipe and atmospheric pressure head 10·3 m of water. Neglect minor losses.

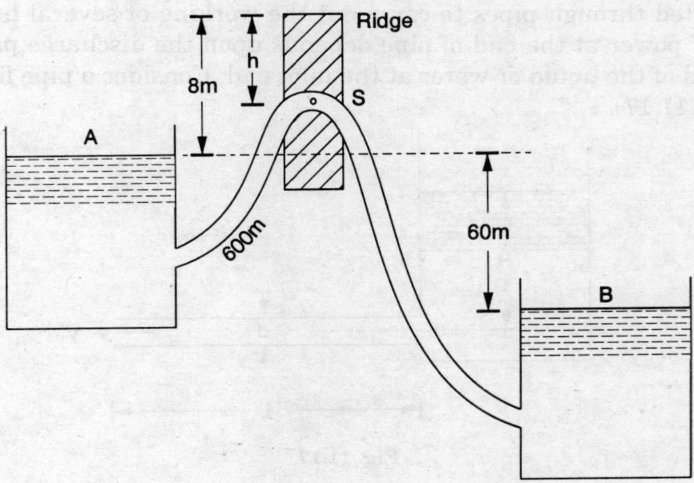

Fig. 11.16

Solution Applying Bernoulli's equation at A and B (datum at B),

$$\frac{p_A}{w}+\frac{V_A{}^2}{2g}+z_A=\frac{p_B}{w}+\frac{V_B{}^2}{2g}+z_B+h_{AB}$$

$$0+0+60=0+0+0+\frac{4fLV^2}{2gd}$$

$$60=\frac{4\times 0{\cdot}007\times 6000\times V^2}{2\times 9{\cdot}81\times 0{\cdot}24}$$

$$=35{\cdot}678V^2$$

$$V=1{\cdot}297\text{ m/s}$$

Let the minimum depth of the pipe below the summit of the ridge = h

Applying Bernoulli's equation at A and S (datum at A and considering absolute pressures),

$$\frac{p_A}{w}+\frac{V_A{}^2}{2g}+z_A=\frac{p_S}{w}+\frac{V_S{}^2}{2g}+z_S+h_{AS}$$

$$10{\cdot}3+0+0=2{\cdot}8+\frac{V_S{}^2}{2g}+(8-h)+\frac{4fLV^2}{2gd}$$

$$10{\cdot}3=2{\cdot}8+\frac{1{\cdot}297^2}{2\times 9{\cdot}81}+(8-h)+\frac{4\times 0{\cdot}007\times 600\times 1{\cdot}297^2}{2\times 9{\cdot}81\times 0{\cdot}24}$$

$$=2{\cdot}8+0{\cdot}0857+(8-h)+6$$

$$=16{\cdot}886-h$$

$$h=6{\cdot}586\text{ m}$$

$$Q=AV=\frac{\pi}{4}\times 0{\cdot}24^2\times 1{\cdot}297=0{\cdot}0587\text{ m}^3/\text{s}$$

11.10 POWER TRANSMISSION THROUGH PIPES

Power is transmitted through pipes to carry out the working of several hydraulic machines. The availability of power at the end of pipe depends upon the discharge passing through the pipe and total head of the liquid or water at the inlet end. Consider a pipe fitted to a high level storage tank (Fig. 11.17).

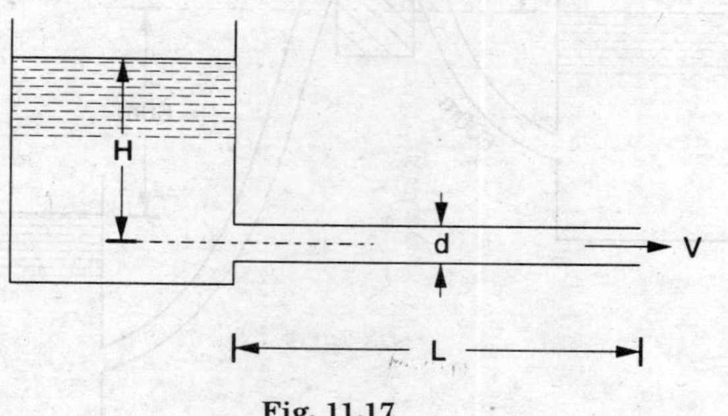

Fig. 11.17

Let H = head of water at the inlet of pipe
L = length of the pipe
d = diamter of the pipe
V = velocity of flow in the pipe
h_f = frictional head loss in the pipe
f = coefficient of friction

Neglecting minor head losses,

Head available at the outlet of pipe = Head at the inlet of pipe – Frictional head loss

$$= H - \frac{4fLV^2}{2gd} \quad \text{...(11.15)}$$

Power transmitted = Weight of liquid flowing per second × Head at the outlet

$$= w \times \text{Volume flow rate} \times \text{Head}$$

$$= w \times Q \times (H - h)$$

$$= wAV\left(H - \frac{4fLV^2}{2gd}\right) \quad \text{...(11.16)}$$

If other parameters are constant, the power transmitted depnds upon the velocity.

Power transmitted is maximum if

$$\frac{d\,(P)}{dV} = 0$$

$$\frac{d}{dV}\left[wAV\left(H - \frac{4fLV^2}{2gd}\right)\right] = 0$$

$$\frac{d}{dV}\left[wA\left(HV - \frac{4fLV^3}{2gd}\right)\right] = 0$$

$$wA\left(H - \frac{4 \times 3fLV^2}{2gd}\right) = 0$$

$$\left(H - 3 \times \frac{4fLV^2}{2gd}\right) = 0$$

$$H - 3h_f = 0$$

$$H = 3h_f$$

$$h_f = \frac{H}{3} \quad \text{...(11.17)}$$

The power transmitted through a pipe is maximum when the head loss due to friction is equal to one third of the total head at the inlet.

Efficiency of transmission,

$$\eta = \frac{\text{Power available at the oulet}}{\text{Power supplied at the inlet}}$$

$$= \frac{H - h_f}{H}$$

Maximum $\eta = \dfrac{H - \dfrac{H}{3}}{H} = 1 - \dfrac{1}{3} = \dfrac{2}{3}$ or 66·7%

Example 11.15 150 kW power is to be transmitted through a 3000 m long pipe. The pressure at the inlet is 5000 kN/m^2, the pressure drop over the entire length of pipe is 1000 kN/m^2 and the coefficient of friction 0·007. Determine the diameter of the pipe and the efficienty of transmission.

Solution $P = 150$ kW $\qquad H = (5000/9{\cdot}81)$ m

$L = 3000$ m $\qquad h_f = (1000/9{\cdot}81)$ m

$f = 0{\cdot}007$

Power transmitted is given by

$$P = wQ\,(H - h_f)$$

$$150 = 9{\cdot}81\left(\frac{\pi}{4}d^2{\cdot}V\right)\left(\frac{5000}{9{\cdot}81} - \frac{1000}{9{\cdot}81}\right)$$

$$= \left(\frac{\pi}{4}d^2{\cdot}V\right) \times 4000$$

$$V = \frac{0{\cdot}0477}{d^2}\ \text{m/s}$$

$$h_f = \frac{4fLV^2}{2gd}$$

$$\frac{1000}{9{\cdot}81} = \frac{4 \times 0{\cdot}007 \times 3000 \times (0{\cdot}0477/d^2)^2}{2 \times 9{\cdot}81 \times d}$$

$$1000 = \frac{0{\cdot}09556}{d^5}$$

$$d = 0{\cdot}157\ \text{m}$$

$$\eta = \frac{H - h_f}{H} = \frac{\dfrac{5000}{9{\cdot}81} - \dfrac{1000}{9{\cdot}81}}{\dfrac{500}{9{\cdot}81}} = \frac{5000 - 1000}{5000} = 0{\cdot}8 \text{ or } 80\%$$

11.11 FLOW THROUGH NOZZLES

A nozzle is a mouthpiece with narrowing area fitted at the outlet end of a pipe (Fig. 11.18). By fitting a nozzle at the end of the pipe, the total head of the liquid is converted into kinetic head which may be used in a Pelton turbine or in some other equipment.

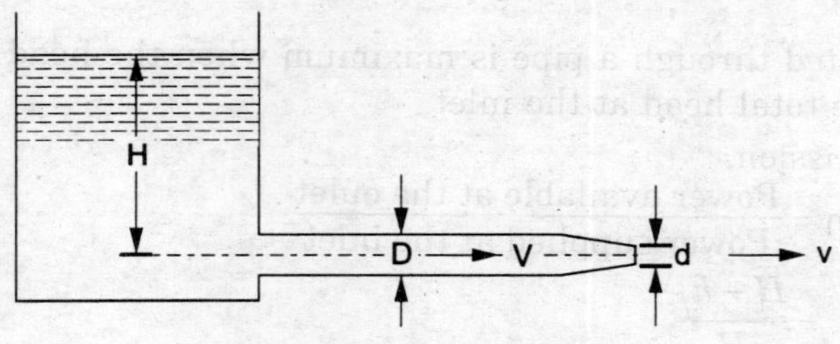

Fig. 11.18

Let $\quad H$ = head at the inlet of a pipe

D = diameter of the pipe

d = diameter of the nozzle
a = cross-sectional area of the nozzle
V = velocity of flow in the pipe
v = velocity of flow at the outlet of nozzle
f = coefficient of friction

Assuming no losses in the nozzle,

Head at the outlet of nozzle (or head at the end of pipe) = $\dfrac{v^2}{2g}$

Now, Head at the inlet of pipe = Head at the end of pipe + Losses in the pipe

$$H = \frac{v^2}{2g} + \frac{4fLV^2}{2gd} \qquad \text{...(i)}$$

$$= \frac{v^2}{2g} + \frac{4fL}{2gd}\left(\frac{av}{A}\right)^2 \qquad (\because AV = av)$$

$$= \frac{v^2}{2g}\left(1 + \frac{4fLa^2}{dA^2}\right)$$

$$v^2 = \frac{2gH}{1 + \dfrac{4fLa^2}{dA^2}}$$

$$v = \sqrt{\frac{2gH}{1 + \dfrac{4fLa^2}{dA^2}}} \qquad \text{...(11.18)}$$

Discharge through the nozzle = $a \cdot v$
Power available at the outlet of nozzle

$$= \frac{1}{2} mv^2$$

$$= \frac{1}{2} \rho Q \cdot v^2$$

$$= \frac{1}{2} \rho\, (av) \cdot v^2$$

$$= \frac{1}{2} \rho a \cdot v^3 \qquad \text{...(11.19)}$$

This power available can be shown to be equal to the power transmitted through the pipe as in Section 11.10.

Power available $= \dfrac{1}{2} \rho av \cdot v^2$

$$= \frac{1}{2} \rho\, AV \cdot 2g \cdot \frac{v^2}{2g}$$

$$= \frac{1}{2} \rho\, AV \cdot 2g \left(H - 4\frac{fLV^2}{2gd}\right) \qquad \left(\text{from (i) } \frac{v^2}{2g} = H - \frac{4fLV^2}{2gd}\right)$$

$$= wAV\left(H - 4\frac{fLV^2}{2gd}\right) \qquad \text{...(11.20)}$$

= Power transmitted through the pipe

The power transmitted in this case is also the same as in the previous section. Thus for mxiumum power transmission,

$$\frac{dP}{dv} = 0$$

$$\frac{d}{dv}\left[wAV\left(H - \frac{4fLV^2}{2gd}\right)\right] = 0$$

Thus the required condition for maximum power transmission is the same *ie* $h_f = \dfrac{H}{3}$

$$\eta = \frac{\text{Power at the outlet of nozzle}}{\text{Power at the inlet of pipe}}$$

$$= \frac{\frac{1}{2}\rho a v^3}{wQH}$$

$$= \frac{\frac{1}{2}\rho a v^3}{\rho g \times av \times H}$$

$$= \frac{v^2}{2gH}$$

$$= \frac{1}{2gH}\,\frac{2gH}{1 + \dfrac{4fLa^2}{dA^2}} \qquad \text{(using Eq. 11.18)}$$

$$= \frac{1}{1 + \dfrac{4fLa^2}{dA^2}}$$

For maximum power transmission,

$$H = 3h_f = 3 \times \frac{4fLV^2}{2gd}$$

Equating this value of H with that given in (i),

$$3 \times \frac{4fLV^2}{2gd} = \frac{v^2}{2g} + \frac{4fLV^2}{2gd}$$

$$v^2 = \frac{8fLV^2}{d}$$

$$\left(\frac{AV}{a}\right)^2 = \frac{8fLV^2}{d}$$

$$\left(\frac{(\pi/4)\,D^2}{(\pi/4)\,d^2}\right)V^2 = \frac{8fLV^2}{d}$$

$$\frac{D^4}{d^4} = \frac{8fL}{D}$$

$$d^4 = \frac{D^5}{8fL}$$

$$d = \left(\frac{D^5}{8fL}\right)^{1/4} \qquad \text{...(11.21)}$$

Example 11.16 A nozzle 150 mm in diameter and 300 m long is fitted to a pipe and discharges freely into the atmosphere. The available head at the nozzle is 120 m. Assuming the coefficient of friction for the nozzle as 0.008, determine the diameter of the nozzle and the maximum power transmitted by the jet.

Solution $D = 0{\cdot}15$ m $\qquad L = 300$ m

$f = 0{\cdot}008 \qquad H' = 120$ m

For maximum power transmitted,

$$d = \left(\frac{D^5}{8fL}\right)^{1/4} = \left(\frac{0{\cdot}15^5}{8 \times 0{\cdot}008 \times 300}\right)^{1/4} = 0{\cdot}0446 \text{ m}$$

For maximum efficiency,

The head loss due to friction = 120/3 = 40 m

Therefore, the useful head = 120 − 40 = 80 m

$$P = wQH = wavH = w\cdot\left(\frac{\pi}{d^2}\right)(\sqrt{2gH})\cdot H$$

$$= 9{\cdot}81\left(\frac{\pi}{4} \times 0{\cdot}0446^2\right)(\sqrt{2 \times 9{\cdot}81 \times 80 \times 80})$$

$$= 48{\cdot}58 \text{ kW}$$

Example 11.17 The head of water at the entrance of a pipe 3 km long and 750 mm in diameter is 80 m. At the outlet of pipe, a nozzle of 150 mm diameter is fitted. Determine the velocity of water at the outlet of nozzle assuming a coefficient of friction for the pipe to be 0.01.

Solution $L = 3000$ m $\qquad H = 80$ m

$d = 0{\cdot}15$ m $\qquad D = 0{\cdot}75$ m

$f = 0{\cdot}01$

$$v = \sqrt{\frac{2gH}{1 + \frac{4fL}{d}\left(\frac{a}{A}\right)^2}}$$

$$= \sqrt{\frac{2gH}{1 + \frac{4fL}{d}\left(\frac{d^2}{D^2}\right)^2}}$$

$$= \sqrt{\frac{2 \times 9{\cdot}81 \times 80}{1 + \frac{4 \times 0{\cdot}01 \times 3000}{0{\cdot}15}\left(\frac{0{\cdot}15^2}{0{\cdot}75^2}\right)^2}}$$

$$= \sqrt{\frac{1569{\cdot}6}{1 + 1{\cdot}28}}$$

$$= 26{\cdot}24 \text{ m/s}$$

11.12 WATER HAMMER

The water flowing in a pipe has momentum due to its motion. If this flowing water is suddenly brought to rest by closing the valve, the momentum is destroyed which causes a high pressure on the valve. This is followed by a series of pressure vibration accompanied by noises in the pipe, known as knocking. The sudden rise of pressure also cause hammering action on the wall of the pipe known as **water hammer** or **hammer blow**. Sometimes, the water hammer can be so high as to burst the pipe.

Gradual Closing of Valve

Consider a uniform pipe line through which water is flowing at a uniform velocity and the valve provided at the end of the pipe is closed gradually (Fig. 11.19).

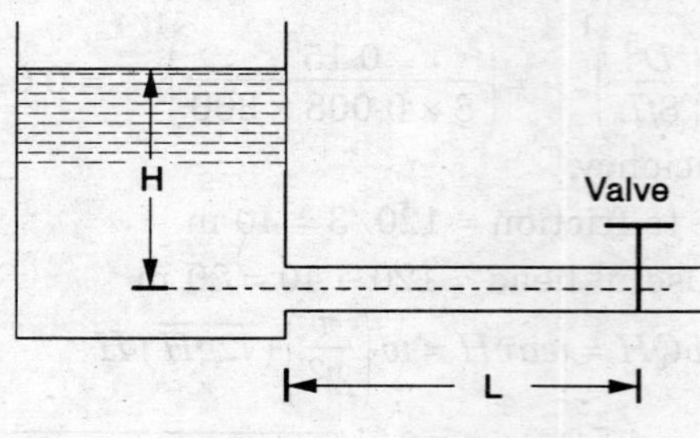

Fig. 11.19

Let L = length of the pipe

A = area of cross-section of the pipe

V = velocity of water in the pipe

p = increase in pressure

t = time required to close the valve (in seconds)

As the valve is closed gradually, the velocity of water decreases from V to zero in time t

$$\text{Retardation of water} = \frac{V-0}{t} = \frac{V}{t}$$

$$\text{Retarding force} = \text{Mass} \times \frac{V}{t} = \rho AL \cdot \frac{V}{t}$$

The force due to increase in the pressure = $p \times A$

Equating the two forces,

$$pA = \rho AL \frac{V}{t}$$

$$p = \frac{\rho LV}{t} \qquad \ldots(11.22)$$

$$\frac{p}{w} = \frac{\rho LV}{wt} = \frac{\rho LV}{\rho gt} = \frac{LV}{gt} \qquad \ldots(11.23)$$

The valve closure is known as gradual if $t > \frac{2L}{C}$ ...(11.24)

The valve closure is known as sudden if $t > \frac{2L}{C}$ ...(11.25)

where C = Velocity of the pressure wave

Sudden Closure of Valve (Rigid Pipe)

Equation 11.22 indicates that if the valve is closed instantly ($t = 0$), the pressure increase is infinite. However, since it takes a little time to close the valve, the pressure rise is finite. Also at very high pressures, the liquids get compressed to some extent and no longer remain incompressible.

Consider a pipe of uniform cross-section through which water is flowing with uniform velocity. As the valve is closed suddenly, the kinetic energy of the flowing water converts to strain energy of water (neglectng the effect of friction). Thus

Loss of kinectic energy = Gain in the strain energy

$$\frac{1}{2} mV^2 = \frac{1}{2}\left(\frac{p^2}{K}\right) \times \text{Volume}$$

where K = Bulk modulus of water

$$\frac{1}{2} (\rho AL)V^2 = \frac{1}{2}\left(\frac{p^2}{K}\right) \times AL$$

$$p = \sqrt{\rho K V^2}$$

$$= V\sqrt{\rho K}$$

$$= V\sqrt{\rho^2 \frac{K}{\rho}}$$

$$= \rho V\sqrt{\frac{K}{\rho}}$$

$$= \rho VC \qquad \ldots(11.26)$$

where C = Velocity of pressure wave

Sudden Closure of Valve (Elastic Pipe)

If the pipe through which the water flows is elastic, then circumferential and longitudinal stresses are produced in the pipe wall as the rise of pressure takes place on closing the valve instantly.

Consider a pipe of length L, diameter D and thickness t through which water flows.

Let p = increase in pressure due to hammer blow

E = modulus of elasticity for the pipe material

$\frac{1}{m}$ = possion's ratio for the pipe meterial

From the study of strength of materials,

$$f_c = \frac{pD}{2t} \quad \text{and} \quad f_t = \frac{pD}{4t}$$

where f_c and f_t are the circumferential and tangential stresses respectively.

Also, strain energy stored in the pipe material per unit volume

$$= \frac{1}{2E}\left(f_c^2 + f_t^2 - \frac{2f_c f_t}{m}\right)$$

$$= \frac{1}{2E}\left(\frac{p^2D^2}{4t^2} + \frac{p^2D^2}{16t^2} - \frac{p^2D^2}{4mt^2}\right)$$

Let $\dfrac{1}{m} = \dfrac{1}{4}$

Strain energy per unit volume

$$= \frac{1}{2E}\left(\frac{p^2D^2}{4t^2} + \frac{p^2D^2}{16t^2} - \frac{p^2D^2}{16t^2}\right)$$

$$= \frac{p^2D^2}{8Et^2}$$

Total strain energy $= \dfrac{p^2D^2}{8Et^2} \times \text{Volume}$

$$= \frac{p^2D^2}{8Et^2} \times \pi D \times t \times L$$

$$= \frac{p^2 \pi D^3 L}{8Et}$$

$$= \frac{1}{2}\frac{p^2}{Et}\left(\frac{\pi}{4}D^2\right)DL$$

$$= \frac{1}{2}\frac{p^2 ADL}{Et}$$

Gain of strain energy in water $= \dfrac{1}{2}\dfrac{p^2}{K} \times AL$

Now loss of kinetic energy of water = Gain of strain energy of water + Strain energy stored in the material

$$\frac{1}{2}mV^2 = \frac{1}{2}\frac{p^2}{K} \times AL + \frac{1}{2}\frac{p^2ADL}{Et}$$

$$\frac{1}{2}\rho ALV^2 = \frac{1}{2}\frac{p^2}{K}AL + \frac{p^2 ADL}{2Et}$$

$$\rho V^2 = \frac{p^2}{K} + \frac{p^2 D}{Et} = p^2\left(\frac{1}{K} + \frac{D}{Et}\right)$$

$$p^2 = \frac{\rho V^2}{\frac{1}{K} + \frac{D}{Et}}$$

$$p = \sqrt{\frac{\rho V^2}{\frac{1}{K} + \frac{D}{Et}}}$$

$$= V\sqrt{\frac{\rho}{\frac{1}{K} + \frac{D}{Et}}} \qquad \text{...(11.27)}$$

Time taken by the pressure wave to travel from the valve to the tank and from tank to the valve,

$$t = \frac{\text{Distance travelled from the valve to the tank and back}}{\text{Velocity of pressure wave}}$$

$$= \frac{L + L}{C}$$

$$= \frac{2L}{C}$$

Example 11.18 A pipe 4000 m long and 800 mm in diameter is provided with a valve at its end. The velocity of water flowing through the pipe is 3 m/s. Assuming the velocity of pressure wave to be 1480 m/s, determine the rise in pressure if

(i) the valve is closed in 24 seconds

(ii) the valve is closed in 3 seconds and the pipe is rigid. The bulk modulus of water is 2 GN/m^2.

Solution $L = 4000$ m $\quad C = 1480$ m/s

$V = 3$ m/s $\quad K = 2 \times 10^9$ m^2/s

(i) The ratio,

$$\frac{2L}{C} = \frac{2 \times 4000}{1480} = 5{\cdot}405$$

Now $t = 24$ s,

Thus $$t > \frac{2L}{C}$$

and therefore the valve is closed gradually. The rise of pressure is given by

$$p = \frac{\rho LV}{t} = \frac{1000 \times 4000 \times 3}{24} = 500 \times 10^3 \text{ N/m}^2 = 500 \text{ kN/m}^2$$

(ii) The ratio,

$$\frac{2L}{C} = \frac{2L}{\sqrt{\frac{K}{\rho}}} = \frac{2 \times 4000}{\sqrt{\frac{2 \times 10^9}{1000}}} = \frac{8000}{1414{\cdot}2} = 5{\cdot}657$$

Now $t = 3$ s, thus

$$t < \frac{2L}{C}$$

The valve is closed instantaneously. The rise of pressure is given by

$$p = V\rho C = 3 \times 1000 \times 1414{\cdot}2 = 4242{\cdot}6 \times 10^3 \text{ N/m}^2 = 4242{\cdot}6 \text{ kN/m}^2$$

PROBLEMS

1. What is the major energy loss in a pipe ? How is it determined ?
2. What are minor energy lossed in pipes ? Name them.
3. Describe Darcy-Weisback and Chezy's formulae to find the major energy losses in a pipe.
4. Derive the relations to calculate loss of head due to (a) sudden enlargement and (b) sudden contraction.

5. In what way are the following losses taken into account :
   (a) entrance to a pipe (b) exit of a pipe
   (c) bend in a pipe (d) obstruction in a pipe ?
6. What do you mean by hdraulic gradient line and total energy line ?
7. How and why are the pipes connected in (i) series (ii) parallel ? Explain.
8. What is an equivalent pipe ?
9. What is a syphon ? Where do we use it ?
10. Deduce the condition for maximum transmission of power through pipes. Also find the corresponding efficiency of transmission.
11. Prove that the diameter of the nozzle for maximum power transmission is given by

$$d = \left(\frac{D^5}{8fL}\right)^{\frac{1}{4}}$$

where D = diameter of pipe

L = length of pipe

12. What is the phenomenon of water hammer ? Obtain the expressions for the rise of pressure when the flow of water in a pipe is brought to rest by closing the valve (i) gradually (ii) suddenly.
13. Water flows through a 200 mm diameter pipe at a velocity of 3 m/s. The coefficient of friction is given by

$$f = 0{\cdot}002 + \frac{0{\cdot}09}{\text{Re}^{0{\cdot}03}}$$

where Re is the Reynolds number. Determine the head lost due to friction for a length of 5 m. The kinematic viscosity of water is 0·01 stoke. (0·993 m)
14. The diameter of a pipe used to carry water enlarges from 240 mm to 480 mm suddenly. As a result hydraulic gradient rises by 10 mm. Determine the rate of flow. (32·72 l/s)
15. Two reservoirs are connected by a pipeline 20 m long. The diameter of the pipe is 150 mm for the first 5 m and 250 mm for the remaining length. The water level in the upper reservoir is 7·5 m above that in the lower reservoir. Entry to and exit from the pipe are sharp. Assuming the coefficient of friction to be 0·01 for both the pipes, find the discharge. Draw the hydraulic gradient and energy gradient lines also. ($0{\cdot}131\ m^3/s$)
16. The difference in the water levels of two reservoirs is 180 m. They are connected by a 500 mm diameter pipe, 64 km long. Determine the discharge through the pipe. Take the value of frictin factor as 0·015.

    To increase the discharge by 50 percent, it is proposed to add another pipe of the same diameter from the lower reservoir for some part of the length and connected to the first pipe. What will be the length of the additional pipe required ? ($0{\cdot}42\ m^3/s$, 47·417 km)
17. Three pipes of the same diameter d, same length l and of the same value of coefficient of friction f are connected in parallel. If the pipes are to be replaced by a single pipe

of the same length and of the same value of coefficient of friction, determine the diameter D of the new pipe in terms of d which will give the same discharge for the same head loss. ($D = 1{\cdot}5518\ d$)

18. A penstock is used to supply water from a reservoir to a turbine which is 200 m below the level of the reservoir. Determine the smallest diameter of the penstock which produces 530 kW of power with a turbine of 80% efficiency. Take the friction factor to be 0·025. (412·8 mm)
19. A nozzle is fitted at the outlet of a 75 mm diameter pipe 500 m long. The head at the inlet of pipe is 60 m. Find the diameter of the pipe and the power transmitted if the power transmitted is maximum. Take 4f = 0·025. (17·55 mm, 2·651 kW)
20. two reservoirs with 10 m difference in their levels are connected by two pipes joined in series. The first pipe is 20 m long and 100 mm in diameter and of coefficient of friction 0·005. The second pipe is 25 m long, 160 mm in diameter and of coefficient of friction 0·0045. Determine the discharge through the pipes. What difference in the elevations of two reservoirs is necessary to have a discharge of 15 l/s ? (0·0438 m^3/s, 1·171 m)
21. A pipeline whose diameter varies from 200 mm to 100 mm in a length of 10 m is used to discharge water through it. Assuming a coefficient of friction for the pipe to be 0·005, determine the head loss in friction if the discharge through the pipe is 50 l/s. (0·968 m)
22. Two pipes are used in parallel to connect two reservoirs. One pipe is 400 m long, 300 mm in diameter and of friction factor 0·02. The second pipe is 500 m long, 200 m in diameter and of friction factor 0·015. Determine the necessary difference in the elevations of two reservoirs to produce a total flow of 0·025 m^3/s. (9 m)
23. Two reservoirs are connected by three pipes 1, 2, and 3 laid in parallel. All the pipes are of the same length, their diameters being d, 2d and 3d respectively. Find the discharge through pipes 2 and 3 if the discharge through pipe 1 is 1 m^3/s. Assume coefficient of friction for each pipe to be the same. (5·657 m^3/s , 15·588m^3/s)
24. Water available at a head of 150 m is to be carried through a 500 m long pipe. What is the minimum diameter of the pipe that conveys the water for an output of 1000 kW at 90 percent efficiency ? Take friction factor to be 0·025. (484 mm)
25. A nozzle is to be fitted at the end of a 300 m long pipe to get maximum power. The level of the nozzle is 30 m below the level of reservoir. Determine the maximum power available and the nozzle diameter. Assume coefficient of friction to be 0·005. (116·5 kW, 197 mm)
26. Water flows through a 150 mm diameter and 5 mm thick pipe at a velocity of 2·5 m/s. If the valve is closed suddenly, determine the rise in pressure assuming the pipe to be elastic. Take the value of Poisson's ratio 0·25, the bulk modulus of water 2·06 GN/m^2, modulus of elasticity for the pipe material 206 GN/m^2. (3148 kN/m^2)

Chapter **12**

FLOW IN OPEN CHANNELS

12.1 INTRODUCTION

Flow in canals, rivers and drainage channels in which the liquid surface is open to atmosphere is known as **open channel flow**. This type of flow is caused due to gravity. Whereas in a pipe, the flow can be forced in the upward direction also by pressure, a downward slope of the bed is necessary to maintain the flow in a channel. However, if a pipe does not run full, the flow is similar to that in an open channel.

12.2 CLASSIFICATION OF FLOW IN OPEN CHANNELS

The flow in open channels can be

1. Steady and unsteady
2. Uniform and non-uniform
3. Laminar and turbulent
4. Subcritical, critical and super critical

1. Steady and Unsteady Flow

The flow is said to be **steady** if the flow characteristics such as depth of flow, velocity of flow, rate of flow etc. at any point in the flow do not change with respect to time. Mathematically,

$$\frac{\partial y}{\partial t} = 0, \frac{\partial V}{\partial t} = 0 \quad \text{and} \quad \frac{\partial Q}{\partial t} = 0$$

where y, V and Q are the depth, velocity and the rate of flow respectively. However, if these parameters change with time, the flow is said to be **unsteady**. Mathematically,

$$\frac{\partial y}{\partial t} \neq 0, \frac{\partial y}{\partial t} \neq 0 \quad \text{and} \quad \frac{\partial Q}{\partial t} \neq 0$$

2. Uniform and Non-Uniform Flow

For a given length of a channel, if the depth, slope, cross-section and the velocity of the flow are constant, the flow is said to be **uniform.** Otherwise, the flow is called **non-uniform.** Mathematically,

$$\frac{\partial y}{\partial s} = 0 \, , \frac{\partial V}{\partial s} = 0 \text{ etc.} \qquad \text{for a uniform flow}$$

$$\frac{\partial y}{\partial s} \neq 0 \, , \frac{\partial V}{\partial s} \neq 0 \text{ etc.} \qquad \text{for a non uniform flow}$$

A non-uniform flow (also known as varied flow) is classified into the following types :

- RVF (Rapidly varied flow)
- GVF (Gradually varied flow)

If the change in the depth of flow is rapid in a small length of channel, it is known as **rapidly varied flow** and if the change is gradual over a long length, it is **gradually varied flow** (Fig. 12.1).

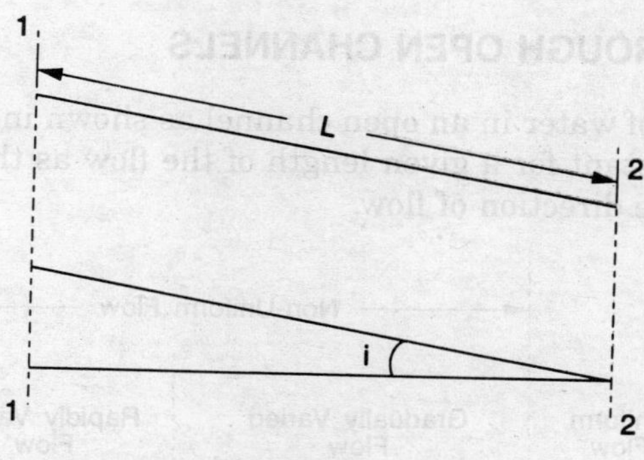

Fig. 12.1

3. Laminar and Turbulent Flow

Reynolds number in an open channel is defined as

$$\text{Re} = \frac{\rho V m}{\mu} \qquad \text{...(12.1)}$$

where ρ = density of the liquid

μ = viscosity of the liquid

V = mean velocity of the flow

m = hydraulic mean depth or hydraulic radius

$= \dfrac{\text{cross–sectional area of flow}}{\text{wetted perimeter}}$

If the Reynolds number is below 500, the flow is **laminar** and if it is more than 2000, it is **turbulent**. A flow with Reynolds number between 500 and 2000 is known as **transitional** flow.

4. Subcritical, Critical and Super Critical Flow

Froude number in a channel flow is an important parameter since gravitational force is a predominant force in such a flow. Froude number is defined as

$$\mathrm{Fr} = \frac{V}{\sqrt{2gD}} \qquad \text{...(12.2)}$$

where V = mean velocity of flow

D = hydraulic depth

$= \dfrac{\text{Wetted area}}{\text{Top width}}$

In case of rectangular channel,

$$D = \frac{\text{width} \times \text{depth}}{\text{width}} = \text{depth}(y)$$

If $\mathrm{Fr} > 1$, the flow is **subscritical** or **stream** or **tranquil**

$\mathrm{Fr} = 1$, the flow is **critical**

$\mathrm{Fr} > 1$, the flow is **supercritical** or **rapid** or **shooting** or **torrential.**

12.3 DISCHARGE THROUGH OPEN CHANNELS

Coinsider a uniform flow of water in an open channel as shown in Fig. 12.2. The velocity, area and depth of flow are constant for a given length of the flow as the flow is uniform. Consider sections 1-1 and 2-2 in the direction of flow.

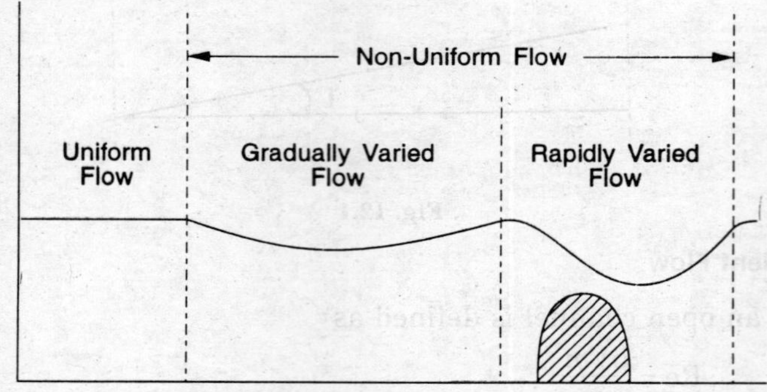

Fig. 12.2

Let A = area of flow of water

V = mean velocity of flow

L = length of channel

i = slope of the bed

P = wetted perimeter

f = frictional resistance/unit velocity/unit area

The pressure forces on the two sections are equal and acting in the opposite directions as the depths of water at the two sections are the same. Therefore, these forces cancel each other.

The velocity of flow is also constant for the given length as the flow is uniform. Thus no acceleration acts on the water and the resultant force acting in the direction of flow is zero.

Now, the weight of water between sections 1-1 and 2-2,

$$W = w \times \text{Volume} = wAL$$

Frictional resistance against the motion of water offered by the surface of the channel is experimently found to vary as square of velocity.

Thus $F = f \times \text{Surface area} \times V^2$

$$= f \times P.L.V^2$$

Resolving all forces in the direction of flow,

$$wAL \sin i = fPLV^2$$

$$V^2 = \frac{wAL \sin i}{fPL} = \frac{w}{f}\frac{A}{P} \sin i$$

$$V = \sqrt{\frac{w}{f}} \times \sqrt{\frac{A}{P} \sin i} = C.\sqrt{m.\sin i}$$

where $C = \sqrt{\frac{w}{f}}$ Chezy's constant (depends on the roughness of channel surface and Reynolds number)

and $\frac{A}{P} = m$, hydraulic mean depth or hydraulic radius

Also for small values of i,

$$\sin i \approx \tan i \approx i$$

$$V = C\sqrt{mi} \qquad ...(12.3)$$

This is known as **Chezy's formula**. It was developed by a French engineer Antoine Chezy in 1975.

The Chezy's constant C is not a dimensionless coefficient. The dimensions of C are

$$C = \frac{V}{\sqrt{m.i}} = \frac{L/T}{\sqrt{L}} = \frac{\sqrt{L}}{T} = L^{1/2}.T^{-1} \qquad (i \text{ is dimensionless})$$

Hence its value depends upon the system of units.

Discharge, $Q = A \times V$

$$= AC\sqrt{mi} \qquad ...(12.4)$$

$$= K\sqrt{i} \qquad ...(12.5)$$

where $K = AC\sqrt{m}$

The factor K is known as the **conveyance** of the channel section and is a measure of the carrying capacity of the channel.

The Chezy's constant C is given by the following empirical relations :

1. Bazin Formula

According to this formula,

$$C = \frac{157.6}{181 + \frac{k}{\sqrt{m}}} \qquad ...(12.6)$$

where K = Bazin's constant, the value of which depends upon the surface roughness and m is the hydraulic mean depth. The following table provides the values of K :

S.N.	Surface of Channel	K
1.	Smooth cement plaster or planed wood	0.11
2.	Unplaned wood or concrete or brick work	0.21
3.	Poor brickwork or rubble stone	0.83
4.	Earth channel in good condition	1.54
5.	Earth channel in ordinary condition	2.35
6.	Earth channel in poor condition	3.17

2. Kutter's Formula

Swiss engineers Ganguillet and Kutter gave the following formula in 1869 :

$$C = \frac{23 + \dfrac{0.00155}{i} + \dfrac{1}{N}}{1 + \left(23 + \dfrac{0.00155}{i}\right)\dfrac{N}{\sqrt{m}}} \qquad \text{...(12.7)}$$

where N is Kutter's constant, the values of which are given in the following table :

S.N.	Surface of Channel	N
1.	Smooth cement plaster or planed wood	0.010
2.	Smooth concrete	0.012
3.	Brick surface lined with cement mortar	0.015
4.	Poor brick work or rubble stone	0.017
5.	Unlined earth in good condition	0.020
6.	Earth channel of ordinary surface	0.025
7.	Earth channel of rough surface	0.030

3. Manning Formula

Rober Manning prescribed the following formula :

$$C = \frac{1}{N} m^{1/6} \qquad \text{...(12.8)}$$

where N is the Manning's constant having the same values as of Kutter's constant.

Example 12.1 The bed slope of a rectangular channel 10 m wide is 1 in 1600. Determine the rate of flow and the convyance of the channel if the depth of water is 4 m. The Chezy's constant C is 60.

Also state the type of flow (tranquil or rapid).

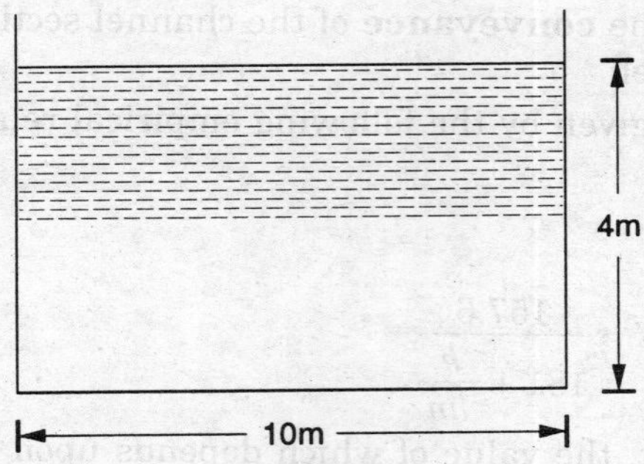

Fig. 12.3

Solution $A = b.y = 10 \times 4 = 40 \text{ m}^2$

$$P = b + 2y = 10 + 2 \times 4 = 18$$

$$m = \frac{A}{P} = \frac{40}{18} = 2.222$$

$$V = C\sqrt{mi} = 60\sqrt{2.222 \times \frac{1}{1600}} = 2.236 \text{ m/s}$$

$$Q = AV = 40 \times 2.236 = 89.44 m^3/s$$

Conveyance,

$$K = AC\sqrt{m} = 40 \times 60 \times \sqrt{2.222} = 3577{\cdot}5$$

$$\text{Fr} = \frac{V}{\sqrt{gy}} = \frac{2.236}{\sqrt{9.81 \times 4}} = 0.357$$

As Fr is less than 1, the flow is tranquil in nature.

Example 12.2 Determine the discharge through the channel shown in Fig. 12.4. The slope of the bed is 1 in 800 and the value of the Chezy's constant 55.

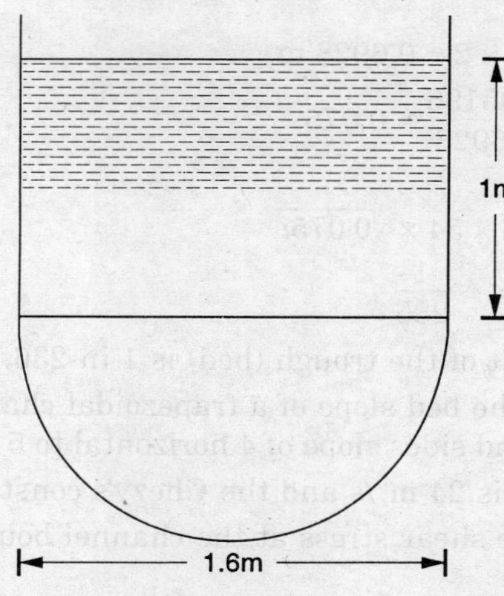

Fig. 12.4

Solution $A = 16 \times 1 + \frac{1}{2}\left(\frac{\pi}{4} \times 1.6^2\right) = 2.605 \text{ m}^2$

$$P = 1 \times 2 + (\pi \times 1.6/2) = 4.515 \text{ m}$$

$$m = \frac{A}{P} = \frac{2.605}{4.515} = 0.577 \text{ m}$$

$$Q = AC\sqrt{mi} = 2.605 \times 55 \times \sqrt{0.577 \times \frac{1}{800}} = 3.848 \text{ m}^3/\text{s}$$

Example 12.3 A triangular channel with included angle as 60° conveys water at a uniform depth of 300 mm. The discharge is 50 litres per second. Find the gradient of the trough. Take Chezy's constant C as 54.

Solution As the included angle is 60°, the water cross-section is an equilateral triangle each side of which is 300/cos 30° = 346.4 mm.

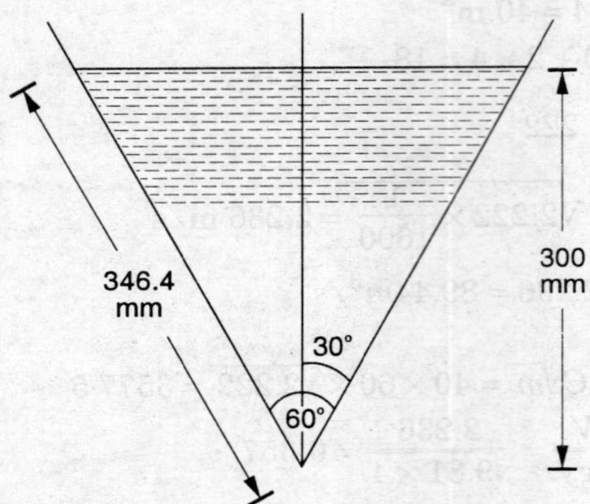

Fig. 12.5

$$A = \frac{1}{2} \times 0.3464 \times 0.3 = 0.05196 \text{ m}^2$$

$$P = 0.3464 \times 2 = 0.6928 \text{ m}$$

$$m = \frac{A}{P} = \frac{0.05196}{0.6928} = 0.075$$

$$Q = AC\sqrt{mi}$$

$$0{\cdot}05 = 0{\cdot}05196 \times 54 \times \sqrt{0{\cdot}075i}$$

$$i = 0{\cdot}004234 = \frac{1}{236}$$

Thus the gradient of the trough (bed) is 1 in 236.

Example 12.4 Determine the bed slope of a trapezoidal channel of bed width 5 m, depth of water 2 m and sides slope of 4 horizontal to 5 vertical. The discharge through the channel is 24 m^3/s and the Chezy's constant C is 60.

Also find the shear stress at the channel boundary.

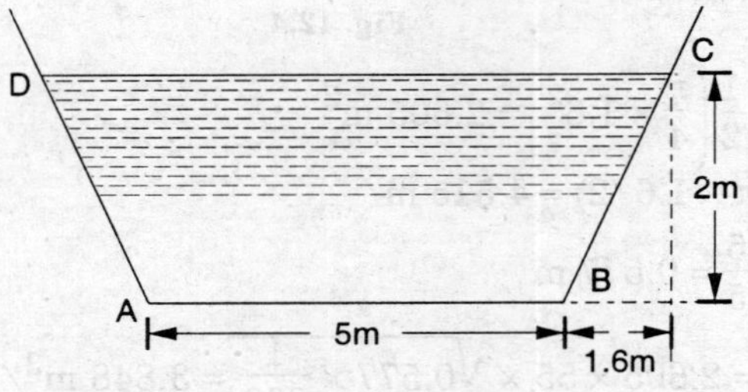

Fig. 12.6

Solution $BK = 2 \times \frac{4}{5} = 16$ m

$$CD = 5 + 16 \times 2 = 8.2 \text{ m}$$

$$AD = BC = \sqrt{1{\cdot}6^2 + 2^2} = 2{\cdot}561 \text{ m}$$

$$A = \frac{5 + 8 \cdot 2}{2} \times 2 = 13 \cdot 2 \text{ m}^2$$

$$P = 5 + 2 \times 2 \cdot 561 = 10 \cdot 122 \text{ m}$$

$$m = \frac{A}{P} = \frac{13 \cdot 2}{10 \cdot 122} = 1 \cdot 304 \text{ m}$$

$$Q = AC\sqrt{mi}$$

$$24 = 13 \cdot 2 \times 60 \times \sqrt{1 \cdot 304 i}$$

$$i = 0 \cdot 000704 = \frac{1}{1420}$$

Shear stress × Wetted perimeter × Length

= Component of weight parallel to the bed surface

$$\tau_o \cdot PL = wAL \sin \theta$$

$$\tau_o = w \cdot \frac{A}{p} \cdot \sin \theta = w \cdot m \cdot \sin \theta = 9 \cdot 81 \times 1 \cdot 304 \times \frac{1}{1420} = 0 \cdot 009 \text{kN/m}^2 = 9 \text{ N/m}^2$$

Example 12.5 What is the discharge through a rectangular channel 3 m wide and having depth of water 2 m and bed slope 1 in 2000 ? The value of N in Kutter's formula is 0·025.

Solution

$$m = \frac{A}{P} = \frac{3 \times 2}{3 + 2 \times 2} = 0 \cdot 857 \text{ m}$$

$$C = \frac{23 + \frac{0 \cdot 00155}{i} + \frac{1}{N}}{1 + \left(23 + \frac{0 \cdot 00155}{i}\right)\frac{N}{\sqrt{m}}} = \frac{23 + \frac{0 \cdot 00155}{(1/2000)} + \frac{1}{0 \cdot 025}}{1 + \left(23 + \frac{0 \cdot 00155}{(1/2000)}\right)\frac{0 \cdot 025}{\sqrt{0 \cdot 857}}}$$

$$= \frac{23 + 3 \cdot 1 + 40}{1 + (23 + 3 \cdot 1) \times 0 \cdot 027} = \frac{66 \cdot 1}{1 \cdot 7047} = 38 \cdot 78$$

$$Q = AC\sqrt{mi} = 6 \times 38 \cdot 78 \times \sqrt{0 \cdot 857 \times (1/2000)} = 4 \cdot 817 \text{ m}^3/\text{s}$$

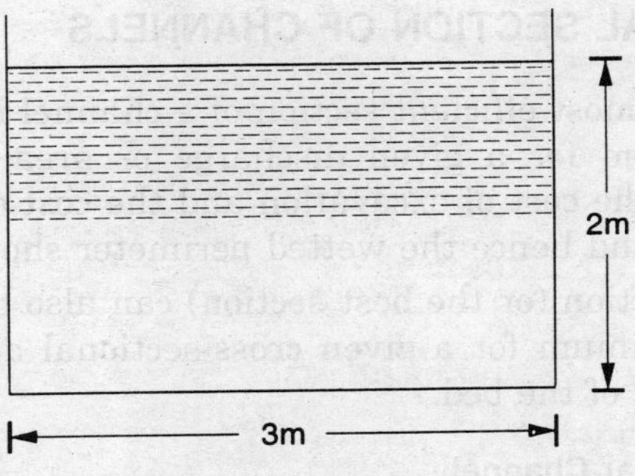

Fig. 12.7

Example 12.6 A circular sewer pipe is laid at a slope of 1 in 6000 and carries a discharge of 600 litres of water per second when flowing half full. Find its diameter by taking the value of N in Manning's formula as 0·015.

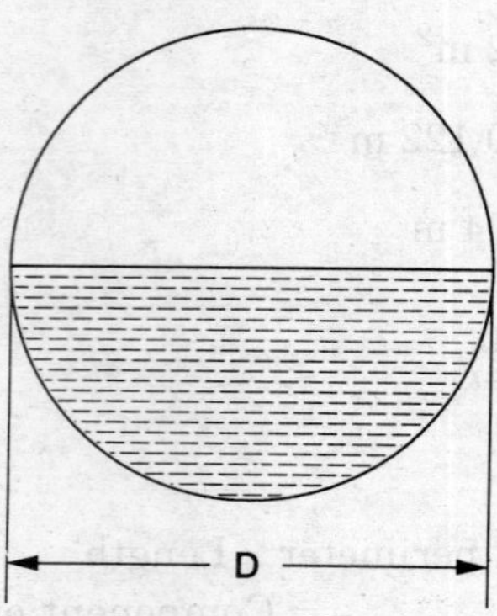

Fig. 12.8

Solution
$$A = \frac{1}{2} \times \frac{\pi}{4} D^2 = \frac{\pi D^2}{8}$$

$$m = \frac{A}{P} = \frac{\frac{\pi D^2}{8}}{\frac{\pi D}{2}} = \frac{D}{4}$$

$$Q = AC\sqrt{mi} = A \cdot \left(\frac{1}{N} m^{1/6}\right) \sqrt{mi} = A \cdot \frac{1}{N} m^{2/3} \sqrt{i}$$

$$0 \cdot 6 = \frac{\pi D^2}{8} \times \frac{1}{0 \cdot 015} \times \left(\frac{D}{4}\right)^{2/3} \times \sqrt{\frac{1}{6000}}$$

$$= 0 \cdot 1341 D^{8/3}$$

$$D^{8/3} = 4 \cdot 473$$

$$D = 1 \cdot 754 \text{ m}$$

12.4 MOST ECONOMICAL SECTION OF CHANNELS

The most economical or the most efficient section of a channel is the one in which the cost of construction is the minimum for a given discharge or area of cross-section. The cost of construction depends upon the cost of excavation and the cost of lining. To keep these to the minimum, the wetted area and hence the wetted perimeter should be minimum.

The most economical section (or the best section) can also be defined as the one through which the discharge is maximum for a given cross-sectional area, slope of the bed and the surface resistance coefficient of the bed.

1. Most Economical Rectangular Channel

Figure 12.9 shows a rectangular channel. Let b be the width and y the depth of the flow.

The wetted perimeter, $P = b + 2y$

$$= \frac{A}{y} + 2y$$

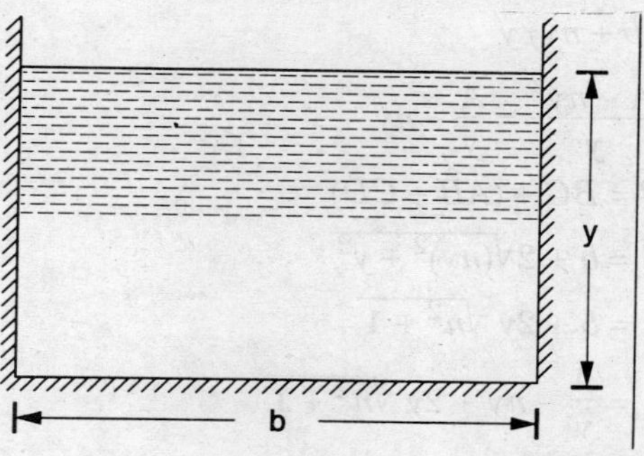

Fig. 12.9

It is minimum, if

$$\frac{dP}{dy} = 0$$

$$\frac{d}{dy}\left(\frac{A}{y} + 2y\right) = 0$$

$$\frac{d}{dy}(Ay^{-1} + 2y) = 0$$

$$-\frac{A}{y^2} + 2 = 0$$

$$A = 2y^2$$

$$b \cdot y = 2y^2$$

$$y = \frac{b}{2} \quad \text{or} \quad b = 2y \qquad \text{...(12.9)}$$

Hydraulic radius (or hydraulic mean depth),

$$m = \frac{A}{p} = \frac{b \times y}{b + 2y} = \frac{2y \cdot y}{2y + 2y} = \frac{2y^2}{4y} = \frac{y}{2} \qquad \text{...(12.10)}$$

Thus the most economical rectangular channel will be the one having its depth equal to half of the width of the channel or the hydraulic mean depth equal to half of the depth.

2. Most Economical Trapezoidal Channel

Let θ be the angle made by the sides with the horizontal and the side slope as 1 vertical to n horizontal (Fig. 12.10).

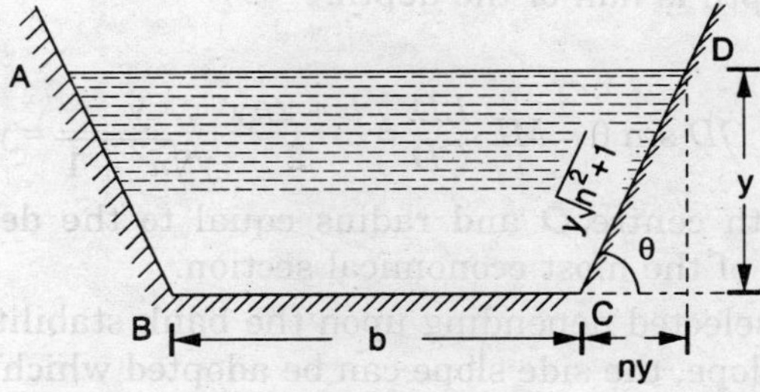

Fig. 12.10

Area, $A = (b + ny)\,y$

$$b = \frac{A - ny^2}{y} = \frac{A}{y} - ny$$

The wetted perimeter, $P = BC + (AB + CD)$

$$= b + 2\sqrt{(ny)^2 + y^2}$$

$$= b + 2y\sqrt{n^2 + 1}$$

$$= \frac{A}{y} - ny + 2y\sqrt{n^2 + 1}$$

For minimum P, y is found by treating A and n as constants.

$$\frac{d}{dy}(P) = 0$$

$$\frac{d}{dy}\left[\frac{A}{y} - ny + 2y\sqrt{n^2 + 1}\right] = 0$$

$$-\frac{A}{y^2} - n + 2\sqrt{n^2 + 1} = 0$$

$$\frac{A}{y^2} + n = 2\sqrt{n^2 + 1}$$

$$\frac{(b + ny)y}{y^2} + n = 2\sqrt{n^2 + 1}$$

$$\frac{b + ny}{y} + n = 2\sqrt{n^2 + 1}$$

$$\frac{b + 2ny}{y} = 2\sqrt{n^2 + 1}$$

$$\frac{b + 2ny}{2} = y\sqrt{n^2 + 1} \quad \text{...(12.11)}$$

or half of the top width = One of the sloping sides

Hydraulic mean depth,

$$m = \frac{A}{P} = \frac{(b + ny)y}{b + 2y\sqrt{n^2 + 1}} = \frac{(b + ny)y}{b + (b + 2ny)} = \frac{(b + ny)y}{2(b + ny)} = \frac{y}{2} \quad \text{...(12.12)}$$

as for a rectangular section.

Thus hydraulic mean depth is half of the depth.

In Fig. 12.11,

$$OH = OD \cdot \sin\theta = ML \cdot \frac{LD}{CD} = \frac{b + 2ny}{2} \cdot \frac{y}{y\sqrt{n^2 + 1}} = y\sqrt{n^2 + 1} \cdot \frac{1}{\sqrt{n^2 + 1}} = y$$

Thus a circle drawn with centre O and radius equal to the depth of the flow will be tangential to the three sides of the most economical section.

Side slopes are usually selected depending upon the bank stability. However, if the bank can be made stable for any slope, the side slope can be adopted which is the most economical.

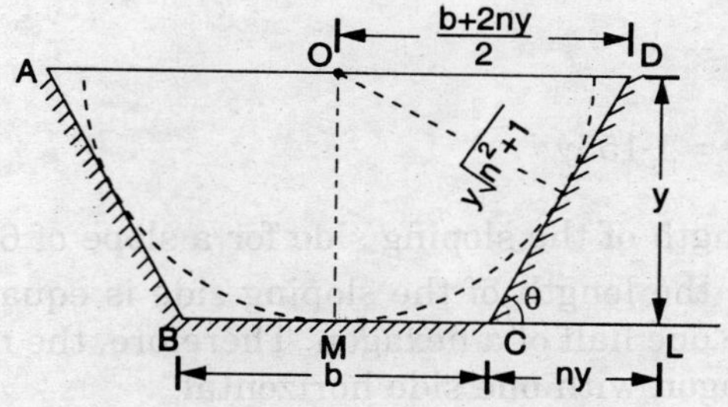

Fig. 12.11

For minimum P, n can be found from $dp/dn = 0$ treating A and y constants.

$$\frac{d}{dn}(P) = 0$$

$$\frac{d}{dn}\left[\frac{A}{y} - ny + 2y\sqrt{n^2+1}\right] = 0$$

$$\frac{d}{dn}\left[\frac{A}{y} - ny + 2y(n^2+1)^{1/2}\right] = 0$$

$$0 - y + 2y\left(\frac{1}{2}\right)(n^2+1)^{-1/2} \times 2n = 0$$

$$-1 + \frac{2n}{\sqrt{n^2+1}} = 0$$

$$2n = \sqrt{n^2+1}$$

$$4n^2 = n^2 + 1$$

$$3n^2 = 1$$

$$n = \frac{1}{\sqrt{3}} \qquad \text{...(12.13)}$$

$$\tan\theta = \frac{y}{ny} = \sqrt{3} = \tan 60^\circ$$

$$\theta = 60^\circ \qquad \text{...(12.14)}$$

Hence the best slope is at 60° to the horizontal.

As half of top width = Length of sloping side (Eq. 12.11)

$$\frac{b+2ny}{2} = y\sqrt{n^2+1}$$

$$b + 2ny = 2y\sqrt{n^2+1}$$

$$b + 2\frac{y}{\sqrt{3}} = 2y\sqrt{\frac{1}{3}+1}$$

$$b + \frac{2}{\sqrt{3}}y = \frac{4y}{\sqrt{3}}$$

$$\sqrt{3}b + 2y = 4y$$

$$\sqrt{3}b = 2y$$

$$b = \frac{2y}{\sqrt{3}} = 1{\cdot}155y \quad \text{...(12.15)}$$

= Length of the sloping side for a slope of 60°

Thus for a slope of 60°, the length of the sloping side is equal to base width or the base and the two sides constitute one half of a hexagon. Therefore, the most economical trapezoidal section is one-half of a hexagon with one side horizontal.

3. Triangular Section

Refer Fig. 12.12,

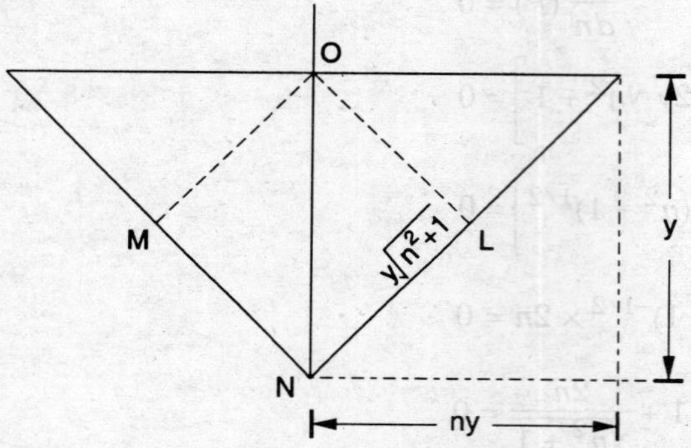

Fig. 12.12

$$A = y \times ny = ny^2$$

or
$$y = \sqrt{\frac{A}{n}}$$

$$P = 2\sqrt{n^2y^2 + y^2} = 2y\sqrt{n^2+1} = 2\sqrt{\frac{A}{n}}\sqrt{n^2+1} = 2\sqrt{A}\sqrt{n + \frac{1}{n}}$$

For the most economical section, P should be minimum for a given value of A, *i.e.*

$$\frac{dP}{dn} = 0$$

$$\frac{d}{dn}\left[2\sqrt{A}\sqrt{n + \frac{1}{n}}\right] = 0$$

$$\frac{d}{dn}\left[\sqrt{n + \frac{1}{n}}\right] = 0$$

$$\frac{d}{dn}\left[n + \frac{1}{n}\right] = 0$$

$$1 - \frac{1}{n^2} = 0$$

$$n^2 = 1$$

$$n = 1 \qquad ...(12.16)$$

$$\tan \theta = \frac{y}{ny} = 1 = \tan 45^\circ \ \ or \ \ \theta = 45^\circ \qquad ...(12.17)$$

Thus the most economical triangular section is half of a square with diagonal horizontal.

$$OL = y \sin \theta = y \sin 45^\circ = \frac{y}{\sqrt{2}} = OM$$

i.e. a semi-circle with O as centre and radius equal to $\frac{y}{\sqrt{2}}$ is tangential to the two sides.

$$m = \frac{A}{P} = \frac{ny^2}{2y\sqrt{n^2+1}} = \frac{ny}{2\sqrt{n^2+1}} = \frac{y}{2\sqrt{2}} \qquad (n = 1) \qquad ...(12.18)$$

4. Circular Section

In the earlier cases, for the most economical section, the perimeter is minimised for a required value of area needed to have the requisite discharge. In case of circular sections, however, it is not possible because two different perimeters with the same area in a circular section of any radius connot be obtained. With the change in the depth of flow, the wetted area as well as the perimeter both are changed. Thus for a most economical circular section, two separate conditions are obtained,

- Condition for maximum velocity and
- Condition for maximum discharge.

Fig. 12.13 shows a circular channel with y as the depth of flow.

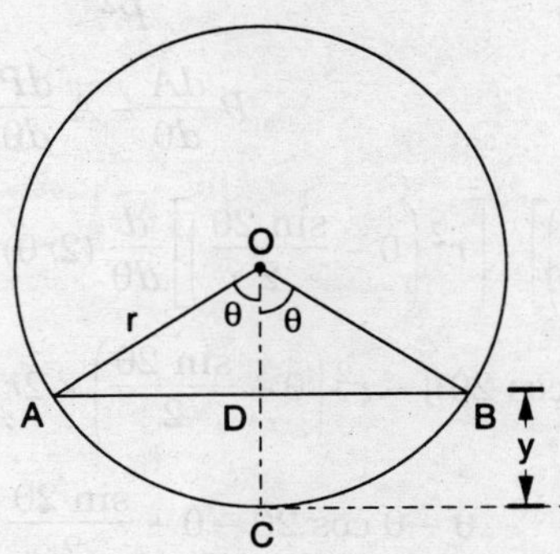

Fig. 12.13

Let $\quad r$ = radius of the section

2θ = angle subtended by the water surface at the centre

y = depth of flow

A = area of the channel

= area of the sector $ACBO$ – Area of two triangles

$$= \pi r^2 \times \frac{2\theta}{2\pi} - \frac{1}{2} AD \times DO$$

$$= r^2 \theta - \frac{1}{2} \times 2\, DB \times DO$$

$$= r^2 \theta - r \sin\theta \times r\cos\theta$$

$$= r^2 \left(\theta - \frac{\sin 2\theta}{2}\right) = \frac{r^2}{2}(2\theta - \sin 2\theta) \qquad \ldots(12.19)$$

(i) Condition for Maximum Velocity

The velocity of flow according to Chezy's formula,

$$V = C\sqrt{mi} = C\sqrt{\frac{A}{P}\, i}$$

For maximum velocity for a given value of C and i,

$$\frac{dV}{d\theta} = 0$$

$$\frac{d}{d\theta}\left(\sqrt{\frac{A}{P}}\right) = 0$$

$$\frac{d}{d\theta}\left(\frac{A}{P}\right) = 0 \qquad \ldots(i)$$

A and P, both are functions of θ,

Differentiating (i) with respect to θ ,

$$\frac{p\dfrac{dA}{d\theta} - A\dfrac{dP}{d\theta}}{P^2} = 0$$

$$P\frac{dA}{d\theta} - A\frac{dP}{d\theta} = 0$$

$$(2r\theta)\frac{d}{d\theta}\left[r^2\left(\theta - \frac{\sin 2\theta}{2}\right)\right] - \left[r^2\left(\theta - \frac{\sin 2\theta}{2}\right)\right]\frac{d}{d\theta}(2r\theta) = 0$$

$$(2r\theta)\,[r^2(1 - \cos 2\theta)] - r^2\left(\theta - \frac{\sin 2\theta}{2}\right) \times 2r = 0$$

$$\theta - \theta\cos 2\theta - \theta + \frac{\sin 2\theta}{2} = 0 \qquad \text{(cancelling } 2r^3)$$

$$\theta\cos 2\theta = \frac{\sin 2\theta}{2}$$

$$\tan 2\theta = 2\theta$$

By trial and error, $2\theta = 257{\cdot}5°$ ...(12.20)

or $\theta = 128{\cdot}75°$ ...(12.20a)

Depth of flow will be

$$\begin{aligned} y &= r - r\cos\theta \\ &= r\,(1 - \cos 128{\cdot}75^\circ) \\ &= r\,(1 + 0{\cdot}62) \\ &= 1{\cdot}62\,r \\ &= 0{\cdot}81\,d \qquad \ldots(12.21) \end{aligned}$$

where d is the diameter of the circular section.

Thus the maximum velocity in a circular cross-section is when the depth of flow is 0·81 times the diameter of the section.

$$m = \frac{A}{P} = \frac{r^2\left(\theta - \dfrac{\sin 2\theta}{2}\right)}{2r\theta} = \frac{r}{2\theta}\left(\theta - \frac{\sin 2\theta}{2}\right)$$

$$= \frac{r}{2\times\left(128{\cdot}75^\circ \times \dfrac{\pi}{180}\right)}\left(128{\cdot}75^\circ \times \frac{\pi}{180} - \frac{\sin 257{\cdot}5^\circ}{2}\right)$$

$$= 0{\cdot}61r$$

$$= 0{\cdot}305d \qquad \ldots(12.12)$$

(ii) Condition for Maximum Discharge

The discharge is given by,

$$Q = AC\sqrt{mi} = AC\sqrt{\frac{A}{P}i} = C\sqrt{\frac{A^3}{P}i}$$

For maximum discharge for a given value of C and i ,

$$\frac{dQ}{d\theta} = 0$$

$$\frac{d}{d\theta}\left[C\sqrt{\frac{A^3}{P}i}\right] = 0$$

$$\frac{d}{d\theta}\left[\frac{A^3}{P}\right] = 0$$

$$\frac{\rho\,.\,3A^2\,\dfrac{dA}{d\theta} - A^3\,\dfrac{dP}{d\theta}}{P^2} = 0$$

$$3P\frac{dA}{d\theta} - A\frac{dP}{d\theta} = 0$$

$$3\times(2r\theta)\,\frac{d}{d\theta}\left[r^2\left(\theta - \frac{\sin 2\theta}{2}\right)\right] - \left[r^2\left(\theta - \frac{\sin 2\theta}{2}\right)\right]\frac{d}{d\theta}(2r\theta) = 0$$

$$(6r\theta)[r^2 (1 - \cos 2\theta)] - r^2 \left(\theta - \frac{\sin 2\theta}{2}\right) \times 2r = 0$$

$$3\theta - 3\theta\cos 2\theta - \theta + \frac{\sin 2\theta}{2} = 0 \qquad \text{(cancelling } 2r^3\text{)}$$

$$2\theta - 3\theta \cos 2\theta + \frac{\sin 2\theta}{2} = 0$$

$$4\theta - 6\theta \cos 2\theta + \sin 2\theta = 0$$

By trial and error,

$$2\theta = 308° \qquad \text{...(12.23)}$$

$$\theta = 154° \qquad \text{...(12.23a)}$$

Depth of flow,

$$y = r - r \cos \theta = r(1 - \cos 154°) = r(1 + 0{\cdot}9) = 1{\cdot}9r = 0{\cdot}95d$$

Hydraulic mean depth,

$$m = \frac{A}{P} = \frac{r^2\left(\theta - \frac{\sin 2\theta}{2}\right)}{2r\theta} = \frac{r}{2\theta}\left(\theta - \frac{\sin 2\theta}{2}\right)$$

$$= \frac{r}{2 \times \left(154° \times \frac{\pi}{180}\right)}\left(154 \times \frac{\pi}{180} - \frac{\sin 308°}{2}\right)$$

$$= 0{\cdot}574r$$

$$= 0{\cdot}287d \qquad \text{...(12.24)}$$

Example 12.7 Find the most economical section of a rectangular channel carrying water, the bed slope of which is 1 in 1800 and the rate of flow 600 l/s. Take Chezy's constant C as 50.

Solution For the most economical rectangular channel,

$b = 2y$ and $m = y/2$

$$A = b{\cdot}y = 2y{\cdot}y = 2y^2$$

$$Q = AC\sqrt{mi}$$

$$0{\cdot}6 = 2y^2 \times 50 \times \sqrt{\frac{y}{2} \times \frac{1}{1800}}$$

$$= 1{\cdot}6667y^{5/2}$$

$$y^{5/2} = 0{\cdot}36$$

$$y = 0{\cdot}665 \text{ m}$$

$$b = 2y = 2 \times 0{\cdot}665 = 1{\cdot}33 \text{ m}$$

Example 12.8 The slope of the bed of a rectangular channel is 1 in 800 and the cross-section of the bed is 5 m wide and 2 m deep. It is proposed to increase the discharge to a maximum for the same area of cross-section, slope of the bed and roughness of the channel. What will be the new dimensions and the increase in discharge if the value of the Chezy's constant is 55.

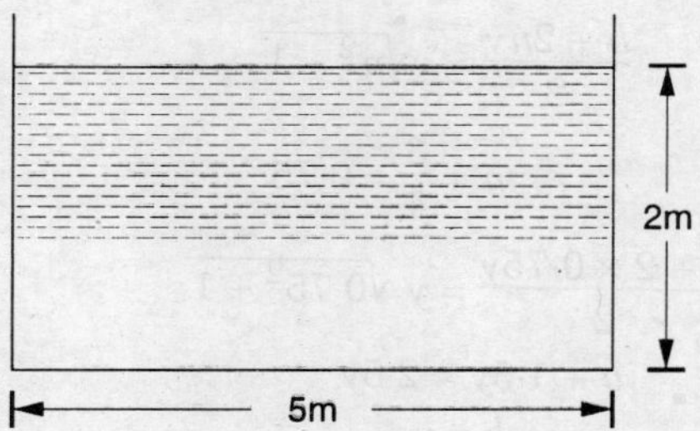

Fig. 12.14

Solution $A = by = 5 \times 2 = 10 \text{ m}^2$

$P = b + 2y = 5 + 2 \times 2 = 9 \text{ m}$

$m = \frac{A}{P} = \frac{10}{9} - 1{\cdot}111 \text{ m}$

$Q = AC\sqrt{mi} = 10 \times 55 \times \sqrt{\frac{10}{9} \times \frac{1}{800}} = 20{\cdot}5 \text{ m}^3/\text{s}$

For maximum flow or the most economical rectangular channle, let b' and y' be the new width and depth of water channel respectively.

Then, $b' = 2y'$

and $b' \times y' = \text{Area} = 10 \text{ m}^2$

or $2y'^2 = 10$

$y' = 2{\cdot}236 \text{ m}$

and $b' = 4{\cdot}472 \text{ m}$

$m = \frac{y}{2} = \frac{2{\cdot}236}{2} = 1{\cdot}118$

$Q = AC\sqrt{mi} = 10 \times 55 \times \sqrt{1{\cdot}118 \times \frac{1}{800}} = 20{\cdot}56 \text{m}^3/\text{s}$

Increase in discharge $= 2{\cdot}56 - 20{\cdot}5 = 0{\cdot}06 \text{ m}^3/\text{s}$

Example 12.9 In a trapezoidal channel, the side slopes are 3 horizontal to 4 vertical and the slope of the bed is 1 in 2500. The flow of water through the channel is 600 litres per second. Find the optimum dimensions of the channel taking Chezy's constant C as 80.

Solution For the most economical section,

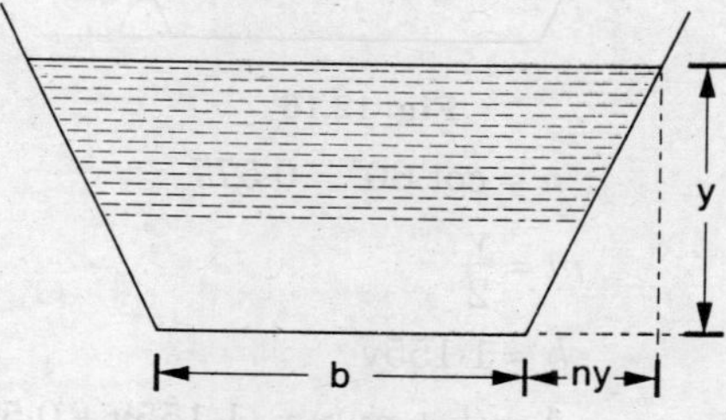

Fig. 12.15

$$\frac{b + 2ny}{2} = y\sqrt{n^2 + 1}$$

where $$n = \frac{3}{4} = 0{\cdot}75$$

$$\frac{b + 2 \times 0{\cdot}75y}{2} = y\sqrt{0{\cdot}75^2 + 1}$$

$$b + 1{\cdot}5y = 2{\cdot}5y$$

$$b = y$$

$$Q = AC\sqrt{mi}$$

$$0{\cdot}6 = (b + ny)y \times C\sqrt{\frac{y}{2}\,i}$$

$$\left(m = \frac{y}{2} \text{ for the most economical section}\right)$$

$$= (y + 0{\cdot}75y)y \times 80\sqrt{\frac{y}{2} \times \frac{1}{2500}}$$

$$= (1 + 0{\cdot}75)y^2 \times 80 \times \frac{\sqrt{y}}{50\sqrt{2}}$$

$$= 1{\cdot}9799y^{5/2}$$

$$y^{5/2} = 0{\cdot}303$$

$$y = 0{\cdot}62 \text{ m}$$

$$b = y = 0{\cdot}62 \text{ m}$$

Example 12.10 An open channel is of the most economical section and is in the form of a half hexagon with one side being the horizontal bottom. It is to provide 25000 litres of water per second. The slope of the bottom is 1 in 2000. Determine the dimensions of the channel taking the Chezy's constant C as 60.

Solution $Q = 25 \text{ m}^3/\text{s}$

As the channel has the form of half hexagon, the angle made by the sloping side is 60°.

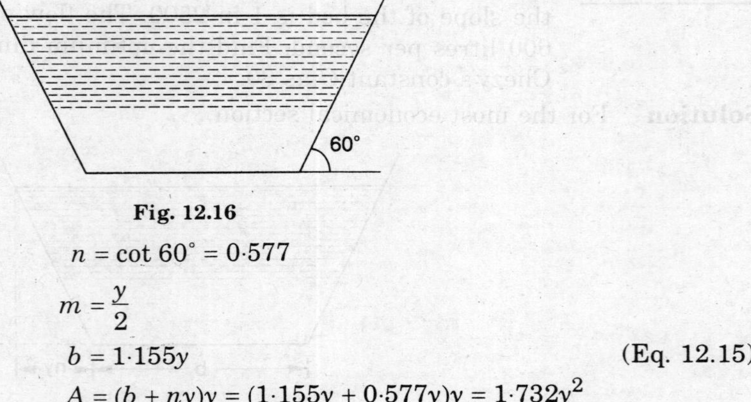

Fig. 12.16

$$n = \cot 60° = 0{\cdot}577$$

$$m = \frac{y}{2}$$

$$b = 1{\cdot}155y \qquad \text{(Eq. 12.15)}$$

$$A = (b + ny)y = (1{\cdot}155y + 0{\cdot}577y)y = 1{\cdot}732y^2$$

$$Q = AC\sqrt{mi}$$

$$25 = 1{\cdot}732y^2 \times 60 \times \sqrt{\frac{y}{2} \times \frac{1}{2000}}$$

$$= 1{\cdot}643y^{5/2}$$

$$y = 2{\cdot}971 \text{ m}$$

$$b = 1{\cdot}155 \times 2{\cdot}971 = 3{\cdot}43 \text{ m}$$

Example 12.11 A canal of a trapezoidal section is to be excavated to irrigate the land, the cross-section of which is to be such that a discharge of 20 m^3/s is obtained with the least cost. The slope of the bed is 1 in 1600. Take Manning's constant $N = 0.025$. Find the dimensions of the section.

Solution For the most economical trapezoidal cross-section,

$$b = \frac{2y}{\sqrt{3}} = 1{\cdot}155y$$

$$A = (b + ny)y = (1{\cdot}155y + 0{\cdot}577y)y = 1{\cdot}732y^2$$

$$m = \frac{y}{2}$$

$$Q = AC\sqrt{mi} = A\frac{1}{N}m^{1/6}\sqrt{mi} = A\frac{1}{N}m^{2/3}\sqrt{i}$$

$$20 = 1{\cdot}732y^2 \times \frac{1}{0{\cdot}025}\left(\frac{y}{2}\right)^{2/3} \times \sqrt{\frac{1}{1600}}$$

$$= 1{\cdot}091y^{8/3}$$

$$y^{8/3} = 18{\cdot}33$$

$$y = 2{\cdot}977 \text{ m}$$

$$b = 1{\cdot}155 \times 2{\cdot}977 = 3{\cdot}438 \text{ m}$$

Angle of the slopping side = 60°

Example 12.12 A trapezoidal channel with side slpes of 1 : 1 is to discharge water at the rate of 20 m^3/s. The slope of the channel is 1 in 800. If the channel is unlined the value of Chezy's constant C is 45 and if lined with concrete it is 70. The cost per m^3 of excavation is three times the cost per m^2 of lining.

Which of the two options is economical ? Lined or unlined.

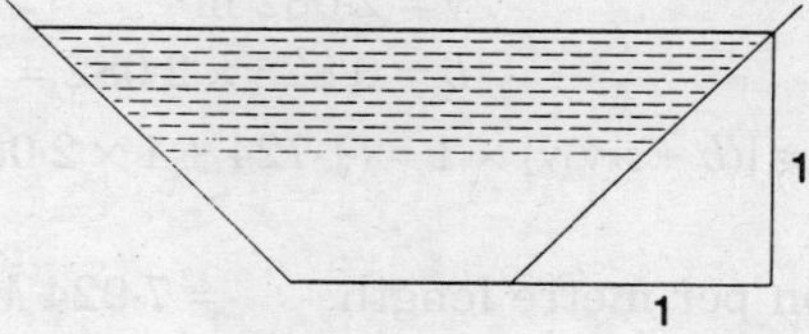

Fig. 12.17

Solution $$\frac{b+2ny}{2} = y\sqrt{n^2+1}$$

where $$n = \frac{1}{1} = 1$$

$$\frac{b+2\times 1\times y}{2} = y\sqrt{1^2+1}$$

$$b + 2y = 2\times\sqrt{2}y$$

$$b = 0{\cdot}828y$$

For unlined channel,

$$Q = AC\sqrt{mi}$$

$$= (b+ny)y\times C\sqrt{\frac{y}{2}i} \qquad \left(m = \frac{y}{2}\right)$$

$$20 = (0{\cdot}828y + y)y\times 45\sqrt{\frac{y}{2}\times\frac{1}{800}}.$$

$$= 2{\cdot}057\, y^{5/2}$$

$$y^{5/2} = 9{\cdot}72$$

$$y = 2{\cdot}484 \text{ m}$$

$$b = 0{\cdot}828\times 2{\cdot}484 = 2{\cdot}057 \text{ m}$$

Let the cost of excavation per unit volume be M, then

Cost per metre length $= \text{Area}\times 1\times M$

$$= [(b+ny)\,y]\times 1\times M$$

$$= [(2{\cdot}057 + 1\times 2{\cdot}484)\ 2{\cdot}484]\ M$$

$$= 11{\cdot}28\ M$$

For lined channel,

$$Q = AC\sqrt{mi}$$

$$= (b = ny)y\times C\sqrt{\frac{y}{2}i}\ \left(m = \frac{y}{2}\right)$$

$$20 = (0{\cdot}828y + y)y\times 70\sqrt{\frac{y}{2}\times\frac{1}{800}}$$

$$= 3{\cdot}199\, y^{5/2}$$

$$y^{5/2} = 6{\cdot}252$$

$$y = 2{\cdot}082 \text{ m}$$

$$b = 0{\cdot}828\times 2{\cdot}082 = 1{\cdot}724 \text{ m}$$

$$A = [(b+ny)y]\times 1 = (1{\cdot}724 + 1\times 2{\cdot}082)2{\cdot}082 = 7{\cdot}924 \text{ m}^2$$

Cost of excavation per metre length $= 7{\cdot}924\ M$

$$m = \frac{A}{P} = \frac{y}{2}$$

$$p = \frac{2A}{y} = \frac{2\times 7{\cdot}924}{2{\cdot}082} = 7{\cdot}612$$

Cost of lining $= \text{Perimeter} \times \text{length} \times \text{Cost/m}^2$

$= 7{\cdot}612 \times 1 \times (M/3) = 2{\cdot}537\ M$

Total cost per metre length $= (7{\cdot}612 + 2{\cdot}537)\ M = 10{\cdot}149\ M$

As the cost of lined channel per m length is less than the unlined channel per m length, the lined channel is economical.

Example 12.13 The bed slope of a concrete lined circular channel of 4 m diameter is 1 in 500. Find the velocity and the discharge for the condition of

(i) maximum velocity

(ii) maximum discharge

The Chezy's constant C is 50.

Solution $d = 4$ m

$r = 2$ m

For maximum velocity condition,

$$2\theta = 257{\cdot}5^\circ = 257{\cdot}5 \times \frac{\pi}{180} = 4{\cdot}494 \text{ radians}$$

$$A = \frac{r^2}{2}(2\theta - \sin 2\theta) = \frac{2^2}{2}(4{\cdot}49 - \sin 257{\cdot}5^\circ) = 10{\cdot}93 \text{ m}^2$$

$$m = 0{\cdot}61r = 0{\cdot}61 \times 2 = 1{\cdot}22 \text{ m}$$

$$V = C\sqrt{mi} = 50 \times \sqrt{1{\cdot}22 \times \frac{1}{500}} = 2{\cdot}47 \text{ m/s}$$

$$Q = AV = 10{\cdot}93 \times 2{\cdot}47 = 26{\cdot}996 \text{ m}^3/\text{s}$$

For maximum discharge condition,

$$2\theta = 308^\circ = 308 \times \frac{\pi}{180} = 5{\cdot}375 \text{ radians}$$

$$A = \frac{r^2}{2}(2\theta - \sin 2\theta) = \frac{2^2}{2}(5{\cdot}375 - \sin 308^\circ) = 12{\cdot}326 \text{ m}^2$$

$$m = 0{\cdot}574r = 0{\cdot}574 \times 2 = 1{\cdot}148 \text{ m}$$

$$V = C\sqrt{mi} = 60 \times \sqrt{1{\cdot}148 \times \frac{1}{500}} = 2{\cdot}396 \text{ m/s}$$

$$Q = AV = 12{\cdot}326 \times 2{\cdot}396 = 29{\cdot}53 \text{ m}^3/\text{s}$$

12.5 NON-UNIFORM FLOW THROUGH OPEN CHANNELS

In open channels, due to change of bed slope, cross-sectional areas and obstructions to the flow (like spillway, dam, sluice gate etc.), the water surface does not remain parallel to the bed and consequently the velocity varies from section to section. Such type of flow is known as non-uniform flow. As mentioned earlier, a non-uniform flow may be RVF (rapidly varied flow) or GVF (gradually varied flow).

Specific Energy

Specific energy is the total energy of flow per unit weight with the channel bed taken as the datum (Fig. 12.18). If y is the depth of flow and V the mean velocity of flow, then the specific energy,

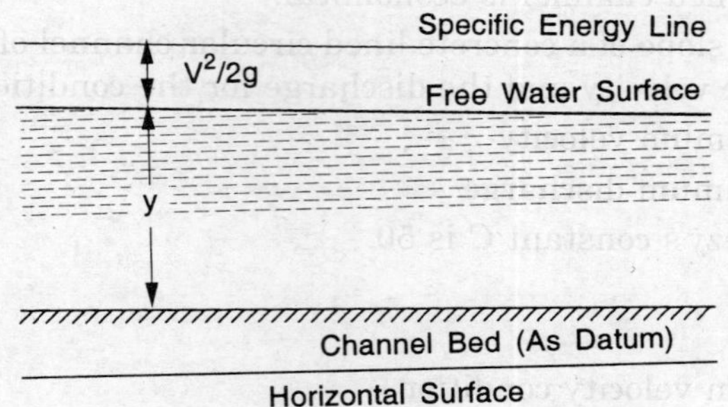

Fig. 12.18

$$E = y + \frac{V^2}{2g} = E_p + E_k \qquad \ldots(12.25)$$

= Potential energy of flow + Kinetic energy of flow

The curve which shows the variation of specific energy with depth of flow is known as the **specific energy curve.**

Consider a channel of rectangular cross-section through which a steady but non-uniform flow takes place.

Let Q = discharge through the channel

q = discharge per unit width

y = depth of flow

b = width of channel

Q = area × velcoity = $(b \times y) \,.\, V$

$$q = \frac{b \cdot y \cdot V}{b} = y \cdot V$$

Velocity of flow,

$$V = \frac{Q}{b.y} = \frac{q}{y} \qquad \ldots(12.26)$$

$$E = y + \frac{(q/y)^2}{2g} = y + \frac{q^2}{2gy^2} = E_p + E_k \qquad \ldots(12.27)$$

The above equation relates the specific energy with the depth of flow. For a given discharge, a curve may be potted for specific energy E against the depth of flow. The specific energy curve (Fig. 12.19) can also be spotted by first drawing a curve for potential energy which is a straight line passing through the origin at 45° with the x-axis (because the potential energy $E = y$ and the scales are the same) and then drawing another curve for the kinetic energy which is a parabola. The resultant of the two curves is the specific energy curve. It may be noted that the specific energy is asympotic to the horizontal axis for small values of y and to 45° line for higher values of y. In general, there are two values of the depth for a particular value of specific

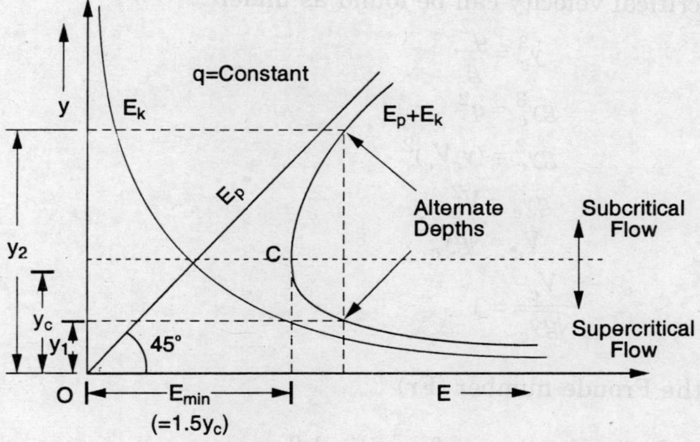

Fig. 12.19

energy. These two depths for a particular value of specific energy are referred as **alternate** or **conjugate depths.** There is a certain depth y at which the specific energy is minimum. This depth is known as **critical depth** and the corresponding velocity as the **critical velocity.** Mathematically, the critical depth can be obtained by differentiating the specific energy equation with respect to depth of flow and equating the sum to zero.

$$\frac{dE}{dy} = 0$$

$$\frac{d}{dy}\left[y + \frac{q^2}{2gy^2}\right] = 0$$

$$\frac{d}{dy}\left[y + \frac{q^2}{2g}(y^{-2})\right] = 0$$

$$1 + \frac{q^2}{2g}\left(\frac{-2}{y^3}\right) = 0$$

$$1 - \frac{q^2}{gy^3} = 0$$

$$1 = \frac{q^2}{gy^3}$$

$$y^3 = \frac{q^2}{g}$$

$$y = \left(\frac{q^2}{g}\right)^{1/3}$$

The critical depth is denoted by y_c.

Thus the critical depth

$$y_c = \left(\frac{q^2}{g}\right)^{1/3} \quad \text{...(12.28)}$$

Further the critical velocity can be found as under :

$$y_c^3 = \frac{q^2}{g}$$

$$gy_c^3 = q^2$$

$$gy_c^3 = (y_c V_c)^2$$

$$gy_c = V_c^2$$

$$V_c = \sqrt{gy_c}$$

Also
$$\frac{V_c}{\sqrt{gy_c}} = 1 \quad \text{...(12.29)}$$

But $\frac{V_c}{\sqrt{gy_c}}$ is the Froude number (Fr)

Thus the Froude number is one for critical flow.

The minimum specific energy can be expressed in terms of critical depth as follows :

$$E_{min} = y_c + \frac{q^2}{2gy_c^2}$$

$$= y_c + \frac{q^2}{g}\frac{1}{2y_c^2}$$

$$= y_c + y_c^3 \frac{1}{2y_c^2} \qquad \left[y_c = \left(\frac{q^2}{g}\right)^{1/3} \right]$$

$$= y_c + \frac{y_c}{2} = \frac{3y_c}{2} = 1 \cdot 5y_c \quad \text{...(12.30)}$$

Critical Flow

The flow corresponding to critical depth is known as critical flow.

Fr = 1 for critical flow

Subcritical Flow (Streaming or Tranquil Flow)

When Fr < 1 and the depth of flow is more than the critical depth in a channel.

Supercritical Flow (Shooting or Torrential Flow)

When Fr > 1 and the depth of flow is less than the critical flow.

Condition for Maximum Discharge for a Given *E*

$$E = y_c + \frac{q^2}{2gy^2}$$

$$= y + \left(\frac{Q}{b}\right)^2 \frac{1}{2gy^2}$$

$$= y + \frac{Q^2}{2gb^2y^2}$$

$$Q = \sqrt{(E-y)\,2gb^2y^2}$$
$$= b\sqrt{2g\,(Ey^2 - y^3)}$$

For maximum discharge, the expression $(Ey^2 - y^3)$ is to be maximum.

$$\frac{d}{dy}(Ey^2 - y^2) = 0$$
$$2Ey - 3y^2 = 0$$
$$2E - 3y = 0$$
$$y = \frac{2}{3}E \qquad \text{...(12.31)}$$
$$E = \frac{3}{2}y = 1{\cdot}5y \qquad \text{...(12.31a)}$$

i.e. the condition to have the maximum discharge for a given E is that the specific energy is 1.5 times the depth. But from Eq. 12.31a, this condition is satisfied when the depth is critical and E is minimum. Thus for maximum discharge through a channel the depth should be critical. Fig. 12.20 shows a discharge curve.

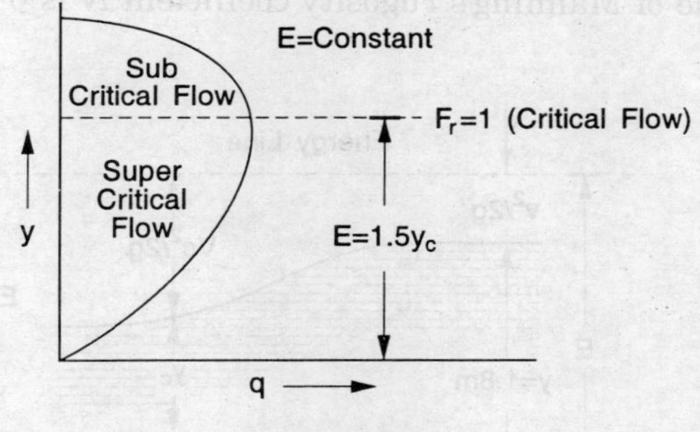

Fig. 12.20

Example 12.14 The flow of water through a 10 m wide channel is $18\text{m}^3/\text{s}$ and the depth of water in the channel is 800 mm. Determine

(i) specific energy of the flow

(ii) critical depth, critical velocity and minimum specific energy

State whether the flow is subcritical or supercritical.

Solution Discharge per unit width.

$$q = \frac{Q}{b} = \frac{18}{10} = 1{\cdot}8 \text{ m}^3/\text{s}$$

$$V = \frac{q}{y} = \frac{1{\cdot}8}{0{\cdot}8} = 2{\cdot}25 \text{ m/s}$$

Specific energy,

$$E = y + \frac{V^2}{2g} = 0{\cdot}8 + \frac{2{\cdot}25^2}{2 \times 9{\cdot}81} = 1{\cdot}058 \text{ m}$$

Critical depth, $y_c = \left(\frac{q^2}{g}\right)^{1/3} = \left(\frac{1\cdot8^2}{9\cdot81}\right)^{1/3} = 0\cdot691 \text{ m}$

Critical Velocity, $V_c = \sqrt{gy_c} = \sqrt{9\cdot81 \times 0\cdot568} = 2\cdot604 \text{ m/s}$

Minimum Specific energy $= \frac{3}{2} y_c = 1\cdot5 \times 0\cdot691 = 10365 \text{ m}$

or

$$E_{\min} = y_c + \frac{V_c^2}{2g} = 0\cdot691 + \frac{2\cdot604^2}{2 \times 9\cdot81} = 1\cdot037 \text{ m}$$

Froude number,

$$\text{Fr} = \frac{V}{\sqrt{gy}} = \frac{2\cdot25}{\sqrt{9\cdot81 \times 0\cdot8}} = 0\cdot803$$

As Froude number is less than 1, the flow is subcritical.

Example 12.15 An open channel of rectangular cross-section is 4m wide and is laid at a slope of 1 in 800. The depth of water in the channel is 1·8 m. It is proposed to have a hump in its bed to obtain critical flow. Determine the height of the hump. The value of Manning's rugosity coefficient N is 0·02.

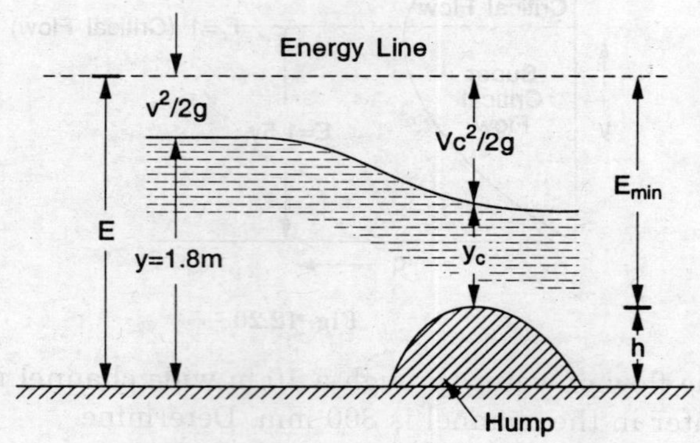

Fig. 12.21

Solution $m = \frac{A}{P} = \frac{4 \times 1\cdot8}{4 + 2 \times 1\cdot8} = 0\cdot947$

$$Q = AC\sqrt{mi} = A \cdot \frac{1}{N} \cdot m^{1/6}\sqrt{mi} = \frac{A}{N} \cdot m^{2/3}\sqrt{i}$$

$$= \frac{(4 \times 1\cdot8)}{0\cdot02} \times 0\cdot947^{2/3} \times \sqrt{\frac{1}{800}}$$

$$= 12\cdot274 \text{ m}^3/\text{s}$$

Specific energy in the normal flow,

$$V = \frac{Q}{A} = \frac{12\cdot274}{4 \times 1.8} = 1\cdot705 \text{ m/s}$$

$$E = y + \frac{V^2}{2g} = 1{\cdot}8 + \frac{1{\cdot}705^2}{2 \times 9{\cdot}81} = 1{\cdot}948 \text{ m}$$

Specific energy in the critical flow (or minimum specific energy),

$$q = \frac{Q}{b} = \frac{12{\cdot}274}{4} = 3{\cdot}0685 \text{ m}^3/\text{s}$$

$$y_c = \left(\frac{q^2}{g}\right)^{1/3} = \left(\frac{3{\cdot}0685^2}{9{\cdot}81}\right)^{1/3} = 0{\cdot}986 \text{ m}$$

$$E_{\min} = \frac{3}{2} y_c = 1{\cdot}5 \times 0{\cdot}986 = 1{\cdot}48 \text{ m}$$

Height of the hump to be provided = 1·948 – 1·48 = 0·468 m

Example 12.16 The depth of water flowing in a 4 m wide rectangular channel is 2m. The slope of the channel is 0·001. Taking Manning's N as 0·015, find the maximum height of hump on the floor to obtain critical depth.

What will be the width of contraction which produces critical depth without increasing the upstream depth of flow ?

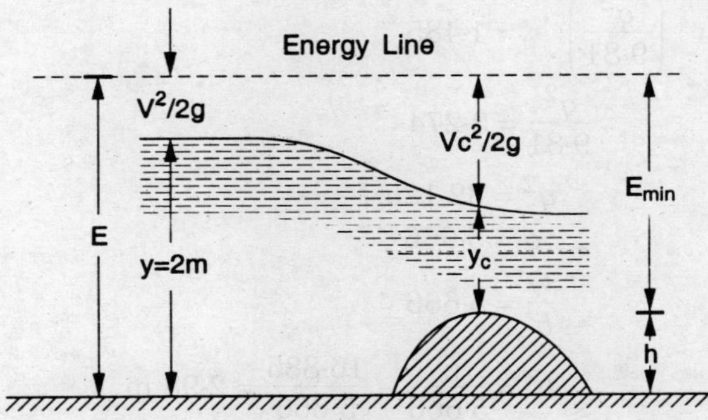

Fig. 12.22

Solution

$$m = \frac{A}{P} = \frac{4 \times 2}{4 + 2 \times 2} = 1$$

$$Q = AC\sqrt{mi} = A.\frac{1}{N}.m^{1/6}\sqrt{mi} = \frac{A}{N}.m^{2/3}\sqrt{i}$$

$$= \frac{(4 \times 2)}{0{\cdot}015} \times 1^{2/3} \times \sqrt{0{\cdot}001}$$

$$= 16{\cdot}865 \text{ m}^3/\text{s}$$

Specific energy in the normal flow,

$$V = \frac{Q}{A} = \frac{16{\cdot}865}{4 \times 2} = 2{\cdot}108 \text{ m/s}$$

$$E = y + \frac{V^2}{2g} = 2 + \frac{2{\cdot}108^2}{2 \times 9{\cdot}81} = 2{\cdot}227 \text{ m}$$

$$q = \frac{Q}{b} = \frac{16{\cdot}865}{4} = 4{\cdot}216 \text{ m}^3/\text{s}$$

$$y_c = \left(\frac{q^2}{g}\right)^{1/3} = \left(\frac{4{\cdot}216^2}{9{\cdot}81}\right)^{1/3} = 1{\cdot}219 \text{ m}$$

$$E_{\min} = \frac{3}{2}y_c = 1{\cdot}5 \times 1{\cdot}219 = 1{\cdot}829 \text{ m}$$

Height of hump $= 2{\cdot}227 - 1{\cdot}829 = 0{\cdot}398$ m

Let b' be the width at the contracted portion to cause the critical depth of water. Then specific energy of the upstream is to be the specific energy in the critical flow on the downstream.

$$E_{\min} = \frac{3}{2}y_c = 2{\cdot}227$$

$$\frac{3}{2}\left(\frac{q^2}{g}\right)^{1/3} = 2{\cdot}227$$

$$\left(\frac{q^2}{9{\cdot}81}\right)^{1/3} = 1{\cdot}485$$

$$\frac{q^2}{9{\cdot}81} = 3{\cdot}274$$

$$q^2 = 32{\cdot}125$$

$$q = 5{\cdot}666$$

$$\frac{Q}{b'} = 5{\cdot}666$$

$$b' = \frac{Q}{5{\cdot}666} = \frac{16{\cdot}885}{5{\cdot}666} = 2{\cdot}98 \text{ m}$$

12.6 HYDRAULIC JUMP (STANDING WAVE)

In open channels, when the depth of flow is more than the critical depth, it is known as subcritical flow whereas if the depth is less than the critical, it is supercritical flow. It is observed that a supercritical (shooting) flow is an unstable flow and soon it transforms into a subcritical (streaming) flow. In doing so, the depth of the flow increases. This increase in the depth is known as **hydraulic jump** or **standing wave**. This is analogous to shock wave in compressible fluids. The hydraulic jump converts kinetic energy of the rapidly flowing stream into potential energy, thereby increasing the depth. During this phenomenon, a loss of energy due to eddy formation and turbulence occurs.

A hydraulic jump occurs in practice at the toe of spillways or below a sluice gate.

To analyse such a flow, it is assumed that the losses due to friction on the surface of the bed of the channel are small and hence neglected. Also, the slope of the bed is small so that the component of the weight of the water in the direction of flow is negligible.

Consider a uniform flow of water over a dam as shown in Fig. 12.23. The flow from section 1-1 to 2-2 depicts a hydraulic jump.

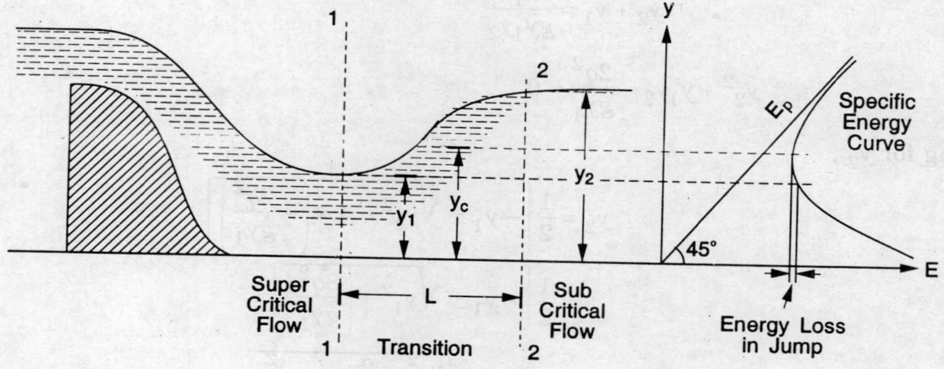

Fig. 12.23

Let y_1, y_2 = depths at section 1-1 and 2-2 respectively

V_1, V_2 = velocities at sections 1-1 and 2-2 respectively

Consider unit width of the channel,

Pressure force on section 1–1 = wAy

$$= w \times (y_1 \times 1) \frac{y_1}{2}$$

$$= w \frac{y_1^2}{2}$$

Pressure force on section 2-2 = $w \dfrac{y_2^2}{2}$

Net force on the mass of water = $\dfrac{w}{2} y_2^2 - \dfrac{w}{2} y_1^2 = \dfrac{w}{2}(y_2^2 - y_1^2) = \dfrac{\rho g}{2}(y_2^2 - y_1^2)$

According to momentum principle,

Net force acting = Rate of change of momentum in the direction of force

= Mass flow rate × Change in velocity

= Density × Discharge/unit width × Change in velocity

$$\frac{\rho g}{2}(y_2^2 - y_1^2) = \rho \times q\,(V_1 - V_2)$$

$$\frac{g}{2}(y_2^2 - y_1^2) = q\left(\frac{q}{y_1 \times 1} - \frac{q}{y_2 \times 1}\right)$$

$$\frac{g}{2}(y_2^2 - y_1^2) = q^2\left(\frac{1}{y_1} - \frac{1}{y_2}\right)$$

$$\frac{g}{2}(y_2 + y_1)(y_2 - y_1) = q^2\left(\frac{y_2 - y_1}{y_1 y_2}\right)$$

$$y_2 + y_1 = \frac{2q^2}{gy_1y_2} \quad \text{...(12.32)}$$

$$y_2^{\,2} + y_1y_2 - \frac{2q^2}{gy_1} = 0$$

Solving for y_2,

$$y_2 = \frac{1}{2}\left[-y_1 \pm \sqrt{y_1^{\,2} - 4 \times \left(-\frac{2q^2}{gy_1}\right)}\right]$$

$$= \frac{1}{2}\left[-y_1 \pm \sqrt{y_1^{\,2} + \left(\frac{8q^2}{gy_1}\right)}\right]$$

$$= -\frac{y_1}{2} \pm \sqrt{\frac{y_1^{\,2}}{4} + \frac{2y_1}{g}\left(\frac{q}{y_1}\right)^2}$$

$$= -\frac{y_1}{2} \pm \sqrt{\frac{y_1^{\,2}}{4} + \frac{2y_1V_1^{\,2}}{g}} \qquad (q = y_1V_1)$$

y_2 cannot be negative,

$$\therefore \qquad y_2 = -\frac{y_1}{2} + \sqrt{\frac{y_1^{\,2}}{4} + \frac{2y_1V_1^{\,2}}{g}} \quad \text{...(12.33)}$$

Depth of hydraulic jump = $y_2 - y_1$

Also,

$$y_2 = -\frac{y_1}{2}\left[-1 + \sqrt{1 + 8\frac{V_1^{\,2}}{gy_1}}\right] = \frac{y_2}{2}\left[\sqrt{1 + 8\left(\frac{V_1}{\sqrt{gy_1}}\right)^2} - 1\right] = \frac{y_1}{2}\,[\sqrt{1 + 8\,(\text{Fr}_1)^2} - 1] \quad \text{...(12.34)}$$

Loss of Energy Due to Hydraulic Jump

The loss of energy is equal to the difference of specific energies at sections 1-1 and 2-2.

$$h_L = E_1 - E_2 = \left(y_1 + \frac{V_1^{\,2}}{2g}\right) - \left(y_2 + \frac{V_2^{\,2}}{2g}\right)$$

$$= \left(\frac{V_1^2}{2g} - \frac{V_2^{\,2}}{2g}\right) - (y_2 - y_1)$$

$$= \left(\frac{q^2}{2gy_1^{\,2}} - \frac{q^2}{2gy_2^{\,2}}\right) - (y_2 - y_1)$$

$$= \frac{q^2}{2g}\left(\frac{1}{y_1^{\,2}} - \frac{1}{y_2^{\,2}}\right) - (y_2 - y_1)$$

$$= \frac{q^2}{2g}\left(\frac{y_2^2 - y_1^2}{y_1^2 y_2^2}\right) - (y_2 - y_1)$$

$$= (y_2 - y_1)\left[\frac{q^2}{2g}\frac{y_2 + y_1}{y_1^2 y_2^2} - 1\right]$$

But $$q^2 = \frac{(y_2 + y_1)\, g y_1 y_2}{2} \quad \text{(from Eq. 12.32)}$$

$$h_L = (y_2 - y_1)\left[\frac{(y_2 + y_1)\, g y_1 y_2}{2}\,\frac{1}{2g}\,\frac{(y_2 + y_1)}{y_1^2 y_2^2} - 1\right]$$

$$= (y_2 - y_1)\left[\frac{(y_2 + y_1)^2}{4y_1 y_2} - 1\right]$$

$$= (y_2 - y_1)\left[\frac{(y_2 + y_1)^2 - 4y_1 y_2}{4y_1 y_2}\right]$$

$$= (y_2 - y_1)\left[\frac{y_2^2 + y_1^2 + 2y_2 y_1 - 4y_2 y_1}{4y_1 y_2}\right]$$

$$= (y_2 - y_1)\left[\frac{y_2^2 + y_1^2 - 2y_2 y_1}{4y_1 y_2}\right]$$

$$= (y_2 - y_1)\left[\frac{(y_2 - y_1)^2}{4y_1 y_2}\right]$$

$$= \frac{(y_2 - y_1)^3}{4y_1 y_2} \quad \text{...(12.35)}$$

Height of Hydraulic Jump

It is the increase in depth of water after the jump.

$$H = y_2 - y_1 \quad \text{...(12.36)}$$

Length of Hydraulic Jump

It is the length of the flow during which the jump occurs. The experiments have shown that this length is equal to 5 to 7 times the height in case of rectangular channels.

i.e. $$L = 6\,H = 6\,(y_2 - y_1) \quad \text{(average value)}$$

Example 12.17 A sluice gate discharges water into a rectangular channel with a velocity of 12 m/s. The depth of the flow is 1·2 m. Find the depth of flow after the jump and the loss in the total head.

Solution $q = V \cdot y = 12 \times 1{\cdot}2 = 14{\cdot}4 \text{ m}^3/\text{s}$

$$y_2 = -\frac{y_1}{2} + \sqrt{\frac{{y_1}^2}{4} + \frac{2y_1{V_1}^2}{g}}$$

$$= -\frac{1{\cdot}2}{2} + \sqrt{\frac{1{\cdot}2^2}{4} + \frac{2 \times 1{\cdot}2 \times 12^2}{9{\cdot}81}}$$

$$= -\,0{\cdot}6 + \sqrt{0{\cdot}36 + 35{\cdot}23}$$

$$= -\,0{\cdot}6 + 5{\cdot}966$$

$$= 5{\cdot}366 \text{ m}$$

$$E_L = \frac{(y_2 - y_1)^3}{4y_1y_2} = \frac{(5{\cdot}366 - 1{\cdot}2)^3}{4 \times 1{\cdot}2 \times 5{\cdot}366} = 2{\cdot}807 \text{ m}$$

Example 12.18 A 4 m wide rectangular channel conveys 12 m^3/s of water with a velocity of 8 m/s. Find out if there is a condition for hydraulic jump to occur. If so determine the height, length and the strength of the jump. Also find the loss of energy per unit mass and the power dissipated.

Solution $q = \frac{Q}{b} = \frac{12}{4} = 3 \text{ m}^3/\text{s per m}$

$$y_1 = \frac{q}{V_1} = \frac{3}{8} = 0{\cdot}375 \text{ m}$$

$$y_c = \left(\frac{q^2}{g}\right)^{1/3} = \left(\frac{3^2}{9{\cdot}81}\right)^{1/3} = 0{\cdot}972 \text{ m}$$

As the depth of water in the channel is less than the critical depth, the jump will occur.

$$y_2 = -\frac{y_1}{2} + \sqrt{\frac{{y_1}^2}{4} + \frac{2y_1{V_1}^2}{g}}$$

$$= -\frac{0{\cdot}375}{2} + \sqrt{\frac{0{\cdot}375^2}{4} + \frac{2 \times 0{\cdot}375 \times 8^2}{9{\cdot}81}}$$

$$= -\,0{\cdot}188 + 2{\cdot}22$$

$$= 2{\cdot}032 \text{ m}$$

Alternately,

$$\text{Fr} = \frac{V_1}{\sqrt{gy_1}} = \frac{8}{\sqrt{9{\cdot}81 \times 0{\cdot}375}} = 4{\cdot}171$$

$$y_2 = \frac{y_1}{2}\left[\sqrt{1 + 8\,(\text{Fr}_1)^2} - 1\right] = \frac{0{\cdot}375}{2}\left[\sqrt{1 + 8\,(4{\cdot}171)^2} - 1\right] = 2{\cdot}032 \text{ m}$$

Height of jump = $y_2 - y_1 = 2{\cdot}032 - 0{\cdot}375 = 1{\cdot}657$ m

Length of jump = 6 × height of jump = 6 × 1·657 = 9·942 m

Strength of jump,

$$= \frac{y_2}{y_1} = \frac{2{\cdot}032}{0{\cdot}375} = 5{\cdot}42$$

Loss of energy,

$$E_L = \frac{(y_2 - y_1)^3}{4y_1y_2} = \frac{(2{\cdot}032 - 0{\cdot}375)^3}{4 \times 2{\cdot}032 \times 0{\cdot}375} = 1{\cdot}493 \text{ m}$$

Power dissipated,

$$P = wQE_L = 9{\cdot}81 \times 12 \times 1{\cdot}493 = 175{\cdot}7 \text{ kW}$$

Example 12.19 Derive an equation for a hydraulic jump occuring in a *V*-shaped channel with side sloping at 45° relating the depths before and after the jump and the flow rate.

If the depths before and after the jump are 1 m and 2 m, find the flow rate. also the Froude numbers before and after the jump.

Solution Let the flow from section 1-1 to 2-2 depicts a hydraulic jump.

Let V_1 = velocity of flow before the hydraulic jump

V_2 = velocity of flow after the hydraulic jump

y_1 = depth of flow before the hydraulic jump

y_2 = depth of flow after the hydraulic jump

From the geometry of the *V*-shaped channel,

Top width before the hydraulic jump = $2y_1$

Top width after the hydraulic jump = $2y_2$

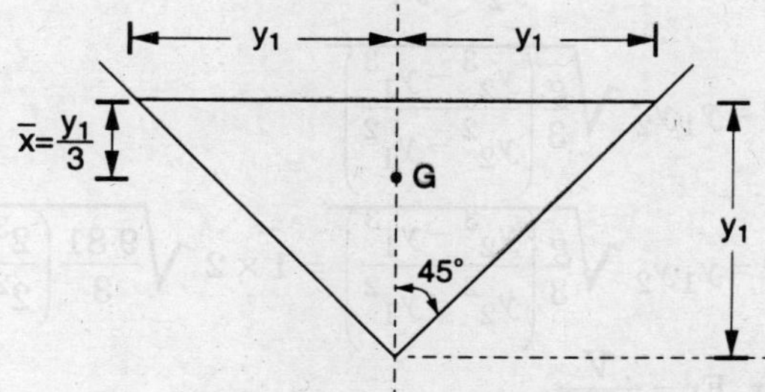

Fig. 12.24

Area $A_1 = \dfrac{2y_1 \cdot y_1}{2} = y_1^{\,2}$

$$A_2 = \frac{2y_2 \cdot y_2}{2} = y_2^{\,2}$$

$$V_1 = \frac{Q}{A_1} = \frac{Q}{y_1^{\,2}} \quad \text{and} \quad V_2 = \frac{Q}{A_2} = \frac{Q}{y_2^{\,2}}$$

Pressure force on section 1-1

$$= wA\bar{x} = w \cdot y_1^2 \cdot \frac{y_1}{3} = \frac{wy_1^{\,3}}{3}$$

Pressure force on section 2-2

$$= wA\bar{x} = w \cdot y_2 \cdot \frac{y_1}{3} = \frac{wy_2^3}{3}$$

Net force on the mass of water

$$= \frac{wy_2^3}{3} - \frac{wy_1^3}{3} = \frac{w}{3}(y_2^3 - y_1^3)$$

According to momentum principle,

Net force acting = Rate of change of momentum in the direction of force
= Mass flow rate × Change in velocity
= Density × Discharge × Change in velocity

$$\frac{w}{3}(y_2^3 - y_1^3) = \frac{w}{g} Q (V_1 - V_2)$$

$$\frac{1}{3}(y_2^3 - y_1^3) = \frac{1}{g} \cdot Q \left(\frac{Q}{y_1^2} - \frac{Q}{y_2^2} \right)$$

$$= \frac{Q^2}{g} \left(\frac{y_2^2 - y_1^2}{y_1^2 \cdot y_2^2} \right)$$

$$Q^2 = \frac{g\,(y_1^2 \cdot y_2^2)}{3} \cdot \frac{y_2^3 - y_1^3}{y_2^2 - y_1^2}$$

$$Q = y_1 \cdot y_2 \sqrt{\frac{g}{3} \left(\frac{y_2^3 - y_1^3}{y_2^2 - y_1^2} \right)}$$ the required relation

$$Q = y_1 \cdot y_2 \sqrt{\frac{g}{3} \left(\frac{y_2^3 - y_1^3}{y_2^2 - y_1^2} \right)} = 1 \times 2 \sqrt{\frac{9.81}{3} \left(\frac{2^3 - 1^3}{2^2 - 1^2} \right)} = 5.524 \text{ m}^3/\text{s}$$

Froude number, $\text{Fr} = \dfrac{V}{\sqrt{gD}}$

where D = hydraulic depth = $\dfrac{\text{Area of cross–section}}{\text{Top width of channel}}$

$$F_{r1} = \frac{V_1}{\sqrt{gD_1}} = \frac{Q}{A_1} \cdot \frac{1}{\sqrt{g\,(y_1^2/2y_1)}} = \frac{Q}{y_1^2} \cdot \frac{1}{\sqrt{gy_1/2}} = \frac{5.524}{1^2} \frac{1}{\sqrt{9.81 \times 1/2}} = 2.5$$

$$F_{r2} = \frac{Q}{y_2^2} \cdot \frac{1}{\sqrt{gy_2/2}} = \frac{5.524}{2^2} \frac{1}{\sqrt{9.81 \times 2/2}} = 0.44$$

12.7 GRADUALLY VARIED FLOW

In a channel flow, if the depth of flow varies gradually over a long length, the flow is said to be gradually varies flow (GVF). The depth of flow in a channel may vary due to change in the shape, size or slope of the channel cross-section, change in the frictional force at the boundaries or presence of some obstruction.

For the analysis of a gradually varied flow, the following assumptions are made :

1. The slope of the bed is small.
2. The channel is prismatic *i.e.* with constant cross-section and alignment.
3. Steady flow with constant discharge.
4. The energy correction factor is unity.
5. The roughness coefficient is constant and does not depend on the depth of the flow.
6. The formulae which are applicable to uniform flow are also applicable to gradually varied flow to determine the slope of the energy line *e.g.* Chezy's and Manning's formulae.
7. The pressure distribution is hydrostatic over the channel cross-section and acceleration effects are negligible.

Consider a channel with rectangular cross-section having a gradually varied flow (Fig. 12.25). The depth of flow gradually decreases in the direction of flow.

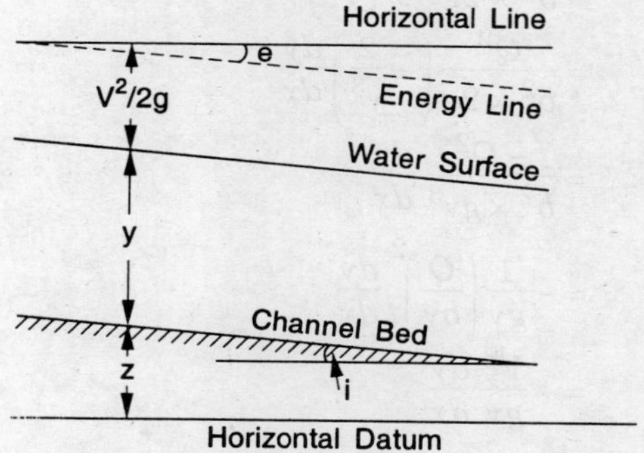

Fig. 12.25

Let Q = discharge through the channel

b = width of the channel

y = depth of the channel

V = mean velocity of flow

z = height of the channel bed above the datum

i = slope of the channel bed

e = slope of the energy line

According to the Bernoulli's equation, the energy at any section is

$$E = z + y + \frac{V^2}{2g}$$

Differentiating this equation with respect to x, where x-axis is taken along the bottom of the channel.

$$\frac{dE}{dx} = \frac{dz}{dx} + \frac{dy}{dx} + \frac{d}{dx}\left(\frac{V^2}{2g}\right)$$

$$-e = -i + \frac{dy}{dx} + \frac{d}{dx}\left(\frac{V^2}{2g}\right) \qquad \text{...(i)}$$

e and i, both are negative as E and z both decrease with increase in x. Also the term

$$\frac{d}{dx}\left(\frac{V^2}{2g}\right) = \frac{d}{dx}\left[\left(\frac{Q}{A}\right)^2 \frac{1}{2g}\right]$$

$$= \frac{d}{dx}\left[\frac{Q^2}{(by)^2}\frac{1}{2g}\right]$$

$$= \frac{Q^2}{b^2 \times 2g}\frac{d}{dx}\left(\frac{1}{y^2}\right)$$

$$= \frac{Q^2}{b^2 \times 2g}\frac{d}{dy}\left(\frac{1}{y^2}\right)\frac{dy}{dx}$$

$$= \frac{Q^2}{b^2 \times 2g}\left(-\frac{2}{y^3}\right)\frac{dy}{dx}$$

$$= \frac{-Q^2}{b^2 \times gy^3}\frac{dy}{dx}$$

$$= -\frac{1}{gy}\left(\frac{Q}{by}\right)^2\frac{dy}{dx}$$

$$= -\frac{V^2}{gy}\frac{dy}{dx}$$

Substituting this value in (i),

$$-e = -i + \frac{dy}{dx} - \frac{dy}{dx}\frac{V^2}{gy}$$

$$-e = -i + \frac{dy}{dx}\left(1 - \frac{V^2}{gy}\right)$$

$$\frac{dy}{dx} = \frac{i-e}{1-\frac{V^2}{gy}} = \frac{i-e}{1-F_r^2} \qquad \text{...(12.37)}$$

$\frac{dy}{dx}$ is the variation of depth along the bed of the channel and represents the slope of the free water surface.

- When $dy/dx = 0$, $i = e$ (constant slope) or the depth of the water above the bottom is constant or the free water surface is parallel to the bed of the channel.
- When $dy/dx > 0$ or the slope is positive which means that the depth of water increases in the direction of flow. This type of profile is also known as back water curve.
- When $dy/dx < 0$ or the slope is negative, the depth of water decreases in the direction of flow and the profile is known as drop down curve.

Back Water Curve and Afflux

As stated earlier, if $dy/dx > 0$, a back water curve is obtained. This happens when there comes some obstruction in the path of flow. Consider the flow over a dam as shown in Fig. 12.26. On the upstream side of the dam, the depth of the water increases as the water approaches the dam and it has maximum depth from the bed at some section.

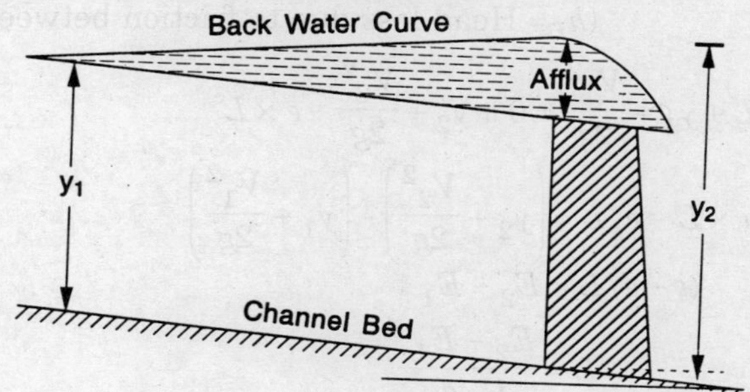

Fig. 12.26

Let $\quad y_1$ = depth of water at the point where the water starts rising up

y_2 = maximum height attained by the water from the bed.

Then $y_2 - y_1$ is the increase in depth of water and is known as **afflux**. The curved surface of water on the upstream side is known as **back water curve**. The distance measured along the bed of the channel between the section from where the water starts rising to the section where it has maximum height is called the length of back water curve.

Refer Fig. 12.27,

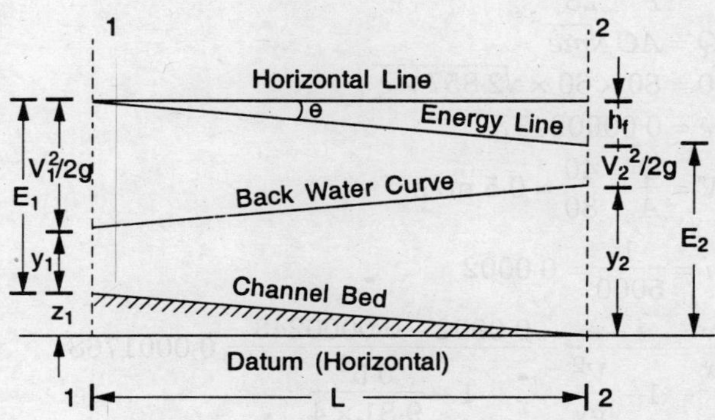

Fig. 12.27

Let $\quad y_1$ = depth of flow at section 1-1

y_2 = depth of flow at section 2-2

V_1 = velocity of flow at section 1-1

V_2 = velocity of flow at section 2-2

i = slope of the bed

e = slope of the energy line

L = length of the back water curve

Applying Bernoulli's equation at sections 1-1 and 2-2,

$$(z_1 + y_1) + \frac{V_1^2}{2g} = (z_2 + y_2) + \frac{V_2^2}{2g} + h_f$$

where (h_f = Head loss due to friction between the two sections)

$$i \times L + y_1 + \frac{V_1^2}{2g} = 0 + y_2 + \frac{V_2^2}{2g} + e \times L$$

$$i \times L - e.\, L = \left(y_2 + \frac{V_2^2}{2g}\right) - \left(y_1 + \frac{V_1^2}{2g}\right)$$

$$(i - e)\, L = E_2 - E_1$$

$$L = \frac{E_2 - E_1}{i - e} \quad \text{...(12.38)}$$

The value of e is found either by Chezy's formula or by Manning's formula corresponding to flow conditions at the mean depth of flow.

Example 12.20 A rectangular channel 20 m wide has a depth of water 4 m and the rate of flow 40 m^3/s. The slope of the bed is 1 in 5000. Taking Chezy's constant C as 60, determine the slope of the free water surface.

Solution

$$A = 20 \times 4 = 80 \text{ m}^2$$

$$p = 20 + 2 \times 4 = 28 \text{ m}$$

$$m = \frac{A}{P} = \frac{80}{28} = 2{\cdot}857 \text{ m}$$

$$Q = AC\sqrt{me}$$

$$40 = 80 \times 60 \times \sqrt{2{\cdot}857 \times e}$$

$$e = 0{\cdot}0000243$$

$$V = \frac{Q}{A} = \frac{40}{80} = 0{\cdot}5 \text{ m/s}$$

$$i = \frac{1}{5000} = 0{\cdot}0002$$

$$\frac{dy}{dx} = \frac{i - e}{1 - \frac{V^2}{gy}} = \frac{0{\cdot}0002 - 0{\cdot}0000243}{1 - \frac{0{\cdot}5^2}{9{\cdot}81 \times 4}} = 0{\cdot}0001768$$

Example 12.21 A weir installed across a rectangular open channel raises the depth of flow from 2 m in a normal flow to 3·5 m at the weir. The channel is 30 m wide and is laid at a slope of 1 in 8000. Considering the average velocity, average depth and average slope midway between the two sections, determine the apporximate length of the backwater curve. Take N in Manning's formula as 0·03.

Solution $i = \dfrac{1}{8000} = 0{\cdot}000125$

length of the back water curve is given by,

$$L = \frac{E_2 - E_1}{i - e}$$

E_1 and E_2 are found as follows :

$$A_1 = 30 \times 2 = 60 \text{ m}^2$$

$$P_1 = 30 + 2 \times 2 = 34 \text{ m}$$

$$m = \frac{A}{P} = \frac{60}{34} = 1{\cdot}765 \text{ m}$$

$$V_1 = C\sqrt{mi} = \frac{1}{N} m^{1/6} \sqrt{mi} = \frac{1}{N} m^{2/3} \sqrt{i} = \frac{1}{0{\cdot}03} 1{\cdot}765^{2/3} \sqrt{\frac{1}{8000}} = 0{\cdot}544$$

$$E_1 = y_1 + \frac{V_1^2}{2g} = 2 + \frac{0{\cdot}544^2}{2 \times 9{\cdot}81} = 2{\cdot}015 \text{ m}$$

$$A_2 = 30 \times 3{\cdot}5 = 105 \text{ m}^2$$

$$P_2 = 30 + 2 \times 3{\cdot}5 = 37 \text{ m}$$

$$m = \frac{A}{P} = \frac{30 \times 3{\cdot}5}{30 + 2 \times 3{\cdot}5} = 2{\cdot}838 \text{ m}$$

$$A_1 V_1 = A_2 V_2$$

$$V_2 = \frac{A_1}{A_2} \cdot V_1 = \frac{60}{105} \times 0{\cdot}544 = 0{\cdot}311 \text{ m/s}$$

$$E_2 = y_2 + \frac{V_2^2}{2g} = 3{\cdot}5 + \frac{0{\cdot}311^2}{2 \times 9{\cdot}81} = 3{\cdot}505 \text{ m}$$

The value of e is found by considering the conditions at the average depth of flow,

At average depth of flow,

$$y = \frac{y_1 + y_2}{2} = \frac{2 + 3{\cdot}5}{2} = 2{\cdot}75 \text{ m}$$

$$A = b \cdot y = 30 \times 2{\cdot}75 = 82{\cdot}5 \text{ m}^2$$

$$P = 30 + 2 \times 2{\cdot}75 = 35{\cdot}5 \text{ m}$$

$$m = \frac{A}{P} = \frac{82{\cdot}5}{35{\cdot}5} = 2{\cdot}324 \text{ m}$$

$$AV = A_1 V_1$$

$$V = \frac{A_1 V_1}{A} = \frac{60 \times 0{\cdot}544}{82{\cdot}5} = 0{\cdot}3956 \text{ m/s}$$

Also $V = C\sqrt{me}$

$$0{\cdot}3956 = \frac{1}{N} m^{2/3} \sqrt{e} = \frac{1}{0{\cdot}03} 2{\cdot}324^{2/3} \sqrt{e}$$

$$e = 0{\cdot}00004576$$

$$L = \frac{3{\cdot}505 - 2{\cdot}015}{0{\cdot}000125 - 0{\cdot}00004576} = 18789 \text{ m or } 18{\cdot}789 \text{ km}$$

PROBLEMS

1. Define an open channel. How is the flow classified in an open channel ?
2. Why is the bed slope essential in open channels ?
3. In case of open channels, mention the difference between
   (a) steady and unsteady flow
   (b) uniform and non-uniform flow
   (c) laminar and turbulent flow.
4. What is the criterion of defining subcritical, critical and super critical flow in open channels ?
5. Derive the Chezy's formula from Darcy's equation. In what way does the roughness of channel effect the Chezy's constant ?
6. What are Bazin's and Kutter's formulae and how are they used ?
7. State and derive the conditions under which a rectangular section of an open channel is the most economical.
8. Why should the trapezoidal shaped open channels be preferred to the rectangular ones ?
9. Prove that the hydraulic mean depth is half of the depth of flow for the most economical trapezoidal channel.
10. Show that for the most economical circular cross-section, the depth of liquid is 0·305 *d* for the maximum velocity and 0·287 *d* to have the maximum discharge where *d* is the diameter of the open channel.
11. What is specific energy of flow in an open channel ? Write it in the form of an equation.
12. What do you mean by critical velocity of flow in an open channel ?
13. What is the critical depth of an open channel if the flow is not uniform ?
14. Draw a specific energy diagram and describe it. How is it useful for the explanation of flow phenomenon in an open channel ?
15. What do you mean by hydraulic jump in an open channel flow ? What is the condition required for its formation in case of a rectangular channel of constant width. Determine the loss of head in terms of depths just before and after the jump.
16. Discuss the formation of a standing wave.
17. Deduce an equation for the back water curve of an afflux caused in a river by some obstruction.
18. The bed slope of a 2·5 m wide rectangular channel is 0·0036. Determine the average shear stress on the boundary if the depth of water in the channel is 1·2 m. (21·58 n/m^2)
19. A rectangular channel 3·5 m wide discharges 5 m^3/s of water. The channel is laid on a slope of 0·0005. Determine the normal depth of flow. The Manning's coefficient N is to be taken as 0·02. (1·48 m)

20. A triangular channel has side slope of 3 horizontal to 2 vertical. Calculate the normal depth required for a discharge of 0·3 m^3/s. The longitudinal slope of the channel is 1 in 1650 and the Manning's N 0·013. (0·54 m)

21. A long triangular flume has side slopes 1 : 1 and is laid at a slope of 1 in 200. If it discharges 0·4 cumec of water, determine the type of flow. What is the possible minimum specific energy ? (supercritical, 0·628 m)

22. The specific energy for a rectangular channel 3 m in width is 2·7 m. Determine the maximum possible discharge. (22·69 m^3/s)

23. The discharge of water in a very wide rectangular channel having a bed slope of 1 in 2500 is 2 m^3/s. Assuming Manning's $N = 0{\cdot}025$, determine the depth of flow. (1·732 m)

24. A trapezoidal channel has side slope 3 horizontal to 2 vertical and a bed slope of 1 in 2000. The discharge is at the rate of 20 cumecs. Taking Manning's N to be 0·01, design the channel for the best form. ($b = 1{\cdot}23$ m, $d = 2{\cdot}05$ m)

25. A circular pipe of 2 m diameter is laid at an inclination of 50 with the horizontal. If the depth of water in the pipe is 750 mm, determine the discharge through the pipe. Take Chezy's constant C as 65. (13·24 m^3/s)

26. The diameter of a circular channel with a bed slope of 1 in 1500 is 1 m. Taking Chezy's constant C as 50, find the maximum discharge through the channel. (532m^3/s)

27. A hydraulic jump takes place in a rectangular channel 500 mm wide at a point where the depth of water is 150 mm and the Froude number 2·5. Determine
   (a) specific energy
   (b) critical depth y_c and sequent depth y_2
   (c) loss of head
   (d) energy dissipated. (0·618 m, 0·277 m, 9·461 m, 0·108 m; 240 w)

28. The discharge through a 3 m wide rectangular channel having a depth of 2 m is 12m^3/s. Determine
   (a) specific energy and the alternate depth
   (b) critical depth and critical velocity
   (c) minimum specific energy
   (d) Froude number.
   What is the nature of flow ? Subcritical or supercritical ?
   (2·204m, 0·33m, 1·177m, 3·398m/s, 1·763m, 0·453 subcritical)

29. The bed slope of a concrete lined circular channel of 3 m diameter is 1 in 500. Determine the rate of flow for the conditions of maximum velocity as well as for the maximum discharge. Chezy's constant $C = 50$. (13·11 m^3/s , 14·37 m^3/s)

30. The discharge through a 5 m wide rectangular channel is 20 m^3/s. The depth of flow is 2 m. Determine
   (a) the depth of flow over a hump of 0.25 m on the bed
   (b) the minimum rise in the bed level required to obtain the critical depth over the hump

(c) the water depth upstream and over the hump if the bed level is raised by 600 mm.

(1·657m, 0·438m, 2·197m, 1·177m)

31. The discharge through a 5 m wide and 2 m deep rectangular channel is 20 m^3/s. Determine

(a) the width to which the channel should be contracted to obtain the critical depth

(b) the depth at the contracted section if the width is 4m

(c) the depth on the upstream and in the contracted section if the width is reduced to 2·5 m. (3·586m, 1·82m, 2·694m, 1·869m)

32. In a hydraulic jump, if the discharge is 10m^3/s per unit width and the head loss 20 m, find the depths of flow before and after the jump.

[**Note** : Use trial and error method assuming some value of y, in steps]

(0·44m, 6·59m)

33. The velocity of water in a rectangular channel 3·6 m deep and 12 m wide is 1·2 m/s. The bed slope of the channel is 0·00025. The flow of water through the channel is regulated in such a way that the energy line has a slope of 1 in 25000. Determine the rate of change of depth of water in the channel. (0·0002189)

34. A channel is in the shape of an isosceles triangle such that the sides make an angle of 45° with the bed. Find the ratio of depth of flow to the bed width for the maximum velocity condition and maximum discharge condition. Depth of flow is less than half the bed width. Use Manning's formula. (0·338, 0·438)

Chapter 13

BOUNDARY LAYER THEORY

13.1 INTRODUCTION

Boundary layer theory pertains to the flow in the vicinity of the boundary of the flow surface. When a real fluid flows past a body with a solid stationary surface, the velocity of the fluid particles which come in contact with the solid surface reduces to zero. This is known as the condition of no slip. As the velocity of the main flow is V, there is a region of retarded flow in which the velocity gradient du/dy exists. The velocity of the fluid varies from zero at the stationary boundary to the maximum at a short distance from the boundary surface. This narrow region in the vicinity of solid boundary is known as **boundary layer**. In the boundary layer, due to the velocity gradient du/dy, the fluid exerts a shear stress on the boundary surface in the direction of flow. The boundary exerts an equal and opposite force on the fluid known as shear resistance.

13.2 BOUNDARY LAYER CHARACTERISTICS

Figure 13.1 shows a continuous flow of fluid with a free stream velocity v over the surface of a thin flat plate parallel to the direction of the free stream. The boundary layer region starts

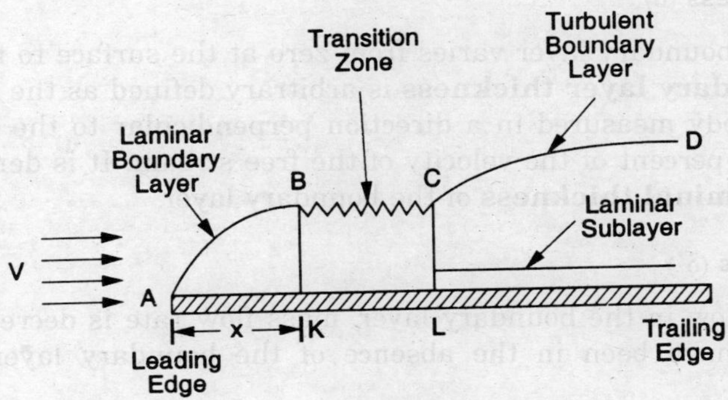

Fig. 13.1

at the leading edge and its thickness increases with the distance from the leading edge as more fluid is slowed down by the viscous boundary. This is referred as the growth of the boundary layer. It is observed that irrespective of the type of the main flow, the flow in the boundary layer is always laminar near the leading edge and is known as the **laminar boundary layer** (*AB*). The length along the plate upto which the laminar boundary layer exists is called the **laminar zone** (*AK*). The length of the laminar zone is obtained from the fact that for laminar flow, the maximum value of Reynolds number is 5×10^5.

As Reynolds number is given by

$$Re = \frac{V \, . \, x}{\nu}$$

where V = free stream velocity

ν = kinematic viscosity of the fluid

x = distance of the leading edge

$$\frac{V \, . \, x}{\nu} = 5 \times 10^5$$

Thus if the values of V and ν are known, x can be calculated.

In case, the length of the plate is less than x, the boundary layer is wholly laminar. However, if the length of the plate is more than x, the thickness of the boundary layer further increases in the downstream direction. For some of the distance along the plate length, the flow becomes irregular indicating a transition zone (*KL*). Further downstream, the flow is turbulent and its thickness continues to increase. This is known as turbulent boundary layer (*CD*). It is also observed that under this turbulent boundary layer, there is a small region of laminar sublayer over the solid surface of the plate. Though the velocity variation in the laminar sublayer is parabolic, it is assumed linear due to very small thickness of the sublayer. Thus the velocity gradient is reasonably considered to be constant and the shear stress near the boundary is given by,

$$\tau_0 = \mu \left(\frac{\partial u}{\partial y} \right) = \mu \frac{u}{y} \qquad \text{...(13.1)}$$

Also note that, in general the thickness of the boundary layer decreases with increase in the velocity of the main flow and decreases with increase in the value of the kinematic viscosity.

13.3 BOUNDARY LAYER PARAMETERS

1. Boundary Layer Thickness (δ)

The velocity within the boundary layer varies from zero at the surface to the velocity of the main stream. The **boundary layer thickness** is arbitrary defined as the distance from the boundary of the solid body measured in a direction perpendicular to the flow upto a point where the velocity is 99 percent of the velocity of the free stream. It is denoted by symbol δ and is termed as the **nominal thickness** of the boundary layer.

2. Displacement Thickness (δ^*)

Owing to the retarded flow in the boundary layer, mass flow rate is decreased compared to the flow which would have been in the absence of the boundary layer. **Displacement**

thickness is defined as the imaginary displacement of the boundary by an amount δ^* such that the actual flow rate is same when moving with the main stream velocity.

Consider the flow of a free stream of a fluid of density ρ, with a velocity V over a thin smooth plate as shown in Fig. 13.2. Consider an elementary strip of thickness dy at a distance y from the surface of the plate.

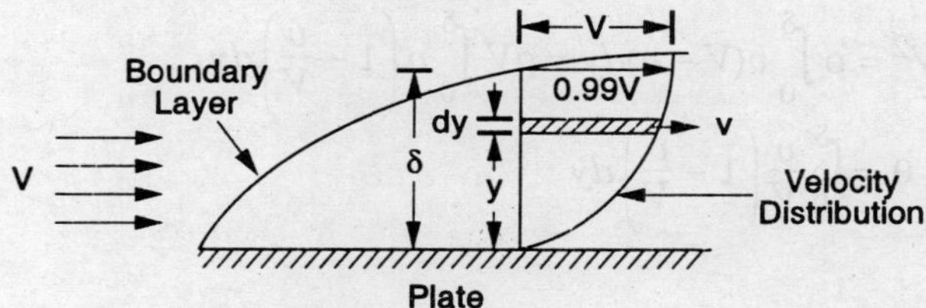

Fig. 13.2

Assuming unit width of the plate.

Mass of fluid/s flowing through the strip = Density × Velocity × Area

$$= \rho \times v \times dy \times 1$$

Mass of fluid/s flowing through the strip with the velocity of the main stream

$$= \rho \cdot V \cdot dy$$

Reduction in mass flow rate $= \rho \cdot (V - v) \cdot dy$

[$(V - v)$ is known as the **velocity of defect**]

Total reduction in mass flow rate (for an incompressible fluid)

$$= \int_0^{\delta} \rho (V - v) \cdot dy$$

$$= \rho \int_0^{\delta} (V - v) \cdot dy$$

Let the plate be displaced by an amount δ^*. Then the flow through the displaced volume must be equal to the reduction in the mass flow rate.

ie $$\rho V \delta^* = \rho \int_0^{\delta} (V - v) dy = \rho V \int_0^{\delta} \left(1 - \frac{v}{V}\right) dy \quad (V \text{ is constant})$$

$$\delta^* = \int_0^{\delta} \left(1 - \frac{v}{V}\right) dy \qquad \ldots(13.2)$$

Momentum Thickness (θ)

There is a reduction of momentum due to the retardation of flow in the boundary layer. The **momentum thickness** is defined as the thickness of an imaginary layer in the free stream, the momentum of which is equal to the deficiency of momentum caused to the flow due to retarded velocity in the boundary layer.

Reduction of mass flow rate through the elementary strip $= \rho \cdot (V - v) \cdot dy$

Reduction in momentum/s $= \rho \cdot (V - v) \cdot dy \times v$

Total reduction in momentum/s

$$= \int_0^{\delta} \rho \cdot v \, (V - v) \cdot dy$$

Momentum of a layer of thickness θ in the free stream (whose momentum is equal to this deficiency of momentum) = Mass × Velocity

$$= (\rho V\theta)\cdot V$$

$$= \rho \,.\, \theta V^2$$

Equating the two,

$$\rho\theta V^2 = \rho \int_0^\delta v(V - v)dy = \rho V \int_0^\delta v\left(1 - \frac{v}{V}\right)dy$$

$$\theta = \int_0^\delta \frac{v}{V}\left(1 - \frac{v}{V}\right)dy \qquad \ldots(13.3)$$

Energy Thickness (δ^{**})

Energy thickness is defined as the thickness of an imaginary layer in the free stream whose kinetic energy is equal to the reduction of kinetic energy caused to the actual mass flowing in the boundary layer.

Kinetic energy of the fluid through elementary strip

$$= \frac{1}{2}(\rho V\cdot dy)v^2$$

Kinetic energy of the same mass in the absence of boundary layer

$$= \frac{1}{2}(\rho V\cdot dy)V^2$$

Loss of kinetic energy of elementary strip

$$= \frac{1}{2}\rho v(V^2 - v^2)dy$$

Total loss of kinetic energy of fluid

$$= \frac{1}{2}\int_0^\delta \rho v(V^2 - v^2)\cdot dy$$

Kinetic energy of a layer of thickness δ^{**} in the free stream (whose kinetic energy is equal to this deficiency of kinetic energy)

$$= \frac{1}{2} \times \text{Mass} \times (\text{Velocity})^2$$

$$= \frac{1}{2}(\rho V\delta^{**})V^2$$

Equating,

$$\frac{1}{2}\rho V\delta^{**}V^2 = \frac{1}{2}\rho\int_0^\delta v(V^2 - v^2)dy$$

$$= \frac{1}{2}\rho V^2 \int_0^\delta v\left(1 - \frac{v^2}{V^2}\right)dy$$

$$\delta^{**} = \int_0^\delta \frac{v}{V}\left(1 - \frac{v^2}{V^2}\right)dy \qquad \ldots(13.4)$$

Example 13.1 The velocity distribution in the boundary layer of a flow is given by

(i) $\frac{v}{V} = \frac{y}{\delta}$

(ii) $\frac{v}{V} = 2\left(\frac{y}{\delta}\right) - \left(\frac{y}{\delta}\right)^2$

where V is the velocity at a distance y from the plate and $v = V$ at $y = \delta$, δ being the boundary layer thickness. Determine the displacement thickness, the momentum thickness, the energy thickness and the value of δ^*/θ.

Solution (i)

$$\delta^* = \int_0^\delta \left(1 - \frac{v}{V}\right) dy = \int_0^\delta \left(1 - \frac{y}{\delta}\right) dy = \left(y - \frac{y^2}{2\delta}\right)_0^\delta = \delta - \frac{\delta^2}{2\delta} = \frac{\delta}{2}$$

$$\theta = \int_0^\delta \frac{v}{V}\left(1 - \frac{v}{V}\right) dy = \int_0^\delta \frac{y}{\delta}\left(1 - \frac{y}{\delta}\right) dy = \int_0^\delta \left(\frac{y}{\delta} - \frac{y^2}{\delta^2}\right) dy$$

$$= \left(\frac{y^2}{2\delta} - \frac{y^3}{3\delta^2}\right)_0^\delta = \frac{\delta^2}{2\delta} - \frac{\delta^3}{3\delta^2} = \frac{\delta}{6}$$

$$\delta^{**} = \int_0^\delta \frac{v}{V}\left(1 - \frac{v^2}{V^2}\right) dy = \int_0^\delta \frac{y}{\delta}\left(1 - \frac{y^2}{\delta^2}\right) dy = \int_0^\delta \left(\frac{y}{\delta} - \frac{y^3}{\delta^3}\right) dy$$

$$= \left(\frac{y^2}{2\delta} - \frac{y^4}{4\delta^3}\right)_0^\delta = \frac{\delta^2}{2\delta} - \frac{\delta^4}{4\delta^3} = \frac{\delta}{4}$$

$$\frac{\delta^*}{\theta} = \frac{\delta/2}{\delta/6} = 3$$

(ii)

$$\delta^* = \int_0^\delta \left(1 - \frac{v}{V}\right) dy$$

$$= \int_0^\delta \left[1 - \left\{2\left(\frac{y}{\delta}\right) - \left(\frac{y}{\delta}\right)^2\right\}\right] dy$$

$$= \int_0^\delta \left[1 - \frac{2y}{\delta} + \frac{y^2}{\delta^2}\right] dy$$

$$= \left(y - \frac{y^2}{\delta} + \frac{y^3}{3\delta^2}\right)_0^\delta$$

$$= \delta - \frac{\delta^2}{\delta} + \frac{\delta^3}{3\delta^2}$$

$$= \frac{\delta}{3}$$

$$\theta = \int_0^\delta \frac{v}{V}\left(1 - \frac{v}{V}\right) dy$$

$$= \int_0^\delta \left\{2\left(\frac{y}{\delta}\right) - \left(\frac{y}{\delta}\right)^2\right\}\left[1 - \left\{2\left(\frac{y}{\delta}\right) - \left(\frac{y}{\delta}\right)^2\right\}\right] dy$$

$$= \int_0^\delta \left\{\frac{2y}{\delta} - \frac{y^2}{\delta^2}\right\}\left[1 - \frac{2y}{\delta} + \frac{y^2}{\delta^2}\right] dy$$

$$= \int_0^\delta \left[\frac{2y}{\delta} - \frac{4y^2}{\delta^2} + \frac{2y^3}{\delta^3} - \frac{y^2}{\delta^2} + \frac{2y^3}{\delta^3} - \frac{y^4}{\delta^4}\right]$$

$$= \int_0^\delta \left[\frac{2y}{\delta} - \frac{5y^2}{\delta^2} + \frac{4y^3}{\delta^3} - \frac{y^4}{\delta^4}\right]$$

$$= \left[\frac{y^2}{\delta} - \frac{5y^3}{3\delta^2} + \frac{y^4}{\delta^3} - \frac{y^5}{5\delta^4}\right]_0^\delta$$

$$= \delta - \frac{5}{3}\delta + \delta - \frac{1}{5}\delta$$

$$= \frac{2}{15}\delta$$

$$\delta^{**} = \int_0^\delta \frac{v}{V}\left(1 - \frac{v^2}{V^2}\right) dy$$

$$= \int_0^\delta \left\{2\left(\frac{y}{\delta}\right) - \left(\frac{y}{\delta}\right)^2\right\}\left[1 - \left\{2\left(\frac{y}{\delta}\right) - \left(\frac{y}{\delta}\right)^2\right\}^2\right]$$

$$= \int_0^\delta \left(\frac{2y}{\delta} - \frac{y^2}{\delta^2}\right)\left[1 - \left(\frac{2y}{\delta} - \frac{y^2}{\delta^2}\right)^2\right] dy$$

$$= \int_0^\delta \left(\frac{2y}{\delta} - \frac{y^2}{\delta^2}\right)\left[1 - \frac{4y^2}{\delta^2} - \frac{y^4}{\delta^4} + \frac{4y^3}{\delta^3}\right] dy$$

$$= \int_0^\delta \left(\frac{2y}{\delta} - \frac{8y^3}{\delta^3} - \frac{2y^5}{\delta^5} + \frac{8y^4}{\delta^4} - \frac{y^2}{\delta^2} + \frac{4y^4}{\delta^4} + \frac{y^6}{\delta^6} - \frac{4y^5}{\delta^5}\right) dy$$

$$= \int_0^\delta \left(\frac{2y}{\delta} - \frac{y^2}{\delta^2} - \frac{8y^3}{\delta^3} + \frac{14y^4}{\delta^4} - \frac{6y^5}{\delta^5} + \frac{y^6}{\delta^6}\right) dy$$

$$= \left[\frac{y^2}{\delta} - \frac{y^3}{3\delta^2} - \frac{2y^4}{\delta^3} + \frac{12y^5}{5\delta^4} - \frac{y^6}{\delta^5} + \frac{y^7}{7\delta^5}\right]_0^\delta$$

$$= \frac{22}{105}\delta$$

$$\frac{\delta^*}{\theta} = \frac{\delta/3}{\frac{2}{15}\delta} = 2{\cdot}5$$

Example 13.2 The velocity distribution in the boundary layer over the face of a spillway is in the form

$$\frac{v}{V} = \left(\frac{y}{\delta}\right)^{0.22}$$

At a section, the free stream velocity is found to be 40 m/s and the boundary layer thickness 80 mm estimated from the velocity distribution measured at the section. If the discharge passing over the spillway is 8 m^3/s, determine the displacement thickness, the energy thickness and the loss of energy upto the section under consideration.

Solution

$$\delta^* = \int_0^\delta \left(1 - \frac{v}{V}\right) dy$$

$$= \int_0^\delta \left[1 - \left(\frac{y}{\delta}\right)^{0.22}\right] dy$$

$$= \left(y - \frac{y^{1.22}}{1.22 \times \delta^{0.22}}\right)_0^\delta$$

$$= \left(\delta - \frac{\delta^{1.22}}{1.22 \times \delta^{0.22}}\right)$$

$$= \delta\left(1 - \frac{1}{1.22}\right)$$

$$= 0.18\delta$$

$$= 0.18 \times 80 = 14.4 \text{ mm}$$

$$\theta = \int_0^\delta \frac{v}{V}\left[1 - \left(\frac{v}{V}\right)^2\right] dy$$

$$= \int_0^\delta \left(\frac{y}{\delta}\right)^{0.22} \left[1 - \left(\frac{y}{\delta}\right)^{0.44}\right] dy$$

$$= \int_0^\delta \left[\left(\frac{y}{\delta}\right)^{0.22} - \left(\frac{y}{\delta}\right)^{0.66}\right] dy$$

$$= \left(\frac{y^{1.22}}{1.22\delta^{0.22}} - \frac{y^{1.66}}{1.66\delta^{0.66}}\right)_0^\delta$$

$$= \left(\frac{\delta}{1.22} - \frac{\delta}{1.66}\right)$$

$$= 0.217\delta$$

$$= 0.217 \times 80$$

$$= 17.36 \text{ mm or } 0.01736 \text{ m}$$

The energy loss/m length of spillway

$$= \frac{1}{2} \times \text{Mass} \times V^2$$

$$= \frac{1}{2} \times (\text{Density} \times \text{Volume}) \times V^2$$

$$= \frac{1}{2} \times (\text{Density} \times \text{Area} \times V)\, V^2$$

$$= \frac{1}{2} \rho \times (\text{Energy thickness} \times 1) \times V^3$$

$$= \frac{1}{2} \times 1000 \times 0{\cdot}01736 \times 40^3$$

$$= 555{\cdot}5 \times 10^3 \text{ N.m/s or } 555{\cdot}5 \text{kN.m/s}$$

The energy loss/m length of spillway = Weight of water × Head = wQH

$$555{\cdot}5 \times 10^3 = (1000 \times 9{\cdot}81) \times 8 \times H$$

$$H = 7{\cdot}078 \text{ m}$$

13.4 DRAG ON A FLAT PLATE

Let a fluid flowing with a velocity V move over a thin plate kept in a free stream (Fig. 13.3). ABCD is an elementary strip of the boundary layer at a distance x from the leading edge. Assume unit width of the plate perpendicular to the direction of flow.

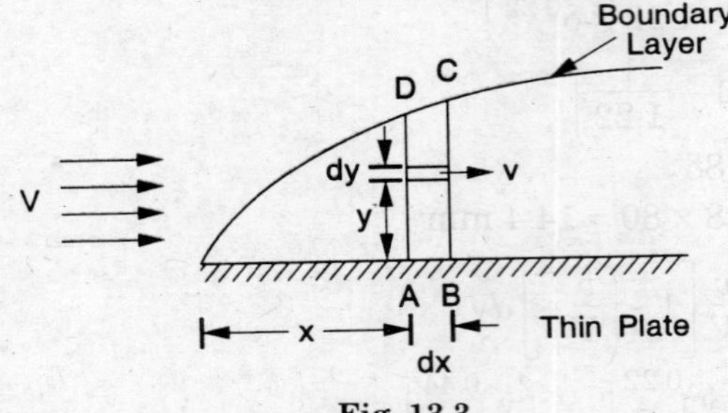

Fig. 13.3

Mass flow rate entering AD + Mass flow rate enetering DC = Mass flow rate leaving BC

Mass flow rate entering AD

$$= \int_0^\delta \rho \cdot v \cdot dy$$

Mass flow rate leaving BC

$$= m + \frac{\partial m}{\partial x} \cdot dx$$

$$= \int_0^\delta \rho v \cdot dy + \frac{\partial}{\partial x}\left[\int_0^\delta \rho v \cdot dy\right] dx$$

Mass flow rate entering DC

$$= \int_0^\delta \rho v \cdot dy + \frac{\partial}{\partial x}\left[\int_0^\delta \rho v \cdot dy\right] dx - \int_0^\delta \rho v \cdot dy$$

$$= \frac{\partial}{\partial x}\left[\int_0^\delta \rho v \cdot dy\right] dx$$

Now rate of change of momentum of the control volume

= Momentum flux leaving through BC – Momentum flux entering through AB – Momentum flux entering through DC

$$= \int_0^\delta \rho v^2 \cdot dy + \frac{d}{dx}\left[\int_0^\delta \rho v^2 \cdot dy\right] dx - \int_0^\delta \rho v^2 \cdot dy - \frac{d}{dx}\left[\int_0^\delta \rho v V \cdot dy\right] dx$$

$$= \frac{d}{dx}\left[\int_0^\delta \rho v^2 \cdot dy - \int_0^\delta \rho v V \cdot dy\right] dx$$

$$= \frac{d}{dx}\left[\rho \int_0^\delta (v^2 - vV) \cdot dy\right] dx$$

$$= \rho \frac{d}{dx}\left[\int_0^\delta (v^2 - vV) \cdot dy\right] dx \qquad \ldots(13.5)$$

According to the principle of momentum, the total force on the control volume must be equal to the rate of change of momentum in the same direction. The only external force on the control volume is the shear force on the boundary surface *AB* (in the opposite direction).

External force in the direction of flow $= -\tau_0 \times (dx \times 1)$ ...(13.6)

(*-ve* sign due to its opposite direction)

From Eqs. (13.5) and (13.6),

$$-\tau_0 \cdot dx = \rho \frac{d}{dx}\left[\int_0^\delta (v^2 - vV) \cdot dy\right] dx$$

$$-\tau_0 = -\rho \frac{d}{dx}\left[\int_0^\delta (v^2 - vV) \cdot dy\right]$$

$$= -\rho \frac{d}{dx}\left[\int_0^\delta V^2 \left(\frac{v^2}{V^2} - \frac{v}{V}\right) \cdot dy\right]$$

$$= \rho \frac{d}{dx}\left[\int_0^\delta V^2 \left(\frac{v}{V} - \frac{v^2}{V^2}\right) \cdot dy\right]$$

$$= \rho V^2 \frac{d}{dx}\left[\int_0^\delta \frac{v}{V}\left(1 - \frac{v}{V}\right) \cdot dy\right]$$

$$\frac{\tau_0}{\rho V^2} = \frac{d}{dx}\left[\int_0^\delta \frac{v}{V}\left(1 - \frac{v}{V}\right) \cdot dy\right] = \frac{d}{dx}(\theta) = \frac{d\theta}{dx} \qquad \ldots(13.7)$$

It is known as Von Karman momentum equation for boundary layer flow. This equation is applicable to both laminar as well as turbulent flow.

The shear stress τ_0 can be obtained from the above equation.

The drag force on the elementary length

$= \tau_0 \times (dx \cdot 1)$

Total drag force on the plate of length L

$$= \int_0^L \tau_0 \cdot dx$$

Total drag force on the plate of length L and width b

$$= \int_0^L \tau_0 \cdot b \cdot dx \qquad \ldots(13.8)$$

Local Coefficient of Drag (C_D^*)

It is the ratio of the shear stress τ_0 to the dynamic pressure of the uniform flow stream $\left(\frac{1}{2}\rho V^2\right)$.

$$C_D^* = \frac{\tau_0}{\frac{1}{2}\rho V^2} \qquad \ldots(13.9)$$

C_D^* is also known as the **skin friction drag.**

The total drag force F_D on the plate is given by

$$F_D = C_D \times \frac{\rho V^2}{2} \times \text{Area}$$

where C_D is the **average coefficient of drag.**

$$C_D = \frac{F_D}{\frac{\rho V^2}{2} \times Area} \qquad \ldots(13.10)$$

Boundary Conditions

The following boundary conditions must be satisfied by any velocity profile :

At $y = 0$, $v = 0$ and dv/dt = Some finite value

At $y = \delta$, $v = V$ and $dv/dy = 0$

13.5 LAMINAR BOUNDARY LAYER

Assume the velocity distribution to be

$$\frac{v}{V} = 2\left(\frac{y}{\delta}\right) - \left(\frac{y}{\delta}\right)^2$$

The values of the boundary layer thickness, shear stress, local drag coefficient and the coefficient of drag are worked out as follows :

Boundary Layer Thickness

From Von Karman momentum equation

$$\frac{\tau_0}{\rho V^2} = \frac{d\theta}{dx} = \frac{d}{dx}\left[\int_0^\delta \frac{v}{V}\left(1 - \frac{v}{V}\right) \cdot dy\right]$$

$$= \frac{d}{dx}\left[\int_0^\delta \left(\frac{2y}{\delta} - \frac{y^2}{\delta^2}\right)\left\{1 - \left(\frac{2y}{\delta} - \frac{y^2}{\delta^2}\right)\right\} \cdot dy\right]$$

$$= \frac{d}{dx}\left[\int_0^\delta \left(\frac{2y}{\delta} - \frac{y^2}{\delta^2}\right)\left(1 - \frac{2y}{\delta} + \frac{y^2}{\delta^2}\right) \cdot dy\right]$$

$$= \frac{d}{dx}\left[\int_0^\delta \left(\frac{2y}{\delta} - \frac{4y^2}{\delta^2} + \frac{2y^3}{\delta^3} - \frac{y^2}{\delta^2} + \frac{2y^3}{\delta^3} - \frac{y^4}{\delta^4}\right) \cdot dy\right]$$

$$= \frac{d}{dx}\left[\int_0^\delta \left(\frac{2y}{\delta} - \frac{5y^2}{\delta^2} + \frac{4y^3}{\delta^3} - \frac{y^4}{\delta^4}\right) dy\right]$$

$$= \frac{d}{dx}\left(\frac{y^2}{\delta} - \frac{5y^3}{3\delta^2} + \frac{y^4}{\delta^3} - \frac{y^5}{5\delta^4}\right)_0^\delta$$

$$= \frac{d}{dx}\left(\delta - \frac{5\delta}{3} + \delta - \frac{\delta}{5}\right)$$

$$= \frac{d}{dx}\left(\frac{2\delta}{15}\right)$$

$$\tau_0 = \frac{2}{15}\rho V^2 \frac{d\delta}{dx} \qquad \ldots\text{(i)}$$

Also

$$\tau_0 = \mu\left(\frac{dv}{dy}\right)_{y=0}$$

$$= \mu \frac{d}{dy}\left[V\left\{\frac{2y}{\delta} - \left(\frac{y}{\delta}\right)^2\right\}\right]_{y=0}$$

$$= \mu V\left[\frac{d}{dy}\left(\frac{2y}{\delta} - \frac{y^2}{\delta^2}\right)\right]_{y=0}$$

$$= \mu V\left(\frac{2}{\delta} - \frac{2y}{\delta^2}\right)_{y=0}$$

$$= \mu V\left(\frac{2}{\delta} - 0\right)$$

$$= \frac{2\mu V}{\delta} \qquad \ldots\text{(ii)}$$

Equating (i) and (ii),

$$\frac{2}{15}\rho V^2 \frac{d\delta}{dx} = \frac{2\mu V}{\delta}$$

$$\delta d\delta = \frac{15\mu}{\rho V}\, dx$$

$$\frac{\delta^2}{2} = \frac{15\mu}{\rho V}\, dx + C$$

where C is a constant of integration.

Applying the boundary conditions, at $x = 0$, $\delta = 0$

$$0 = 0 + C$$

$$C = 0$$

$$\frac{\delta^2}{2} = \frac{15\mu}{\rho V} . x$$

$$\delta = \sqrt{\frac{30\mu x}{\rho V}} = 5{\cdot}48\sqrt{\frac{x^2}{\rho V x/\mu}} = 5{\cdot}48\,\frac{x}{\sqrt{R_{ex}}} \qquad \text{...(13.11)}$$

Shear Stress

From (ii),

$$\tau_0 = \frac{2\mu V}{\delta} = \frac{2\mu V}{5{\cdot}48\,\dfrac{x}{\sqrt{R_{ex}}}} = 0{\cdot}365\,\frac{\mu V}{x}\sqrt{R_{ex}} \qquad \text{...(13.12)}$$

Local Coefficient of Drag

$$C_D^* = \frac{\tau_0}{\frac{1}{2}\rho V^2} = \frac{0{\cdot}365\,\dfrac{\mu V}{x}\sqrt{R_{ex}}}{\frac{1}{2}\rho V^2} = \frac{0{\cdot}73\sqrt{R_{ex}}}{\dfrac{\rho V x}{\mu}} = \frac{0{\cdot}73\sqrt{R_{ex}}}{R_{ex}} = \frac{0{\cdot}73}{\sqrt{R_{ex}}} \qquad \text{...(13.13)}$$

Coefficient of Drag

$$C_D = \frac{\int_0^L \tau_0 \cdot b \cdot dx}{\dfrac{\rho A V^2}{2}}$$

$$= \frac{\int_0^L 0{\cdot}365\,\dfrac{\mu V}{x}\sqrt{R_{ex}} \cdot b \cdot dx}{\dfrac{\rho L \cdot b \cdot V^2}{2}}$$

$$= \frac{0{\cdot}73\int_0^L \dfrac{1}{x}\sqrt{\dfrac{\rho V x}{\mu}} \cdot dx}{\dfrac{\rho V L}{\mu}}$$

$$= \frac{0{\cdot}73\sqrt{\dfrac{\rho V}{\mu}}\int_0^L x^{-1} \cdot dx}{R_{el}} \qquad \text{...(13.14)}$$

$$= \frac{0.73\sqrt{\frac{\rho V}{\mu}}\left(\frac{x^{1/2}}{1/2}\right)_0^L}{R_{el}}$$

$$= \frac{1.46\sqrt{\frac{\rho VL}{\mu}}}{R_{el}}$$

$$= \frac{1.46\sqrt{R_{el}}}{R_{el}}$$

$$= \frac{1.46}{\sqrt{R_{el}}}$$

The results obtained above are not changed much if some other velocity distribution is assumed for the laminar boundary layer. Blasius gave the following numerical values after giving rigorous mathematical treatment to the general equations of motion :

$$\delta = \frac{5x}{\sqrt{R_{ex}}}, \quad C_D^* = \frac{0.664}{\sqrt{R_{ex}}}, \quad C_D = \frac{1.328}{\sqrt{R_{el}}} \qquad \ldots(13.15)$$

These results are commonly used to work out the solutions to numerical problems.

Example 13.3 For the following velocity distributions, determine the boundary layer thickness, shear stress, and local coefficient of drag and the coefficient of drag :

(i) $\frac{v}{V} = \frac{3}{2}\left(\frac{y}{\delta}\right) - \frac{1}{2}\left(\frac{y}{\delta}\right)^3$

(ii) $\frac{v}{V} = 2\left(\frac{y}{\delta}\right) - 2\left(\frac{y}{\delta}\right)^3 + \left(\frac{y}{\delta}\right)^4$

(iii) $\frac{v}{V} = \sin\left(\frac{\pi}{2}\frac{y}{\delta}\right)$

Solution (i) **Boundary Layer Thickness**

$$\frac{\tau_0}{\rho V^2} = \frac{d\theta}{dx} = \frac{d}{dx}\left[\int_0^\delta \frac{v}{V}\left(1 - \frac{v}{V}\right)\cdot dy\right]$$

$$= \frac{d}{dx}\left[\int_0^\delta \left(\frac{3}{2}\frac{y}{\delta} - \frac{1}{2}\frac{y^3}{\delta^3}\right)\left\{1 - \left(\frac{3}{2}\frac{y}{\delta} - \frac{1}{2}\frac{y^3}{\delta^3}\right)\right\}\cdot dy\right]$$

$$= \frac{d}{dx}\left[\int_0^\delta \left(\frac{3}{2}\frac{y}{\delta} - \frac{1}{2}\frac{y^3}{\delta^3}\right)\left(1 - \frac{3}{2}\frac{y}{\delta} + \frac{1}{2}\frac{y^3}{\delta^3}\right)\cdot dy\right]$$

$$= \frac{d}{dx}\left[\int_0^\delta \left(\frac{3}{2}\frac{y}{\delta} - \frac{9}{4}\frac{y^2}{\delta^2} + \frac{3}{4}\frac{y^4}{\delta^4} - \frac{1}{2}\frac{y^3}{\delta^3} + \frac{3}{4}\frac{y^4}{\delta^4} - \frac{1}{4}\frac{y^6}{\delta^6}\right)\cdot dy\right]$$

$$= \frac{d}{dx}\left(\frac{3^2}{4\delta} - \frac{9y^3}{12\delta^2} + \frac{3}{20}\frac{y^5}{\delta^4} - \frac{1}{8}\frac{y^4}{\delta^3} + \frac{3}{20}\frac{y^5}{\delta^4} - \frac{1}{28}\frac{y^7}{\delta^6}\right)_0^\delta$$

$$= \frac{d}{dx}\left(\frac{3\delta}{4} - \frac{9\delta}{12} + \frac{3\delta}{20} - \frac{\delta}{8} + \frac{3\delta}{20} - \frac{\delta}{28}\right)$$

$$= \frac{d}{dx}\left(\frac{39\delta}{280}\right)$$

$$\tau_0 = \frac{39}{280}\rho V^2 \frac{d\delta}{dx} \quad \text{...(i)}$$

Also

$$\tau_0 = \mu\left(\frac{dv}{dy}\right)_{y=0}$$

$$= \mu\frac{d}{dy}\left[V\left\{\frac{3}{2}\frac{y}{\delta} - \frac{1}{2}\left(\frac{y}{\delta}\right)^3\right\}\right]_{y=0}$$

$$= \mu V\left[\frac{d}{dy}\left\{\frac{3}{2}\frac{y}{\delta} - \frac{y^3}{2\delta^3}\right\}\right]_{y=0}$$

$$= \mu V\left(\frac{3}{2\delta} - \frac{3y^2}{2\delta^3}\right)_{y=0}$$

$$= \mu V\left(\frac{3}{2\delta} - 0\right)$$

$$= \frac{3\mu V}{2\delta} \quad \text{...(ii)}$$

Equation (i) and (ii),

$$\frac{39}{280}\rho V^2 \frac{d\delta}{dx} = \frac{3\mu V}{2\delta}$$

$$\delta d\delta = \frac{140\mu}{13\rho V}dx$$

$$\frac{\delta^2}{2} = \frac{140\mu}{13\rho V}x + C$$

where C is a constant of integration.

Applying the boundary conditions, at $x = 0$, $\delta = 0$

$$0 = 0 + C$$

$$C = 0$$

$$\frac{\delta^2}{2} = \frac{140\mu}{13\rho V}.x$$

$$\delta = \sqrt{\frac{280\mu x}{13\rho V}} = 4{\cdot}64\sqrt{\frac{x^2}{\rho V x/\mu}} = 4{\cdot}64\frac{x}{\sqrt{R_{ex}}}$$

Shear Stress

$$\tau_0 = \frac{3\mu V}{2\delta} = \frac{3\mu V}{2 \times 4{\cdot}64 \dfrac{x}{\sqrt{R_{ex}}}} = 0{\cdot}323 \frac{\mu V}{x} \sqrt{R_{ex}}$$

Local Coefficient of Drag

$$C_D^* = \frac{\tau_0}{\frac{1}{2}\rho V^2} = \frac{0{\cdot}323 \dfrac{\mu V}{x} \sqrt{R_{ex}}}{\frac{1}{2}\rho V^2} = \frac{0{\cdot}646 \sqrt{R_{ex}}}{\dfrac{\rho V x}{\mu}} = \frac{0{\cdot}646 \sqrt{R_{ex}}}{R_{ex}} = \frac{0{\cdot}646}{\sqrt{R_{ex}}}$$

Coefficient of Drag

$$C_D = \frac{\int_0^L \tau_0 {\cdot} b {\cdot} dx}{\dfrac{\rho A V^2}{2}}$$

$$= \frac{\int_0^L 0{\cdot}332 \dfrac{\mu V}{x} \sqrt{R_{ex}} {\cdot} b {\cdot} dx}{\dfrac{\rho\, L {\cdot} b {\cdot} V^2}{2}}$$

$$= \frac{0{\cdot}646 \int_0^L \dfrac{1}{x} \sqrt{\dfrac{\rho V x}{\mu}} {\cdot} dx}{\dfrac{\rho V L}{\mu}}$$

$$= \frac{0{\cdot}646 \sqrt{\dfrac{\rho V}{\mu}} \int_0^L x^{-1} {\cdot} dx}{R_{el}}$$

$$= \frac{0{\cdot}646 \sqrt{\dfrac{\rho V}{\mu}} \int_0^L \left(\dfrac{x^{1/2}}{1/2}\right)_0^L}{R_{el}}$$

$$= \frac{1{\cdot}292 \sqrt{\dfrac{\rho V L}{\mu}}}{R_{el}}$$

$$= \frac{1{\cdot}292 \sqrt{R_{el}}}{R_{el}}$$

$$= \frac{1{\cdot}292}{\sqrt{R_{el}}}$$

(ii) **Boundary Layer Thickness**

$$\frac{\tau_0}{\rho V^2} = \frac{d\theta}{dx} = \frac{d}{dx}\left[\int_0^\delta \frac{v}{V}\left(1 - \frac{v}{V}\right) {\cdot} dy\right]$$

$$= \frac{d}{dx}\left[\int_0^\delta \left(2\frac{y}{\delta} - 2\frac{y^3}{\delta^3} + \frac{y^4}{\delta^4}\right)\left\{1 - \left(2\frac{y}{\delta} - 2\frac{y^3}{\delta^3} + \frac{y^4}{\delta^4}\right)\right\} {\cdot} dy\right]$$

$$= \frac{d}{dx}\left[\int_0^\delta \left(2\frac{y}{\delta} - 2\frac{y^3}{\delta^3} + \frac{y^4}{\delta^4}\right)\left(1 - 2\frac{y}{\delta} + 2\frac{y^3}{\delta^3} - \frac{y^4}{\delta^4}\right)\cdot dy\right]$$

$$= \frac{d}{dx}\left[\int_0^\delta \left(\begin{array}{l} 2\frac{y}{\delta} - 4\frac{y^2}{\delta^2} + 4\frac{y^4}{\delta^4} - 2\frac{y^5}{\delta^5} - 2\frac{y^3}{\delta^3} + 4\frac{y^4}{\delta^4} \\ -4\frac{y^6}{\delta^6} + 2\frac{y^7}{\delta^7} + \frac{y^4}{\delta^4} - 2\frac{y^5}{\delta^5} + 2\frac{y^7}{\delta^7} - \frac{y^8}{\delta^8}\end{array}\right)\cdot dy\right]$$

$$= \frac{d}{dx}\left[\int_0^\delta \left(2\frac{y}{\delta} - 4\frac{y^2}{\delta^2} - 2\frac{y^3}{\delta^3} + 9\frac{y^4}{\delta^4} - 4\frac{y^5}{\delta^5} - 4\frac{y^6}{\delta^6} + 4\frac{y^7}{\delta^7} - \frac{y^8}{\delta^8}\right)\cdot dy\right]$$

$$= \frac{d}{dx}\left(\frac{y^2}{\delta} - \frac{4y^3}{3\delta^2} - \frac{y^4}{2\delta^3} + \frac{9y^5}{5\delta^4} - \frac{2y^6}{3\delta^5} - \frac{4}{7}\frac{y^7}{\delta^6} + \frac{1}{2}\frac{y^8}{\delta^7} - \frac{y^9}{9\delta^8}\right)_0^\delta$$

$$= \frac{d}{dx}\left(\delta - \frac{4\delta}{3} - \frac{\delta}{2} + \frac{9\delta}{5} - \frac{2\delta}{3} - \frac{4\delta}{7} + \frac{\delta}{2} - \frac{\delta}{9}\right)$$

$$= \frac{d}{dx}\left(\frac{37\delta}{315}\right)$$

$$\tau_0 = \frac{37}{315}\rho V^2 \frac{d\delta}{dx} \qquad \ldots\text{(i)}$$

Also

$$\tau_0 = \mu\left(\frac{dv}{dy}\right)_{y=0}$$

$$= \mu\frac{d}{dy}\left[V\left\{\frac{2y}{\delta} - \frac{2y^3}{\delta^3} + \frac{y^4}{\delta^4}\right\}\right]_{y=0}$$

$$= \mu V\left[\frac{d}{dy}\left\{\frac{2y}{\delta} - \frac{2y^3}{\delta^3} + \frac{y^4}{\delta^4}\right\}\right]_{y=0}$$

$$= \mu V\left(\frac{2}{\delta} - \frac{6y^2}{\delta^3} + 4\frac{y^3}{\delta^4}\right)_{y=0}$$

$$= \mu V\left(\frac{2}{\delta} - 0\right)$$

$$= \frac{2\mu V}{\delta} \qquad \ldots\text{(ii)}$$

Equating (i) and (ii),

$$\frac{37}{315}\rho V^2\frac{d\delta}{dx} = \frac{2\mu V}{\delta}$$

$$\delta d\delta = \frac{630\mu}{37\rho V}dx$$

$$\frac{\delta^2}{2} = \frac{630\mu}{37\rho V}x + C$$

where C is a constant of integration.

Applying the boundary conditions, at $x = 0$, $\delta = 0$

$$0 = 0 + C$$

$$C = 0$$

$$\frac{\delta^2}{2} = \frac{630\mu}{37\rho V}.x$$

$$\delta = \sqrt{\frac{1260\mu x}{37\rho V}} = 5{\cdot}836\sqrt{\frac{x^2}{\rho Vx/\mu}} = 5{\cdot}836\frac{x}{\sqrt{R_{ex}}}$$

Shear stress

$$\tau_0 = \frac{2\mu V}{\delta} = \frac{2\mu V}{5{\cdot}836\dfrac{x}{\sqrt{R_{ex}}}} = 0{\cdot}343\frac{\mu V}{x}\sqrt{R_{ex}}$$

Local Coefficient of Drag

$$C_D^* = \frac{\tau_0}{\frac{1}{2}\rho V^2} = \frac{0{\cdot}343\dfrac{\mu V}{x}\sqrt{R_{ex}}}{\frac{1}{2}\rho V^2} = \frac{0{\cdot}686\sqrt{R_{ex}}}{\dfrac{\rho Vx}{\mu}} = \frac{0{\cdot}686\sqrt{R_{ex}}}{R_{ex}} = \frac{0{\cdot}686}{\sqrt{R_{ex}}}$$

Coefficient of Drag

$$C_D = \frac{\int_0^L \tau_0{\cdot}b{\cdot}dx}{\dfrac{\rho AV^2}{2}} = \frac{\int_0^L 0{\cdot}343\dfrac{\mu V}{x}\sqrt{R_{ex}}{\cdot}b{\cdot}dx}{\dfrac{\rho L{\cdot}b{\cdot}V^2}{2}} = \frac{0{\cdot}686\int_0^L \dfrac{1}{x}\sqrt{\dfrac{\rho Vx}{\mu}}{\cdot}dx}{\dfrac{\rho VL}{\mu}}$$

$$= \frac{0{\cdot}686\sqrt{\dfrac{\rho V}{\mu}}\int_0^L x^{-1}{\cdot}dx}{R_{el}} = \frac{0{\cdot}686\sqrt{\dfrac{\rho V}{\mu}}\left(\dfrac{x^{1/2}}{1/2}\right)_0^L}{R_{el}}$$

$$= \frac{1{\cdot}372\sqrt{\dfrac{\rho VL}{\mu}}}{R_{el}} = \frac{1{\cdot}372\sqrt{R_{el}}}{R_{el}}$$

$$= \frac{1{\cdot}372}{\sqrt{R_{el}}}$$

(iii) Boundary Layer Thickness

$$\frac{\tau_0}{\rho V^2} = \frac{d\theta}{dx} = \frac{d}{dx}\left[\int_0^\delta \frac{v}{V}\left(1 - \frac{v}{V}\right){\cdot}dy\right]$$

$$= \frac{d}{dx}\left[\int_0^\delta \sin\left(\frac{\pi}{2}\frac{y}{\delta}\right)\left\{1 - \sin\left(\frac{\pi}{2}\frac{y}{\delta}\right)\right\}{\cdot}dy\right]$$

$$= \frac{d}{dx}\left[\int_0^\delta \left\{\sin\left\{\frac{\pi}{2}\frac{y}{\delta}\right\} - \sin^2\left(\frac{\pi}{2}\frac{y}{\delta}\right)\right\}{\cdot}dy\right]$$

$$= \frac{d}{dx}\left[\int_0^\delta \left\{\sin\left(\frac{\pi}{2}\frac{y}{\delta}\right) - \frac{1-\cos(\pi y/\delta)}{2}\right\}\cdot dy\right]$$

$$= \frac{d}{dx}\left[\int_0^\delta \left\{\sin\left(\frac{\pi}{2}\frac{y}{\delta}\right) - \frac{1}{2} - \frac{\cos(\pi y/\delta)}{2}\right\}\cdot dy\right]$$

$$= \frac{d}{dx}\left[-\frac{2\delta}{\pi}\cos\left(\frac{\pi}{2}\frac{y}{\delta}\right) - \frac{y}{2} + \cdot\frac{\delta}{\pi}\sin\frac{\pi y}{\delta}\right]_0^\delta$$

$$= \frac{d}{dx}\left[0 - \frac{\delta}{2} + 0 - \left(-\frac{2\delta}{\pi} - 0 + 0\right)\right]$$

$$= \frac{d\delta}{dx}\left(\frac{4-\pi}{2\pi}\right)$$

$$\tau_0 = \frac{2}{15}\rho V^2\frac{d\delta}{dx}$$

Also $\quad \tau_0 = \mu\left(\frac{dv}{dy}\right)_{y=0}$

$$= \mu\frac{d}{dy}\left[V\left\{\sin\left(\frac{\pi}{2}\frac{y}{\delta}\right)\right\}\right]_{y=0}$$

$$= \mu V\left[\frac{d}{dy}\left\{\sin\left(\frac{\pi}{2}\frac{y}{\delta}\right)\right\}\right]_{y=0}$$

$$= \mu V\left[\frac{\pi}{2\delta}\cos\left(\frac{\pi}{2}\frac{y}{\delta}\right)\right]_{y=0}$$

$$= \mu V\left(\frac{\pi}{2\delta}\right)$$

$$= \frac{3\mu V\pi}{2\delta}$$

$$\left(\frac{4-\pi}{2\pi}\right)\rho V^2\frac{d\delta}{dx} = \frac{\mu V\pi}{2\delta}$$

$$\delta d\delta = \frac{\pi^2}{(4-\pi)}\frac{\mu}{\rho V}dx = 11{\cdot}4975$$

$$\frac{\delta^2}{2} = 11{\cdot}4975\frac{\mu}{\rho V}x + C$$

where C is a constant of integration.

Applying the boundary conditions, at $x = 0$, $\delta = 0$

$$0 = 0 + C$$

$$C = 0$$

$$\frac{\delta^2}{2} = 11{\cdot}4975\frac{\mu}{\rho V}\cdot x$$

$$\delta = \sqrt{2\times 11{\cdot}4975\frac{\mu x}{\rho V}} = 4{\cdot}795\sqrt{\frac{x^2}{\rho Vx/\mu}} = 4{\cdot}795\frac{x}{\sqrt{R_{ex}}}$$

Shear Stress

$$\tau_0 = \frac{\mu V\pi}{2\delta} = \frac{\mu V\pi}{2 \times 4{\cdot}795 \dfrac{x}{\sqrt{R_{ex}}}} = 0{\cdot}3275 \frac{\mu V}{x} \sqrt{R_{ex}}$$

Local Coefficent of Drag

$$C_D^* = \frac{\tau_0}{\frac{1}{2}\rho V^2} = \frac{0{\cdot}3275 \dfrac{\mu V}{x} \sqrt{R_{ex}}}{\frac{1}{2}\rho V^2} = \frac{0{\cdot}655 \sqrt{R_{ex}}}{\dfrac{\rho V x}{\mu}} = \frac{0{\cdot}655 \sqrt{R_{ex}}}{R_{ex}} = \frac{0{\cdot}655}{\sqrt{R_{ex}}}$$

Coefficient of Drag

$$C_D = \frac{\int_0^L \tau_0 {\cdot} b {\cdot} dx}{\dfrac{\rho A V^2}{2}} = \frac{\int_0^L 0{\cdot}3275 \dfrac{\mu V}{x} \sqrt{R_{ex}} {\cdot} b {\cdot} dx}{\dfrac{\rho L.b{\cdot}V^2}{2}} = \frac{0{\cdot}655 \int_0^L \dfrac{1}{x} \sqrt{\dfrac{\rho V x}{\mu}} {\cdot} dx}{\dfrac{\rho V L}{\mu}}$$

$$= \frac{0{\cdot}655 \sqrt{\dfrac{\rho V}{\mu}} \int_0^L x^{-1}{\cdot}dx}{R_{el}} = \frac{0{\cdot}655 \sqrt{\dfrac{\rho V}{\mu}} \left(\dfrac{x^{1/2}}{1/2} \right)_0^L}{R_{el}}$$

$$= \frac{1{\cdot}31 \sqrt{\dfrac{\rho V L}{\mu}}}{R_{el}} = \frac{1{\cdot}31 \sqrt{R_{el}}}{R_{el}}$$

$$= \frac{1{\cdot}31}{\sqrt{R_{el}}}$$

Table 13.1 shows the values of boundary layer thickness, local coefficient of drag and the average coefficient of drag in terms of Reynolds number for various velocity distributions.

Table 13.1

S.N.	Velocity distribution	δ	C_D^*	C_D
1.	$\frac{v}{V} = 2\left(\frac{y}{\delta}\right) - \left(\frac{y}{\delta}\right)^2$	$5{\cdot}48 \frac{x}{\sqrt{R_{ex}}}$	$\frac{0{\cdot}73}{\sqrt{R_{ex}}}$	$\frac{1{\cdot}46}{\sqrt{R_{el}}}$
2.	$\frac{v}{V} = \frac{3}{2}\left(\frac{y}{\delta}\right) - \frac{1}{2}\left(\frac{y}{\delta}\right)^3$	$4{\cdot}64 \frac{x}{\sqrt{R_{ex}}}$	$\frac{0{\cdot}646}{\sqrt{R_{ex}}}$	$\frac{1{\cdot}292}{\sqrt{R_{el}}}$
3.	$\frac{v}{V} = 2\left(\frac{y}{\delta}\right) - 2\left(\frac{y}{\delta}\right)^3 + \left(\frac{y}{\delta}\right)^4$	$5{\cdot}84 \frac{x}{\sqrt{R_{ex}}}$	$\frac{0{\cdot}686}{\sqrt{R_{ex}}}$	$\frac{1{\cdot}372}{\sqrt{R_{el}}}$
4.	$\frac{v}{V} = \sin\left(\frac{\pi}{2}\frac{y}{\delta}\right)$	$4{\cdot}795 \frac{x}{\sqrt{R_{ex}}}$	$\frac{0{\cdot}654}{\sqrt{R_{ex}}}$	$\frac{1{\cdot}31}{\sqrt{R_{el}}}$
5.	Blasius results	$5 \frac{x}{\sqrt{R_{ex}}}$	$\frac{0{\cdot}664}{\sqrt{R_{ex}}}$	$\frac{1{\cdot}328}{\sqrt{R_{el}}}$

Example 13.4 Air flows over a smooth plate with a velocity of 15 m/s. The velocity profile is given by

$$\frac{v}{V} = 2\left(\frac{y}{\delta}\right) - \left(\frac{y}{\delta}\right)^2$$

The plate is 1 m long and 800 mm wide. The kinematic viscosity of air is 0·15 stoke. Assuming that the laminar boundary layer exists upto a value of Re = 2×10^5, determine the maximum distance from the leading edge upto which the laminar boundary layer exists and the maximum thickness of the laminar boundary layer.

Solution $V = 15$ m/s $\qquad L = 1$ m

$\upsilon = 0{\cdot}15 \times 10^{-4}\ \text{m}^2/\text{s} \qquad B = 0{\cdot}8$ m

The distance from the leading edge is given by,

$$R_{ex} = \frac{V{\cdot}x}{\nu}$$

$$2 \times 10^5 = \frac{15x}{0{\cdot}15 \times 10^{-4}}$$

$$x = 0{\cdot}2\ \text{m}$$

The maximum thickness of boundary layer for the given profile,

$$\delta = \frac{5{\cdot}48x}{\sqrt{R_{ex}}} = \frac{5{\cdot}48 \times 0{\cdot}2}{\sqrt{2 \times 10^5}} = 0{\cdot}00245\ \text{m or } 2{\cdot}45\ \text{mm}$$

Example 13.5 Air flows over a plate of 600 mm length and 800 mm width with a velocity of 5 m/s. The velocity profile is assumed to be of the form

$$\frac{\nu}{V} = \sin\left(\frac{\pi}{2}\frac{y}{\delta}\right)$$

The density of air is 1·24 kg/m^3 and the kinematic viscosity $0{\cdot}15 \times 10^{-4}$ m^2/s. Find

(i) Boundary layer thickness at the end of the plate

(ii) Shear stress at 300 mm from the leading edge

(iii) Drag force on one side of the plate.

Solution (i)

$$R_{el} = \frac{V{\cdot}L}{\nu} = \frac{5 \times 0{\cdot}6}{0{\cdot}15 \times 10^{-4}} = 2 \times 10^5$$

As R_{el} is less than 5×10^5, the boundary layer is laminar throughout the length of the plate. For the given profile, the boundary layer thickness is

$$\delta = \frac{4{\cdot}795x}{\sqrt{R_{ex}}} = \frac{4{\cdot}795 \times 0{\cdot}6}{\sqrt{2 \times 10^5}} = 0{\cdot}00643\ \text{m or } 6{\cdot}43\ \text{mm}$$

(ii)

$$C_D^* = \frac{0{\cdot}654}{\sqrt{R_{ex}}} = \frac{0{\cdot}654}{\sqrt{\frac{V.x}{\nu}}} = \frac{0{\cdot}654}{\sqrt{\frac{5 \times 0{\cdot}3}{0{\cdot}15 \times 10^{-4}}}} = 0{\cdot}002068$$

$$\tau_0 = C_D^* \times \frac{1}{2}\rho V^2 = 0{\cdot}002068 \times \frac{1}{2} \times 1{\cdot}24 \times 5^2 = 0{\cdot}032\ \text{N/m}^2$$

(iii) $$F_D = C_D \times \frac{1}{2}\rho V^2 \times A$$

where $$C_D = \frac{1{\cdot}31}{\sqrt{R_{el}}} = \frac{1{\cdot}31}{\sqrt{2 \times 10^5}} = 0{\cdot}002929$$

$$F_D = 0{\cdot}002929 \times \frac{1}{2} \times 1{\cdot}24 \times 5^2 \times 0{\cdot}6 \times 0{\cdot}8 = 0{\cdot}02179 \text{ N}$$

Example 13.6 A stream of crude oil flows over a flat plate of length 800 mm and width 300 mm placed longitudinally with a velocity of 4 m/s. The kinematic viscosity of the oil is 1 stoke and the specific gravity 0·9. Determine the boundary layer thickness and the shear stress at the middle of the plate and the friction drag on one side of the plate.

Solution At the middle of the plate,

$$x = \frac{L}{2} = \frac{0{\cdot}8}{2} = 0{\cdot}4 \text{ m}$$

$$\nu = 1 \times 10^{-4} \text{ N·s/m}^2$$

$$R_{ex} = \frac{V{\cdot}x}{\nu} = \frac{4 \times 0{\cdot}4}{1 \times 10^{-4}} = 16000$$

As Re is less than 5×10^5, the boundary layer is laminar upto the middle of the plate. As the velocity profile is not given, Blasius solution should be used.

$$\delta = \frac{5x}{\sqrt{R_{ex}}} = \frac{5 \times 0{\cdot}4}{\sqrt{16000}} = 0{\cdot}01158 \text{ m or } 15{\cdot}8 \text{ mm}$$

$$C_D^* = \frac{0{\cdot}664}{\sqrt{R_{ex}}} = \frac{0{\cdot}664}{\sqrt{16000}} = 0{\cdot}00525$$

$$\tau_0 = C_D^* \times \frac{1}{2}\rho V^2 = 0{\cdot}00525 \times \frac{1}{2} \times 0{\cdot}9 \times 1000 \times 4^2 = 37{\cdot}79 \text{ N/m}^2$$

$$F_D = C_D \times \frac{1}{2}\rho V^2 \times A$$

where $$C_D = \frac{1{\cdot}328}{\sqrt{R_{el}}} = \frac{1{\cdot}328}{\sqrt{\frac{V{\cdot}x}{\nu}}} = \frac{1{\cdot}328}{\sqrt{\frac{4 \times 0{\cdot}8}{1 \times 10^{-4}\ \nu}}} = 0{\cdot}007424$$

$$F_D = 0{\cdot}007424 \times \frac{1}{2} \times (0{\cdot}9 \times 1000) \times 4^2 \times (0{\cdot}3 \times 0{\cdot}8) = 12{\cdot}83 \text{ N}$$

Example 13.7 Assuming the boundary layer to be laminar over the entire length of a flat plate kept at zero incidence in a stream of uniform velocity, determine the ratio of friction drag on the front half and the rear half of the plate.

Solution Friction drag on the entire plate

$$F_D = C_D \times \frac{1}{2}\rho V^2 \times A = \frac{1{\cdot}328}{\sqrt{\frac{V{\cdot}L}{\nu}}} \times \frac{1}{2}\rho V^2 \times L{\cdot}B = 0{\cdot}664\rho V^2\, LB \times \sqrt{\frac{\nu}{V{\cdot}L}}$$

Friction drag on the first half

$$F_{D1} = C_D \times \frac{1}{2}\rho V^2 \times A = \frac{1.328}{\sqrt{\dfrac{V \cdot L/2}{\nu}}} \times \frac{1}{2}\rho V^2 \times \frac{L}{2} \cdot B = 0.664 \rho V^2 L \cdot B \sqrt{\frac{\nu}{2V \cdot L}}$$

Friction drag on the rear half

$$F_{D2} = 0.664 \rho V^2 LB \left(\sqrt{\frac{\nu}{VL}} - \sqrt{\frac{\nu}{2VL}}\right)$$

$$\frac{F_{D1}}{F_{D2}} = \frac{\sqrt{\dfrac{\nu}{2VL}}}{\left(\sqrt{\dfrac{\nu}{VL}} - \sqrt{\dfrac{\nu}{2VL}}\right)} = \frac{\sqrt{\dfrac{\nu}{2VL}}}{\sqrt{\dfrac{\nu}{2VL}}\,[\sqrt{2} - 1]} = \frac{1}{[\sqrt{2} - 1]} = 2.414$$

13.6 TURBULENT BOUNDARY LAYER

The velocity profile for a turbulent boundary layer is given by

$$\frac{v}{V} = \left(\frac{y}{\delta}\right)^{\frac{1}{n}} \qquad \text{...(13.16)}$$

where $n = 7$ for Re $> 5 \times 10^5$ and $< 1 \times 10^7$

This is known as one seventh power law. However, this equation cannot be applied at the boundary itself as in that case

$$v = V\left(\frac{y}{\delta}\right)^{\frac{1}{7}}$$

$$\frac{dv}{dy} = \frac{V}{\delta^{1/7}}\frac{1}{7} \cdot y^{-6/7} = \frac{1}{7}\frac{V}{\delta^{1/7}} \cdot \frac{1}{y^{6/7}}$$

So when $y = 0$, dv/dx or the velocity gradient is infinity. This difficulty is obviated by assuming the velocity in the viscous laminar sublayer to be linear and tangent to the seventh root profile at the point where the laminar sublayer and the turbulent part of the boundary layer merge.

Blasius gave the following relation for the viscous shear stress :

$$\tau_0 = 0.0225 \rho V^2 \left(\frac{\mu}{\rho V \delta}\right)^{\frac{1}{4}} \qquad \text{...(13.17)}$$

Assuming the velocity profile according to one seventh law in the turbulent boundary layer, the analysis is made as under :

Boundary Layer Thickness

$$\frac{\tau_0}{\rho V^2} = \frac{d\theta}{dx} = \frac{d}{dx}\left[\int_0^\delta \frac{v}{V}\left(1 - \frac{v}{V}\right) \cdot dy\right]$$

$$= \frac{d}{dx}\left[\int_0^\delta \left(\frac{y}{\delta}\right)^{1/7}\left\{1 - \left(\frac{y}{\delta}\right)^{1/7}\right\} \cdot dy\right]$$

$$= \frac{d}{dx}\left[\int_0^\delta \left(\frac{y^{1/7}}{\delta^{1/7}} - \frac{y^{2/7}}{\delta^{2/7}}\right) dy\right]$$

$$= \frac{d}{dx}\left(\frac{7}{8}\frac{y^{8/7}}{\delta^{1/7}} - \frac{7y^{9/7}}{9\delta^{2/7}}\right)_0^\delta$$

$$= \frac{d}{dx}\left(\frac{7\delta}{8} + \frac{7\delta}{9}\right)$$

$$= \frac{7}{72}\frac{d\,\delta}{dx}$$

$$\tau_0 = \frac{7}{72}\rho V^2 \frac{d\delta}{dx} \qquad \text{...(13.18)}$$

Also

$$\tau_0 = 0{\cdot}0225\rho V^2 \left(\frac{\mu}{\rho V\delta}\right)^{1/4} \qquad \text{...(Eq. 13.17)}$$

Equating Eqs. (13.17) and (13.18),

$$\frac{7}{72}\rho V^2 \frac{d\delta}{dx} = 0{\cdot}0225\,\rho V^2 \left(\frac{\mu}{\rho V\delta}\right)^{1/4}$$

$$\frac{7}{72}\frac{d\delta}{dx} = 0{\cdot}0225\left(\frac{\mu}{\rho V\delta}\right)^{1/4}$$

$$\delta^{1/4}{\cdot}d\delta = 0{\cdot}231\left(\frac{\mu}{\rho V}\right)^{1/4}{\cdot}dx$$

$$\frac{4}{5}\delta^{5/4} = 0{\cdot}231\left(\frac{\mu}{\rho V}\right)^{1/4}{\cdot}x + C$$

where C is a constant of integration.

At

$$x = 0\,,\ \delta = 0$$

$$0 = 0 + C$$

$$C = 0$$

$$\frac{4}{5}\delta^{5/4} = 0{\cdot}231\left(\frac{\mu}{\rho V}\right)^{1/4}{\cdot}x$$

$$\delta = \left[\frac{5}{4}\times 0{\cdot}231\left(\frac{\mu}{\rho V}\right)^{1/4}{\cdot}x\right]^{\frac{4}{5}}$$

$$= 0{\cdot}37\left(\frac{\mu}{\rho V}\right)^{1/5}{\cdot}x^{4/5}$$

$$= 0{\cdot}37 \times \frac{1}{(\rho Vx/\mu)^5}{\cdot}x^{1/5}{\cdot}x^{4/5}$$

$$= \frac{0{\cdot}37x}{(R_{ex})^{1/5}} \qquad \text{...(13.19)}$$

Shear Stress

$$\tau_0 = 0{\cdot}0225\rho V^2 \left(\frac{\mu}{\rho V\delta}\right)^{\frac{1}{4}} \qquad \text{(Blausius Equation)}$$

$$= 0{\cdot}0225\rho V^2 \left(\frac{\mu}{\rho V \dfrac{0{\cdot}37\,x}{(R_{ex})^{1/5}}}\right)^{\frac{1}{4}}$$

$$= \rho V^2 \times 0{\cdot}0288 \left[\frac{(R_{ex})^{1/5}}{\rho Vx/\mu}\right]^{\frac{1}{4}}$$

$$= \frac{\rho V^2}{2} \times 0{\cdot}0576 \left[\frac{(R_{ex})^{1/5}}{R_{ex}}\right]^{\frac{1}{4}}$$

$$= \frac{\rho V^2}{2} \times 0{\cdot}0576 \left[\frac{1}{(R_{ex})^{4/5}}\right]^{\frac{1}{2}}$$

$$= \frac{\rho V^2}{2} \times \frac{0{\cdot}0576}{(R_{ex})^{1/5}} \qquad \text{...(13.20)}$$

Local Coefficient of Drag

$$C_D^* = \frac{\tau_0}{\frac{1}{2}\rho V^2} = \frac{\dfrac{\rho V^2}{2} \times \dfrac{0{\cdot}576}{(R_{ex})^{1/5}}}{\frac{1}{2}\rho V^2} = \frac{0{\cdot}0576}{(R_{ex})^{1/5}} \qquad \text{...(13.21)}$$

Coefficient of Drag

$$C_D = \frac{\int_0^L \tau_0 {\cdot} b {\cdot} dx}{\dfrac{\rho A V^2}{2}}$$

$$= \frac{\int_0^L \dfrac{\rho V^2}{2} \dfrac{0{\cdot}576}{R_{ex}^{1/5}} {\cdot} b {\cdot} dx}{\dfrac{\rho L {\cdot} b {\cdot} V^2}{2}}$$

$$= \frac{\dfrac{\rho V^2}{2} \int_0^L \dfrac{0{\cdot}576}{R_{ex}^{1/5}} {\cdot} b {\cdot} dx}{\dfrac{\rho L {\cdot} b {\cdot} V^2}{2}}$$

$$= \frac{0{\cdot}576}{L}\int_0^L \frac{1}{(\rho Vx/\mu)^{1/5}}\,dx$$

$$= \frac{0{\cdot}576 \int_0^L x^{-1/5}{\cdot}dx}{L^{4/5}{\cdot}\left(\dfrac{\rho VL}{\mu}\right)^{1/5}}$$

$$= \frac{0{\cdot}576 \times \dfrac{5}{4}\left(\dfrac{x^{4/5}}{1/2}\right)_0^L}{L^{4/5}{\cdot}R_{eL}^{1/5}}$$

$$= \frac{0{\cdot}072\, L^{4/5}}{L^{4/5}{\cdot}R_{eL}^{1/5}}$$

$$= \frac{0{\cdot}072}{R_{eL}^{1/5}} \qquad \ldots(13.22)$$

These relations are valid for Re = 5×10^5 to 1×10^7

For Reynolds number between 10^7 and 10^9, Prandtl and Schlichting suggested the following relation :

$$C_D = \frac{0{\cdot}455}{(\log_{10} R_{eL})^{2{\cdot}58}} \qquad \ldots(13.23)$$

13.7 TOTAL DRAG ON A FLAT PLATE

The above relations derived for the turbulent boundary layer are on the assumption that the boundary layer starts from the leading edge of the plate whereas in fact turbulent boundary layer is preceded by a laminar boundary layer upto a length x_1 (Fig. 13.4). Thus the total drag on a flat plate will be given by (assuming Re > 1×10^7)

F_D = Drag for laminar boundary layer for length l

+ (Drag for turbulent boundary layer for length L

– Drag for turbulent boundary layer for length l)

$$= C_{Dl}\text{ (Blausius sol)} \times \frac{\rho V^2}{2}{\cdot}b{\cdot}l + C_{DL} \times \frac{\rho V^2}{2}{\cdot}b{\cdot}L - C_{Dt} \times \frac{\rho V^2}{2}{\cdot}b{\cdot}l$$

$$= \frac{1{\cdot}328}{\sqrt{R_{ex}}} \times \frac{\rho V^2}{2}{\cdot}b{\cdot}l + \frac{0{\cdot}455L}{(\log_{10} R_{eL})^{2{\cdot}58}} \times \frac{\rho V^2}{2}{\cdot}b{\cdot}L - \frac{0{\cdot}072x_l}{(R_{ex})^{1/5}} \times \frac{\rho V^2}{2}{\cdot}b{\cdot}l$$

$$= b{\cdot}\frac{\rho V^2}{2}\left[\frac{1{\cdot}328l}{\sqrt{R_{el}}} + \frac{0{\cdot}455L}{(\log_{10} R_{eL})^{2{\cdot}58}} - \frac{0{\cdot}072l}{(R_{el})^{1/5}}\right]$$

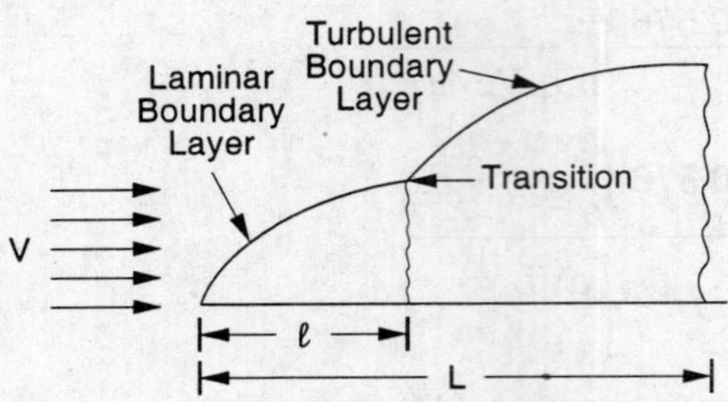

Fig. 13.4

Since $\frac{R_{el}}{R_{eL}} = \frac{\rho Vl/\mu}{\rho VL/\mu} = \frac{l}{L}$ *or* $l = \frac{R_{el}}{R_{eL}} \cdot L$

$$F_D = \frac{bL\rho V^2}{2}\left[\frac{1.328\sqrt{R_{el}}}{R_{eL}} + \frac{0.455}{(\log_{10} R_{eL})^{2.58}} - \frac{0.072\, R_{el}^{4/5}}{R_{eL}}\right]$$

Assuming that transition from the laminar boundary layer occurs at Re = 5×10^5

$$F_D = \frac{bL\rho V^2}{2}\left[\frac{0.455}{(\log_{10} R_{eL})^{2.58}} - \frac{1700}{R_{eL}}\right]$$

Also, $$C_D = \frac{F_D}{\frac{1}{2}\rho V^2 \cdot bL} = \frac{0.455}{(\log_{10} R_{eL})^{2.58}} - \frac{1700}{R_{eL}} \qquad \text{...(13.24)}$$

Example 13.8 Assuming the boundary layer to be turbulent over the entire length of a flat plate kept at zero incidence in a stream of uniform velocity, determine the ratio of friction drag on the front half and the rear half of the plate.

Soltuion Friction drag on the entire plate

$$F_D = C_D \times \frac{1}{2}\rho V^2 \times A = \frac{0.072}{\left(\frac{V \cdot L}{\nu}\right)^{1/5}} \times \frac{1}{2}\rho V^2 \times L \cdot B = 0.036 \times \rho V^2 BL \cdot \left(\frac{\nu}{VL}\right)^{1/5}$$

Friction drag on the first half

$$F_{D1} = C_D \times \frac{1}{2}\rho V^2 \times A = \frac{0.072}{\left(\frac{V \cdot L}{2\nu}\right)^{1/5}} \times \frac{1}{2}\rho V^2 \times \frac{L}{2} \cdot B = 0.036 \times \rho V^2 BL \times \frac{1}{2}\left(\frac{2\nu}{VL}\right)^{1/5}$$

Friction drag on the rear half

$$F_{D2} = 0.036 \times \rho V^2\, BL\left[\left(\frac{\nu}{VL}\right)^{1/5} - \frac{1}{2}\left(\frac{2\nu}{VL}\right)^{1/5}\right]$$

$$\frac{F_{D1}}{F_{D2}} = \frac{\frac{1}{2}\left(\frac{2\nu}{VL}\right)^{1/5}}{\left(\frac{\nu}{VL}\right)^{1/5} - \frac{1}{2}\left(\frac{2\nu}{VL}\right)^{1/5}} = \frac{\frac{1}{2}\left(\frac{2\nu}{VL}\right)^{1/5}}{\frac{1}{2}\left(\frac{2\nu}{VL}\right)^{1/5}\left[\frac{2}{(2)^{1/5}} - 1\right]} = \frac{1}{[1{\cdot}7413 - 1]} = 1{\cdot}349$$

Example 13.9 Water flows over a thin smooth plate of length 4·8 m and width 2·4 m with a velocity of 2 m/s. The boundary layer flow changes from laminar to turbulent at a Reynolds number 5×10^5, determine

(i) the distance from the leading edge upto which the boundary layer is laminar

(ii) the thickness of the boundary layer at the transition point

(iii) the drag on one side of the plate.

The viscosity of water is 0·001 N·s/m².

Solution (i) Let x be the distance from the leading edge upto which the boundary layer is laminar. Then

$$5 \times 10^5 = \frac{\rho V l}{\mu} = \frac{1000 \times 2 \times l}{0{\cdot}001}$$

$$l = 0{\cdot}25 \text{ m}$$

(ii) Thickness of boundary layer at this point is given by Blasius' solution.

$$\delta = \frac{4{\cdot}91x}{\sqrt{R_{ex}}} = \frac{5 \times 0{\cdot}25}{\sqrt{5 \times 10^5}} = 0{\cdot}001768 \text{ m or } 1{\cdot}768 \text{ mm}$$

(iii) Drag force due to laminar boundary layer is given by Blasius' solution,

$$F_D = C_D \times \frac{1}{2}\rho V^2 \times A$$

where
$$C_D = \frac{1{\cdot}328}{\sqrt{R_{el}}} = \frac{1{\cdot}328}{\sqrt{5 \times 10^5}} = 0{\cdot}001878$$

$$F_D = 0{\cdot}001878 \times \frac{1}{2} \times 1000 \times 4^2 \times (0{\cdot}25 \times 2{\cdot}4) = 2{\cdot}2536 \text{ N}$$

Drag force due to turbulent boundary layer

= Drag force due to turbulent boundary layer over entire length

– Drag force due to turbulent boundary layer upto 0·25 m

$$R_{el} = \frac{\rho V l}{\mu} = \frac{1000 \times 2 \times 4{\cdot}8}{0{\cdot}001} = 9{\cdot}6 \times 10^6$$

$$F_D = \frac{1}{2}\rho V^2 (C_{DL} \times B{\cdot}L - C_{Dl} \times B{\cdot}l)$$

$$= \frac{1}{2}\rho V^2 \left(\frac{0{\cdot}072}{R_{eL}^{1/5}} \times B{\cdot}L - \frac{0{\cdot}072}{R_{el}^{1/5}} \times B{\cdot}l\right)$$

$$F_D = \frac{1}{2} \times 1000 \times 2^2 \times \left(\frac{0{\cdot}072}{(96 \times 10^5)^{1/5}} \times 2{\cdot}4 \times 4{\cdot}8 - \frac{0{\cdot}072}{(5 \times 10^5)^{1/5}} \times 2{\cdot}4 \times 0{\cdot}25\right)$$

$= 2000\,(0{\cdot}03329 - 0{\cdot}00313)$

$= 60{\cdot}326$ N

Total drag force = Drag force due to laminar boundary layer + Drag force due to turbulent boundary layer

$= 2{\cdot}2536 + 60{\cdot}326$

$= 62{\cdot}5796$ N

Example 13.10 Determine the approximate friction drag of a streamlined train 250 m long running at 108 km/h. The cross-section of the train is of 10 m perimeter above the wheels. Assume the kinematic viscosity of the air 1.5×10^{-5} m^2/s and density $1{\cdot}24$ kg/m^3.

Solution

$$V = \frac{108 \times 1000}{3600} = 30 \text{ m/s}$$

$$R_{eL} = \frac{VL}{\nu} = \frac{30 \times 250}{15 \times 10^{-5}} = 5 \times 10^{8}$$

$$F_D = \frac{PL\rho V^2}{2}\left[\frac{0{\cdot}455}{(\log_{10} R_{eL})^{2{\cdot}58}} - \frac{1700}{R_{eL}}\right]$$

$$= \frac{10 \times 250 \times 1{\cdot}24 \times 30^2}{2}\left[\frac{0{\cdot}455}{(\log_{10}(5 \times 10^8))^{2{\cdot}58}} - \frac{1700}{5 \times 10^8}\right]$$

$$= \frac{10 \times 250 \times 1{\cdot}24 \times 30^2}{2}(0{\cdot}0017147 - 0{\cdot}0000034)$$

$$= \frac{10 \times 250 \times 1{\cdot}24 \times 30^2}{2} \times 0{\cdot}001711$$

$$= 2387 \text{ N}$$

13.8 BOUNDARY LAYER SEPARATION

Consider the flow past a converging-diverging boundary surface as shown in Fig. 13.5. In the region of converging boundary AB, the flow is accelerated *ie* the free stream velocity V increases and becomes maximum at B. As a result, the pressure decreases in the direction of flow or the pressure gradient dp/dx is negative. The negative pressure gradient is said to be favorable as the fluid is pushed high pressure region to the low pressure region which is according to the natural law. The boundary layer remain thin and is stabilised. Losses are low and the efficiency is high. Flow around a nozzle and around the blades of a turbine conforms to this condition.

In the diverging portion BC, the flow is decelerated and pressure increases in the direction of flow or the pressure gradient dp/dx is positive. This is considered an unfavorable pressure gradient. The net pressure force is in the backward direction and magnifies the effect of the shear stress near the boundary surface. The continuous reduction in the flow momentum causes the boundary layer to become thicker rapidly. If these effects persist over a large distance, the fluid becomes unable to move forward in the region of increased pressure and at C the velocity as well as the velocity gradient dV/dy become zero. If the deceleration continues,

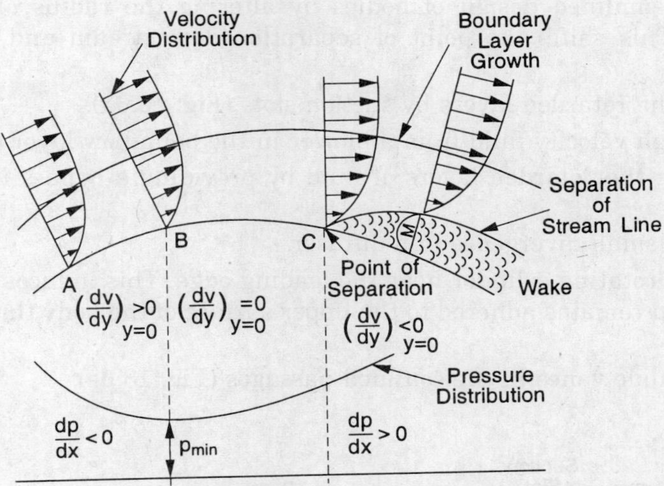

Fig. 13.5

the fluid separates at C and tends to flow in a direction opposite to the main flow in the separated region known as wake. Eddies and turbulence are developed in the wake resulting in the loss of energy and decrease in efficiency. The separation line is obtained by joining all the zero velocity points by a smooth curve. Such a flow is observed around the blades of compressor and through a diffuser.

The point of separation depends upon

- Shape of the boundary surface
- Reynolds number of the flow
- Roughness of the boundary surface

For a given velocity profile, the following conditions determine whether the boundary layer separates or not :

– if $\left(\dfrac{\partial v}{\partial y}\right)_{y=0}$ is positive, there is no separation and the flow is attached

– if $\left(\dfrac{\partial v}{\partial y}\right)_{y=0}$ is zero, the flow is on the verge of separation

– if $\left(\dfrac{\partial v}{\partial y}\right)_{y=0}$ is negative, the flow is separated.

As the separation of flow is associated with loss of energy and decrease in efficiency, efforts are made to delay or prevent its occurance by using any of the following methods :

1. Artificial roughening of the boundary surface by sand coating, strips or bars to produce an early onset of turbulence. Turbulent boundary layers resist the separation in a more effective way.

2. Making streamlined design of bodies by altering the radius of curvature of the boundary. This shifts the point of separation downstream and reduces the wake region.
3. Suction of the retarded layers by suction slots (Fig. 13.6a).
4. Injecting high velocity fluid from a blower in the boundary layer (Fig. 6b).
5. Accelerating the retarded layers of fluid by providing a bypass in the slotted wing (Fig. 13.6c).
6. Providing a small divergence in a diffuser.
7. Providing a rotating cylinder near the leading edge. This induces the Magnus effect and the fluid remains adhered to the upper surface of the body throughout its length (Fig. 13.6d).
8. Providing guide vanes in the confined passages (Fig. 13.6e).

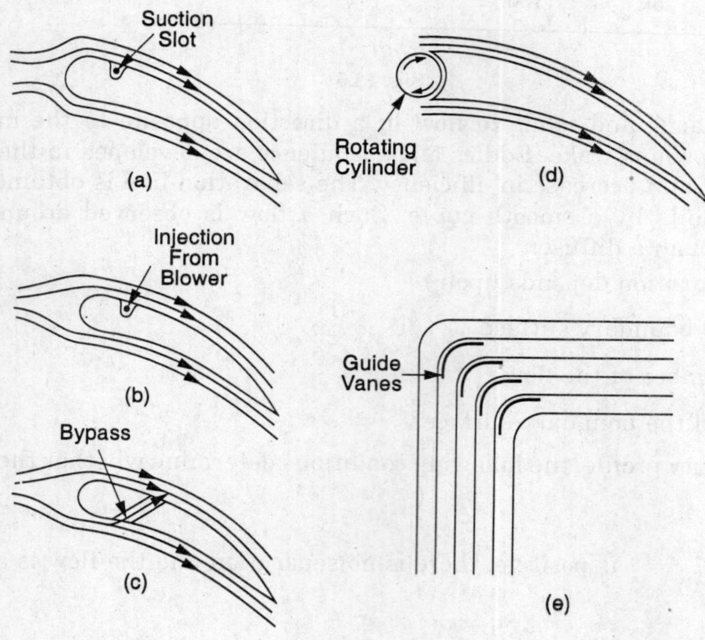

Fig. 13.6

Example 13.11 For the following velocity profiles, find whether the flow is attached or detached or on the verge of separation :

(i) $\dfrac{v}{V} = \dfrac{3}{2}\left(\dfrac{y}{\delta}\right) - \dfrac{1}{2}\left(\dfrac{y}{\delta}\right)^3$

(ii) $\frac{v}{V} = 2\left(\frac{y}{\delta}\right)^2 + \left(\frac{y}{\delta}\right)^3 - 2\left(\frac{y}{\delta}\right)^4$

(iii) $\frac{v}{V} = -2\left(\frac{y}{\delta}\right) + \left(\frac{y}{\delta}\right)^2$

Solution (i) $\frac{v}{V} = \frac{3}{2}\left(\frac{y}{\delta}\right) - \frac{1}{2}\left(\frac{y}{\delta}\right)^3$

$$v = \frac{3V}{2}\left(\frac{y}{\delta}\right) - \frac{V}{2}\left(\frac{y}{\delta}\right)^3$$

$$\frac{\partial v}{\partial y} = \frac{3V}{2}\frac{1}{\delta} - \frac{3V}{2}\left(\frac{y}{\delta}\right)^2\frac{1}{\delta}$$

At $y = 0$, $\left(\frac{\partial v}{\partial y}\right)_{y=0} = \frac{3V}{2\delta}$

As ∂v/∂y is +ve, the flow is attached.

(ii) $\frac{v}{V} = 2\left(\frac{y}{\delta}\right)^2 + \left(\frac{y}{\delta}\right)^3 - 2\left(\frac{y}{\delta}\right)^4$

$$v = 2V\left(\frac{y}{\delta}\right)^2 + V\left(\frac{y}{\delta}\right)^3 - 2V\left(\frac{y}{\delta}\right)^4$$

$$\frac{\partial v}{\partial y} = 4V\left(\frac{y}{\delta}\right)\frac{1}{\delta} + 3V\left(\frac{y}{\delta}\right)^2\frac{1}{\delta} - 8V\left(\frac{y}{\delta}\right)^3\frac{1}{\delta}$$

At $y = 0$, $\left(\frac{\partial v}{\partial y}\right)_{y=0} = 0$

As $\partial v/\partial y$ is zero, the flow is on the verge of separation.

(iii) $\frac{v}{V} = -2\left(\frac{y}{\delta}\right) + \left(\frac{y}{\delta}\right)^2$

$$v = -2V\left(\frac{y}{\delta}\right) + \left(\frac{y}{\delta}\right)^2$$

$$\frac{\partial v}{\partial y} = -2V\left(\frac{1}{\delta}\right) + 2V\left(\frac{y}{\delta}\right)\frac{1}{\delta}$$

At $y = 0$, $\left(\frac{\partial v}{\partial y}\right)_{y=0} = \frac{-2V}{\delta}$

As $\partial v/\partial y$ is –ve, the flow has separated.

PROBLEMS

1. Explain the term boundary layer.
2. What are characteristics of laminar and turbulent boundary layer.
3. Define and explain the following terms :
   (a) displacement thickness
   (b) momentum thickness
   (c) energy thickness.
4. What is Von-Karman momentum equation ? How is it helpful in determining the thickness of boundary layer, shear stress and the drag along a flat plate ?
5. Derive an expression for the boundary shear stress in terms of momentum thickness.
6. How do you find the drag on a flat plate due to laminar and turbulent boundary layers ?
7. What is meant by separation of boundary layer ? In what way does the pressure gradient affect it ?
8. How do you determine whether a boundary layer is attached, detached or on the verge of separation ?
9. Enumerate the methods to prevent the separation of boundary layer.
10. A 5 m long and 1·8 m wide plate is moved through stationary air at a velocity of 1·75 m/s parallel to its length. The density of air is 1·22 kg/m^3 and the viscosity $1{\cdot}8 \times 10^{-5}$ m^2/s. Calculate the drag force on one side of the plate assuming the flow to be (i) laminar (ii) turbulent. (0·029 N, 0·0871N)
11. A submarine is assumed to have a cylindrical shape with rounded nose. It is 55 m long and 6 m in diameter. Determine the power required to overcome boundary friction if it cruises at a velocity of 8 m/s in sea water at 20°C· $\rho = 1030$ kg/m^3, $\nu = 1 \times 10^{-6}$ m^2/s. (476·54 kW)
12. A smooth rectangular plate 1 m wide and 20 m long is towed through water at 20°C lengthwise. It experiences a drag of 1·44 kN on both the sides. Determine
   (a) average drag coefficient
   (b) velocity of the plate
   (c) boundary layer thickness at the edge of the plate.
   (0·00208, 5·889m/s, 198·6 mm)
13. In laminar boundary layer over a flat plate, the velocity distribution is given by a second order polynomial $v = a + by + cy^2$. Determine its form using the necessary boundary conditions.
   $$\left[\frac{v}{V} = 2\left(\frac{y}{\delta}\right) - \left(\frac{y}{\delta}\right)^2\right]$$
14. Air flows over a flat plate 5 m long and 2·5 m wide with a velocity of 4 m/s at 15°C. Assuming that laminar boundary layer exists upto Re = 5×10^5 , determine
   (a) length of plate over which the boundary layer is laminar
   (b) thickness of the laminar boundary layer

(c) shear stress at the section where boundary layer ceases to be laminar

(d) total drag on that portion of the plate where boundary layer is laminar.

(0·837m , 12·99mm , 0·0091N/m^2 , 0·0835N)

15. Velocity profile for air flowing at a velocity of 10 m/s over a smooth plate is given by

$$\frac{v}{V} = 2\left(\frac{y}{\delta}\right) - \left(\frac{y}{\delta}\right)^2$$

The plate is 1.2 m long and 800 mm wide. Determine the maximum distance from the leading edge upto which the boundary layer is laminar and the maximum thickness. Assume that the laminar boundary layer exists upto a value of Re = 2×10^5. (300 mm, 3·67 mm)

16. Free stream velocity of oil over a thin plate 2 m × 2 m size is 2 m/s. Find the boundary layer thickness and the shear stress at the trailing end point. Also determine the total surface resistance of the plate. The kinematic viscosity of the oil is 1×10^{-5} m^2/s and the specific gravity 0·86. (15·5 mm, 1·805 N/m^2, 28·88 N)

17. A plate 450 mm long and 150 mm wide is placed longitudinally in a stream of crude oil which flows with a velocity of 6 m/s. Determine the friction drag on the plate, thickness of the boundary layer and the shear stress at the trailing edge of the plate. The specific gravity of the oil is 0·925 and the kinematic viscosity 0·9 stoke.

(8·62 N, 13 mm, 63·8N/m^2)

18. Assuming the boundary layer to be turbulent and Reynolds number to be of the order 10^6, determine the ratio of skin friction drag on the front two thirds and rear one third of a smooth flat plate placed in a uniform stream at zero incidence. (2·61)

19. The velocity of wind blowing over a 6 m flat plate is 20 km per hour. Determine the force per metre width of the plate and the thickness of the boundary layer at the trailing edge. The density and the kinematic viscosity of the air are 1·2 kg/m^3 and $1{\cdot}5 \times 10^{-5}$ m^2/s respectively. (0·3578 N, 121·7 mm)

Chapter 14

IMMERSED BODIES

14.1 INTRODUCTION

The bodies which are completely submerged in a fluid and have motion relative to the fluid are known as **immersed** or **submerged bodies.** It is observed that such bodies are subjected to a force exerted by the fluid if either the bodies move in stationary fluids or fluids move over stationary bodies. Such situations are encountered in the designs of vehicles such as aircrafts, submarines and bridges or buildings which are submerged either in air or water.

14.2 LIFT AND DRAG

In general, if a body is placed in the passage of flowing stream of a viscous fluid (or when a body moves through a viscous fluid at rest), two types of surface forces act on the body; one due to pressure and the other due to viscous shear. Over any infinitely small area of the body surface, pressure force is always normal to that surface area, whereas the viscous shear is parallel or tangent to the surface area. Sum of the components of the pressure and viscous forces over the entire body surface, in the direction of motion of the fluid, is called the **drag (or profile drag)** and the sum of the forces in a direction normal to the fluid flow is called the lift.

It is to be noted that the lift force is produced only if the axis of the body is inclined to the direction of flow or the body is asymmetrical. In case, a flow approaches a body along the axis of symmetry, the force acting on the body is only the drag force and no lift force.

Consider a body held stationary in a stream of real fluid moving at a uniform velocity V (Fig. 14.1). Let p and τ_0 be the static pressure and shear stress respectively on an elementary area dA on the surface of the body. Let θ be the inclination of the perpendicular to the surface with the direction of flow. Then the forces acting on the area dA are

Pressure force, perpendicular to the surface $= p \cdot dA$

Shear force, acting along tangential direction $= \tau_0 \cdot dA$

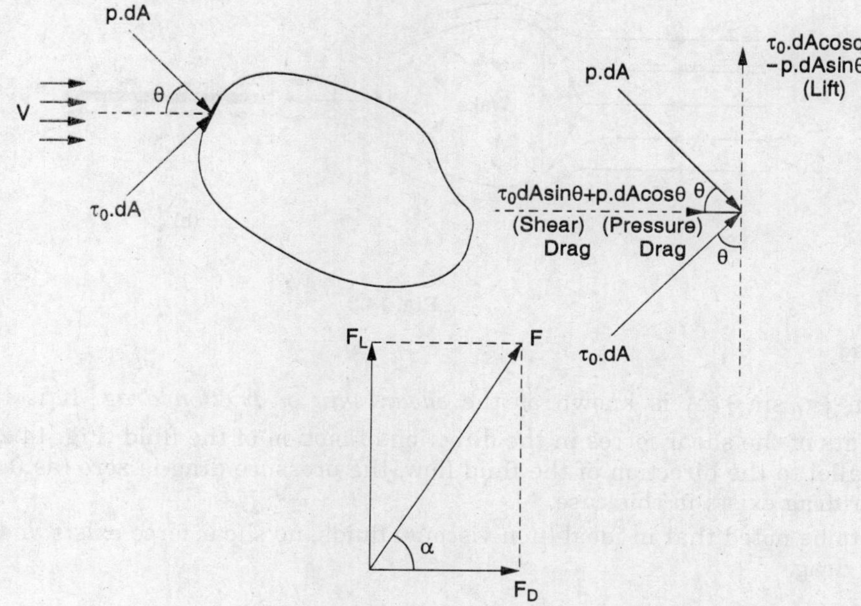

Fig. 14.1

Thus

$$\text{Drag force} = \tau_0 dA \sin\theta + p\, dA \cos\theta$$

$$\text{Lift force} = \tau_0 dA \cos\theta + p\, dA \sin\theta$$

The total drag and lift forces are found by integrating over the entire body surface

$$F_D = \int \tau_0 \sin\theta \,.\, dA + \int p \cos\theta \,.\, dA \qquad \text{...(14.1)}$$

$$F_L = \int \tau_0 \cos\theta \,.\, dA - \int p \sin\theta \,.\, dA \qquad \text{...(14.2)}$$

Total force acting on the body = $\sqrt{F_L{}^2 + F_D{}^2}$

and the angle made by the resultant is given by,

$$\tan\alpha = \frac{F_L}{F_D}$$

Pressure Drag

The term $\int p \cos\theta\, dA$ is known as the **pressure drag** *i.e.* It is the sum of the components of the pressure force in the direction of motion of the fluid. If a flat plate is held normal to a fluid stream (Fig. 14.2a), shear force acts along the surface or normal to the incoming stream and thus has no effect on the drag force directly. The total drag is only due to the pressure drag in this case and depends on the shape of the surface and Reynolds number of the fluid flow.

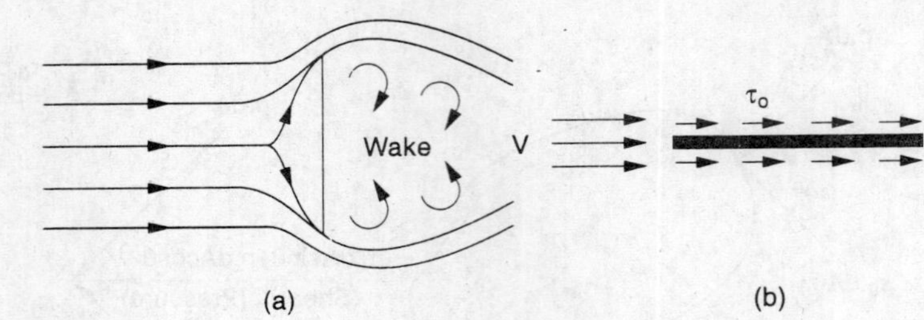

Fig. 14.2

Shear Drag

The term $\int \tau_0 \sin \theta \, dA$ is known as the *shear, skin or friction drag*. It is the sum of the components of the shear forces in the direction of motion of the fluid (Fig. 14.2b). If a plate is held parallel to the direction of the fluid flow, the pressure drag is zero (as $\theta = 90°$) and only the shear drag exists in this case.

It is to be noted that in ideal (non-viscous) fluids, no shear force exists and hence there is no shear drag.

14.3 LIFT AND DRAG COEFFICIENTS

Usually, it is not possible to determine the magnitude of the lift or drag forces acting on a body since both the pressure and the shear stress distribution depend on the geometrical configuraton of the body, the fluid properties such as density (ρ), viscosity (μ), modulus of elasticity (E), flow velocity (V). In general, the practice is to find out the forces experimently. To carry out the experiments in an effective way and to reduce the number of experiments, the method of dimensional analysis is adopted.

$$F = f(V, d, \rho, E, \mu, g)$$

By adopting the method of dimensional analysis, the following functional relationship is obtained :

$$\frac{F}{\rho V^2 d^2} = f\left(\frac{\rho V d}{\mu}, \frac{V}{\sqrt{E/\rho}}, \frac{V}{\sqrt{gd}}\right)$$

$$= f(\text{Re, M, Fr}) \qquad \text{...(14.3)}$$

where Re , M and Fr denote the Reynolds number, Mach number and the Froude number respectively.

- At low speeds ($M < 0{\cdot}3$) the change in the density of the fluid is negligible and thus the effect of Mach number is neglected.
- The Froude number has significance only when the bodies are partially submerged. For fully immersed bodies, it has no effect and is neglected.

Thus the above equation for low speeds and fully submerged bodies can be written as

$$\frac{F}{\rho V^2 d^2} = f(\text{Re}) \qquad \text{...(14.4)}$$

or

$$F = \rho V^2 d^2 f(\text{Re})$$

The parameter d^2 represents the area of the body.

Thus $F = C_1 \times \rho V^2 \times \text{Area}$ (C_1 is a force coefficient and a function of Reynolds number)

$$= 2 \times C_1 \times \frac{\rho V^2}{2} \times \text{Area}$$

Let the coefficient $C = 2C_1$

Then $$C = \frac{F}{\frac{\rho V^2}{2} \times \text{Area}} = \frac{F}{\text{Dynamic pressure} \times \text{Area}}$$

This equation is applicable to the lift as well the drag *i.e.*

Lift coefficient, $$C_L = \frac{F_L}{\frac{1}{2}\rho V^2 . A} \quad \text{...(14.5)}$$

Drag coefficient,

$$C_D = \frac{F_D}{\frac{1}{2}\rho V^2 . A} \quad \text{...(14.6)}$$

The area taken is either the largest possible projected area on a plane or projected area perpendicular to the direction of flow. For example, the projected area of a cylinder is dL (diameter d and length L) and of an airfoil, the product of the span and the mean chord.

Example 14.1 A vehicle travelling at a speed of 72 km/h has a projected area of 8 m^2. It has a total resistance of 2400 kN out of which 18 percent is due to the rolling friction and 10 percent due to surface friction and the rest due to the form drag. Determine the coefficient of form drag. The density of air is 1·24 kg/m^3.

Solution $V = 72 \text{ km/h} = \frac{72 \times 1000}{60 \times 60} = 20 \text{ m/s}$

Resistance due to rolling friction $= 0{\cdot}18 \times 2400 = 432$ kN

Resistance due to surface friction $= 0{\cdot}1 \times 2400 = 240$ kN

Resistance due to form drag, $F_D = 2400 - 432 - 240 = 1728$ kN

$$F_D = C_D \times \frac{\rho V^2}{2} \times A$$

$$1728 = C_D \times \frac{1{\cdot}24 \times 20^2}{2} \times 8$$

$$C_D = 0{\cdot}871$$

Example 14.2 Experiments were conducted on a flat plate of 2·5 m in length and 0·8 m in width in a wind tunnel with a wind speed of 54 km/h. The coefficient of drag and lift were found to be 0·2 and 0·8 respectively for the plate when kept at a particular angle. Taking the density of air as 12 kg/m^3, determine

(i) the drag, the lift and the resultant force

(ii) power exerted by the air stream on the plate.

Solution $V = \dfrac{54 \times 1000}{60 \times 60} = 15 \text{ m/s}$

$$A = 2{\cdot}5 \times 0{\cdot}8 = 2 \text{ m}^2$$

$$F_D = C_D \times \frac{\rho V^2}{2} \times A = 0{\cdot}2 \times \frac{1{\cdot}2 \times 15^2}{2} \times 2 = 54 \text{ N}$$

$$F_L = C_L \times \frac{\rho V^2}{2} \times A = 0{\cdot}8 \times \frac{1{\cdot}2 \times 15^2}{2} \times 2 = 216 \text{ N}$$

$$F_R = \sqrt{F_D{}^2 + F_L{}^2} = \sqrt{54^2 + 216^2} = 222{\cdot}65 \text{ N}$$

$$\tan \theta = \frac{F_L}{F_D} = \frac{216}{54} = 4$$

$$\theta = 75{\cdot}96^\circ$$

$$P = F_D \times V = 54 \times 15 = 810 \text{ W}$$

Example 14.3 A kite flying in a wind of 54 km/h makes an angle of 6° to the horizontal. The kite weighs 10 N and has dimensions 0·7 m × 0·7 m. The pull on the string attached to the kite and inclined to the horizontal at 45° is 40 N. If the density of air is 1·24 kg/m^3, determine the coefficients of drag and lift.

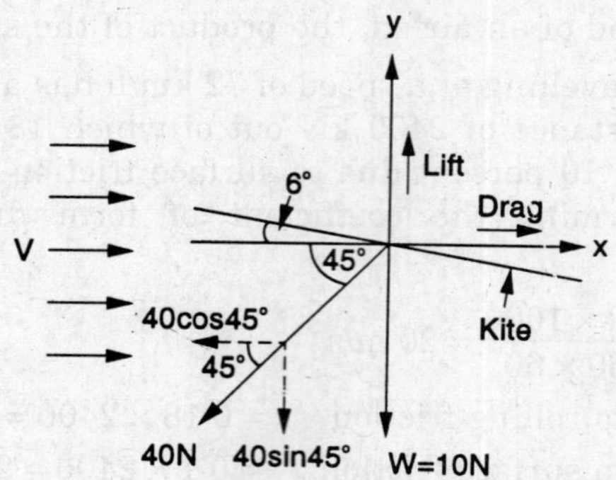

Fig. 14.3

Solution $V = \dfrac{54 \times 1000}{60 \times 60} = 15 \text{ m/s}$

$$A = 0{\cdot}7 \times 0{\cdot}7 = 0{\cdot}49 \text{ m}^2$$

$$F_D = \text{Component of pull along } X\text{-direction} = F \,.\, \cos \theta$$

$$F \,.\, \cos \theta = C_D \times \frac{\rho V^2}{2} \times A$$

$$40 \times \cos 45^\circ = C_D \times \frac{1{\cdot}24 \times 15^2}{2} \times 0{\cdot}49$$

$$C_D = 0{\cdot}414$$

F_L = Component of pull along Y-direction + Weight of the kite

$$40 \times \sin 45^\circ + 10 = C_L \times \frac{\rho V^2}{2} \times A$$

$$28{\cdot}28 + 10 = C_L \times \frac{1{\cdot}24 \times 15^2}{2} \times 0{\cdot}49$$

$$C_L = 0{\cdot}56$$

(The angle of kite with the horizontal is superfluous data).

Example 14.4 A kite weighing 5 N has an area of 0·84 m^2. It is maintained in the air at an angle of 7° to the horizontal by a string attached to the kite which makes an angle of 45° to the horizontal. For this position, the coefficients of drag and lift are found to be 0·5 and 0·75 respectively. Determine the wind speed and the tension in the string. Density of the air is 1·22 kg/m^3.

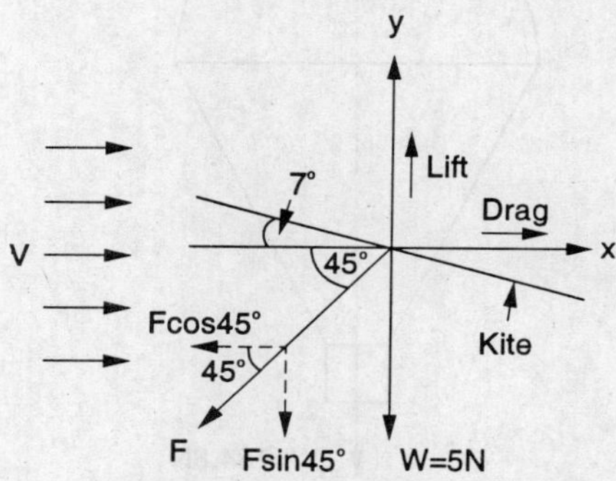

Fig. 14.4

Solution F_D = Component of pull along X-direction = F . cos θ

$$F \,.\, \cos\theta = C_D \times \frac{\rho V^2}{2} \times A$$

$$F \cos 45^\circ = 0{\cdot}5 \times \frac{1{\cdot}22 V^2}{2} \times 0{\cdot}84$$

$$F \cos 45^\circ = 0{\cdot}2562 V^2 \qquad \ldots(i)$$

F_L = Component of pull along Y-direction + Weight of the kite

$$F \sin 45^\circ + 5 = 0{\cdot}75 \times \frac{1{\cdot}22 V^2}{2} \times 0{\cdot}84$$

$$F \cos 45^\circ = 0{\cdot}3843 V^2 - 5 \qquad \ldots(ii)$$

From (i) and (ii),

$$0{\cdot}2562 V^2 = 0{\cdot}3843 V^2 - 5$$

$$V^2 = 39{\cdot}032$$

$$V = 6{\cdot}248 \text{ m/s}$$

$$= \frac{6{\cdot}248 \times 3600}{1000} = 22{\cdot}49 \text{ km/h}$$

$$F \cos 45^\circ = 0{\cdot}2562V^2$$

$$= 0{\cdot}2562 \times 39{\cdot}032$$

$$F = 14{\cdot}14 \text{ N} = (T)$$

Example 14.5 A parachute is assumed in the form of an open hemisphere. The component of its landing speed in the vertical direction is 10 m/s. The total weight carried by the parachute is 80 kg. If the density of air is 1·2 kg/m^3 and the coefficient of drag 0·8, determine the diameter of the parachute.

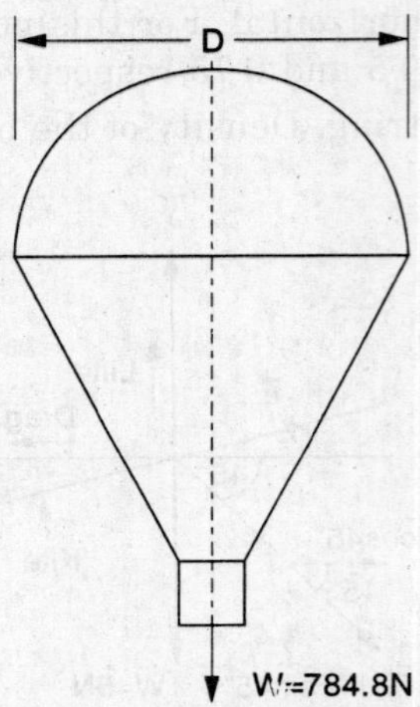

Fig. 14.5

Solution $W = mg = (80 \times 9{\cdot}81) \text{ N} = 784{\cdot}8 \text{ N}$

$$F_D = C_D \times \frac{\rho V^2}{2} \times A$$

$$784{\cdot}8 = 0{\cdot}8 \times \frac{1{\cdot}2 \times 10^2}{2} \times \frac{\pi}{4} \times d^2$$

$$d^2 = 20{\cdot}82$$

$$d = 4{\cdot}563 \text{ m}$$

14.4 STREAMLINED AND BLUFF BODIES

Drag is an energy loss and thus efforts are made to minimise the drag coefficient as far as possible. For this purpose, a body may be provided with a rounded leading edge and along the flow direction a gentle curvature may be given.

A streamlined body is one in which the surface coincides with the streamlines when placed in a flow and the flow separation occurs at the trailing edge only (Fig. 14.6). The wave formation zone at the back of the body is very small, thus reducing the pressure drag. However, in streamlined bodies, the skin friction is higher due to the greater surface area. Still, the net

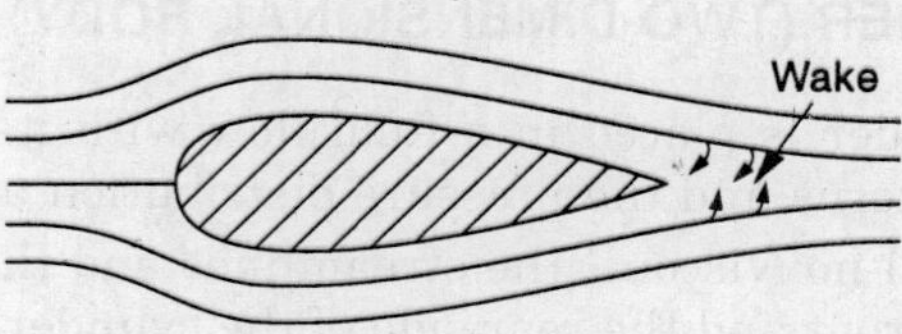

Fig. 14.6

effect is a great reduction of the total drag. Owing to this reason, the wings of an aeroplane or the blades of a marine propeller are made streamlined. It may be noted that the orientation of a body and the velocity of flow also determine the effect of streamlining of the body. A body may behave as a streamlined body in a particular position but may not be so if placed in another position or at low velocities a body may act as a streamlined body whereas at high velocities it may not.

On the other hand, the surface of a bluff body does not coincide with the streamlines. There is extreme boundary layer sepration resulting in a large wake formation and large scale eddies (Fig. 14.2a). This leads to a very large pressure drag as compared to the skin or friction drag.

14.5 FLOW PAST AN INCLINED PLATE

Figure 14.7 shows a flat plate held in the path of a fluid stream flowing with velocity V. The plate is inclined at an angle α to the direction of flow. The fluid streams which strike the underside of the plate are deviated in a direction parallel to the surface by the pressure of the streams. This causes a tangential friction drag on the plate. The impact of the fluid produces a normal pressure on the plate.

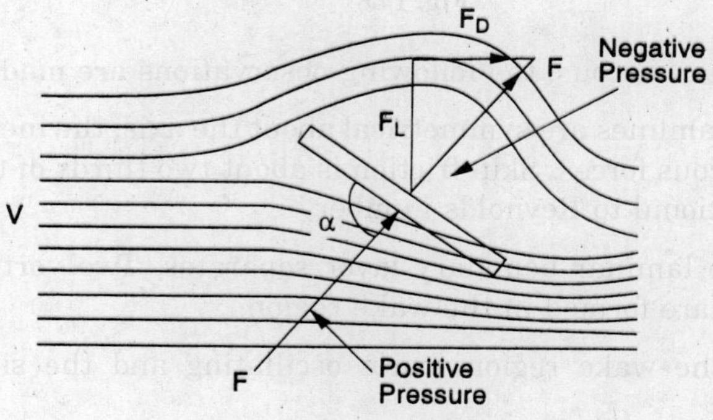

Fig. 14.7

Fluid streams that strike the upper surfaces are deflected by the front edge of plate. But soon they are forced down by the pressure of the fluid streams above them and finally all are merged in the main stream. However, due to the deflection of the stream lines, a wake region with negative pressure is produced on the upper surface of the plate.

The resultant pressure force F is, thus partly due to the positive pressure on the lower surface and partly due to the effect of the negative pressure on the upper surface of the plate. The component of F perpendicular to the direction of flow is the lift and the component along the direction of flow is the drag.

14.6 DRAG ON A CYLINDER (TWO-DIMENSIONAL BODY)

When an infinitely long cylinder is placed in a fluid flow with its axis perpendicular to the direction of flow, the flow patterns and the pressure distribution are as follows :

1. If the flow is an ideal-nonviscous, the streamlines and the pressure distribution are symmetrical on the front and the rear side of the cylinder (Fig. 14.8). The flow which comes in contact with the cylinder first stagnates at S_1, accelerates to a maximum velocity at the top and bottom and decelerates to a second stagnation point S_2. Again the flow from S_2 accelerates upto the approach velocity several diameters away from the cylinder and merges with the main flow. Pressure at the corresponding points on the front and rear of the cylinder are equal and no unbalanced force acts on the cylinder and thus the drag is zero.

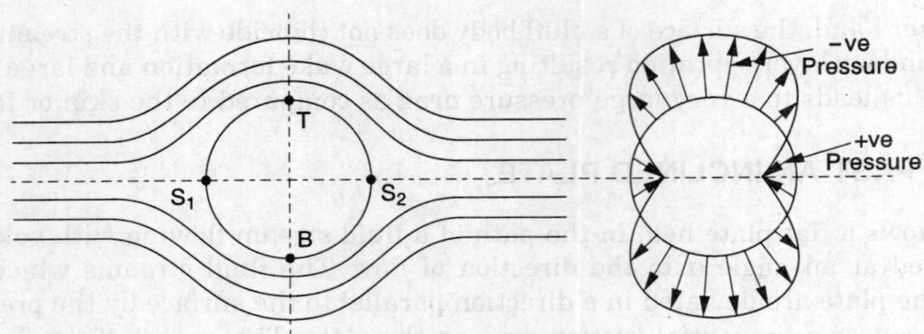

Fig. 14.8

2. If the flow is real or viscous, the following observations are made (Fig. 14.9) :

- For Re < 1, the streamlines are symmetrical about the axis, the inertia forces are small as compared to viscous forces. Skin friction is about two thirds of the total drag which is inversely proportional to Reynolds number.
- For Re = 1– 30, the laminar boundary layer separates. Two vortices rotating in the opposite directions are formed in the wake region.
- For Re = 30 – 60, the wake region starts oscillating and the size of the eddies is enlarged.
- For Re = 60 – 120, the vortices from the two sides of the cylinder detach alternately and moves downstream forming two rows of vortices in the wake which rotate in the opposite directions. The rows of the vortices are known as **Karman vortex street** or **trail** after Von-Karman. He also deduced the following relation for the spacing of the vortices :

$$a/b = 0{\cdot}281$$

where a and b are the vertical and the horizontal pitch of vortices.

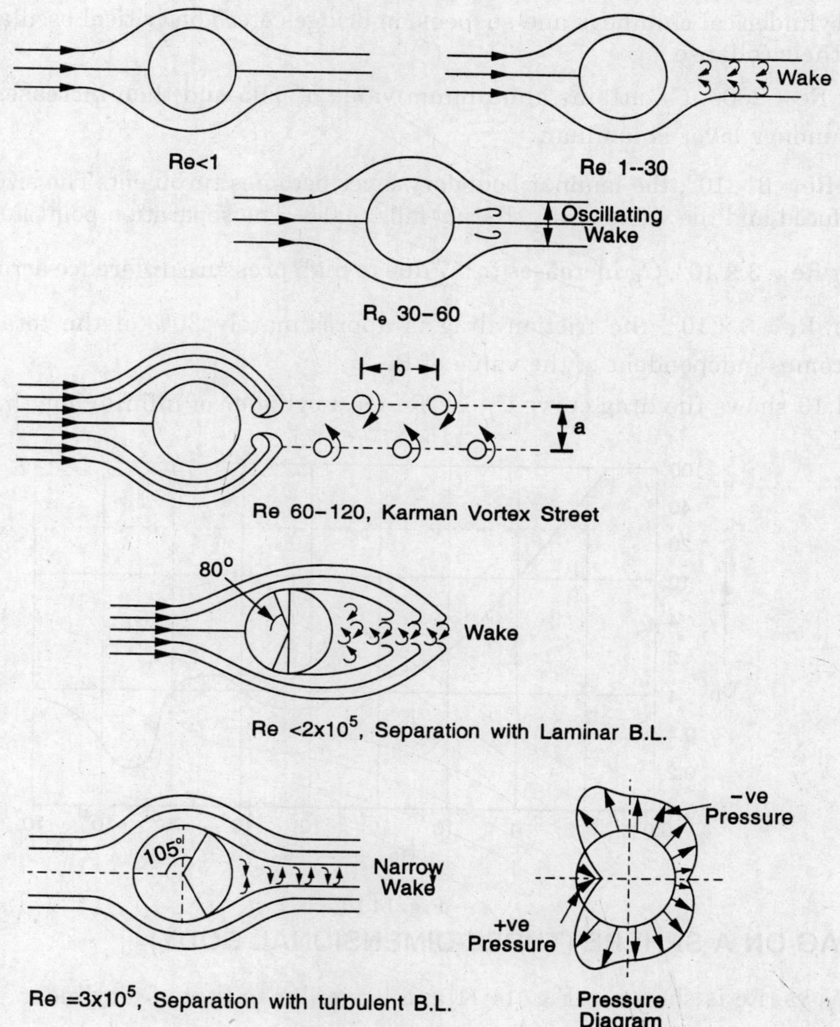

Fig. 14.9

- For Re = 120 – 250, the Karman vortex street is fully established. Form drag is about two-third of the total drag.
- For Re = 250 – 2×10^5, there is alternate shedding of the vortices from the two sides. This results in a periodic shifting of the low pressure zone in the wake and a fluctuating drag alongwith a transverse force. The dynamic transverse or lateral force makes the body to vibrate with frequency equal to that of the vortex shedding. In case, the speed of the wind is such that the frequency of the vortex shedding is near to the natural frequency of the cylinder, a state of resonance is reached. This can give rise to the phenomenon of 'singing' of telephone wires in the wind, fluttering of poles, vibration

of cylinderical chimneys and suspension bridges etc. The critical oscillations may lead to their collapse.

At Re = 2000, C_D attains a minimum value of 0·95 and then increases to 1·2 till the boundary layer is laminar.

- At Re = 3×10^5, the laminar boundary layer becomes turbulent. The size of the wake is reduced and the value of C_D sharply falls to 0·3. The separation point lies at $\theta = 105°$.
- For Re > 3×10^5, C_D increases to 0·7 due to high pressure difference across the cylinder.
- For Re > 3×10^6, the friction drag is approximately 30% of the total drag and C_D becomes independent of the value of Re.

Fig. 14.10 shows the drag curve C_D vs. Re for a cylinder of infinite length.

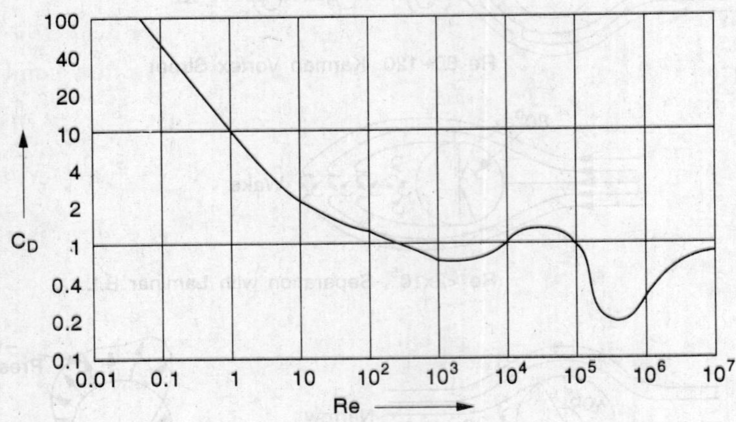

Fig. 14.10

14.7 DRAG ON A SPHERE (THREE-DIMENSIONAL BODY)

A plot of C_D vs. Re is shown in Fig. 14.11 and is similar to that of a cylinder.

- For Re < 0·1, the viscous forces are predominant as compared to inertia forces. C.G. Stokes (1980) gave the following equation by solving Navier Stokes equation for the flow around a sphere under very low velocities.

$$F_D = 3\pi\mu VD \qquad \text{...(14.7)}$$

Also since

$$F_D = C_D \times \frac{\rho V^2}{2} \times A$$

$$C_D = \frac{F_D}{A\dfrac{\rho V^2}{2}} = \frac{3\pi\mu VD}{\dfrac{\pi}{4}D^2\dfrac{\rho V^2}{2}} = \frac{24}{\rho VD/\mu} = \frac{24}{\text{Re}} \qquad \text{...(14.8)}$$

This equation is known as 'Stokes law'.

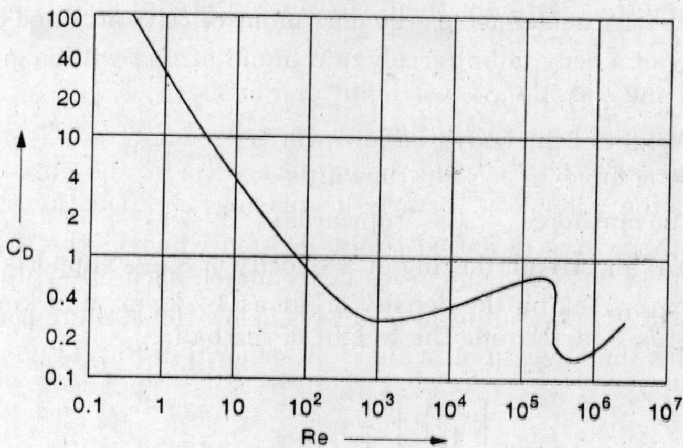

Fig. 14.11

He also found that the skin friction drag is about two-third of the total drag and the rest is pressure drag.

- For Re = 0·1 – 1, inertia effects increase. Carl *W.* Oseen (1910) improved the Stokes equation by taking into account the inertia terms,

$$C_D = \frac{24}{\text{Re}}\left(1 + \frac{3}{16}\,\text{Re}\right) = \frac{24}{\text{Re}} + 4{\cdot}5 \qquad \text{...(14.9)}$$

- For Re ≈ 1 – 1000, the laminar boundary layer exists. Separation takes place at $\theta \approx 80°$ at Re ≈ 1000.

- For Re = 1000 – 2×10^5, C_D is approximately 0·4 to 0·5, practically independent of Re. Laminar boundary layer exists.

- For Re > 2×10^5, C_D drops approximately to a value of 0·2. Turbulent boundary layer is formed. The point of separation moves downstream and the wake becomes smaller.

It was observed that for a flow around a cylinder, a vortex street is formed. However, as the flow around a sphere is axisymmetrical, a vortex ring is formed instead of a vortex street. When a vortex ring becomes unstable, it is washed away and the next ring is formed. Transverse vibrations as seen in a 2-*D* body (cylinder) are absent.

14.8 TERMINAL VELOCITY OF A BODY

When a body falls from rest in the atmosphere, its velocity increases because of acceleration due to gravity. As the velocity increases the drag force opposing the motion of the body also increases. At the moment when the drag force becomes equal to the weight of the body, the acceleration reduces to zero and the net external force acting on the body becomes zero. Onwards, the body moves at a constant speed which is known as the **terminal velocity.**

This terminal velocity may be defined as the maximum velocity attained by a falling body.

The terminal velocity of a body falling freely in a liquid at rest will be given by

$$W = F_D + F_B \quad \ldots(14.10)$$

where W = Weight of the body (downwards)

F_D = Drag force (upwards)

F_B = Buoyant force (upwards)

Example 14.6 A vertical air stream flowing at a velocity of 8 m/s supports a ball of 60 mm in diameter. Taking the density of air as 1·2 kg/m^3 and kinematic viscosity as 1·6 stokes, determine the weight of the ball.

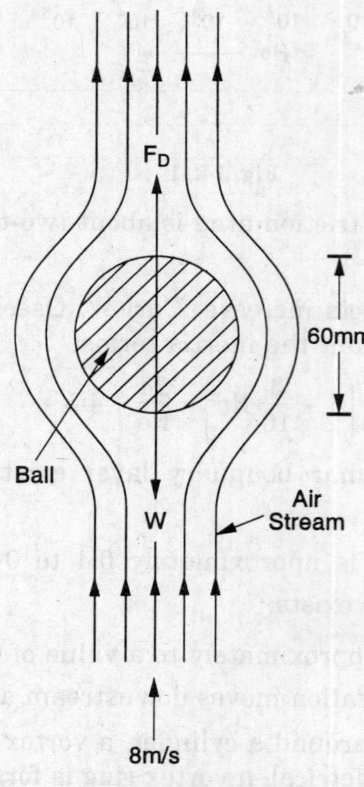

Fig. 14.12

Solution Reynolds number is given by

$$\text{Re} = \frac{VD}{\nu} = \frac{8 \times 0{\cdot}06}{1{\cdot}6 \times 10^{-4}} = 3000$$

Since the Reynolds number lies between 1000 and 10000, C_D may be taken as 0·5.

As the ball is supported by the stream, its weight must be equal to the drag force (neglecting the bouyant force, being very small). Thus

$$F_D = C_D \times \frac{\rho V^2}{2} \times A$$

$$= 0{\cdot}5 \times \frac{1{\cdot}2 \times 8^2}{2} \times \frac{\pi}{4} \times 0{\cdot}06^2$$

$$= 0{\cdot}0543 \text{ N}$$

Thus the weight of the ball = 0·0543 N

Example 14.7 A metallic ball 3 mm in diameter having a specific gravity 12 falls in a fluid of specific gravity 0·84 and viscosity 1·5 N.s/m^2. Determine

(i) the drag on the ball

(ii) the pressure drag and the skin drag

(iii) terminal velocity of the ball in the fluid.

Solution Weight of the ball = Volume × Specific weight of ball

$$= \frac{\pi}{6} d^3 \times w_b$$

$$= \frac{\pi}{6} \times 0{\cdot}0003^3 \times (12 \times 9{\cdot}81 \times 10^3)$$

$$= 1{\cdot}664 \times 10^{-3} \text{ N}$$

Buoyant force, F_B = Volume × Specific weight of fluid

$$= \frac{\pi}{6} d^3 \times w_f$$

$$= \frac{\pi}{6} \times 0{\cdot}0003^3 \times (0{\cdot}84 \times 9{\cdot}81 \times 10^3)$$

$$= 1{\cdot}165 \times 10^{-4} \text{ N}$$

$$W = F_D + F_B$$

$$F_D = W - F_B = (1{\cdot}664 - 0{\cdot}1165) \times 10^{-3} = 1{\cdot}548 \times 10^{-3} \text{ N}$$

$$\text{Pressure drag} = \frac{1}{3} F_D = \frac{1}{3} \times 1{\cdot}548 \times 10^{-3} = 5{\cdot}156 \times 10^{-4} \text{ N}$$

$$\text{Skin friction drag} = \frac{2}{3} F_D = \frac{2}{3} \times 1{\cdot}548 \times 10^{-3} = 1{\cdot}0319 \times 10^{-3} \text{ N}$$

$$F_D = 3\pi\mu DV$$

$$1{\cdot}548 \times 10^{-3} = 3\pi \times 1{\cdot}5 \times 0{\cdot}003 \times V$$

$$V = 0{\cdot}0365 \text{ m/s}$$

Now Reynolds number can be checked,

$$\text{Re} = \frac{\rho V d}{\mu} = \frac{(0{\cdot}84 \times 1000) \times 0{\cdot}0365 \times 0{\cdot}003}{1{\cdot}5} = 0{\cdot}0613$$

As Re < 0·2, the expression used above ($F_D = 3\pi\mu DV$) to calculate the terminal velocity is valid.

14.9 CIRCULATION AND LIFT ON A CYLINDER

When an ideal fluid moving with a velocity V flows past a stationary cylinder of radius R (Fig. 14.13a), the flow pattern is symmetrical about the axis and the circumferential velocity u_c at any point on the surface of the cylinder is

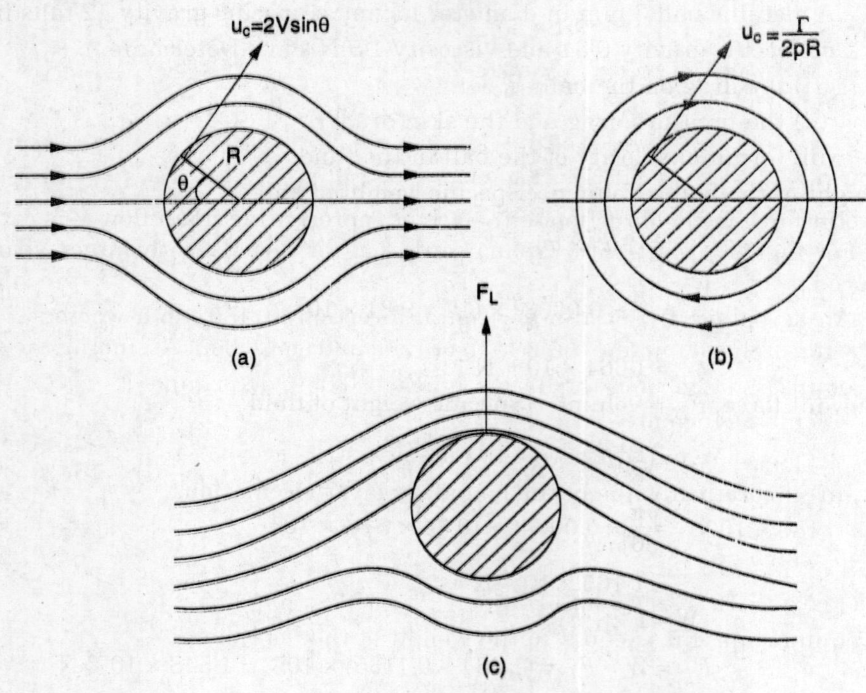

Fig. 14.13

$$u_c = 2V \sin \theta$$

where

θ = angular distance of the point from the forward stagnation point.

In this case, no lift force acts on the cylinder as the flow pattern is symmetrical about the horizontal axis and the pressure distribution is identical on the upper and lower halves of the cylinder.

- When the same cylinder is subjected to a constant circulation Γ, streamlines are in the form of concentric circles (Fig. 14.13b). The peripheral velocity on the surface of cylinder is given by

$$u_c = \frac{\Gamma}{2\pi R} \qquad \text{...(14.11)}$$

- By superimposing the above two patterns, a composite flow pattern as shown in Fig. 14.13c is obtained and the resultant velocity at any point on the surface of the cylinder becomes

$$u = 2V \sin \theta + \frac{\Gamma}{2\pi R} \quad ...(14.12)$$

This flow pattern is unsymmetrical about the horizontal axis. In the upper half, θ varies from 0° to 180° and therefore $2V \sin \theta$ is positive, the maximum value being at 90° indicating that at the top point of the cylinder, the velocity is maximum. In the lower half, θ varies from 180° to 360° which makes $2V \sin \theta$ to be negative. This means that the summation of the two terms will provide a low value of u. The minimum value of velocity in the lower half is at 270°. This is apparent from the flow pattern indicating crowding of streamlines around the upper half and widening of streamlines on the lower half. According to Bernoulli's principle, the pressure on the higher velocity side will be lower as compared to the other side. Thus on the lower half of the cylinder, the pressure will be more than the pressure on the upper half. Due to this pressure difference on the two halves of the cylinder, a force acts on the cylinder in a direction perpendicular to the flow direction. This upward force is known as the lift force. However, as the flow is symmetrical on the front and the rear side of the cylinder, the pressures at the corresponding points on the front and the rear of the cylinder are equal and no unbalanced forces act on the cylinder and therefore, the drag is zero.

The phenomenon of generation of lift by a spinning cylinder in a fluid stream is known as **magnus Effect**.

The magnitude of the lift force is changed by varying either the speed of rotation of the cylinder or the velocity of the stream or both. These effects are observed in bullets fired from a rifle or firing of shells from guns.

Position of Stagnation Points

The velocity at any point on the surface of rotating cylinder is

$$u = 2V \sin \theta + \frac{\Gamma}{2\pi R}$$

At the stagnation points, $u = \theta$

$$0 = 2V \sin \theta + \frac{\Gamma}{2\pi R}$$

$$\sin \theta = \frac{\Gamma}{4\pi RV} \quad ...(14.13)$$

- When $\Gamma = 0$, $\sin \theta = 0$ or $\theta = 0°$ and 180°, stagnation points are at the leading and trailing edges (Fig. 14.14a)
- When $\Gamma < 4\pi RV$, $\sin \theta < -1$, θ is $< -90°$ and $> 180°$. S_1 and S_2 are as shown in Fig. 14.14b.
- When $\Gamma = 4\pi RV$, $\sin \theta = -1$, $\theta = -90°$ and 270° which is the same point at the bottom of the cylinder (Fig. 14.14c).
- When $\Gamma > 4\pi RV$, $\sin \theta > -1$ which is not feasible. In other words, the stagnation points do not occur on the surface of the rotating cylinder (Fig. 14.14d).

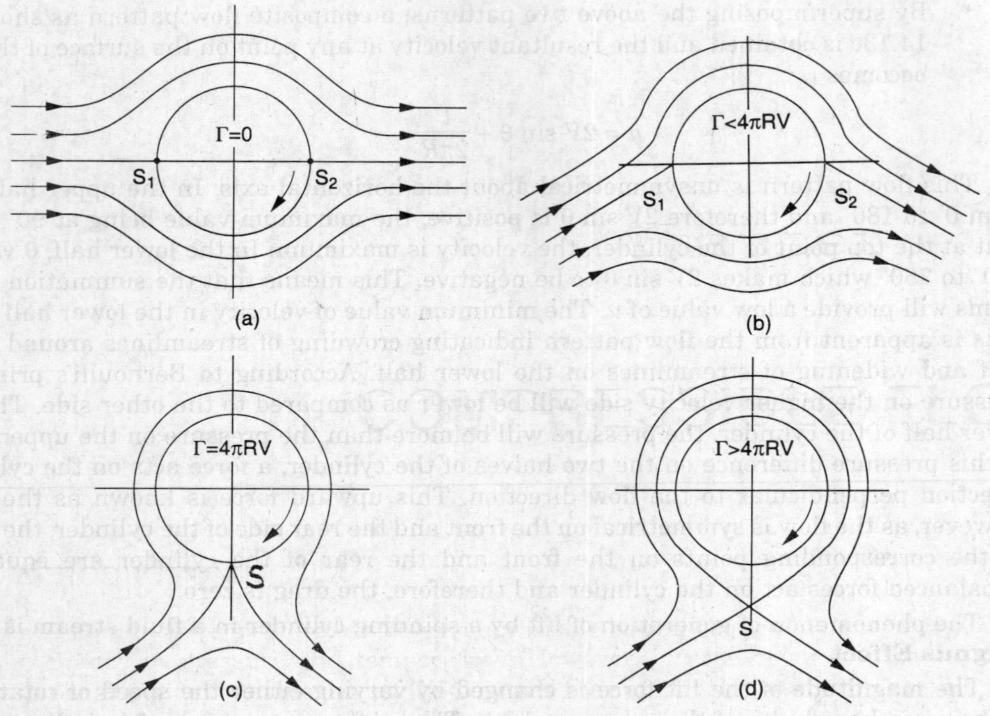

Fig. 14.14

14.10 EXPRESSION FOR LIFT ON A ROTATING CYLINDER

Let a cylinder rotate in a uniform flow field. Consider a small elemental length on the surface of cylinder (Fig. 14.15).

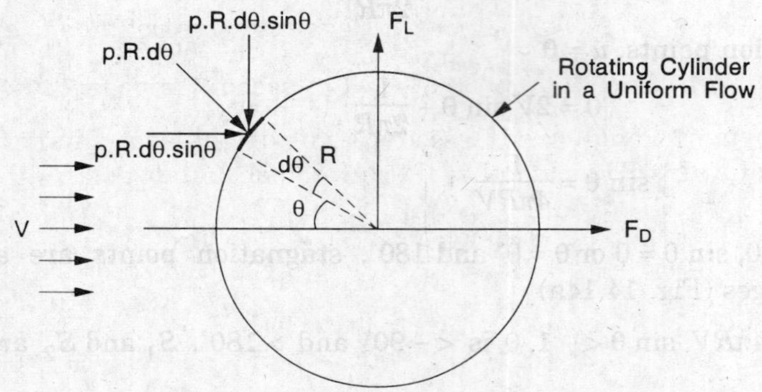

Fig. 14.15

Let p_0 = pressure of the fluid far away from the cylinder

V = velocity of the fluid far away from the cylinder

p = pressure on the surface of the element

u = velocity of the fluid on the surface of the element

$d\theta$ = angle made by the length of the element at the centre of cylinder

R = radius of the cylinder

Applying Bernoulli's equation to points 1 (far away from the cylinder) and 2 (on the surface of cylinder), both the points being at the same horizontal line,

$$\frac{p_0}{\rho} + \frac{V^2}{2} = \frac{p}{\rho} + \frac{u^2}{2}$$

$$p_0 + \frac{\rho V^2}{2} = p + \frac{\rho u^2}{2}$$

$$p - p_0 = \frac{1}{2}\rho\,(V^2 - u^2)$$

$$= \frac{1}{2}\rho V^2\left[1 - \left(\frac{u}{V}\right)^2\right]$$

$$p = p_0 + \frac{1}{2}\rho V^2\left[1 - \left(2\sin\theta + \frac{\Gamma}{2\pi RV}\right)\right]$$

$$= p_0 + \frac{1}{2}\rho V^2 - \frac{1}{2}\rho V^2\left(2\sin\theta + \frac{\Gamma}{2\pi RV}\right) \qquad \text{...(14.14)}$$

Assuming unit length of the cylinder,

Lift force on the elemental area $= -p\,.\,(R\,.\,d\theta)\sin\theta$

(–ve sign is taken as the component of p is in the downward direction)

The total lift force on the entire surface of the cylinder is obtained by integrating the above expression *i.e.*

$$F_L = -\int_0^{2\pi} p\,.\,(R\,.\,d\theta)\sin\theta$$

$$= -R\int_0^{2\pi} p\,.\,\sin\theta\,.\,d\theta$$

$$= -R\left[\int_0^{2\pi} p_0\,.\,\sin\theta\,.\,d\theta + \int_0^{2\pi}\frac{1}{2}\rho V^2\sin\theta\,.\,d\theta - \int_0^{2\pi}\frac{1}{2}\rho V^2\left(2\sin\theta + \frac{\Gamma}{2\pi RV}\right)^2\sin\theta\,.\,d\theta\right]$$

$$= -R\left[\int_0^{2\pi} p_0\,.\,\sin\theta\,.\,d\theta + \int_0^{2\pi}\frac{1}{2}\rho V^2\sin\theta\,.\,d\theta\right]$$

$$= -\frac{1}{2}\int_0^{2\pi}\rho V^2\left\{\int_0^{2\pi} 4\,.\,\sin^3\theta\,d\theta + \int_0^{2\pi}\frac{2\Gamma}{\pi RV}\sin^2\theta + \int_0^{2\pi}\frac{\Gamma^2}{4\pi^2R^2V^2}\sin\theta\,.\,d\theta\right\}$$

As $\sin^n\theta\,.\,d\theta = 0$ for n = odd number. Therefore, all the terms in the above expression are zero except

$$F_L = R\,.\,\frac{1}{2}\rho V^2 \times \frac{2\Gamma}{\pi RV}\int_0^{2\pi}\sin^2\theta\,d\theta$$

$$= \frac{\rho V\Gamma}{\pi}\int_0^{2\pi}\left(\frac{1 - \cos 2\theta}{2}\right)d\theta$$

$$= \frac{\rho V\Gamma}{\pi}\left(\frac{\theta}{2} - \frac{\sin 2\theta}{4}\right)^2$$

$$= \frac{\rho V\Gamma}{\pi}(\pi - 0)$$

$$= \rho V\Gamma \text{ per m length of cylinder} \qquad \ldots(14.15)$$

$$\text{Total lift on the cylinder} = \rho V\Gamma L \qquad \ldots(14.16)$$

This equation is known as **Kutta-Joukowski equation.** It is applicable to all types of bodies including airfoils.

14.11 LIFT COEFFICIENT FOR A ROTATING CYLINDER

Lift coefficient

$$C_L = \frac{F_L}{\frac{1}{2}\rho V^2 A}$$

$$C_L = \frac{\rho V L\Gamma}{\frac{1}{2}\rho V^2 2RL} \qquad (F_L = \rho V L\Gamma)$$

$$= \frac{\Gamma}{VR}$$

$$= \frac{2\pi R u_c}{VR} \qquad \left(u_c = \frac{\Gamma}{2\pi R}\right)$$

$$= \frac{2\pi u_c}{V} \qquad \ldots(14.17)$$

This shows that the lift coefficient depends upon the ratio u_c/V or the ratio of tangential velocity of the cylinder due to circulation to the velocity of the main flow. Fig. 14.16 shows

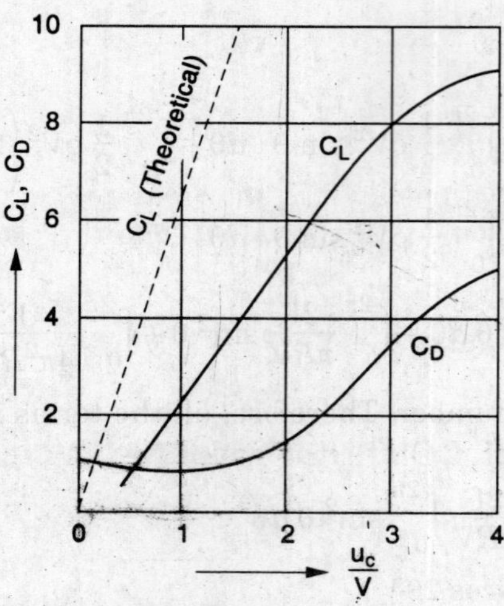

Fig. 14.16

theoretical variation of C_L with u_c/V from Eq. 14.17 alongwith the experimental lift and drag coefficients in real fluids with the variation of u_c/V.

For the theoretical case, when $u_c/V = 2$

$C_L = 4\pi = 12{\cdot}56$ which is the theoretical maximum for lift coefficient and for this case the stagnation points meet at the bottom ($\because\ \Gamma = 2\pi R u_c = 4\pi RV$, refer section 14.9)

The experimental values of C_L differ from the theoretical values due to the effect of viscosity, length and circulation around the rotating cylinder differing from an irrotational vortex.

Example 14.8 A cylinder 15 m long and 2 m in diameter rotates at 200 rpm in a stream of water flowing with a velocity of 20 m/s. The axis of the cylinder is perpendicular to the stream of water. Assuming that there is no slip between the cylinder and the circulating flow, determine the circulation, theoretical lift, the position of the stagnation points and the speed of the cylinder for a single stagnation point.

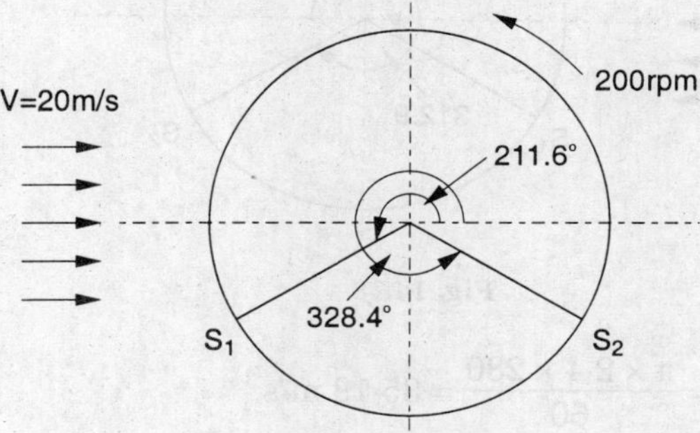

Fig. 14.17

Solution $u_c = \dfrac{\pi DN}{60} = \dfrac{\pi \times 2 \times 200}{60} = 20{\cdot}94$ m/s

$$\Gamma = 2\pi R u_c = \pi D u_c = \pi \times 2 \times 20{\cdot}94 = 131{\cdot}6 \text{ m}^2/\text{s}$$

Theoretical lift,

$$F_L = \rho L V \Gamma = 1000 \times 15 \times 20 \times 131{\cdot}6 = 39{\cdot}48 \times 10^6 \text{ N}$$

The position of the stagnation points is given by,

$$\sin\theta = -\frac{\Gamma}{4\pi RV} = \frac{131{\cdot}6}{4\pi \times 1 \times 20} = -\,0{\cdot}5236$$

$$\theta = 211{\cdot}6^\circ \text{ and } 328{\cdot}4^\circ$$

To have a single stagnation point,

$$\Gamma = 4\pi VR = 4\pi \times 20 \times 1 = 251{\cdot}3 \text{ m}^2/\text{s}$$

$$u_c = \frac{\Gamma}{2\pi R}$$

$$\frac{\pi DN}{60} = \frac{\pi \times 2N}{60} = \frac{251{\cdot}3}{2\pi \times 1} = 40 \text{ m/s}$$

$$N = 382 \text{ rpm}$$

Example 14.9 A cylinder having a diameter of 2·4 m and 12 m long has its axis perpendicular to the stream of air. It rotates at 280 rpm in the air stream, the velocity of which is 24 m/s. Determine

(i) the circulation and the theoretical lift per unit length

(ii) the position of stagnation points

(iii) the actual lift, drag and direction of resultant force, assuming coefficients of drag and lift 0·6 and 3·5 respectively

(iv) the speed of the cylinder which provides only one stagnation point.

The density of air is 1·2 kg/m^3.

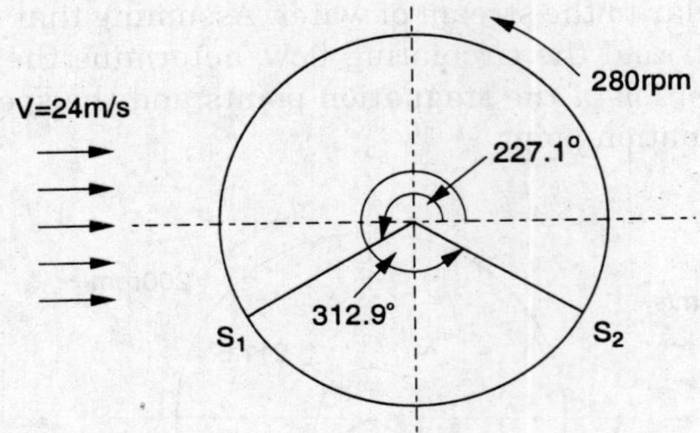

Fig. 14.18

Solution (i)

$$u_c = \frac{\pi DN}{60} = \frac{\pi \times 2{\cdot}4 \times 280}{60} = 35{\cdot}19 \text{ m/s}$$

$$\Gamma = 2\pi R u_c = 2\pi\left(\frac{2{\cdot}4}{2}\right) \times 35{\cdot}19 = 265{\cdot}3 \text{ m}^2/\text{s}$$

Theoretical lift per unit length is given by

$$F_L = \rho V \Gamma = 1{\cdot}2 \times 24 \times 265{\cdot}3 = 7640{\cdot}5 \text{ N/m}$$

(ii) The position of the stagnation points is given by,

$$\sin\theta = -\frac{\Gamma}{4\pi RV} = \frac{265{\cdot}3}{4\pi \times \left(\frac{2{\cdot}4}{2}\right) \times 24} = -\,0{\cdot}733$$

$$\theta = 227{\cdot}1^\circ \text{ and } 312{\cdot}9^\circ$$

(iii) The actual drag and lift,

$$F_D = C_D \times \frac{\rho V^2}{2} \times A = 0{\cdot}6 \times \frac{1{\cdot}2 \times 24^2}{2} \times 2{\cdot}4 \times 12 = 5972 \text{ N}$$

$$F_L = C_L \times \frac{\rho V^2}{2} \times A = 3{\cdot}5 \times \frac{1{\cdot}2 \times 24^2}{2} \times 2{\cdot}4 \times 12 = 34836 \text{ N}$$

$$F_R = \sqrt{F_D{}^2 + F_L{}^2} = \sqrt{5972^2 + 34836^2} = 35344 \text{ N}$$

$$\tan\alpha = \frac{F_L}{F_D} = \frac{34836}{5972} = 5{\cdot}833$$

$$\alpha = 80{\cdot}27^\circ$$

(iv) To have a single stagnation point,

$$\Gamma = 4\pi VR = 4\pi \times 20 \times \left(\frac{2{\cdot}4}{2}\right) = 351{\cdot}9 \text{ m}^2/\text{s}$$

$$u_c = \frac{\Gamma}{2\pi R}$$

$$\frac{\pi DN}{60} = \frac{361{\cdot}9}{2\pi \times \left(\frac{2{\cdot}4}{2}\right)}$$

$$\frac{\pi \times 2{\cdot}4 \times N}{60} = 48$$

$$N = 381{\cdot}97 \text{ rpm}$$

Example 14.10 60% of power of the engine of a jet plane which delivers 6000 kW is consumed in overcoming the drag resistance of the wing. The plane weighs 20 kN and has a wing area of 18 m^2. The speed of the jet plane is 720 km/h. Determine the coefficients of lift and drag for the wings. The density of the air is 1·2 kg/m^3.

Solution

$$V = \frac{720 \times 1000}{3600} = 200 \text{ m/s}$$

Lift force = weight of the jet plane

$$C_L \times \frac{\rho V^2}{2} \times A = 20 \times 10^3$$

$$C_L \times \frac{1{\cdot}2 \times 200^2}{2} \times 18 = 20 \times 10^3$$

$$C_L = 0{\cdot}0463$$

Power required to overcome the friction = Drag force × Velocity

$$0{\cdot}6 \times (6000 \times 10^3) = \left(C_D \times \frac{\rho V^2}{2} \times A\right) \times V$$

$$= C_D \times \frac{1{\cdot}2 \times 200^2}{2} \times 18 \times 200$$

$$C_D = \frac{0{\cdot}6 \times (6000 \times 10^3) \times 2}{1{\cdot}2 \times 200^2 \times 18 \times 200} = 0{\cdot}0417$$

Example 14.11 Figure 14.19 shows a cup anemometer which rotates freely without air friction. The coefficient of drag for the hollow upstream hemisphere is 1·6 and for the hollow downstream hemisphere 0·4. Determine the speed of rotation against a wind speed of 72 km/h.

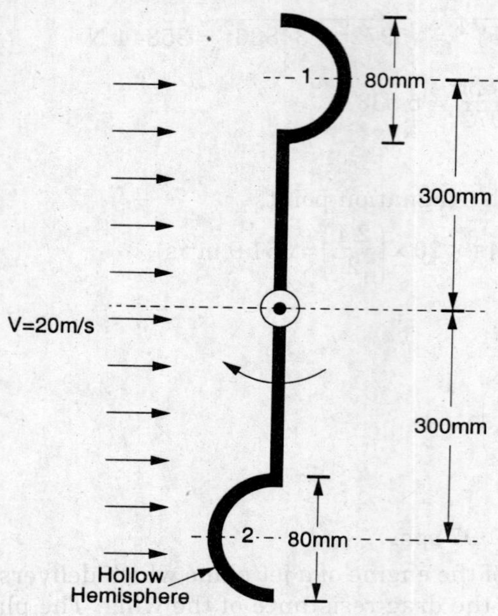

Fig. 14.19

Solution $V = \dfrac{72 \times 1000}{60 \times 60} = 20$ m/s

Let the anemometer rotate at ω rad/s in the clockwise direction.

Fluid velocity relative to hemisphre 1 = 20 – 0·3ω

Fluid velocity relative to hemisphere 2 = 20 + 0·3ω

Drag forces on the hemispheres are

$$F_{D1} = C_{D1} \times \frac{\rho V_1^2}{2} \times A = \frac{1}{2} \times 1{\cdot}6 \times \rho \,.\, A \times (20 - 0{\cdot}3\omega)^2$$

$$F_{D2} = C_{D2} \times \frac{\rho V_2^2}{2} \times A = \frac{1}{2} \times 0{\cdot}4 \times \rho \,.\, A \times (20 + 0{\cdot}3\omega)^2$$

If the anemometer rotates at a steady uniform angular velocity, the net torque about the axis of rotation is zero.

$$\text{Torque} = (F_{D1} - F_{D2}) \times 0{\cdot}3 = 0$$

$$F_{D1} = F_{D2}$$

$$\frac{1}{2} \times 1{\cdot}6 \times \rho \,.\, A\,(20 - 0{\cdot}3\omega)^2 = \frac{1}{2} \times 0{\cdot}4 \times \rho \,.\, A\,(20 + 0{\cdot}3\omega)^2$$

$$1{\cdot}6 \times (20 - 0{\cdot}3\omega)^2 = 0{\cdot}4 \times (20 + 0{\cdot}3\omega)^2$$

$$\left(\frac{20 - 0{\cdot}3\omega}{20 + 0{\cdot}3\omega}\right)^2 = \frac{0{\cdot}4}{1{\cdot}6}$$

$$\frac{20 - 0{\cdot}3\omega}{20 + 0{\cdot}3\omega} = 0{\cdot}5$$

$$20 - 0{\cdot}3\omega = 10 + 0{\cdot}15\omega$$

$$0{\cdot}45\omega = 10$$

$$\omega = 22{\cdot}22 \text{ rad/s}$$

$$\frac{2\pi N}{60} = 22{\cdot}22$$

$$N = 212{\cdot}2 \text{ rpm}$$

14.12 AIRFOIL

Some of the terms related to an airfoil (Fig. 14.20) are defined below :

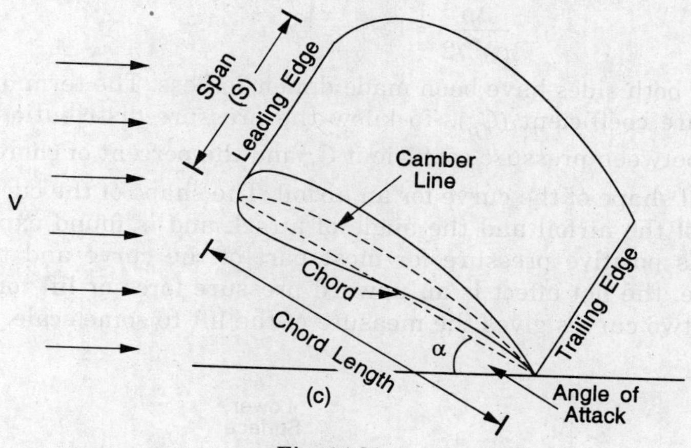

Fig. 14.20

Leading Edge	It is the front edge of the blade on thicker side.
Trailing Edge	It is the back edge of the blade on thinner side.
Chord (c)	It is the line joining the leading and the trailing edges in a direction perpendicular to the two edges.
Span (s)	It is the length of the airfoil blade.
Angle of Attack (α)	It is the inclination of the chord to the direction of motion of the fluid stream. It is also known as the angle of incidence.
Camber Line	A line joining the mid-points of the profile of the airfoil is known as the camber line.
Aspect Ratio	Aspect ratio is the ratio of the span to the chord length (s/c).

The airfoil sections are standardised by National Advisory Committee for Aeronautics (NACA) in the form of tables containing percentage thickness against percentage length along x- and y-directions.

Let P_x = pressure on the surface at a distance x from the leading edge of the airfoil

V_x = Velocity on the surface at a distance x from the leading edge of the airfoil

p = pressure of the free stream

V = Velocity of the free stream

Applying Bernoulli's equation to a point in the free stream and a point on the surface of the airfoil at a distance x from the leading edge of the airfoil. Neglecting any datum and friction effects,

$$\frac{p_x}{\rho} + \frac{{V_x}^2}{2} = \frac{p}{\rho} + \frac{V^2}{2}$$

$$\frac{p_x - p}{\rho} = \frac{V^2 - {V_x}^2}{2}$$

$$= \frac{V^2}{2}\left[1 - \left(\frac{V_x}{V}\right)^2\right]$$

$$\frac{\Delta p}{\rho V^2/2} = 1 - \left(\frac{V_x}{V}\right)^2 \qquad ...(14.18)$$

The terms on both sides have been made dimensionless. The term on the left hand side is called the pressure coefficient (C_p). To know the pressure distribution around an airfoil, a graph is plotted between pressure coefficient C_p and the percent of chord length. Figure 14.21 shows the general shape of the curve for an airfoil. The shape of the curve changes depending upon the shape of the airfoil and the angle of attack and is found experimentaly. Since the lower surface has positive pressure for most part of the curve and the upper surface the negative pressure, the net effect is an upward pressure force or lift force on the airfoil. The difference of the two curves gives the measure of the lift to some scale.

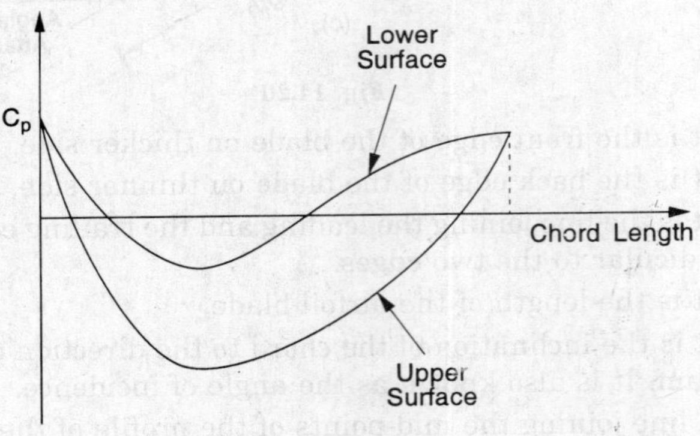

Fig. 14.21

The rise in pressure at any point from the free stream value is called the **dynamic pressure.** At a point on the surface of the airfoil at a distance x from the leading edge, it is $(p_x - p)$. At a stagnation point, the value of the coefficient is one. Fig. 14.22 shows the stagnation point at the leading edge. The point is shifted depending upon the shape of the airfoil and the angle of attack.

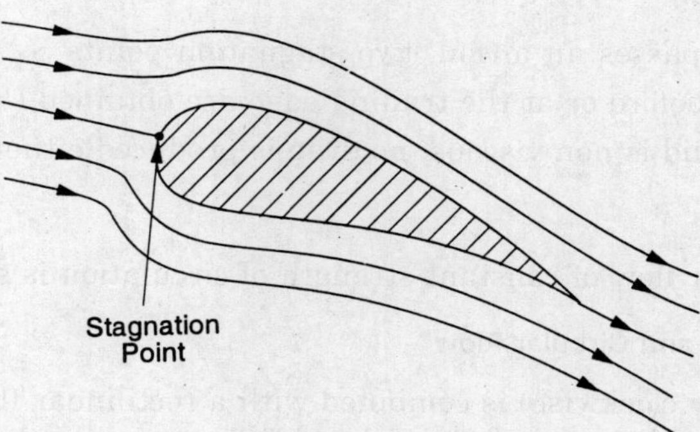

Fig. 14.22

14.13 LIFT AND DRAG ON AN AIRFOIL

First consider an airfoil in a non-viscous (ideal) fluid under following flow conditions :

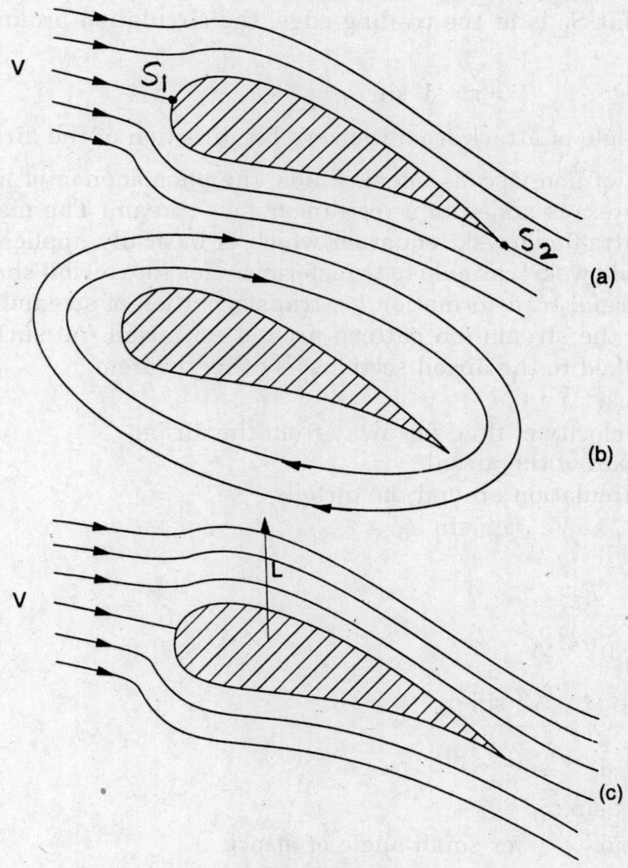

Fig. 14.23

Flow Rectilinear (Irrotational)

When a rectilinear flow passes an airfoil, two stagnation points S_1 at the leading edge and S_2 on the upper surface before or at the trailing edge are obtained (Fig. 14.23a). No lift force is produced. Since the fluid is non-viscous, no drag is produced either.

Flow Circular

Flow pattern of a circular flow of constant strength of circulation is shown in Fig. 14.23b.

Combination of Rectilinear and Circular Flow

When a circular flow (say clockwise) is combined with a rectilinear flow, a change in the flow velocities on the upper and lower surfaces of the airfoil occurs. On the upper surface, the flow velocity is added to the velocity of the circulation whereas on the lower surface, flow velocity decreases by the velocity of circulation (Fig. 14.23c). This results in the crowding of streamlines on the upper surface and moving apart of streamlines under the lower surface. Thus according to the law of Bernoulli's equation, the pressure on the upper surface is lower than on the lower surface.

When the flow past an airfoil is such that the trailing edge streamlines are tangent to it *i.e.* the stagnation point S_2 is at the trailing edge, the circulation around the airfoil is given by

$$\Gamma = \pi c \, . \, V \sin \alpha_0 \quad \ldots(14.19)$$

where α_0 is the angle of attack from the zero lift direction of the airfoil.

Therefore, in case of non-viscous (ideal) fluids, the phenomenon of lift is developed when a circulation (free vortex) is added to a rectilinear flow stream. The magnitude of the lift is obtained from the Kutta-Joukowski equation which is basically applicable to flow around a cylinder. However, Joukowski was able to transform circles into airfoil shapes by a mathematical technique of conformal transformation *i.e.* transformation of streamline pattern around a circular cylinder into the streamline pattern around an airfoil. As such, theorem of Kutta-Joukowski is also applied to the airfoil sections. By this theorem,

Lift, $F_L = \rho \, . \, s \, . \, V \, . \, \Gamma$

where V = Velocity of fluid far away from the airfoil

s = span of the airfoil

Γ = circulation around the airfoil

$$F_L = \rho \, . \, s \, . \, V \, . \, (\pi c V \sin \alpha_0)$$

Lift coefficient

$$C_L = \frac{F_L}{\frac{1}{2} \rho V^2 \, . \, A}$$

$$= \frac{\pi \rho \, . \, c \, . \, s V^2 \sin \alpha_0}{\frac{1}{2} \rho V^2 \, . \, (c \, . \, s)}$$

$$= 2\pi \sin \alpha_0 \quad \ldots(14.20)$$

$$\approx 2\pi\alpha_0 \quad \text{for small angle of attack} \quad \ldots(14.20a)$$

In case of a real (viscous) fluid, a circulation is also necessary in addition to a rectilinear flow. However, in this case the linear flow itself produces a circulation around the body to affect the phenomenon of lift as explained below :

When the flow passes an airfoil, the velocity on the upper surface increases whereas the velocity on the lower surface decreases. Since the two streams of the upper and the lower surfaces are to mix again at the trailing edge, there is a tendency of the fluid to flow from the lower surface to the upper surface at the trailing edge. This also emanates from the fact that the pressure at the lower surface is higher than the pressure on the upper surface and the fluid has a tendency to travel from region of higher pressure to that of a lower pressure. Thus vortices are formed at the trailing edge having anticlockwise circulation. Production of vortices at the trailing edge is called starting vortices (Fig. 14.24). As the main flow is without circulation, a circulation in the opposite direction is produced in the flow to wash away the vortices at the trailing edge in accordance with the law of circulation. Continuous production of vortices at the trailing edge gives rise to a continuous uniform circulation around the airfoil which is necessary to cause the lift.

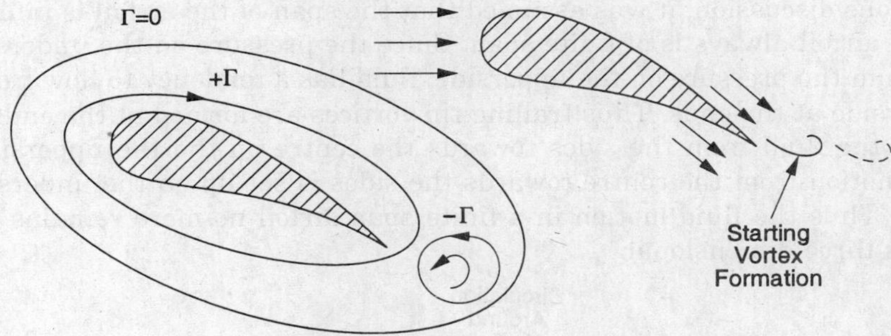

Fig. 14.24

Variation of C_L and C_D with angle of attack

Fig. 14.25a shows a graph plotted for the coefficients of lift found experimently for different values of angle of attack. It is observed from the plot that for zero value of coefficient of lift,

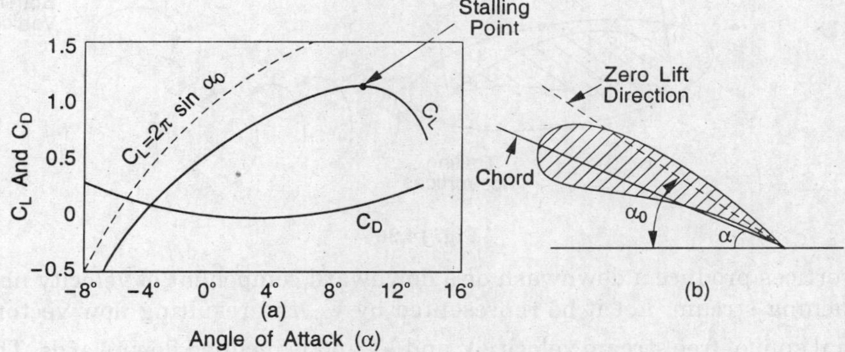

Fig. 14.25

the value of angle of attack is negative. This is true when the airfoil has camber or curvature. As mentioned earlier, the angle of attack measured from the zero lift direction is referred as α_0 (Fig. 14.25b). For a symmetrical airfoil without curvature, zero lift direction coincides with the chord and $\alpha_0 = \alpha$.

The curve between C_L and α is almost a straight line for most of the part. At point S, it reaches the maximum height indicating the maximum value of C_L, after which its value suddenly drops. This is attributed to a wider wake formed after this value. The point of maximum lift after which the value of C_L starts decreasing is called the **stalling point.** Alongwith this curve, another curve in dotted is also plotted which is the theoretical curve $C_L = 2\pi \sin \alpha_0$ given by Eq. 14.20. The two curves fairly match each other upto the stalling point. In practice, the operating point of airfoil is always below the stalling point.

The variation of C_D with the angle of attack α is also shown in Fig. 14.25a.

Induced Drag

In the previous discussion, it was assumed that the span of the airfoil is infinite. However, in practice the airfoil always is of finite span. Since the pressure on the underside of the airfoil is greater than the pressure on the upperside, fluid has a tendency to flow from the underside to the upperside at the ends. Thus trailing tip vortices are formed at the ends. This results in a motion of the fluid from the sides towards the centre on the the upperside of the airfoil, whereas a motion from the centre towards the sides is set up on the underside of the airfoil (Fig. 14.26). Thus the fluid motion in a finite span airfoil no more remains two-dimensional but becomes three dimensional.

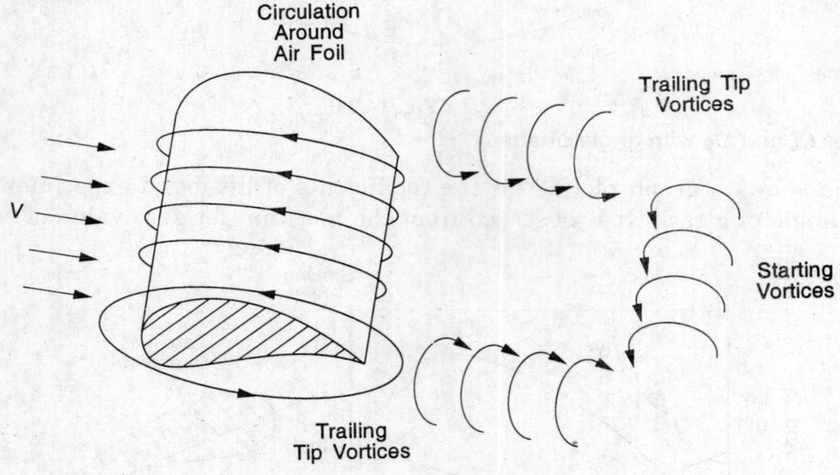

Fig. 14.26

The tip vortices produce a downwash or a downward component of velocity normal to that of the approaching stream. Let it be represented by V_1, the resulting flow vector V_r is given by the vectorial sum of free stream velocity V and V_1 and is inclined downwards. The directions of the true drag and the true lift is also affected and they become inclined (Fig. 14.27). The

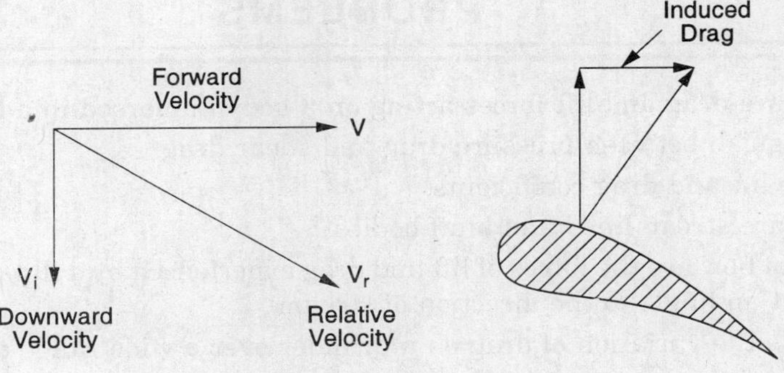

Fig. 14.27

component of the lift force in the direction of the vector representing main flow V is called the **induced drag.** For higher aspect ratio (span/chord), *i.e.* for airfoils with larger spans, the value of the induced drag is lower.

Example 14.12 An aeroplane flying in a horizontal direction at 180 km/h weighs 35 kN. A wing of the aeroplane has a span of 12 m and chord 3 m. Assuming the density of air 1·2 kg/m^3 and the drag coefficient 0·032, determine

(i) lift coefficient

(ii) power required to drive the plane

(iii) theoretical value of the boundary layer circulation.

Solution $V = \dfrac{180 \times 1000}{60 \times 60} = 50$ m/s

Area $= s \times c = 12 \times 3 = 36$ m^2

For equilibrium, lift is equal to the weight of the aeroplane,

$$W = F_L = C_L \times \frac{\rho V^2}{2} \times A$$

$$35 \times 10^3 = C_L \times \frac{1{\cdot}2 \times 50^2}{2} \times 36$$

$$C_L = 0{\cdot}648$$

Power required = Drag force × Velocity

$$= F_D \times V$$

$$= \left(C_D \times \frac{\rho V^2}{2} \times A\right) \times V$$

$$= 0{\cdot}032 \times \frac{1{\cdot}2 \times 50^2}{2} \times 36 \times 50$$

$$= 86400 \text{ N} \quad \text{or} \quad 86{\cdot}4 \text{ kN}$$

Theoretical value of boundary layer circulation is found from the relation,

$$F_L = \rho L V \Gamma$$

$$35 \times 10^3 = 1{\cdot}2 \times 12 \times 50 \times \Gamma$$

$$\Gamma = 48{\cdot}6 \text{ m}^2/\text{s}$$

PROBLEMS

1. What are drag and lift forces acting on a body immersed in a fluid stream ?
2. Distinguish between pressure drag and shear drag.
3. Define lift and drag coefficients.
4. What are streamlined and bluff bodies ?
5. Explain how are the forces of lift and drag experienced by a flat plate held in a moving fluid at an angle to the direction of stream.
6. Discuss the variation of drag on a cylinder over a wide range of Reynolds number.
7. In what way is the drag on a sphere differ from that on a cylinder ?
8. What is Stoke's law as applicable to drag of a sphere ? Under what condition is it applicable ?
9. What is meant by terminal velocity of a body ?
10. What is Magnus effect ? Explain.
11. What do you mean by stagnation point ? How do you determine the position of stagnation point for a rotating cylinder in a uniform flow ? What is the condition for single stagnation point ?
12. Determine an expression for the lift on a rotating cylinder placed in a uniform flow such that the axis of cylinder is perpendicular to the direction of low.
13. What is an airfoil ? Define the terms with reference to an airfoil : leading edge, trailing edge, span, angle of attack, camber line and aspect ratio.
14. Give the pressure distribution around an airfoil placed in the way of a moving fluid stream.
15. How can the force of lift be experienced by an airfoil placed in an ideal flow ?
16. How is the force of lift produced in an airfoil placed iin a stream of real fluid ? Is a circulation necessary to produce the force of lift ?
17. Explain the term induced drag.
18. The cross-sectional area of a car weighing 30 kN is 2·7 m^2. If the drag coefficient is 0·6, determine the power requirement at a speed of 60 km/h. The density of air is 1·208 kg/m^3 and the coefficient of friction 0·012. (10·533 kW)
19. Experiments were conducted on a flat plate 2 m in length and 1 m wide at a speed of 50 km/h. The plate is kept at such an angle that the coefficient of lift and drag are 0·75 and 0·15 respectively. Determine the lift force, the drag force, the resultant force and the power exerted by the airstream on the plate. (166·3 N, 33·3 N, 259·6 N, 462 W)
20. A parachute is treated as an open hemisphere. The vertical component of the landing speed of the parachute is 6 m/s. Determine its diameter if the total weight to be carried is 1·5 kN. Assume the density of air to be 1·208 kg/m^3 and the coefficient of drag as 1·33. (8·126 m)
21. A ship is propelled by two cylindrical rotors each of diameter 1·5 m and length 6 m revolving at 250 rpm about their axes which are horizontal. Determine the force

exerted upon the rotors in the direction of motion when the relative wind velocity is 40 km/h at an angle of 30° to the horizontal. Assume density of air as 1·22 kg/m^3.
(4·148 kN)

22. A kite of dimensions 600 mm × 450 mm weighs 0·6 N. It is maintained in air at an angle 10° to the horizontal by a string attached to the kite making an angle of 30° to the vertical. The drag coefficient is estimated to be 0·25. The wind speed is 15 km/h. Assuming the density of air to be 1·2 kg/m^3, determine tension in the string and the lift coefficient.
(1·406 N, 0·717)

23. A kite weighing 0·8 N has an area of 0·6 m^2. It is found to have an angle of inclination of 8° to the horizontal while flying in a wind of 10 km/h. Determine the tension in the chord and the inclination of the chord to the horizontal at the kite assuming the coefficients of lift and drag to be 0·6 and 0·2 respectively. ρ for air = 1·2 kg/m^3.
(1·032 N, 57·4°)

24. Assuming that Stokes law holds good, find the velocity of fall of rain drops of 0·3 mm diameter in the atmospheric air having a density 1·2 kg/m^3 and kinematic viscosity 0·15 stoke.
(2·72 m/s)

25. A rotary mixer has two circular discs each 100 mm in diameter. The discs are spaced 1 m apart on the two ends of a horizontal rod. The centre of the rod has a vertical shaft attached to it. The mixer is rotated at 90 rpm in a large vessel containing liquid of density 930 kg/m^3 and kinematic viscosity 0·8 stokes. Neglecting the resistance of the rod and the shaft, determine the torque and energy required to drive the mixer.
(71·55N.m, 374·6 W)

Chapter 15

COMPRESSIBLE FLOW

15.1 INTRODUCTION

In general, liquids are considered to be incompressible as the change in density is negligible with the change in pressure and temperature. However, in case of air and gases, relative change in density is appreciable during the flow and they are treated as **compressible fluids.** Thus in cases such as flow of gases through orifices, nozzles, compressors and aeroplanes and projectiles flying at high altitudes with high velocities, the change in the density of the fluids varies by the change in pressure and temperature and therefore, the thermodynamic behaviour of the flow is taken into account.

15.2 BASIC THERMODYNAMIC RELATIONS

The following are the important relations :

1. Isobaric (Constant Pressure) Process

If the pressure remains constant during a process, the volume of a given mass of perfect gas varies directly with its absolute temperature. It is know as **Charle's law.**

$$V \propto T$$

$$\frac{V}{T} = \text{constant}$$

$$\frac{V_1}{T_1} = \frac{V_2}{T_2} \quad \text{or} \quad \frac{V_1}{V_2} = \frac{T_1}{T_2} \qquad \ldots(15.1)$$

2. Isochoric (Constant Volume) Process

If the volume is kept constant during a process, the absolute pressure varies directly with the change in its absolute temperature and the process is known as **isochoric process.**

$$p \propto T$$

$$\frac{p}{T} = \text{constant}$$

$$\frac{p_1}{T_1} = \frac{p_2}{T_2} \quad \text{or} \quad \frac{p_1}{p_2} = \frac{T_1}{T_2} \qquad \text{...(15.2)}$$

3. Isothermal (Constant Temperature) Process

In a process in which temperature is kept constant by adding or subtracting heat, the absolute pressure varies inversely as its volume. It is known as **Boyle's law.**

$$p \propto \frac{1}{V}$$

$$pV = \text{constant}$$

$$p_1V_1 = p_2V_2 \quad \text{or} \quad \frac{p_1}{p_2} = \frac{V_2}{V_1} \qquad \text{...(15.3)}$$

Since Density = Mass/Volume

or Mass = Density × Volume

$$m = \rho_1 V_1 = \rho_2 V_2 \quad \text{or} \quad \frac{V_2}{V_1} = \frac{\rho_1}{\rho_2}$$

and Eq. 15.3 can be written as

$$\frac{p_1}{p_2} = \frac{\rho_1}{\rho_2} \qquad \text{...(15.4)}$$

4. Adiabatic Process

If no heat is transferred from the gas during a process, it is known as an **adiabatic process.** According to this process

$$pV^{\gamma} = \text{constant}$$

$$p_1V_1^{\gamma} = p_2V_2^{\gamma}$$

or

$$\frac{p_1}{p_2} = \left(\frac{V_2}{V_1}\right)^{\gamma} = \left(\frac{\rho_1}{\rho_2}\right)^{\gamma} \qquad \text{...(15.5)}$$

If the adiabatic process is frictionless, it is known as a reversible (isentropic) process. Also, in case $\gamma \neq C_p/C_v$, but equal to some positive value, then the process is known as **polytropic** and the relation

$$pV^{\chi} = \frac{p}{\rho^{\chi}} = \text{constant} \qquad \text{...(15.6)}$$

15.3 BASIC EQUATIONS OF COMPRESSIBLE FLOW

The following are the basic equations of compressible flow :

1. Equation of State
2. Continuity Equation

3. Momentum Equation
4. Energy Equation (or Bernoulli's Equation)

1. Equation of State

If a perfect gas having a mass m, absolute pressure p, absolute temperature T occupies a volume V, the following relation is known as the **characteristic equation of state** :

$$pV = mRT$$

where R is gas constant. The equation is also written as

$$\frac{p}{\rho} = RT \qquad \ldots\left(\because \frac{m}{V} = \rho\right) \qquad \ldots(15.7)$$

Units of R are

$$R = \frac{pV}{mT} = \frac{(\text{N/m}^2)\,\text{m}^3}{\text{kg K}} = \text{N.m/kg.K or J/kg.K}$$

In practice, no gas follows this law rigidly, but many gases tend towards it.

2. Continuity Equation

In a one-dimensional flow, mass flow rate is to be constant according to the conservation of mass,

$$\rho AV = \text{constant}$$

Differentiating it, $\quad d\,(AV) = 0$

$$\rho d\,(AV) + AV \,.\, d\rho = 0$$

$$\rho\,(A \,.\, dV + V \,.\, dA) + AV \,.\, d\rho = 0$$

$$\rho A dV + \rho V dA + AV \,.\, d\rho = 0$$

Dividing by ρAV throughout,

$$\frac{dV}{V} + \frac{dA}{A} + \frac{d\rho}{\rho} = 0 \qquad \ldots(15.8)$$

It is known as **continuity equation** in differential form.

3. Momentum Equation

The momentum per second of flowing fluid is the product of the mass per second and the velocity of the flow. As the mass (or $\rho A.V$) is constant at every section of the flow due to continuity equation, the momentum per second is independent of the compressible effects and the **momentum equation** is same as for incompressible fluids.

i.e. Net force = Rate of change of momentum

$$F = m\,(V_1 - V_2)$$

where V_1 and V_2 are the initial and final velocities of flow in the given direction.

4. Energy Equation (Bernoulli's equation)

In section 6.3, the energy equation was derived for incompressible fluids. Following the same procedure, for compressible fluids also the same **Euler energy equation** is obtained

$$\frac{dp}{\rho} + V \,.\, dV + g \,.\, dz = 0 \qquad \ldots(15.9)$$

Integrating it,

$$\int \frac{dp}{\rho} + \int V \,.\, dV + \int g \,.\, dz = \text{constant}$$

$$\int \frac{dp}{\rho} + \frac{V^2}{2} + gz = \text{constant} \qquad \text{...(15.10)}$$

As p and ρ are both variables in this case, hence any of these cannot be taken outside integration sign. Moreover, the variation of p and ρ is governed by the process (isothermal or adiabatic).

Bernoulii's Equation for Isothermal Process

In an isothermal process,

$$\frac{p}{\rho} = \text{constant} = C_1 \quad \text{or} \quad \rho = \frac{p}{C_1}$$

Inserting this value in Eq. 15.10,

$$C_1 \int \frac{dp}{p} + \frac{V^2}{2} + gz = \text{constant}$$

$$C_1 \log p + \frac{V^2}{2} + gz = \text{constant}$$

$$\frac{p}{\rho} \log p + \frac{V^2}{2} + gz = \text{constant}$$

$$\frac{p}{\rho g} \log p + \frac{V^2}{2g} + z = \text{constant} \qquad \text{...(15.11)}$$

Bernoulli's Equation for Adiabatic Process

In an adiabatic process,

$$\frac{p}{\rho^\gamma} = \text{constant} = C_1 \quad \text{or} \quad \rho = \left(\frac{p}{C_1}\right)^{\frac{1}{\rho}}$$

Now $$\int \frac{dp}{\rho} = \int \frac{dp}{(p/C_1)^{1/\gamma}} = C_1{}^{1/\gamma} \int \frac{dp}{p^{1/\gamma}} = C_1{}^{1/\gamma} \,.\, \int p^{-1/\gamma} \,.\, dp$$

$$= C_1{}^{1/\gamma} \frac{p^{-1/\gamma + 1}}{-\frac{1}{\gamma} + 1} = \frac{C_1{}^{1/\gamma} \,.\, p^{(\gamma - 1)/\gamma}}{\frac{\gamma - 1}{\gamma}} = \frac{\gamma}{\gamma - 1} C_2{}^{1/\gamma} p^{(\gamma - 1)/\gamma} = \frac{\gamma}{\gamma - 1} \left(\frac{p}{\rho^\gamma}\right)^{\frac{1}{\gamma}} \,.\, p^{(\gamma - 1)/\gamma}$$

$$= \frac{\gamma}{\gamma - 1} \frac{p^{1/\gamma + (\gamma - 1)/\gamma}}{\rho} = \frac{\gamma - 1}{\gamma} \frac{p}{\rho}$$

Thus the Bernoulli's equation becomes,

$$\frac{\gamma}{\gamma - 1} \frac{p}{\rho} + \frac{V^2}{2} + gz = \text{constant}$$

$$\frac{\gamma}{\gamma - 1} \frac{p}{\rho g} + \frac{V^2}{2g} + z = \text{constant} \qquad \text{...(15.12)}$$

Example 15.1 A gas at a temperature of 5° is flowing through a horizontal pipe (Fig. 15.1). The gauge pressures at sections 1 and 2 are 3·5 bar and 2·6 bar respectively. Find the velocity of the gas at these sections if the atmospheric pressure is 1 bar and $R = 287$ J/kg.K.

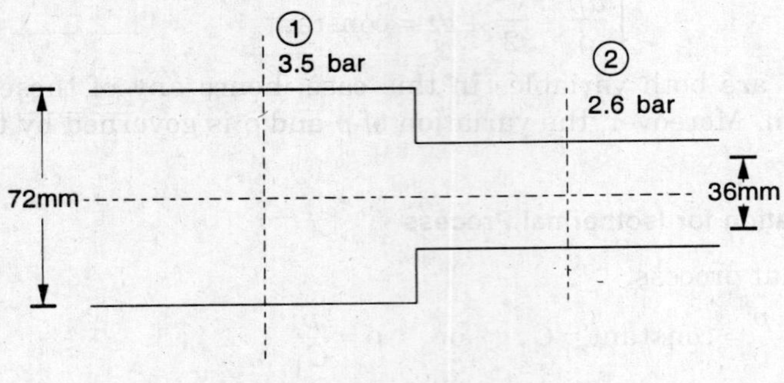

Fig. 15.1

Solution $p_1 = 3{\cdot}5 + 1 = 4{\cdot}5$ bar (absolute)

$p_2 = 2{\cdot}6 + 1 = 3{\cdot}6$ bar (absolute)

$T_1 = 273 + 5 = 278°\text{K}$

$$A_1 = \frac{\pi}{4} \times 0{\cdot}072^2 = 0{\cdot}004072 \text{ m}^2$$

$$A_2 = \frac{\pi}{4} \times 0{\cdot}036^2 = 0{\cdot}001018 \text{ m}^2$$

Applying Bernoulli's equation at sections 1 and 2 for isothermal process,

$$\frac{p_1}{\rho_1 g} \log p_1 + \frac{V_1^2}{2g} + z_1 = \frac{p_2}{\rho_2 g} \log p_2 + \frac{V_2^2}{2g} + z_2$$

The pipe is horizontal, $z_1 = z_2$, and $\dfrac{p_1}{\rho_1} = \dfrac{p_2}{\rho_2}$ for isothermal process,

$$\frac{p_1}{\rho_1 g} \log p_1 + \frac{V_1^2}{2g} = \frac{p_1}{\rho_1 g} \log p_2 + \frac{V_2^2}{2g}$$

$$\frac{p_1}{\rho_1} \log p_1 - \frac{p_1}{\rho_1} \log p_2 = \frac{V_2^2 - V_1^2}{2} \quad \text{(cancelling } g\text{)}$$

$$\frac{p_1}{\rho_1} \log \frac{p_1}{p_2} = \frac{V_2^2 - V_1^2}{2} \qquad \ldots\text{(i)}$$

Also by applying continuity equation at section 1 and 2,

$$\rho_1 A_1 V_1 = \rho_2 A_2 V_2$$

$$V_2 = \frac{\rho_1 A_1 V_1}{\rho_2 A_2}$$

$$= \frac{p_1 A_1}{p_2 A_2} V_1 \qquad \left(\frac{\rho_1}{\rho_2} = \frac{p_1}{p_2}\right)$$

$$= \frac{4{\cdot}5 \times 0{\cdot}004072}{3{\cdot}6 \times 0{\cdot}001018} V_1$$

$$= 5V_1$$

From (i),

$$\frac{p_1}{\rho_1} \log\left(\frac{4{\cdot}5}{3{\cdot}6}\right) = \frac{(5V_1)^2 - V_1^2}{2} = 12V_1^2$$

$$0{\cdot}223 \frac{p_1}{\rho_1} = 12V_1^2$$

$$0{\cdot}223RT_1 = 12V_1^2$$

$$V_1^2 = \frac{0{\cdot}223 \times 287 \times 278}{12} = 1483{\cdot}6$$

$$V_1 = 38{\cdot}52 \text{ m/s}$$

$$V_2 = 5V_1 = 5 \times 38{\cdot}51 = 192{\cdot}59 \text{ m/s}$$

Example 15.2 A gas flows through a horizontal pipe and at a section its velocity is 400 m/s where its pressure is 80 kN/m^2 absolute and temperature 50°. The diameter of the pipe change after this section and the pressure increases to 120 kN/m^2 absolute. Assuming the flow to be adiabatic and the value of R = 287 J/kg.K and γ = 1·4, find the velocity of the gas at the new section.

Solution $T_1 = 273 + 50 = 323$°K

$V_1 = 400$ m/s

$p_1 = 80$ kN/m^2

$p_2 = 120$ kN/m^2

Applying Bernoulli's equation at sections 1 and 2 for adiabatic process,

$$\left(\frac{\gamma}{\gamma - 1}\right)\frac{p_1}{\rho_1 g} + \frac{V_1^2}{2g} + z_1 = \left(\frac{\gamma}{\gamma - 1}\right)\frac{p_2}{\rho_2 g} + \frac{V_2^2}{2g} + z_2$$

The pipe is horizontal, $z_1 = z_2$, and cancelling g,

$$\left(\frac{\gamma}{\gamma - 1}\right)\frac{p_1}{\rho_1} + \frac{V_1^2}{2} = \left(\frac{\gamma}{\gamma - 1}\right)\frac{p_2}{\rho_2} + \frac{V_2^2}{2}$$

$$\left(\frac{\gamma}{\gamma - 1}\right)\left(\frac{p_1}{\rho_1} - \frac{p_2}{\rho_2}\right) = \frac{V_2^2 - V_1^2}{2}$$

$$\left(\frac{\gamma}{\gamma - 1}\right)\frac{p_1}{\rho_1}\left(1 - \frac{p_2}{\rho_2}\frac{\rho_1}{p_1}\right) = \frac{V_2^2 - V_1^2}{2}$$

$$\left(\frac{\gamma}{\gamma-1}\right)\frac{p_1}{\rho_1}\left[1-\frac{p_2}{p_1}\left(\frac{p_1}{p_2}\right)^{\frac{1}{\gamma}}\right]=\frac{V_2^{\,2}-V_1^{\,2}}{2} \qquad \left[\frac{\rho_1}{\rho_2}=\left(\frac{p_1}{p_2}\right)^{\frac{1}{\gamma}}\right]$$

$$\left(\frac{\gamma}{\gamma-1}\right)\frac{p_1}{\rho_1}\left[1-\left(\frac{p_2}{p_1}\right)^{1-\frac{1}{\gamma}}\right]=\frac{V_2^{\,2}-V_1^{\,2}}{2}$$

$$\left(\frac{\gamma}{\gamma-1}\right)RT_1\left[1-\left(\frac{p_2}{p_1}\right)^{\frac{\gamma-1}{\gamma}}\right]=\frac{V_2^{\,2}-V_1^{\,2}}{2g}$$

$$\left(\frac{14}{14-1}\right)287\times 323\left[1-\left(\frac{120}{80}\right)^{\frac{1\cdot 4-1}{1\cdot 4}}\right]=\frac{V_2^{\,2}-400^2}{2}$$

$$3\cdot 5\times 92701\times(-0\cdot 1228)\times 2=V_2^{\,2}-160000$$

$$-79685\cdot 8+160000=V_2^{\,2}$$

$$V_2^{\,2}=80314\cdot 2$$

$$V_2=283\cdot 4\text{ m/s}$$

15.4 WAVE PROPAGATION

A minor disturbance on one end of a rigid body is instantaneously transmitted to the other end as the molecules are close together. But in fluids, the molecules are relatively apart and each molecule has to travel through a small distance to transmit the disturbance to the next molecule. The distance between the molecules depends upon the density and pressure of the fluids *i.e.* on the elastic properties of the fluid.

The propagation of disturbance through a medium is similar to the propagation of sound. Thus an infinitesimal disturbance at a point in a stationary fluid propagates in a spherical form *i.e.* radially in all directions at the velocity of sound.

The speed of propagation of sound in a media is known as **sonic acoustic velocity.**

Consider a long rigid pipe of uniform cross-sectional area fitted with a piston and having a compressible fluid initially at rest (Fig. 15.2). Now, if the piston is suddenly moved towards right, a pressure wave is propagated through the fluid with the sonic velocity.

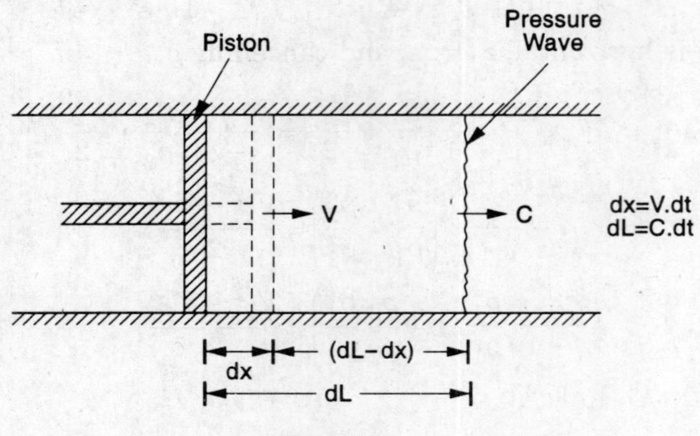

Fig. 15.2

Let A = cross-sectional area of the pipe

V = velocity of the piston

ρ = density of the fluid at rest

p = pressure of the fluid at rest

dt = a small period of time during which the piston moves

C = velocity of the pressure (sound) wave travelling in the fluid

Consider a controlled volume of the fluid of length dL. Let the piston move through a distance dx in time dt so that during this interval of time, the pressure wave travels a distance dL. As the volume of the fluid is decreased from $A \,.\, dL$ to $A\,(dL - dx)$, the pressure and the density of the fluid are increased.

Let $p + dp$ = pressure of the fluid after compression

$\rho + d\rho$ = density of the fluid after compression

From continuity equation,

Mass of the fluid before compression = Mass of the fluid after compression

$$\rho \,.\, A \,.\, dL = (\rho + d\rho) \,.\, A \,.\, (dL - dx)$$

$$\rho \,.\, A \,.\, C \,.\, dt = (\rho + d\rho) \,.\, A \,.\, (C \,.\, dt - V \,.\, dt)$$

$$= (\rho + d\rho) \,.\, A \,.\, dt\,(C - V)$$

$$\rho \,.\, C = \rho \,.\, C - \rho \,.\, V + C \,.\, d\rho - V \,.\, d\rho$$

$$C \,.\, d\rho = \rho \,.\, V + V \,.\, d\rho$$

As the velocity of the piston (V) is very small as compared to the velocity of the pressure wave (C), neglecting the term $V \,.\, d\rho$,

$$\rho \,.\, V = C \,.\, d\rho \qquad \text{...(15.13)}$$

When the piston is moved with a velocity V, the fluid also moves with a velocity equal to the velocity of the piston and the force on the piston increases from $p \,.\, A$ to $(p + dp) \,.\, A$.

Applying impulse momentum equation,

Force on the fluid = Rate of change of momentum

= Mass flow rate $\times$ Change in velocity

$$(p + dp)\,A - pA = \frac{p \,.\, A \,.\, dL}{dt}\,(V - 0)$$

$$dp \,.\, A = \frac{p \,.\, A \,.\, C \,.\, dt}{dt}\,V = \rho \,.\, A \,.\, CV$$

$$C = \frac{dp}{\rho \,.\, V}$$

$$= \frac{dp}{C \,.\, d\rho} \qquad \text{(from Eq. 15.13)}$$

or
$$C^2 = \frac{dp}{d\rho}$$

$$C = \sqrt{\frac{dp}{d\rho}} \qquad \text{...(15.14)}$$

Also as the mass of a fluid is constant

Density $\times$ Volume = Constant

or $\rho \,.\, V$ = Constant

Differentiating both sides,

$$\rho \,.\, dV + V \,.\, d\rho = 0$$

$$\rho \,.\, dV = -V \,.\, d\rho$$

$$\frac{dV}{V} = -\frac{d\rho}{\rho} \qquad \text{...(i)}$$

Now, bulk modulus, K is defined as

$$K = \frac{\text{Increase in pressure}}{\text{Decrease in volume/Original volume}}$$

$$= \frac{dp}{-(dV/V)}$$

(Negative sign indicates that with increase in pressure, the volume decreases).

$$K = \frac{dp}{d\rho/\rho} \qquad \left(\because \frac{dV}{V} = -\frac{d\rho}{\rho} \text{ from (i)}\right)$$

$$= \frac{\rho \,.\, dp}{d\rho}$$

$$\frac{dp}{d\rho} = \frac{K}{\rho}$$

Thus Eq. 15.14 is also written as

$$C = \sqrt{\frac{dp}{d\rho}} = \sqrt{\frac{K}{\rho}} \qquad \text{...(15.15)}$$

In case of isothermal process,

$$\frac{p}{\rho} = \text{constant or } p\rho^{-1} = \text{Constant}$$

On differentiating,

$$p\,(-1)\,\rho^{-2} \,.\, d\rho + \rho^{-1} \,.\, dp = 0$$

$$\frac{p}{\rho^2} \cdot d\rho = \frac{dp}{\rho}$$

$$\frac{dp}{d\rho} = \frac{p}{\rho} = RT$$

and

$$C = \sqrt{\frac{dp}{d\rho}} = \sqrt{\frac{p}{\rho}} = \sqrt{RT} \qquad \text{...(15.16)}$$

If the process is adiabatic,

$$\frac{p}{\rho^{\gamma}} = \text{Constant}$$

$$p \,.\, \rho^{-\gamma} = \text{Constant}$$

Differentiating both sides,

$$p\,(-\gamma)\,\rho^{-\gamma-1} \,.\, d\rho + \rho^{-\gamma} \,.\, dp = 0$$

$$\frac{p \,.\, \gamma}{\rho^{\gamma+1}} \cdot d\rho = \frac{dp}{\rho^{\gamma}}$$

$$\frac{p \,.\, \gamma}{\rho} \cdot d\rho = dp$$

$$\therefore \frac{dp}{d\rho} = \frac{p}{\rho} \cdot \gamma = \gamma RT$$

$$C = \sqrt{\frac{dp}{d\rho}} = \sqrt{\frac{\gamma p}{\rho}} = \sqrt{\gamma RT} \qquad ...(15.17)$$

If the disturbances are propagated through air, the process is assumed as adiabatic as the velocity of the pressure wave is very high and there is no appreciable heat transfer.

15.5 MACH NUMBER

Mach number is defined as the square root of the ratio of the inertia force of a fluid to the elastic force.

Mach number,

$$M = \sqrt{\frac{\text{Inertia force}}{\text{Elastic force}}} = \sqrt{\frac{\rho L^2 V^2}{KL^2}} = \sqrt{\frac{V^2}{K/\rho}} = \frac{V}{\sqrt{K/\rho}} = \frac{V}{C} \qquad ...(15.18)$$

$$= \frac{\text{Velocity of a body moving in a fluid}}{\text{Velocity of sound in the fluid}}$$

For a compressible flow, Mach number is an important parameter. Depending upon its value, the flow is classified as

1. Subsonic flow ($M < 1$ or $V < C$)
2. Sonic flow ($M = 1$ or $V = C$)
3. Supersonic flow ($M > 1$ or $V > C$)

Example 15.3 Determine the sonic velocity of (i) mercury with a bulk modulus of 27 GN/m^2 (ii) crude oil with a specific gravity 0·9 and bulk modulus 1·6 GN/m^2.

Solution $C_{mer} = \sqrt{\frac{K}{\rho}} = \sqrt{\frac{27 \times 10^9}{13.6 \times 1000}} = 1409$ m/s

$$C_{\text{oil}} = \sqrt{\frac{1.6 \times 10^9}{0.9 \times 1000}} = 1333.3 \text{ m/s}$$

Example 15.4 The speed of an aeroplane corresponds to $M = 1{\cdot}8$. The aeroplane flies at a height of 10 km where the temperature is – 30°C. If the value of $R = 287$ J/kg.K and $\gamma = 1{\cdot}4$, determine the speed of the aeroplane.

Solution $T = 273 - 30 = 243$°K

$$M = \frac{V}{C}$$

$$M = \frac{V}{\sqrt{\gamma RT}}$$

$$1.8 = \frac{V}{\sqrt{1.4 \times 287 \times 243}}$$

$$V = 562.45 \text{ m/s} = 562.45 \times \frac{3600}{1000} = 2024.8 \text{ km/h}$$

15.6 PRESSURE WAVES IN COMPRESSIBLE FLUIDS

Any disturbance in a stationary compressible fluid propagates radially in all directions with the velocity of sound (Fig. 15.3a). Let us now consider a small projectile moving in a straight line with velocity V through a stationary compressible fluid. Initially, the projectile is at A when time $t = 0$.

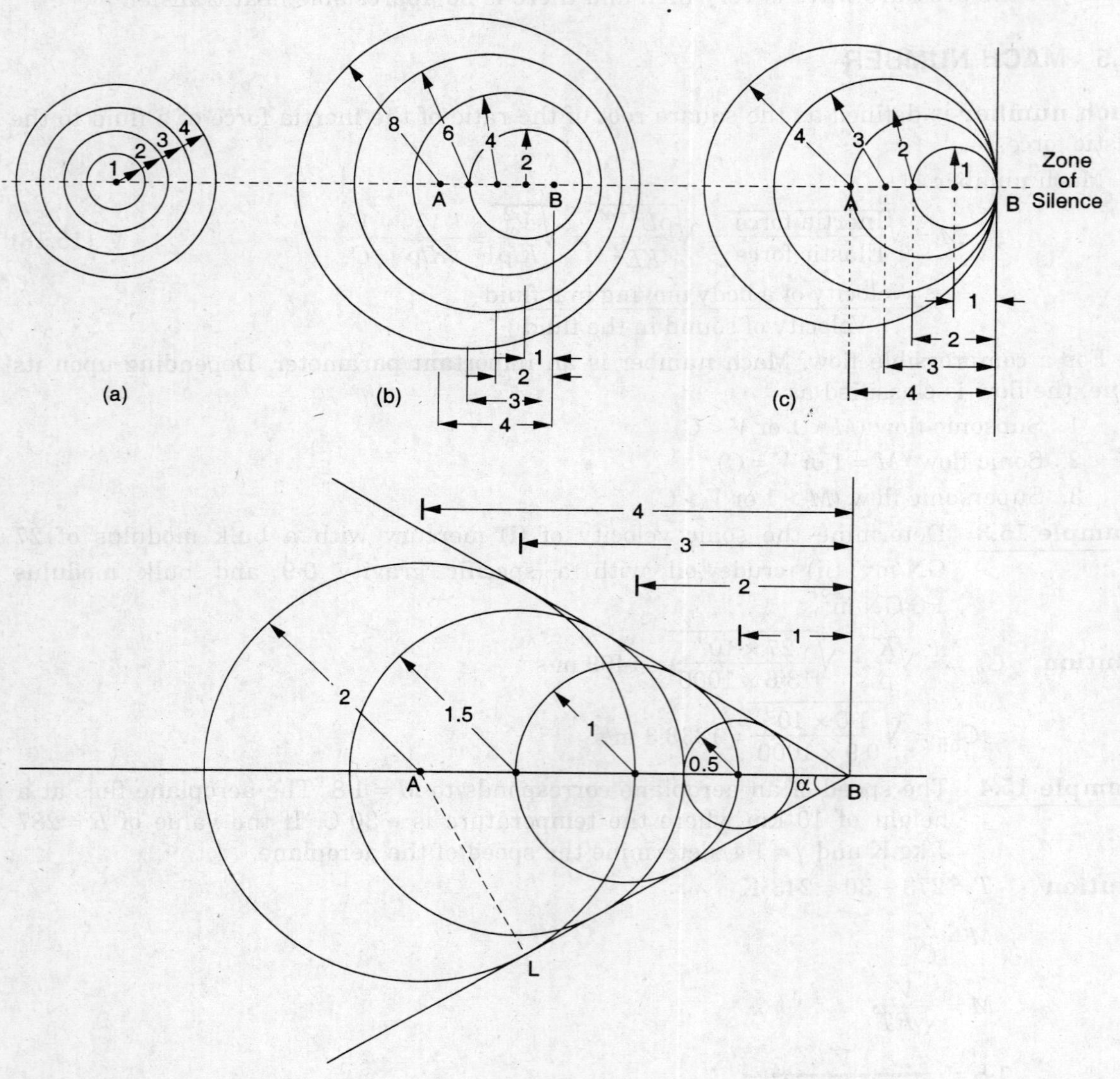

Fig. 15.3

Case 1.

$$M < 1 \text{ or } V < C \left(\text{say } V = \frac{1}{2} C \right)$$

Assume that a projectile travels at a uniform velocity of one unit distance per second. In 4 seconds, the projectile travels from point A to point B, a distance of 4 units (Fig. 15.3b). During this time, the pressure wave propagates radially in all directions at velocity $C = 2V$ and covers a distance of 8 units. Since V is less than C, the point B lies inside the sphere of the pressure wave. If the projectile is considered at any intermediate point, say at D from where it takes 3 seconds to reach the point B, the pressure wave will propagate in a sphere of radius of 6 units during the same period of time. Thus again the point B lies inside the sphere of the pressure wave. Similarly, a pressure wave originated at midpoint E (from where the point B is reached in 2 seconds) travels in a sphere of radius of 4 units and so on. Thus the projectile at B lies inside the sphere formed by the wave started at A or at intermediate point.

Case 2.

$$M = 1 \text{ or } V = C$$

In this case, the pressure wave always travels with the projectile. Any pressure wave propagated at any point during the path of the projectile reaches point B alongwith the projectile (Fig. 15.3c). Therefore, in case of sonic flow, the zone of action is always behind the projectile and ahead of that there is zone of silence.

Case 3.

$$M > 1 \text{ or } V > C \text{ (say } V = 2C)$$

As the velocity of the projectile is more than the pressure wave, the projectile always outside the sphere of pressure wave. If the projectile travels from A to B (say 4 units in 4 secnds), the pressure wave travels 2 units in the the same time in all directions (Fig. 15.3d). Similarly, a projectile at D reaches B in 3 seconds whereas the pressure wave travels only 3/2 units and so on. Thus in all cases, the sphere of propagation of disturbance or pressure wave always lags behind the projectile. If tangents are drawn to the different circles that represent the propagated spherical waves on both sides, a cone is obtained with vertex at B. This cone is known as Mach cone. The semi-vertex angle of the Mach cone is known as **Mach angle.**

$$\sin \alpha = \frac{AL}{AB} = \frac{C \cdot t}{V \cdot t} = \frac{C}{V} = \frac{1}{M} \quad \text{...(15.19)}$$

As the effect of the disturbance is felt only in the inside region of the Mach cone, it is known as the zone of action. The region outside the Mach cone is zone of silence. Thus when an aeroplane moves with supersonic speed, its noise is heard only after the plane has already passed over us.

Example 15.5 Determine the velocity of a bullet fired in standard air if the Mach angle is 36°. R = 287 J/kg.K and γ = 1·4. The air temperature is 18°C.

Solution $T = 273 + 18 = 291°\text{K}$

$$\sin \alpha = \frac{C}{V} = \frac{\sqrt{\gamma RT}}{V}$$

$$\sin 36^\circ = \frac{\sqrt{1{\cdot}4 \times 287 \times 291}}{V}$$

$$= \frac{341{\cdot}94}{V}$$

$$V = 581{\cdot}75 \text{ m/s}$$

15.7 STAGNATION PROPERTIES

In a flowing fluid, a point on the surface of an immersed body where the velocity is zero, is called stagnation point. The values of pressure, temperature and density at that point are known as **stagnation properties.**

Stagnation Pressure

Consider a compressible fluid flowing past an immersed body. Let the conditions be adiabatic frictionless. Consider two points O and S on a streamline. Point O is far away from the body and S is the stagnation point where the velocity of flow is zero (Fig. 15.4).

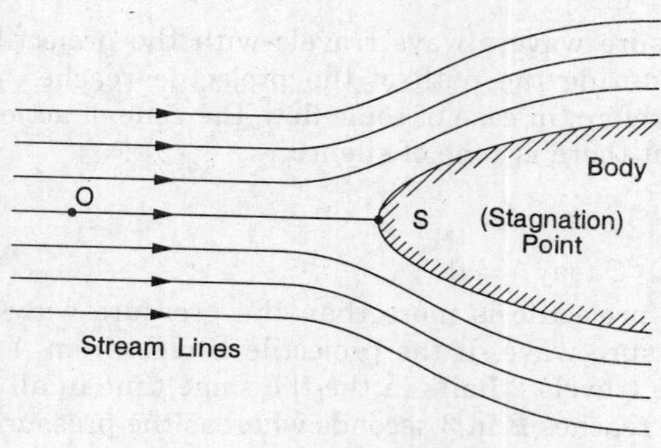

Fig. 15.4

Let p, V, ρ and T represent the pressure, velocity, density and temperature respectively of the fluid with subscripts at respective points.

Applying Bernoulli's equation for adiabatic conditions,

$$\left(\frac{\gamma}{\gamma-1}\right)\frac{p_0}{\rho_0 g} + \frac{V_0^{\,2}}{2g} + z_0 = \left(\frac{\gamma}{\gamma-1}\right)\frac{p_s}{\rho_s g} + \frac{V_s^{\,2}}{2g} + z_s$$

Now $z_0 = z_s$, $V_s = 0$

$$\left(\frac{\gamma}{\gamma-1}\right)\frac{p_0}{\rho_0} + \frac{V_0^{\,2}}{2} = \left(\frac{\gamma}{\gamma-1}\right)\frac{p_s}{\rho_s}$$

$$\left(\frac{\gamma}{\gamma-1}\right)\left(\frac{p_0}{\rho_0} - \frac{p_s}{\rho_s}\right) = \frac{V_0^{\,2}}{2}$$

$$\left(\frac{\gamma}{\gamma-1}\right)\frac{p_0}{\rho_0}\left(1 - \frac{p_s}{\rho_s}\frac{\rho_0}{p_0}\right) = -\frac{V_0^{\,2}}{2}$$

$$\left(\frac{\gamma}{\gamma-1}\right)\frac{p_0}{\rho_0}\left(1-\frac{p_s}{p_0}\frac{\rho_0}{\rho_s}\right)=-\frac{V_0{}^2}{2}$$

$$\left(\frac{\gamma}{\gamma-1}\right)\frac{p_0}{\rho_0}\left[1-\frac{p_s}{p_0}\left(\frac{p_0}{p_s}\right)^{\frac{1}{\gamma}}\right]=-\frac{V_0{}^2}{2} \qquad \text{(using Eq. 15.5)}$$

$$\left(\frac{\gamma}{\gamma-1}\right)\frac{p_0}{\rho_0}\left[1-\left(\frac{p_s}{p_0}\right)^{1-\frac{1}{\gamma}}\right]=-\frac{V_0{}^2}{2}$$

$$1-\left(\frac{p_s}{p_0}\right)^{\frac{\gamma-1}{r}}=-\frac{V_0{}^2}{2}\times\frac{\gamma-1}{\gamma}\cdot\frac{\rho_0}{p_0}$$

$$\left(\frac{p_s}{p_0}\right)^{\frac{\gamma-1}{\gamma}}=1+\frac{V_0{}^2}{2}\frac{\gamma-1}{\gamma}\cdot\frac{\rho_0}{p_0}$$

$$=1+\frac{V_0{}^2}{2}(\gamma-1)\cdot\frac{1}{\gamma p_0/\rho_0}$$

$$=1+\frac{V_0{}^2}{2}(\gamma-1)\cdot\frac{1}{C_0^2} \qquad \text{(using Eq. 15.17)}$$

$$=1+\frac{V_0{}^2}{2C_0{}^2}(\gamma-1)$$

$$=1+\frac{M_0{}^2}{2}(\gamma-1)$$

$$(p_s)^{(\gamma-1)/\gamma}=(p_0)^{(\gamma-1)/\gamma}\left[1+\frac{M_0{}^2}{2}(\gamma-1)\right]$$

$$p_s=p_0\left[1+\frac{(\gamma-1)}{2}M_0^2\right]^{\frac{\gamma}{\gamma-1}} \qquad \text{...(15.20)}$$

$$\frac{p_s}{p_0}=\left[1+\frac{(\gamma-1)}{2}M_0{}^2\right]^{\frac{\gamma}{\gamma-1}}$$

$$\left(\frac{\rho_s}{\rho_0}\right)^{\gamma}=\left[1+\frac{(\gamma-1)}{2}M_0{}^2\right]^{\frac{\gamma}{\gamma-1}}$$

$$\rho_s=\rho_0\left[1+\frac{(\gamma-1)}{2}M_0{}^2\right]^{\frac{1}{\gamma-1}} \qquad \text{...(15.21)}$$

As $\dfrac{p_s}{\rho_s}=RT_s$

$$T_s = \frac{1}{R}\frac{p_s}{\rho_s} = \frac{1}{R}\frac{p_0\left[1+\frac{(\gamma-1)}{2}M_0^2\right]^{\frac{\gamma}{\gamma-1}}}{\rho_0\left[1+\frac{(\gamma-1)}{2}M_0^2\right]^{\frac{1}{\gamma-1}}}$$

$$= \frac{1}{R}\frac{p_0}{\rho_0}\left[1+\frac{(\gamma-1)}{2}M_0^2\right]^{\frac{\gamma}{\gamma-1}-\frac{1}{\gamma-1}}$$

$$= \frac{1}{R}RT_0\left[1+\frac{(\gamma-1)}{2}M_0^2\right]$$

$$T_s = T_0\left[1+\frac{(\gamma-1)}{2}M_0^2\right] \qquad \ldots(15.22)$$

Example 15.6 An aeroplane is flying at 1080 km/h through still air having a pressure of 80 kN/m^2 absolute and temperature – 6°C. Determine the stagnation pressure, temperature and density on the stagnation point on the nose of the plane. R = 287 J/kg.K and γ = 1·4.

Solution T_s = 273 – 6 = 267°C

$$p_0 = 80 \text{ kN/m}^2$$

$$V_0 = \frac{1080 \times 1000}{3600} = 300 \text{ m/s}$$

$$M_0 = \frac{V_0}{C_0} = \frac{V_0}{\sqrt{\gamma R T_0}} = \frac{300}{\sqrt{1{\cdot}4 \times 287 \times 267}} = 0{\cdot}916$$

$$p_s = p_0\left[1+\frac{(\gamma-1)}{2}M_0^2\right]^{\frac{\gamma}{\gamma-1}}$$

$$= 80\left[1+\frac{(14-1)}{2}\times 0{\cdot}916^2\right]^{\frac{1{\cdot}4}{1{\cdot}4-1}}$$

$$= 80 \times (1{\cdot}1678)^{35}$$

$$= 80 \times 1{\cdot}7211$$

$$= 137{\cdot}69 \text{ kN/m}^2$$

$$T_s = T_0\left[1+\frac{(\gamma-1)}{2}M_0^2\right]$$

$$= 267\left[1+\frac{(1{\cdot}4-1)}{2}\times 0{\cdot}916^2\right]$$

$$= 267 \times 1{\cdot}1678$$

$$= 311{\cdot}8\text{°K} \quad \text{or} \quad (311{\cdot}8 - 273) = 38{\cdot}8\text{°C}$$

$$\frac{p_s}{\rho_s} = RT_s$$

$$\rho_s = \frac{p_s}{RT_s} = \frac{137{\cdot}69}{287 \times 311{\cdot}8} = 1{\cdot}539 \text{ kg/m}^2$$

15.8 AREA VELOCITY RELATIONSHIPS

The continuity equation for an compressible flow is

$$\frac{dV}{V} + \frac{dA}{A} + \frac{d\rho}{\rho} = 0 \qquad \text{(Eq. 15.8)}$$

$$\frac{dA}{A} = -\frac{dV}{V} - \frac{d\rho}{\rho} \qquad \text{...(15.23)}$$

The Euler's equation for compressible flow (neglecting z-terms),

$$\frac{dp}{\rho} + V \,.\, dV = 0 \qquad \text{(Eq. 15.9)}$$

$$\frac{dp}{d\rho} \cdot \frac{d\rho}{\rho} + V \,.\, dV = 0$$

$$C^2 \cdot \frac{d\rho}{\rho} + V \,.\, dV = 0$$

$$\frac{d\rho}{\rho} = -\frac{V \,.\, dV}{C^2}$$

From Eq. 15.23,

$$\frac{dA}{A} = -\frac{dV}{V} + \frac{V \,.\, dV}{C^2}$$

$$= \frac{dV}{V}\left(\frac{V^2}{C^2} - 1\right)$$

$$= \frac{dV}{V}(M^2 - 1) \qquad \text{...(15.24)}$$

The following cases are worth consideration :

Subsonic Flow ($M < 1$)

Eq. 15.24 will be of the form $dA/A = -C \,.\, dV/V$ where C denotes some numerical value. dA and dV are of opposite signs, indicating that if the area increases, velocity decreases and vice-versa. This is also true for incompressible fluids (Fig. 15.5).

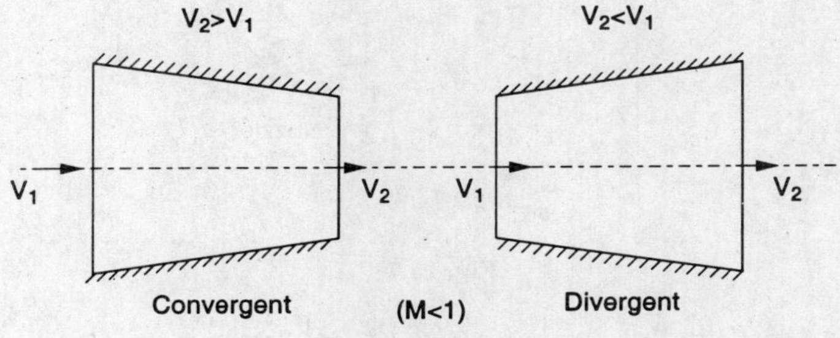

Fig. 15.5

Supersonic Flow ($M > 1$)

Eq. 15.24 will be of the form $dA/A = C \cdot dV/V$. dA and dV are of same signs which means that when the area increases, the velocity also increases (Fig. 15.6).

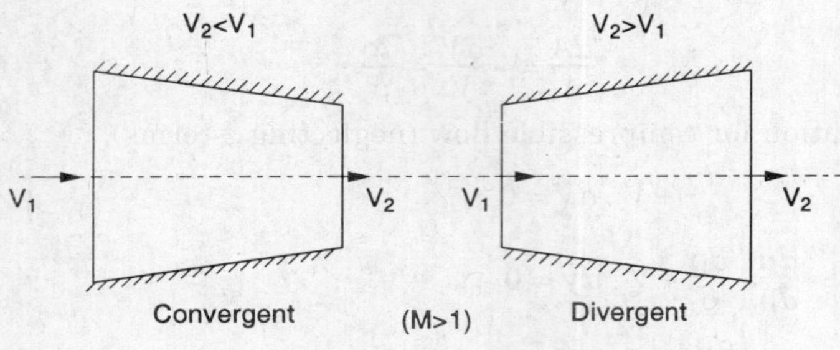

Fig. 15.6

Sonic Flow ($M = 1$)

$dA/A = 0$ which means area is constant.

Thus if a fluid is to accelerate in a subsonic region, the area decreases whereas in supersonic region, it increases. This means a convergent-divergent nozzle can be used to accelerate the flow throughout such that in the convergent portion, the flow accelerates from subsonic velocity to sonic velocity (at the throat) and in the divergent portion, the velocity accelerates from sonic to supersnoic.

15.9 DISCHARGE THROUGH A CONVERGENT NOZZLE

Consider a compressible fluid in a large tank fitted with a convergent nozzle (Fig. 15.7). Let the flow be adiabatic. Then applying the Bernoulli's equation for adiabatic flow to two points, 1 inside the tank and 2 at the exit of nozzle

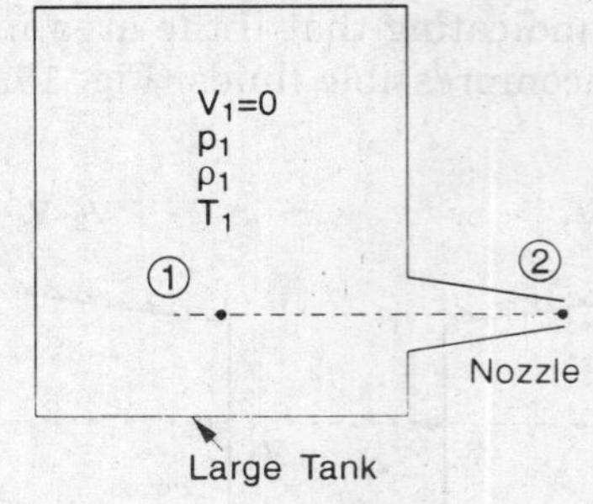

Fig. 15.7

$$\left(\frac{\gamma}{\gamma-1}\right)\frac{p_1}{\rho_1 g}+\frac{V_1^2}{2g}+z_1=\left(\frac{\gamma}{\gamma-1}\right)\frac{p_2}{\rho_2 g}+\frac{V_2^2}{2g}+z_2$$

Now $z_1 = z_2$ and $V_1 = 0$

$$\left(\frac{\gamma}{\gamma-1}\right)\frac{p_1}{\rho_1 g}=\left(\frac{\gamma}{\gamma-1}\right)\frac{p_2}{\rho_2 g}+\frac{V_2^2}{2g}$$

$$\frac{\gamma}{\gamma-1}\left(\frac{p_1}{\rho_1}-\frac{p_2}{\rho_2}\right)=\frac{V_2^2}{2}$$

$$V_2=\sqrt{\frac{2\gamma}{\gamma-1}\frac{p_1}{\rho_1}\left(1-\frac{p_2}{\rho_2}\frac{\rho_1}{p_1}\right)}$$

$$=\sqrt{\frac{2\gamma}{\gamma-1}\frac{p_1}{\rho_1}\left(1-\frac{p_2}{p_1}\frac{\rho_1}{\rho_2}\right)}$$

$$=\sqrt{\frac{2\gamma}{\gamma-1}\frac{p_1}{\rho_1}\left[1-\frac{p_2}{p_1}\left(\frac{p_1}{p_2}\right)^{\frac{1}{\gamma}}\right]}$$

$$=\sqrt{\frac{2\gamma}{\gamma-1}\frac{p_1}{\rho_1}\left[1-\left(\frac{p_2}{p_1}\right)^{1-\frac{1}{\gamma}}\right]}$$

$$=\sqrt{\frac{2\gamma}{\gamma-1}\frac{p_1}{\rho_1}\left[1-\left(\frac{p_2}{p_1}\right)^{\frac{\gamma-1}{\gamma}}\right]} \quad \ldots(15.25)$$

Mass flow rate,

$$m=\rho_2 A_2 V_2$$

$$=\rho_2 A_2\sqrt{\frac{2\gamma}{\gamma-1}\frac{p_1}{\rho_1}\left[1-\left(\frac{p_2}{p_1}\right)^{\frac{\gamma-1}{\gamma}}\right]}$$

$$=A_2\sqrt{\frac{2\gamma}{\gamma-1}p_1\frac{\rho_2^2}{\rho_1}\left[1-\left(\frac{p_2}{p_1}\right)^{\frac{\gamma-1}{\gamma}}\right]}$$

But from Eq. 15.5,

$$\frac{p_1}{p_2}=\left(\frac{\rho_1}{\rho_2}\right)^{\gamma}$$

or

$$\frac{\rho_2}{\rho_1}=\left(\frac{p_2}{p_1}\right)^{\frac{1}{\gamma}}$$

or

$$\rho_2^2=\rho_1^2\left(\frac{p_2}{p_1}\right)^{\frac{2}{\gamma}}$$

$$m=A_2\sqrt{\frac{2\gamma}{\gamma-1}p_1\frac{\rho_1^2}{\rho_1}\left(\frac{p_2}{p_1}\right)^{\frac{2}{\gamma}}\left[1-\left(\frac{p_2}{p_1}\right)^{\frac{\gamma-1}{\gamma}}\right]}$$

$$=A_2\sqrt{\frac{2\gamma}{\gamma-1}p_1\rho_1\left[\left(\frac{p_2}{p_1}\right)^{\frac{2}{\gamma}}-\left(\frac{p_2}{p_1}\right)^{\frac{\gamma+1}{\gamma}}\right]} \quad \ldots(15.26)$$

Thus the mass flow rate depends upon the ratio of p_2/p_1 for the given values of p_1 and ρ_1.

for maximum mass flow rate,

$$\frac{dm}{d\left(\frac{p_2}{p_1}\right)} = 0$$

$$\frac{d}{d\left(\frac{p_2}{p_1}\right)}\left[\left(\frac{p_2}{p_1}\right)^{\frac{2}{\gamma}} - \left(\frac{p_2}{p_1}\right)^{\frac{\gamma+1}{\gamma}}\right] = 0$$

$$\frac{2}{\gamma}\left(\frac{p_2}{p_1}\right)^{\frac{2}{\gamma}-1} = \frac{\gamma+1}{\gamma}\left(\frac{p_2}{p_1}\right)^{\frac{\gamma+1}{\gamma}-1}$$

$$\frac{2}{\gamma}\left(\frac{p_2}{p_1}\right)^{\frac{2-\gamma}{\gamma}} = \frac{\gamma+1}{\gamma}\left(\frac{p_2}{p_1}\right)^{\frac{1}{\gamma}}$$

$$\frac{2}{\gamma}\,\frac{\gamma}{\gamma+1} = \left(\frac{p_2}{p_1}\right)^{\frac{1}{\gamma}-\frac{2-\gamma}{\gamma}}$$

$$\frac{2}{\gamma+1} = \left(\frac{p_2}{p_1}\right)^{\frac{\gamma-1}{\gamma}}$$

$$\frac{p_2}{p_1} = \left(\frac{2}{\gamma+1}\right)^{\frac{\gamma}{\gamma-1}} \quad \text{...(15.27)}$$

$$m_{max} = A_2\sqrt{\frac{2\gamma}{\gamma-1}\,p_1\cdot\rho_1\left[\left(\frac{2}{\gamma+1}\right)^{\frac{\gamma}{\gamma-1}\cdot\frac{2}{\gamma}} - \left(\frac{2}{\gamma+1}\right)^{\frac{\gamma}{\gamma-1}\cdot\frac{\gamma+1}{\gamma}}\right]}$$

$$= A_2\sqrt{\frac{2\gamma}{\gamma-1}\,p_1\rho_1\left[\left(\frac{2}{\gamma+1}\right)^{\frac{2}{\gamma-1}} - \left(\frac{2}{\gamma+1}\right)^{\frac{\gamma+1}{\gamma-1}}\right]}$$

For $\gamma = 1{\cdot}4$,

$$m_{max} = A_2\sqrt{7p_1\rho_1\,(0{\cdot}402 - 0{\cdot}335)}$$

$$= 0{\cdot}685A_2\sqrt{p_1\cdot\rho_1} \quad \text{...(15.28)}$$

and

$$\frac{p_2}{p_1} = \left(\frac{2}{1{\cdot}4+1}\right)^{\frac{1{\cdot}4}{1{\cdot}4-1}} = 0{\cdot}528 \quad \text{...(15.29)}$$

Also from Eq. 15.25,

$$V_{2max} = \sqrt{\frac{2\gamma}{\gamma-1}\,\frac{p_1}{\rho_1}\left[1 - \left(\frac{2}{\gamma+1}\right)^{\frac{\gamma}{\gamma-1}\cdot\frac{\gamma-1}{\gamma}}\right]}$$

$$= \sqrt{\frac{2\gamma}{\gamma-1}\frac{p_1}{\rho_1}\left[1-\frac{2}{\gamma+1}\right]}$$

$$= \sqrt{\frac{2\gamma}{\gamma-1}\frac{p_1}{\rho_1}\left[\frac{\gamma-1}{\gamma+1}\right]}$$

$$= \sqrt{\frac{2\gamma}{\gamma+1}\frac{p_1}{\rho_1}}$$

$$= \sqrt{\frac{2\gamma}{\gamma+1}\cdot\frac{p_1}{p_2}\cdot\frac{p_2}{\rho_2}\cdot\frac{\rho_2}{\rho_1}}$$

$$= \sqrt{\frac{2\gamma}{\gamma+1}\frac{p_2}{\rho_2}\left(\frac{p_2}{p_1}\right)^{-1}\left(\frac{p_2}{p_1}\right)^{\frac{1}{\gamma}}} \qquad \left[\because \frac{\rho_2}{\rho_1} = \left(\frac{p_2}{p_1}\right)^{\frac{1}{\gamma}}\right]$$

$$= \sqrt{\frac{2\gamma}{\gamma+1}\frac{p_2}{\rho_1}\left(\frac{p_2}{p_1}\right)^{\frac{1-\gamma}{\gamma}}}$$

$$= \sqrt{\frac{2\gamma}{\gamma+1}\frac{p_2}{\rho_2}\left(\frac{2}{\gamma+1}\right)^{\frac{\gamma}{\gamma-1}\cdot\frac{1-\gamma}{\gamma}}}$$

$$= \sqrt{\frac{2\gamma}{\gamma+1}\frac{p_2}{\rho_2}\left(\frac{2}{\gamma+1}\right)^{-1}}$$

$$= \sqrt{\gamma\frac{p_2}{\rho_2}}$$

$$= C_2 \qquad \text{...(15.30)}$$

Thus the velocity at the outlet of the nozzle is equal to sound velocity for maximum flow rate.

A passage in which sonic velocity is reached and the flow is maximum, is known as a choked passage and is said to operate under choked conditions. Equation 15.26 shows that the mass flow rate depends upon the ratio of p_2/p_1. The plot of mass flow rate vs. p_2/p_1 is shown in Fig. 15.8 which shows that as the pressure ratio decreases from 1, the mass flow rate increases in the subsonic flow till the sonic velocity is attained at $p_2/p_1 = 0{\cdot}528$ for air. With

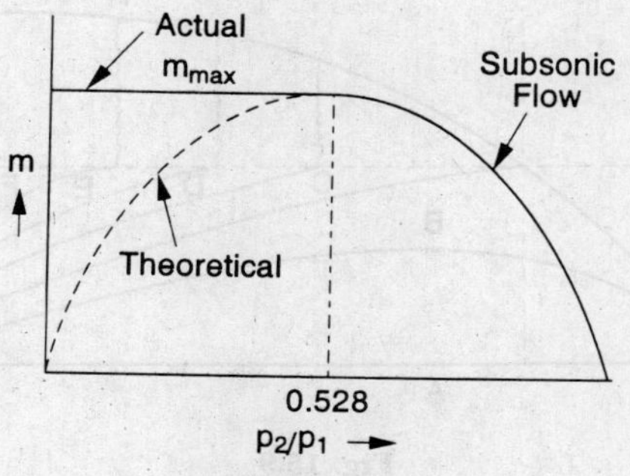

Fig. 15.8

further decrease in exit pressure, the theoretical mass flow rate decreases. However, in actual practice this does not happen. This is because once the velocity at the throat reaches the sonic value, it cannot reduce as long as the pressure is below $0{\cdot}528\ p_1$. The convergent portion of the nozzle continues to behave independently of the conditions beyond the throat.

15.10 FLOW THROUGH A CONVERGENT-DIVERGENT NOZZLE

For different values of pressure at the exit, the flow is as follows :

- When the pressure at the exit (point 3) is equal to that at the inlet (point 1), there is no flow.
- When the pressure at the exit is reduced, the flow starts (Fig. 15.9). As long as the pressure at 3 is such that the pressure at the throat is above the critical value ($0{\cdot}528p_1$), the flow in the convergent as well as in the divergent portions will be

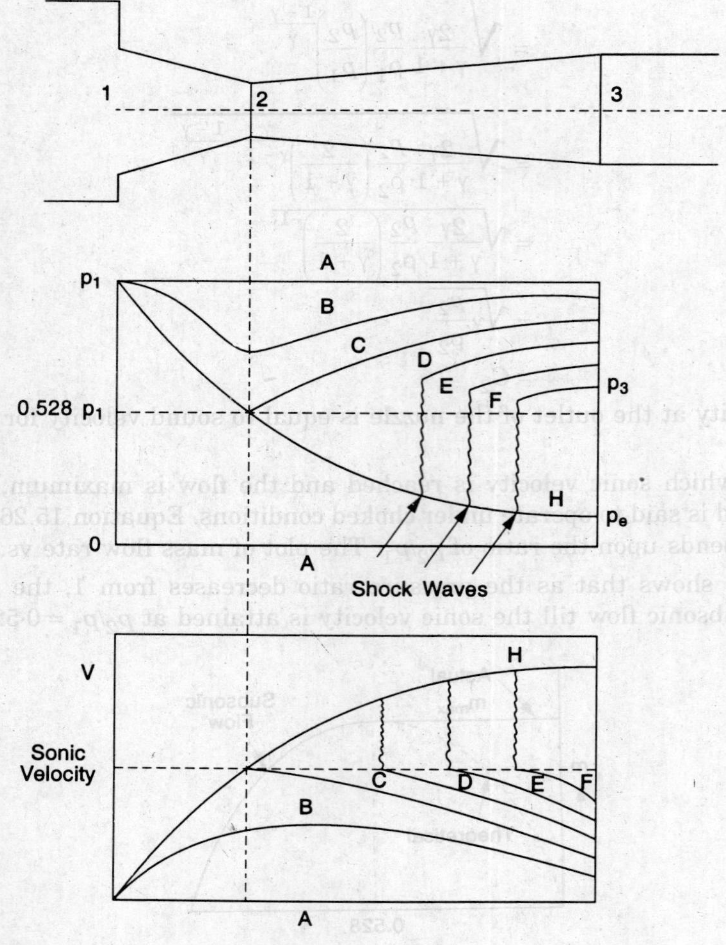

Fig. 15.9

subsonic (line B). The flow accelerates in the convergent portion and decelerates in the divergent portion. This will happen till the velocity at the throat becomes equal to sonic velocity (curve C).

- As p_3 is further reduced, the flow in the convergent portion is not affected because in a supersonic flow, the velocity can increase only in a divergent portion. Beyond the throat, thus there will be supersonic flow. The velocity will continue to increase. In case the pressure is not decreased enough to justify the increase in velocity, a shock wave will occur in the divergent portion and the flow changes to subsonic. The flow across the shock wave is non isentropic. Downstream the shock wave, the flow is subsonic and decelerating. (Curves D, E and F).
- When the discharge pressure equals the nozzle design pressure (p_e), there is acceleration throughout the diverging passage and no shock wave occurs.
- On further decrease of the outlet pressure, oblique shock waves are formed.

Example 15.7 A nozzle fitted to a large vessel containing air at a pressure of 3200 kN/m^2 absolute and at a temperature of 24°C has outlet pressure as 1880 kN/m^2 absolute. Determine the velocity of air at the outlet of nozzle. $R = 287$ J/kg.K and $\gamma = 1{\cdot}4$ for air.

Solution $T_1 = 273 + 24 = 297°\text{K}$

$$p_1 = 3200 \text{ kN/m}^2$$

$$p_2 = 1880 \text{ kN/m}^2$$

$$\frac{p_1}{\rho_1} = RT_1 \quad \text{or} \quad \rho_1 = \frac{p_1}{RT_1} = \frac{3200 \times 10^3}{287 \times 297} = 37{\cdot}54 \text{ kg/m}^3$$

$$V_2 = \sqrt{\frac{2\gamma}{\gamma - 1}\frac{p_1}{\rho_1}\left[1 - \left(\frac{p_2}{p_1}\right)^{\frac{\gamma-1}{\gamma}}\right]} = \sqrt{\frac{2 \times 1{\cdot}4}{1{\cdot}4 - 1} \times \frac{3200 \times 10^3}{37{\cdot}54}\left[1 - \left(\frac{1880}{3200}\right)^{\frac{1{\cdot}4-1}{1{\cdot}4}}\right]}$$

$$= \sqrt{7 \times 85242 \times (1 - 0{\cdot}8593)} = 289{\cdot}7 \text{ m/s}$$

Example 15.8 A tank containing air at a temperature of 24°C is fitted with a convergent nozzle having a diameter of 20 mm at its outlet. Determine the mass flow rate of the air through the nozzle into the atmosphere if the gauge pressure in the tank is (i) 50 kN/m^2 (ii) 250 kN/m^2. Barometric pressure is 100 kN/m^2.

$R = 287$ J/kg.K and $\gamma = 1{\cdot}4$ for air.

Solution $T_1 = 273 + 24 = 297°\text{K}$

$$A_2 = \frac{\pi}{4} \times (0{\cdot}02)^2 = 0{\cdot}0003142 \text{ m}^2$$

(i) $p_1 = 50 \text{ kN/m}^2 \text{ gauge} = 100 + 50 = 150 \text{ kN/m}^2 \text{ absolute}$

$$\frac{p_1}{\rho_1} = RT_1 \quad \text{or} \quad \rho_1 = \frac{p_1}{RT_1} = \frac{150 \times 10^3}{287 \times 297} = 1{\cdot}7598 \text{ kg/m}^3$$

$$m = A_2 \sqrt{\frac{2\gamma}{\gamma - 1} p_1 \rho_1 \left[\left(\frac{p_2}{p_1}\right)^{\frac{2}{\gamma}} - \left(\frac{p_2}{p_1}\right)^{\frac{\gamma+1}{\gamma}}\right]}$$

$$= 0{\cdot}0003142 \sqrt{\frac{2 \times 1{\cdot}4}{1{\cdot}4 - 1} \times 150 \times 10^3 \times 1{\cdot}7598 \left[\left(\frac{100}{150}\right)^{\frac{2}{\gamma}} - \left(\frac{100}{150}\right)^{\frac{1{\cdot}4+1}{1{\cdot}4}}\right]}$$

$$= 0{\cdot}0003142 \sqrt{7 \times 263970 \times (0{\cdot}5603 - 0{\cdot}499)}$$

$$= 0{\cdot}0003142 \times 336{\cdot}5$$

$$= 0{\cdot}1057 \text{ kg/s}$$

(ii) $p_1 = 250$ kN/m^2 gauge $= 100 + 250 = 350$ kN/m^2 absolute

$$\text{Pressure ratio} = \frac{p_2}{p_1} = \frac{100}{350} = 0{\cdot}286$$

As the pressure ratio is less than the critical value of 0·528, the flow is sonic and the flow rate is maximum.

$$\frac{p_1}{\rho_1} = RT_1 \quad \text{or} \quad \rho_1 = \frac{p_1}{RT_1} = \frac{350 \times 10^3}{287 \times 297} = 4{\cdot}106 \text{ kg/m}^3$$

$$m_{\max} = 0{\cdot}685 A_2 \sqrt{p_1 \rho_1} \qquad \text{(Eq. 15.28)}$$

$$= 0{\cdot}685 \times 0{\cdot}0003142 \times \sqrt{350 \times 10^3 \times 4{\cdot}106}$$

$$= 0{\cdot}258 \text{ kg/s}$$

15.11 FLOW THROUGH A VENTURIMETER

Consider the flow of compressible fluid through a horizontal venturimeter. Let the suffixes 1 and 2 denote the main and the throat of the venturimeter.

Applying Bernoulli's equation (for adiabatic flow),

$$\left(\frac{\gamma}{\gamma - 1}\right)\frac{p_1}{\rho_1 g} + \frac{V_1^2}{2g} + z_1 = \left(\frac{\gamma}{\gamma - 1}\right)\frac{p_2}{\rho_2 g} + \frac{V_2^2}{2g} + z_2$$

Taking $z_1 = z_2$

$$\left(\frac{\gamma}{\gamma - 1}\right)\frac{p_1}{\rho_1} + \frac{V_1^2}{2g} = \left(\frac{\gamma}{\gamma - 1}\right)\frac{p_2}{\rho_2} + \frac{V_2^2}{2}$$

$$\frac{1}{2}(V_2^2 - V_1^2) = \frac{\gamma}{\gamma - 1}\frac{p_1}{\rho_1}\left(1 - \frac{p_2}{p_1}\frac{\rho_1}{\rho_2}\right)$$

$$\frac{V_2^2}{2}\left[1 - \left(\frac{V_1}{V_2}\right)^2\right] = \frac{\gamma}{\gamma - 1}\frac{p_1}{\rho_1}\left[1 - \frac{p_2}{p_1}\left(\frac{p_1}{p_2}\right)^{\frac{1}{\gamma}}\right]$$

$$= \frac{\gamma}{\gamma - 1}\frac{p_1}{\rho_1}\left[1 - \left(\frac{p_2}{p_1}\right)^{1 - \frac{1}{\gamma}}\right]$$

$$= \frac{\gamma}{\gamma - 1} \frac{p_1}{\rho_1} \left[1 - \left(\frac{p_2}{p_1} \right)^{\frac{\gamma - 1}{\gamma}} \right]$$

Since mass flow rate, $m = \rho_1 . A_1 . V_1 = \rho_2 . A_2 . V_2$,

$$\frac{V_1}{V_2} = \frac{\rho_2 A_2}{\rho_1 A_1}$$

or

$$\left(\frac{V_1}{V_2} \right)^2 = \left(\frac{\rho_2}{\rho_1} \right)^2 \left(\frac{A_2}{A_1} \right)^2$$

Substituting this value of $(V_1/V_2)^2$ in the above relation,

$$\frac{V_2^2}{2} \left[1 - \left(\frac{\rho_2}{\rho_1} \right)^2 \left(\frac{A_2}{A_1} \right)^2 \right] = \frac{\gamma}{\gamma - 1} \frac{p_1}{\rho_1} \left[1 - \left(\frac{p_2}{p_1} \right)^{\frac{\gamma - 1}{\gamma}} \right]$$

$$\frac{V_2^2}{2} \left[1 - \left(\frac{p_2}{p_1} \right)^{\frac{2}{\gamma}} \left(\frac{A_2}{A_1} \right)^2 \right] = \frac{\gamma}{\gamma - 1} \frac{p_1}{\rho_1} \left[1 - \left(\frac{p_2}{p_1} \right)^{\frac{\gamma - 1}{\gamma}} \right]$$

$$V_2 = \sqrt{\frac{\frac{2\gamma}{\gamma - 1} \frac{p_1}{\rho_1} \left[1 - \left(\frac{p_2}{p_1} \right)^{\frac{\gamma - 1}{\gamma}} \right]}{1 - \left(\frac{p_2}{p_1} \right)^{\frac{2}{\gamma}} \left(\frac{A_2}{A_1} \right)^2}}$$

Mass flow rate through venturimeter,

$$m = \rho_2 A_2 V_2$$

$$= \rho_2 A_2 \sqrt{\frac{\frac{2\gamma}{\gamma - 1} \frac{p_1}{\rho_1} \left[1 - \left(\frac{p_2}{p_1} \right)^{\frac{\gamma - 1}{\gamma}} \right]}{1 - \left(\frac{p_2}{p_1} \right)^{\frac{2}{\gamma}} \left(\frac{A_2}{A_1} \right)^2}}$$

$$= \rho_2 A_2 \sqrt{\frac{\frac{2\gamma}{\gamma - 1} \frac{p_1}{\rho_1} \left[1 - \left(\frac{p_2}{p_1} \right)^{\frac{\gamma - 1}{\gamma}} \right]}{1 - \left(\frac{p_2}{p_1} \right)^{\frac{2}{\gamma}} \left(\frac{D_2}{D_1} \right)^4}} \qquad ...(15.31)$$

The pressure ratio p_2/p_1 must be greater than the critical value (*i.e.* 0·528) for the flow.

Example 15.9 A venturimeter has inlet diameter 360 mm and throat diameter 180 mm. The temperature and the pressure at the inlet of the venturimeter are 18°C and 150 kN/m^2 absolute respectively and the pressure at the throat 140 kN/m^2 absolute. Determine the mass flow rate of air through the venturimeter. R = 287 J/kg.K and γ = 1·4.

Solution $T = 273 + 18 = 291^\circ\text{C}$

$$p_1 = 150 \text{ kN/m}^2$$

$$p_2 = 140 \text{ kN/m}^2$$

$$\rho_1 = \frac{p_1}{RT} = \frac{150 \times 10^3}{287 \times 291} = 1{\cdot}796 \text{ kg/m}^3$$

$$\frac{p_1}{p_2} = \left(\frac{\rho_1}{\rho_2}\right)^{\gamma} \text{ or } \rho_2 = \rho_1 \left(\frac{p_2}{p_1}\right)^{\frac{1}{\gamma}} = 1{\cdot}796 \times \left(\frac{140}{150}\right)^{\frac{1}{1\cdot4}} = 1{\cdot}7096 \text{ kg/m}^3$$

$$m = \rho_2 A_2 \sqrt{\frac{\frac{2\gamma}{\gamma-1}\frac{p_1}{\rho_1}\left[1-\left(\frac{p_2}{p_1}\right)^{\frac{\gamma-1}{\gamma}}\right]}{1-\left(\frac{p_2}{p_1}\right)^{\frac{2}{\gamma}}\left(\frac{D_2}{D_1}\right)^4}}$$

$$= 1{\cdot}7096 \times \frac{\pi}{4} \times 0{\cdot}18^2 \sqrt{\frac{\frac{2 \times 1\cdot4}{1\cdot4 - 1}\left(\frac{150 \times 10^3}{1\cdot796}\right)\left[1-\left(\frac{140}{150}\right)^{\frac{1\cdot4-1}{1\cdot4}}\right]}{1-\left(\frac{140}{150}\right)^{\frac{2}{1\cdot4}}\left(\frac{0\cdot18}{0\cdot36}\right)^4}}$$

$$= 0{\cdot}0435 \sqrt{\frac{7 \times 83520 \times (1 - 0{\cdot}98046)}{1 - 0{\cdot}9061 \times 0{\cdot}0625}}$$

$$= 0{\cdot}0435 \sqrt{\frac{7 \times 83520 \times 0{\cdot}01954}{0{\cdot}94337}}$$

$$= 0{\cdot}0435 \times 110{\cdot}044$$

$$= 4{\cdot}787 \text{ kg/s}$$

15.12 FLOW THROUGH A PITOT TUBE

A pitot tube used to determine the velocity at a point gives the difference between the stagnation head and the static head. In case of compressible fluids (section 5.5),

$$\frac{p_s}{p_0} = \left[1 + \frac{\gamma - 1}{2} M_0^{\,2}\right]^{\frac{\gamma}{\gamma-1}}$$

Assuming $M_0 < 1$ and expanding the terms on the right hand side by Bionomial theorem,

$$p_s = p_0 \left[1 + \frac{\gamma}{2} M_0^{\,2} + \frac{\gamma}{8} M_0^{\,4} + \frac{\gamma(2-\gamma)}{48} M_0^{\,6} + \ldots\ldots\right]$$

$$= p_0 \left[1 + \frac{\gamma M_0^{\,2}}{2}\left(1 + \frac{M_0^{\,2}}{4} + \frac{2-\gamma}{24} M_0^{\,4} + \ldots\ldots\right)\right]$$

$$= p_0 \left[1 + \frac{\gamma}{2} \left(\frac{V_0}{C_0} \right)^2 \left(1 + \frac{M_0^2}{4} + \frac{2-\gamma}{24} M_0^4 + \ldots\ldots \right) \right]$$

$$= p_0 + \frac{p_0 \gamma}{2} \frac{V_0^2}{\gamma p_0 / \rho_0} \left[1 + \frac{M_0^2}{4} + \frac{2-\gamma}{24} M_0^4 + \ldots\ldots \right] \quad \text{(using Eq. 15.17)}$$

$$= p_0 + \frac{\rho_0 V_0^2}{2} \left[1 + \frac{M_0^2}{4} + \frac{2-\gamma}{24} M_0^4 + \ldots\ldots \right] \quad \ldots(15.32)$$

For incompressible fluid, applying Bernoulli's equation to points S and O,

$$\frac{p_s}{\rho} + 0 = \frac{p_0}{\rho} + \frac{V_0^2}{2} \quad \text{or} \quad p_s = p_0 + \frac{\rho_0 V_0^2}{2}$$

Comparing this with Eq. 15.32, it is observed that in case of compressible fluids, the stagnant pressure becomes greater for higher approaching velocity (or higher values of M). The quantity

$$\left[1 + \frac{M_0^2}{4} + \frac{2-\gamma}{24} M_0^4 + \ldots\ldots \right]$$

represents the compressibility effect and is known as **compressibility factor.**

Example 15.10 Air has a velocity of 900 km/h at a temperature of 42°C and at a pressure of 10 kN/m^2 vacuum. Find the stagnation properties and the local Mach number. Assume atmospheric pressure to be 100 kN/m^2, R = 287 J/kg.K and $\gamma = 1{\cdot}4$.

Also determine the compressibility correction factor for a pitot static tube to measure the velocity at a Mach number of 0·9.

Solution $T_0 = 273 + 42 = 315$°K

$$p_0 = 100 - 10 = 90 \text{ kN/m}^2 \text{ (absolute)}$$

$$C_0 = \sqrt{\gamma R T_0} = \sqrt{1{\cdot}4 \times 287 \times 315} = 355{\cdot}8 \text{ m/s}$$

$$V_0 = \frac{900 \times 1000}{3600} = 250 \text{ m/s}$$

$$M_0 = \frac{V_0}{C_0} = \frac{250}{355{\cdot}8} = 0{\cdot}7027$$

$$p_s = p_0 \left[1 + \frac{(\gamma - 1)}{2} M_0^2 \right]^{\frac{\gamma}{\gamma - 1}}$$

$$= 90 \left[1 + \frac{(1{\cdot}4 - 1)}{2} \times 0{\cdot}7027^2 \right]^{\frac{1{\cdot}4}{1{\cdot}4 - 1}}$$

$$= 90 \times (1{\cdot}09876)^{3{\cdot}5}$$

$$= 125{\cdot}14 \text{ kN/m}^2$$

$$T_s = T_0 \left[1 + \frac{(\gamma - 1)}{2} M_0^2 \right]$$

$$= 315\left[1 + \frac{(1.4 - 1)}{2} \times 0.7027^2\right]$$

$$= 346.1^\circ K \quad \text{or} \quad 73.1^\circ C$$

$$\frac{p_s}{\rho_s} = RT_s$$

$$\rho_2 = \frac{p_s}{RT_s} = \frac{125.14 \times 10^3}{287 \times 346.1} = 1.26 \text{ kg/m}^2$$

Compressibility factor,

$$= 1 + \frac{M_0^2}{4} + \frac{2-\gamma}{24} M_0^4 + \ldots\ldots$$

$$= 1 + \frac{0.7027^2}{4} + \frac{2-1.4}{24} 0.7027^4 + \ldots\ldots$$

$$= 1 + 0.1234 + 0.006096$$

$$= 1.1295$$

15.13 SHOCK WAVES

A shock wave occurs where the compressible flow changes abruptly from supersonic to subsonic state with a sudden rise in pressure, density, temperature and entropy. The Mach number of the approaching flow is more than 1. The phenomenon is similar to the formation of a hydraulic jump in an open channel when the flow changes from supercritical to subcritical for which the Froude number of the approaching flow is more than 1.

A shock wave occurs in pipes, diffusers, diverging section of a nozzle, throat of a supersonic wind tunnel etc. There are two types of shock waves; the **normal shock waves** which are perpendicular to the flow and the **oblique shock waves** which are inclined to the direction of flow.

Normal Shock Waves

Consider a control surface around the wave (Fig. 15.10). The flow is supersonic upstream and subsonic on the downstream of the shock.

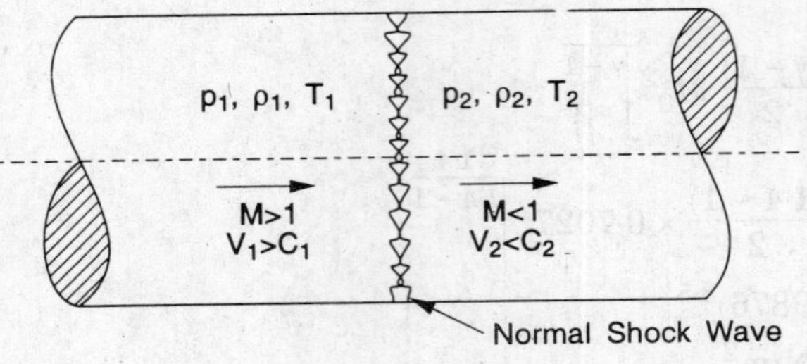

Fig. 15.10

Let p_1 = fluid pressure on upstream of the shock

V_1 = velocity of fluid on upstream of the shock

ρ_1 = density of the fluid on upstream

T_1 = temperature of the fluid on upstream

p_2, V_2, ρ_2 and T_2 are the corresponding values on the downstream side of the shock. The changes in the flow variables across the shock wave are irreversible and, therefore, are not isentropic.

Considering $A_1 = A_2 = A$

Continuity equation :

$$m = \rho_1 A_1 V_1 = \rho_2 A_2 V_2 \quad \text{or} \quad m = \rho_1 A V_1 = \rho_2 A V_2$$

Momentum equation :

$$\Sigma F = p_1 A_1 - p_2 A_2 = m\,(V_2 - V_1)$$

$$A\,(p_1 - p_2) = m\,(V_2 - V_1)$$

$$V_2 - V_1 = \frac{A}{m}\,(p_1 - p_2)$$

$$(V_2 - V_1)\,(V_2 + V_1) = \frac{A}{m}\,(p_1 - p_2)\,(V_2 + V_1)$$

[multiplying both sides by $(V_2 + V_1)$]

$$V_2^{\,2} - V_1^{\,2} = (p_1 - p_2)\left(\frac{AV_2}{m} + \frac{AV_1}{m}\right)$$

$$= (p_1 - p_2)\left(\frac{AV_2}{\rho_2 AV_2} + \frac{AV_1}{\rho_1 AV_1}\right)$$

$$= (p_1 - p_2)\left(\frac{1}{\rho_2} + \frac{1}{\rho_1}\right) \qquad \text{...(i)}$$

Energy equation,

$$\left(\frac{\gamma}{\gamma - 1}\right)\frac{p_1}{\rho_1} + \frac{V_1^{\,2}}{2} = \left(\frac{\gamma}{\gamma - 1}\right)\frac{p_2}{\rho_2} + \frac{V_2^{\,2}}{2}$$

$$V_2^{\,2} - V_1^{\,2} = \frac{2\gamma}{\gamma - 1}\left(\frac{p_1}{\rho_1} - \frac{p_2}{\rho_2}\right) \qquad \text{...(ii)}$$

Comparing (i) and (ii)

$$(p_1 - p_2)\left(\frac{1}{\rho_2} + \frac{1}{\rho_1}\right) = \frac{2\gamma}{\gamma - 1}\left(\frac{p_1}{\rho_1} - \frac{p_2}{\rho_2}\right)$$

$$\frac{p_1}{\rho_2} + \frac{p_1}{\rho_1} - \frac{p_2}{\rho_2} - \frac{p_2}{\rho_1} = \frac{2\gamma}{\gamma - 1}\frac{p_1}{\rho_1} - \frac{2\gamma}{\gamma - 1}\frac{p_2}{\rho_2}$$

Multiplying throughout by ρ_2/p_1

$$1 + \frac{\rho_2}{\rho_1} - \frac{p_2}{\rho_1} - \frac{p_2}{p_1}\frac{\rho_2}{\rho_1} = \frac{2\gamma}{\gamma - 1}\frac{\rho_2}{\rho_1} - \frac{2\gamma}{\gamma - 1}\frac{p_2}{p_1}$$

$$\frac{p_2}{p_1}\left(\frac{2\gamma}{\gamma - 1} - 1 - \frac{\rho_2}{\rho_1}\right) = \frac{\rho_2}{\rho_1}\left(\frac{2\gamma}{\gamma - 1} - 1\right) - 1$$

$$\frac{p_2}{p_1}\left(\frac{\gamma + 1}{\gamma - 1} - \frac{\rho_2}{\rho_1}\right) = \frac{\rho_2}{\rho_1}\left(\frac{\gamma + 1}{\gamma - 1}\right) - 1 \qquad \text{...(15.33)}$$

$$\frac{p_2}{p_1} = \frac{\frac{\rho_2}{\rho_1}\left(\frac{\gamma + 1}{\gamma - 1}\right) - 1}{\frac{\gamma + 1}{\gamma - 1} - \frac{\rho_2}{\rho_1}} \qquad \text{...(15.34)}$$

Solving Eq. 15.33 for ρ_2/ρ_1,

$$\frac{p_2}{p_1}\frac{\gamma + 1}{\gamma - 1} - \frac{p_2}{p_1}\frac{\rho_2}{\rho_1} = \frac{\rho_2}{\rho_1}\left(\frac{\gamma + 1}{\gamma - 1}\right) - 1$$

$$\frac{\rho_2}{\rho_1}\left(\frac{\gamma + 1}{\gamma - 1} + \frac{p_2}{p_1}\right) = \frac{p_2}{p_1}\left(\frac{\gamma + 1}{\gamma - 1}\right) + 1$$

$$\frac{\rho_2}{\rho_1} = \frac{\left(\frac{\gamma + 1}{\gamma - 1}\right)\frac{p_2}{p_1} + 1}{\frac{\gamma + 1}{\gamma - 1} + \frac{p_2}{p_1}} = \frac{V_1}{V_1} \qquad \text{...(15.35)}$$

Eqs. 15.34 and 15.35 are known as **Rankine-Hugoniot Equations.**

The following relation between the Mach numbers upstream and downstream may be derived :

$$M_2^2 = \frac{(\gamma - 1)\,M_1^2 + 2}{2\gamma M_1^2 - (\gamma - 1)} \qquad \text{...(15.36)}$$

The ratio of fluid properties in terms of Mach numbers,

$$\frac{p_2}{p_1} = \frac{1 + \gamma M_1^2}{1 + \gamma M_2^2} \qquad \text{...(15.37)}$$

$$\frac{T_2}{T_1} = \frac{1 + \frac{1}{2}(\gamma - 1)\,M_1^2}{1 + \frac{1}{2}(\gamma - 1)\,M_2^2} \qquad \text{...(15.38)}$$

Also the ratio of fluid properties may be expressed in terms of Mach number of the upstream,

$$\frac{p_2}{p_1} = \frac{2\gamma M_1^2 - (\gamma - 1)}{\gamma + 1} \qquad \text{...(15.39)}$$

$$\frac{\rho_2}{\rho_1} = \frac{V_1}{V_2} = \frac{(\gamma + 1)\, M_1^{\,2}}{(\gamma - 1)\, M_1^{\,2} + 2} \qquad \ldots(15.40)$$

$$\frac{T_2}{T_1} = \frac{[2\gamma M_1^{\,2} - (\gamma - 1)]\,[(\gamma - 1)\, M_1^{\,2} + 2]}{(\gamma + 1)^2\, M_1^{\,2}} \qquad \ldots(15.41)$$

Shock Strength

It is the ratio of the rise of pressure across the shock to the upstream pressure.

$$\text{Shock strength} = \frac{p_2 - p_1}{p_1}$$

$$= \frac{p_1 \cdot \dfrac{2\gamma M_1^{\,2} - (\gamma - 1)}{\gamma + 1} - p_1}{p_1}$$

$$= \frac{2\gamma M_1^{\,2} - (\gamma - 1)}{\gamma + 1} - 1$$

$$= \frac{2\gamma M_1^{\,2} - \gamma + 1 - \gamma - 1}{\gamma + 1}$$

$$= \frac{2\gamma}{\gamma + 1}\,(M_1^{\,2} - 1) \qquad \ldots(15.42)$$

Oblique Shock Wave

An oblique shock wave is the result of a supersonic flow taking a sudden turn through a small angle δ as shown in Fig. 15.11. During the shock, mechanical energy is converted into thermal energy and thus is a loss. Therefore, the formation of shock waves should be avoidable or made weak. Oblique shock waves are weaker as compared to normal shock waves.

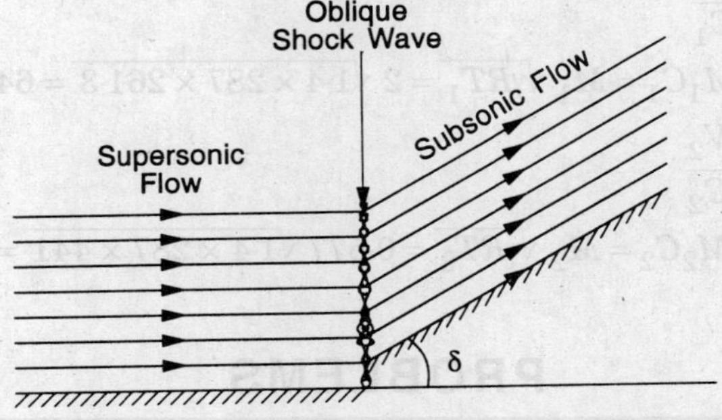

Fig. 15.11

Example 15.11 The Mach number for a normal shock wave in air is 2. The atmospheric pressure and density are 30 kN/m^2 and 0·4 kg/m^3. Taking $\gamma = 1{\cdot}4$, determine the flow conditions before and after the shock wave.

Solution $p_1 = 30 \text{ kN/m}^2$

$\rho_1 = 0.4 \text{ kg/m}^3$

$M_1 = 2$

$$M_2^2 = \frac{(\gamma - 1) M_1^2 + 2}{2\gamma M_1^2 - (\gamma - 1)} = \frac{(1.4 - 1) 2^2 + 2}{2 \times 1.4 \times 2^2 - (1.4 - 1)} = 0.3333$$

$$M_2 = 0.577$$

$$\frac{p_2}{p_1} = \frac{2\gamma M_1^2 - (\gamma - 1)}{\gamma + 1} = \frac{2 \times 1.4 \times 2^2 - (1.4 - 1)}{1.4 + 1} = 4.5$$

$$p_2 = 4.5 \times 30 = 135 \text{ kN/m}^2$$

$$\frac{\rho_2}{\rho_1} = \frac{(\gamma + 1) M_1^2}{(\gamma - 1) M_1^2 + 2} = \frac{(1.4 + 1) 2^2}{(1.4 - 1) \times 2^2 + 2} = 2.667$$

$$\rho_2 = 0.4 \times 2.667 = 1.0668 \text{ kg/m}^3$$

$$\frac{p_1}{\rho_1} = RT_1 \quad \text{or} \quad T_1 = \frac{30 \times 10^3}{0.4 \times 287} = 261.3^\circ\text{K} = -11.7^\circ\text{C}$$

$$\frac{T_2}{T_1} = \frac{[2\gamma M_1^2 - (\gamma - 1)]\,[(\gamma - 1) M_1^2 + 2]}{(\gamma + 1)^2 M_1^2}$$

$$= \frac{[2 \times 1.4 \times 2^2 - (1.4 - 1)]\,[(1.4 - 1) 2^2 + 2]}{(1.4 + 1)^2 2^2}$$

$$= 1.688$$

$$T_2 = 261.3 \times 1.688 = 441^\circ\text{K} = 168^\circ\text{C}$$

$$M_1 = \frac{V_1}{C_1}$$

$$V_1 = M_1 C_1 = M_1 \sqrt{\gamma R T_1} = 2\sqrt{1.4 \times 287 \times 261.3} = 648 \text{ m/s}$$

$$M_2 = \frac{V_2}{C_2}$$

$$V_2 = M_2 C_2 = M_2 \sqrt{\gamma R T_2} = 0.577\sqrt{1.4 \times 287 \times 441} = 242.88 \text{ m/s}$$

PROBLEMS

1. Distinguish between compressible and incompressible flows.
2. Define the following processes :
   (a) Isobaric (b) Isochoric
   (c) Isothermal (d) Isentropic.
3. What are the basic equations of compressible flow ?

4. In what way is Bernoulli's equation for a compressible flow different from that for an incompressible flow ?
5. Derive an expression for continuity equation, for one-dimensional compressible flow in the differential form.
6. In what way is the disturbance in a compressible flow propagated ?
7. Derive an expression for the sound wave in a compressible flow in terms of change in pressure and in density.
8. Define Mach number.
9. Explain the terms : sonic flow, subsonic flow, supersonic flow, Mach angle and Mach cone.
10. What do understand by the terms : stagnation point, stagnation pressure, stagnation temperature.
11. Describe the flow of a compressible fluid through a convergent-divergent nozzle.
12. What is the compressibility correction factor ? Find an expression for the same.
13. Deduce an expression for mass flow rate through a venturimeter.
14. How are shock waves produced in a compressible flow ?
15. Determine the velocity of a bullet fired in the atmosphere at 15°C, if the Mach angle is 30°. Take $\gamma = 1{\cdot}4$ and $R = 287$ J/kg.K. (2449·3 km/h)
16. A man standing on the ground hears the sonic boom of a supersonic fighter plane 7·5 seconds after passing of the plane over his head. The altitude of the plane is 3000 m. Determine the speed of the plane and the Mach number. Assume $\gamma = 1{\cdot}4$, $R = 287$ J/kg.K and the average temperature 11°C. (2271 km/h)
17. Calculate the sonic velocity and the maximum possible flow velocity for air corresponding to stagnation temperature of 477°C. Also find the sonic velocity when the flow takes place at half the maximum velocity. $\gamma = 1{\cdot}4$ and $R = 287$ J/kg.K for air. (549 m/s, 1227·6 m/s, 475·45 m/s)
18. A jet plane flies at a speed of 960 km/h in air where the pressure is 80 kN/m^2 and temperature – 20°C. Determine the pressure, temperature and the density at the stagnation point on the nose of the plane. $R = 287$ J/kg.K and $\gamma = 1{\cdot}4$. (126 kN/m^2, 15°C, 1·523 kg/m^3)
19. In a conduit flow Mach number is 1·5. Determine the percentage of the original area to increase the velocity by 25 percent. What would be the area if M is reduced to 0·5. (131·25%, 81·25%)
20. Air flows through a convergent-divergent nozzle. At particular section where the area of cross-section is 1000 m^2, the values of pressure, temperature and the velocity are 200 kN/m^2, 200°C and 170 m/s respectively. Considering the flow conditions to be isentropic, determine

    (a) stagnation temperature, stagnation pressure, sonic velocity and Mach number at this section.

    (b) Mach number and the flow area at the outlet section where pressure is 110 kN/m^2.

(c) Pressure temperature and the flow area at the throat of the nozzle.

(214·5°C, 222·2 kN/m^2, 436 m/s, 0·39; 1·05, 620 mm^2; 117·8 kN/m^2, 133·2°C, 616 mm^2).

21. A venturimeter used to measure the rate of flow of a gas is 25 mm in diameter and is fixed in a 75 mm diameter pipe. The absolute pressures at the inlet and the throat of venturimeter are equivalent to 1·01 m and 0·91 m of mercury. Assuming the flow to be isentropic, determine the volumetric flow rate. Take $\gamma = 1{\cdot}4$ and density of the gas in the main pipe 1·6 kg/m^3. (0·0599 m^3/s)

22. For a normal shock in air, Mach number is 2. The atmospheric pressure and density are 26·5 kN/m^2 and 0·413 kg/m^3 respectively. Taking $\gamma = 1{\cdot}4$ and $R = 290$ J/kg.K determine the flow conditions before and after the shock wave.

(p_2 = 119·3 kN/m^2, p_2 = 1·101 kN/m^2, T_1 = – 51·71°C, T_2 = 100·43°C, V_1 = 599·4 m/s, V_2 = 224·6 m/s).

23. A normal shock wave occurs in a diverging section where the speed of the air is 420 m/s, pressure 100 kN/m^2 and temperature 10°C. Determine the Mach number before and after the shock and the pressure rise. (1·25, 0·91, 66 kN/m^2)

24. Air moving at a velocity of 200 m/s has a pressure of 220 kN/m^2 and temperature 27°C. Determine the stagnation pressure. What is the stagnation pressure if the compressibility is also taken into account? For air $\gamma = 1{\cdot}4$ and $R = 287$ J/kg.K.

(271 kN/m^2, 275·5 kN/m^2)

Chapter **16**

IMPACT OF JETS

16.1 INTRODUCTION

When a steady jet of liquid issued from a nozzle impinges on a solid smooth surface, a thin layer of the liquid is formed which glides over the surface up to the boundaries. This layer of liquid then leaves the surface approximately tangent to the surface. In doing so, the jet exerts a force on the surface which is known as **impact of jet**. As the motion of the liquid produces this force, it is dynamic in nature and is called **hydrodynamic force.** It can be evaluated by applying the impulse-momentum principle *i.e.* the force of a jet in that direction. The surface on which the jet strikes may be stationary or moving.

16.2 FORCE ON STATIONARY PLATES

The following cases of stationary surfaces are considered :

- Flat plate normal to jet
- Flat plate inclined to jet
- Curved plate, jet impinging at the centre
- Curved plate, jet impinging at one end.

Flat Plate Normal to Jet

Let a jet of liquid issued from a nozzle impinges normally on the flat surface of a fixed plate (Fig. 16.1). The jet is deflected through 90° after impinging on the plate. It leaves the plate tangentially and the component of velocity in the original direction of the jet is zero. If the plate is quite smooth, the frictional resistance may be neglected.

As the jet is exposed to atmosphere, the atmospheric pressure prevails throughout. If the change in elevation as well as the energy loss due to impact are neglected, then by Bernoulli's

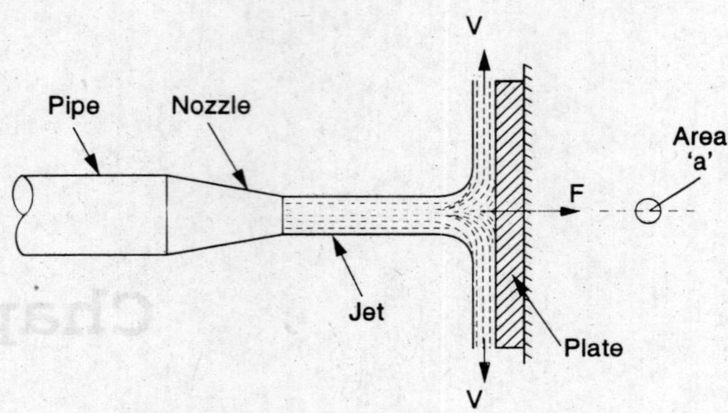

Fig. 16.1

equation, the velocity of the leaving jet is equal to that of the incoming jet. Any energy loss reduces the velocity of the liquid leaving the plate.

Let Q = volumetric flow rate of liquid

a = cross-sectional area of the jet

ρ = density of the liquid

V = velocity of the jet

Force, F = Mass flow rate × Change in velocity along direction of jet

$$= \rho \cdot Q \times (V - 0)$$

$$= \rho \cdot (a \cdot V) \cdot V$$

$$= \rho a V^2 \quad \text{...(16.1)}$$

Force exerted by plate on the jet is equal and opposite to this force.

Flat Plate Inclined to Jet

Fig. 16.2 shows a smooth plate held against the flow and inclined at an angle θ to the direction of the jet. Neglecting any frictional loss and the change in the elevation, the velocity of the leaving jet is equal to that of the incoming jet.

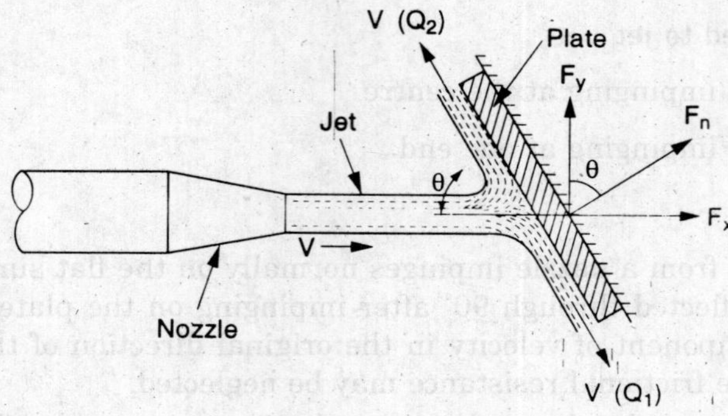

Fig. 16.2

Normal force on the plate

$$F_n = \text{Mass flow rate} \times \text{Change in velocity normal to the plate}$$
$$= \rho \cdot Q \cdot (V \sin\theta - 0)$$
$$= \rho \cdot (a \cdot V) \cdot V \sin\theta$$
$$= \rho a V^2 \sin\theta \quad \ldots(16.2)$$

As no force acts on the plate along its surface, the normal force on the plate is also the resultant or the maximum force.

Component of force along the jet direction $= F_n \times \sin\theta$

$$= (\rho a V^2 \sin\theta) \sin\theta$$
$$= \rho a V^2 \sin^2\theta \quad \ldots(16.3)$$

Component of force perpendicular to the jet direction

$$= F_n \cos\theta$$
$$= (\rho a V^2 \sin\theta) \cos\theta$$
$$= \rho a V^2 \sin\theta \cos\theta \quad \ldots(16.4)$$

If the jet is prevented from spreading sideways by providing flanges, a part of the stream escapes from each end of the plate. If Q_1 and Q_2 are the discharges and m_1 and m_2 the mass flow rates in the two directions respectively, then

The momentum parallel to the plate before impact

= The momentum parallel to the plate after impact

$$mV\cos\theta = m_1 V - m_2 V$$
$$m\cos\theta = m_1 - m_2$$
$$\rho Q\cos\theta = \rho Q_1 - \rho Q_2$$
$$Q\cos\theta = Q_1 - Q_2$$

But $\qquad Q = Q_1 + Q_2$

On solving the two equations,

$$Q_1 = \frac{Q}{2}(1 + \cos\theta)$$
$$Q_2 = \frac{Q}{2}(1 - \cos\theta)$$

Curved Plate, Jet Impinging at the Centre

Fig. 16.3 shows a jet of liquid striking at the centre of a smooth and symmetrical curved plate or vane. After striking, the jet glides over the plate in the outward directions and leaves at the ends tangentially. The velocity of the jet leaving the vane is the same as that of the incoming jet as before.

Let $\qquad \theta$ = angle between the directions of the incoming and the leaving jets

= vane angles at the tip of each side

V = velocity of the jet

Velocity of the jet at the leaving tip of the vane may be resolved into two components.

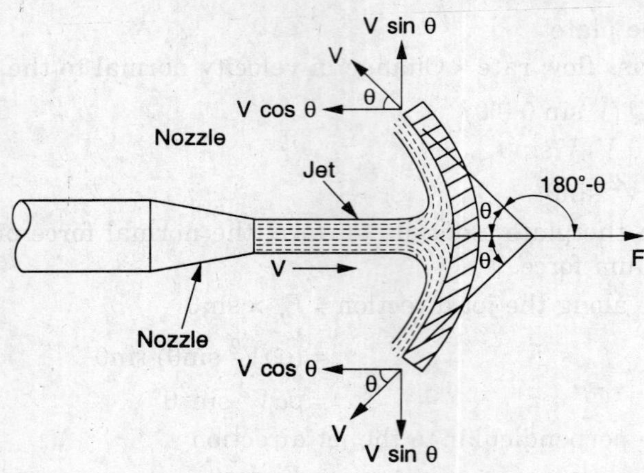

Fig. 16.3

Velocity along the direction of the jet $= -V\cos\theta$

(Negative as the direction is opposite to that of the jet)

Velocity perpendicular to the direction of the jet $= V\sin\theta$

Force on the vane in the direction of the jet

$$F_x = \text{Mass flow rate} \times \text{Change in velocity}$$
$$= \rho \cdot Q \cdot [V - (-V\cos\theta)]$$
$$= \rho \cdot a \cdot V \cdot V\,(1 + \cos\theta)$$
$$= \rho \cdot a \cdot V^2\,(1 + \cos\theta) \quad \ldots(16.5)$$

The net force in a direction perpendicular to the direction of the jet is zero due to symmetry.

- If $\theta = 90^\circ$, it becomes a case of flat plate held perpendicular to the direction of the jet and the force $F = \rho a V^2$ as before.
- If $\theta = 0^\circ$ *i.e.* if the plate is semi-circular and the jet is deflected back in the incoming direction,

$$F = \rho a V^2\,(1 + \cos 0^\circ) = 2\,\rho a V^2$$

Thus, the force exerted on a semi-circular plate is twice that of exerted on a flat plate.

Curved Plate, Jet Impinging at One End

A jet of liquid impinges at one end of a fixed curved plate, glides over it and then leaves it tangentially (Fig. 16.4). The velocity at the outlet is equal to the velocity at the inlet as before.

Let θ_1 = Angle of vane at the inlet with the x-direction
= Angle of jet at the inlet with the x-direction
θ_2 = Angle of vane at the outlet with the x-direction
= Angle of jet at the outlet with the x-direction

Components of velocity V,

At inlet, in x-direction $= V\cos\theta_1$

At inlet, in y-direction $= -V\sin\theta_1$

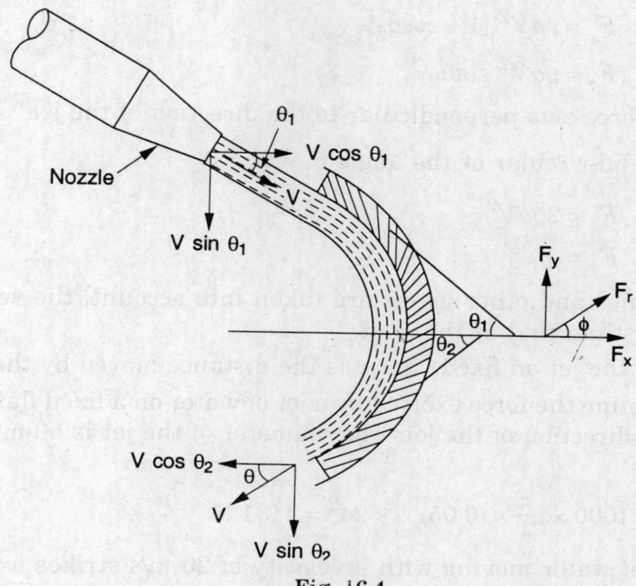

Fig. 16.4

At outlet, in x-direction $= V \cos\theta_2$

At outlet, in y-direction $= - V \sin\theta_2$

Force in the x-direction, $F_x = \rho a V [V \cos\theta_1 - (- V \cos\theta_2)]$

$$= \rho a V^2 [\cos\theta_1 + \cos\theta_2] \quad \text{...(16.6)}$$

Force in the y-direction, $F_y = \rho a V [- V \sin\theta_1 - (- V \sin\theta_2)]$

$$= \rho a V^2 [\sin\theta_2 - \sin\theta_1] \quad \text{...(16.7)}$$

Resultant force

$$\sqrt{F_x^2 + F_y^2} \quad \text{...(16.8)}$$

and its direction is given by

$$\phi = \tan^{-1} \frac{F_y}{F_x} \text{ with } x\text{-direction} \quad \text{...(16.9)}$$

Total angle of deflection of the jet $= 180° - (\theta_1 + \theta_2)$

- If the vane is symmetrical, $\theta_1 = \theta_2 = \theta$

and
$$F_x = 2\rho a V^2 \cos\theta$$
$$F_y = 0$$

i.e. no force in the the y-direction and thus F_x is the resultant force.

- If the inlet is parallel to the x-axis,

$$F_x = \rho a V^2 [1 + \cos\theta_2]$$

$$F_y = \rho a V^2 \sin\theta_2$$

i.e. an upward force acts perpendicular to the direction of the jet.

- If the vane is semi-circular or the angle $\theta_1 = \theta_2 = 0$

$$F_x = 2\rho a V^2$$

$$F_y = 0$$

If friction of the vanes and other losses are taken into account, the velocity of the jet at the outlet will be lesser than that at the inlet.

No work is done by the jet on fixed vanes as the distance moved by the vanes is zero.

Example 16.1 Determine the force exerted by a jet of water on a fixed flat plate held normal to the direction of the jet. The diameter of the jet is 50 mm and the velocity 24 m/s.

Solution $F = \rho a V^2 = 1000 \times \left[\frac{\pi}{4} \times (0.05)^2\right] \times 24^2 = 1131 \text{ N}$

Example 16.2 A jet of water moving with a velocity of 20 m/s strikes a fixed plate in such a way that the angle between the jet and the plate is 40°. The diameter of the jet is 60 mm. Determine

(i) Maximum force of the jet on the plate

(ii) Force along the direction of the jet

(iii) Force normal to the direction of the jet.

Solution (i) $F_n = \rho a V^2 \sin\theta = 1000 \times \left[\frac{\pi}{4} \times (0.06)^2\right] \times 20^2 \times \sin 40° = 727 \text{ N}$

(ii) $F_x = F_n \sin\theta = 727 \sin 40° = 467.3 \text{ N}$

(iii) $F_y = F_n \cos\theta = 727 \cos 40° = 556.9 \text{ N}$

Example 16.3 A jet of water 25 mm in diameter and moving with a velocity of 10 m/s strikes horizontally at the centre of square plate of edge 250 mm. The plate is of uniform thickness and has a mass of 7 kg. It is suspended vertically by a hinge on its top horizontal edge. Find

(i) the force to be applied at the lower edge to keep the plate vertical

(ii) the angle of inclination of the plate if it is allowed to swing freely.

Solution (i) Refer Fig. 16.5a,

$$F = \rho a V^2 = 1000 \times \left[\frac{\pi}{4} \times (0.025)^2\right] \times 10^2 = 49.1 \text{ N}$$

Let f = Force to be applied at the lower edge to keep the plate in the vertical position

Then taking moments about the hinge,

$$f \times 0.25 = F \times \frac{0.25}{2}$$

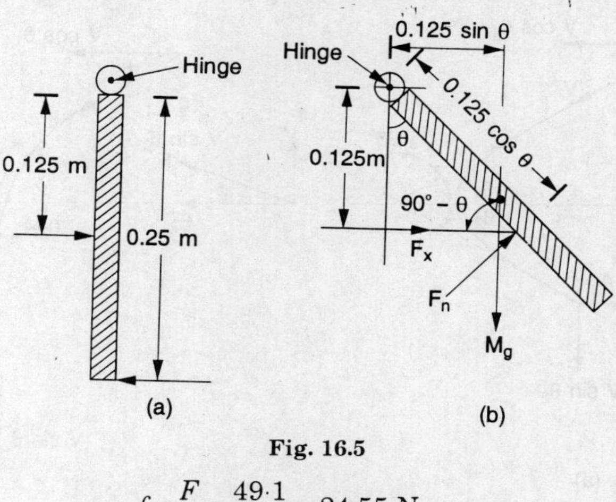

Fig. 16.5

$$f = \frac{F}{2} = \frac{49.1}{2} = 24.55 \text{ N}$$

(ii) Let θ be the angle of inclination when the plate is allowed to swing freely (Fig. 16.5b).

Normal force on the plate, $F_n = \rho a V^2 \sin(90° - \theta)$

$$= 49.1 \times \sin(90° - \theta)$$

$$= 49.1 \cos\theta$$

Weight of the plate acts downwards through the centre of mass of the plate.

Taking moments of the forces about the hinge,

$$7 \times 9.81 \times 0.125 \sin\theta = 49.1 \cos\theta \times \frac{0.125}{\cos\theta}$$

$$\sin\theta = 0.715$$

$$\theta = 45.6°$$

Example 16.4 A jet of water 80 mm in diameter and moving with a velocity of 60 m/s strikes at the centre of a fixed symmetrical curved plate. If the jet is deflected through an angle of 130°, determine the force exerted by the jet on the plate.

Solution $\theta = 180° - 130° = 50°$

$$F = \rho a V^2 (1 + \cos\theta) = 1000 \times \left[\frac{\pi}{4} \times (0.08)^2\right] \times 60^2 \times (1 + \cos 50°) = 29727 \text{ N}$$

Example 16.5 A jet of water impinges at one end of a fixed curved plate tangentially. The diameter of the jet is 60 mm and its velocity 30 m/s. The angle of the plate at the inlet is 30° to the horizontal. Determine the force exerted by the jet on the plate in the horizontal and in the vertical direction if the jet is deflected through an angle of

(i) 125° (ii) 50°

Also, find the magnitude and the direction of resultant force in each case.

Solution (i) Angle of deflection = $180° - (\theta_1 + \theta_2)$

$$125° = 180° - (30° + \theta_2)$$

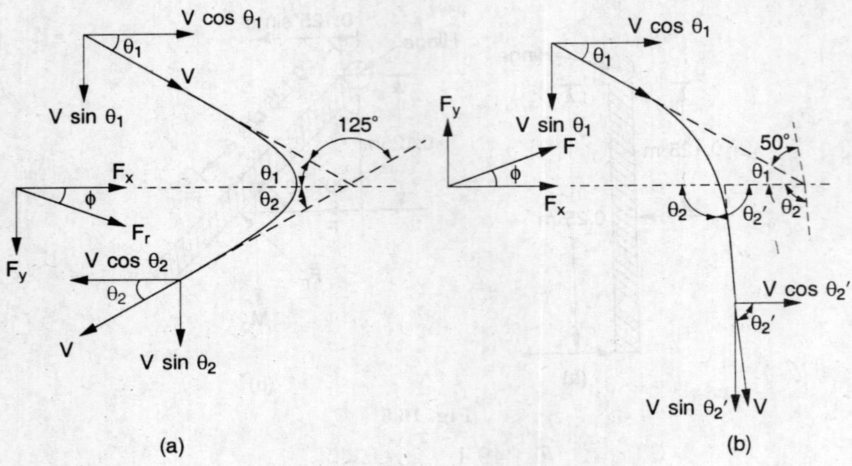

Fig. 16.6

$$\theta_2 = 25^\circ$$

$$F_x = \rho a V^2 (\cos \theta_1 + \cos \theta_{21})$$

$$= 1000 \times \left[\frac{\pi}{4} \times (0{\cdot}05)^2\right] \times 30^2 \times (\cos 30^\circ + \cos 25^\circ)$$

$$= 2544{\cdot}7 \times 1{\cdot}7723$$

$$= 4510 \text{ N}$$

$$F_y = \rho a V^2 [\sin \theta_2 - \sin \theta_1]$$

$$= 2544{\cdot}7 \times (\sin 25^\circ - \sin 30^\circ)$$

$$= 2544{\cdot}7 \times (- {\cdot}0774)$$

$$= -\ 196{\cdot}9 \text{ N}$$

Negative sign indicates that the force in the vertical direction is downwards.

Resultant force

$$= \sqrt{F_x^{\ 2} + F_y^{\ 2}} = \sqrt{4510^2 + 196{\cdot}9^2} = 4514{\cdot}3 \text{ N}$$

$$\tan \phi = \frac{196{\cdot}9}{4510} = 0{\cdot}0437$$

$$\phi = 2{\cdot}5^\circ$$

(ii) Angle of deflection $= 180^\circ - (\theta_1 + \theta_2)$

$$50^\circ = 180^\circ - (30^\circ + \theta_2)$$

$$\theta_2 = 100^\circ$$

$$F_x = \rho a V^2 (\cos \theta_1 + \cos \theta_2)$$

$$= 2544{\cdot}7 \times (\cos 30^\circ + \cos 100^\circ)$$

$$= 2544{\cdot}7 \times 0{\cdot}6924$$

$$= 1761{\cdot}9 \text{ N}$$

$$F_y = \rho a V^2 \left[\sin \theta_2 - \sin \theta_1\right]$$
$$= 2544{\cdot}7 \times (\sin 100^\circ - \sin 30^\circ)$$
$$= 2544{\cdot}7 \times 0{\cdot}4848$$
$$= 1233{\cdot}7 \text{ N}$$

i.e. the force in the vertical direction is upwards.

Resultant force

$$= \sqrt{F_x^{\,2} + F_y^{\,2}}$$
$$= \sqrt{1761{\cdot}9^2 + 1233{\cdot}7^2}$$
$$= 2150{\cdot}9N$$
$$\tan \phi = \frac{1233{\cdot}7}{1761{\cdot}9} = 0{\cdot}7$$
$$\phi = 35^\circ$$

16.3 FORCE ON MOVING PLATES

In case of moving plates, the change of momentum is found by considering the change in the relative velocities.

The cases, which were considered for fixed plates, are again taken for moving plates.

Flat Plate Normal to Jet

Consider a smooth flat plate moving in the same direction as the jet (Fig. 16.7). Let V and u are the absolute velocities of the jet and the plate respectively. The velocity of the plate u must be smaller than the velocity V of the jet, otherwise, the jet is not able to strike the plate. The velocity with which the jet impinges on the plate is its velocity relative to the plate *i.e.* velocity of jet assuming the plate to be stationary. Thus, the jet strikes the plate with a velocity equal to $(V - u)$.

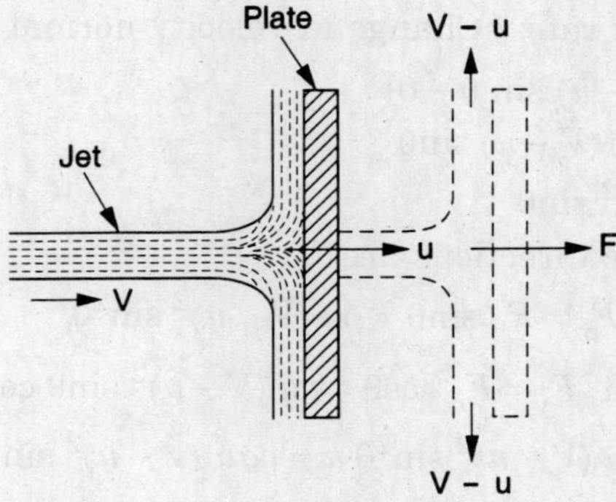

Fig. 16.7

The velocity of the leaving jet is equal to the impinging velocity of the jet *i.e.* equal to $(V - u)$.

In case of moving plates, the mass of liquid per second striking the plate depends upon the relative velocity.

Force on the plate, F = Mass flow rate × Change in relative velocity

$$= [\rho \cdot a \cdot (V - u)]\,[(V - u) - 0]$$

$$= \rho a\,(V - u)^2 \quad ...(16.10)$$

In case of a moving plate, the work is done on the plate.

Work done per unit time,

W = Force × Distance moved by the plate per second

$$= F \cdot u$$

$$= \rho a\,(V - u)^2 \cdot u \quad ...(16.11)$$

Flat Plate Inclined to Jet

Consider a smooth flat plate inclined at an angle θ to the direction of the jet moving in the same direction as the jet (Fig. 16.8). The velocity of the leaving jet is equal to the impinging velocity of the jet as before. Since the jet leaves the plate tangentially, the velocity of the leaving jet perpendicular to the plate is zero.

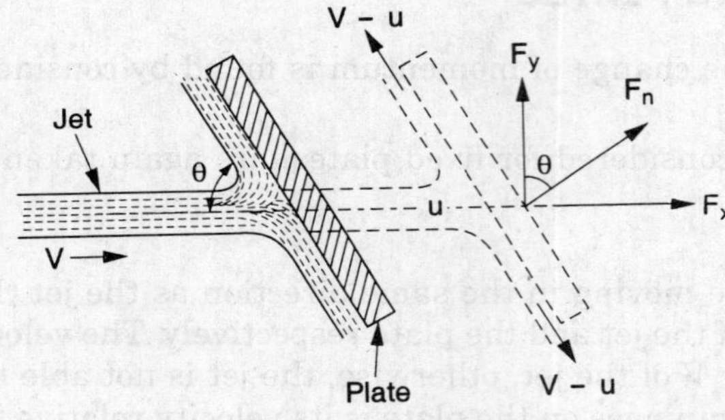

Fig. 16.8

Normal force on the plate

F_n = Mass flow rate × Change in velocity normal to the plate

$$= \rho \cdot Q \times [(V - u)\sin\theta - 0]$$

$$= \rho \cdot a \cdot (V - u) \cdot (V - u)\sin\theta$$

$$= \rho a\,(V - u)^2 \sin\theta \quad ...(16.12)$$

The components in the two directions are

Along the direction of jet, $F_x = F_n \sin\theta = \rho a\,(V - u)^2 \sin^2\theta$

In a direction normal to jet, $F_y = F_n \cos\theta = \rho a\,(V - u)^2 \sin\theta \cos\theta$

Work done, $W = F_x \times u = \rho a\,(V - u)^2 \sin^2\theta \cdot u = \rho a u\,(V - u)^2 \sin^2\theta$ ...(16.13)

If θ = 90°, the plate is normal to the jet and the results for that case are obtained.

Curved Plate, Jet Impinging at the Centre

Let a jet of liquid impinge at the centre of a moving symmetrical smooth curved vane (Fig. 16.9). The liquid glides over the vane and leaves it tangentially.

Let θ = angle of the vane at the tips on each side

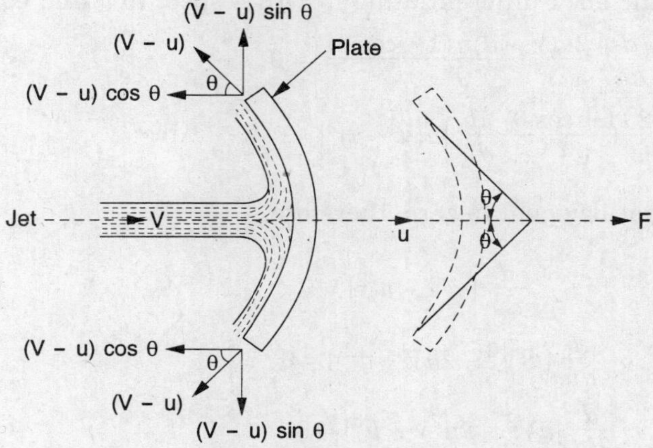

Fig. 16.9

Force on the vane in the direction of the jet,

$$F_x = \text{Mass flow rate} \times \text{Change in velocity}$$
$$= \rho \cdot Q \cdot [(V - u) - \{-(V - u)\cos\theta\}]$$
$$= \rho \cdot a \cdot (V - u)\,[(V - u)\,(1 + \cos\theta)]$$
$$= \rho \cdot a \cdot (V - u)^2\,(1 + \cos\theta) \qquad \text{...(16.14)}$$

Work done, $W = F_x \times u = \rho a\,(V - u)^2\,(1 + \cos\theta) \cdot u$

- When $\theta = 90^\circ$, it becomes the case of a flat plate

  and $\quad F_x = \rho a\,(V - u)^2$

- When $\theta = 0^\circ$ or the vane is semicircular,

  $F_x = 2\rho a\,(V - u)^2$ *i.e.* twice the force on a flat plate.

As the vane is symmetrical, the net force in a direction perpendicular to that of jet is zero.

Efficiency of the system,

$$\eta = \frac{\text{Output}}{\text{Input}} = \frac{\text{Work done}}{\text{Kinetic energy of jet}}$$

Kinetic energy of jet coming out of the nozzle

$$= \frac{1}{2} \times \text{Mass of liquid} \times (\text{Velocity of the jet})^2$$
$$= \frac{1}{2} \times (\rho a V) \cdot V^2$$
$$= \frac{1}{2}\,\rho a V^3$$

Thus efficiency,

$$\eta = \frac{\rho a (V - u)^2\,(1 + \cos\theta) \cdot u}{\frac{1}{2}\,\rho a V^3} = \frac{2u(V - u)^2\,(1 + \cos\theta)}{V^3} \qquad \text{...(16.15)}$$

For maximum efficiency, differentiating it witch respect to u and equating to zero.

$$\frac{d\eta}{du} = \frac{d}{du}\left[\frac{2u(V-u)^2(1+\cos\theta)}{V^3}\right] = 0$$

$$\frac{2(1+\cos\theta)}{V^3}\cdot\frac{d}{du}[u(V-u)^2] = 0$$

$\frac{2(1+\cos\theta)}{V^3}$ cannot be equal to zero, therefore,

$$\frac{d}{du}[u(V-u)^2] = 0$$

$$\frac{d}{du}[u(V^2-2uV+u^2)] = 0$$

$$\frac{d}{du}[uV^2-2u^2V+u^3] = 0$$

$$V^2-4uV+3u^2 = 0$$

$$(V-u)(V-3u) = 0$$

Either $V = u$ or $V = 3u$.

If $V = u$, no liquid can strike the vane and no force acts.

Thus $V = 3u$

and $$\eta_{max} = \frac{2u(2u)^2(1+\cos\theta)}{(3u)^3} = \frac{8(1+\cos\theta)}{27} = \frac{8}{27}\times 2\cos^2\frac{\theta}{2} = \frac{16}{27}\cos^2\frac{\theta}{2}$$

If the vane is semicircular, $\theta = 0°$

$$\eta_{max} = \frac{16}{27} = 0{\cdot}592$$

Curved Plate, Jet Impinging at One End.

Consider a jet of liquid impinging on a curved plate or vane tangentially at one end. In Fig. 16.10, ab represents the absolute velocity V_1 of the jet and ac is the linear velocity u_1 of the vane at the inlet. cb is the vector difference of the two and represents the velocity of the jet relative to vane and is denoted by V_{r1}. The vector cb also indicates the direction of the jet with which it strikes the inlet tip of the vane. For smooth entry of the liquid, the inlet angle of the vane must conform to this direction. Let α_1 and β_1 be the angles made by the absolute velocity ab and relative velocity cb respectively with the direction of the motion of the vane. It may be observed that α_1 is the angle of the nozzle and β_1 the inlet angle of the vane.

Taking x-axis along the direction of jet, the compnent of absolute velocity ab in the x-direction is V_{w1} and is known as **velocity of whirl** at the inlet. The component in the y-direction is V_{f1} and is known as the **velocity of flow** at the inlet.

After gliding over the vane, the liquid leaves it with a relative velocity V_{r2} at an angle β_2, which is the angle of the tip of the vane at the outlet. V_{r2} is equal to V_{r1} if friction of the vane is neglected. As V_{r2} is the vector difference of the absolute velocity of the jet V_2 and that of the vane tip at the outlet u_2, the magnitudes and directions of these must be as shown in

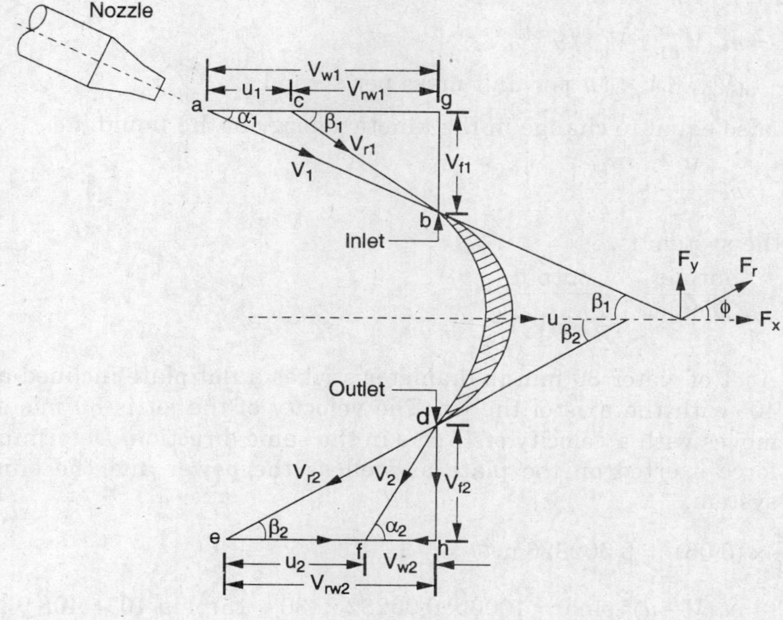

Fig. 16.10

the figure. The vectors indicating V_2 and u_2 must be towards or away from a point to obtain the difference. α_2 is the angle of the absolute velocity with the direction of motion of the vane tip at the outlet.

As before, V_{w2} and V_{f2} are the components of the absolute velocity in the x- and y-directions and are known as **velocity of whirl** and **velocity of flow** respectively at the outlet.

Let V_{rw1} and V_{rw2} be the components of relative velocities at the inlet and the outlet respectively in the x-direction. Then,

Now, $V_{rw1} = V_{w1} - u_1$ (directions of the two are the same)

$V_{rw2} = V_{w2} + u_2$ (directions of the two are opposite)

Force exerted by jet in the direction of motion of the vane,

$$F_x = \text{Mass flow rate} \times \text{Change in relative velocities in the } x\text{-direction}$$

$$= m \times (V_{rw1} - V_{rw2})$$

$$= m \times [(V_{w1} - u_1) - \{-(V_{w2} + u_2)\}]$$

$$= m\,(V_{w1} + V_{w2}) \qquad \ldots(16.17)$$

Mass of the liquid striking the vane, $m = \rho \cdot a \cdot V_{r1}$

If $\beta_2 > 90^\circ$, then V_{rw2} is positive, and

$$F_x = m\,(V_{w1} - V_{w2})$$

Thus in general,

$$F_x = m\,(V_{w1} \pm V_{w2})$$

Work done, $W = F_x \times u$

$= m\ (V_{w1} \pm V_{w2})\cdot u$

$= (V_{w1} \pm V_{w2})\cdot u$ per unit mass per second ...(16.18)

Work done is also equal to change in the kinetic energy of the liquid, *i.e.*

$$W = m\left(\frac{V_1^2}{2} - \frac{V_2^2}{2}\right) \qquad ...(16.19)$$

Efficiency of the system is,

$$\eta = \frac{\text{output}}{KE \text{ of jet}} = \frac{\text{output}}{\frac{1}{2}(\rho a V)V^2}$$

Example 16.6 A jet of water 60 mm in diameter strikes a flat plate inclined at an angle of 40° with the axis of the jet. The velocity of the jet is 30 m/s and the plate moves with a velocity of 15 m/s in the same direction. Determine the normal force exerted on the plate as well as the power and the efficiency of the system.

Solution $a = \frac{\pi}{4} \times (0{\cdot}06)^2 = 0{\cdot}002826 \text{ m}^2$

$F_n = \rho a(V - u)^2 \cdot \sin\theta = 1000 \times 0{\cdot}00282 \times (30 - 15)^2 \cdot \sin 40° = 408{\cdot}9$ N

Power is the work done by the jet per second,

$P = F_x \times u = F_n \sin 40° \times 15 = 408{\cdot}9 \sin 40° \times 15 = 3942{\cdot}5$ W or $3{\cdot}9425$ kW

$$\eta = \frac{\text{output}}{KE \text{ of jet}} = \frac{\text{output}}{\frac{1}{2}(\rho a V)V^2} = \frac{3942{\cdot}5}{\frac{1}{2} \times 1000 \times 0{\cdot}002826 \times 30^3} = 0{\cdot}1033 \quad \text{or} \quad 10{\cdot}33\%$$

Example 16.7 A jet of water 60 mm in diameter impinges on a curved vane at its centre. The velocity of the jet is 24 m/s and is deflected through an angle of 160°. The curved vane moves with a velocity of 10 m/s in the same direction as the jet. Determine

(i) Force exerted on the vane

(ii) Power and efficiency of the jet.

Solution $a = \frac{\pi}{4} \times (0{\cdot}06)^2 = 0{\cdot}002826 \text{ m}^2$

$\theta = 180° - 160° = 20°$

$F = \rho \cdot a \cdot (V - u)^2 (1 + \cos\theta) = 1000 \times 0{\cdot}002826 \times (24 - 10)^2 \cdot (1 + \cos 20°) = 1075$ N

$P = F \times u = 1072 \times 10 = 10720$ W or $10{\cdot}72$ kW

$$\eta = \frac{\text{output}}{KE \text{ of jet}} = \frac{\text{output}}{\frac{1}{2}(\rho a V)V^2} = \frac{10720}{\frac{1}{2} \times 1000 \times 0{\cdot}002826 \times 24^3} = 0{\cdot}549 \quad \text{or} \quad 54{\cdot}9\%$$

Example 16.8 A jet of water moving with a velocity of 22 m/s impinges on a curved vane at the one end tangentially. The jet leaves the vane at an angle of 120° to the direction of motion of the vane. The velocity of the vane is 10 m/s and the angle of the nozzle is 20°. Determine

(i) Vane angles at the inlet and the outlet and

(ii) Work done per second per unit mass of water.

Solution $\alpha_1 = 20^\circ$ $\alpha_2 = 180^\circ - 120^\circ = 60^\circ$

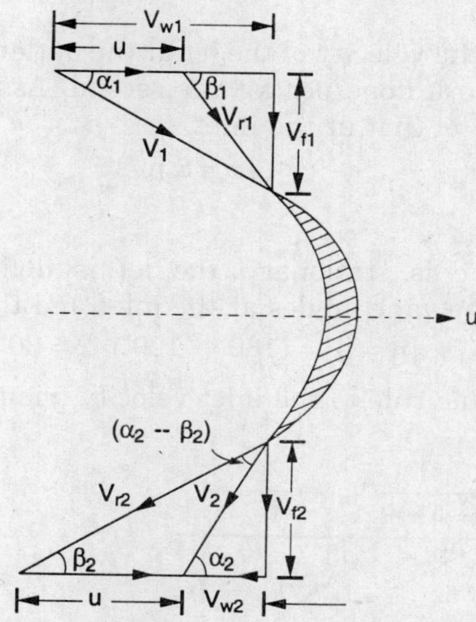

Fig. 16.11

$$V_{f1} = V_1 \sin \alpha_1 = 22 \sin 20^\circ = 7{\cdot}52 \text{ m/s}$$

$$V_{w1} = V_1 \cos \alpha_1 = 22 \cos 20^\circ = 20{\cdot}67 \text{ m/s}$$

$$\tan \beta_1 = \frac{V_{f1}}{V_{w1} - u} = \frac{7{\cdot}52}{20{\cdot}67 - 10} = 0{\cdot}705$$

$$\beta_1 = 35{\cdot}19^\circ$$

$$V_{r2} = V_{r1} = \frac{V_{f1}}{\sin \beta_1} = \frac{7{\cdot}52}{\sin 35{\cdot}19^\circ} = 13{\cdot}05 \text{ m/s}$$

Applying sine rule,

$$\frac{V_{r2}}{\sin (180^\circ - \alpha_2)} = \frac{u}{\sin (\alpha_2 - \beta_2)}$$

$$\frac{13{\cdot}05}{\sin (180^\circ - 60^\circ)} = \frac{10}{\sin (60^\circ - \beta_2)}$$

$$\sin (60^\circ - \beta_2) = 0{\cdot}664$$

$$60^\circ - \beta_2 = 41{\cdot}58^\circ$$

$$\beta_2 = 18{\cdot}42^\circ$$

(ii) $V_{w2} = V_{r2} \cos \beta_2 - u = 13{\cdot}05 \cos 18{\cdot}42^\circ - 10 = 2{\cdot}38$ m/s

$W/kg/s = (V_{w1} + V_{w2}){\cdot}u = (20{\cdot}67 + 2{\cdot}38) \times 10 = 230{\cdot}5$ J

Example 16.9 A jet of water moving at a velocity of 16 m/s impinges on a concave shaped vane. When the vane is stationary, it is deflected through an angle of 120°. Assuming the vane to be symmetrical, determine the angle of the nozzle to avoid the shock at the inlet when the vane starts moving at a velocity of 8 m/s.

Also, find the velocity of the jet at the outlet both in magnitude and direction and the work done per kg per second. Assume the relative velocity at the outlet 90% of that at the inlet.

Solution $V_1 = 16$ m/s $\qquad u = 8$ m/s

$V_{r2} = 0{\cdot}9 V_{r1}$

When the vane is stationary, the jet is deflected by 120°. As the vane is symmetrical, the vane angles at the inlet and the outlet are

$$\beta_1 = \beta_2 = (180^\circ - 120^\circ)/2 = 30^\circ$$

Applying the sine rule to the inlet velocity triangle,

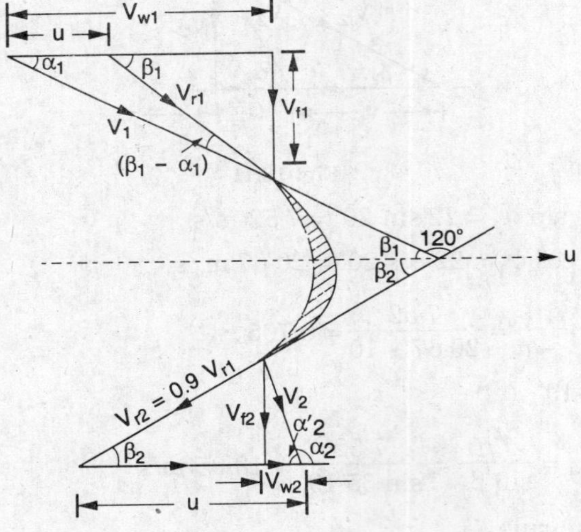

Fig. 16.12

$$\frac{V_1}{\sin(180^\circ - \beta_1)} = \frac{u}{\sin(\beta_1 - \alpha_1)}$$

$$\frac{16}{\sin(180^\circ - 30^\circ)} = \frac{8}{\sin(30^\circ - \alpha_1)}$$

$$\sin(30^\circ - \alpha_1) = 0{\cdot}25$$

$$30^\circ - \alpha_1 = 14{\cdot}48^\circ$$

$$\alpha_1 = 15{\cdot}52^\circ$$

$$V_{w1} = V_1 \cos \alpha_1 = 16 \cos 15{\cdot}52° = 15{\cdot}42 \text{ m/s}$$

$$V_{f1} = V_1 \sin \alpha_1 = 16 \sin 15{\cdot}52° = 4{\cdot}28 \text{ m/s}$$

$$V_{r1} = \frac{V_{f1}}{\sin \beta_1} = \frac{4{\cdot}28}{\sin 30°} = 8{\cdot}56 \text{ m/s}$$

At the outlet,

$$V_{r2} = 0{\cdot}9V_{r1} = 0{\cdot}9 \times 8{\cdot}56 = 7{\cdot}7 \text{ m/s}$$

$$V_{w2} = V_{r2} \cos\beta_2 - u = 7{\cdot}7 \cos 30° - 8 = -1{\cdot}328 \text{ m/s}$$

Nagative sign indicates that the direction of the whirl velocity is opposite to the usual assumed direction. In other words, it is in the direction of the jet velocity.

$$V_{f2} = V_{r2} \sin\beta_2 = 7{\cdot}7 \sin 30° = 3{\cdot}85 \text{ m/s}$$

$$\tan \alpha_2' = \frac{V_{f2}}{V_{w2}} = \frac{3{\cdot}85}{1{\cdot}328} = 2{\cdot}899$$

$$\alpha_2' = 78°$$

$$V_2 = \sqrt{V_{f2}^2 + V_{w2}^2} = \sqrt{3{\cdot}85^2 + 1{\cdot}328^2} = 4{\cdot}073 \text{ m/s}$$

Work done per kg per second = $(V_{w1} - V_{w2}) \cdot u = (15{\cdot}42 - 1{\cdot}328) \times 8 = 112{\cdot}7$ J

16.4 FORCE ON SERIES OF VANES

In case of a single moving plate or vane, the distance between the nozzle and the vane constantly increases. This makes it impractical for any useful output. However, if a number of vanes are mounted on a wheel so that when the wheel is rotated about its axis, one vane after the other will come before the jet. As the variation in the distance will be small, it may be assumed that the average distance between the jet and the vane is constant.

The following cases are being considered :

- Series of Flat Moving Plates
- Series of Curved Moving Vanes
- Series of Radial Curved Vanes.

Series of Flat Moving Plates

Figure 16.13 shows a number of smooth flat plates mounted on a rotating wheel all around its periphery. As the plates come before the jet one after the other, it may be assumed that there remains a plate before the jet all the time and at the same distance.

Mass of liquid impinging on the plates per second,

$$m = \rho \cdot a \cdot V$$

Force, F = Mass flow rate × Change in relative velocity

$$= m\,[(V - u) - 0]$$

$$= \rho a V (V - u) \quad \text{...(16.20)}$$

work,

$$W = \rho a V (V - u) \cdot u \quad \text{...(16.21)}$$

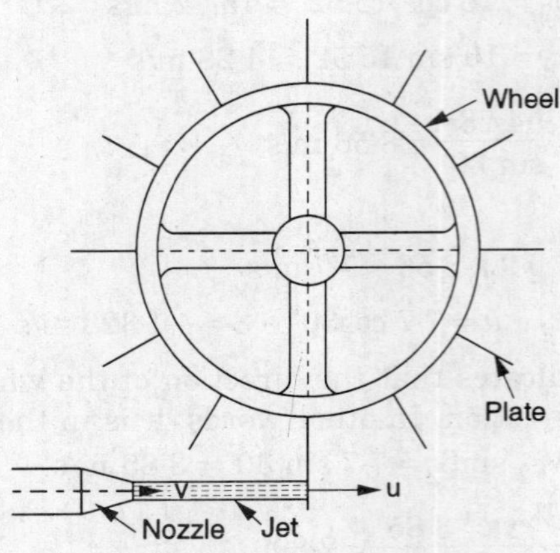

Fig. 16.13

Kinetic energy of the issuing jet $= \frac{1}{2}(\rho a V)\, V^2$

Efficiency,

$$\eta = \frac{\rho a V\,(V-u)u}{\frac{1}{2}(\rho a V)V^2} = \frac{2u(V-u)}{V^2} \qquad \text{...(16.22)}$$

For maximum efficiency, differentiating it with respect to u and equating to zero.

$$\frac{d\eta}{du} = \frac{d}{du}\left[\frac{2(uV-u^2)}{V^2}\right] = 0$$

$$\frac{2}{V^2}(V-2u) = 0$$

$$u = \frac{V}{2}$$

$$\eta_{max} = \frac{2u(2u-u)}{4u^2} = \frac{2u^2}{4u^2} = \frac{1}{2} = 0{\cdot}5$$

2. Series of Curved Moving Vanes

A number of curved vanes are mounted on a wheel rotating about its axis. The jet impinges at the centre of the moving vanes.

Force, F = Mass flow rate × Change in relative velocity

$= \rho \cdot a \cdot V \times [(V-u) - \{-(V-u)\cos\theta\}]$

$= \rho \cdot a \cdot V\,[(V-u)(1+\cos\theta)]$ ...(16.23)

Work done, $W = F_x \times u = \rho \cdot a \cdot V\,[(V-u)(1+\cos\theta)] \cdot u$ ...(16.24)

Kinetic energy of jet coming out of the nozzle

$= \frac{1}{2}$ × Mass of liquid × (Velocity of the jet)^2

$$= \frac{1}{2} \times (\rho a V) \cdot V^2$$

$$= \rho a V^3$$

Thus efficiency,

$$\eta = \frac{\rho a V (V - u)(1 + \cos\theta) \cdot u}{\frac{1}{2} \rho a V^3} = \frac{2u (V - u)(1 + \cos\theta)}{V^2} \quad ...(16.25)$$

For maximum efficiency, differentiating it with respect to u and equating to zero.

$$\frac{d\eta}{du} = \frac{d}{du}\left[\frac{2u (V - u)(1 + \cos\theta)}{V^2}\right] = 0$$

$$\frac{2(1 + \cos\theta)}{V^2} \cdot \frac{d}{du}[uV - u^2] = 0$$

$\frac{2(1 + \cos\theta)}{V^2}$ cannot be equal to zero.

$$\therefore \quad \frac{d}{du}[uV - u^2] = 0$$

$$V - 2u = 0$$

$$u = \frac{V}{2}$$

$$\eta_{max} = \frac{2u(2u - u)(1 + \cos)}{4u^2} = \frac{1 + \cos\theta}{2} \quad ...(16.26)$$

which is more than the efficiency of a series of flat moving plates.

When $\theta = 90°$, it becomes the case of flat plates mounted on the periphery of a wheel and

$$\eta_{max} = \frac{1}{2}(1 + \cos 90°) = \frac{1}{2}$$

If the vanes are semicircular, $\theta = 0°$

$$\eta_{max} = 1$$

The principle is applicable to Pelton turbines with double hemispherical shape of buckets. The jet impinges the vanes at the centre and turns through 160° to 170°.

3. Series of Radial Curved Vanes

A number of radial curved vanes mounted on the circumference of a wheel is a practical case. The flow is admitted at one tip of a vane and is discharged at the other. The liquid particles travel in a radial direction inwards or outwards. All along the travel, the liquid particles ramain in a plane perpendicular to the axis. Such type of flow is known as **radial flow** (Fig. 16.14). The torque produced in case of radial vanes is given by the rate of change of angular momentum.

Let R_1 = radius of the wheel at the inlet of vane

R_2 = radius of the wheel at the outlet of vane

Mass of liquid impinging on the vanes per second $= \rho a V_1$

From the inlet and outlet triangles :

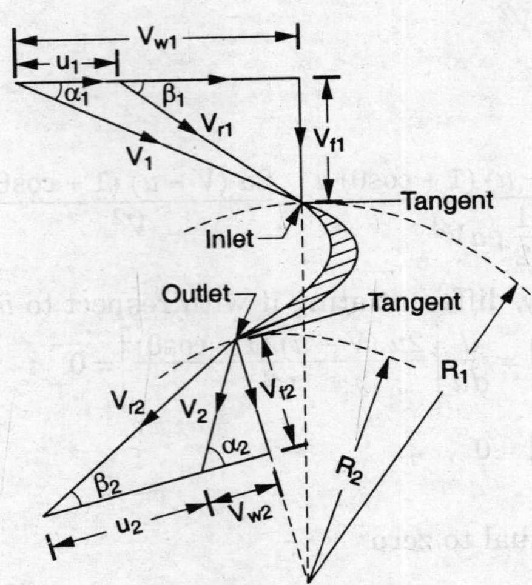

Fig. 16.14

At the inlet :

Momentum of liquid in the tangential direction $= m \cdot V_{w1}$

Angular momentum = Momentum × Radius $= m \cdot V_{w1} \cdot R_1$

At the outlet :

Momentum of liquid in the tangential direction $= m \cdot (-V_{w2})$

Angular momentum = Momentum × Radius $= m \cdot (-V_{m2}) \cdot R_2 = -m \cdot R_{w2} \cdot R_2$

Torque produced by the liquid on the wheel per second,

$$
\begin{aligned}
T &= \text{Change of angular momentum per second} \\
&= m \cdot V_{w1} \cdot R_1 - (-m \cdot V_{w2} \cdot R_2) \\
&= m \cdot V_{w1} \cdot R_1 + m \cdot V_{w2} \cdot R_2 \\
&= m\,(V_{w1} \cdot R_1 + V_{w2} \cdot R_2) \\
&= \rho a V_1\,(V_{w1} \cdot R_1 + V_{w2} \cdot R_2) \qquad \ldots(16.27)
\end{aligned}
$$

Work done on the wheel per second,

$$
\begin{aligned}
W &= \text{Torque} \times \text{Angular velocity} \\
&= T \times \omega \\
&= \rho a V_1\,(V_{w1} \cdot R_1 + V_{w2} \cdot R_2) \cdot \omega \\
&= \rho a V_1\,(V_{w1} \cdot R_1 \cdot \omega + V_{w2} \cdot R_2 \cdot \omega) \\
&= \rho a V_1\,(V_{w1} \cdot u_1 + V_{w2} \cdot u_2) \qquad \ldots(16.28)
\end{aligned}
$$

or Work done/unit mass/s $= V_{w1} \cdot u_1 + V_{w2} \cdot u_2$

- If $\alpha_2 > 90^\circ$, $W = \rho a V_1\,(V_{w1} \cdot u_1 - V_{w2} \cdot u_2)$

  In general, $W = \rho a V_1\,(V_{w1} \cdot u_1 \pm V_{w2} \cdot u_2)$ ...(16.29)

- If $\alpha_2 = 90^\circ$ *ie* the discharge is radial,

$$W = \rho a V_1 (V_{w1} \cdot u_1) \qquad ...(V_{w2} = 0)$$

Efficiency,

$$\eta = \frac{\text{Work done per second}}{\text{KE of the jet per second}}$$

$$= \frac{\rho a V_1 (V_{w1} \cdot u_1 \pm V_{w1} \cdot u_2)}{\frac{1}{2}(\rho a V_1) V_1^2}$$

$$= \frac{2(V_{w1} \cdot u_1 \pm V_{w1} \cdot u_2)}{V_1^2} \qquad ...(16.30)$$

Work done on the wheel is also equal to the change in kinetic energy of the jet provided there is no loss of energy on the vane surface.

$$W = \frac{1}{2} \text{m} (V_1^2 - V_2^2) \qquad ...(16.31)$$

$$\eta = \frac{\frac{1}{2} \text{m}(V_1^2 - V_2^2)}{\frac{1}{2} \text{m} \cdot V_1^2} = \frac{V_1^2 - V_2^2}{V_1^2} = 1 - \left(\frac{V_2}{V_1}\right)^2 \qquad ...(16.32)$$

For a given value of the inlet velocity, efficiency is maximum if the absolute velocity at the outlet V_2 is minimum. It cannot be zero, as in that case, there is no flow. However, it may be reduced by keeping the angle of the vane at the outlet minimum. If the energy supplied to the wheel is given in terms of head (H), then the efficiency can also be written as

$$\eta = \frac{V_{w1} \cdot u_1 \pm V_{w1} \cdot u_2}{gH} \qquad ...(16.33)$$

Example 16.10 A jet of water impinges on a curved vane with the inlet angle being zero degree. The diameter of the jet is 50 mm and it is deflected through an angle of 160°. The rate of flow through the nozzle is 200 l/s. Determine

(i) Magnitude and direction of the force on the fixed vane.

(ii) Magnitude and direction of the force on the fixed vane if the vane moves with a velocity of 30 m/s in the same direction as the jet. Also, find the power developed and the efficiency.

(iii) Power developed and the efficiency if a series of vanes are mounted on a wheel, the tangential velocity being 30 m/s.

Solution $Q = 0 \cdot 2 \text{ m}^3/\text{s}$ $\qquad d = 0 \cdot 5 \text{ m}$

$\theta_1 = 0^\circ$ $\qquad \theta_2 = 180^\circ - 160^\circ = 20^\circ$

$$a = \frac{\pi}{4} \times d^2 = \frac{\pi}{4} \times 0 \cdot 05^2 = 0 \cdot 001962 \text{ m}^2$$

$$V = \frac{Q}{a} = \frac{0 \cdot 2}{0 \cdot 001962} = 101 \cdot 94 \text{ m/s}$$

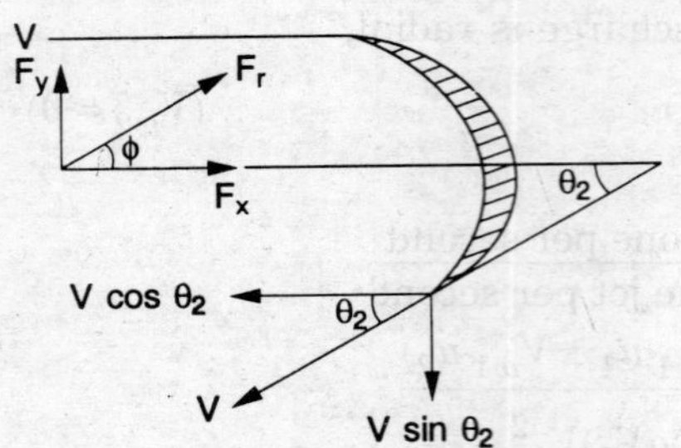

Fig. 16.15

(i) $F_x = \rho a V^2 \, [\cos\theta_1 + \cos\theta_2]$

$$= 1000 \times 0{\cdot}001962 \times 101{\cdot}94^2 \, [\cos 0^\circ + \cos 20^\circ]$$

$$= 39545 \text{ N}$$

$$F_y = \rho a V^2 \, [\sin\theta_2 - \sin\theta_1] = 1000 \times 0{\cdot}001962 \times 101{\cdot}94^2$$

$$[\sin 0^\circ - (-\sin 20^\circ)] = 6973 \text{ N}$$

Resultant force,

$$F = \sqrt{F_x^{\,2} + F_x^{\,2}} = \sqrt{39545^2 + 6973^2} = 32515 \text{ N}$$

$$\tan\phi = \frac{6973}{39545} = 0{\cdot}176$$

$$\phi = 10^\circ$$

(ii) $F_x = m\,(V_{w1} + V_{w2})$

where $m = \rho a\,(V - u) = 1000 \times 0{\cdot}001962 \times (101{\cdot}94 - 30) = 141{\cdot}15$ kg/s

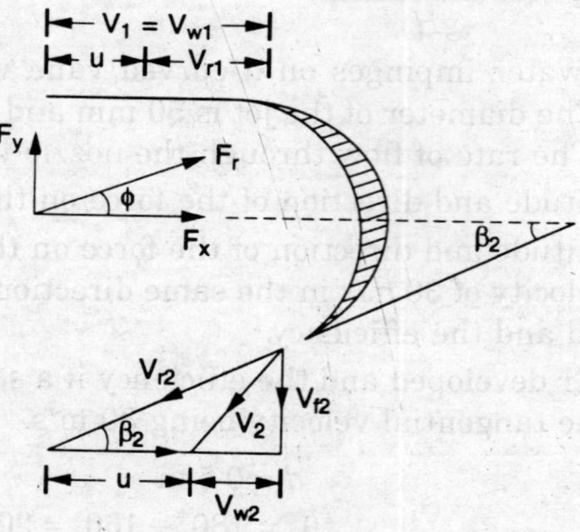

Fig. 16.16

$$V_{w1} = V_1 = 101{\cdot}94 \text{ m/s}$$

$$V_{w2} = V_{r2}\cos\beta_2 - u$$

$$= V_{r1}\cos\beta_2 - u$$

$= (V_1 - u)\cos\beta_2 - u$

$= (101{\cdot}94 - 30)\cos 20^\circ - 30$

$= 37{\cdot}6$ m/s

$F_x = 121{\cdot}05\,(91{\cdot}7 + 27{\cdot}98) = 14488$ N

$F_y = m\,(V_{f1} + V_{f2})$

where $V_{f1} = V_{r2}\sin\beta_2$

$= V_{r1}\sin\beta_2$

$= (V_1 - u)\sin\beta_2$

$= (101{\cdot}94 - 30)\sin 20^\circ$

$= 24{\cdot}6$ m/s

$F_y = 141{\cdot}15\,[0 - (-24{\cdot}6)] = 3473$ N

ie F_y acts in the upward direction.

Resultant force,

$$F = \sqrt{F_x^{\,2} + F_x^{\,2}} = \sqrt{19696^2 + 3473^2} = 20000 \text{ N}$$

$$\tan\phi = \frac{3473}{19696} = 0{\cdot}176$$

$$\phi = 10^\circ$$

The direction of the resultant force remains same as in the previous case.

$$P = F_x{\cdot}u = 19696 \times 30 = 590880 \text{ W} \quad \text{or} \quad 590{\cdot}88 \text{ kW}$$

$$\eta = \frac{P}{\frac{1}{2}(\rho a V)\,V^2} = \frac{590880}{\frac{1}{2} \times 1000 \times 0{\cdot}001962 \times 101{\cdot}94^3} = 0{\cdot}5686 \quad \text{or} \quad 56{\cdot}86\%$$

(iii) $m = \rho a V = 1000 \times 0{\cdot}001962 \times 101{\cdot}94 = 200$ kg/s

$$F_x = m\,(V_{w1} + V_{w2}) = 200 \times (101{\cdot}94 + 37{\cdot}6) = 27908 \text{ N}$$

$$P = F_x{\cdot}u = 27908 \times 30 = 837240 \text{ W} \quad \text{or} \quad 837{\cdot}24 \text{ kW}$$

$$\eta = \frac{P}{\frac{1}{2}(\rho a V)V^2} = \frac{837240}{\frac{1}{2} \times 1000 \times 0{\cdot}001962 \times 101{\cdot}94^3} = 0{\cdot}8057 \quad \text{or} \quad 80{\cdot}57\%$$

Example 16.11 A jet of water impinges on a series of radial curved vanes mounted on a wheel which rotates at 220 rpm. The inlet and the outlet radii of the wheel are 400 mm and 200 mm respectively. The velocity of the jet is 25 m/s and it makes an angle of 30° with the tangent to the wheel at the inlet tip. Water leaves the wheel with a velocity of 4 m/s at an angle of 130° to the tangent to the wheel at the outlet. Find

(i) vane angles at the inlet and the outlet

(ii) work done per unit mass per second

(iii) efficiency

Also determine the above quantities for the radial outlet velocity.

Solution $V_1 = 25$ m/s $\alpha_1 = 30°$

$V_2 = 4$ m/s $\alpha_2 = 180° - 130° = 50°$

$N = 220$ rpm $r_1 = 0{\cdot}4$ m

$r_2 = 0{\cdot}2$ m

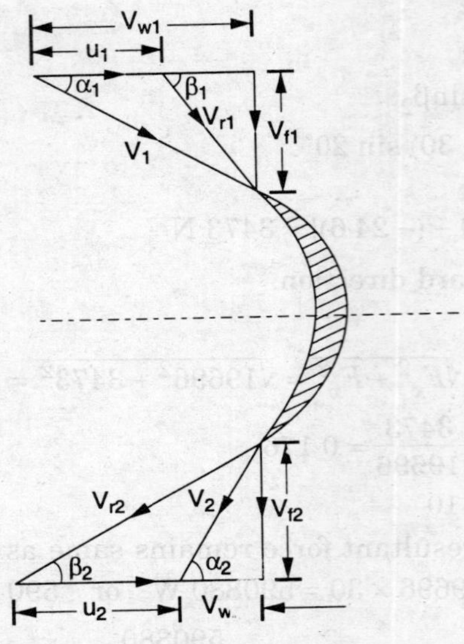

Fig. 16.17

$$\omega = \frac{2\pi N}{60} = \frac{2\pi \times 220}{60} = 23{\cdot}04 \text{ rad/s}$$

$$u_1 = w{\cdot}r_1 = 23{\cdot}04 \times 0{\cdot}4 = 9{\cdot}22 \text{ m/s}$$

$$u_2 = w{\cdot}r_2 = 23{\cdot}04 \times 0{\cdot}2 = 4{\cdot}61 \text{ m/s}$$

(i) At inlet

$$\tan \beta_1 = \frac{V_{f1}}{V_{w1} - u_1} = \frac{V_1 \sin \alpha_1}{V_1 \cos \alpha_1 - u_1} = \frac{25 \sin 30°}{25 \cos 30° - 9{\cdot}22} = \frac{12{\cdot}5}{21{\cdot}65 - 9{\cdot}22} = 1$$

$$\beta_1 = 45°$$

At outlet

$$\tan \beta_2 = \frac{V_{f2}}{V_{w2} + u_2} = \frac{V_2 \sin \alpha_2}{V_2 \cos \alpha_2 + u_2} = \frac{4 \sin 50°}{4 \cos 50° + 4{\cdot}61} = \frac{3{\cdot}06}{2{\cdot}57 + 4{\cdot}61} = 0{\cdot}426$$

$$\beta_2 = 23{\cdot}07°$$

$$W = V_{w1}{\cdot}u_1 + V_{w2}{\cdot}u_2 = 21{\cdot}65 \times 9{\cdot}22 + 2{\cdot}57 \times 4{\cdot}61 = 211{\cdot}5 \text{ J/s}$$

$$\eta = \frac{2(V_{w1}{\cdot}u_1 \pm V_{w1}{\cdot}u_2)}{V_1^{\,2}} = \frac{2 \times 211{\cdot}5}{25^2} = 0{\cdot}677 \text{ or } 67{\cdot}7\%$$

When the velocity at the outlet is radial, V_{w0} is zero and angle $\alpha_0 = 90°$

Thus at the outlet,

$$V_{f0} = V_0 = 4 \text{ m/s}$$

$$\tan \beta_2 = \frac{V_{f2}}{u_2} = \frac{4}{4{\cdot}61} = 0{\cdot}868$$

$$\beta_2 = 40{\cdot}96°$$

$$W = V_{w1} \cdot u_1 = 21{\cdot}65 \times 9{\cdot}22 = 199{\cdot}6 \text{ J/s}$$

$$\eta = \frac{W}{\frac{1}{2} V_1^2} = \frac{199{\cdot}6}{\frac{1}{2} \times 25^2} = 0{\cdot}639 \quad \text{or} \quad 63{\cdot}9\%$$

16.5 JET PROPULSION OF SHIPS

The principle of jet is also applicable to propulsion of ships. A ship is steered through water due to reaction of the force of the jet which is issued out at the back of the ship. The water for the purpose is taken in through the inlet orifices by centrifugal pumps.

Let u = velocity of the ship

V = Absolute velocity of the issuing jet (opposite to u)

a = Area of cross-section of the jet

Considering the velocities relative to the ship :

Issuing velocity of the jet, $V_r = V + u$ in the backward direction

The water after coming out of the ship becomes stationary, but as the ship moves with velocity u , the final velocity of water is u relative to the ship in the backward direction.

Mass flow rate of the issuing jet,

$$m = \text{Density} \times \text{Discharge} = \text{Density} \times \text{Relative velocity} \times \text{Area} = \rho \cdot V_r \cdot a$$

Propulsive force,

$$F = \text{Mass} \times \text{Change in velocity}$$
$$= \text{Mass} \times (\text{Initial velocity} - \text{Final velocity})$$
$$= \rho \cdot V_r \cdot a\ (V_r - u)$$

Work done per second, $W = \rho \cdot V_r \cdot a\ (V_r - u) \cdot u$

Efficiency,

$$\eta = \frac{\text{Work done per second}}{\text{Energy supplied per second}}$$

The energy supplied depends upon the way in which the water is supplied to the ship. There are two ways :

Inlet Orifices at Right Angle

If the inlet orifices are at right angles to the motion of the ship (Fig. 16.18), then

Velocity at the inlet in the direction of the jet = 0

Velocity at the outlet in the direction of the jet = V_r

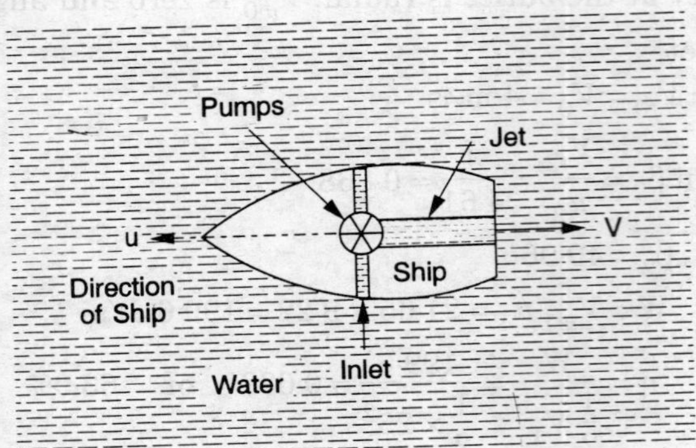

Fig. 16.18

$\therefore$ Energy supplied/s = Kinetic energy at outlet – kinetic energy at inlet

$$= \frac{1}{2}(\rho \cdot a \cdot V_r) \cdot V_r^2 - 0$$

$$= \frac{1}{2}\rho \cdot a \cdot V_r^3$$

$$\eta = \frac{\rho \cdot a \cdot V_r (V_r - u)u}{\frac{1}{2}\rho \cdot a \cdot V_r^3} = \frac{2(V_r - u)u}{V_r^2} \quad \text{...(16.34)}$$

For maximum efficiency,

$$\frac{d\eta}{du} = 0$$

$$\frac{d}{du}\left[\frac{2(V_r - u)u}{V_r^2}\right] = 0$$

$$\frac{d}{du}(V_r \cdot u - u^2) = 0$$

$$V_r - 2u = 0$$

$$u = \frac{V_r}{2}$$

$$\eta_{max} = \frac{2(2u - u)u}{(2u)^2} = 0{\cdot}5$$

Inlet Orifices in the Direction of Ship Motion

If the inlet orifices face the direction of the ship motion (Fig. 16.19), then

Velocity at the inlet in the direction of the jet = u

Velocity at the outlet in the direction of the jet = V_r

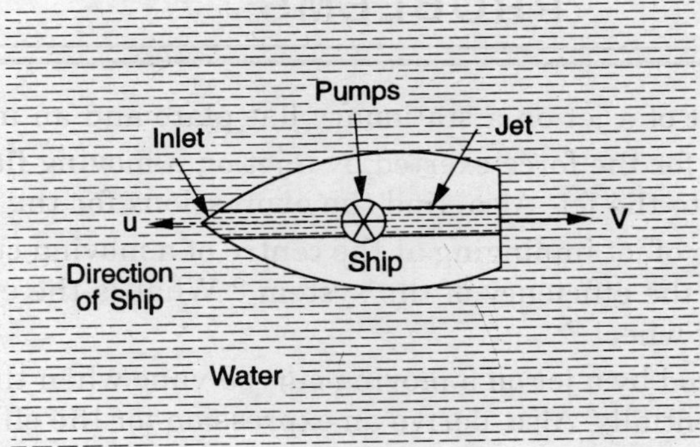

Fig. 16.19

$\therefore$ Energy supplied/s = Kinetic energy at outlet – Kinetic energy at inlet

$$= \frac{1}{2}(\rho \cdot a \cdot V_r) \cdot V_r^2 - \frac{1}{2}(\rho \cdot a \cdot V_r) \cdot u^2$$

$$= \frac{1}{2}\rho \cdot a \cdot V_r \cdot (V_r^2 - u^2)$$

$$\eta = \frac{\rho \cdot a \cdot V_r \cdot (V_r - u)u}{\frac{1}{2}\rho \cdot a \cdot V_r \cdot (V_r^2 + u^2)} = \frac{2u}{V_r + u} \qquad \text{...(16.35)}$$

Example 16.12 A jet propelled boat moves at 18 km per hour. The jet has a cross-sectional area of 0·02 m^2 and is discharged back with a relative velocity of 12 m/s. Find the propelling force, power and the efficiency of the jet propulsion if

(i) the inlet orifices are at right angles to the direction of the motion of the boat.

(ii) the inlet orifices face the direction of motion of the boat.

Solution $a = 0{\cdot}02\ \text{m}^2$ $\qquad V = 12\ \text{m/s}$

$$u = \frac{18 \times 1000}{3600} = 5\ \text{m/s}$$

(i) $F = \rho \cdot V_r \cdot a\ (V_r - a) = 1000 \times 12 \times 0{\cdot}02 \times (12 - 5) = 1680\ \text{N}$

$P = F \times u = 1680 \times 5 = 8400\ \text{W}$ or $8{\cdot}4\ \text{kW}$

$$\eta = \frac{2(V_r - u)u}{V_r^2} = \frac{2(12-5) \times 5}{12^2} = 0{\cdot}486 \quad \text{or} \quad 48{\cdot}6\%$$

(ii) F and P remain same as in (i)

$$\eta = \frac{2u}{V_r + u} = \frac{2 \times 5}{12 + 5} = 0{\cdot}588 \quad \text{or} \quad 58{\cdot}8\%$$

PROBLEMS

1. Describe the impact of a jet on a stationary flat plate and an inclined plate.
2. Derive expressions for the force exerted by a jet on a moving flat plate (a) normal to the jet (b) inclined to the jet. Also, find the expressions for the work done.
3. Determine the force of jet impinging at the centre of a moving curved vane. What are the work done and the efficiency of the system ? What is the effect of mounting the vanes on a rotating wheel ?
4. Find the force exerted by a jet on a moving curved vane when the jet strikes the vane at its one end tangentially. Also, derive an expression for the efficiency of the system.
5. What is meant by velocity of whirl and the velocity of flow ?
6. Describe the velocity diagrams at the inlet and the outlet of the moving vanes.
7. Derive an expression for the change in the angular momentum when the radial distance of the liquid particles change while passing over the vanes mounted on a wheel. What is the work done and the efficiency of such a wheel ?
8. A jet of water exerts a force of 1420 N on a flat plate held normal to the jet. Determine the velocity of the jet if the diameter of the jet is 60 mm. (22·4 m/s)
9. A jet of water 40 mm in diameter moving with a velocity of 12 m/s impinges on a fixed plate inclined at 30° to the direction of the jet. What is the normal force of the jet on the plate ? What are the components of the force in the direction of the jet and perpendicular to it ? (90·5 N , 45·24 N , 78·4 N)
10. A jet of water impinges on a symmetrically curved vane at the centre. The velocity of the jet is 60 m/s and the diameter 120 mm. The jet is deflected through an angle of 120°. Calculate the force on the vane if the vane is fixed. Also, determine the force if the vane moves with a velocity of 25 m/s in the direction of the jet. What will be the power and the efficiency ? (61·04 kN , 20·76 kN , 519 kW , 42·5%)
11. A jet of water 80 mm in diameter moves with a velocity of 20 m/s and impinges on a flat plate inclined at 45° to the direction of the jet. If the power developed is 2·2 kW, find the velocity of the plate in the direction of the jet. Also, find the efficiency of the system. (11·1 m/s , 10·96%)
12. A horizontal jet of the water 20 mm in diameter impinges normally on a rectangular plate of 6 kg mass. The plate is suspended vertically by a hinge on the top horizontal edge. The centre of gravity of the plate is 100 mm from the hinge and the axis of the jet is 150 mm below the hinge. If the velocity of the jet is 5 m/s, calculate the horizontal force applied at the centre of gravity to maintain the plate in the vertical position. If the plate is deflected through 30° under the same force, determine the corresponding velocity of the jet. (11·77 N , 9·18 m/s)
13. A jet of water moving at a velocity of 12 m/s impinges on a concave shaped vane to deflect the jet through 120° when stationary. If the vane moves at 5 m/s, determine the angle of the jet so that there is no shock at the inlet. What is the absolute velocity of water at the exit in magnitude and direction ? Also, find the work done per nit mass of water. Assume the vane smooth. (18° , 3·96 m/s , 69·12° , 64· J/s)

14. A wheel mounted with curved vanes revolves at a speed of 250 rpm. The velocity of the jet is 25 m/s and the wheel diameter 1·2 m. The tips of the vanes make angles of 20° and 30° to the wheel tangent at the inlet and the outlet respectively. If the water enters the vanes without shock, determine the tangential force on the wheel per unit mass of water. What are the specific work done and the hydraulic efficiency ?
(17·45 N/kg , 274 J/s , 87·7%)

15. A jet propelled boat moving at 30 km/h draws water through inlet openings facing the direction of motion of the boat. The absolute velocity of the jet issued at the back is 22 m/s and the area of the jet is 0.025 m^2. Determine the power and the efficiency of propulsion.
(139 kW , 43·1%)

Chapter 17

HYDRAULIC TURBINES

17.1 INTRODUCTION

Turbines and pumps are machines that use fluid as medium. If the fluid happens to be water, they are known as **hydraulic machines.** In a hydraulic turbine, the pressure energy of water is converted into mechanical energy, whereas a pump converts the mechanical energy into pressure energy. The mechanical energy obtained from a turbine is further used to generate electric power. Pressure energy imparted to water by a pump is used either to increase the pressure of water to do some useful work or to raise it to a higher level.

Since early times, man started tapping energy from natural sources of water by converting it into mechanical energy. Initially, this was done in a limited way by using water wheels which did not give high efficiency. They were used as grinding wheels in India. They were also developed in China and Egypt and later in European countries. The famous Italian, Leonardo Da Vinci (1452-1519) provided the initial design and construction of a water wheel. Some more developments were made by Borda and Poncelet. Afterwards, a Swiss scientist Bernoulli wrote a book on the conversion of hydro power into other forms of energy. In 1970, Euler forwarded the theory of hydraulic machines which is still in use.

17.2 WATER WHEELS

The following are the main types of water wheels :

Overshot Water Wheel

This type of wheel rotates entirely by the weight of water. Water from the headrace falls on a series of buckets fixed to a wheel on a horizontal axis (Fig. 17.1). Due to the weight of water, the buckets move down and the wheel rotates. The flow is regulated by sluice gates which may be hand driven or governed by some mechanism.

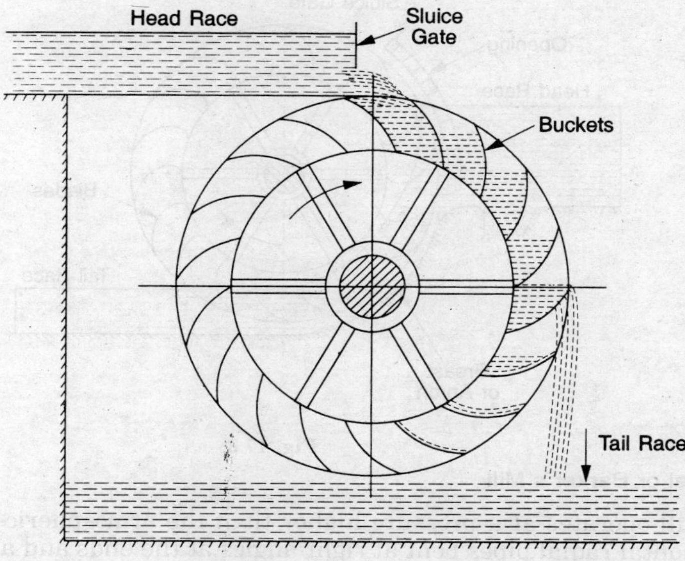

Fig. 17.1

Undershot Water Wheel

It is a wheel which rotates due to the impulse action of water (Fig. 17.2a). The energy of the available water is converted into kinetic energy and the wheel rotates due to change in momentum. An improvement in its design is made in the Poncelet water wheel in which the blades are curved and losses are reduced (Fig. 17.2b). It is known as a forerunner of Pelton turbine, a modern turbine for high heads as it works on the same principle.

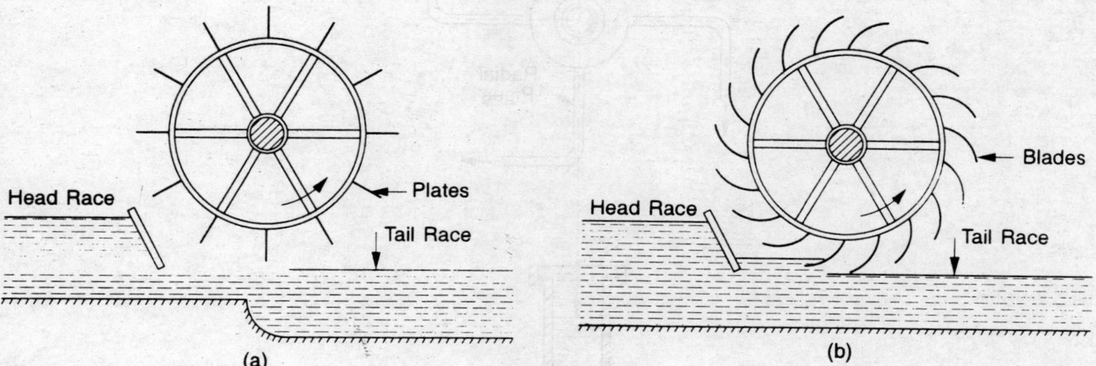

Fig. 17.2

Breast Water Wheel

This type of wheel rotates partly due to the weight of water and partly due to its impulse action. It is provided with a breast or apron to keep the buckets full from headrace to tailrace (Fig. 17.3).

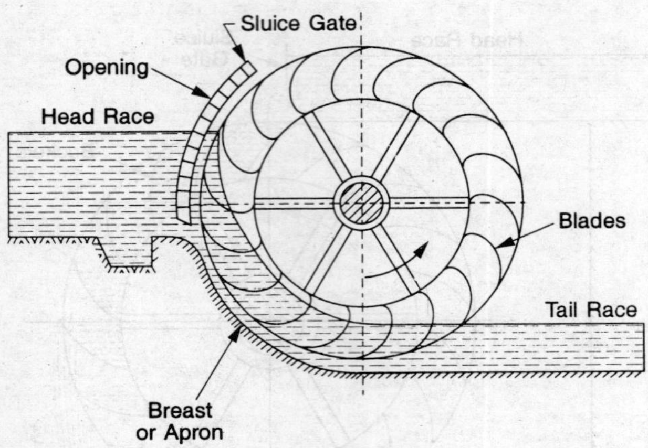

Fig. 17.3

Reaction Wheel or Barker's Mill

A barker's mill operates at a pressure higher than the atmospheric pressure. There are two or more horizontal radial pipes bent at right angles at the ends and attached to a vertical pipe (Fig. 17.4). The central pipe is provided with water at high pressure and out of each horizontal pipe, a jet of water comes out. Due to change in momentum of water, the wheel starts rotating in the opposite direction.

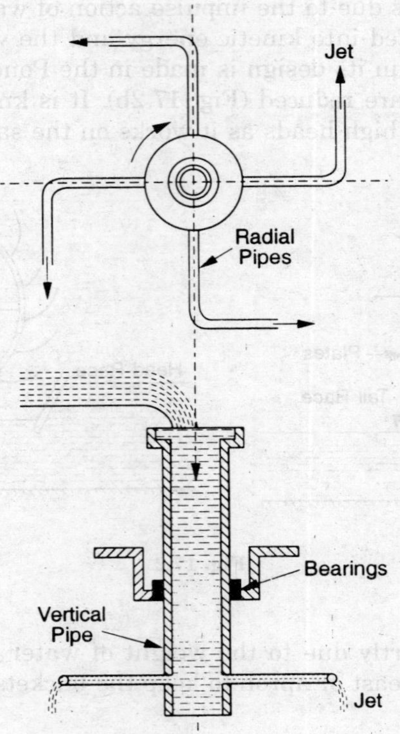

Fig. 17.4

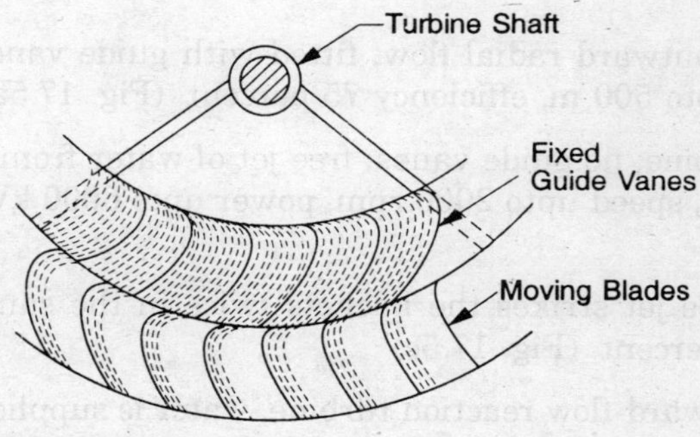

(a) Outward Radial flow Girard Turbine

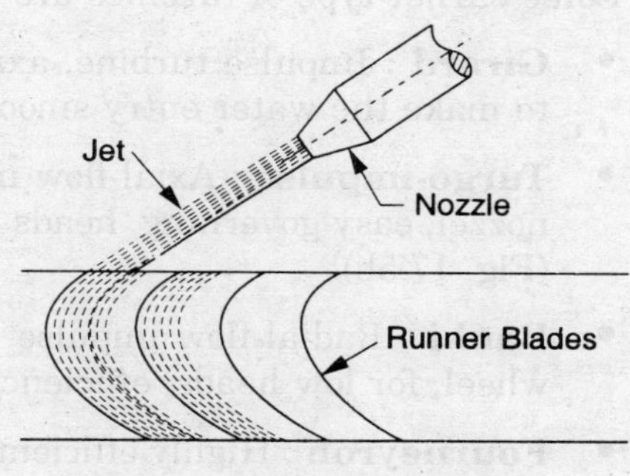

(b) Turgo--impulse Turbine

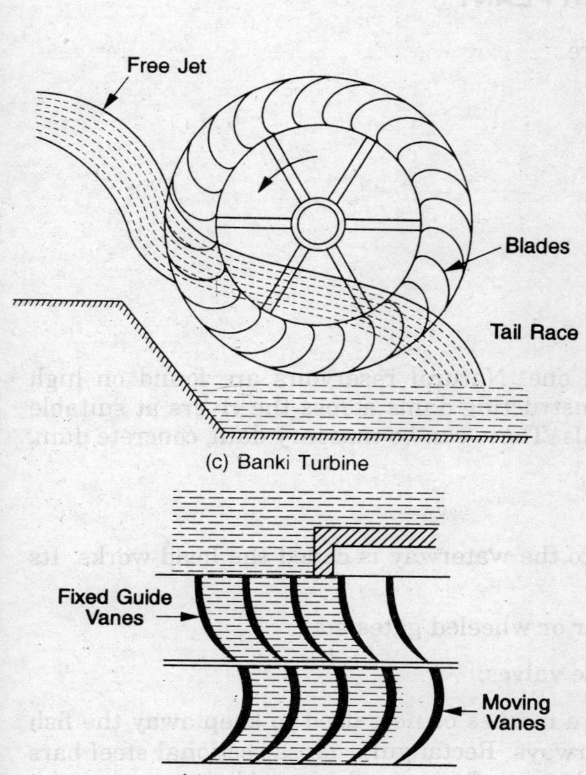

(c) Banki Turbine

(e) Jonval Turbine

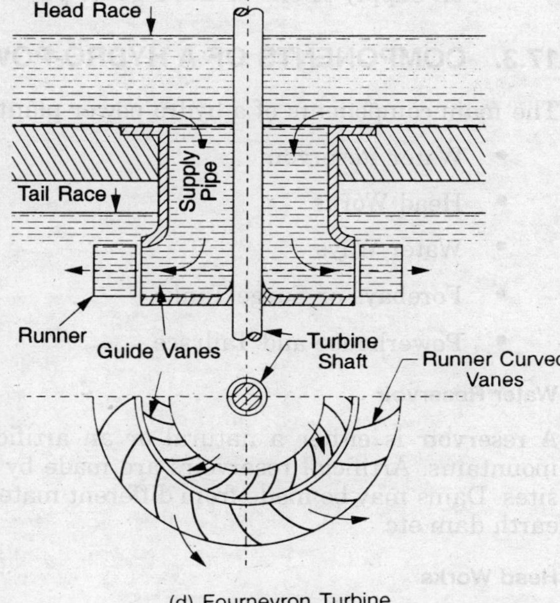

(d) Fourneyron Turbine

Fig. 17.5

Some earlier type of turbines are :

- **Girard :** Impulse turbine, axial-flow or outward radial flow; fitted with guide vanes to make the water entry smooth; head upto 500 m, efficiency 75 percent. (Fig. 17.5a)
- **Turgo-impulse** : Axial-flow impulse turbine; no guide vanes, free jet of water from a nozzel, easy governing, heads upto 300 m, speed upto 2000 rpm, power upto 6000 kW. (Fig. 17.5b)
- **Banki** : Radial-flow impulse turbine, free jet strikes the buckets twice in the same wheel, for low heads, efficiency upto 70 percent. (Fig. 17.5c)
- **Fourneyron** : Highly efficient radial-outward-flow reaction turbine, water is supplied at centre, a series of guide vanes, runner with curved vanes, heads upto 100 m, efficiency upto 75 percent. (Fig. 17.5d)
- **Jonval** : Axial-flow reaction turbine, series of guide vanes, easy governing by cutting off supply to one or more guide passages. (Fig. 17.5e)

17.3. COMPONENTS OF A HYDRO-POWER PLANT

The main components of a hydro-power plant are :

- Water Reservoir
- Head Works
- Water Ways
- Forebays or Surge Tanks
- Powerhouse and Tailrace.

Water Reservoir

A reservoir is either a natural or an artificial one. Natural reservoirs are found on high mountains. Artificial reservoirs are made by constructing dams across the rivers at suitable sites. Dams may be made from different materials. There can be masonry dam, concrete dam, earth dam etc.

Head Works

The equipment used for the control of water into the waterway is called the head-works. Its components are

- **Gates** : These can be plain sliding, roller or wheeled gates etc.
- **Valves** : These can be butterfly or needle valves.
- **Fish-traps and trash-racks** : These are meshes or nets used to keep away the fish and the debris from going into the waterways. Rectangular cross-sectional steel bars are used for the purpose. Some type of cleaning device is also provided to remove the debris from the trash racks.
- In cold countries, some heating arrangement is also provided to melt the ice at the inlet.

Water Ways

The passages, through which water is brought from the reservoir to the power house, are known as waterways. They can be tunnels, channels, penstocks etc.

- A tunnel is cut through a hill if it comes between the reservoir and the power house.
- Open channels are used when the distance of the water reservoir and the power house is considerable.
- Penstocks are the steel or reinforced concrete pipes and are used to bring the water from the reservoir level to the power house level. If the distance of the power house from the reservoir is not large, the penstock alone can take the water from the reservoir to the power house.

Forebays and Surge Tanks

Small water storage, to meet the load fluctuations for small periods as for a day, is known as a forebay (Fig. 17.6). This may be situated at the end of a tunnel or the open channel. In case

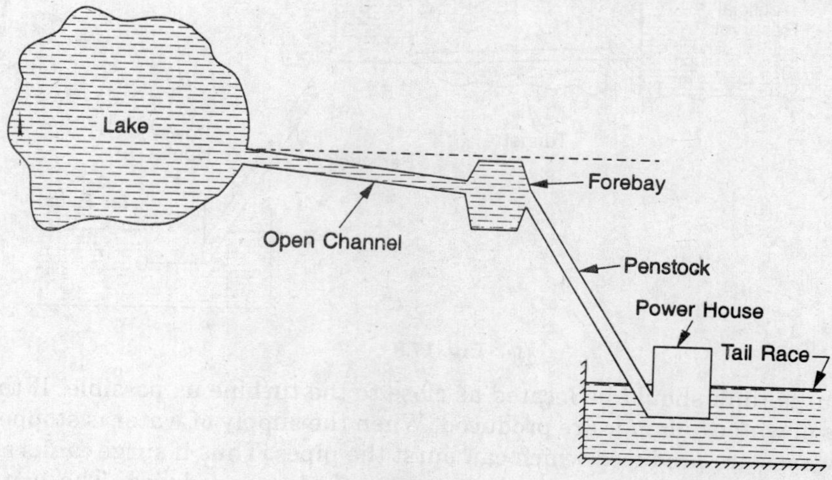

Fig. 17.6

Fig. 17.7

of open channels, the channel is enlarged just before the penstock which acts as a forebay. However, if the distance between the reservoir and the power house is small and only penstocks are used, the reservoir itself acts as a forebay (Fig. 17.7). Forebays are used in medium and low-head plants where the length of the penstock is small.

A surge tank is a cylindrical open-topped tank and it also acts as a small reservoir (Fig. 17.8). The normal level of water in it is at the level of reservoir minus the head losses in transition between the reservoir and the surge tank. When the load on the turbine is reduced, water has to be retarded. Since it takes some time to do so, the excess flow of water is temporarily stored in the surge tank and the level in the surge tank becomes higher than the average. Similarly, when the load on the turbine increases and the supply of water is to be increased, some of the supply is supplemented from the surge tank.

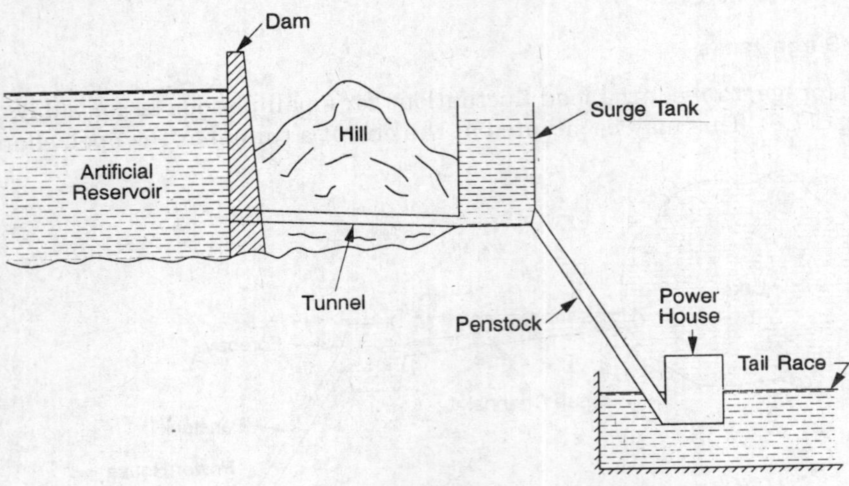

Fig. 17.8

Ideally, a surge tank should be located as close to the turbine as possible. If the pipes are long, the effects of water hammer are produced. When the supply of water is stopped suddenly, high-pressure waves are produced which can burst the pipes. Thus if surge tanks are provided in between the reservoir and the powerhouse, the effects are reduced. The water hammer effects due to the flow downstream of the surge tank are always there.

Powerhouse and Tailrace

A powerhouse consists of turbines, generators, governing mechanism and some other equipment. Tailrace is a waterway which takes the water from the turbine outlet into some channel or river.

17.4 TYPES OF TURBINES

Turbines are classified according to

1. head and quantity of water available
2. hydraulic action of water

3. direction of flow of water in the runner
4. disposition of shaft of the runner
5. Specific speed of turbines

1. Based on Head and Quantity of Water Available

According to head and quantity of water available, the turbines are classified into

- High-head turbines
- Medium-head turbines
- Low-head turbines

High-head turbines : The turbines that work under heads more than 250 m are known as **high-head turbines.** The quantity of water in such turbines is usually small, as the power produced by a turbine is proportional to QH and H is high. Heads up to 1800 m have been utilised. Pelton turbines are the usual choice for high heads.

Medium-head turbines : Medium-head turbines are the turbines which work under a head of 45 m to 250 m. They require medium flow of water. Francis turbines are used for the medium heads.

Low-head turbines : Turbines that work under heads below 45 m are known as low-head turbines. Due to low head, the quantity of water required is generally large. Kaplan and propeller are the low-head turbines.

2. Based on Hydraulic Action of Water

According to this criterion, the turbines are classified into

- Impulse turbines
- Reaction turbines

Impulse turbines : If the runner of a turbine revolves by impact or impulse action of water, it is known as an impulse turbine. To increase the impact of water, the available pressure energy of water is converted into kinetic energy by passing it through a nozzle. The jet of water smoothly impinges on the blades or buckets of the runner. An impulse turbine functions in free atmosphere and so the atmospheric pressure prevails at all point in the runner. No pressure difference exists between the inlet and the outlet of runner. Owing to this reason, the casing of an impulse turbine does not perform any hydraulic function. However, a casing has to be provided to avoid accidents and splashing of water. A Pelton turbine is an impulse type of turbine.

Reaction turbines : These are the turbines the runner of which rotate due to reaction of the pressure difference between the inlet and the outlet of the runner. The runners of reaction turbines have curved blades.

Reaction turbines may be inward-flow or outward-flow types. In an inward-flow type of turbine, the water is fed to the turbine all around the periphery of the runner. An outlet is provided along the shaft of the runner. A draft tube is provided at the outlet to direct the water from the turbine to the tailrace. The pressure at the tailrace is atmospheric. Due to the difference in pressure at the inlet and the outlet, the water flows from inlet towards outlet

There is continuous decrease in pressure as the water flows through the runner. The continuous decrease in the pressure energy during the flow through the runner results in transformation into kinetic energy which is transmitted to the runner in the form of its rotary motion. In an outward-flow turbine, water is admitted along the shaft which comes out along the periphery of the runner.

A casing is necessary in a reaction turbine to maintain different pressures at the inlet and the outlet and to insulate it from the atmosphere. In reaction turbines, the runner must remain full of water. Francis, Kaplan and propeller are the commonly used reaction turbines.

3. Based on Direction of Flow of Water in the Runner

According to the direction of flow of water through the runner, the turbines are classified as

- Tangential-flow turbines
- Radial-flow turbines
- Axial-flow turbines
- Mixed-flow turbines

Tangential-flow turbines : When the flow is tangential to the wheel circle, it is called a **tangential-flow turbine.** A Pelton turbine is a tangential-flow turbine.

Radial-flow turbines : When the path of flow of water remains in the radial direction and in a plane normal to the runner shaft, it is known as a **radial-flow turbine.** If the inlet of water is at the outer periphery of the runner and it flows towards the centre, it is an inward-radial flow turbine. Similarly, there can be outward- radial flow turbine. These days, no pure radial-flow turbine is in use.

Axial-flow turbines : In an axial-flow turbine, the path of flow of water remains parallel to the axis of the shaft. Kaplan and propeller types of turbines are axial-flow turbines.

Mixed-flow turbines : In a **mixed-flow turbine,** the water enters the runner in a radial direction and comes out in the axial direction. There is gradual change of flow from radial to axial in the runner. Francis turbine is a mixed-flow turbine.

4. Based on Disposition of Shaft of Runner

Generally, Pelton turbines are set up with horizontal shafts whereas the others have vertical shafts.

5. Based on Specific Speed of Turbines

Specific speed of a turbine is defined as the speed of a geometrically similar turbine which produces a unit power while working under a unit head. The range of specific speed for various types of turbines is as under :

Type of Turbine	Specific Speed
Pelton : With one nozzle	8–30
more than one nozzle	up to 50
Francis	50–250
Kaplan or propeller	250–850

17.5 EULER'S EQUATION AND DEGREE OF REACTION

Equation for work done or energy transfer in case of series of radial curved vanes is derived as

$$E/\text{unit mass}/s = V_{w1}\, u_1 \pm V_{w2}\, u_2$$

This is the fundamental equation of hydraulic machines. This is known as Euler's equation and is applicable to turbines as well as centrifugal pumps. A positive value of E indicates that the energy is transferred from water to the wheel which gives motion to that and is the principle of motion of the turbines. Similarly, a negative value indicates that the energy is transferred by the wheel to the liquid which can raise its pressure or velocity. This is the principle applicable to centrifugal pumps.

The above equation can be transformed into the following form also :

From inlet velocity traingle (Fig. 16.14).

$$V_{rw1}^{\ 2} = (V_{w1} - u_1)^2$$

$$V_{r1}^{\ 2} - V_{f1}^{\ 2} = V_{w1}^{\ 2} + u_1^{\ 2} - 2\, V_{w1}u_1$$

$$V_{r1}^{\ 2} - (V_1^{\ 2} - V_{w1}^{\ 2}) = V_{w1}^{\ 2} + u_1^{\ 2} - 2\, V_{w1}u_1$$

$$2\, V_{w1}u_1 = V_1^{\ 2} + u_1^{\ 2} - V_{r1}^{\ 2}$$

From outlet velocity triangle,

$$V_{rw2}^{\ 2} = (V_{w2} + u_2)^2$$

$$V_{r2}^{\ 2} - V_{f2}^{\ 2} = V_{w2}^{\ 2} + u_2^{\ 2} + 2\, V_{w2}u_2$$

$$V_{r2}^{\ 2} - (V_2^{\ 2} - V_{w2}^{\ 2}) = V_{w2}^{\ 2} + u_2^{\ 2} + 2\, V_{w2}u_2$$

$$2\, V_{w2}u_2 = V_{r2}^{\ 2} - u_2^{\ 2} - V_2^{\ 2}$$

Thus energy transfer,

$$E = \frac{V_1^{\ 2} + u_1^{\ 2} - V_{r1}^{\ 2}}{2} + \frac{V_{r2}^{\ 2} - u_2^{\ 2} - V_2^{\ 2}}{2}$$

$$= \frac{V_1^{\ 2} - V_2^{\ 2}}{2} + \frac{u_1^{\ 2} - u_2^{\ 2}}{2} + \frac{V_{r2}^{\ 2} - V_{r1}^{\ 2}}{2}$$

The equation shows that the energy transfer in a machine is the sum of the following :

- Difference in squares of absolute liquid velocities or the change in the kinetic energy of the liquid.
- Difference in squares of tangential rotor velocities or the change in the centrifugal head and the pressure change due to that.
- Difference in squares of relative liquid velocities or the pressure change due to change in relative kinetic energy.

In case of impulse turbines, the flow is tangential ($u_1 = u_2$) and the pressure is atmospheric. This means that the relative velocity in the rotor does not vary if friction of the buckets is neglected. Thus, the energy transfer is only due to change in the kinetic energy of the liquid.

In reaction turbines, if the flow is axial, the second term is zero as in Kaplan turbines. If the flow is radial or mixed, all the three terms are effective as in Francis turbines.

Generally, the hydraulic machines are designated based on relative importance of the second and the third terms compared to all the three terms. The conversion of energy of the first part is the dynamic energy change and does not take place in the rotor of the turbine. Energy transfer due to the second and the third parts takes place inside the rotor. It is known as the reactionary part and the pressure change due to it is known as reaction pressure. All types of hydraulic machines have reaction pressure to various degrees and there is always some part of impulse action.

Degree of reaction of a hydraulic turbine is defined as the ratio of the second and the third terms to all the three terms. It is the energy conversion inside the runner to the total energy conversion.

17.6 HEADS, LOSSES AND EFFICIENCIES OF HYDRAULIC TURBINES

The heads, losses and efficiencies are explained below.

1. Heads

Heads are defined as

Gross head : Total or gross head is the difference between the headrace level and the tailrace level when there is no flow.

Net head : The head available at the turbine is known as the net or the effective head. This takes into account the losses occurring in the flow passage from the reservoir to the turbine inlet or in the penstock.

2. Losses

Various types of head losses are

Head loss in the penstock : It is the friction loss in the pipe of the penstock.

Head loss in the nozzle : In case of impulse turbines, there is nead loss due to friction in the nozzle.

Hydraulic losses :

- In impulse turbines, these losses occur due to blade fricition, eddy formation and kinetic energy of the leaving water.
- In reaction turbines, apart from above losses, disc friction loss (friction encountered by a rotating wheel in a fluid) and the fricition in the draft tube also occur.

Leakage losses : Loss due to leakage or wastage of water is known as leakage loss.

- In impulse turbines, whole of the water may not strike the buckets and some of the water may go waste.
- In reaction turbines, some of the water may pass through the clearance between the casing and the runner without imparting any energy to the runner.

Mechanical losses : Loss due to friction in the bearings is known as the mechanical loss. Thus, useful power available is less than the power generated in the turbine shaft.

Generator losses : Owing to the losses that occur in the generator, the useful power available is less than what is generated by the turbine.

3. Efficiencies

The efficiencies are defined as

Hydraulic efficiency : It takes into account the hydraulic losses and is the ratio of the runner output to the power supplied to the runner.

$$\eta_h = \frac{\text{Runner output}}{\text{Runner input}} = \frac{\text{Runner output}}{wQH} \quad \text{...(17.1)}$$

where Q = quantity of water actually impinging on the runner vanes

H = Net head available at the turbine inlet

Volumetric efficiency : It takes into account the volumetric losses and is the ratio of the actual quantity of water impinging on the runner to the quantity of water supplied to the turbine. If δQ is the quantity of water leaking or not striking the runner vanes, then

$$\eta_v = \frac{Q}{Q + \delta Q} \quad \text{...(17.2)}$$

Mechanical efficiency : It takes into account the mechanical loss. It is defined as the ratio of the shaft output to the runner output.

$$\eta_m = \frac{\text{Shaft output}}{\text{Runner output}} \quad \text{...(17.3)}$$

Overall efficiency : It takes into account the hydraulic, volumetric and mechanical losses and is the ratio of the shaft output to the net power available at the turbine inlet.

$$\eta_0 = \frac{\text{Shaft output}}{\text{Net power available}}$$

$$= \frac{\text{Shaft outut}}{W(Q + \delta Q)H}$$

$$= \frac{\text{Shaft outut}}{\text{Runner output}} \times \frac{\text{Runner output}}{\text{wQH}} \times \frac{Q}{Q + \delta Q}$$

$$= \eta_m \times \eta_h \times \eta_v \quad \text{...(17.4)}$$

17.7 PELTON TURBINE

A Pelton turbine is a free-jet turbine since it uses a free jet of water under atmospheric pressure (Fig. 17.9). It is more efficient under high heads. Thus, the power plants with heads more than 250 m are fitted with Pelton turbines. A Pelton turbine uses small quantity of water. Water from the reservoir is brought to the turbine inlet through a steel or concrete pipe, known as penstock. At the end of penstock, a nozzle is fitted. The nozzle converts the whole of the available pressure head into the kinetic head in the form of a high velocity jet. The jet impinges on the buckets mounted on the periphery of a wheel, called **runner.** The force of jet causes the runner and the shaft to revolve. Thus, mechanical power is produced. In the end, the water goes into a discharge channel known as TAILRACE. The number of nozzles can be upto 6 depending upon specific speed.

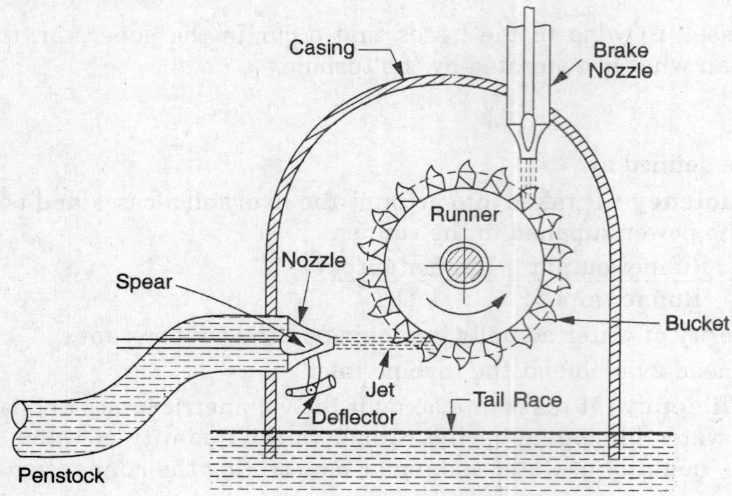

Fig. 17.9

1. Components

The main components of Pelton turbine are as follows :

(a) Runner with buckets

(b) Nozzle with guide mechanism

(c) Casing

(a) **Runner with buckets** : The runner of a Pelton turbine consists of double-cupped or hemispherical buckets fixed to the periphery of the wheel called runner. Each bucket has a sharp edge at the centre, known as **splitter.** The jet impinges on the splitter and is divided into two portions to avoid any unbalanced thrust on the shaft (Fig. 17.10). As the splitter is to bear the full impact of the jet, it is made very strong. Therefore, it is not practical to have a sharp edge and zero angle at the centre of bucket. Usually, this angle is made 3 to 6 degrees. However, for simplicity sake, the inlet angle β_1 may be assumed zero degree.

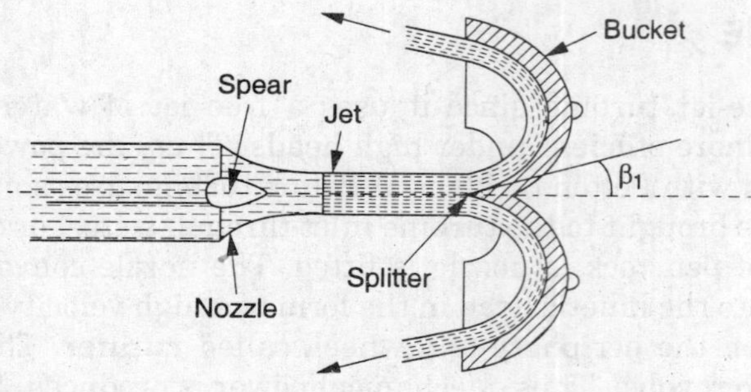

Fig. 17.10

To have the maximum impact of the jet, the buckets should turn the jet through 180° However, in that case, the jet may strike the incoming buckets, thereby opposing the motion of the runner. Usually, the angle turned by the jet is kept as 160° to 170°. Average value is taken as 165°. Since the surface of the buckets is spherical, the outlet angle is different at all points on the outer edge. At the bottom portion of the buckets, surface is raised and water is not able to turn through the proper angle. The outgoing water may obstruct the incoming bucket. Therefore, the bottom portion of the buckets is generally cut which also facilitates the jet to pass through the succeeding buckets.

The water impinging on the buckets may contain some impurities that can produce erosion of the buckets. Thus, the material used is cast iron for low-head turbines and cast-steel, stainless steel and bronze for medium and high head turbines. The buckets are either cast integral with the wheel or bolted to its rim. Bolted arrangement facilitates the easy replacement of the damaged buckets.

(b) **Nozzle with guide mechanism :** The function of the nozzle is to convert the available pressure into high velocity jet. The quantity of water to pass through the nozzle is to be proportional to the load on the turbine. To control the flow through the turbine, some governing mechanism is incorporated.

- To vary the discharge through the nozzle, either the velocity or the area of flow is to be changed. The velocity of jet does not vary as it is dependent upon the head which normally does not change during short periods. Thus, the variation in discharge is affected by a change in area. Usually, the movement of a spear inside the nozzle changes the area. A governor controls the movement of the spear through a servomechanism to be discussed later. If the load on the turbine reduces and the runner tends to speed up, the governor is activated which moves the spear forward inside the nozzle thus reducing the area of flow. This reduces the discharge. If the load is increased, the spear moves backwards and the area is increased thus increasing the discharge.
- In large turbines, if the flow is suddenly stopped, high pressures are produced in the penstock pipes. To avoid this, a bypass nozzle is incorporated through which the flow can pass and directly go to the tailrace instead of impinging on the buckets.
- A deflector may be used in between the nozzle and the buckets operated by a lever arrangement. It is withdrawn when the flow through the nozzle settles.
- A brake nozzle may also be used to stop the runner. The jet from a brake nozzle is directed on the back of the buckets apart from cutting the supply of water through the main nozzle, thus reducing the speed of the runner or stopping it completely.

The jet issued by the nozzle must be solid, uniform and free from surface spray at different rates of flow.

The material of the nozzle is either cast-iron or cast-steel. The mouth ring of the nozzle and the tip of the spear are made from any non-abrasive material such as stainless steel or bronze and can easily be replaced.

(c) **Casing :** The casing of a Pelton turbine does not perform any hydraulic function. It is required to

- avoid accidents
- avoid splashing of water
- lead the water into the tailrace
- support the housing for the bearing and the nozzle.

The casing is usually made in two parts to facilitate erection and assembling. Normally, the material for the casing is cast- iron.

2. Number of Buckets

The main considerations to select the number of buckets are

- Friction loss is less which means the number of buckets should be small.
- Deflected water from one bucket does not impinge back on the incoming bucket and the path of the jet is not disturbed. This implies that the number of buckets should be small.
- The jet is fully utilised *i.e.* whole of the water impinges the buckets. This means that the number of buckets should be large.

The attempts were made to determine the number of buckets analytically. However, the following empirical relation forwarded by Tygun gives satisfactory results :

$$Z = 0{\cdot}5\ \text{m} + 15$$

$$= \frac{D}{2d} + 15 \qquad \ldots(17.5)$$

where $m = \frac{D}{d}$ = Jet ratio

The following table may be used for the number of buckets based on the satisfactory performance of running units :

$m = D/d$	6	8	10	15	20	25
Z	17—21	18—22	19—24	22—27	24—30	26—33

3. Dimensions of Buckets

The dimensions of the buckets are based on the following considerations (Fig. 17.11) :

- The jet is turned through an angle of 165°. Therefore, if d is the diameter of the jet, the mininim possible width of a bucket is approximately 2 d. Since the water is to glide on the bucket, the width B is usually taken as 2·8 d to 3·2 d. Some prefer to take it as 3 d to 4 d.
- The height H of the bucket is kept as 2·4 d to 2·8 d. The shape of the bucket is hemispherical due to which the proper angle can be attained only in the middle portion of the bucket.

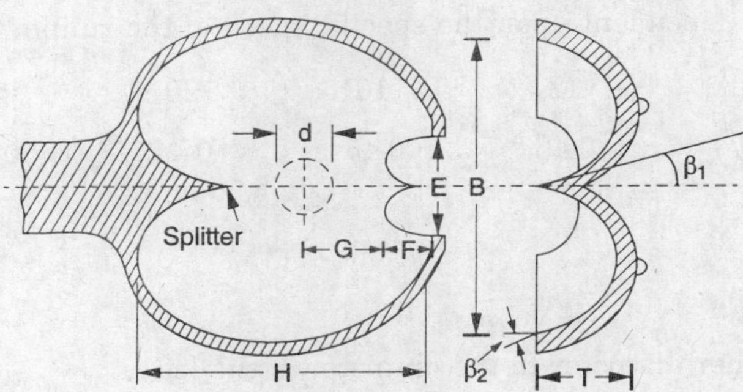

Fig. 17.11

- Depth T of the bucket is betwen 0·8 d to 1·2 d as this much depth is found satisfactory for proper gliding of water and turn through the required angle.
- Other dimensions based on the experimental results are as under :

$E = 1{\cdot}2\, d$
$F = 0{\cdot}4\, d$
$G = 0{\cdot}9\, d - 1{\cdot}0d$
$\beta_1 = 3^\circ - 6^\circ$
$\beta_2 = 10^\circ - 20^\circ$

4. Jet diameter

The velocity of jet through a nozzle is given by

$$V_1 = C_v \sqrt{2gH}$$

where V_1 = Velocity of the jet
C_v = Coefficient of velocity of the nozzle (0·96 – 0·99)

Let d = diameter of the jet

Flow through the nozzle,

$$Q = \frac{\pi}{4} d^2 \cdot V_1 = \frac{\pi}{4} d^2 \cdot C_v \sqrt{2gH}$$

$$d = \sqrt{\frac{4Q}{\pi C_v \sqrt{2gH}}}$$

$$= 0{\cdot}54 \sqrt{\frac{Q}{\sqrt{H}}} \quad \text{(for } C_v = 0{\cdot}98\text{)} \qquad \ldots(17.6)$$

5. Speed Ratio (K_u)

SPEED RATIO is defined as the ratio of the tangential velocity of the runner to the theoretical velocity of the jet.

$$K_u = \frac{u}{\sqrt{2gH}} \qquad \ldots(17.7)$$

Theoretical value of the speed ratio for maximum efficiency is 0·5. However, practically the speed ratio is found dependent upon the specific speed of the runner as follows :

N_s	8	12	16	20	24	28
K_u	0·47	0·46	0·45	0·44	0·43	0·42

6. Jet Ratio (m)

$$m = \frac{D}{d}$$

It is the ratio of runner diameter to the diameter of the jet

It can be shown that the jet ratio is a function of the specific speed of the turbine as follows :

Let N = Speed of turbine in rpm

P = power of the turbine in kW

H = head in m

Specific speed is defined as

$$N_s = \frac{N\sqrt{P}}{H^{5/4}}$$

Now

$$N = \frac{60u}{\pi D} \qquad \left(\because u = \frac{\pi DN}{50}\right)$$

$$= \frac{60 \cdot K_u \sqrt{2gH}}{\pi D}$$

$$P = (wQH \cdot \eta_0)\text{ W} = (WQH \cdot \eta_0 \times 10^{-3})\text{ kW} = W\left(\frac{\pi}{4} d^2 \cdot C_v \sqrt{2gH}\right) H \cdot \eta_0 \times 10^{-3}\text{ kW}$$

$$N_s = \frac{60 \cdot K_u \sqrt{2gH}}{\pi D H^{5/4}} \sqrt{W\left(\frac{\pi}{4} d^2 \cdot C_v \sqrt{2gH}\right) H \cdot \eta_0 \times 10^{-3}}$$

Assuming

$\eta_0 = 0{\cdot}87$

$K_u = 0{\cdot}46$

$C_v = 0{\cdot}98$

$$N_s = 210 \frac{D}{d} \qquad \ldots(17.8)$$

For various values of N_s , ratio D/d or the jet ratio changes as follows :

N_s	8	16	24	32
$m = D/d$	26	13	9	7

7. Number of Jets

Power generated by a turbine = Power available at the inlet × Efficiency

$$\propto wQH$$

$$\propto QH$$

Discharge through a nozzle, Q = Area × Velocity of jet $= \frac{\pi}{4} d^2 \cdot V_1$

$$\propto d^2 V_1$$

- The velocity of the jet is given by $V_1 = C_v \sqrt{2gH}$ *i.e.* it is dependent upon the head or $V_1 \propto \sqrt{H}$.
- Since

$$m = \frac{D}{d}$$

or

$$d = \frac{D}{m} = \frac{60u/\pi N}{m} \qquad \left(\because u = \frac{\pi DN}{60}\right)$$

$$= \frac{60K_u \sqrt{2gH}}{\pi Nm}$$

$$= \frac{60 \times 0{\cdot}46 \times \sqrt{2 \times 9{\cdot}81 \times H}}{\pi Nm}$$

$$\propto \frac{\sqrt{H}}{Nm}$$

Therefore,

$$Q \propto \frac{H}{N^2 m^2} \cdot \sqrt{H} \propto \frac{H^3/2}{N^2 m^2}$$

Thus power generated $\propto \frac{H^3/2}{N^2 m^2} \times H \propto \frac{H^5/2}{N^2 m^2}$

H is constant for a power plant, the jet ratio m is a function of the specific speed and speed N of the turbine is the speed dictated by the speed of the generator and is usually high. This shows that the power generated by the turbine has limited value though we may have large quantity of water at our disposal.

To increase the discharge or power, the following options are available :

(a) Number of jets can be increased. Usually, the number can be easily increased to 4, but never more than 6 as a higher number of jets make the governing mechanism complicated (Fig. 17.12).

(b) More than one runner can be mounted on one shaft. Sometimes, two turbines are put on two sides of a generator which is located in the middle. It becomes a double-over-hung system.

(c) Number of turbine can be increased, each operating individually. However, this may prove costly.

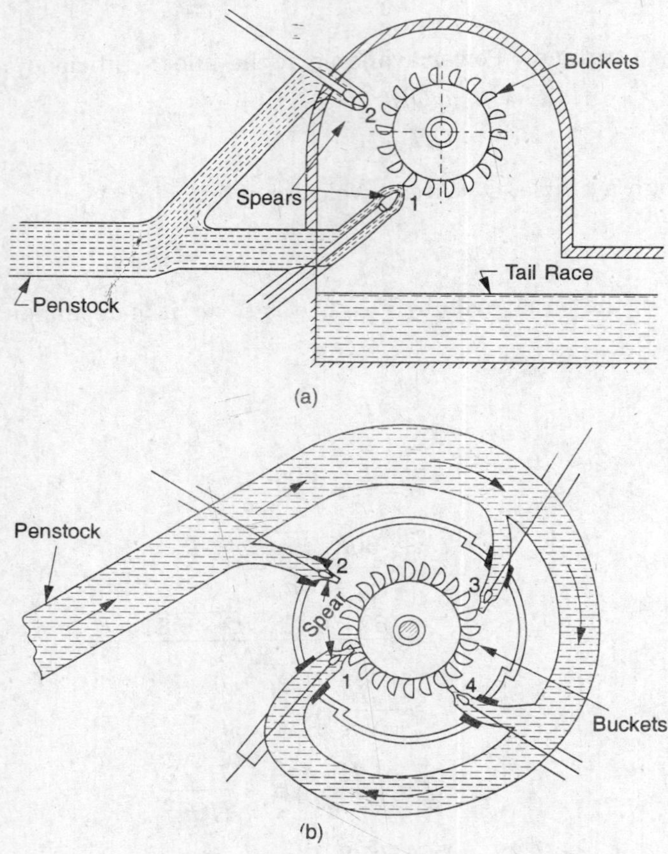

Fig. 17.12

8. Power and efficiency

The work done by the jet on the buckets is found by the change in momentum of the jet.

Let V_1 = velocity of jet before impinging the buckets

V_2 = velocity of jet leaving the buckets

u = velocity of the buckets or tangential velocity of the runner

V_{w1} = velocity of whirl at the inlet

V_{w2} = velocity of whirl at the outlet

V_{f2} = velocity of flow at the outlet

α_2 = angle of absolute velocity at the outlet

β_2 = angle of relative velocity at the outlet

Since angle at the inlet is assumed zero, velocity V_1 is collinear with velocity of the bucket. At the outlet, the shape of the velocity triangle depends upon the speed of the runner *i.e.* whether it is fast, slow or medium.

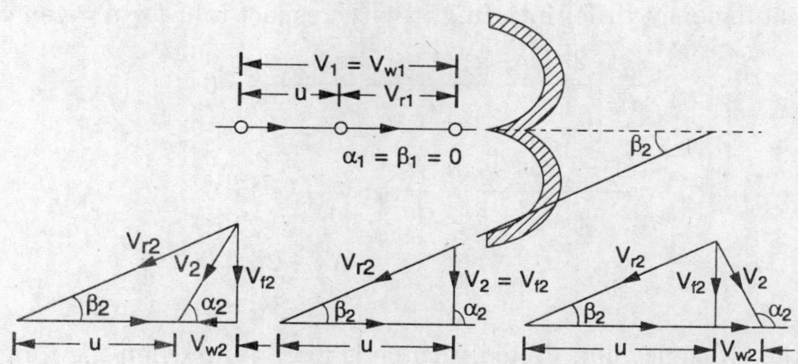

Fig. 17.13

Force on the runner per unit mass = $V_{w1} \pm V_{w2}$

- V_{w2} is positive for Fig. Fig. 17.13a
- V_{w2} is zero for Fig. 17.13b
- V_{w2} is negative for Fig. 17.13c.

Work done per unit mass $= (V_{w1} \pm V_{w2}) \cdot u$

Kenetic energy per unit mass $= V_1^2$

Hydraulic efficiency $= \dfrac{(V_{w1} \pm V_{w2})\, u}{\frac{1}{2} V_1^2}$

For Fig. 17.13a,

Let relative velocity at the outlet be k times of that at the inlet due to friction of the surface of the buckets ($k < 1$).

- $V_{w1} = V_1$
- $V_{w2} = V_{r2} \cos \beta_2 - u = k \cdot V_{r1} \cos \beta_2 - u = k\,(V_{w1} - u) \cos \beta_2 - u = k\,(V_1 - u) \cos \beta_2 - u$

Thus efficiency,

$$\eta_h = \frac{2\,(V_{w1} \pm V_{w2})\, u}{V_1^2}$$

$$= \frac{2\,[V_1 + k\,(V_1 - u) \cos \beta_2 - u]\, u}{V_1^2}$$

$$= \frac{2\,[(V_1 - u) + k\,(V_1 - u) \cos \beta_2]\, u}{V_1^2}$$

$$= \frac{2\,(V_1 - u)\,(1 + k \cos \beta_2)\, u}{V_1^2} \qquad \ldots(17.9)$$

For maximum efficiency, differentiating it with respect to u for a given value of V_1.

$$\frac{d}{du}\left[\frac{2\,(V_1 - u)\,(1 + k\cos\beta_2)\,u}{V_1^{\,2}}\right] = 0$$

$$\frac{(1 + k\cos\beta_2)\,u}{V_1^{\,2}}\,\frac{d}{du}\,(u \cdot V_1 - u^2) = 0$$

$$V_1 - 2u = 0$$

$$V_1 = 2u \qquad \text{...(17.10)}$$

Thus, hydraulic efficiency of a Pelton turbine is maximum when the tangential velocity of the runner is half the velocity of the jet at the inlet. If the surface of the buckets is smooth and polished and there is no friction, then $k = 1$ or the relative velocity at the outlet is equal to the relative velocity at the inlet. However, friction has no effect on the condition of maximum efficiency.

Maximum hydraulic efficiency,

$$\eta_h = \frac{2\left[\left(V_1 - \frac{V_1}{2}\right)(1 + k\cos\beta_2)\right]\frac{V_1}{2}}{V_1^{\,2}}$$

$$= \frac{2 \times \frac{V_1}{2}\,(1 + k\cos\beta_2)\,\frac{V_1}{2}}{V_1^{\,2}}$$

$$= \frac{(1 + k\cos\beta_2)}{2} \qquad \text{...(17.11)}$$

Example 7.1 Each turbine of a hydropower plant operates under a head of 860 m and produced 6000 kW. The speed of the turbine is 600 rpm. Determine

(i) the least diameter of the jet

(ii) the mean diameter of the wheel

(iii) the jet ratio

(iv) the number of jets.

The coefficient of velocity is 0·98, speed ratio 0·46 and the overall efficiency 0·88.

Solution $P = 6000$ kW $\qquad C_v = 0{\cdot}98$

$N = 600$ rpm $\qquad K_u = 0{\cdot}46$

$H = 860$ rpm $\qquad \eta_0 = 0{\cdot}88$

(i) We have,

$$V_1 = C_v\,\sqrt{2gH} = 0{\cdot}98\,\sqrt{2 \times 9{\cdot}81 \times 860} = 127{\cdot}3 \text{ m/s}$$

$$\eta_0 = \frac{\text{Power produced}}{\text{Available power}} = \frac{P}{wQH}$$

$$0.88 = \frac{6000}{9.81 \times Q \times 860}$$

$$Q = 0.808 \text{ m}^3/\text{s}$$

$$\frac{\pi}{4} d^2 \times V_1 = 0.808$$

$$\frac{\pi}{4} d^2 \times 127.3 = 0.808$$

$$d = 0.0899 \text{ m} \quad \text{or} \quad 89.9 \text{ mm}$$

(ii)

$$K_u = \frac{u}{\sqrt{2gH}} = \frac{\pi DN}{60} \cdot \frac{1}{\sqrt{2gH}}$$

$$0.46 = \frac{\pi D \times 600}{60} \cdot \frac{1}{\sqrt{2 \times 9.81 \times 860}}$$

$$D = 1.902 \text{ m}$$

(iii) Jet ratio $= \frac{D}{d} = \frac{1.902}{0.0899} = 21.2$

(iv) Number of buckets $= 0.5 \text{ m} + 15 = 0.5 \times 21.2 + 15 = 25.6 \approx 26$

Example 17.2 Flow through a Pelton turbine is 250 l/s under a head of 320 m. Determine the power developed and the efficiency if the mean bucket speed is 35 m/s. The buckets deflect the jet through 160°. Take $C_v = 0.97$.

Solution $Q = 0.25 \text{ m}^3/\text{s}$ $\quad C_v = 0.97$

$u = 35 \text{ m/s}$ $\quad \beta_0 = 180° - 160° = 20°$

$H = 320$ rpm

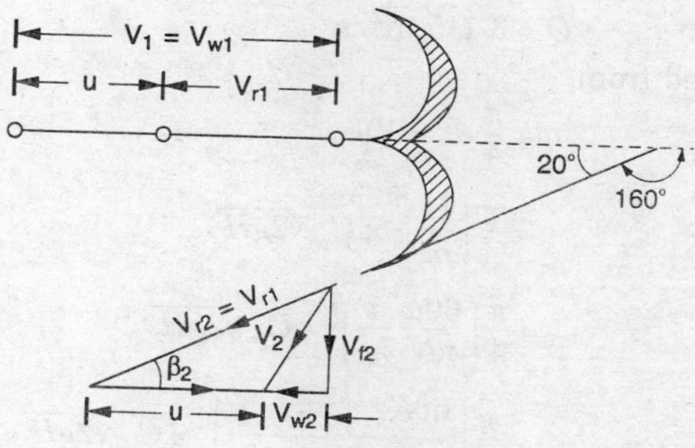

Fig. 17.14

$$P = pQ\,(V_{w1} \pm V_{w2}) \cdot u$$

$$V_{w1} = V_1 = C_v\sqrt{2gH} = 0.97\sqrt{2 \times 9.81 \times 320} = 76.86 \text{ m/s}$$

$$V_{w2} = V_{r2} \cos \beta_2 - u$$

$$= V_{r1} \cos \beta_2 - u$$

$$= (V_1 - u) \cos \beta_2 - u$$

$$= (76{\cdot}86 - 35) \cos 20^\circ - 35$$

$$= 4{\cdot}33 \text{ m/s}$$

$$P = \rho Q\,(V_{w1} \pm V_{w2}) \cdot u = 1000 \times 0{\cdot}25 \times (76{\cdot}86 + 4{\cdot}33) \times 35$$

$$= 710{\cdot}45 \times 10^3 \text{ W} = 710{\cdot}45 \text{ kW}$$

$$\eta_h = \frac{(V_{w1} \pm V_{w2})\, u}{\frac{1}{2} V_1^{\,2}} = \frac{(76{\cdot}86 + 4{\cdot}33)\, 35}{\frac{1}{2} \times 76{\cdot}86^2} = 0{\cdot}962 \quad \text{or} \quad 96{\cdot}2\%$$

Example 17.3 Determine the number of jets required for a Pelton turbine to develop 10000 kW of power under a head of 380 m when running at 500 rpm. Ratio of wheel diameter to jet diameter is 12. Assume suitable values of speed ratio, coefficient of velocity and the overall efficiency.

Solution $P = 10000$ kW $\quad H = 380$ m $\quad N = 500$ rpm

Assume $\eta_0 = 0{\cdot}85$

$C_v = 0{\cdot}97$

$K_u = 0{\cdot}47$

Number of jets,

$$n = \frac{\text{Total discharge}}{\text{discharge of one jet}} = \frac{Q}{q}$$

Q is found from the relation,

$$\eta_0 = \frac{\text{Power produced}}{\text{Available power}} = \frac{P}{wQH}$$

$$0{\cdot}85 = \frac{10000}{9{\cdot}81 \times Q \times 380}$$

$$Q = 3{\cdot}115 \text{ m}^2/\text{s}$$

and q is evaluated from,

$$q = \frac{\pi}{4} d^2 \times V_1$$

$$= \frac{\pi}{4} \left(\frac{D}{m}\right)^2 \times C_v \sqrt{2gH} \qquad \left(m = \frac{D}{d}\right)$$

$$= \frac{\pi}{4} \left(\frac{60u}{\pi N} \frac{1}{m}\right)^2 \times C_v \sqrt{2gH} \qquad \left(u = \frac{\pi D N}{60}\right)$$

$$= \frac{\pi}{4} \left(\frac{60 K_u \sqrt{2gH}}{\pi N} \frac{1}{m}\right)^2 \times C_v \sqrt{2gH}$$

$$= \frac{\pi}{4} \left(\frac{60 \times 0{\cdot}47 \times \sqrt{2 \times 9{\cdot}81 \times 380}}{\pi \times 500} \times \frac{1}{12}\right)^2 \times 0{\cdot}97 \sqrt{2 \times 9{\cdot}81 \times 380}$$

$$= 0{\cdot}0131 \times 83{\cdot}766$$

$$= 1{\cdot}098 \text{ m}^3/\text{s}$$

$$n = \frac{3{\cdot}115}{1{\cdot}098} = 2{\cdot}84 \approx 3$$

Example 17.4 The runner of a Pelton turbine has tangential velocity of 20 m/s and works under a head of 60 m. The jet is turned through 165°. The discharge through the nozzle is 100 l/s. Determine the power developed by the runner and the efficiency. Take $C_v = 0{\cdot}97$.

Solution $Q = 0{\cdot}1\ \text{m}^3/\text{s}$ $\qquad u = 20$ m/s

$H = 60$ m $\qquad \beta_2 = 180° - 165° = 15°$

$C_v = 0{\cdot}97$

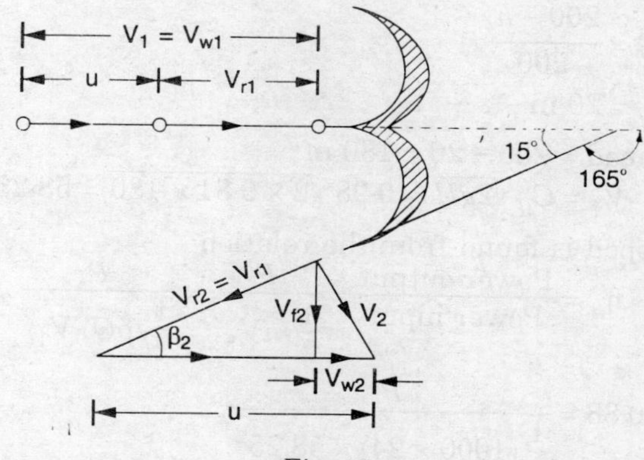

Fig. 17.15

$$P = \rho Q\,(V_{w1} \pm V_{w2}) \cdot u$$

$$V_{w1} = V_1 = C_v\sqrt{2gH} = 0{\cdot}97\sqrt{2 \times 9{\cdot}81 \times 60} = 33{\cdot}29 \text{ m/s}$$

and

$$\begin{aligned} V_{w2} &= V_{r2}\cos\beta_2 - u \\ &= V_{r1}\cos\beta_2 - u \\ &= (V_1 - u)\cos\beta_2 - u \\ &= (33{\cdot}29 - 20)\cos 15° - 20 \\ &= -\,7{\cdot}16 \text{ m/s} \end{aligned}$$

The negative sign indicates that the direction of the absolute velocity at the outlet is the same as that of the vane.

Thus

$$P = \rho Q\,(V_{w1} - V_{w2}) \cdot u = 1000 \times 0{\cdot}1 \times (33{\cdot}29 - 7{\cdot}16) \times 20 = 52254 \text{ W} = 52{\cdot}254 \text{ kW}$$

Hydraulic efficiency

$$= \frac{(V_{w1} - V_{w2})\,u}{\frac{1}{2}V_1^2} = \frac{(33{\cdot}29 - 7{\cdot}16) \times 20}{\frac{1}{2} \times 33{\cdot}29^2} = 0{\cdot}943 = 94{\cdot}3\%$$

Example 17.5 A power house has five Pelton turbines. Each turbine has two runners. Each runner is fitted with two nozzles. Total discharge through the turbines is 24 m²/s. The head on the turbine is 200 m and the length of the penstock pipe is 2500 m. Assuming the efficiency of transmission through the penstock pipe and the nozzles as 90%, coefficient of friction 0·009, coefficient of velocity 0·98 and the hydraulic efficiency 88%, determine

(a) the power developed

(b) diameter of the jet and

(c) the diameter of the penstock pipe.

Solution Efficiency of transmission

$$= \frac{\text{Total head} - \text{Friction head loss}}{\text{Total head}}$$

$$0{\cdot}9 = \frac{200 - h_f}{200}$$

$$h_f = 20 \text{ m}$$

Therefore, Net head = 200 − 20 = 180 m

$$V_1 = C_v \sqrt{2gH} = 0{\cdot}98\sqrt{2 \times 9{\cdot}81 \times 180} = 58{\cdot}25 \text{ m/s}$$

(a) Power developed is found from the relation

$$\eta_h = \frac{\text{Power output}}{\text{Power input}} = \frac{P}{\frac{1}{2} m V_1^2} = \frac{P}{\frac{1}{2}(\rho Q) V_1^2}$$

$$0{\cdot}88 = \frac{P}{\frac{1}{2}(1000 \times 24) \times 58{\cdot}25^2}$$

$$P = 35831 \times 10^3 \text{ W} \quad \text{or} \quad 35831 \text{ kW}$$

Discharge per jet,

$$q = \frac{\text{Total discharge}}{\text{Number of turbines} \times \text{Number of runners} \times \text{Number of jets}}$$

$$\frac{\pi}{4} \times d^2 \cdot V_1 = \frac{24}{5 \times 2 \times 2}$$

$$\frac{\pi}{4} \times d^2 \times 58{\cdot}25 = 1{\cdot}2 \text{ m}^3/\text{s}$$

$$d = 0{\cdot}162 \text{ m} \quad \text{or} \quad 162 \text{ mm}$$

Let V be the velocity of water in the penstock pipe. Also let D and L represent the diameter and the length of the penstock pipe.

$$h_f = \frac{4fLV^2}{2gD} = \frac{4fL}{2gD}\left(\frac{Q}{\frac{\pi}{4}D^2}\right)^2$$

$$20 = \frac{4 \times 0{\cdot}009 \times 2500}{2 \times 9{\cdot}81 \times D}\left(\frac{24}{\frac{\pi}{4}D^2}\right)^2$$

$$= \frac{4{\cdot}587}{D}\,\frac{933{\cdot}78}{D^4}$$

$$D^5 = 214{\cdot}16$$

$$D = 2{\cdot}925 \text{ m}$$

Example 17.6 The following data relate to a Pelton turbine :

Power developed	= 4000 kW
Head	= 320 m
Speed ratio	= 0·45
Jet ratio	= 12
Angle of jet deflection	= 165°
Coefficient of velocity	= 0·98
Overall efficiency	= 84%
Generator eficiency	= 96%
Friction loss of buckets	= 10%

Determine

(i) Diameter of the jet

(ii) Tangential force on the buckets

(iii) Diameter of the runner

(iv) Synchronous speed.

Solution $P = 4000$ kW $\qquad \eta_0 = 0{\cdot}84$

$H = 320$ m $\qquad \eta_g = 0{\cdot}96$

$m = 12$ $\qquad \beta_0 = 180° - 165° = 15°$

$K_u = 0{\cdot}45$ $\qquad h_f = 10\%$ or $V_{r2} = 0{\cdot}9V_{r1}$

$C_v = 0{\cdot}98$

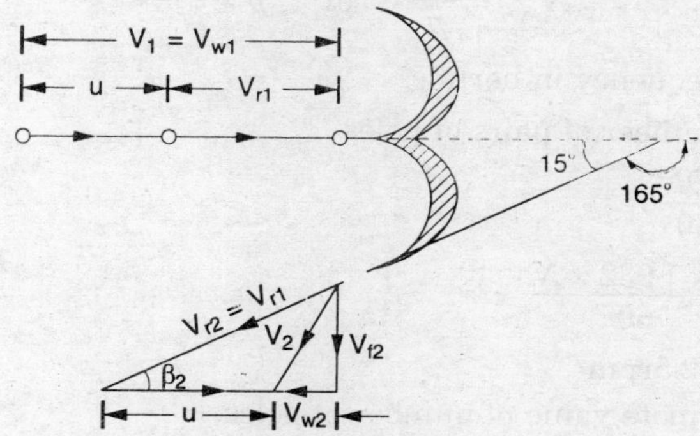

Fig. 17.16

(a) $Q = \frac{\pi}{4} \times d^2 {\cdot} V_1$

Q is found from the relation,

$$P = wQH \times \eta_0 \times \eta_g$$

$$4000 = 9{\cdot}81 \times Q \times 320 \times 0{\cdot}84 \times 0{\cdot}96$$

$$Q = 1{\cdot}58 \text{ m}^3$$

and $$V_1 = C_v \sqrt{2gH} = 0{\cdot}98 \times \sqrt{2 \times 9{\cdot}81 \times 320} = 77{\cdot}65 \text{ m/s}$$

Therefore, $$1{\cdot}58 = \frac{\pi}{4} \times d^2 \times 77{\cdot}65^2$$

$$d = 0{\cdot}161 \text{ m} \quad \text{or} \quad 161 \text{ mm}$$

(b) $F = \rho Q\,(V_{w1} \pm V_{w2})$

$$u = K_u \sqrt{2gH} = 0{\cdot}45 \sqrt{2 \times 9{\cdot}81 \times 320} = 35{\cdot}66 \text{ m/s}$$

$$V_{w1} = V_1 = 77{\cdot}65 \text{ m/s}$$

and
$$\begin{aligned} V_{w2} &= V_{r2} \cos \beta_2 - u \\ &= 0{\cdot}9 V_{r1} \cos \beta_2 - u \\ &= 0{\cdot}9\,(V_1 - u) \cos \beta_2 - u \\ &= 0{\cdot}9\,(77{\cdot}65 - 35{\cdot}66) \cos 15^\circ - 35{\cdot}66 \\ &= 0{\cdot}843 \text{ m/s} \end{aligned}$$

$$F = 1000 \times 1{\cdot}58 \times (77{\cdot}65 + 0{\cdot}843) = (124 \times 10^3) \text{ W} \quad \text{or} \quad 124 \text{ kW}$$

(c) $m = \dfrac{D}{d}$

$$12 = \frac{D}{0{\cdot}161}$$

$$D = 1{\cdot}932 \text{ m}$$

(d) Synchronous speed,

$$N_{syn} = \frac{60f}{p} \quad \text{or} \quad p = \frac{60f}{N_{syn}}.$$

Where f = frequency in hertz

P = Number of pairs of poles

$$u = \frac{\pi D N}{60}$$

$$35{\cdot}66 = \frac{\pi \times 1{\cdot}932 \times N}{60}$$

$$N = 352{\cdot}5 \text{rpm}$$

The approximate value of number of poles,

$$P = \frac{60f}{N_{syn}}$$

$$352{\cdot}5 = \frac{60 \times 50}{p}$$

$$P = 8{\cdot}51$$

Assuming the number of poles = 8

$$N_{syn} = \frac{60 \times 50}{8} = 375 \text{ rpm}$$

With change in speed, diameter of the runner is revised as

$$u = \frac{\pi DN}{60}$$

$$35{\cdot}66 = \frac{\pi D \times 375}{60}$$

$$D = 1{\cdot}816 \text{ m}$$

17.8 REACTION TURBINES

As discussed in the previous section under number of jets,

$$P \propto \frac{H^{5/2}}{N^2 m^2}$$

This expression shows that the power generated by a Pelton turbine is dependent upon H, N and m. If N and m are assumed constant, then for low head, the power generated will also be less and at some stage, setting up of a power plant may not become feasible. This limitation of Pelton or impulse tubines led to the investigation of some other type of turbines. Consequently, reaction turbines came into existence in which water is admitted all around the turbine runner under pressure and the turbine functions under a difference of pressure between the inlet and the outlet.

Neglecting the effect of thickness of runner blades, flow through a reaction turbine at any cross-section of the runner is given by

Q = Area of flow × Velocity of flow

= Circumference × Width × Velocity of flow

$= \pi D \times B \times V_f$

where V_f = Velocity of flow at the cross-section

B = Width of the runner

D = Diameter of the runner at the cross-section

To study the effect of V_f, consider the outlet velocity triangles shown in Fig. 17.17.

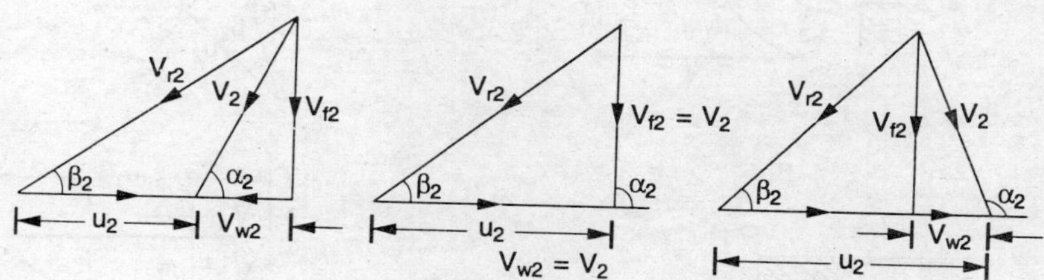

Fig. 17.17

- The kinetic energy of water at the outlet is $V_2^2/2g$ where V_2 is the absolute velocity at the outlet and is a loss. This reduces the efficiency of the turbine. Thus, it is desired to keep this value to the minimum. From the outlet velocity triangles it can be seen that the velocity V_2 is minimum if angle α_2 is 90°. In that case V_2 coincides with V_{f2}

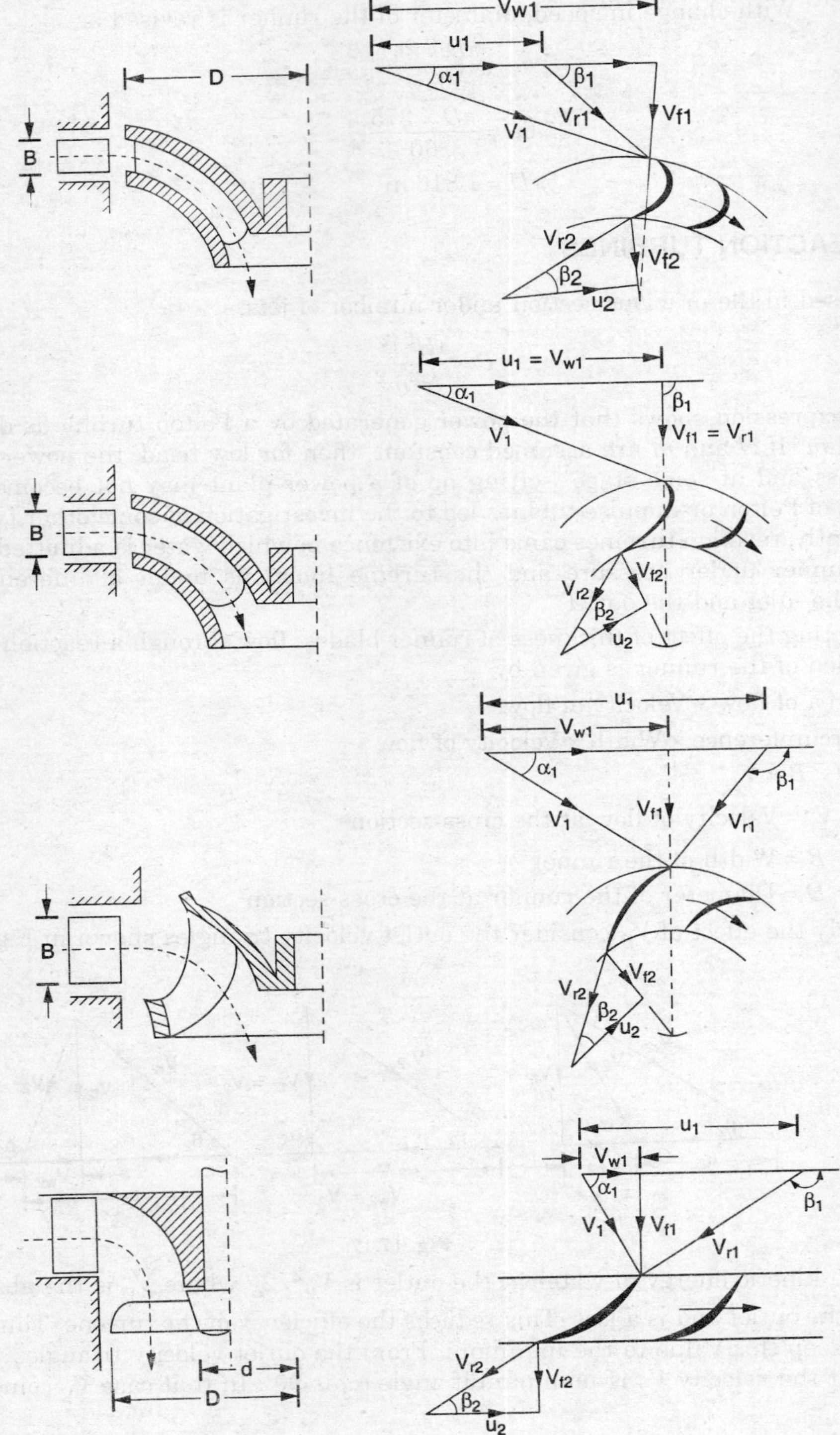

V
w1
u
1
D
α
1
β
1
B
V
r1
V
f1
V
1
V
r2
V
f2
β
2
u
2
u
1
= V
w1
α
1
β
1
V
1
V
f1
= V
r1
B
V
f2
V
r2
β
2
u
2
u
1
V
w1
α
1
β
1
V
f1
V
1
V
r1
B
V
r2
V
f2
β
2
u
2
u
1
V
w1
α
1
β
1
V
1
V
f1
V
r1
V
r2
V
f2
d
β
2
D
u
2

Fig. 17.18

and the velocity of whirl V_{w2} is zero. Owing to this reason, all the reaction turbines are usually designed in this way where $\alpha_2 = 0$. Such a discharge at the outlet is called radial discharge, which implies that the direction of the absolute velocity of water at the outlet is perpendicular to that of tangential velocity of the runner.

- The value of V_{f2} and thus V_2 also depends upon the angle of the vane at the outlet β_2. Therefore, this angle is also kept small (approximately 15°).

The above considerations indicate that the flow velocity V_f can be increased only upto a certain extent. Thus, there is possibility of increasing either D or B or both to get more discharge. Diameter can not be increased to any value due to stress considerations. Thus to increase the discharge, width (or ratio B/D) at the inlet is increased sufficiently so that the area is increased. At the outlet also, sufficient area is required. Such a change in the shape of runner changes the flow from pure radial to a mixed one. Finally, the flow changes to an axial one in which the water enters the runner from one side of the runner and leaves from the other. The shape of the runner vanes and the corresponding velocity triangles are also changed due to change in the direction of flow of water as shown in Fig. 17.18.

To enable large quantity of water to pass through the runner and to reduce the area occupied by the thickness of the vanes, number of vanes must decrease. It can be shown that the shape of a reaction turbine runner depends upon the specific speed of the turbine.

As the specific speed of the turbine is given by $N_s = \dfrac{N\sqrt{P}}{H^{3/4}}$, for the constant speed of a runner, specific speed is a function of P and H. As the head decreases, Q must increase to produce the same power. Thus with a lower value of H, specific speed N_s as well as area of flow of the runner increases and the flow changes from radial to axial. Therefore, Kaplan or propeller turbine has the highest value of specific speed. Table 17.1 gives the approximate values of various parameters for reaction turbines.

Table 17.1

Parameter	**Francis Turbine**			**Kaplan or Propeller turbine**
	Slow runner	**Medium runner**	**Fast runner**	
N_s (SI)	50-100	100-150	150-250	250-850
K_u	0·60-0·69	0·69-0·75	0·75-0·88	0·88-2·20
K_f	0·12	0·12-0·28	0·28-0·35	0·35-0·70
α_1 (deg)	12-25	25-33	33-40	40-74
β_1 (deg)	60-90	90	90-120	120-165
α_2 (deg)	90	90	90	90
β_2 (deg)	15	15	15	15
Z	20-25	15-20	11-15	3-6
$N = B/D$	0·04	0·12-0·24	0·24-0·5	0·35-0·6*

* ratio of hub diameter to runner diameter.

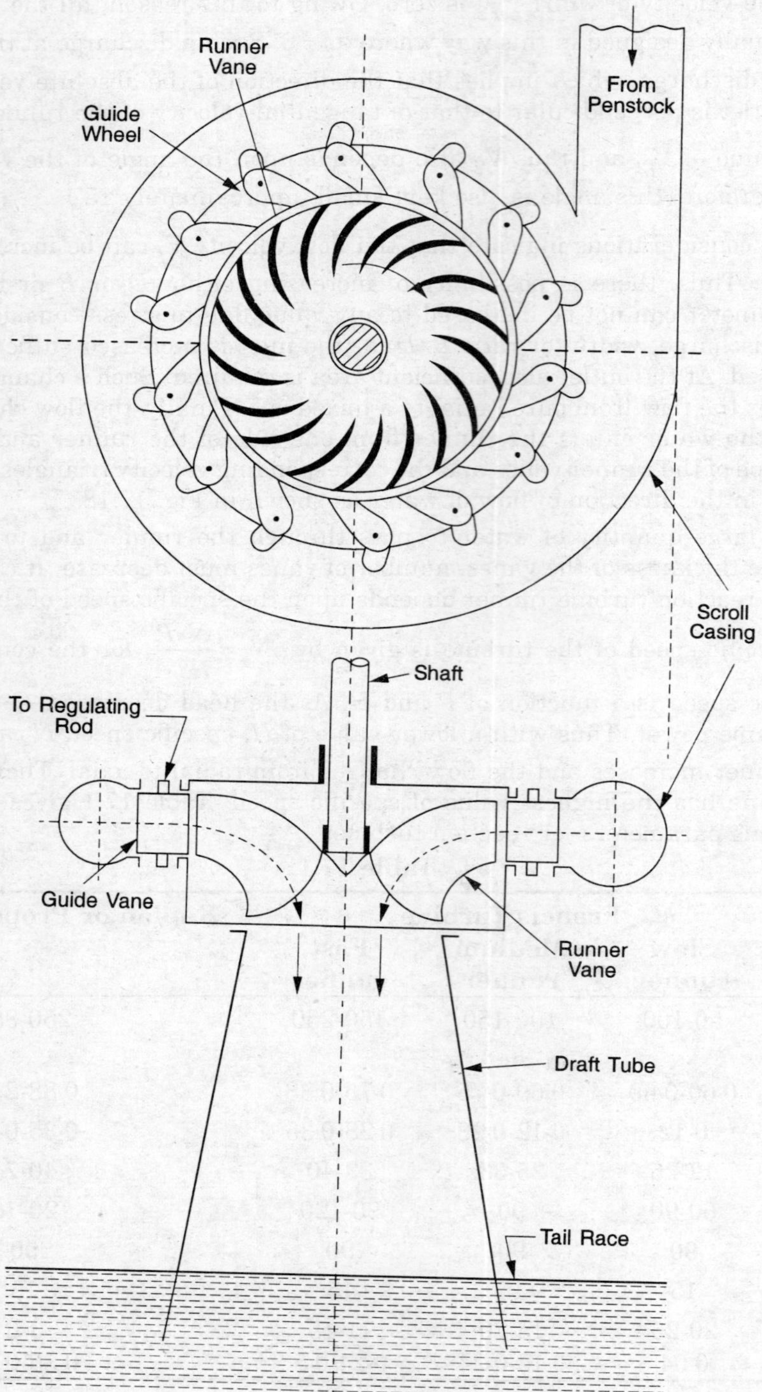
Runner Vane
Guide Wheel
From Penstock
Scroll Casing
Shaft
To Regulating Rod
Guide Vane
Runner Vane
Draft Tube
Tail Race

Fig. 17.19

17.9 FRANCIS TURBINE

Francis turbine is a mixed flow tubine (Fig. 17.19). It operates under medium heads and requires medium quantity of water. The water at high pressure enters at the circumference of the runner and moves towards the axis. It comes out of the runner axially through a passage along the shaft.

The water first passes through the guide vanes which direct the water at the proper angle so that the water enters the runner vanes without shock and glides properly. As the water passed through the runner and proceeds towards the outlet, continuously a part of the pressure energy is converted into kinetic energy which gives motion to the wheel. Finally, the water is discharged into the tailrace through a draft tube.

1. Components

The following are the main components of a Francis turbine :

(a) **Penstock** : A penstock is a waterway used to carry the water from the reservoir to the turbine. At the inlet, trashracks are used to prevent the debris from going into the turbine. Due to difficulty in transportation, a penstock is made in sections that are welded together at the site.

(b) **Scroll casing** : A scroll casing is a casing around the turbine wheel (Fig. 17.20). It distributes the water around the circumference of the wheel evenly.

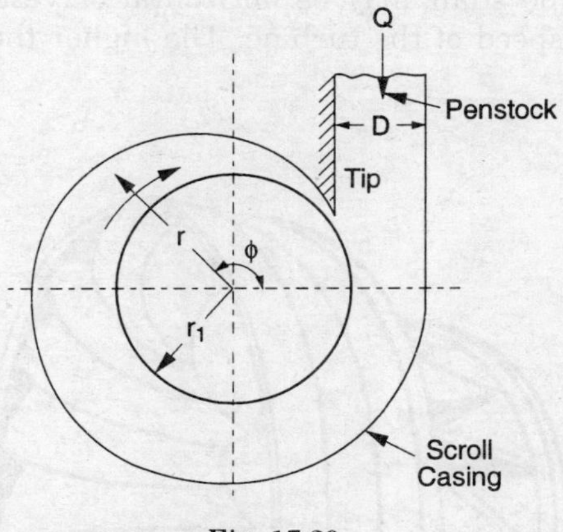

Fig. 17.20

The quantity of water reduces from maximum at the inlet of the casing to zero at the tip. This is because a portion of water enters all along the circumference of the runner. Therefore, the cross-section of a scroll casing reduces uniformly from maximum at the entrance to zero at the tip. Otherwise, it will not be possible to maintain the same speed throughout.

The material of the casing is

- Concrete for low heads
- Rolled steel for medium heads and
- Cast steel for high heads.

(c) **Guide mechanism** : A guide mechanism is necessary to regulate the flow or discharge according to the load on the turbine. It consists of a guide wheel having guide vanes (Fig. 17.21). The guide vanes are of airfoil section to have a smooth and eddiless flow. Each guide vane is able to move on its pivot centre which changes the area of flow. The movement of the guide wheel is regulated by a regulating shaft which in turn is operated by a governor through a servomechanism.

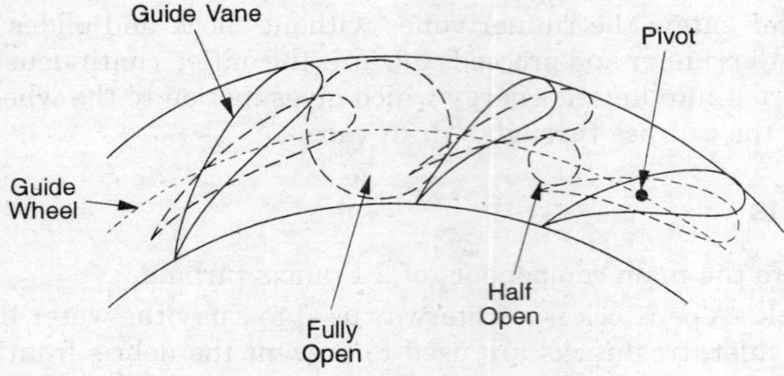

Fig. 17.21

The guide vanes are usually made of cast steel.

(d) **Runner** : Runner of a Francis turbine is keyed to the shaft which is coupled to the generator shaft. The turbine shaft may be horizontal or vertical. The height of the runner depends upon the specific speed of the turbine. The higher the specific speed, the higher the height.

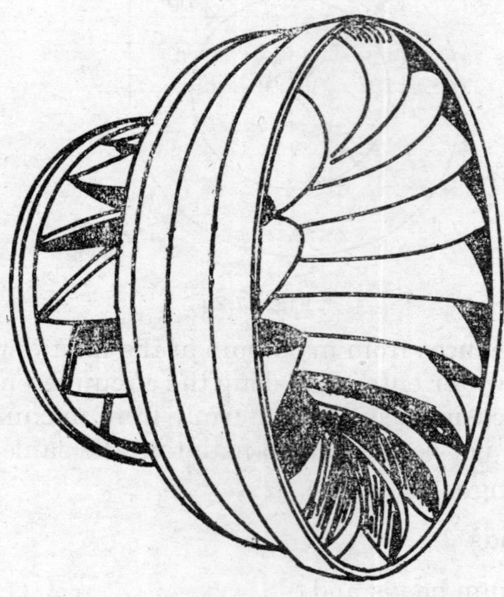

Fig. 17.22

The runner of a Francis turbine is made either as one piece or of separate steel plates welded together to form one piece. Material for the shaft is forged steel and for the runner cast iron for small power output and cast steel for large output. In case, the water is not chemically pure, the runner may be made of stainless steel or bronze to avoid corrosion. Fig. 17.22 shows the runner of a Francis turbine.

(e) **Draft tube** : A tube of enlarging cross-sectional area fixed to the outlet of a turbine that takes the water into the tailrace is known as **draft tube**. It is discussed in the next section in detail.

2. Power and Efficiency

In case of series of radial curved vanes,

Power produced per unit mass = $V_{w1} \cdot u_1 \pm V_{w2} \cdot u_2$

The reaction tubines are usually designed for radial flow at the outlet and the whirl component V_{w2} is zero. Thus,

Power produced per unit mass of water = $V_{w1} \cdot u_1$

Input to the runner per unit mass = gH

Therefore,
$$\eta_h = \frac{V_{w1} u_1}{gH} \qquad \text{...(17.12)}$$

If losses due to friction, eddies etc. are neglected and hydraulic loss is considered only due to kinetic energy loss at the outlet,

$$\eta_h = \frac{V_{w1} u_1}{gH} = \frac{gH - \dfrac{V_2^2}{2}}{gH} = 1 - \frac{V_2^2}{2gH} \qquad \text{...(17.13)}$$

Also,

$$V_{w1} u_1 = gH - \frac{V_2^2}{2} \qquad \text{...(17.14)}$$

This is known as the condition of frictionless vanes. In some problems, the condition of frictionless vanes may be assumed if not stated otherwise.

3. Francis Turbine Proportions

- Ratio of width B to the outside diameter D of the runner is denoted by n *ie* $n = B/D$
- Ratio of tangential velocity at the inlet u_1 to the quantity $\sqrt{2gH}$ is known as **speed ratio** and is denoted by K_u.

$$K_u = \frac{u_1}{\sqrt{2gH}} \qquad \text{...(17.15)}$$

K_u lies between 0·6 and 0·88.

- Ratio of flow velocity at the inlet V_{f1} to the quantity $\sqrt{2gH}$ is known as **flow ratio** and is denoted by K_f.

$$K_f = \frac{V_{f1}}{\sqrt{2gH}} \qquad \text{...(17.16)}$$

K_f lies between 0·12 and 0·36.

Example 17.7 The diameter of the runner of a reaction turbine is 1 m and flow area $0{\cdot}35\ m^2$ at the inlet. The speed of the runner is 360 rpm and the head is 50 m. Angles of the guide vanes and the runner vanes at the inlet are 18° and 60° respectively with the tangential direction of the wheel. Assuming radial flow at the outlet, determine the discharge, power developed and the hydraulic efficiency.

Solution $H = 50$ m $\qquad D_1 = 1$ m

$N = 360$ rpm $\qquad A_1 = 0{\cdot}35\ m^2$

$\alpha_1 = 18^\circ \qquad \beta_1 = 60^\circ$

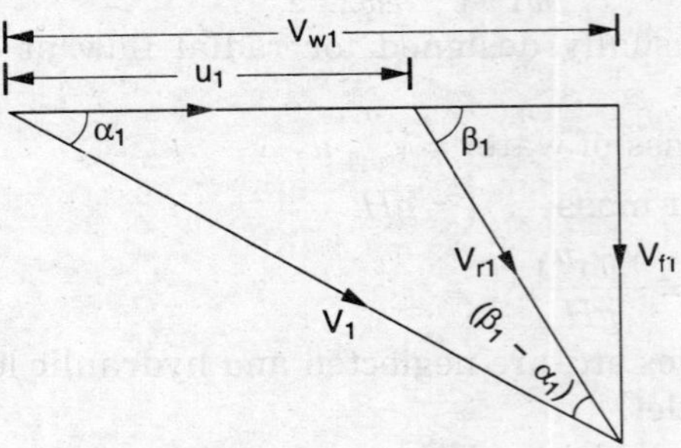

Fig. 17.23

$$u_1 = \frac{\pi D_1 N}{60} = \frac{\pi \times 1 \times 360}{60} = 18{\cdot}85 \text{ m/s}$$

Apply sine rule to the velocity triangle at the inlet,

$$\frac{V_{r1}}{\sin \alpha_1} = \frac{u_1}{\sin (\beta_1 - \alpha_1)}$$

$$\frac{V_{r1}}{\sin 18^\circ} = \frac{18{\cdot}85}{\sin (60^\circ - 18^\circ)}$$

$$V_{r1} = 8{\cdot}71 \text{ m/s}$$

$$V_{f1} = V_{r1} \cdot \sin 60^\circ = 8{\cdot}71 \sin 60^\circ = 7{\cdot}54 \text{ m/s}$$

$$Q = A_1 \cdot V_{f1} = 0{\cdot}35 \times 7{\cdot}54 = 2{\cdot}64 \text{ m}^3/\text{s}$$

Also $\qquad V_{f1} = V_{w1} \cdot \tan\alpha_1$

$$7{\cdot}54 = V_{w1} \cdot \tan 18^\circ$$

$$V_{w1} = 23{\cdot}21 \text{ m/s}$$

$$P = \rho Q \cdot V_{w1} \cdot u_1 = (1000 \times 2{\cdot}64 \times 23{\cdot}21 \times 18{\cdot}85) \text{ W} = 1154{\cdot}8 \text{ kW}$$

$$\eta_h = \frac{V_{w1} \cdot u_1}{gH} = \frac{23{\cdot}21 \times 18{\cdot}85}{9{\cdot}81 \times 60} = 0{\cdot}892 \quad \text{or} \quad 89{\cdot}2\%$$

Example 17.8 Runner diameter of a reaction turbine is 3 m at the inlet and 2 m at the outlet. The width of the wheel is constant. The discharge through the turbine is 75 m^3/s and is radial at the outlet having a velocity of 12 m/s. The angle of the vanes at the inlet is 120°. The working head is 160 m. If the efficiency of the turbine is 90 percent, determine the power produced and the speed of the turbine.

Solution $Q = 75\ \text{m}^3/\text{s}$ $\quad V_{f2} = 12$ m/s

$H = 160$ m $\quad \beta_1 = 120°$

$D_1 = 3$ m $\quad \beta_1' = 180° - 120° = 60°$

$D_2 = 2$ m

$$Q = \pi D_1 \cdot B_1 \cdot V_{f1} = \pi D_2 \cdot B_2 \cdot V_{f2}$$

$$\pi D_1 \cdot V_{f1} = \pi D_2 \cdot V_{f2} \qquad (\because B_1 = B_2)$$

$$\pi \times 3 \times V_{f1} = \pi \times 2 \times 12$$

$$V_{f1} = 8 \text{ m/s}$$

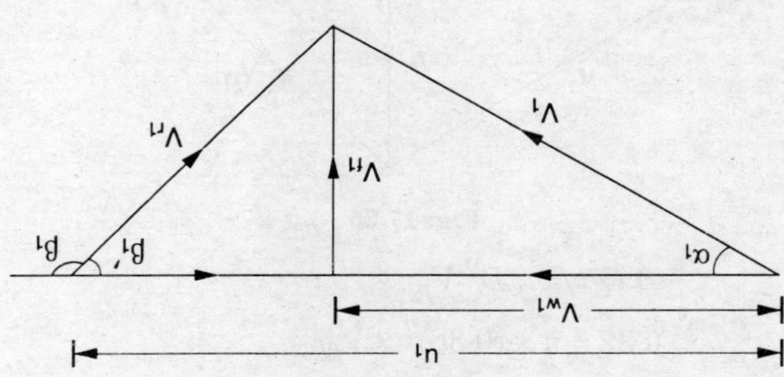

Fig. 17.24

$$\eta_h = \frac{V_{w1} \cdot u_1}{gH} = \frac{(u_1 - V_{f1} \cot \beta_1')\, u_1}{gH}$$

$$0.9 = \frac{(u_1 - 8 \cot 60°)\, u_1}{9.81 \times 160}$$

$$u_1^2 - 4.62 u_1 = 1412.6$$

$$u_1^2 - 4.62 u_1 + \left(\frac{4.62}{2}\right)^2 = 1412.6 + \left(\frac{4.62}{2}\right)^2$$

$$(u_1 - 2.31)^2 = 37.66^2$$

$$u_1 = 39.97 \text{ m/s}$$

$$V_{w1} = u_1 - V_{f1} \cot \beta_1' = 39.97 - 8 \cot 60° = 35.35 \text{ m/s}$$

$$P = \rho Q \cdot V_{w1} \cdot u_1 = (1000 \times 75 \times 35.35 \times 39.97) \text{ W} = 105974 \text{ kW}$$

$$u_1 = \frac{\pi D_1 N}{60}$$

$$39.97 = \frac{\pi \times 3 \times N}{60}$$

$$N = 254.5 \text{ rpm}$$

Example 17.9 An inward-flow reaction turbine is supplied water at the rate of 0·32 m³ per second. Outlet pipe of the turbine is 360 mm in diameter. The turbine operates under a head of 35 m. Radial velocity of flow through the wheel is same as the velocity in the outlet pipe. The tangential velocity of wheel at the inlet is 22 m/s. Determine the guide blade and the runner blade angles at the inlet. Also find the power developed.

Solution $H = 35$ m $\qquad u_1 = 22$ m/s

$D = 0.36$ m $\qquad Q = 0.32$ m³/s

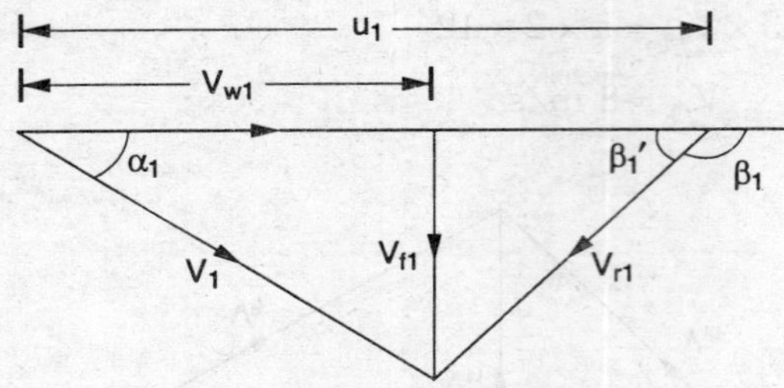

Fig. 17.25

$$Q = \pi \times D^2 . V_{f2}$$

$$0.32 = \pi \times (0.36)^2 \times V_{f2}$$

$$V_{f2} = 3.144 \text{ m/s}$$

Assuming the discharge to be radial, $V_2 = V_{f2} = V_{f1} = 3.144$ m/s

$$V_{w1} u_1 = gH - \frac{V_2^{\,2}}{2}$$

$$V_{w1} \times 22 = 9.81 \times 35 - \frac{(3.144)^2}{2}$$

$$V_{w1} = 15.38 \text{ m/s}$$

$$\tan \alpha_1 = \frac{V_{f1}}{V_{w1}} = \frac{3.144}{15.38} = 0.204$$

$$\alpha_1 = 11.55^\circ$$

$$\tan\beta_1' = \frac{V_{f1}}{u_1 - V_{w1}} = \frac{3.144}{22 - 15.38} = 0.475$$

$$\beta_1' = 25.4^\circ \quad \text{or} \quad \beta_1 = 180^\circ - 25.4^\circ = 154.6^\circ$$

$$P = \rho Q \cdot V_{w1} \cdot u_1 = (1000 \times 0.32 \times 15.38 \times 22) \text{ W} = 108.3 \text{ kW}$$

Example 17.10 An inward-flow reaction turbine has inner diameter of the runner 0·6 times the outer diameter. The water enters the turbine at 12° tc the tangent to the runner wheel. The vane angles are radial at the inlet. Velocity of flow is constant through the turbine and is 2·4 m/s. Speed of the runner is 280 rpm. Runner blades block 5 percent of area of flow. The width of the wheel at the inlet is 100 mm. Determine

(i) the inner and the outer diameters of the wheel

(ii) working head

(iii) blade angles at the outlet

(iv) power produced

Solution $B_1 = 0{\cdot}1$ m $k = 1 - 0{\cdot}05 = 0{\cdot}95$

$V_{f1} = V_{f2} = 2{\cdot}4$ m/s $N = 280$ rpm

$\alpha_1 = 12^\circ$ $D_2 = 0{\cdot}6\, D_1$

$\beta_1 = 90^\circ$

$$\tan \alpha_1 = \frac{V_{f1}}{V_{w1}}$$

$$\tan 12^\circ = \frac{2{\cdot}4}{V_{w1}}$$

$$V_{w1} = 11{\cdot}29 \text{ m/s}$$

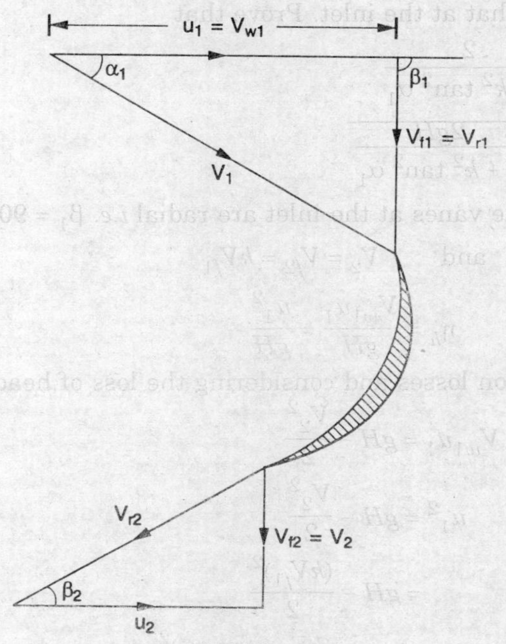

Fig. 17.26

Also $u_1 = 11{\cdot}29$ m/s $(u_1 = V_{w1})$

$$\frac{\pi D_1 N}{60} = 11{\cdot}29$$

$$\frac{\pi D_1 \times 280}{60} = 11{\cdot}29$$

$$D_1 = 0{\cdot}77 \text{ m}$$

$$D_2 = 0{\cdot}77 \times 0{\cdot}6 = 0{\cdot}462 \text{ m}$$

Neglecting friction losses and considering the loss of head only at the exit.

$$gH - \frac{V_2^{\,2}}{2} = V_{w1} u_1$$

$$H = \frac{V_{w1} u_1}{g} + \frac{V_2^{\,2}}{2g} = \frac{11{\cdot}29 \times 11{\cdot}29}{9{\cdot}81} + \frac{2{\cdot}4^2}{2 \times 9{\cdot}81} = 13{\cdot}12 \text{ m}$$

$$\tan \beta_2 = \frac{V_{f2}}{u_2} = \frac{V_{f2}}{(\pi D_2 N/60)} = \frac{2{\cdot}4}{(\pi \times 0{\cdot}462 \times 280/60)} = \frac{2{\cdot}4}{6{\cdot}733} = 0{\cdot}354$$

$$\beta_2 = 19{\cdot}5^\circ$$

$$Q = k\pi D_1 B_1 V_{f1} = 0{\cdot}95 \times \pi \times 0{\cdot}77 \times 0{\cdot}1 \times 2{\cdot}4 = 0{\cdot}552 \text{ m}^3/\text{s}$$

$$P = \rho Q {\cdot} V_{w1} {\cdot} u_1 = 1000 \times 0{\cdot}552 \times 11{\cdot}29 \times 11{\cdot}29 - (70{\cdot}3 \times 10^3) \text{ W} \quad \text{or} \quad 70{\cdot}3 \text{ kW}$$

Example 17.11 In a turbine with radial vanes at the inlet, the velocity of flow at the outlet is k times that at the inlet. Prove that

(i) $\eta_h = \dfrac{2}{2 + k^2 \tan^2 \alpha_1}$

(ii) $u_1 = \sqrt{\dfrac{2gH}{2 + k^2 \tan^2 \alpha_1}}$

Solution The angles of the vanes at the inlet are radial *i.e.* $\beta_1 = 90^\circ$

Also $u_1 = V_{w1}$ and $V_2 = V_{f2} = kV_{f1}$

(i)
$$\eta_h = \frac{V_{w1} u_1}{gH} = \frac{u_1^{\,2}}{gH} \qquad \ldots\text{(i)}$$

Neglecting friction losses and considering the loss of head only at the exit.

$$V_{w1} u_1 = gH - \frac{V_2^{\,2}}{2}$$

$$u_1^{\,2} = gH - \frac{V_2^{\,2}}{2}$$

$$= gH - \frac{(kV_{f1})^2}{2}$$

$$= gH - \frac{(kV_{w1} \tan \alpha_1)^2}{2}$$

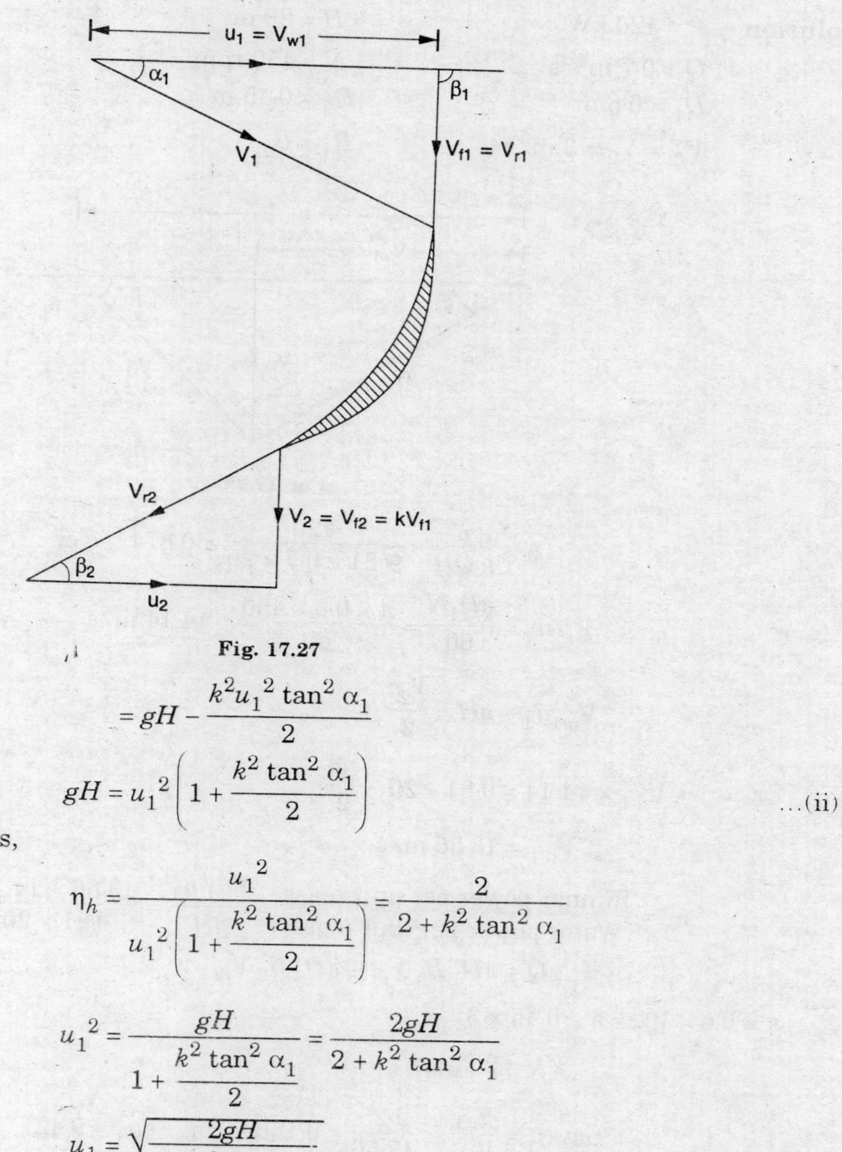

Fig. 17.27

$$= gH - \frac{k^2 {u_1}^2 \tan^2 \alpha_1}{2}$$

$$gH = {u_1}^2 \left(1 + \frac{k^2 \tan^2 \alpha_1}{2}\right) \qquad \ldots\text{(ii)}$$

∴ (i) becomes,

$$\eta_h = \frac{{u_1}^2}{{u_1}^2 \left(1 + \frac{k^2 \tan^2 \alpha_1}{2}\right)} = \frac{2}{2 + k^2 \tan^2 \alpha_1}$$

from (ii),

$${u_1}^2 = \frac{gH}{1 + \frac{k^2 \tan^2 \alpha_1}{2}} = \frac{2gH}{2 + k^2 \tan^2 \alpha_1}$$

$$u_1 = \sqrt{\frac{2gH}{2 + k^2 \tan^2 \alpha_1}}$$

Example 17.12 An inward-flow reaction turbine produces 120 kW of power working under a head of 20 m. The discharge through the turbine in 700 l/s and is radial at the outlet. The diameters at the inlet and the outlet of the turbine are 600 mm and 450 mm respectively. The radial velocity at the outlet is 3 m/s, width of the wheel constant and the speed of the runner 480 rpm. Determine

(i) overall and hydraulic efficiencies

(ii) inlet angles of the guide and runner blades.

Solution $P = 120$ kW $\quad H = 20$ m

$Q = 0.7\ m^3/s \quad N = 450$ rpm

$D_1 = 0.6$ m $\quad D_2 = 0.45$ m

$V_{f2} = V_2 = 3$ m/s $\quad B_1 = B_2$

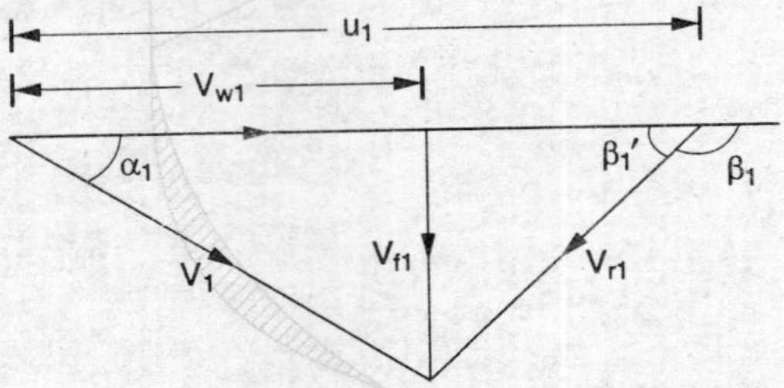

Fig. 17.28

$$\eta = \frac{p}{wQH} = \frac{120}{9.81 \times 0.7 \times 20} = 0.874 \quad \text{or} \quad 87.4\%$$

$$u_1 = \frac{\pi D_1 N}{60} = \frac{\pi \times 0.6 \times 450}{60} = 14.14 \text{ m/s}$$

$$V_{w1} u_1 = gH - \frac{V_2^{\,2}}{2}$$

$$V_{w1} \times 14.14 = 9.81 \times 20 - \frac{3^2}{2}$$

$$V_{w1} = 13.56 \text{ m/s}$$

$$\eta_h = \frac{\text{Runner power per unit mass}}{\text{Water power per unit mass}} = \frac{V_{w1} \cdot u_1}{gH} = \frac{13.56 \times 14.14}{9.81 \times 20} = 0.977 \quad \text{or} \quad 97.7\%$$

$$Q = \pi D_1 \cdot B_1 \cdot V_{f1} = \pi D_2 \cdot B_2 \cdot V_{f2}$$

$$\pi \times 0.6 \times V_{f1} = \pi \times 0.45 \times 3 \qquad (B_1 = B_2 \text{ given})$$

$$V_{f1} = 2.25 \text{ m/s}$$

$$\tan \alpha_1 = \frac{V_{f1}}{V_{w1}} = \frac{2.25}{13.56} = 0.166 \quad \text{or} \quad \alpha_1 = 9.42^\circ$$

$$\tan \beta_1' = \frac{V_{f1}}{u_1 - V_{w1}} = \frac{2.25}{14.14 - 13.56} = 3.88 \quad \text{or} \quad \beta_1' = 75.5$$

Example 17.13 An inward flow reaction turbine has a constant velocity of flow 2·5 m/s through the runner and the discharge is radial. The width of the runner at the inlet is 220 mm and the guide vane angle is 15° to the tangent to the wheel. The diameters of the wheel at the inlet and the outlet are 800 mm and 600 mm respectively. The wheel rotates at a speed of 180 rpm.

(i) runner blade angles
(ii) power developed
(iii) Hydraulic efficiency

Solution $N = 180$ rpm $\quad D_1 = 0{\cdot}8$ m

$B_1 = 0{\cdot}22$ m $\quad D_2 = 0{\cdot}6$ m

$V_{f1} = V_{f2} = 2{\cdot}5$ m $\quad \alpha_1 = 15^\circ$

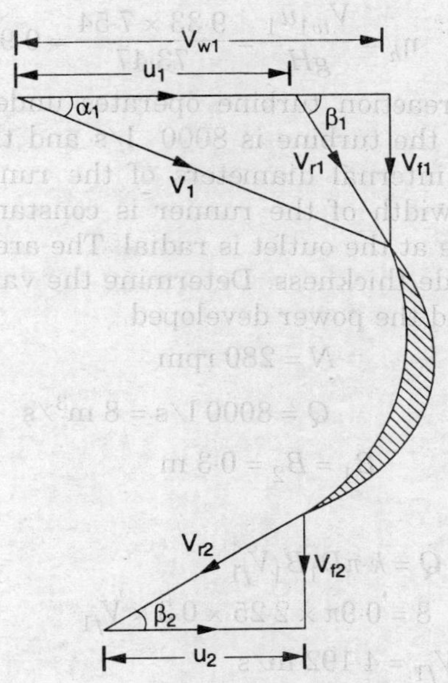

Fig. 17.29

(i)

$$V_{f1} = V_1 \sin \alpha_1$$

$$1{\cdot}5 = V_1 \sin 15^\circ$$

$$V_1 = 9{\cdot}66 \text{ m/s}$$

$$u_1 = \frac{\pi D_1 N}{60} = \frac{\pi \times 0{\cdot}8 \times 180}{60} = 7{\cdot}54 \text{ m/s}$$

$$V_{w1} = V_1 \cos \alpha_1 = 9{\cdot}66 \cos 15^\circ = 9{\cdot}33 \text{ m/s}$$

$$\tan \beta_1 = \frac{V_{f1}}{V_{w1} - u_1} = \frac{2{\cdot}5}{9{\cdot}33 - 7{\cdot}54} = 1{\cdot}397 \quad \text{or} \quad \beta_1 = 54{\cdot}4^\circ$$

$$u_2 = \frac{\pi D_2 N}{60} = \frac{\pi \times 0{\cdot}6 \times 180}{60} = 5{\cdot}65 \text{ m/s}$$

$$\tan \beta_2 = \frac{V_{f2}}{u_2} = \frac{2{\cdot}5}{5{\cdot}65} = 0{\cdot}442$$

$$\beta_2 = 23{\cdot}9^\circ$$

(ii) $Q = \pi D_1 \cdot B_1 \cdot V_{f1} = \pi \times 0{\cdot}8 \times 0{\cdot}22 \times 2{\cdot}5 = 1{\cdot}38 \text{ m/s}$

$P = \rho Q \cdot V_{w1} u_1 = (1000 \times 1{\cdot}38 \times 9{\cdot}33 \times 7{\cdot}54) \text{ W} = 97{\cdot}1 \text{ kW}$

(iii)
$$V_{w1} u_1 = gH - \frac{V_2^2}{2}$$
$$9{\cdot}33 \times 7{\cdot}54 = gH - \frac{2{\cdot}5^2}{2}$$
$$gH = 73{\cdot}47$$
$$\eta_h = \frac{V_{w1} u_1}{gH} = \frac{9{\cdot}33 \times 7{\cdot}54}{73{\cdot}47} = 0{\cdot}957 \quad \text{or} \quad 95{\cdot}7\%$$

Example 17.14 An outward-flow reaction turbine operates under a head of 200 m. The discharge through the turbine is 8000 l/s and the speed is 280 rpm. The external and the internal diameters of the runner are 3 m and 2·25 m respectively. The width of the runner is constant throughout and is 300 mm. The discharge at the outlet is radial. The area of flow is reduced by 10 percent due to blade thickness. Determine the vane angles at the inlet and the outlet. Also find the power developed.

Solution $D_1 = 2{\cdot}25$ m $\qquad N = 280$ rpm

$D_2 = 3$ m $\qquad Q = 8000 \text{ l/s} = 8 \text{ m}^3\text{/s}$

$H = 200$ m $\qquad B_1 = B_2 = 0{\cdot}3$ m

$V_2 = V_{f2}$

$$Q = k \cdot \pi D_1 B_1 V_{f1}$$
$$8 = 0{\cdot}9\pi \times 2{\cdot}25 \times 0{\cdot}3 \times V_{f1}$$
$$V_{f1} = 4{\cdot}192 \text{ m/s}$$
$$u_1 = \frac{\pi D_1 N}{60} = \frac{\pi \times 2{\cdot}25 \times 280}{60} = 32{\cdot}99 \text{m/s}$$
$$Q = k \cdot \pi D_2 B_2 V_{f2}$$
$$8 = 0{\cdot}9 \times \pi \times 3 \times 0{\cdot}3 \times V_{f2}$$
$$V_{f2} = 3{\cdot}144 \text{ m/s}$$
$$V_{w1} u_1 = gH - \frac{V_2^2}{2}$$
$$V_{w1} \times 32{\cdot}99 = 9{\cdot}81 \times 200 - \frac{3{\cdot}144^2}{2}$$
$$V_{w1} = 59{\cdot}32 \text{ m/s}$$
$$\tan \beta_1 = \frac{V_{f1}}{V_{w1} - u_1} = \frac{4{\cdot}192}{59{\cdot}32 - 32{\cdot}99} = 0{\cdot}159$$
$$\beta_1 = 9{\cdot}05^\circ$$

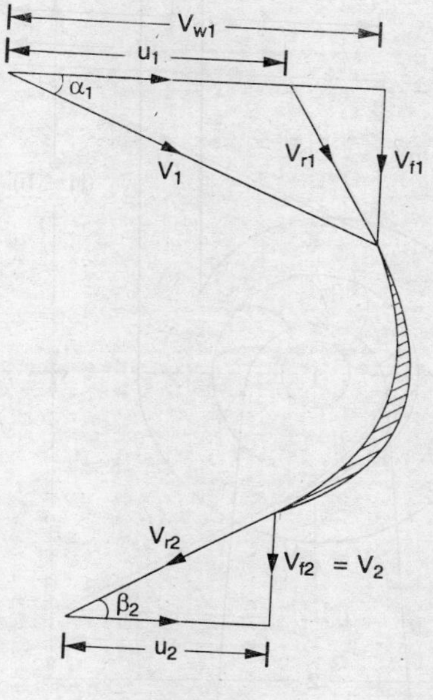

Fig. 17.30

$$u_2 = \frac{\pi D_2 N}{60} = \frac{\pi \times 3 \times 280}{60} = 43{\cdot}98$$

$$\tan \beta_2 = \frac{V_{f2}}{u_2} = \frac{3{\cdot}144}{43{\cdot}98} = 0{\cdot}072$$

$$\beta_2 = 4{\cdot}1^\circ$$

$$P = \rho Q{\cdot}V_{w1}u_1 - (1000 \times 8 \times 59{\cdot}32 \times 32{\cdot}99)\text{ W} = 15656 \times 10^3 \text{ W} \quad \text{or} \quad 15{\cdot}656 \text{ kW}$$

17.10 DRAFT TUBE

A reaction turbine may be installed

- Below the tailrace level. The head utilised is maximum. However, it is not convenient to repair the turbine if it goes out of order.
- At the tailrace level. The head utilised is maximum. As the level of the tailrace does not ramain constant throughout the year, sometimes the turbine will be submerged in water and sometimes exposed to atmosphere. The same problem of repair may occur.
- Above the tailrace level. The head utilised is not maximum. It is reduced by the height of the turbine above the tailrace level.

It may be observed that out of the three options available, the last one is more practical though head utilisation is reduced. It will be shown below that when a draft tube is connected to the outlet of a reaction turbine, this loss is retrieved.

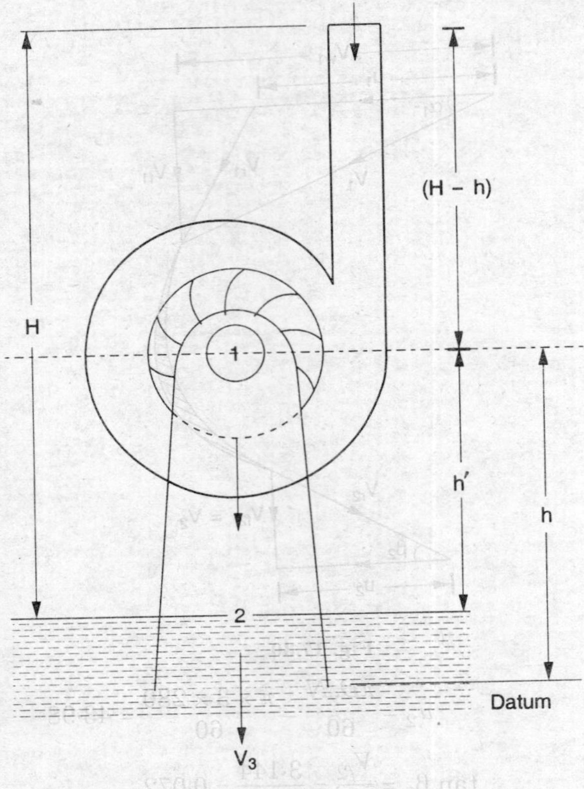

Fig. 17.31

Consider a reaction turbine installed above the tailrace level and fixed with a draft tube (Fig. 17.31). Apply Bernoulli's equation to points 1 and 3 taking point 3 as the datum,

$$\frac{p_1}{w} + \frac{V_1^2}{2g} + h = \frac{p_3}{w} + \frac{V_3^2}{2g} + 0$$

$$= \left[\frac{p_2}{w} + (h - h')\right] + \frac{V_3^2}{2g} + 0$$

$$\frac{p_1}{w} = \frac{p_2}{w} - \left(h' + \frac{V_1^2 - V_3^2}{2g}\right) \quad ...(17.17)$$

where p_2 is the atmospheric pressure and is the datum for the pressure. Thus the pressure head at the point 1 or at the outlet of the turbine is below the atmospheric pressure head. It is negative by an amount equal to the height of turbine above the tailrace level and the difference of the kinetic heads at the inlet and the outlet of the draft tube. Had the turbine been discharging into the atmosphere, the pressure at the outlet of turbine would have been atmospheric. As the head utilised in the turbine is the difference between the inlet and the outlet heads, the head recovered is not only equal to the height of the turbine above the tailrace level but also the difference of kinetic head at the inlet and outlet of the draft tube. Thus, a turbine is conveniently installed above the tailrace level without the loss of any head.

Height of a Draft tube

The maximum height of the turbine above tailrace level and the length of the draft tube are influenced by the following considerations :

- A draft tube is a diverging passage tube in which the velocity of flow reduces. In diverging passages, there is always danger of flow separation from the walls of the draft tube which may result in the formation of eddies. The phenomenon is known as **separation** which is associated with energy loss. From experience it is found that the total angle of divergence should not exceed 9-10 degrees.
- The pressure at the inlet of a draft tube is negative. Lowering of pressure of a fluid also lowers its boiling point. Therefore, if the level of a turbine is raised from a certain value, at some point, the pressure may be decreasing to such a value that the water starts boiling at the room temperature and bubbles are formed. When these bubbles reach some region of higher pressure, they collapse and with a high pressure collide with the surface in contact. In due course of time pits or cavities are formed at the surface. This phenomenon is known as **cavitation**. Thus, effort is made to avoid cavitation.

Efficiency of a Draft Tube

The efficiency of a draft tube takes into account the friction losses in the tube. It is defined as

$$\eta_d = \frac{\dfrac{V_1^2 - V_3^2}{2g} - h_L}{\dfrac{V_1^2 - V_3^2}{2g}} \qquad \ldots(17.18)$$

where h_L represents the friction head loss in the draft tube and V_1 and V_3 are the inlet and outlet velocities in the draft tube.

Types of Draft Tubes

The usual types of draft tubes are the following (Fig. 17.32) :

Conical Tube

It is a simple divergent tube of increased area (Fig. 17.32a). It should sufficiently be immersed in water so that its outlet end remains in water even if the water level in the tailrace falls by a small height. It should be able to decrease the velocity at the outlet to about 1 m/s. It is used where sufficient depth of water in the tailrace is there. The angle of divergence should not be more than 9 to 10 degrees as discussed earlier. Its efficiency is around 90 percent.

Spreading Tube or Hydracone

'Moody' introduced the design of a spreading draft tube. Annular area is provided by a solid conical section that gives increase in area in a small length of tube. Thus, it is suitable where conical tube is not able to give sufficient increase in area due to separation problems. There are two kinds of spreading tube : high cone and low cone as shown in Figs. 17.32 b and c.

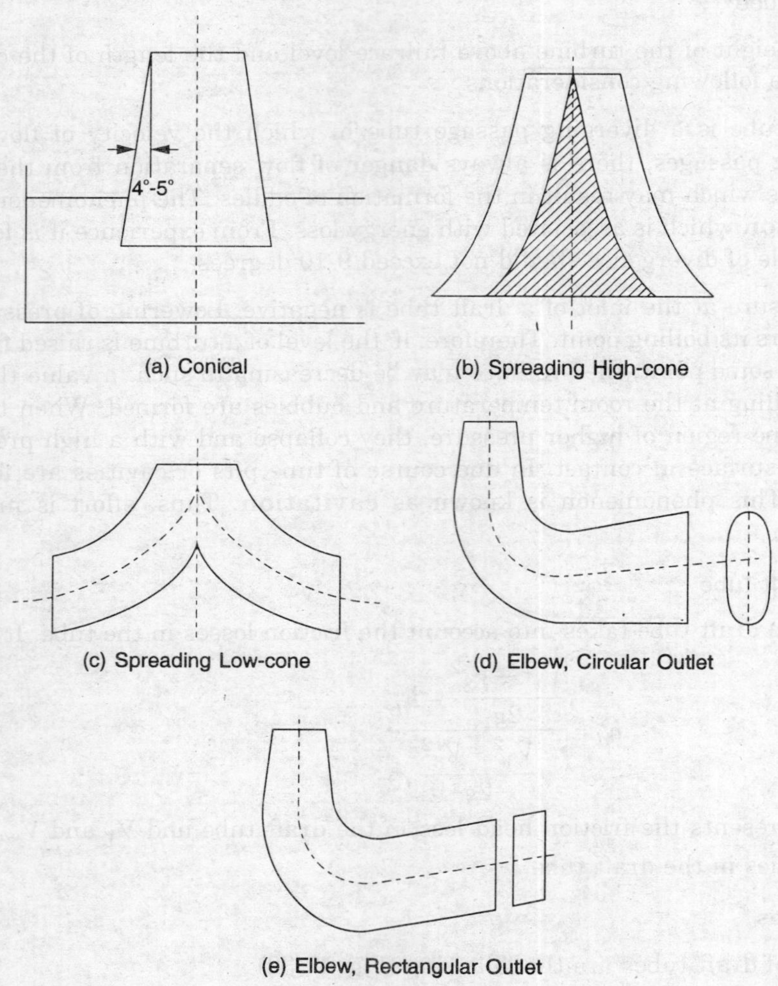

Fig. 17.32

When the turbine runs at part loads, the velocity at the outlet no longer remains radial and there is a component of whirl that may produce eddies in the flow. A spreading draft tube helps in reducing the whirl component. The efficiency of the tube is around 88 percent.

Elbow Tube

If a bend is given to the tube, sufficient increase in area can be made in lesser depth. It is useful where it is not possible to immerse the tube into the tailrace sufficiently below the level. It also provides the outgoing flow a direction consistent with the flow in the tailrace. The outlet of the draft tube may be made rectangular or circular as shown in Figs. 17.32d and e. The efficiency of such tubes is around 60 percent.

Example 17.15 A reaction turbine pruduces 500 kW under a head of 6 m. The discharge through the turbine is 10 m^3/s. Determine the increase in power and efficiency, if its draft tube of cylindrical cross-section of diameter 1·8 m is replaced by a tapered one having an outlet diameter of 2 m. The efficiency of the draft tube is 92 percent.

Solution $Q = 10\ m^3/s$ $\quad H = 6$ m

$D_2 = 1{\cdot}8$ m $\quad D_d = 2$ m

$P = 500$ kW $\quad \eta_d = 0{\cdot}92$

Velocity at the inlet of draft tube,

$$V_2 = \frac{10}{\frac{\pi}{4}(1{\cdot}8)^2} = 3{\cdot}93 \text{ m/s}$$

Velocity at the outlet of draft tube,

$$V = \frac{10}{\frac{\pi}{4}(2)^2} = 3{\cdot}18 \text{ m/s}$$

$$\text{Head gained} = \left(\frac{V_2^{\,2} - V^2}{2g}\right) \times \eta_d = \left(\frac{3{\cdot}93^2 - 3{\cdot}18^2}{2 \times 9{\cdot}81}\right) \times 0{\cdot}92 = 0{\cdot}25 \text{ m}$$

$$\text{Increase in Power} = \frac{\text{Inchrease in head}}{\text{Original head}} \times P$$

$$= \frac{0{\cdot}25}{6} \times 500 = 20{\cdot}34 \text{ kW}$$

Original efficiency,

$$\eta = \frac{P}{wQH} = \frac{500}{9{\cdot}81 \times 10 \times 6} = 0{\cdot}85 \quad \text{or} \quad 85\%$$

$$\text{Gain in efficiency} = \text{original efficiency} \times \frac{\text{Increase in power}}{\text{Original power}}$$

$$= 0{\cdot}85 \times \frac{20{\cdot}84}{500} = 0{\cdot}0354 \quad \text{or} \quad 3{\cdot}54\%$$

17.11 OUTWARD VS. INWARD-FLOW REACTION TURBINES

Outward-flow and inward-flow reaction turbines are compared as follows :

S. No.	Outward-flow Turbines	Inward-flow Turbines
1.	The centrifugal head impressed on water due to the rotation of runner is in the direction of flow. If due to a decrease in the load, there is a slight increase in the speed, the centrifugal head increases. This tends to increase the relative velocity of water. Consequently, the quantity of water passing through the runner tends to increase and the runner tends to race.	The centrifugal head acting on the water is in the opposite direction of flow of water. An increase in the runner speed due to decrease in the load will tend to reduce the quantity of flow of water through the runner. This will tend to reduce the power. This is advantageous, as it tends to prevent racing. Thus, an inward-flow turbine becomes self-regulating to some extent.
2.	The flow is divergent. There is likelihood of eddy formation and flow separation resulting in loss of energy.	The flow is convergent. There is no chance of eddy formation and flow separation. No energy loss.
3.	Guide vanes are not easily accessible and thus regulation of flow is not easy.	The guide vanes are more accessible as they are on the outer periphery of the runner. It is possible to use improved methods of regulating the flow into the runner.
4.	Water with its maximum velocity admitted at the most slowly moving part of the wheel and leaves it at points where its tangential velocity is maximum. It is not convenient to adopt the vane angles designed to give the entry without shock.	Water with its maximum velocity is admitted around the circumference of the wheel where the tangential velocity is maximum and leaves where it is minimum. It is more convenient to adopt the vane angles to made the water enter the runner without shock. Outlet velocity is also easily reduced to give low kinetic energy loss.

17.12 DERIAZ TURBINE

A fixed runner vane turbine is designed for particlar values of head, speed, discharge and power output. The runner vanes are fixed angle blades and have specific values of inlet and outlet angles corresponding to the maximum efficiency. Guide vanes give proper direction to the water to enter into the runner without shock. At part loads, the flow is reduced by partially shutting off the passage of the guide vanes by the governor through a servomechanism. Since the orientation of the guide vanes is changed to reduce the inlet area, the angle of the absolute velocity also changes, which in turn varies the angle of the relative velocity. As the runner vanes remain at the same fixed angle, the water does not enter the runner smoothly and there are shock losses at the entrance. This loss is considerable as the load varies from the designed value. It can be reduced to a large extent if the runner vanes are also rotated according to the guide vanes to make a smooth entry of water. Then, the turbine will be able to give best efficiency over a wide range of load.

In axial flow turbines, there already exists a turbine known as Kaplan turbine in which runner vanes are able to rotate on their axes fixed to the hub. The movement of the vanes is controlled by a mechanism inside the runner shaft. However, Kaplan turbines are efficient only at low heads which can be upto 60 m. For higher heads, mixed flow turbines like Francis turbines which give low efficiency at part loads have to be used,.

Sometime back, P. Deriaz of English Electric Co., Rugby, London, invented a Francis turbine with feathering vanes to utilise higher heads effectively. It is known as **Deriaz turbine** after the name of its inventor. The vanes of a Deriaz turbine are pivoted to the hub like that of a Kaplan turbine, but are inclined to the axis of the shaft at an angle of 45°. Guide vanes are provided before the runner vanes to regulate and direct the flow. The flow through the runner is similar to a Francis turbine. The casing of a Deriaz turbine is of such shape that there is only a small clearance between the runner vanes and the casing to mininise the loss due to leakage. Thus a Deriaz turbine has high efficiency at part and overload conditions.

The other advantages of a Deriaz turbine are

- It can be used as a reversible hydraulic machine very conveniently *i.e.* it can be used as a pump turbine unit without any loss of energy.
- The runner vanes are designed to fit each other closely and form a smooth cone when closed (Fig. 17.33). A Francis turbine needs high initial torque to run it as a pump as the water inside the runner must be rotated. However, a Deriaz turbine does not require high initial torque for starting since the closed runner can be rotated in the water with a small resistance. When the speed increases, the runner vanes are opened completely. The use of guide vanes becomes optional for this pump.

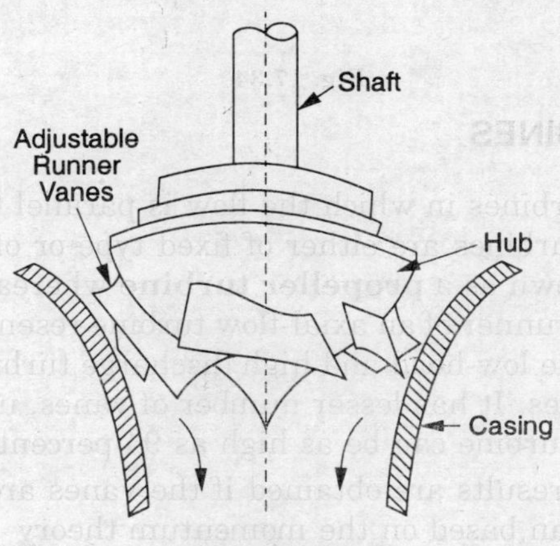

Fig. 17.33

- Since the vanes of a Deriaz turbine are inclined, the load on the outer trunnion journal bearing is reduced.
- In an axial flow turbine, airfoil section is mostly used for the designing of vanes. But the vanes of a Deriaz turbine are simple and can be designed and manufactured easily. Adjusting the runner vanes regulates flow through the machine, whether it is being used as a turbine or a pump. Even overloading is possible when it is being used as a pump.

In Canada, six high capacity units of Deriaz reversible hydraulic machines are installed at the Niagara pumped storage plant. Each produces 40000 kW at a speed of 92·5 rpm. When it is used as a pump, it discharges 113 m^3/s with a delivery head 26 m. The diameter of the runner is 6·4 m. The runner is fitted with cast steel vanes, the outer edge of which is coated with stainless steel. The runner of the turbine is designed by English Electric Company, London and is manufactured by the firm John Inglis, Toronto, Canada (Fig. 17.34).

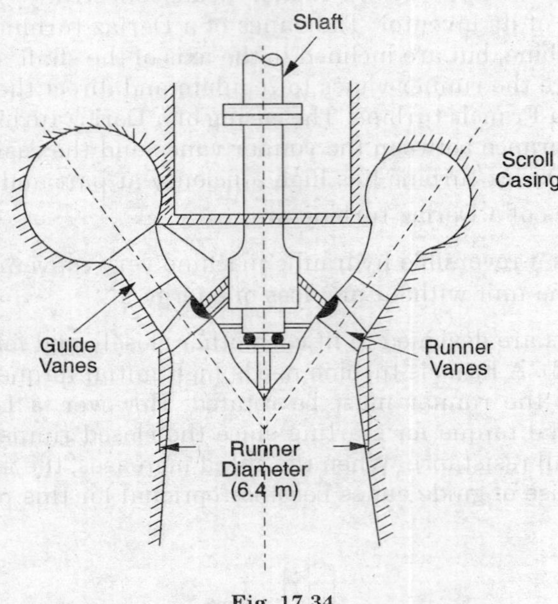

Fig. 17.34

17.13 AXIAL-FLOW TURBINES

Axial-flow turbines are the turbines in which the flow is parallel to the axis of the shaft. The vanes of the runner of such turbines are either of fixed type or of adjustable type. A turbine with fixed type of vanes is known as a **propeller turbine** whereas a turbine with adjustable type is **Kaplan turbine.** The runner of an axial-flow turbine resembles the propeller of a ship. The axial- flow turbines are the low-head and high discharge turbines. In general, heads upto 30 *m* require axial-flow turbines. It has lesser number of vanes, usually 3 to 4 or at the most 6. The efficiency of a Kaplan turbine can be as high as 94 percent.

It is observed that better results are obtained if the vanes are designed according to the aerodynamic theory rather than based on the momentum theory.

17.14 KAPLAN TURBINES

1. Components

The components of a Kaplan or propeller turbine are similar to that of a Francis turbine. Guide vanes or wicket gates impart a tangential and a radial-inward velocity to the liquid. The runner vanes are adjustable just like the guide vanes of a Francis turbine (Fig. 17.35). A space,

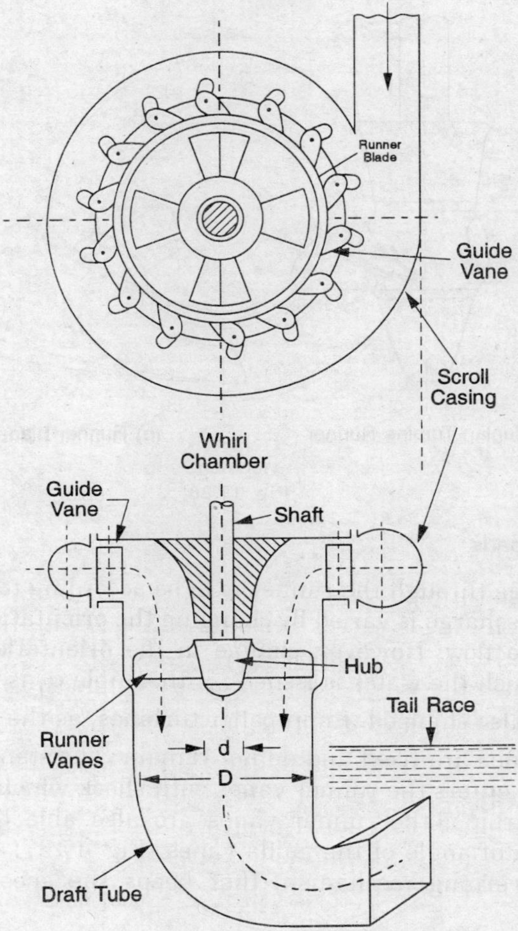

Fig. 17.35

called whirl chamber, is provided between the guide vanes and the runner. In this chamber, the flow is turned through 90 degrees and moves as free vortex, *i.e.* without the aid of any external torque. The radial component is changed into axial component due to the guidance from the fixed housing. The tangential component of velocity increases as the radius is reduced. Angular momentum remains constant.

The discharge through a Kaplan turbine is high, as the head is low. The number of vanes is also small so that the resistance of vanes to the flow is the least. The vanes are of relatively flat shape with little curvature (Fig. 17.36). The vanes are fixed to a hub or boss mounted on a shaft. In propeller turbines, the vanes are fixed. They are fitted at such an angle that at the designed value of flow, water enters the runner tangential to the leading edge of the vanes. However, in Kaplan turbine the vanes are able to move on their axes.

The material of the runner vanes is **stainless steel**. They are directly attached to the hub and there is no bend like that in a Francis turbine. This reduces the frictional losses.

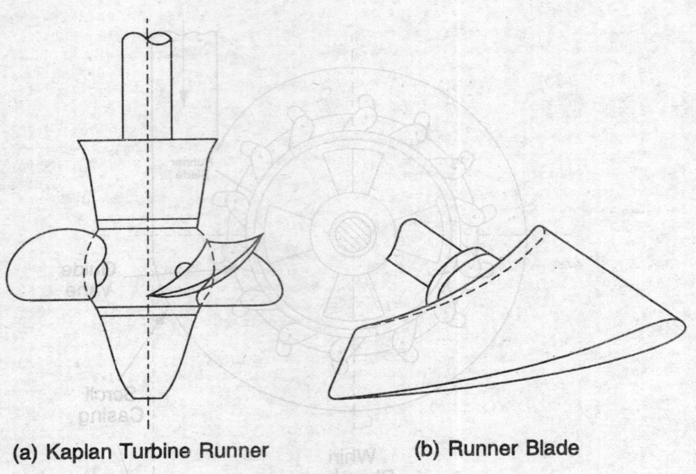

Fig. 17.36

2. Performance At Part Loads

In turbines, the discharge through the runner is to be according to the load on the turbine. In reaction turbines, the discharge is varied by changing the orientation of the guide vanes which changes the area of the flow. However, change in the orientation of the guide vanes also changes the angle at which the water is issued *i.e.* the angle α_1 is changed. With this change, the relative angle β_1 is also changed. In propeller turbines, as the runner vanes are fixed, the angles of the runner vanes are fixed and do not remain consistent with the angle of relative velocity. Thus the water enters the runner vanes with shock which means the losses are high. However, in Kaplan turbine, the runner vanes are also able to change their orientation according to the change of angle of the guide vanes (Fig. 17.37). This is done automatically with the help of a governing mechanism that keeps the speed of the runner constant.

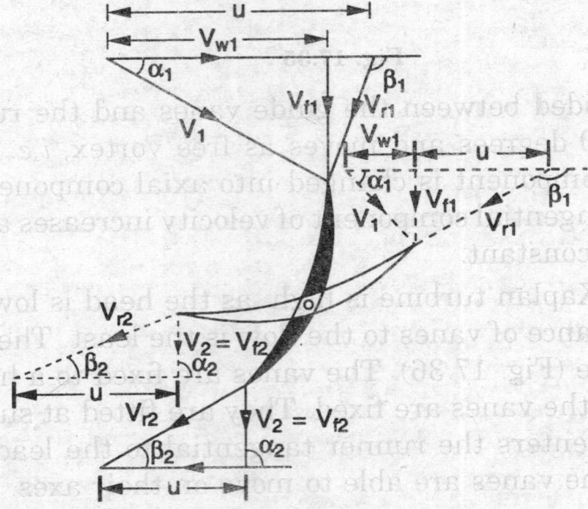

Fig. 17.37

Therefore, the Kaplan turbines are able to give high efficiency over a wide range of load by checking the eddy losses at the inlet.

The governing mechanism consists of a servomotor mechanism which operates inside the hollow coupling of the generator and the turbine shafts (Fig. 17.38). The servomotor mechanism consists of a piston operating inside a cylinder and is under oil pressure. The supply of the pressurised oil is regulated by the governing mechanism.

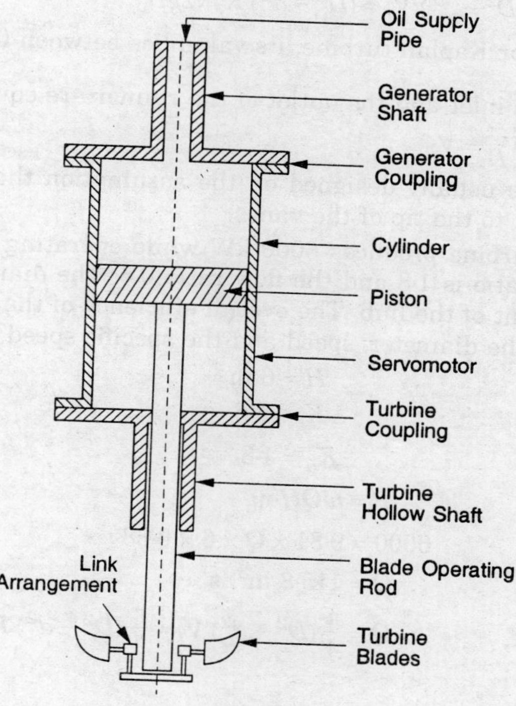

Fig. 17.38

3. Proportions

Ratio of hub diameter to the outer diameter of the runner depends upon the specific speed of the runner. As the specific speed increases, this ratio decreases. The following table indicates the variation of hub diameter to outer diameter, head and the number of vanes with the change in the specific speed of a Kaplan turbine.

N_s	250	325	500	650	725	850
d/D	0·7	0·6	0·5	0·42	0·37	0·3
H (m)	70	55	40	25	15	5
z	10	7	5	4–5	4	3

- The peripheral velocity u of the runner varies with the radius. Thus in axial-flow runners, the tangential velocity varies from section to section at the inlet and the outlet. Depending upon this variation, the inlet and the outlet angles of the vanes are different at different sections. The angles are greater at the tips and smaller at the hub. This gives a twisted shape to the vanes.
- The discharge through the runner is given by

$$Q = (D^2 - d^2)\, V_f = (D^2 - d^2)\, K_f \sqrt{2gH} \quad \text{...(17.19)}$$

K_f is the flow ratio. For Kaplan turbine, its value lies between 0·35 to 0·7.

- Since the areas at the inlet and the outlet of the runner are equal

$$V_f = V_{f1} = V_{f2}$$

The runner vanes are usually designed on the assumption that the flow velocity is uniform from the root to the tip of the vanes.

Example 17.16 A Kaplan turbine produces 6000 kW while operating under a head of 6 m. The speed ratio is 1·8 and the flow ratio 0·6. The diameter of the runner is 2·5 times that of the hub. The overall efficiency of the turbine is 88 percent. Determine the diameter, speed and the specific speed of the runner.

Solution $P = 6000$ kW $\qquad H = 6$ m

$D = 2{\cdot}5\, d \qquad K_f = 0{\cdot}6$

$\eta_0 = 0{\cdot}88 \qquad K_u = 1{\cdot}8$

$$P = wQH{\cdot}\eta_0$$

$$6000 = 9{\cdot}81 \times Q \times 6 \times 0{\cdot}88$$

$$Q = 115{\cdot}8 \text{ m}^3/\text{s}$$

$$Q = \frac{\pi}{4}(D^2 - d^2)\, V_f = \frac{\pi}{4}(D^2 - d^2)\, K_f \cdot \sqrt{2gH}$$

$$115{\cdot}8 = \frac{\pi}{4}[(2{\cdot}5d)^2 - d^2] \times 0{\cdot}6 \times \sqrt{2 \times 9{\cdot}81 \times 6}$$

$$= \frac{\pi d^2}{4}(6{\cdot}25 - 1) \times 6{\cdot}51$$

$$d = 2{\cdot}077 \text{ m}$$

$$D = 2{\cdot}5 \times 2{\cdot}077 = 5{\cdot}193 \text{ m}$$

$$u = \frac{\pi D N}{60} = K_u \sqrt{2gH}$$

$$\frac{\pi \times 5{\cdot}193 \times N}{60} = 1{\cdot}8 \times \sqrt{2 \times 9{\cdot}81 \times 6}$$

$$N = 71{\cdot}8 \text{ rpm}$$

$$N_s = \frac{N\sqrt{P}}{H^{5/4}} = \frac{71{\cdot}8 \times \sqrt{6000}}{(6)^{5/4}} = 592{\cdot}3$$

Example 17.17 The diameter of the hub of a Kaplan turbine is one third that of its runner. The speed of the runner is 120 rpm and the head on the turbine is 12 m. The vane angle at the outlet of the extreme edge of the runner is 15°. Assuming the discharge at the outlet as radial and the flow ratio 0·65, determine the diameters of the hub and the runner. Also, find the discharge through the runner.

Solution $N = 120$ rpm $\quad H = 12$ m

$d = 3\,D \quad K_f = 0{\cdot}65$

$\beta_2 = 15^\circ \quad u_1 = u_2 = u$

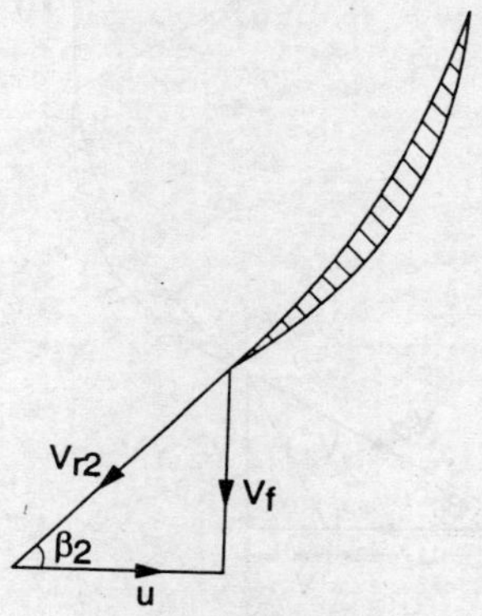

Fig. 17.39

$$V_f = K_f\sqrt{2gH} = 0{\cdot}65\sqrt{2\times 9{\cdot}81\times 12} = 9{\cdot}974 \text{ m/s}$$

$$\tan\beta_2 = \frac{V_f}{u}$$

$$\tan 15^\circ = \frac{9{\cdot}974}{u}$$

$$u = 37{\cdot}22 \text{ m/s}$$

$$\frac{\pi D\times 120}{60} = 37{\cdot}22$$

$$D = 5{\cdot}92 \text{ m}$$

$$d = \frac{D}{3} = \frac{5{\cdot}92}{3} = 1{\cdot}97 \text{ m}$$

$$Q = \frac{\pi}{4}(D^2 - d^2)\,V_f = \frac{\pi}{4}(5{\cdot}92^2 - 1{\cdot}97^2)\times 9{\cdot}974 = 244{\cdot}1 \text{ m/s}$$

Example 17.18 The inlet and the outlet runner blade angles of a propeller turbine are 90 and 25° respectively to the tangential direction of the runner. The inlet guide vane angle is 30°. The speed of the turbine is 30 rpm. The mean diameter of the runner blades is 3·6 m and the area of flow 30 m^2. Assuming that the velocity of flow is constant throughout, find

(i) Discharge

(ii) Power developed
(iii) Hydraulic efficiency
(iv) Specific speed

Solution $\alpha_1 = 30°$ $D = 3{\cdot}6$ m
$\beta_1 = 90°$ $A = 30$ m^2
$\beta_2 = 25°$ $N = 30$ rpm

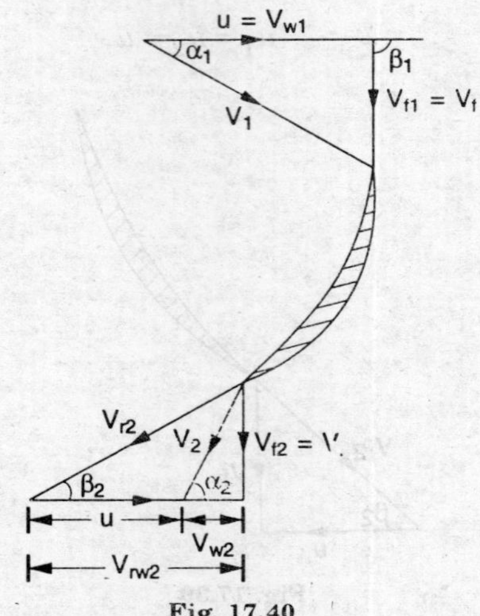

Fig. 17.40

$$u = \frac{\pi DN}{60} = \frac{\pi \times 3{\cdot}6 \times 30}{60} = 5{\cdot}655 \text{ m/s}$$

(i) From the inlet velocity triangle,

$$V_{w1} = u = 5{\cdot}655 \text{ m/s}$$

$$V_f = V_{w1} \cdot \tan\alpha_1 = 5{\cdot}655 \times \tan 30° = 3{\cdot}265 \text{ m/s}$$

$$Q = AV_f = 30 \times 3{\cdot}265 = 97{\cdot}95 \text{ m}^3/\text{s}$$

(ii) $P = \rho Q\ (V_{w1} \cdot u + V_{w2} \cdot u) = \rho Q\ (V_{w1} + V_{w2}) \cdot u$

From outlet velocity triangle,

$$\tan \beta_2 = \frac{V_f}{V_{w2} + u}$$

$$\tan 25° = \frac{3{\cdot}265}{V_{w2} + 5{\cdot}655}$$

$$V_{w2} = 1{\cdot}347 \text{ m/s}$$

$\therefore P = 1000 \times 97{\cdot}95 \times (5{\cdot}655 + 1{\cdot}347)\ 5{\cdot}655 = (3878 \times 10^3)$ W $= 3878$ kW

(iii) $\eta_h = \dfrac{V_{w1} \cdot u + V_{w2} \cdot u}{gH}$ or $\dfrac{P}{wQH}$

But $V_{w1} \cdot u + V_{w2} \cdot u = gH - \frac{1}{2} V_2^2$

Where $V_2 = \sqrt{V_f^2 + V_{w2}^2} = \sqrt{3 \cdot 265^2 + 1 \cdot 347^2} = 3 \cdot 532$ m/s

$\therefore\ 5 \cdot 655 \times 5 \cdot 655 + 1 \cdot 347 \times 5 \cdot 655 = 9 \cdot 81 \times H - \frac{1}{2} V_2^2$

$\therefore\ 39 \cdot 596 = 9 \cdot 81 \times H - 6 \cdot 238$

$$H = 4 \cdot 67 \text{ m}$$

$$\eta_h = \frac{39 \cdot 596}{9 \cdot 81 \times 4 \cdot 67} \quad \text{or} \quad \frac{3878}{9 \cdot 81 \times 97 \cdot 95 \times 4 \cdot 67}$$

$$= 0 \cdot 864 \quad \text{or} \quad 86 \cdot 4\%$$

(iv) $N_s = \frac{N\sqrt{P}}{H^{5/4}} = \frac{30\sqrt{3878}}{(4 \cdot 67)^{5/4}} = 272 \cdot 1$

Example 17.19 The discharge through the runner of a propeller turbine is 100 m³/s and is radial at the outlet. Guide vanes are set at 30°. The diameters of the runner and the hub are 5·5 m and 2·5 m respectively. The runner speed is 100 rpm. The flow velocity is constant throughout the runner. Determine

(i) the blade angles at the tip, the midpoint and the root of the runner

(ii) power produced assuming the midpoint velocities as the average velocities.

Solution $N = 100$ rpm $\qquad D_t = 5 \cdot 5$ m

$Q = 100$ m³/s $\qquad D_r = 2 \cdot 5$ m

$\alpha_1 = 30°$

$$D_m = 2 \cdot 5 + \frac{5 \cdot 5 - 2 \cdot 5}{2} = 4 \text{ m}$$

$$u_t = \frac{\pi D_t N}{60} = \frac{\pi \times 5 \cdot 5 \times 100}{60} = 28 \cdot 8 \text{ m/s}$$

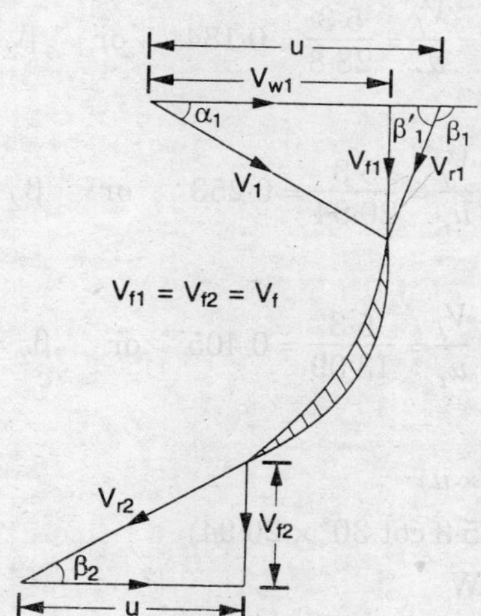

Fig. 17.41

$$u_m = \frac{\pi D_m N}{60} = \frac{\pi \times 4 \times 100}{60} = 20{\cdot}94 \text{ m/s}$$

$$u_r = \frac{\pi D_r N}{60} = \frac{\pi \times 2{\cdot}5 \times 100}{60} = 13{\cdot}09 \text{ m/s}$$

$$Q = \frac{\pi}{4}(D_t^2 - D_r^2)\, V_f$$

$$100 = \frac{\pi}{4}(5{\cdot}5^2 - 2{\cdot}5^2)\, V_f$$

$$V_f = 5{\cdot}3 \text{ m/s}$$

(i) Inlet blade angles,

At the tip :

$$\tan \beta_1' = \frac{V_f}{u_1 - V_{w1}} = \frac{V_f}{u_1 - V_f \cot \alpha_1} = \frac{5{\cdot}3}{28{\cdot}8 - 5{\cdot}3 \cot 30^\circ} = \frac{5{\cdot}3}{28{\cdot}8 - 9{\cdot}18} = 0{\cdot}27$$

$$\beta_1' = 15{\cdot}1^\circ \quad \text{or} \quad \beta_1 = 180^\circ - 15{\cdot}1^\circ = 164{\cdot}9^\circ$$

At the midpoint :

$$\tan \beta_1' = \frac{V_f}{u_1 - V_{w1}} = \frac{5{\cdot}3}{20{\cdot}94 - 9{\cdot}18} = 9{\cdot}451$$

$$\beta_1' = 24{\cdot}3^\circ \quad \text{or} \quad \beta_1 = 180^\circ - 24{\cdot}3^\circ = 155{\cdot}7^\circ$$

At the root :

$$\tan \beta_1' = \frac{V_f}{u_1 - V_{w1}} = \frac{5{\cdot}3}{13{\cdot}09 - 9{\cdot}18} = 1{\cdot}355$$

$$\beta_1' = 53{\cdot}6^\circ \quad \text{or} \quad \beta_1 = 180^\circ - 53{\cdot}6^\circ = 126{\cdot}4^\circ$$

Outlet blade angles,

At the tip :

$$\tan \beta_2 = \frac{V_f}{u_t} = \frac{5{\cdot}3}{28{\cdot}8} = 0{\cdot}184 \quad \text{or} \quad \beta_2 = 10{\cdot}4^\circ$$

At the midpoint :

$$\tan \beta_2 = \frac{V_f}{u_t} = \frac{5{\cdot}3}{20{\cdot}94} = 0{\cdot}253 \quad \text{or} \quad \beta_2 = 14{\cdot}2^\circ$$

At the root :

$$\tan \beta_2 = \frac{V_f}{u_t} = \frac{5{\cdot}3}{13{\cdot}09} = 0{\cdot}405 \quad \text{or} \quad \beta_2 = 22^\circ$$

(ii) $P = \rho Q \cdot (V_{w1} \cdot u)$

$= \rho Q \cdot (V_f \cot 30 \times u)$

$= 1000 \times 100 \times (5{\cdot}3 \cot 30^\circ \times 20{\cdot}94)$

$= (19223 \times 10^3)$ W

$= 19223$ kW

17.15 GOVERNING OF TURBINES

Hydraulic turbines are used to generate electric power by coupling them with the electric alternators. To maintain the frequency of the electric supply constant, the alternators are required to rotate at synchronous speed. This means that the runners of the turbines must also rotate at constant speeds. The speed of a turbine is affected by the flow of water as well as the load on the turbine. If any of these is changed, the other has to be changed to keep the speed constant. Thus if the load on the turbine decreases and no change is made in the flow, the turbine tends to gain in speed and if the load increases, it tends to decrease. The regulation of speed under varying loads is called **governing.** The arrangement by which the speed of a turbine is kept constant under varying load conditions is called governing mechanism or governing system.

1. Governing of Impulse Turbines

Figure 17.42 shows the governing mechanism for an impulse turbine. It consists of a centrifugal governor having flyballs that respond to the speed quickly. The movement of the sleeve of the governor is taken by the main lever and is transmitted to a relay or control valve. The control valve has cylinder in which a double piston moves. It receives supply of pressurised oil in the centre portion b from an oil sump with the help of a rotary gear pump. The pressure of the oil sent to the relay valve is usually between 1200 kN/m^2 and 2000 kN/m^2. There are also pipes p_2 and p_3 to take the extra oil from the relay valve back into the oil sump. The supply of oil received by the relay valve is further sent to the servomotor through pipe p_4 or p_5. Thus, relay valve is a sort of distributing valve which supplies the oil to the servomotor on the required side.

The servomotor consists of a cylinder inside which a piston moves due to oil pressure. It receives the oil on one side of the piston and the oil from the other side goes back to the relay valve and finally to the oil sump. Motion of the piston moves the spear rod that controls the discharge through the nozzle.

When the load on the turbine decreases, the speed of the runner tends to increase. This makes the flyballs of the governor to rotate at higher speed and the sleeve of the governor moves up. The motion of the sleeve and of the main lever lowers the piston in the relay valve. This opens the passage of pipe P_5 to the portion b. Thus oil flows to the left-hand side of the servomotor. The excess oil from the right-hand side flows back to the oil sump through pipe P_4 and pipe P_2. As the piston of the servomotor moves forward in the nozzle, it reduces the area of flow thus reducing the discharge through the nozzle. The reduced discharge reduces the speed of the turbine and the radius of flyballs also reduces. This makes the relay valve to move up and the servomotor starts receiving the oil supply on the right hand side. The excess oil from the left-hand side goes back to sump via pipes P_5 and P_3. The backward motion of the spear increases the area of flow and the discharge increases again increasing the speed. After some time the system stabilises itself according to the required speed. In a similar way the system would respond when the load on the turbine increases.

Deflector arrangement : If the load on the turbine decreases suddenly, there may be a sudden increase in the speed of the turbine while the spear regulation system is still responding. The high speed of the runner sometimes may be dangerous. To obviate such a

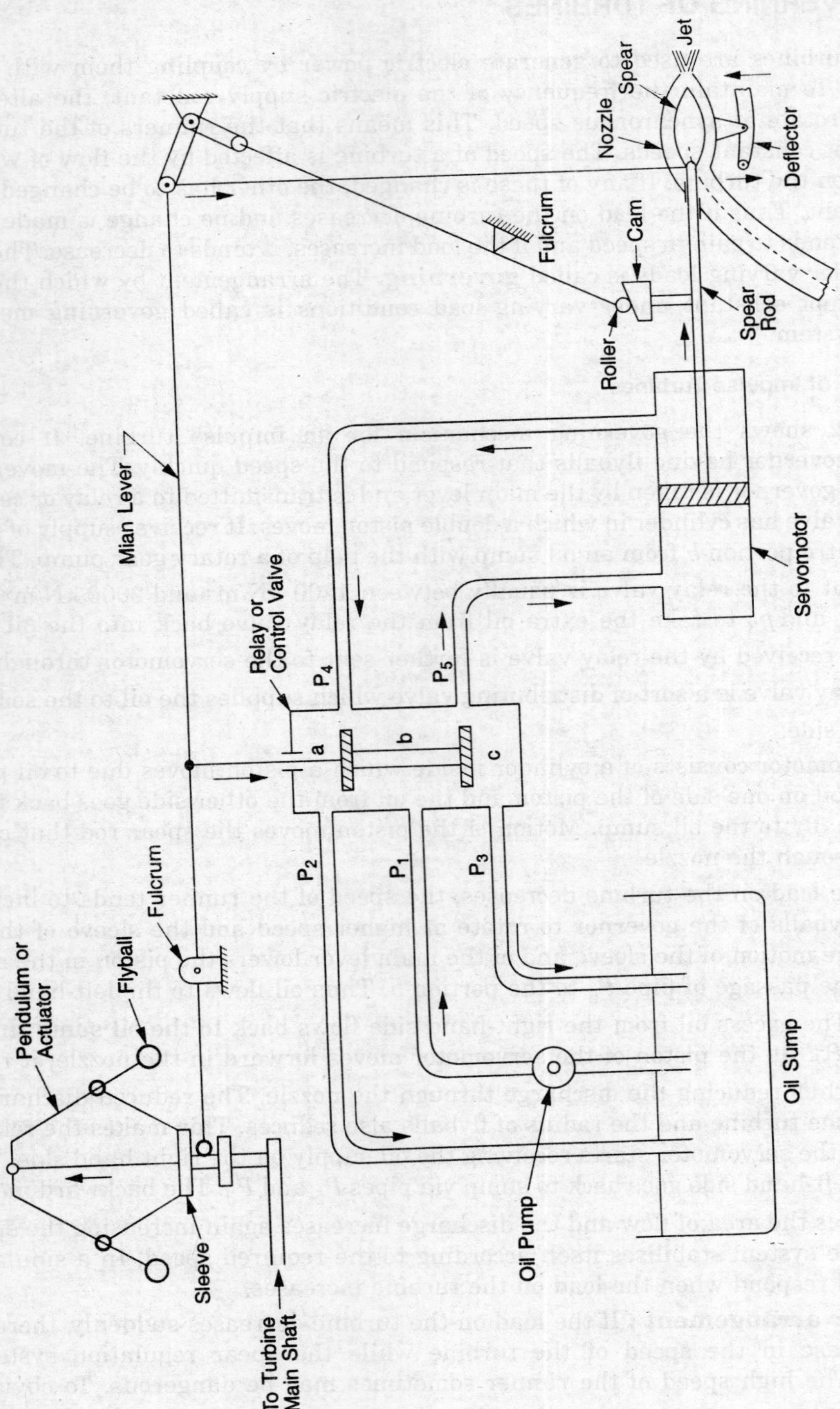
Jet
Spear
Nozzle
Deflector
Fulcrum
Cam
Roller
Spear Rod
Mian Lever
Relay or Control Valve
P_4
P_5
a
b
c
P_2
P_1
P_3
Servomotor
Oil Sump
Oil Pump
Fulcrum
Flyball
Pendulum or Actuator
Sleeve
To Turbine Main Shaft

Fig. 17.42

situation, a deflector arrangement may be used. When the load on the turbine decreases, the deflector immediately comes between the jet of water and the buckets, thus deflecting away a part of the jet. As the speed decreases, the deflector moves backwards. A bell-crank lever, the motion of which is controlled by the main lever controls the movement of the deflector. As the sleeve of the governor moves upwards due to increase in speed, the bell-crank lever moves counter-clockwise and the pin of the deflector is lowered. By its motion about the pivot, the deflector comes in between the jet and the buckets.

Approximate time taken for the whole process is 60-70 seconds.

These days, an electrical mechanism is preferred to a flyball governor. The electrical mechanism responds quickly to any variation of frequency that results from the change in speed. The electrical impulses received from the system are fed to the hydraulic circuit.

2. Governing of Reaction Turbines

In reaction turbines the area of flow is changed by changing the orientation of the guide vanes instead of moving a spear in the nozzle. The movement of the piston of the servomotor moves a regulating lever fixed to a shaft as shown in Fig. 17.43. Two regulating rods are attached to the regulating lever which in turn are attached to a regulating ring. The regulating ring consists of number of guide vanes through levers and links, which are able to move around their axes, when the ring makes some angular movement.

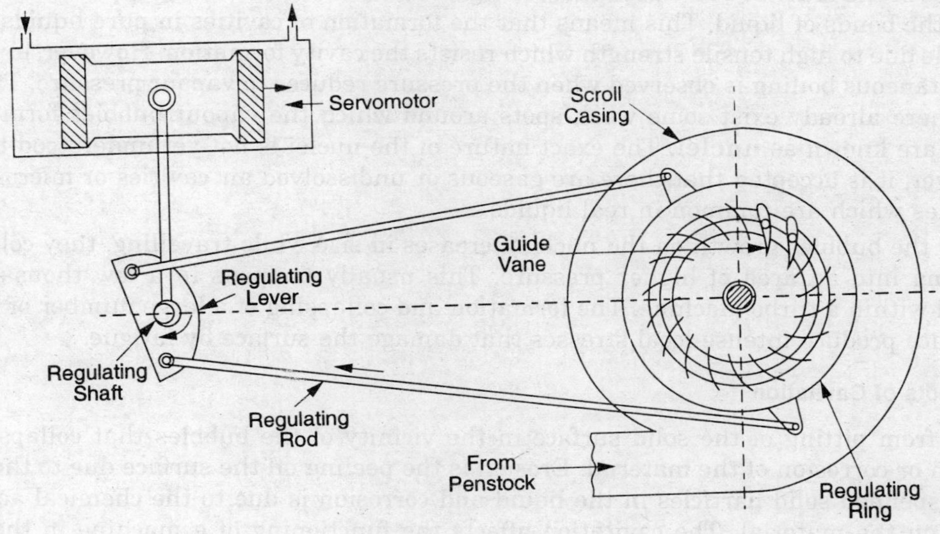

Fig. 17.43

When the load decreases and the speed increases, the piston in the servomotor moves right, thus rotating the regulating lever clockwise. This motion further rotates the regulating ring through the regulating rods. The rotation of the ring makes the guide vanes to rotate about their pivots which reduces the area of flow. Similarly, a left hand movement of the piston in the servomotor increases the area of flow and thus the discharge.

Double regulation of reaction turbines : In case the guide vanes are to be closed suddenly, a relief valve or a pressure regulator may be used along with the governing mechanism. The relief valve opens and let the water pass from the penstock direct to the tialrace and the turbine is saved from racing. When the flow is adjusted to the new value, the relief valve closes. The operation of the relief valve is also controlled by the governor.

17.16 CAVITATION IN TURBINES

While passing through hydraulic turbines or other passages, water comes in a region where its pressure falls below its vapour pressure at the prevailing temperature. At this point, it starts boiling and vapour bubbles are formed. This can happen at the outlet of a turbine, the inlet of a pump, the bend of a pipe or the convex surface of curved vanes. The vapour bubbles travel along with the liquid and on reaching in a region of higher pressure, suddenly collapse creating a vacuum in the place. The liquid in the vicinity of vapour bubbles rushes into the vacuum or the cavities creating very huge local pressures. If this happens near a solid surface, the forces exerted by the rushing liquid cause pitting of the surface. This phenomenon is known as **cavitation**. Thus cavitation is a phenomenon of the formation, the growth and the collapsing of the small vapour bubbles in a liquid flow. These days, the basic design of hydraulic machines is extremely influenced by cavitation.

Theoretically, pure liquids have high tensile strengths and can withstand high tensile stresses of the order of 100 – 1100 MN/m^2. Formation of cavities in the liquid involves tearing apart the bonds of liquid. This means that the formation of cavities in pure liquids is just not possible due to high tensile strength which resists the cavity formation. However, in real fluids, a spontaneous boiling is observed when the pressure reduces to vapour pressure. This implies that there already exist some weak spots around which the vapour bubbles form and grow. These are known as **nuclei.** The exact nature of the nuclei is not yet understood thoroughly. However, it is accepted that there are gaseous or undissolved air cavities or microscopic dust particles which are common in real liquids.

As the bubbles growth on the nuclei increases in size while travelling, they collapse after reaching into an area of higher pressure. This usually happens in a few thousandths of a second within a turbo machine. The formation and collapsing of a large number of bubbles on a surface produce intense local stresses that damage the surface by fatigue.

1. Effects of Cavitation

Apart from pitting of the solid surface in the vicinity of the bubbles that collapse, there is erosion or corrosion of the material. Erosion is the peeling off the surface due to the impact of the suspended solid particles in the liquid and corrosion is due to the chemical action of the liquid on the material. The cavitation affects the functioning of a machine in the following ways :

- **Change in flow pattern :** When the surface is pitted by cavitation, the effective size of the passage changes due to which the flow directions change. The pitted surface also increases the turbulence in the flow that increases the number of cavitation bubbles. These bubbles may choke the passage and may cause breakdown of the whole system in due course of time.

- **Noise production :** The noise produced by collapsing of the vapour bubbles and striking of liquid at the solid surface is usually so loud that it may be heard hundred meters away. In some cases such as water missiles or submarines it becomes a serious problem, as the dissipation of noise is not easy.
- **Vibration production :** The runner of a turbo machine does not remain balanced and vibrations are produced. Generally, these vibrations are so powerful that they can be felt hundreds of metres away. The collapsing of the cavitation bubbles also produce shock waves. In due course of time, a great damage may be done to the dam.
- **Energy loss :** In the cavities formed, auxiliary motions are set up which become a source of energy loss.
- **Efficiency loss :** Sharp decrease in efficiency is observed due to alteration in flow pattern and increased resistance to the flow. As a result, the power produced by a turbine is reduced.
- **Increase in drag force :** In axial flow turbines based on airfoil theory, the drag force is increased by cavitation.
- **Emission of light :** Some liquids like alcohol and carbolic acid emit light during cavitation.
- **Material loss :** If the porcess of cavitation continues, after some period of time, the pitted vanes or the casing have to be changed. If this happens frequently, there is loss of material and money.

2. Cavitation Number

It is also known as Thoma's cavitation number or factor, after the name of its inventor *D.* Thoma of Germany. It is defined as

$$\sigma = \frac{H_a - h - H_v}{H} \qquad \ldots(17.20)$$

where H = net available head

H_a = atmospheric pressure head

H_v = vapour pressure head of the liquid

h = height of turbine above the tailrace level

The cavitation number is useful in proper selection of turbines and is helpful in installing the turbines with respect to the tailrace level. Experimentally, it is observed that the Thoma's cavitation factor is a function of specific speed of the turbine. It has been shown that for a particular value of specific speed, cavitation may be reduced upto a certain value without any decrease in the efficiency. However, if the value of cavitation factor is further reduced, a sharp decline in the efficiency is noted. This value of the cavitation factor is known as **critical cavitation factor** σ_c.

Empirical relations for the critical cavitation factors are

$$\text{For Francis turbine} = 0{\cdot}625\left(\frac{N_s}{445}\right)^2 \qquad \ldots(17.21)$$

$$\text{For propeller turbine} = 0{\cdot}28 + 0{\cdot}14\left(\frac{N_s}{445}\right)^{2.75} \quad \ldots(17.22)$$

For Kaplan turbines, the values are increased by 10 percent than that for propeller turbines. The following tables show the values of the critical cavitation factor with change in the specific speed :

Francis turbine

N_s	50	100	150	200	250
σ_c	0·024	0·060	0·082	0·12	0·17

Kaplan turbine

N_s	350	400	500	600	700
σ_c	0·3	0·42	0·72	1·05	1·59

3. Detection of Cavitation

The following methods may be adopted to detect the cavitation :

- By photographing the interior of water turbine, it is easily detected.
- By measuring the noise and the vibration levels that accompany the phenomenon of cavitation.
- A widely used method is by studying the hydraulic performance of turbines. When the suction head is gradually increases by increasing the length of the draft tube at constant operating conditions, the value of cavitation number continuously falls. At some particular value, a sudden drop in efficiency is observed. This value of cavitation number σ corresponds to critical value of the cavitation number (Fig. 17.44).

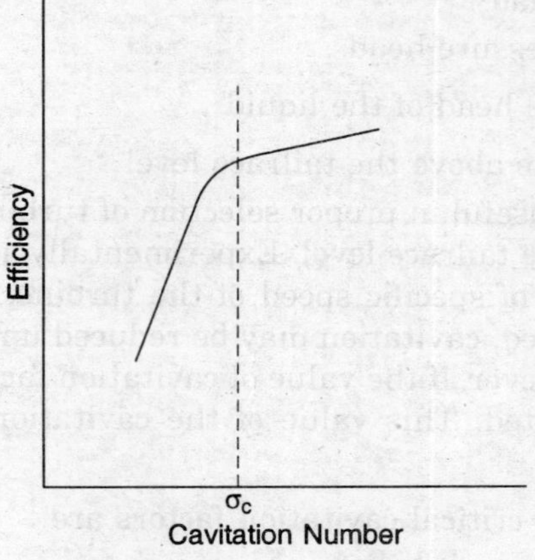

Fig. 17.44

4. Methods to Prevent Cavitation

Though it is not possible to control or prevent the phenomenon of cavitation, yet efforts can be made to minimise it to the least. Some of the methods to do so are

(i) **Design considerations :** To minimise the effect of cavitation, turbines should be suitably designed to prevent the separation of flow. Sharp corners or curvatures that produce vortices or eddies etc. should be avoided as these are the main source of low pressure in a region.

(ii) **Model testing :** Though a design is selected on the basis of theoretical considerations, yet a model of the prototype must be made and thorougly tested in a laboratory. Different possible models based on theoretical considerations with respect to lengths, angles, thickness etc. may be made and tested with the help of stroboscope which can take the photographs during the experiments. Only the best model should be adopted for the prototype.

In a turbine, the most susceptible point where cavitation can occur is the region of low pressure and as such pressure distribution around a blade section may be found by testing it in a wind tunnel.

(iii) **Installation of turbines :** At the oulet of a turbine, the pressure is negative and its value increases with the height of turbine above the tailrace level. Thus to minimise the effect of cavitation, it is always safer to install the turbine at a lesser height than given by the safe value of the suction height. Also velcoity at the outlet should be as small as possible, since high velocity increases the suction pressure and the turbine becomes more prone to cavitation.

(iv) **Choice of materials :** Certain materials offer more resistance to cavitation. The use of such materials helps to a great extent to reduce the chance of cavitation damage. The following table lists various materials in order of their cavitation resistant properties :

S.No.	Material	Relative resistance to cavitation
1.	Stellite	1·000
2.	Bronze	0·330
3.	Cast aluminium brone	0·120
4.	Nickel	0·110
5.	Welded stainless steel, Cr. 17%, Ni 8%	0·100
6.	Welded mild steel	0·006
7.	Aluminium	0·005
8.	Cast iron	0·003

It is seen that Stellite is at the top of the list. However, it cannot be used due to its prohibiting cost and machining difficulty. The material chosen should have adequate stiffness and high fatigue strength and should be able to have fine finish.

(v) **Temperature of liquids :** The evaporation of liquids also depends upon the temperature. The following table provides the values of vapour pressure of water with the change in temperature. As the vapour pressure increases with the rise in temperature, efforts should be made to use as cold a liquid as possible.

Temperature °C	0	3	6	10	15	20	30	40
Vapour pressure (*m* of water)	0·062	0·077	0·095	0·125	0·175	0·239	0·431	0·751

(vi) **Machining :** High finish of the surfaces reduce the chance of cavitation as compared to the relatively rough surfaces which have lesser degree of finish.

(vii) **Admission of air :** If air is admitted into a low pressure region in a turbine, the cavitation can be reduced or prevented. With the admission of air, the pressure at that point is raised and forming of vapour bubbles is prevented. Air admitted forms large bubbles that travel downstream. They do not collapse due to cushioning effect of air.

17.17 SELECTION OF TURBINES

The following considerations govern the selection of turbines :

1. Head

The approximate range for each type of turbine is as given in the following table :

S.No.	Head	Range in *m*	Type of Turbine
1.	Very high head	> 350	Pelton
2.	High head	150–350	Pelton/Francis
3.	Medium	60–150	Francis
4.	Low	30–60	Francis/Propeller/Kaplan
5.	Very low	5–30	Propeller/Kaplan

If the head varies considerably, a variable runner vanes turbine is preferred.

2. Specific speed

The following table gives the approximate values of specific speed for each type of turbine :

S.No.	Specific speed	Type of turbine
1.	10–25	Pelton with single nozzle
2.	25–50	Pelton with more than one nozzle
3.	50–220	Francis
4.	220–350	Francis/Propeller/Kaplan
5.	350–850	Propeller/Kaplan

Care is taken in selecting the specific speed corresponding to the synchronous speed of the generator. As the power produced depends also upon the speed of turbines, a turbine with high rotational speed and thus high specific speed should be preferred. This reduces the size of the turbine, the generators and the powerhouse and thus reducing the overall cost of the plant.

3. Part Load Operation

If the load on the turbines vary frequently, the turbines having fixed vanes are not suitable. In fixed vane runners, the losses are high and efficiency is low at part load conditions due to variation in the inlet angle of the relative velocity. Therefore, while selecting a turbine which is used under variable load conditions, one with a variable vanes should be chosen.

4. Number of Units

Depending upon the availability of head, rotational speed, the specific speed etc. of the turbine, power produced by a single turbine is calculated. Then depending upon the quantity of water available and the power to be generated, the number of turbines is found. In general, the number of units should be as less as possible to reduce the cost. Thus, turbines with high specific speed are preferred.

5. Cavitation

Cavitation depends upon a critical value of the cavitation factor which in turn is a function of specific speed. Cavitation also depends upon the suction height of the turbine above the tailrace level. Cavitation may be reduced or avoided if a turbine is installed at a lower level with respect to the tailrace level.

6. Disposition of Turbine Shaft

A vertical shaft turbine needs deeper foundation and high building. On the other hand, a horizontal shaft turbine covers large floor area. Based on experimental results a Pelton turbine usually has horizontal shaft whereas small size reaction turbines may have horizontal or vertical shafts. Large size reaction turbines are always installed with vertical shafts.

7. Cost

The cost of a power plant always influences the choice of turbines. Initial as well as the running cost has to be considered.

17.18 UNIT QUANTITIES

A turbine is designed for a particular head that is related to other parameters of the turbine such as speed, discharge and power output. To correlate, compare and use experimental data, these parameters are generally reduced according to unit heads. Each parameter is determined for a unit head on the turbine and expressed as a function of head. In case, the head on the turbine does not remain constant, the other parameters are also changed. Unit quantities make an effort to find the other parameters at the changed head on the assumption that the efficiency of the turbine is not changed. This means that the velocity triangles for the actual working head and for the new head are similar and angles are not changed. This also implies that the water enters the turbine without shock at different heads. Fig. 17.45 shows inlet velocity triangles for a Francis turbine. Fig. 17.45a refers to the actual head whereas Fig.

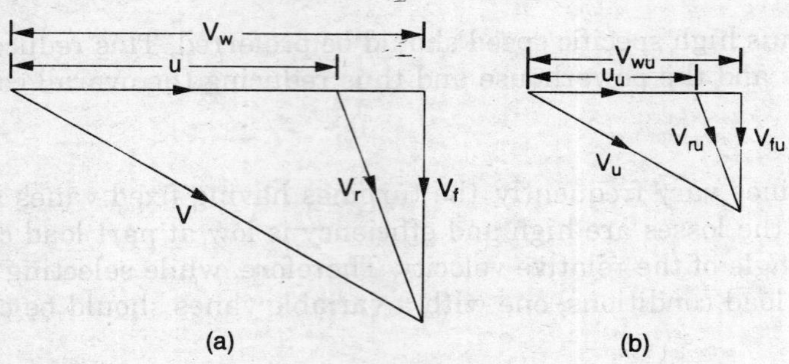

Fig. 17.45

17.45b represents the same turbine under unit head. The two velocity triangles are similar. Let subscript u denote the unit quantities.

Unit Speed

The speed of a turbine working under unit head is known as **unit speed.**

$$\eta_{hu} = \eta_h \qquad \text{(assumed)}$$

$$\frac{V_{wu} \cdot u_u}{g \cdot 1} = \frac{V_w \cdot u}{g \cdot H}$$

or
$$\frac{u_u}{u} = \frac{V_w}{V_{wu}} \cdot \frac{1}{H}$$

But in similar triangles (a) and (b)

$$\frac{u}{V_w} = \frac{u_u}{V_{wu}} \quad \text{or} \quad \frac{V_w}{V_{wu}} = \frac{u}{u_u}$$

$$\therefore \quad \frac{u_u}{u} = \frac{V_w}{V_{wu}} \cdot \frac{1}{H} = \frac{u}{u_u} \cdot \frac{1}{H}$$

$$\frac{u_u^2}{u^2} = \frac{1}{H}$$

$$\frac{u_u}{u} = \frac{1}{\sqrt{H}}$$

or
$$\frac{N_u}{N} = \frac{1}{\sqrt{H}} \qquad ...(17.23)$$

Unit Discharge

It is the discharge through the turbine under unit head.

$$Q_u = k \cdot \pi \cdot D \cdot B \cdot V_{fu} \quad \text{and} \quad Q = k \cdot \pi \cdot D \cdot B \cdot V_f$$

$$\therefore \quad \frac{Q_u}{Q} = \frac{V_{fu}}{V_f} = \frac{u_u}{u} = \frac{1}{\sqrt{H}}$$

or
$$Q_u = \frac{Q}{\sqrt{H}} \qquad ...(17.24)$$

Unit Power

It is the power produced by the turbine under unit head.

$$P_u = (w \cdot Q_u \cdot 1)\,\eta_0 \qquad \text{and} \qquad P = (w \cdot Q \cdot H)\,\eta_0$$

$$\therefore \quad \frac{P_u}{P} = \frac{Q_u}{Q} \cdot \frac{1}{H} = \frac{1}{\sqrt{H}} \cdot \frac{1}{H} = \frac{1}{H^{3/2}}$$

or
$$P_u = \frac{P}{H^{3/2}} \qquad \ldots(17.25)$$

Unit Force

It is the tangential force exerted by water on the runner vanes under unit head.

$$F_u = \rho \cdot Q_u \cdot V_{wu} \qquad \text{and} \qquad F = \rho \cdot Q \cdot V_w$$

$$\therefore \quad \frac{F_u}{F} = \frac{Q_u}{Q} \cdot \frac{V_{wu}}{V_w} = \frac{1}{\sqrt{H}} \cdot \frac{u_u}{u} = \frac{1}{\sqrt{H}} \cdot \frac{1}{\sqrt{H}} = \frac{1}{H}$$

or
$$F_u = \frac{F}{H} \qquad \ldots(17.26)$$

Unit Torque

Unit torque is the torque transmitted to the runner under unit head of water.

$$T_u = F_u \times R \qquad \text{and} \qquad T = F \times R$$

$$\frac{T_u}{T} = \frac{F_u}{F} = \frac{1}{H}$$

or
$$T_u = \frac{T}{H} \qquad \ldots(17.27)$$

The unit quantities help in predicting the performance of a turbine under different heads. Assume that the speed of a turbine operating under head H is N. Let it change to N' at head N'.

$$N_u = \frac{N}{\sqrt{H}}$$

Also when the turbine operates under Head H',

$$N_u = \frac{N'}{\sqrt{H'}}$$

Thus,
$$\frac{N'}{\sqrt{H'}} = \frac{N}{\sqrt{H}}$$

or
$$N' = N\sqrt{\frac{H'}{H}} \qquad \ldots(17.28)$$

Similarly,
$$Q' = Q\sqrt{\frac{H'}{H}} \qquad \ldots(17.29)$$

$$P' = P\left(\frac{H'}{H}\right)^{3/2} \qquad \ldots(17.30)$$

$$F' = F.\frac{H'}{H} \quad \text{...(17.31)}$$

$$T' = T.\frac{H'}{H} \quad \text{...(17.32)}$$

These equations are derived on the assumption that the efficiency of the turbine remains same under different heads which is not true for fixed vanes type of turbines.

Example 17.20 Following data relate to a test on a Francis turbine :

Power produced = 160 kW
Head = 12 m
Discharge = 2 m^3/s
Speed = 240 rpm
Diameter of runner = 1·1 m

Compute the values of speed, discharge and the power produced if the head on the turbine is changed to 5 m.

Solution $P = 160$ kW $\quad N = 240$ rpm

$H = 12$ m $\quad Q = 2\ m^3/s$

$D = 1{\cdot}1$ m $\quad H' = 5$ m

$$N' = N\sqrt{\frac{H'}{H}} = 240\sqrt{\frac{5}{12}} = 154{\cdot}9 \text{ rpm}$$

$$Q' = Q\sqrt{\frac{H'}{H}} = 2\sqrt{\frac{5}{12}} = 1{\cdot}29 \text{ m}^3/\text{s}$$

$$P' = P\left(\frac{H'}{H}\right)^{3/2} = 160\left(\frac{5}{12}\right)^{3/2} = 43 \text{ kW}$$

Example 17.21 A Pelton turbine operates under a head of 300 m and produces 5000 kW. The speed of the runner is 240 rpm and the overall efficiency 0·85. Determine the values of unit speed, unit discharge and unit power. If the head on the turbine falls to 160 m, find the new values of the speed, the discharge and the power. Assume that the efficiency of the turbine remains constant.

Solution $P = 5000$ kW $\quad H = 300$ m

$N = 240$ rpm $\quad \eta_0 = 0{\cdot}85$

$H' = 160$ m

$$\eta_0 = \frac{P}{wQH}$$

$$0{\cdot}85 = \frac{5000}{9{\cdot}81 \times Q \times 300}$$

$$Q = 2 \text{ m}^3/\text{s}$$

$$N_u = \frac{N}{\sqrt{H}} = \frac{240}{\sqrt{300}} = 13{\cdot}86 \text{ rpm}$$

$$Q_u = \frac{Q}{\sqrt{H}} = \frac{2}{\sqrt{300}} = 0{\cdot}116 \text{ m}^3/\text{s}$$

$$P_u = \frac{P}{H^{3/2}} = \frac{5000}{(300)^{3/2}} = 0{\cdot}962 \text{ kW}$$

$$N' = N\sqrt{\frac{H'}{H}} = 240\sqrt{\frac{160}{300}} = 175{\cdot}3 \text{ kW}$$

$$Q' = Q\sqrt{\frac{H'}{H}} = 2\sqrt{\frac{160}{300}} = 1{\cdot}46 \text{ m}^3/\text{s}$$

$$P' = P\left(\frac{H'}{H}\right)^{3/2} = 5000\left(\frac{160}{300}\right)^{3/2} = 1947 \text{ kW}$$

17.19 SPECIFIC SPEED

The unit quantities help in predicting the performance of a turbine at different values of the head. It does not help to compare the performance of different types of turbines. Specific speed helps in distinguishing the perfomance of different types and shapes of turbines.

Specific speed is defined as the speed of an imaginary turbine indentical in shape, blade angles, geometrical proportions etc. to the actual turbine but reduced to such a size that under corresponding conditions, it produces unit power when working under unit head. Such a turbine is known as **specific turbine.**

Let subscript 's' denote the specific quantities whereas subscript 'u' the unit quantities. As the specific turbine is identical to the actual turbine in every respect, its velocity triangle is also identical to that for the actual turbine. Since it also works under unit head, the same velocity triangle holds good for the specific turbine (Fig. 17.46).

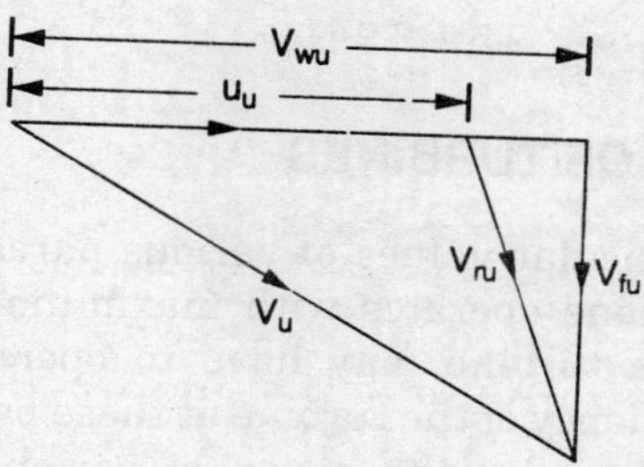

Fig. 17.46

Thus,

$$u_s = u_u \qquad \text{...(i)}$$

$$\frac{\pi D_s N_s}{60} = \frac{\pi D N_u}{60}$$

or

$$N_s = \frac{D}{D_s} \cdot N_u$$

where D is the diameter of the actual turbine.

$$\frac{P_u}{P_s} = \frac{w \cdot Q_u \cdot 1 \cdot \eta_0}{w \cdot Q_s \cdot 1 \cdot \eta_0} = \frac{Q_u}{Q_s} = \frac{k \cdot \pi D \cdot B \cdot V_{fu}}{k \cdot \pi D_s \cdot B_s \cdot V_{fs}} = \frac{k \cdot \pi D \cdot (nD) \cdot V_{fu}}{k \cdot \pi D_s\,(nD_s)\,V_{fs}} = \left(\frac{D}{D_s}\right)^2 \qquad (\because V_{fu} = V_{fs})$$

or $$\frac{D}{D_s} = \sqrt{\frac{P_u}{P_s}} = \sqrt{P_u} \qquad (\because\ P_s = 1\text{, from definition})$$

$$N_s = \sqrt{P_u} \cdot N_u = \sqrt{\frac{P}{H^{3/2}}} \cdot \frac{N}{\sqrt{H}} = \frac{\sqrt{P}}{H^{3/4}} \cdot \frac{N}{H^{1/2}} = \frac{N\sqrt{P}}{H^{5/4}} \qquad ...(17.33)$$

As the specific turbine remains the same for all the actual turbines of different sizes but of the same geometrical shape, the specific speed of all such turbines is the same being the speed of the specific turbine. This also implies that as the specific turbines of actual turbines of different shapes are to be different, they will also have different specific speeds. Thus it is concluded that all turbines of the same geometrical shape, though different in dimensions have the same specific speed. The above expression also indicates that as the head decreases, the specific speed increases. Thus the specific speed of Pelton turbines is the least whereas for the Kaplan turbine it is highest.

17.20 RUNAWAY SPEED

Runaway speed is defined as the maximum speed that a turbine attains under designed head and gate opening, the governing system being disconnected and the load reduced to zero. All the rotating parts of the turbine must be designed to withstand the centrifugal stresses of the runaway speed. The exact value of the runaway speed for a turbine is determined by experimentation. The following are the approximate values of the runaway speed in terms of normal speed :

Pelton turbine	1·8 – 1·9 N
Francis turbine	2·0 – 2·3 N
Kaplan turbine	2·5 – 2·9 N

17.21 CHARACTERISTICS OF TURBINES

A turbine is developed for particular values of various parameters such as head, discharge, speed and power etc. The turbine operates with maximum efficiency at these parameters. However, in actual practice, a turbine may have to operate at different values of these parameters. Obviously, the efficiency of the turbine at these parameters will not be the highest. The behaviour of the turbines under the action of variable parameters is known as the characteristics of the turbines. The characteristics are usually presented by plotting curves among different parameters. The data to plot the curves are obtained by experimentation on models of the turbines. Models are tested in the laboratories and t
actual turbines at actual sites.

The variables in case of a turbine are :

H = head on the turbine

N = speed of the turbine

Q = discharge through the turbine

P = power produced by the turbine

η = efficiency of the turbine

G = Gate opening (nozzle opening or the guide vane opening)

H, Q and N are physically independent quantities. The variation of other quantities with respect to any of these provides a curve or a characteristic of the turbine. For convenience, the curves are plotted in terms of unit quantities.

Types of Characteristic Curves

1. Main or constant head characteristics
2. Operating or constant speed characteristics
3. Constant efficiency or Muschel curves.

1. Main characteristics curves

To draw these curves, the head on the turbine is maintained constant at a particular gate opening and the speed is varied by changing the load on the turbine. Speed regulation system if any is disconnected from the turbine so that the speed is varied. Corresponding values of the discharge are noted. For another gate opening, the same procedure is repeated keeping the head constant at the previous value. The importance of these characteristics lies in the fact that somehow if the governing mechanism goes out of order, the speed of the turbine can increase upto runaway speed. The data obtained is usually reduced to unit quantities and then plotted.

Discharge vs. speed curves :

- In an impulse turbine, the discharge through the nozzle is constant at a particular position of the spear. If the whole of the water impinges on the buckets, the plots of discharge vs. speed curves are horizontal straight lines with different values at different gate openings (Fig. 17.47a).

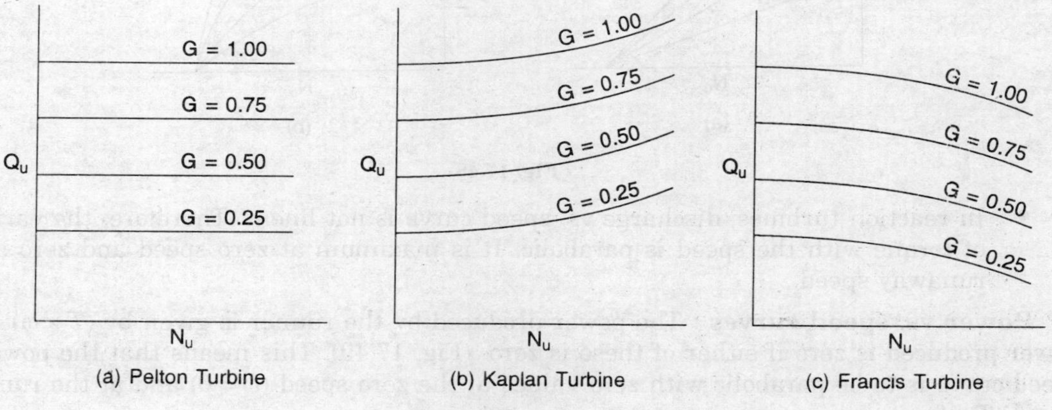

Fig. 17.47

- In reaction turbines, discharge is the product of the area and the velocity of flow. Since the governing mechanism is detached and the gate opening is constant, the area remains constant. Thus, the discharge is proportional to the velocity of flow. As the speed is increased, the tangential velocity increases which in turn proportionally increases the velocity of flow so that the velocity triangle tends to remain similar. Thus, the discharge through the reaction turbines should increase with the increase in speed.

However, in Francis turbines when the speed is increased, centrifugal force on the water acting outwards also increases which has tendency to retard the flow. Therefore, in a Francis turbine the flow is actually reduced. In Kaplan or propeller turbines, centrifugal force does not vary much and thus the flow is increased with speed. The curves do not pass through the origin, as some discharge is necessary to generate a minimum power to overcome the friction at the start of the turbine (Fig. 17.47b and c).

Torque vs. speed curves : Force or torque exerted by a jet of water on a body is always proportional to its velocity relative to the body. If it is linear motion, the force is maximum when the body is stationary and is zero if the body moves with the speed of the jet which means that the jet does not impinge on the body at all. In rotary motions, the speed at which the water does not strike the vanes is called the runaway speed of the turbine. Thus, the torque acting on the runner is maximum when it is stationary and is zero at the runaway speed.

- In Pelton turbines, the discharge is constant for a particular gate opening. Thus, the torque varies linearly with the speed. It is maximum when the runner is stationary and is zero at the runaway speed (Fig. 17.48a).

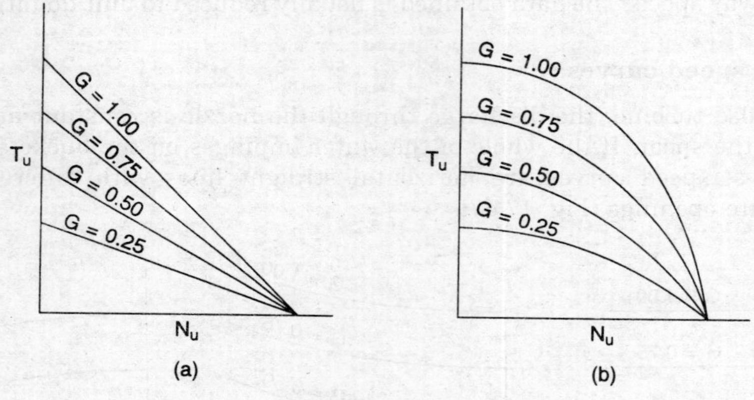

Fig. 17.48

- In reaction turbines, discharge vs. speed curve is not linear. Therefore, the variation of torque with the speed is parabolic. It is maximum at zero speed and zero at the runaway speed.

Power vs. speed curves : The power produced by the runner is given by $(T \times \omega)$. Thus power produced is zero if either of these is zero. (Fig. 17.49). This means that the power vs. speed curve is to be parabolic with zero values at the zero speed ($\omega = 0$) and at the runaway speed ($T = 0$).

Efficiency vs. speed curves : Efficiency is given by $\eta = P/wQH$. If H is constant, the efficiency is proportional to the ratio P/Q. Since the discharge is constant with speed in case of Pelton turbine and does not vary much in case of reaction turbines, the η vs. N curves are similar to those for P vs. N (Fig. 17.50).

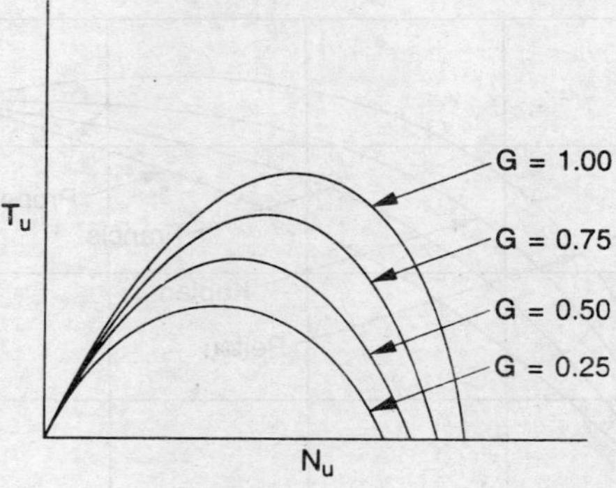

Fig. 17.49

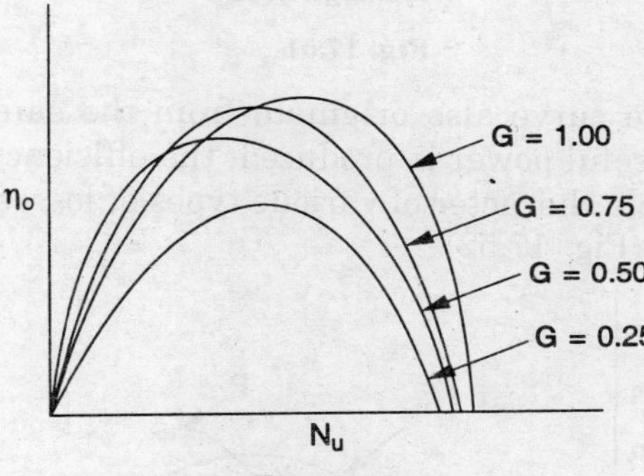

Fig. 17.50

2. Operating characteristic Curves

In practice, the turbines are coupled to generators and are to rotate at the same speed which is always kept constant. In addition, the head on the turbines does not vary much from day to day. The variation in head is usually over longer periods.

If the head and the speed are kept constant and the discharge is varied by varying the gate opening, efficiency vs. load, and power or efficiency vs. discharge curves can be drawn.

- For different values of discharge, power produced and the overall efficiency are calculated. Then curves are drawn showing the efficiency with respect to variation of percentage of full load (Fig. 17.51). It is observed that the Pelton and the Kaplan turbines have high efficiency for larger variation of loads as compared to Francis or propeller turbines. This is because shock losses are prominent in case of fixed blade turbines. The curves pass through origin because if power produced is zero, efficiency is zero.
- Power is given by $T \times \omega$.If speed is constant, the power is proportional to torque. As head is also kept constant, the torque is proportional to the discharge. Therefore, power increases linearly from zero, when the gate is closed to maximum at full gate opening. The curve does not start from zero, as some discharge is necessary to overcome the initial friction before any useful power is obtained. (Fig. 17.52)

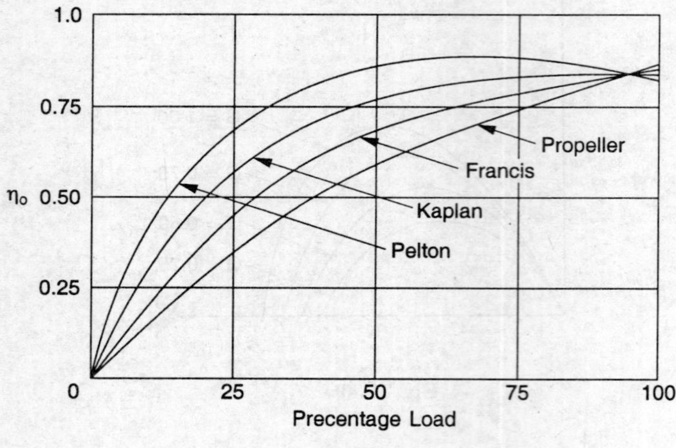

Fig. 17.51

- Efficiency vs. discharge curve also originate from the same point as for the power because unless some useful power is produced, the efficiency is zero. It is not a linear curve because of variable character of various types of losses *i.e.* hydraulic, volumetric and mechanical losses (Fig. 17.52).

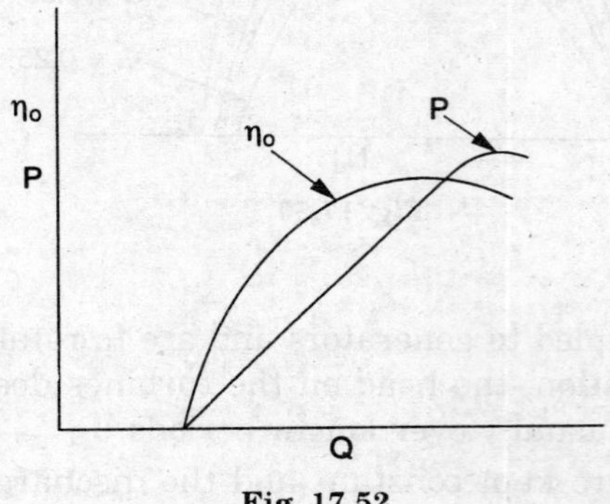

Fig. 17.52

3. Constant Efficiency Curves

These curves are also known as **Muschel, iso-efficiency** or **universal curves.** From the main characteristics, it is observed that there are two values of the speed for a particular value of efficiency for a particular gate opening except for maximum efficiency. This implies that there are two values of the speeds giving the same value of efficiency for the same gate opening. To draw these curves adopt the following procedure :

- Reproduce the discharge vs. speed, and the efficiency vs. speed curves from the main characteristics of the turbines as shown in Fig. 17.53, the former being above the latter one.
- Locate the points corresponding to a certain value of efficiency and a particular gate opening in the discharge curve by drawing the projections as shown in the figure.

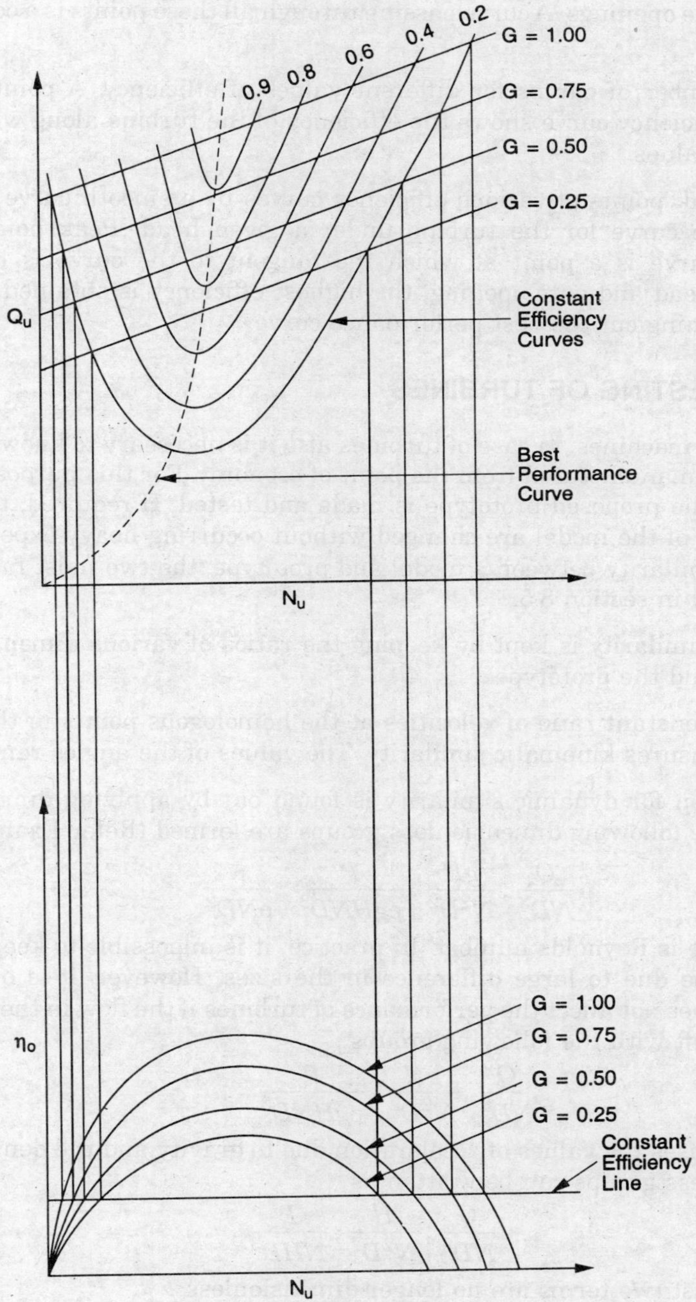
0.9
0.8
0.6
0.4
0.2
G = 1.00
G = 0.75
G = 0.50
G = 0.25
Q_u
Constant Efficiency Curves
Best Performance Curve
N_u
η_o
G = 1.00
G = 0.75
G = 0.50
G = 0.25
Constant Efficiency Line
N_u

Fig. 17.53

- Similarly, locate the points corresponding to the same value of efficiency but for different gate openings. A curve passing through all these points is a constant efficiency curve.
- Draw a number of curves for different values of efficiency. A point anywhere on a constant efficiency curve shows the efficiency of the turbine along with the discharge and speed values.
- Join the peak points of various efficiency curves by a smooth curve. This is the best performance curve for the turbine under a given head. Peak point on a constant efficiency curve is a point at which the tangent to the curve is horizontal. For a particular head and gate opening, the highest efficiency is obtained where the curve for that opening cut the best performance curve.

17.22 MODEL TESTING OF TURBINES

As with many other machines, in case of turbines also it is necessary to know the performance before it is actually manufactured from the point of economy. For this purpose, a model of the turbine similar to the proposed prototype is made and tested. If required, the design, shape and the dimensions of the model are changed without occurring heavy expenditure. In order to have complete similarity between a model and prototype, the two must fulfil the similarity conditions discussed in section 8·5.

- Geometric similarity is kept by keeping the ratios of various dimensions constant in the model and the prototype.
- Keeping a constant ratio of velocities at the homologous points in the model and the prototype ensures kinematic similarity. The values of the angles remain same.
- The condition for dynamic similarity is found out by applying dimensional analysis method. The following dimensionless groups are formed (Refer Example 8·6) :

$$\frac{Q}{ND^3}, \frac{gH}{N^2D^2}, \frac{P}{\rho gHND^3}, \frac{\mu}{\rho ND^2}$$

The last of these is Reynolds number. In practice, it is impossible to keep it constant in a model and prototype due to large difference in the sizes. However, it is observed that the Reynolds number does not affect the performance of turbines if the flow in the two is turbulent. Therefore, we are left with the following groups :

$$\frac{Q}{ND^3}, \frac{gH}{N^2D^2}, \frac{P}{\rho gHND^3}$$

In most of the cases the values of acceleration due to gravity and the density are the same and the dimensionless groups can be written as

$$\frac{Q}{ND^3}, \frac{H}{N^2D^2}, \frac{P}{NHD^3}$$

Note that the last two terms are no longer dimensionless.

Replacing the first term by $\frac{\text{1st term}}{\text{2nd term}}$

or
$$\frac{Q}{ND^3}\sqrt{\frac{N^2D^2}{H}} \quad \text{or} \quad \frac{Q}{D^2\sqrt{H}}$$

The last term may be replaced by $\dfrac{\text{Last term}}{\text{2nd term}}$

$$\frac{P}{HND^3}\sqrt{\frac{N^2D^2}{H}} \quad \text{or} \quad \frac{P}{D^2H^{3/2}}$$

Other important terms, applicable to hydraulic machines, known as the specific speed can also be derived.

$$N_s = \frac{\sqrt{\text{First term}}}{(\text{Second term})^{3/4}} = \frac{\sqrt{Q}}{\sqrt{ND^3}} \cdot \left(\frac{N^2D^2}{H}\right)^{3/4} = \frac{N\sqrt{Q}}{H^{3/4}}$$

$$N_s = \frac{\sqrt{\text{Last term}}}{(\text{Second term})^{3/4}} = \frac{\sqrt{P}}{\sqrt{HND^3}} \cdot \left(\frac{N^2D^2}{H}\right)^{3/4} = \frac{N\sqrt{P}}{H^{5/4}}$$

Thus the necessary conditions for dynamic similarity of turbines are

$$\left(\frac{H}{N^2D^2}\right)_m = \left(\frac{H}{N^2D^2}\right)_p$$

$$\left(\frac{P}{HND^3}\right)_m = \left(\frac{P}{HND^3}\right)_p \quad \text{or} \quad \left(\frac{P}{D^2H^{3/2}}\right)_m = \left(\frac{P}{D^2H^{3/2}}\right)_p$$

$$\left(\frac{N\sqrt{P}}{H^{5/4}}\right)_m = \left(\frac{N\sqrt{P}}{H^{5/4}}\right)_p$$

The following two conditions are applicable to similarity of pumps :

$$\left(\frac{Q}{ND^3}\right)_m = \left(\frac{Q}{ND^3}\right)_p \quad \text{or} \quad \left(\frac{Q}{D^2\sqrt{H}}\right)_m = \left(\frac{Q}{D^2\sqrt{H}}\right)_p$$

$$\left(\frac{N\sqrt{Q}}{H^{3/4}}\right)_m = \left(\frac{N\sqrt{Q}}{H^{3/4}}\right)_p$$

PROBLEMS

1. What is the basic difference between a turbine and a pump?
2. Explain the action of various types of water wheels?
3. What are the various components of a hydropower plant? Describe them.
4. Give plant layout of a typical high-head plant and a lowhead plant.
5. Why are surge tanks necessary in case of high and medium head plants?
6. Classify the turbines on the basis of

(i) Head and quantity of water available

(ii) Hydraulic action of water

(iii) Direction of flow of water in the runner

(iv) Specific speed of turbines

7. What is the difference between an impulse and a reaction turbine?
8. Define hydraulic, volumetric and mechanical efficiencies of a turbine. What is overall efficiency?
9. What is Euler equation of energy transfer in a hydraulic machine? Elaborate.
10. Define degree of reaction in a hydraulic turbine ?
11. What are the main features of a Pelton turbine ? Describe them.
12. Why is the bucket of a Pelton runner made double cupped?
13. Why is the bottom portion of the buckets of a Pelton wheel cut?
14. In case of Pelton turbines, casing is unnecessary. Comment.
15. What is speed ratio? What are its theoretical and actual values for a Pelton turbine?
16. How do you decide the number of jets for a Pelton turbine?
17. Find an expression for the hydraulic efficiency of a Pelton wheel. Find the condition for maximum efficiency.
18. Write a note on the evolution of reaction turbines.
19. What kind of turbine is Francis turbine? Describe its main components.
20. Why is a draft tube necessary in case of reaction turbines? How does it regain the lost head when a turbine is set up above the tailrace level? What are the main types of draft tubes?
21. Why are the inward-flow reaction turbines are preferred over outward-flow tubrines?
22. How does a Kaplan turbine differ from a propeller turbine?
23. What is a whirl chamber of a Kaplan turbine? What is its function?
24. Why is a Kaplan turbine best suited for variable loads?
25. With the help of a sketch, describe the main features of a Kaplan turbine?
26. In what way are the blades of a Kaplan turbine different from that of a Francis turbine?
27. What do you mean by governing of hydraulic turbines? Why is it necessary?
28. Describe various components of the governing mechanism of a Pelton turbine. How does the governing of a reaction turbine differ from that of an impulse turbine ?
29. What are unit quantities? What is their importance?
30. Derive expressions for various unit quantities. What assumptions are made to derive these expressions ?
31. What is specific speed? How is it helpful in comparing the turbines of different types?
32. Derive an expression for the specific speed of turbine? Prove that it can be expressed as $3{\cdot}65\, N_u\, Q_u$ with usual notations.
33. What is run away speed? What are its values for impulse and reaction turbines?
34. What are different types of characteristic curves for turbines? Differentiate between the main and the operating characteristics of hydraulic turbines.
35. What is cavitation? Define Thoma's cavitation number.
36. What are the effects of cavitation in hydraulic machines?
37. Mention some methods to prevent cavitation in hydraulic machines.
38. How will you detect the commencement of cavitation in hydraulic machines?

39. A Pelton turbine generates 12·7 MW of power when working under a head of 510 m. The speed of the runner is 430 rpm. Determine the least diameter of the jet, jet ratio and the number of buckets. Take coefficient of velocity 0·97, speed ratio 0·45 and the overall efficiency 90%. (192·4 mm , 10·39 , 20)
40. The nozzle of a Pelton turbine works under a head of 30 m and has coefficient of velocity 0·96. Its buckets deflect the jet through 165°. The flow rate is 1 m^3/s and the mean bucket speed 10 m/s. Calculate the power and efficiency of the turbine. (261·3 kW , 96·35%)
41. A Pelton turbine produces 4600 kW of power. The head on the turbine is 300 m. the speed of the runner is 430 rpm. Assuming the values of speed ratio 0·45, overall efficiency 85% and the coefficient of velocity 0·97, find the number of jets for the turbine. (3)
42. The discharge through a Pelton turbine is 250 l/s. The head on the turbine is 100 m. and the tangential velocity of the runner is 18 m/s. The jet is turned through 166°. Determine the power produced and the efficiency of the turbine. Take coefficient of velocity as 0·98. (225·3 kW , 95·6%)
43. The runner of a Francis turbine has an outlet diameter 500 mm and the inlet diameter 350 mm. It works under a head of 60 m. The width at the inlet is 75 mm. The angles of the blades at the inlet and the outlet are 93° and 30° respectively. The flow enters the vanes at an angle of 23°. Mechanical losses are 6% and the area taken by the vanes is 7% of the total area. Assume hydraulic efficiency as 90% and the flow velocity constant. Determine the speed of the runner so that the entry of water into the runner is shockless. Also find the power produced. (880 rpm, 524·9 kW)
44. An inward-flow reaction turbine operates under a head of 10 m and produces 150 kW. The speed of the runner is 175 rpm. The diameter of the runner at the outlet is 0·6 times the inlet diameter. The flow velocity is constant and is 2 m/s. The tangential velocity at the inlet is 10 m/s, the hydraulic efficiency 85% and the overall efficiency 80%. Determine
    (i) the diameter of the runner at the inlet and the outlet
    (ii) the angles of the guide vanes
    (iii) the runner vane angles at the inlet and the outlet
    (iv) the discharge. (1·091 m , 0·655 m , 13·5° , 50·3° , 18·43° , 1·911 m^3/s)
45. In an inward flow reaction turbine,

$$V_{w1} = 3{\cdot}2\sqrt{H} \qquad V_{f1} = 1{\cdot}1\sqrt{H}$$

$$V_{w2} = 0{\cdot}24\sqrt{H} \qquad V_{f1} = 0{\cdot}85\sqrt{H}$$

where H is the head in meters. The inner diameter is 0·75 times the outer diameter and the hydraulic efficiency 0·85. Determine the angles of the vanes at the inlet and the outlet. (68·2° , 24·9°)
46. In an inward flow reaction turbine, the flow velocity is constant and the vane angles at the entrance are 90°. Show that the best tangential velocity is given by

$$u_1 = \frac{\sqrt{2gH}}{\sqrt{2 + \tan^2 \alpha_1}}$$

where H is the head on the turbine, u_1 the tengentially velocity of the runner and α_1 the angle of guide vanes at the inlet.

47. The specific speed of a Francis turbine is 140 and it generates 12000 kW of power. The inner and the outer diameters are 0·8 m and 1·6 m respectively. The head on the turbine is 100 m. The water enters the runner at an angle of 12° and the velocity at the outlet is radial. The hydraulic efficiency is 90% and the flow velocity constant. Determine the vane angles at the inlet and the outlet. (35·55°, 18·15°)

48. The inner and outer diameters of an outward flow reaction turbine are 0·75 m and 1 m respectively. The speed of the runner is 300 rpm and the angles at the inlet and the outlet are 90° and 20° respectively. The velocity of flow is 3 m/s and is radial at the outlet. The turbine is placed at 1·2 m below the tailrace level. Determine the guide vane angles, total head, pressure head at the inlet and the efficiency. (14·3°, 15·8 m, 10·27 m. 89·5%)

49. A Kaplan turbine produces 8000 kW of power when working under a head of 18 m. The runner and the hub diameters are 4·5 m and 2·25 m respectively. The angle of the guide vanes at the inlet is 30°. Hydraulic and overall efficiencies are 89% and 84% respectively. The discharge at the outlet is radial. Find the runner vane angles at the tip of the runner and the speed of the runner. (At inlet : 158·1°, outlet 12·7°, 85·2 rpm)

50. The hub and the runner diameters of a Kaplan turbine are 2 m and 4 m at the inlet. The flow velocity is constant from the root to the tip and the discharge is radial at the outlet. Find the vane angles at the root diameter, mean diameter and tip diameter of the runner. (At inlet : 155·7°, 140·83° 103·55°, At outlet : 14·2°, 18·68°, 26·84°)

51. A Kaplan turbine works under a head of 40 m and develops 25 MW. The diameter of the hub is 0·35 times the diameter of the runner. Flow ratio is 0·6 and the speed ratio 2. The overall efficiency is 90%. Determine the diameter, the speed and the specific speed of the runner. (2·75 m, 389 rpm, 611·5)

52. A Kaplan turbine generates 11 MW of power and works under a head of 4 m. The speed ratio is 1·2 and the flow ratio 0·65. The hub diameter is 0·36 times the runner diameter. The overall efficiency is 85%. Find the diameter and the speed of the runner. (9·15 m, 35·1 rpm)

53. A reaction turbine working under a head of 5 m produces 70 kW. The speed of the runner is 180 rpm and the discharge 1·5 m^3/s. If the head increases to 15 m, determine the values of the speed, discharge and the power. (311·8 rpm, 2·6 m^3/s, 363·8 kW)

54. The head on a Pelton turbine is 500 m. It produces 10000 kW of power at a speed of 300 rpm. The efficiency of the turbine is 80%. Determine the values of the unit discharge, unit speed and unit power. In what way the values are changed if the head falls to 350 m ? Assume the efficiency to remain constant. (0·163 m^3/s, 13·4 rpm, 894 W, 3·045 m^3/s, 418·3 rpm, 5857 kW)

Chapter 18

HYDRAULIC PUMPS

18.1 INTRODUCTION

Pumps are used to increase the energy level of water so that it can be raised to higher levels. This can be done by using either dynamic pressure pumps usually known as centrifugal pumps or by reciprocating pumps. The working principle of the two is altogether different.

Centrifugal pumps are the rotodynamic machines in which the mechanical energy of shaft is converted into kinetic and pressure energies of water. The wheel of centrifugal pump is known as impeller. In a centrifugal pump, the energy added by the impeller to the liquid is largely due to centrifugal effect. A centrifugal pump works on the same principle as a turbine, though its operation is opposite to that of a turbine. Recently, reversible type machines have been developed which can work both as a turbine and as a pump. The transfer of energy is given by Euler's fundamental equation of energy transfer in hydraulic machines.

Reciprocating pumps are positive displacement pumps. A small quantity of liquid is taken into a chamber and is physically displaced and forced out with pressure by a moving mechanical element. The moving mechanical element is a piston moving in a cylinder or a gear rotating in a housing.

The use of reciprocating pumps is being limited these days. Owing to their high initial cost as well as maintenance cost as compared to centrifugal pumps, they have become almost obsolete for industrial purposes. However, for small purposes they are still widely used *e.g.* as kerosene pumps, well pumps, in pneumatic pressure systems, in milk shops etc. They are also useful where high head with small discharge is required as in oil drilling.

18.2 CLASSIFICATION OF CENTRIFUGAL PUMPS

Centrifugal pumps are classified according to

1. Working head
2. Type of casing
3. Specific speed

4. Number of entrances to the impeller
5. Direction of flow of water
6. Number of stages
7. Disposition of shaft

1. Working Head

Centrifugal pumps are divided into low, medium and high head pumps.

Low-head centrifugal pumps : These are generally single stage pumps. The heads developed are less than 15 m. The pumps are fitted with volute casing without guide vanes. These are single or double suction pumps and have horizontal shafts.

Medium-head centrifugal pumps : Usually, these are single stage pumps which may or may not use guide vanes. The heads generated are between 15-45 m. These are single or double suction pumps with horizontal shafts.

High-head centrifugal pumps : Generally, these are multistage pumps fitted with guide vanes. The heads developed are more than 45 m. The pumps may have horizontal or vertical shafts. Vertical shafts are useful in deep wells.

2. Type of Casing

Centrifugal pumps are divided into following types depending upon the casing :

- **Volute-chamber pump :** This type of casing is of spiral shape and has a sectional area that increases uniformly from the tongue to the delivery pipe (Fig. 18.1). More area is required to accommodate the increased quantity of water as the water flows towards the delivery pipe. Otherwise, the velocity of whirl will vary in the volute chamber at all cross sections. Generally, the velocity of whirl is kept around 40 percent of the velocity at the impeller outlet. The impact of water entering the volute chamber with higher velocity on the slow moving water in the chamber results in shock losses. Therefore, the casing should be designed carefully to minimize the losses to have the maximum efficiency.

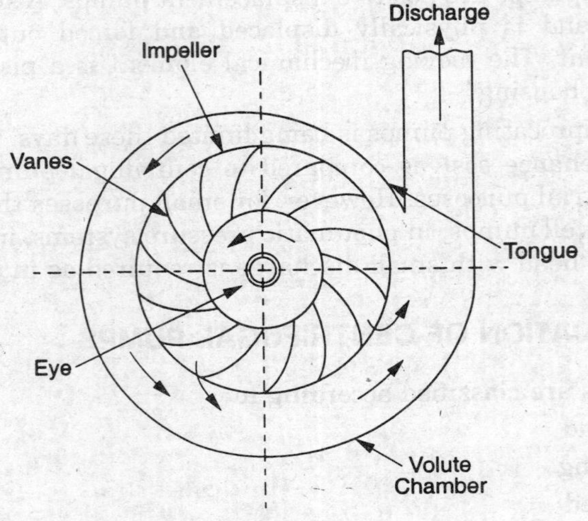

Fig. 18.1

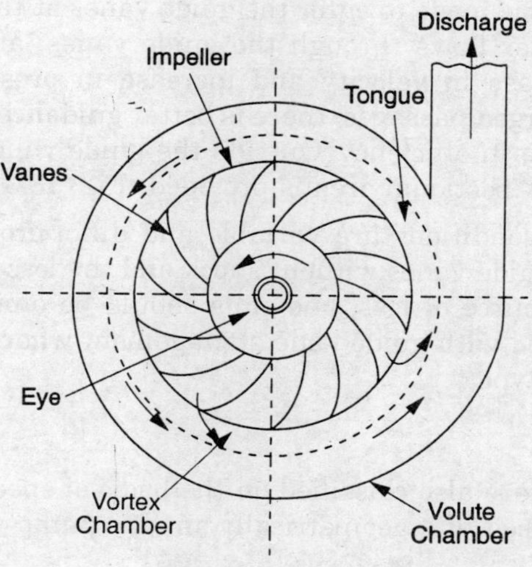

Fig. 18.2

- **Vortex-chamber pump :** This type of chamber is also known as whirlpool chamber. In this type of chamber, an area of uniform radial width is provided between the impeller outer periphery and the volute casing as shown in Fig. 18.2. As the radial distance increases outwards, this provides a uniformly increasing area to the flowing water. The increase in area leads to increase in pressure and decrease in velocity in the outward direction (free vortex). From this chamber, the water is discharged into the volute casing as usual. Though a vortex chamber is more efficient than a volute chamber, still there is instability of flow and eddy formation due to divergent motion of water.

- **Diffuser pump :** This is also knwon as a **turbine pump**. In this type of pump, guide vanes are provided at the outer periphery of the impeller and it resembles a reaction

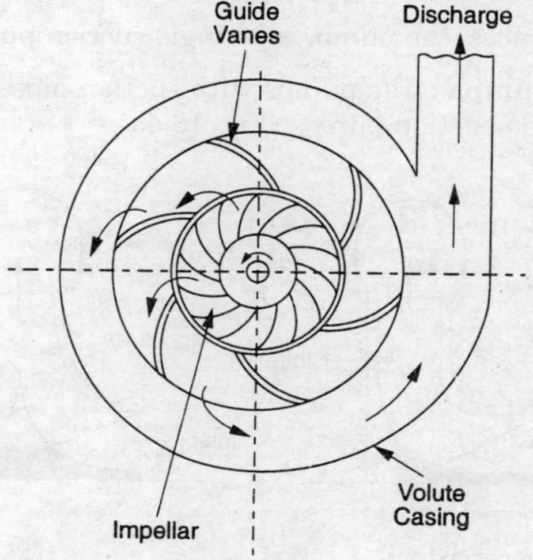

Fig. 18.3

turbine. The water is made to enter the guide vanes at the proper angle to avoid shock losses. As the water flows through the guide vanes, area of cross-section increases resulting in decrease in velocity and increase in pressure. Since the guide vanes provide small enlarged passages, there is better guidance to the flow, reduction in eddy losses and increase in efficiency. Outside the guide vanes, a volute of either uniform or increasing cross- sectional area is provided (Fig. 18.3).

In case, the working conditions are variable and differ from the design conditions, the water does not enter the guide vanes without shock and the losses are high. In such cases, the diffuser ring becomes a source of loss, and thus should be omitted. These days, multistage pumps are usually provided with guide vane arrangement whereas single stage pumps are of volute or vortex-chamber type.

3. Specific Speed

Just like turbines, pumps are also classified on the basis of specific speed. Specific speed of a pump is defined as the speed of a geometrically similar pump which delivers unit discharge at unit head.

$$N_S = \frac{N\sqrt{Q}}{H^{3/4}} \qquad \ldots(18.1)$$

Specific speeds of different types of turbines are given in the following table :

Pump	Speed	Specific Speed
Radial-flow	Slow	10–25
	Normal	25–45
	High	45–70
Mixed flow		70–135
Axial flow, propeller		100–425

4. Number of Entrances to impeller

Depending upon the entrances, the pumps are single suction pumps or double suction pumps.

- **Single-suction pumps :** The pumps with suction pipe only on one side of the pumps are known as single-suction pumps (Fig. 18.4a).

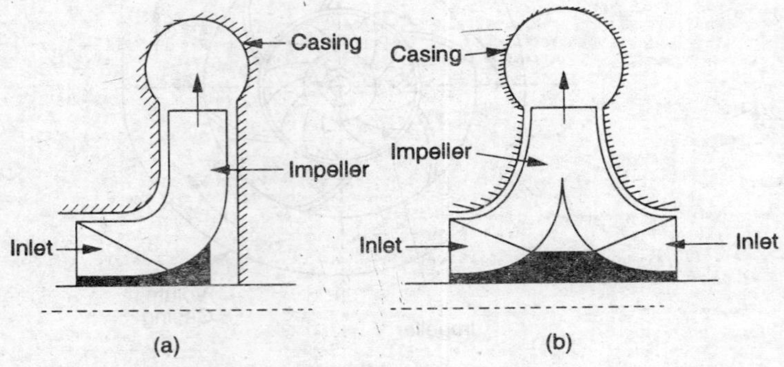

Fig. 18.4

- **Double-suction pumps :** If the suction pipe is provided on both sides of the impeller, it is called a double-suction pump (Fig. 18.4b). The discharge is increased considerably as well as axial thrust is neutralised.

5. Direction of Flow of Water

The flow through the impeller can be radial, mixed or axial similar to that in reaction turbines (Fig. 18.5).

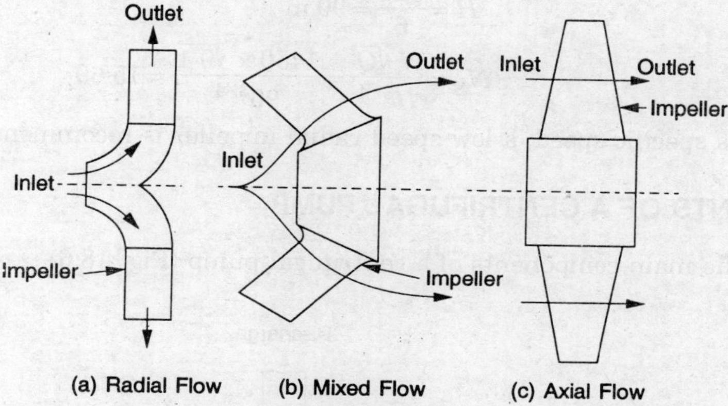

Fig. 18.5

- **Radial flow :** If the flow in the impeller is completely in a radial direction, it is known as a radial flow pump This type is useful when the requirements are high head and low discharge.
- **Mixed flow :** Mixed flow pumps are used when the requirements are medium discharge with medium head. The direction of flow is changed from pure radial to a combination of axial and radial. These are useful mostly for irrigation purposes.
- **Axial flow :** In strict sense, these pumps are not centrifugal pumps as no centrifugal force is developed since the radial distance of liquid particles does not change from the axis of impeller. Usually, the vanes of an axial-flow centrifugal pump are designed on the basis of airfoil theory. The function of an axial-flow pump is opposite to that of an axial- flow turbine. If the blades of an axial-flow pump are adjustable, it is known as **Kaplan pump**. These pumps are useful when high discharge at low head is required.

6. Number of Stages

A centrifugal pump may have single stage or multi stages. In multistage pumps, two or more impellers are mounted on the same shaft and enclosed in the same casing. When high heads are to be raised, these are very helpful.

7. Disposition of Shaft.

Generally, the centrifugal pumps are used with horizontal shafts. However, vertical shafts may be used where there is space limitation such as in deep wells, mines etc.

Example 18.1 A six stage centrifugal pump raises the head of water by 540 m and delivers 100 litres of water per second. Find its specific speed if it rotates at 1450 rpm. What type of impeller should be used for such type of pump?

Solution $N = 1450$ rpm $\quad Q = 0{\cdot}1\ \text{m}^3/\text{s}$

Total head raised = 540 m $\quad n = 6$

Head developed per stage,

$$H = \frac{540}{6} = 90\ \text{m}$$

$$N_S = \frac{N\sqrt{Q}}{H^{3/4}} = \frac{1450 \times \sqrt{0{\cdot}1}}{90^{3/4}} = 15{\cdot}69$$

For this specific speed, a low speed radial impeller is recommended.

18.3 COMPONENTS OF A CENTRIFUGAL PUMP

The following are the main components of a centrifugal pump (Fig. 18.6) :

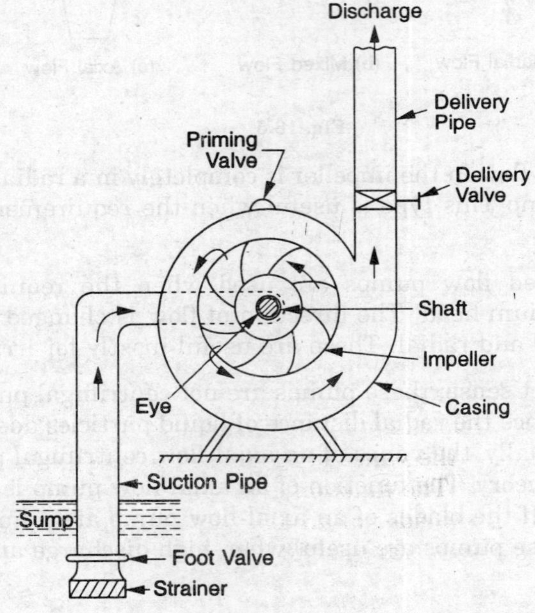

Fig. 18.6

1. Impeller

The wheel of a centrifugal pump is known as impeller. It consists of a number of forward- or backward-curved blades depending upon the speed. It converts the mechanical energy obtained from the external source into pressure and kinetic energy of water. The impellers may be of the following types (Fig. 18.7) :

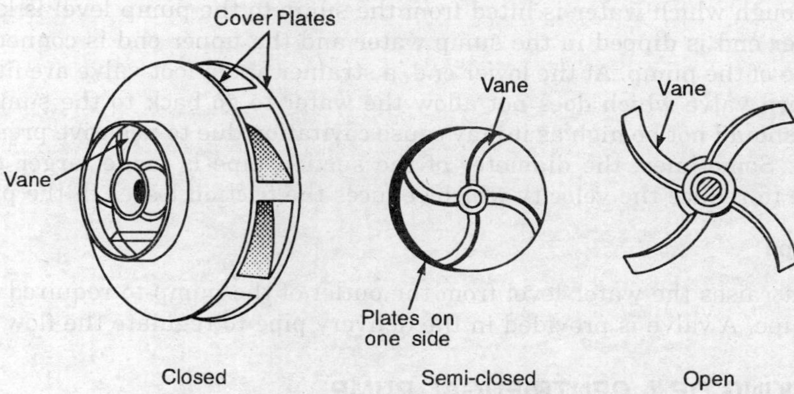

Fig. 18.7

- **Closed impeller :** In case the vanes of the impeller are covered with plates on both sides, it is known as closed impeller. It is used when water is free from external impurities or debris. Otherwise, the impeller may be clogged. It provides better guidance to water; thus it is more efficient.

  The material for the impeller can be cast iron if water to be pumped is fresh. The impeller is cast as one piece and is cheaper. The impeller may be made of stainless steel or gunmetal if it is liable to corrosion due to chemical impurities. If the impellers are to handle hot water with temperature more than 140°C, they can be made of cast steel. Pumps for milk industry are made of stainless steel to avoid contamination.

- **Semi-closed impeller :** If only one side of the impeller vanes is covered with plate, it is called a semi-closed impeller. It is used when the liquid is highly viscous or is a mixture of liquid and solid particles. The impeller has lesser number of vanes.

- **Open impeller :** An impeller without a cover plate is called an open impeller. It is used when the liquid has some solid suspended matter such as sand, paper pulp or molasses etc. There is no chance of its clogging.

2. Casing

The pressures at the inlet and outlet of a centrifgual pump are different. Therefore, the casing has to be airtight. The casing must have the provision to support the bearings, to fix stuffing box and to have packing material to prevent leaking.

The casing is usually split into two halves along a vertical or horizontal plane. Vertical-split casing is used for single-suction single-stage radial-flow pumps whereas the horizontal-split type is used for all other types such as multistage pumps and double suction pumps. It is already mentioned that a volute casing is used for single-stage pumps and a diffuser casing for multistage pumps.

3. Suction Pipe

The pipe through which water is lifted from the sump to the pump level is known as suction pipe. Its lower end is dipped in the sump water and the upper end is connected to the inlet, known as **eye** of the pump. At the lower end, a strainer and a foot-valve are fitted. A foot valve is a non-return valve which does not allow the water to go back to the sump. Losses in the suction pipe should not be high as it may cause cavitation due to negative pressure at the inlet of the pump. Sometimes, the diameter of the suction pipe is made larger than that of the delivery pipe to reduce the velocity which reduces the friction losses in the pipe.

4. Delivery Pipe

The pipe that raises the water level from the outlet of the pump to required height is known as delivery pipe. A valve is provided in the delivery pipe to regulate the flow.

18.4 WORKING OF A CENTRIFUGAL PUMP

To start a pump, first of all its casing is filled with water which ensures that the centrifugal head developed is sufficient to lift the water from the sump. If it is not done, the impeller rotates in air and as the head developed is proportional to the density of air, it is insufficient to lift the water from the sump to the pump level. The process of filling the casing with water before starting the pump is known as **priming**. After completing the priming, the motor is started and the delivery valve is closed to reduce the starting torque. When the motor gains the speed and the water rises to the pump level, the delivery valve is opened and the water starts pouring out of the delivery pipe. As the water flows towards the outlet of the pump, a partial vacuum is created at the eye of the impeller. Due to this suction, water from the sump which is at atmospheric pressure, is raised in the suction pipe. The water leaves the impeller with high pressure and high velocity. The velocity is usually reduced to gain in useful pressure energy in the impeller casing.

18.5 HEADS OF CENTRIFUGAL PUMP

- **Static head :** The vertical distance between water levels in the sump and the reservoir is known as static head. It consists of suction and delivery heads. Suction head is the height of centre line of pump above the sump level and the delivery head is the height of the liquid level in the tank above the centre line of the pump.
- **Total head :** The head which has to be developed by a pump to deliver water from the sump into the tank is called the total head. It takes into account the losses in the pipes and fittings and the loss due to kinetic energy at the delivery outlet. Total head is also known as **gross head, effective head** or **dynamic head.**

$$H = h_s + h_d + h_{fs} + h_{fd} + \frac{V_d^2}{2g} \qquad \ldots(18.2)$$

Where H = total head

h_s = suction head

h_d = delivery head

h_{fs} = losses in the suction pipe

h_{fd} = losses in the delivery pipe

V_d = Velocity of liquid in the delivery pipe.

Losses in the impeller or in the casing are not considered in the total head.

- **Manometric head :** It is the pressure head rise of the liquid in the impeller of the pump. If pressure gauges are installed on the suction and the delivery sides as near to the pump as possible, the difference in their readings gives the change in the pressure head. Let H_{ms} and H_{md} be the readings of the pressure gauges on the suction and delivery sides respectively (Fig. 18.8). Then the manometric head H_m is

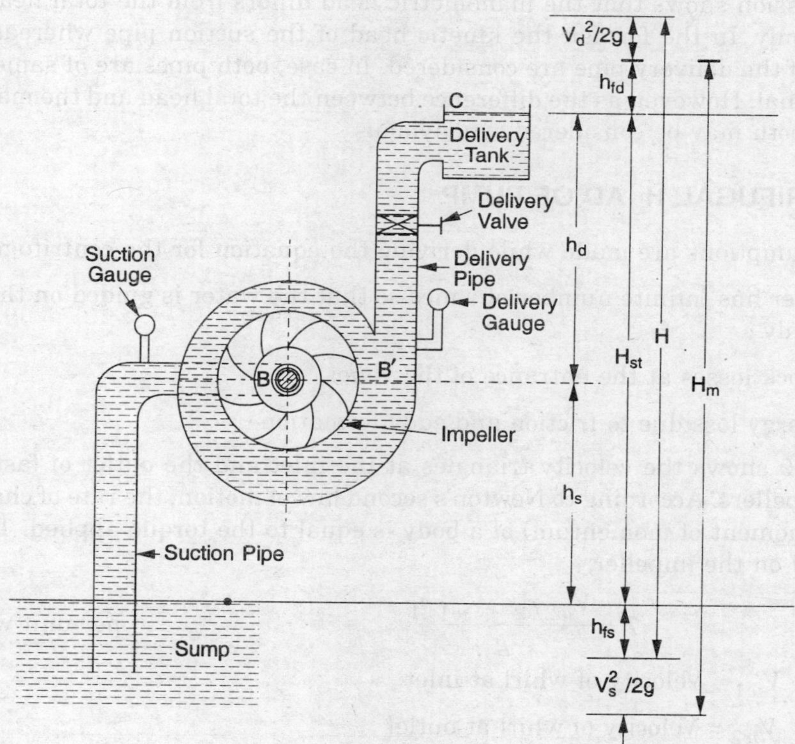

Fig. 18.8

$$H_m = H_{md} - H_{ms}$$

Apply Bernoulli's equation to a point B at the centre line of the pump on the suction side and point A in the sump where water is still. Assuming the sump level as datum,

$$H_{ms} + h_s + \frac{V_s^2}{2g} + h_{fs} = \frac{p_a}{w} + 0 + 0 \qquad \dots (18.3)$$

Similarly, apply Bernoulli's equation to a point B' on the centre line of the pump on the delivery side and to a point C at the discharge end. Neglect the height between the centre line of the pump and the pressure gauge on the delivery side and assume the centre line of the pump as datum.

$$H_{md} + \frac{V_d^2}{2g} + 0 = \frac{p_a}{w} + \frac{V_d^2}{2g} + h_d + h_{fd} \qquad \text{...(18.4)}$$

Subtracting Equation 18.3 from 18.4.

$$H_{md} + \frac{V_d^2}{2g} - H_{ms} - h_s - \frac{V_s^2}{2g} - h_{fs} = \frac{p_a}{w} + \frac{V_d^2}{2g} + h_d + h_{fd} - \frac{p_a}{w}$$

$$H_{md} - H_{ms} = h_s + h_d + h_{fs} + h_{fd} + \frac{V_s^2}{2g} \qquad \text{...(18.5)}$$

The expression shows that the manometric head differs from the total head in respect of kinetic head only. In the former, the kinetic head of the suction pipe whereas in the latter, kinetic head of the delivery pipe are considered. In case, both pipes are of same diameter, the two become equal. However, as the difference between the total head and the manometric head is negligible, both may be considered synonymous.

18.6 CENTRIFUGAL HEAD OF PUMP

Following assumptions are made while deriving the equation for the centrifugal head :

- Impeller has infinite number of vanes so that the water is guided on the vane profiles smoothly.
- No shock losses at the entrance of the vanes.
- No energy loss due to friction and eddy formation.

Figure 18.9 shows the velocity triangles at the inlet and the outlet of fast, medium and slow speed impellers. According to Newton's second law of motion, the rate of change if angular momentum (moment of momentum) of a body is equal to the torque applied. Thus if T is the torque applied on the impeller,

$$T = \frac{V_{w2} \cdot r_2 - V_{w1} \cdot r_1}{g} \quad \text{per unit weight of water}$$

where V_{w1} = Velocity of whirl at inlet

V_{w2} = Velocity of whirl at outlet

r_1 = radius of the impeller at the inlet

r_2 = radius of the impeller at the outlet

Energy transfer per unit weight = $T \cdot \omega$

$$= \frac{(V_{w2} \cdot r_2 - V_{w1} \cdot r_1)\omega}{g}$$

$$= \frac{V_{w2} \cdot u_2 - V_{w1} \cdot u_1}{g}$$

This is the fundamental equation of energy transfer and is known as Euler's energy equation. This also indicates the head developed by the pump which is also called **Euler head** (H_e).

To maximise the Euler head, the absolute velocity at the inlet is made radial with no whirl component *ie* V_{w1} is made zero. Then

$$H_e = \frac{V_{w2} \cdot u_2}{g} \qquad \text{...(18.6)}$$

Further, to obtain a high head V_{w2} *or* u_2 should be increased. u_2 can be increased by increasing the speed of rotation and the diameter whereas V_{w2} can be increased by providing sufficient number of vanes of suitable size and shape.

18.7 FORWARD CURVED AND BACKWARD CURVED VANES

The Euler head is (Fig. 18.9a),

$$H_e = \frac{V_{w2} \cdot u_2}{g} = \frac{u_2}{g}(u_2 - V_{f2} \cot \beta_2') \qquad \text{...(18.7)}$$

where β_2' is the angle of the vane tip with the tangential direction and V_{f2} is the velocity of flow.

Now,

$$Q = \pi D_2 B_2 V_{f2} \qquad \text{or} \qquad V_{f2} = \frac{Q}{\pi D_2 B_2}$$

Therefore,

$$H_e = \frac{u_2}{g}\left(u_2 - \frac{Q}{\pi D_2 B_2} \cot \beta_2'\right) = \frac{u_2^2}{g} - \frac{u_2 Q}{g \pi D_2 B_2} \cot \beta_2'$$

If a pump runs at constant speed, u_2 is constant. Therefore,

$$H_e = C_1 - C_2 Q \cot \beta_2'$$

Where C_1 and C_2 are constants such that

$$C_1 = \frac{u_2^2}{g} \qquad \text{and} \qquad C_2 = \frac{u_2}{g \pi D_2 B_2}$$

- When β_2' is less than 90°, cot β_2' is positive. The vanes are backward curved. The impeller rotates at fast speed. Euler head decreases with discharge (Fig. 18.9a).
- When β_2' is equal to 90°, cot β_2' is zero. The vanes are radial. The impeller is of medium speed. Euler head H_e does not vary with discharge Q and is equal to u_2^2/g (Fig. 18.9b).
- When β_2' is greater than 90°, cot β_2' is negative. The vanes are forward curved. The impeller is of slow speed. Euler head H_e increases with discharge Q (Fig. 18.9c).

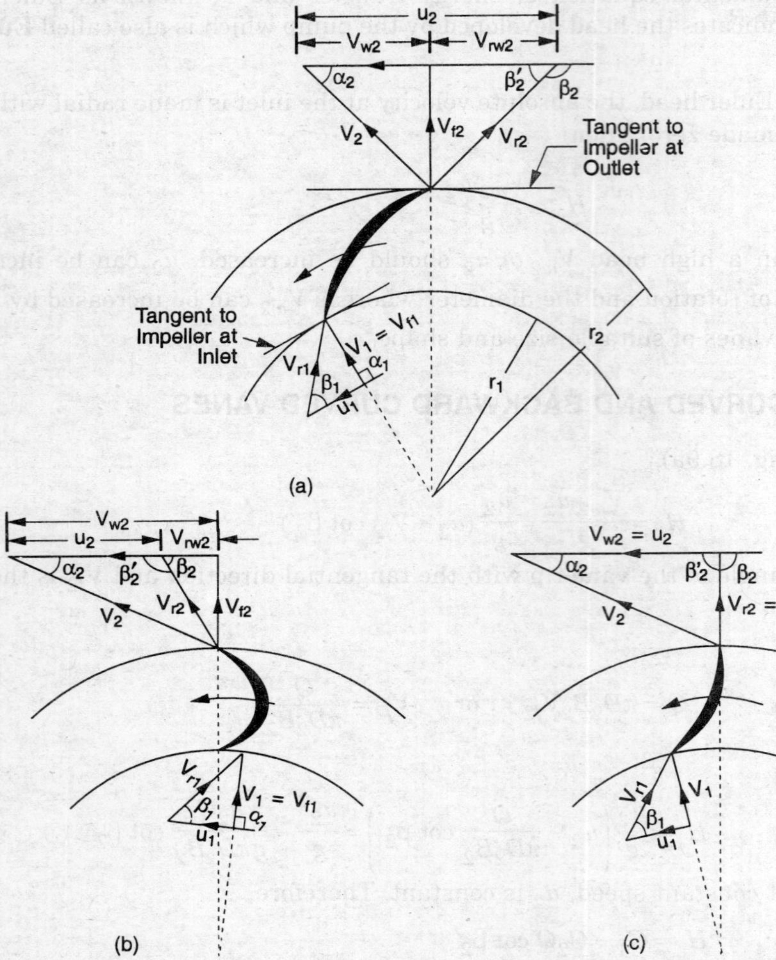

Fig. 18.9

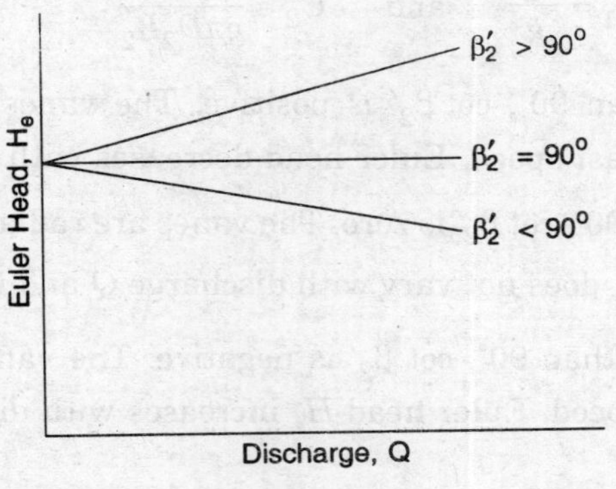

Fig. 18.10

The variation of Euler head with discharge is shown in Fig. 18.10.

From the plots, it is observed that the head developed is more in case of forward curved vanes. However, with the increase in angle β_2', the absolute velocity at the outlet V_2 also increases. A high value of V_2 is not recommended as the losses are increased due to eddies which actually lowers the efficiency. Moreover, these days, as the speed of the impellers is high, the use of backward curved vanes is common.

18.8 EFFECT OF FINITE NUMBER OF VANES

While deriving the equation of Euler head, it is assumed that the impeller has infinite number of vanes so that the water strictly follows the profile of the vanes. But in practice, the impeller has finite number of vanes. Thus the liquid does not follow the prescribed path properly. In the passage between two impeller vanes, the velocity distribution is considered as the result of two flows :

- Uniform velocity distribution, as with infinite number of Vanes (Fig. 18.11a).
- Owing to inertia, liquid particles are reluctant to rotate with the impeller. This results in high or positive pressure on the front face of the vanes and low or negative pressure on the back face. The velocity distribution due to this effect is shown in Fig. 18.11b.

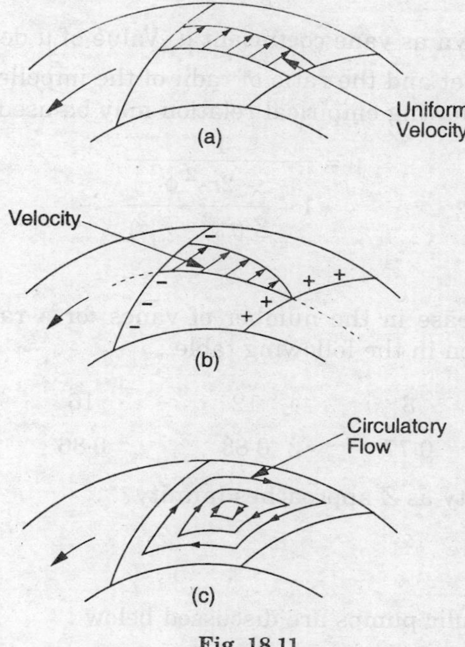

Fig. 18.11

When these two flows are superimposed, a rotational flow is obtained, the direction of which is opposite to that of the impeller (Fig. 18.12). This slows down the flow of liquid and the absolute velocity reduces to V_2'. As the tangential velocity remains same, the liquid leaves

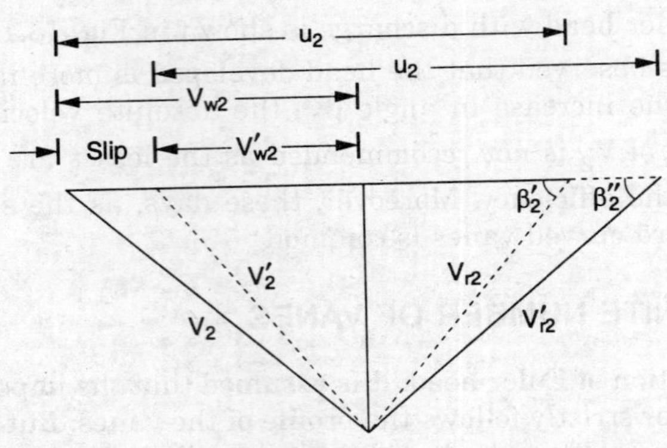

Fig. 18.12

the impeller at an angle less than the vane angle β_2' and the value of whirl velocity V_{w2} reduces to V_{w2}'. Difference between V_{w2} and V_{w2}' is called **slip**. In Fig. 18.7. the firm lines indicate impeller with infinite number of vanes and dotted lines when the number of vanes is finite. Let the reduced head be H_e' which is usually known as input head. Then

$$H_e' = \frac{V_{w2}' u_2}{g} \qquad \text{...(18.8)}$$

Ratio of H_e' to H_e is known as vane coefficient μ. Value of μ depends upon the number of vanes, vane angle at the outlet and the ratio of radii of the impeller. It does not depend upon operating conditions. The following empirical relation may be used to find the value of μ.

$$\mu = \frac{1}{1 + \dfrac{2 r_2^2 \phi}{Z (r_2^2 - r_1^2)}}$$

where $\phi = 0.6\,(1 + \sin \beta_2')$

Variation of μ with increase in the number of vanes for a radial flow pump with ratio $r_2/r_1 = 2$ and $\beta_2' = 30°$ is given in the following table :

Z	4	8	12	16	20	24
μ	0·62	0·77	0·83	0·86	0·89	0·91

Vale of μ approaches unity as Z approaches infinity.

18.9 EFFICIENCIES

Various efficiencies of hydraulic pumps are discussed below :

1. Hydraulic Efficiency

The losses, which occur between the suction and the delivery ends of a pump, are known as hydraulic losses (Fig. 18.13). These are

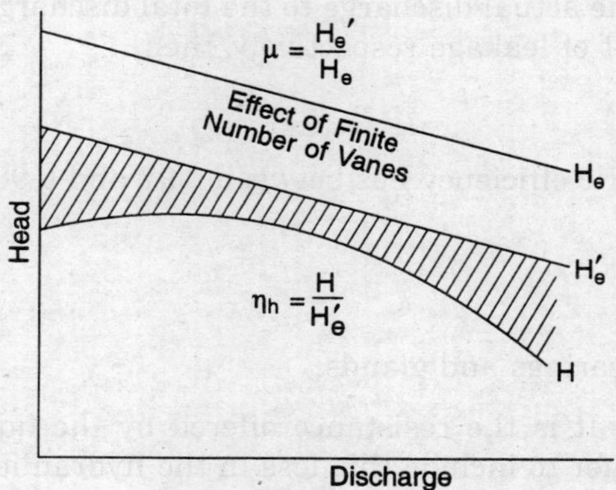

Fig. 18.13

- Shock losses at the inlet of the impeller vanes and in the volute. When a pump is operated at different working conditions than the designed ones, shock losses occur due to the change in angles of the inlet velocity. Also, when the water enters into the volute, losses are there due to the change in velocity as it slows down.
- Friction and eddy losses in the casing and the vanes.
- Friction and eddy losses in the suction and delivery pipes, losses due to sudden change in area and the direction of flow.

Hydraulic efficiency

$$\eta_h = \frac{H}{H_e'} = \frac{gH}{V_{w2}' \cdot u_2} \qquad \text{...(18.9)}$$

Hydraulic efficiency varies from 0·6 to 0·9.

2. Manometric Efficiency

Manometric efficiency is defined as the hydraulic efficiency of an ideal pump *ie* a pump with *i.e.* infinite number of vanes.

$$\eta_{man} = \frac{H}{H_e} = \frac{H}{H_e'/\mu} = \mu\eta_h \qquad \text{...(18.10)}$$

For infinite number of vanes, $\mu = 1$ and manometric efficiency is the same as hydraulic efficiency. In the absence of sufficient data, the problems may be solved assuming infinite number of vanes, and no distinction is made between hydraulic efficiency and manometric efficiency and so

$$\eta_h = \eta_{man} = \frac{gH}{V_{w2} \cdot u_2} \qquad \text{...(18.11)}$$

3. Volumetric Efficiency

As the pressure at the outlet of the impeller is more than that at the Inlet, there is tendency of liquid to slip back through the clearance between the impeller and the casing. Some leakage can also take place from the seals. These losses are known as volumetric losses. Volumetric efficiency is the ratio of the actual discharge to the total discharge. If Q and δQ are the actual discharge and the amount of leakage respectively, then

$$\eta_v = \frac{Q}{Q + \delta Q} \quad \text{...(18.12)}$$

The value of volumetric efficiency lies between 0·97 and 0·98.

4. Mechanical Efficiency

Mechanical losses are

- Friction loss in bearings and glands.
- Disc friction loss. It is the resistance offered by the liquid to the rotating impeller. Some authors prefer to include this loss in the hydraulic losses.

Let P and δP be the total power input to the shaft and the mechanical losses respectively. Then

$$\eta_m = \frac{P - \delta P}{P} \quad \text{...(18.13)}$$

Its value is between 0·95 to 0·98.

5. Overall Efficiency

Ratio of total head developed by a pump to the total power input is known as the overall efficiency. It is the product of the hydraulic, volumetric and mechanical efficiencies.

Range of overall efficiency lies between 0.71 to 0.86.

Example 18.2 A centrifugal pump raises the head of water by 30 m. Its speed is 1000 rpm and discharge 220 litres per second. The outlet angle of the backward curved vanes is 30° and the velocity of flow at the outlet 2.8 m/s. If the hydraulic efficiency of the pump is 79%, find the diameter and the width of the impeller at the outlet.

Solution $Q = 0{\cdot}22\ \text{m}^3/\text{s}$ $\quad N = 1000$ rpm

$H = 30$ m $\quad V_{f2} = 2{\cdot}8$ m/s

$\eta_h = 0{\cdot}79$ $\quad \beta_2' = 30°$

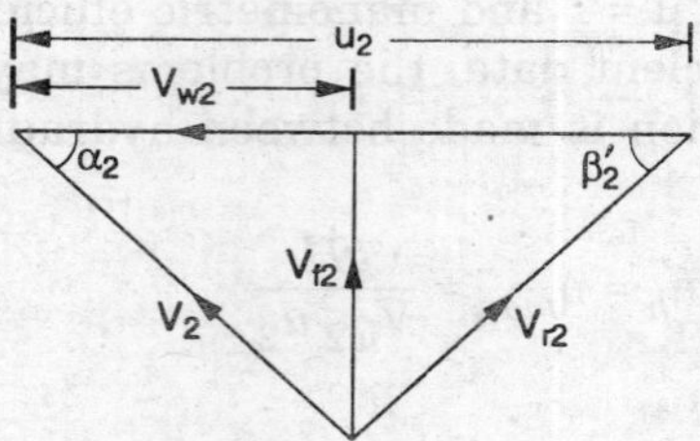

Fig. 18.14

$$\eta_h = \frac{gH}{u_2 \cdot V_{w2}}$$

$$= \frac{gH}{u_2\,(u_2 - V_{f2} \cot \beta_2')}$$

$$0{\cdot}79 = \frac{9{\cdot}81 \times 30}{u_2\,(u_2 - 2{\cdot}8 \cot 30°)}$$

$$= \frac{9{\cdot}81 \times 30}{u_2^2 - 4{\cdot}85\, u_2}$$

$$u_2^2 - 4{\cdot}85 u_2 = 372{\cdot}5$$

$$u_2^2 - 2\left(\frac{4{\cdot}85}{2}\right) u_2 + \left(\frac{4{\cdot}85}{2}\right)^2 = 372{\cdot}5 + \left(\frac{4{\cdot}85}{2}\right)^2$$

$$(u_2 - 2{\cdot}425)^2 = (19{\cdot}3)^2$$

$$u_2 = 19{\cdot}3 + 2{\cdot}425 = 21{\cdot}726 \text{ m/s}$$

$$\frac{\pi D_2 N}{60} = 21{\cdot}726$$

$$\frac{\pi D_2 \times 1000}{60} = 21{\cdot}726$$

$$D_2 = 0{\cdot}415 \text{ m} \quad \text{or} \quad 415 \text{ mm}$$

Also

$$Q = \pi D_2 B_2 V_{f2}$$

$$0{\cdot}22 = \pi \times 0{\cdot}415 \times B_2 \times 2{\cdot}8$$

$$B_2 = 0{\cdot}0603 \text{ m} \quad \text{or} \quad 60{\cdot}3 \text{ mm}$$

Example 18.3 Water flows through a centrifugal pump at the rate of 130 litres per second. The speed of the pump is 1540 rpm and the head developed 30 m. The diameter of the impeller at the outlet is 280 mm and the width 50 mm. Assuming hydraulic efficiency to be 77%, determine the vane angles at the outlet of the impeller.

Solution $Q = 0{\cdot}13 \text{ m}^3/\text{s}$ $\quad H = 30$ m

$D_2 = 0{\cdot}28$ m $\quad B_2 = 0{\cdot}05$ m

$N = 1540$ rpm $\quad \eta_h = 0{\cdot}77$

From the outlet velocity triangle,

$$\tan \beta_2' = \frac{V_{f2}}{u_2 - V_{w2}}$$

u_2, V_{f2} and V_{w2} are found as follows :

$$u_2 = \frac{\pi D_2 N}{60} = \frac{\pi \times 0{\cdot}28 \times 1540}{60} = 22{\cdot}578 \text{ m/s}$$

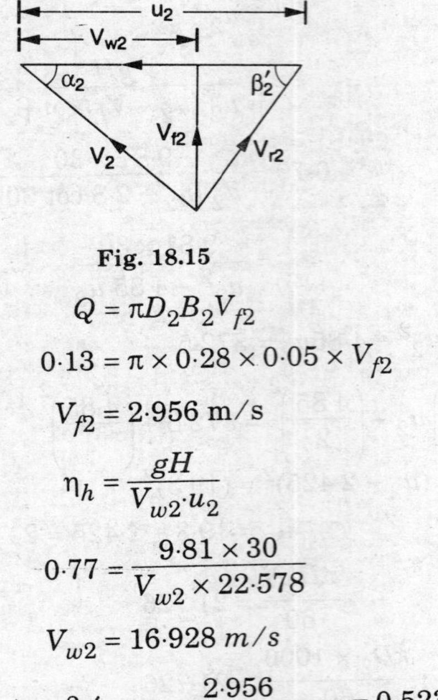

Fig. 18.15

$$Q = \pi D_2 B_2 V_{f2}$$

$$0{\cdot}13 = \pi \times 0{\cdot}28 \times 0{\cdot}05 \times V_{f2}$$

$$V_{f2} = 2{\cdot}956 \text{ m/s}$$

$$\eta_h = \frac{gH}{V_{w2} \cdot u_2}$$

$$0{\cdot}77 = \frac{9{\cdot}81 \times 30}{V_{w2} \times 22{\cdot}578}$$

$$V_{w2} = 16{\cdot}928 \; m/s$$

$$\tan \beta_2' = \frac{2{\cdot}956}{22{\cdot}578 - 16{\cdot}928} = 0{\cdot}523$$

$$\beta_2' = 27{\cdot}6°$$

Example 18.4 The discharge of water through a centrifugal pump is $0{\cdot}05$ m^3/s. The length of the pipe is 22 m and it raises the height through 15 m. The diameter of the pipe is 140 mm. Overall efficiency is 72% and coefficient of friction 0·02. Find the power input of the pump.

Solution $Q = 0{\cdot}05$ m^3/s $\quad H_s = 15$ m

$L = 20$ m $\quad f = 0{\cdot}015$

$d = 0{\cdot}14$ m $\quad \eta_0 = 0{\cdot}72$

$$H = H_s + h_f + \frac{V_2^2}{2g}$$

V_2 and h_f are found as follows :

$$Q = \frac{\pi}{4} d^2 \cdot V_2$$

$$0{\cdot}05 = \frac{\pi}{4} \times 0{\cdot}14^2 \times V_2$$

$$V_2 = 3{\cdot}25 \text{ m/s}$$

$$h_f = \frac{4fLV_2^2}{2gd} = \frac{4 \times 0{\cdot}02 \times 22 \times 3{\cdot}25^2}{2 \times 9{\cdot}81 \times 0{\cdot}14} = 6{\cdot}77 \text{ m}$$

$$H = 15 + 6{\cdot}77 + \frac{3{\cdot}25^2}{2 \times 9{\cdot}81} = 22{\cdot}306 \text{ m}$$

$$\eta_0 = \frac{wQH}{P}$$

$$0{\cdot}72 = \frac{9{\cdot}81 \times 0{\cdot}05 \times 22{\cdot}306}{P}$$

$$P = 15{\cdot}196 \text{ kW}$$

Example 18.5 Show that the pressure head rise in the impeller of a centrifugal pump is given by

$$H = \frac{1}{2g}(V_{f1}^2 + u_2^2 - V_{f2}^2 \operatorname{cosec}^2 \beta_2')$$

where V_{f1} and V_{f2} are the velocities of flow at the inlet and the outlet respectively, u_2 is the vane velocity at the outlet and β_2' the vane angle at the outlet of the backward curved vanes.

Neglect frictional and other losses in the impeller.

Solution Apply Bernoulli's equation at the inlet and the outlet of impeller, (neglecting losses in the impeller),

Head at outlet = Head at inlet + Head added in the impeller

$$\frac{p_2}{w} + \frac{V_2^2}{2g} + z_2 = \frac{p_1}{w} + \frac{V_1^2}{2g} + z_1 + \frac{V_{w2} \cdot u_2}{g}$$

$$\frac{p_2 - p_1}{w} = \frac{V_1^2}{2g} - \frac{V_2^2}{2g} + \frac{V_{w2} \cdot u_2}{g} \qquad (z_1 = z_2)$$

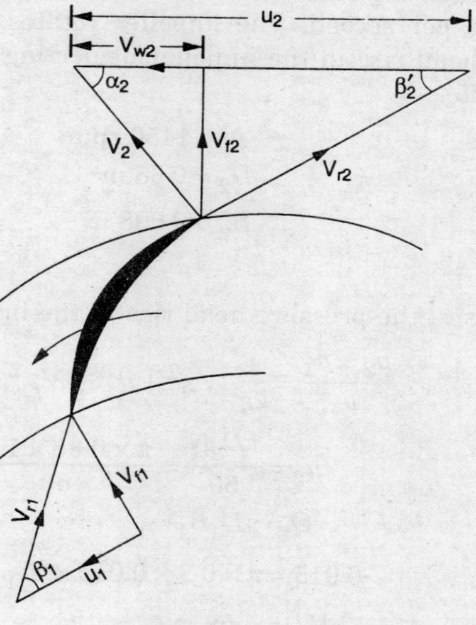

Fig. 18.16

From inlet velocity triangle, $V_1 = V_{f1}$

From outlet velocity triangle,

$$V_{w2} = u_2 - V_{f2} \cot \beta_2'$$

and
$$V_2^2 = V_{f2}^2 + V_{w2}^2$$
$$= V_{f2}^2 + (u_2 - V_{f2} \cot \beta_2')^2$$
$$= V_{f2}^2 + u_2^2 + V_{f2}^2 \cot^2 \beta_2' - 2\, u_2\, V_{f2} \cot \beta_2'$$
$$= V_{f2}^2 (1 + \cot^2\beta_2') + u_2^2 - 2u_2\, V_{f2} \cot \beta_2'$$
$$= V_{f2}^2 \operatorname{cosec}^2\beta_2' + u_2^2 - 2u_2\, V_{f2} \cot \beta_2'$$

Pressure head rise

$$\frac{p_2 - p_1}{w} = \frac{V_1^2}{2g} - \frac{1}{2g}(V_{f2}^2 \operatorname{cosec}^2 \beta_2' + u_2^2 - 2u_2 \cdot V_{f2} \cot \beta_2') + \frac{u_2}{g}(u_2 - V_{f2} \cot \beta_2')$$

$$= \frac{1}{2g}\begin{pmatrix} V_{f1}^2 - V_{f2}^2 \operatorname{cosec}^2 \beta_2' - u_2^2 + 2u_2 \cdot V_{f2} \cot \beta_2' \\ + 2u_2^2 - 2u_2 V_{f2} \cot \beta_2' \end{pmatrix}$$

$$= \frac{1}{2g}(V_{f1}^2 + u_2^2 - V_{f2}^2 \operatorname{cosec}^2 \beta_2')$$

Example 18.6 The internal and external diameters of the impeller of a centrifugal pump are 200 mm and 380 mm respectively. The width at the inlet of impeller is 15 mm and at the outlet 8 mm. The discharge of water through the impeller is 15 litre per second. The impeller rotates at 1450 rpm. Determine the pressure head rise in the impeller neglecting losses. The vane angle at the outlet is 35°.

Solution $Q = 0{\cdot}015 \text{ m}^3/\text{s}$ $\qquad N = 1450$ rpm

$D_1 = 0{\cdot}2$ m $\qquad D_2 = 0{\cdot}38$ m

$B_1 = 0{\cdot}015$ m $\qquad B_2 = 0{\cdot}008$ m

$\beta_2' = 35°$

Neglecting losses, the pressure head rise in the impeller is given by

$$\frac{p_2 - p_1}{w} = \frac{1}{2g}(V_{f1}^2 + u_2^2 - V_{f2}^2 \operatorname{cosec}^2 \beta_2')$$

$$u_2 = \frac{\pi D_2 N}{60} = \frac{\pi \times 0{\cdot}38 \times 1450}{60} = 28{\cdot}85 \text{ m/s}$$

As
$$Q = \pi D_1 B_1 V_{f1}$$
$$0{\cdot}015 = \pi \times 0{\cdot}2 \times 0{\cdot}015 \times V_{f1}$$
$$V_{f1} = 1{\cdot}59 \text{ m/s}$$

Also
$$Q = \pi D_2 B_2 V_{f2}$$
$$0{\cdot}015 = \pi \times 0{\cdot}38 \times 0{\cdot}008 \times V_{f2}$$
$$V_{f2} = 1{\cdot}57 \text{ m/s}$$
$$\frac{p_2 - p_1}{w} = \frac{1}{2 \times 9{\cdot}81}(1{\cdot}59^2 + 28{\cdot}85^2 - 1{\cdot}57^2 \operatorname{cosec}^2 35°)$$
$$= 42{\cdot}17 \text{ m}$$

Example 18.7 The suction lift of a diffuser pump is 3 m and the height of the delivery tank above the pump is 12 m. The radial velocity of water through the impeller is 3 m/s. The vanes of the impeller are curved back at 45° and the water enters the vanes radially. The velocity of water in the delivery pipe is 1·75 m/s. Neglecting the friction and other losses in the pipes and the impeller, find

(i) tangential velocity of the impeller at the outlet
(ii) kinetic head at the outlet of impeller
(iii) pressure head at the outlet of impeller
(iv) angle of the guide vanes.

Solution $H = 3 + 12 = 15$ m $\qquad \beta_2' = 45°$

$V_{f1} = V_{f2} = 3$ m/s $\qquad V_3 = 1{\cdot}75$ m/s

(i) All the losses in the impeller are neglected *i.e.* the hydraulic efficiency is 100%.

or
$$V_{w2} \cdot u_2 = gH$$
$$u_2 (u_2 - V_{f2} \cot \beta_2') = gH$$
$$u_2 (u_2 - 3 \cot 45°) = 9{\cdot}81 \times 15$$
$$u_2^2 - 3u_2 = 147{\cdot}15$$

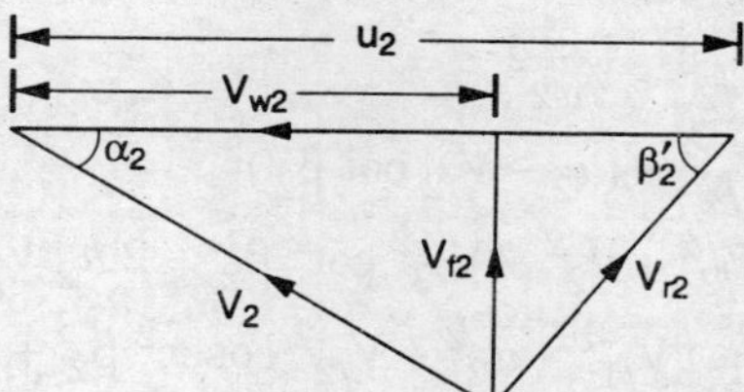

Fig. 18.17

$$u_2^2 - 2\left(\frac{3}{2}\right) u_2 + 1{\cdot}5^2 = 147{\cdot}15 + 1{\cdot}5^2$$
$$(u_2 - 1{\cdot}5)^2 = 12{\cdot}223^2$$
$$u_2 = 13{\cdot}723 \text{ m/s}$$

(ii) $V_2 = \sqrt{V_{f2}^2 + V_{w2}^2}$

we have $V_{w2} = u_2 - V_{f2} \cot \beta_2' = 13{\cdot}723 - 3 \cot 45 = 10{\cdot}723$ m/s
$$V_2 = \sqrt{3^2 + 10{\cdot}723^2} = 11{\cdot}135 \text{ m/s}$$

Kinetic head at the outlet

$$\frac{V_2^2}{2g} = \frac{11{\cdot}135^2}{2 \times 9{\cdot}81} = 6{\cdot}319 \text{ m}$$

(iii) Total head at outlet = Pressure head + kinetic head

$$h_d = \frac{p_2}{w} + \frac{V_2^2}{2g}$$

$$12 = \frac{p_2}{w} + 6{\cdot}319$$

or pressure head = 12 – 6·319 = 5·681 m

(iv) $\tan \alpha_2 = \frac{V_{f2}}{V_{w2}} = \frac{3}{10{\cdot}723} = 0{\cdot}28$

$$\alpha_2 = 15{\cdot}63^\circ$$

Example 18.8 Assuming the flow velocity to be constant, prove that the manometric head of a centrifugal pump running at speed N and having a discharge Q can be expressed as

$$H_{man} = AN^2 + BNQ + CQ^2$$

Solution If losses are ignored, the manometric head is equal to the pressure head rise in the impeller and some fraction of the kinetic head at the impeller which is recovered in the volute chamber.

$$H_{man} = \frac{p_2 - p_1}{w} + \frac{kV_2^2}{2g}$$

$$\frac{p_2 - p_1}{w} = \frac{1}{2g}(V_{f1}^2 + u_2^2 - V_{f2}^2 \operatorname{cosec}^2 \beta_2')$$

$$V_2^2 = V_{f2}^2 + V_{w2}^2$$

$$= V_{f2}^2 + (u_2 - V_{f2} \cot \beta_2')^2$$

$$= V_{f2}^2 + u_2^2 + V_{f2}^2 \cot^2 \beta_2' - 2\, u_2\, V_{f2} \cot \beta_2'$$

$$H_{man} = \frac{1}{2g}\left(\begin{array}{l} V_{f1}^2 + u_2^2 - V_{f2}^2 \operatorname{cosec}^2 \beta_2' + kV_{f2}^2 \\ + kV_{f2}^2 \cot^2 \beta_2' + ku_2^2 - 2ku_2 V_{f2} \cot^2 \beta_2' \end{array}\right)$$

Since the flow velocity is constant, $V_{f1} = V_{f2} = V_f$

$$H_{man} = \frac{1}{2g}((1+k)u_2^2 - 2ku_2 V_f \cot \beta_2' + V_f^2 (1 + k \cot^2 \beta_2' - \operatorname{cosec}^2 \beta_2'))$$

$$= \frac{1}{2g}(A' u_2^2 + B' V_f u_2 + C' V_f^2)$$

where A', B' and C' are constants such that

$$A' = 1 + k$$

$$B' = -2\, k \cot \beta_2'$$

$$C' = 1 + k \cot^2 \beta_2' - \operatorname{cosec}^2 \beta_2'$$

or $$H_{man} = \frac{1}{2g}\left[A'\left(\frac{\pi D_2 N}{60}\right)^2 + B'\frac{Q}{a}\left(\frac{\pi D_2 N}{60}\right) + C'\frac{Q^2}{a^2}\right]$$

$$= \frac{1}{2g}\left(A'\left(\frac{\pi D_2}{60}\right)^2 N^2 + B'\frac{Q}{a}\left(\frac{\pi D_2}{60}\right)N + C'\frac{Q^2}{a^2}\right)$$

$$= AN^2 + BNQ + CQ^2$$

where A, B and C are constants such that

$$A = \frac{A'}{2g}\left(\frac{\pi D_2}{60}\right)^2$$

$$B = \frac{B'}{2ga}\left(\frac{\pi D_2}{60}\right)$$

$$C = \frac{C'}{2ga^2}$$

Example 18.9 The diameter of the impeller of a centrifugal pump is 800 mm. The pump delivers water at the rate of 90 m^3 per minute. The head raised is 72 m and the speed 900 rpm. Peripheral area at the outlet is 0·3 m^2. The vanes of the impeller are curved back at 30° to the tangential direction. The leakage loss is 2·5%, the input head is reduced by 20% of the theoretical head and the external mechanical losses are 30 kW. Find the power input and the overall efficiency of the pump.

Solution $Q = 90/60 = 1{\cdot}5\ m^3/s$ $\quad H = 72$ m

$N = 900$ rpm $\quad D_2 = 0{\cdot}8$ m

$A_2 = 0{\cdot}3\ m^2$ $\quad B_2' = 30°$

$\delta Q = 0{\cdot}025Q$

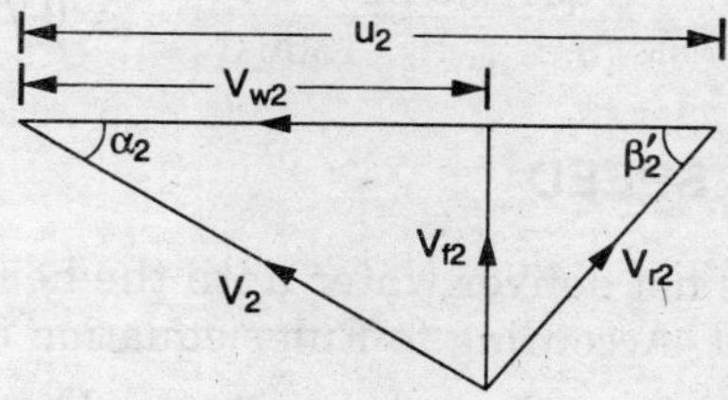

Fig. 18.18

$$\mu = \frac{100 - 20}{100} = 0{\cdot}8$$

$$P = w\,(Q + \delta Q)\,H_e' + \text{Mechanical losses}$$

H_e' is found as below :

$$Q + \delta Q = A_2 V_2$$

$$(1{\cdot}5 + 0{\cdot}025 \times 1{\cdot}5) = 0{\cdot}3 \times V_{f2}$$

or $$V_{f2} = 5{\cdot}125 \text{ m/s}$$

$$V_{w2} = u_2 - V_{f2} \cot \beta_2'$$

$$= \frac{\pi D_2 N}{60} - V_{f2} \cot \beta_2'$$

$$= \frac{\pi \times 0{\cdot}8 \times 900}{60} - 5{\cdot}125 \cot 30^\circ$$

$$= 37{\cdot}7 - 8{\cdot}877$$

$$= 28{\cdot}82 \text{ m/s}$$

$$H_e' = \mu{\cdot}H_e = 0{\cdot}8 \times \frac{V_{w2}{\cdot}u_2}{g} = 0{\cdot}8 \times \frac{28{\cdot}82 \times 37{\cdot}7}{9{\cdot}81} = 88{\cdot}6 \text{ m}$$

Therefore,

$$P = w\,(Q + \delta Q)\, H_e' + \text{Mechanical losses}$$

$$= 9{\cdot}81\,(1{\cdot}5 + 0{\cdot}025 \times 1{\cdot}5) \times 88{\cdot}6 + 30$$

$$= 1336{\cdot}3 + 30 \text{ kW}$$

$$= 1366{\cdot}3 \text{ kW}$$

$$\eta_h = \frac{H}{H_e'} = \frac{72}{88{\cdot}6} = 0{\cdot}813$$

$$\eta_v = \frac{Q}{Q + \delta Q} = \frac{1{\cdot}5}{1{\cdot}5 + 0{\cdot}025 \times 1{\cdot}5} = 0{\cdot}976$$

$$\eta_m = \frac{P - \delta P}{P} = \frac{1366{\cdot}3 - 30}{1366{\cdot}3} = 0{\cdot}978$$

$$\eta_0 = \eta_h \times \eta_v \times \eta_m = 0{\cdot}813 \times 0{\cdot}976 \times 0{\cdot}978 = 0{\cdot}776 \quad \text{or} \quad 77{\cdot}6\%$$

Alternatively,

$$\eta_0 = \frac{wQH}{P} = \frac{9{\cdot}81 \times 1{\cdot}5 \times 72}{1366{\cdot}3} = 0{\cdot}776 = 77{\cdot}6\%$$

18.10 MINIMUM STARTING SPEED

When a pump is started, it does not deliver water until the head generated is enough to raise the water to the required height. According to Euler equation of energy transefer.

$$H_e = \frac{V_2^2 - V_1^2}{2g} + \frac{u_2^2 - u_1^2}{2g} + \frac{V_{r1}^2 - V_{r2}^2}{2g} \quad \text{per unit weight}$$

Till the flow does not start, the velocities are zero. The head developed is only due to the second term in the above equation. As soon as the centrifugal head developed becomes equal to the total head, the flow starts *i.e.* when

$$\frac{u_2^2 - u_1^2}{2g} = H_e = \frac{H}{\eta_h}$$

$$\left[\left(\frac{\pi D_2 N}{60}\right)^2 - \left(\frac{\pi D_1 N}{60}\right)^2\right] = \frac{2gH}{\eta_h}$$

$$\left(\frac{\pi N}{60}\right)^2 (D_2^{\ 2} - D_1^{\ 2}) = \frac{2gH}{\eta_h}$$

$$N_{\min} = \frac{60}{\pi}\sqrt{\frac{2gH}{\eta_h (D_2^{\ 2} - D_1^{\ 2})}} \qquad \text{...(18.14)}$$

Example 18.10 A centrifugal pump raises the head of water through 4 m and delivers 1·5 m^3/s. The speed of the impeller is 180 rpm. The impeller diameter at the outlet is 1·3 m and the area at the outer periphery 0·3 m^2. The ratio of the outlet to the inlet diameter is 2 and the vane angle at the outlet 30°. Determine the hydraulic efficiency, power required and the minimum starting speed.

Solution $Q = 1{\cdot}5\ m^3/s$ $\qquad H = 4$ m

$D_2 = 1{\cdot}3$ m $\qquad D_1 = 1{\cdot}3/2 = 0{\cdot}65$ m

$N = 180$ rpm $\qquad A_2 = 0{\cdot}3\ m^2$

$\beta_2' = 30°$

$$V_{w2} = u_2 - V_{f2} \cot \beta_2'$$

$$= \frac{\pi D_2 N}{60} - \frac{Q}{A_2} \cot \beta_2'$$

$$= \frac{\pi \times 1{\cdot}3 \times 180}{60} - \frac{1{\cdot}5}{0{\cdot}3} \cot 30°$$

$$= 12{\cdot}25 - 5 \cot 30°$$

$$= 3{\cdot}59 \text{ m/s}$$

$$\eta_h = \frac{gH}{V_{w2} \cdot u_2} = \frac{9{\cdot}81 \times 4}{3{\cdot}59 \times 12{\cdot}25} = 0{\cdot}892 \quad \text{or} \quad 89{\cdot}2\%$$

$$P = \frac{wQH}{\eta_h} = \frac{9{\cdot}81 \times 1{\cdot}5 \times 4}{0{\cdot}892} = 66 \text{kW}$$

$$N_{\min} = \frac{60}{\pi}\sqrt{\frac{2gH}{\eta_h (D_2^{\ 2} - D_1^{\ 2})}}$$

$$= \frac{60}{\pi}\sqrt{\frac{2 \times 9{\cdot}81 \times 4}{0{\cdot}892\,(1{\cdot}3^2 - 0{\cdot}65^2)}}$$

$$= \frac{60}{\pi}\sqrt{\frac{78{\cdot}48}{1{\cdot}1306}}$$

$$= 159{\cdot}1 \text{ rpm}$$

18.11 PROPORTIONS

- **Speed Ratio :** It is the ratio of tangential velocity of impeller at the outlet u_2 to the quantity $\sqrt{2gH}$.

$$K_u = \frac{u_2}{\sqrt{2gH}}$$

Its value varies between 0·95 to 1·3.

- **Flow ratio :** The ratio of flow velocity at the outlet V_{f2} to the quantity $\sqrt{2gH}$ is called flow ratio.

$$K_f = \frac{V_{f2}}{\sqrt{2gH}}$$

Its value varies between 0·1 to 0·25.

- **Diameter ratio :** It is the ratio of the inner diameter to the outer diameter of the impeller and its value lies between 0·3 to 0·65.
- **Outlet vane angle :** For best efficiency, the vane angle at the outlet should lie between 20° to 30°.
- **Number of vanes :** The number of vanes depends upon pump size, outlet vane angle, total head and the speed ratio. Usually, 6 to 8 vanes may be sufficient.

18.12 SUCTION LIFT

At the inlet of a centrifugal pump, there exists negative or suction pressure. When the absolute pressure reduces to the level of vapour pressure of water at the room temperature, vapour bubbles are formed. As these vapour bubbles reach in a region of higher pressure, they collapse and vacuum is developed in their place. The surrounding water rushes in to fill the vacuum with high velocity. The impact is so high that the surface in the vicinity is pitted. The phenomenon is known as **cavitation.** To avoid this, it is necessary that the pump is fitted at such a height that the absolute pressure does not fall below the vapour pressure p_v. Rewriting Eq. 18.3.

$$H_{ms} + h_s + \frac{V_s^{\,2}}{2g} + h_{fs} = \frac{p_a}{w}$$

$$\frac{p_s}{w} + h_s + \frac{V_s^{\,2}}{2g} + h_{fs} = \frac{p_a}{w} \qquad \text{...(18.15)}$$

Value of p_s should not fall below the vapour pressure p_v of water at room temperature. Thus at the limit, $p_s = p_v$.

$$\frac{p_v}{w} + h_s + \frac{V_s^{\,2}}{2g} + h_{fs} = \frac{p_a}{w}$$

or

$$h_s = \frac{p_a}{w} - \frac{p_v}{w} - \frac{V_s^{\,2}}{2g} - h_{fs} \qquad \text{...(18.16)}$$

The maximum value of h_s should not exceed the value given by this equation to avoid cavitation.

18.13 NET POSITIVE SUCTION LIFT

Net positive suction head is defined as the total head (velocity head plus pressure head) developed at the centre line of the pump inlet above the vapour pressure head of the liquid *i.e.*

$$NPSH = \left(\frac{V_s^2}{2g} + \frac{p_s}{w}\right) - \frac{p_v}{w}$$

But from equation 18.15.

$$\frac{p_s}{w} = \frac{p_a}{w} - h_s - \frac{V_s^2}{2g} - h_{fs}$$

$$NPSH = \left[\frac{V_s^2}{2g} + \left(\frac{p_a}{w} - h_s - \frac{V_s^2}{2g} - h_{fs}\right)\right] - \frac{p_v}{w}$$

$$= \frac{p_a}{w} - \frac{p_v}{w} - h_s - h_{fs} \quad \text{...(18.17)}$$

18.14 HEAD LOSS DUE TO FLOW VARIATION

If a pump does not operate at the designed values of discharge and speed, shock losses occur which lower the efficiency of the pump. Assume that the value of flow velocity reduces from v_{f1} to v_{f1}' on reducing the discharge (Fig. 18.19a). Since the speed of the impeller remains same, the velocity triangle is changed to $ab'c$. The angle at which the liquid enters the impeller

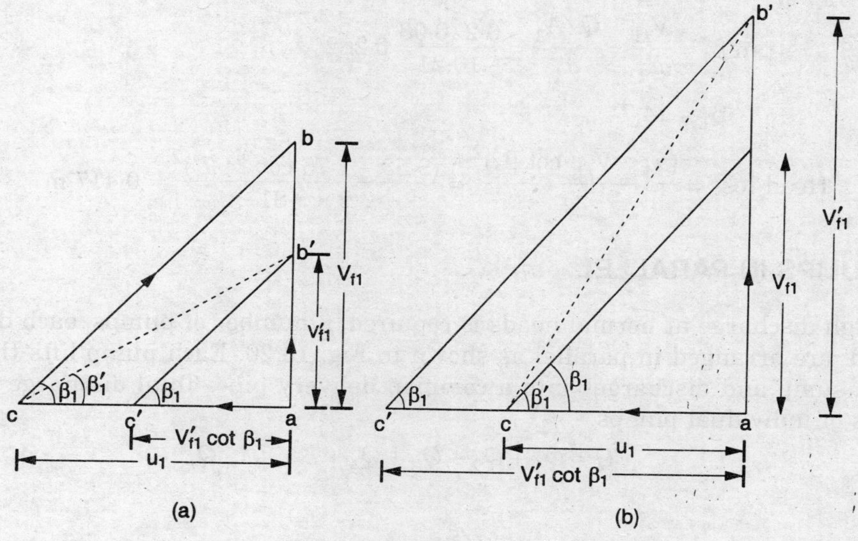

Fig. 18.19

reduces to β_1'. However, the vane angle of the impeller at the inlet remains β_1. Therefore, the water does not enter the impeller smoothly and there are shock losses. Thus there is virtual change in the tangential velocity at the inlet which is a measure of the shock loss.

$$\text{Head loss} = \frac{V_{cc}'^2}{2g} = \frac{(u_1 - V_{f1}' \cot \beta_1)^2}{2g} \qquad \ldots(18.18)$$

The effect of the increased flow is also shown in Fig. 18.19b.

Example 18.11 The flow area at the inlet of a centrifugal pump is $0{\cdot}08\ m^2$ and the discharge of water 200 l/s. The impeller speed is 650 rpm and the diameter at the inlet 280 mm. If the discharge is reduced by 30% keeping the speed of the impeller constant, determine the loss of head at the inlet.

Solution $D_1 = 0{\cdot}28$ m $\qquad A_1 = 0{\cdot}08\ m^2$

$Q = 0{\cdot}2\ m^3/s \qquad Q' = (1 - 0{\cdot}3) \times 0{\cdot}2 = 0{\cdot}14\ m^2$

$N = 650$ rpm

$$\text{Head loss} = \frac{(u_1 - V_{f1}' \cot \beta_1)^2}{2g}$$

$$u_1 = \frac{\pi D_1 N}{60} = \frac{\pi \times 0{\cdot}28 \times 650}{60} = 9{\cdot}53 \text{ m/s}$$

$$Q' = A_1 \times V_f'$$

$$0{\cdot}14 = 0{\cdot}08 \times V_f'$$

$$V_f' = 1{\cdot}75 \text{ m/s}$$

$$\tan \beta_1 = \frac{V_{f1}}{u_1} = \frac{Q/A_1}{u_1} = \frac{0{\cdot}2/0{\cdot}08}{10{\cdot}21}\ 0{\cdot}262$$

$$\beta_1 = 14{\cdot}7^\circ$$

$$\text{Head loss} = \frac{(u_1 - V_{f1}' \cot \beta_1)^2}{2g} = \frac{(9{\cdot}53 - 1{\cdot}75 \cot 14{\cdot}7^\circ)^2}{2 \times 9{\cdot}81} = 0{\cdot}417 \text{ m}$$

18.15 PUMPS IN PARALLEL

When a high discharge at normal heads is required, a number of pumps, each developing the same head, are arranged in parallel as shown in Fig. 18.20. Each pump lifts the water from a common sump and discharges into a common delivery pipe. Total discharge is the sum of discharges of individual pumps.

$$Q = Q_1 + Q_2 + Q_3 + Q_4 + \ldots\ldots\ldots\ldots\ Q_n$$

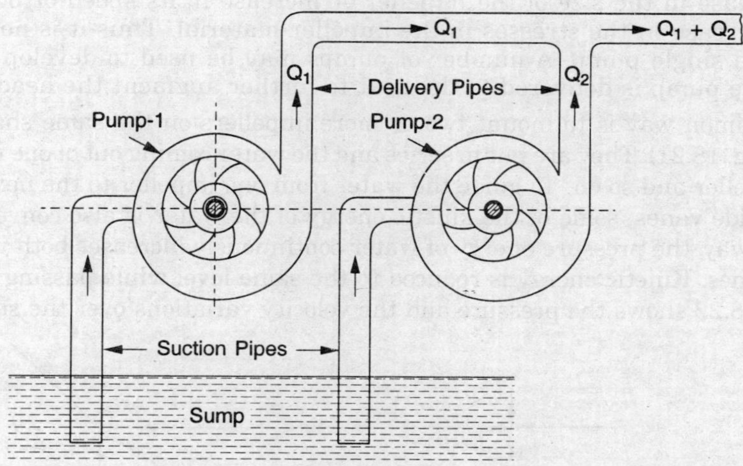

Fig. 18.20

18.16 PUMPS IN SERIES (MULTISTAGE PUMPS)

To obtain high centrifugal heads, it is necessary to increase the tangential velocity u_2 which

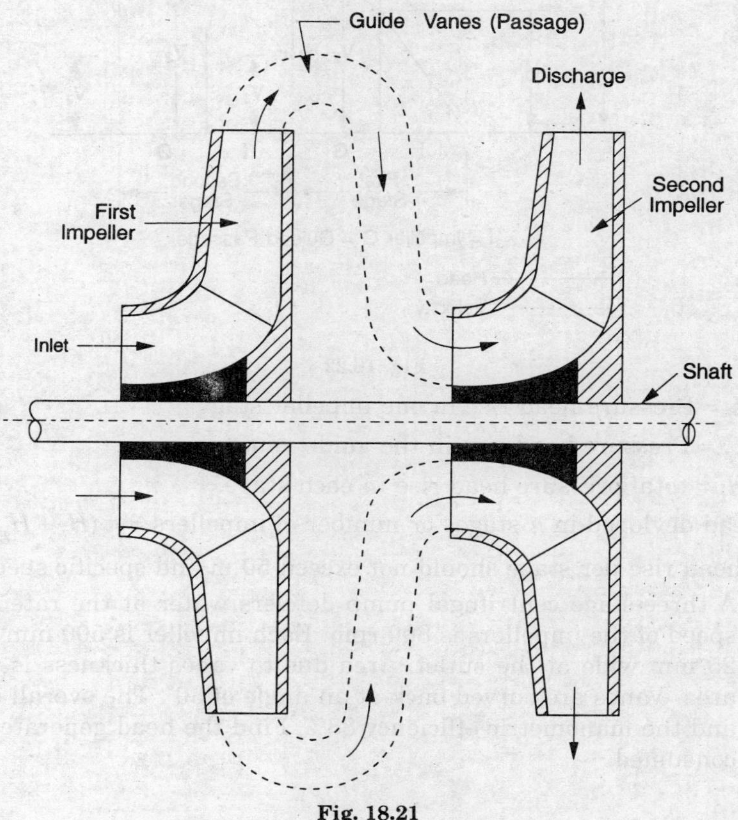

Fig. 18.21

means an increase in the size of the impeller or increase in its speed or both. However, this may lead to increase in the stresses in the impeller material. Thus it is not possible to raise high heads in a single pump. A number of pumps may be used to develop a high head. The discharge of one pump is delivered to the next to further augment the head.

A more common way is to mount two or more impellers on the same shaft enclosed in the same casing (Fig. 18.21). They are put in series and the water coming out of one impeller is directed to the next impeller and so on. To guide the water from one impeller to the next, guide vanes are used. In the guide vanes, some of the kinetic energy of the water is also converted into pressure energy. In this way, the pressure energy of water continuously increases both in the impeller and in the guide vanes. Kinetic energy is reduced to the same level while passing through the guide vanes. Figure 18.22 shows the pressure and the velocity variations over the stages.

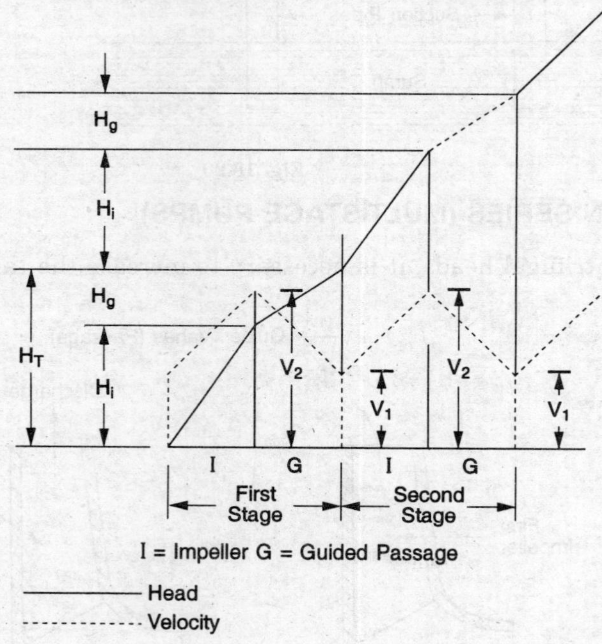

Fig. 18.22

Let H_i = Pressure head rise in one impeller stage

H_g = Pressure head rise in the guide vanes

H_t = total pressure head rise in each stage.

Then total head devloped in n stages or number of impellers $= n\ (H_i + H_g) = n.H_t$

Usually, the head rise per stage should not exceed 50 m and specific speed 600.

Example 18.12 A three-stage centrifugal pump delivers water at the rate of 0·1 m^3/s. The speed of the impellers is 800 rpm. Each impeller is 500 mm in diameter and 25 mm wide at the outlet. Area due to vanes thickness is 10% of the total area. Vanes are curved back at an angle of 30°. The overall efficiency is 80% and the manometric efficiency 88%. Find the head generated and the power consumed.

Solution $Q = 0{\cdot}1 \text{ m}^3/\text{s}$ $\qquad N = 800$ rpm

$D_2 = 0{\cdot}5$ m $\qquad B_2 = 0{\cdot}025$ m

$k = (1 - 0{\cdot}1) = 0{\cdot}9$ $\qquad n = 3$

$\eta_0 = 0{\cdot}8$ $\qquad \eta_{man} = 0{\cdot}88$

$\beta_2' = 30°$

$$A_2 = k{\cdot}\pi D_2 B_2 = 0{\cdot}9 \times \pi \times 0{\cdot}5 \times 0{\cdot}025 = 0{\cdot}03534 \text{ m}^2$$

$$V_{w2} = u_2 - V_{f2} \cot \beta_2'$$

$$= \frac{\pi D_2 N}{60} - \frac{Q}{A_2} \cot \beta_2'$$

$$= \frac{\pi \times 0{\cdot}5 \times 800}{60} - \frac{0{\cdot}1}{0{\cdot}03534} \cot 30°$$

$$= 20{\cdot}94 - 2{\cdot}829 \cot 30°$$

$$= 16{\cdot}04 \text{ m/s}$$

$$\eta_h = \frac{gH}{V_{w2}{\cdot}u_2}$$

$$0{\cdot}88 = \frac{9{\cdot}81 \times H}{16{\cdot}04 \times 20{\cdot}94}$$

$$H = 30{\cdot}13 \text{ m}$$

Total head developed $= n{\cdot}H = 3 \times 30{\cdot}13 = 90{\cdot}4$ m

$$\eta_0 = \frac{WQH}{P}$$

$$0{\cdot}8 = \frac{9{\cdot}81 \times 0{\cdot}1 \times 90{\cdot}4}{P}$$

$$P = 110{\cdot}85 \text{ kW}$$

18.17 SPECIFIC SPEED

Specific speed of pumps is used to compare the performance of different types of pumps. It is defined as the speed of a geometrically similar pump which delivers a unit quantity of water at a unit head under corresponding conditions of working.

$$u = \frac{\pi DN}{60}$$

$$N = \frac{60u}{\pi D} \propto \frac{u}{D}$$

D and u in terms of Q and H are found as below :

$$Q = k{\cdot}\pi DBV_f \propto DBV_f \propto D^2{\cdot}V_f \qquad (B \propto D)$$

$$D \propto \sqrt{\frac{Q}{V_f}} \propto \sqrt{\frac{Q}{K_f\sqrt{2gH}}} \propto \sqrt{\frac{Q}{H^{1/2}}}$$

Also $$u = K_u \sqrt{2gH} \propto \sqrt{H}$$

Therefore, $$N \propto \frac{\sqrt{H}}{\sqrt{Q/H^{1/2}}} \propto \frac{\sqrt{H}}{\sqrt{Q}} H^{1/4} \propto \frac{H^{3/4}}{\sqrt{Q}} = C \frac{H^{3/4}}{\sqrt{Q}}$$

where C is a constant.

From definition of the specific speed, when H and Q are unity each, N becomes the specific speed N_s of the pump. Then

$$C = N = N_s$$

Thus $$N = N_s \frac{H}{\sqrt{Q}}$$

or $$N_s = \frac{N\sqrt{Q}}{H^{3/4}} \qquad \ldots(18.19)$$

18.18 PRIMING DEVICES

To start a pump, first it is filled with water. The process is known as **priming**. Though the suction pipe is provided with a foot valve or a non-return valve so that water does not leak out easily. However, if a pump is being started after a long period, it may be empty and priming may be needed. Thus a check is to be made every time. An air vent is provided in the pump casing so that the air may escape while the water is being poured in. The devices used to prime a centrifugal pump are known as priming devices.

External Priming Devices

- The pump and the suction pipe can be filled with water with the help of a funnel.
- A chamber may be provided on the delivery side to store the water which may be used later on to fill the pump and the suction pipe.
- Some vacuum creating device can be used to generate enough vacuum in the pump to lift the water in the suction pipe.

Self Priming Devices

- **Priming with separator :** In this method, the suction pipe is bent near the pump as shown in Fig. 18.23. Due to this bend, water from the pump does not leak into the sump. Thus the pump always remains filled with water. The delivery side is provided with an air separator chamber made by enlarging the cross- sectional area of the pipe. When the pump is started, the impeller rotates in a mixture of water and air, air being sucked from the suction pipe. The head generated is higher than that developed in a medium of air only. However, the head is not enough to lift the water from the sump upto the level of the pump. As the mixture of water and air rises in the delivery pipe and reaches the enlarged portion of the pipe, its velocity decreases. Water being heavier, falls back into the pump from the other side of the separator and the air goes through the delivery pipe into the atmosphere. Continuous running of the impeller sends more and more air out of the delivery pipe making the remaining mixture richer

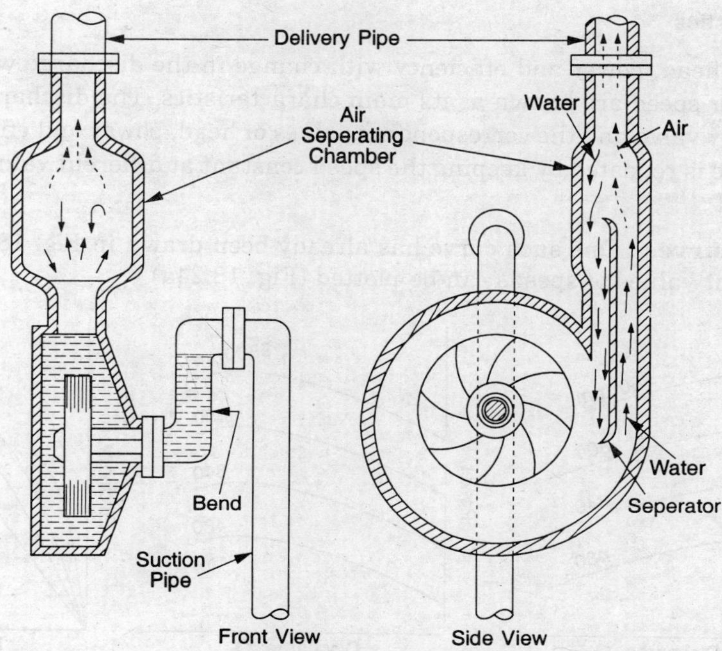

Fig. 18.23

and richer in water. When whole of the air of the suction pipe is pumped out, priming is completed. Soon the water starts coming out of the delivery pipe. This method is widely used in pumping of petroleum products.

- **Priming with vacuum pump :** A vacuum or air pump can be fitted on the suction side of the pump which can be run by the pump shaft. When started, the vacuum pump takes out air from the suction pump discharging it into the atmosphere. When whole of the air in the suction pipe has been taken out, the water starts coming out of the pump.

18.19 PERFORMANCE OF CENTRIFUGAL PUMPS

A pump like any other machine has maximum efficiency when operated at designed values of speed, head and discharge. However, it may have to be operated at different conditions than the designed ones. On changing one parameter, others are affected. Characteristic curves help in predicting the behaviour of a pump under different parameters, Usual characteristics are

1. Main characteristic
2. Operating characteristics
3. Constant head and constant discharge characteristics
4. Constant efficiency characteristics

1. Main Characteristics

The variations of head, power and efficiency with change in the discharge when the pump is run at a particular speed are known as its main characteristics. The discharge is varied with the help of delivery valve and the corresponding values of head, power and efficiency are found out. The procedure is repeated by keeping the speed constant at different values. The following curves are plotted :

- **H vs. Q curves :** One such curve has already been drawn in Fig. 18.10. More curves at different values of speeds can be plotted (Fig. 18.24a)

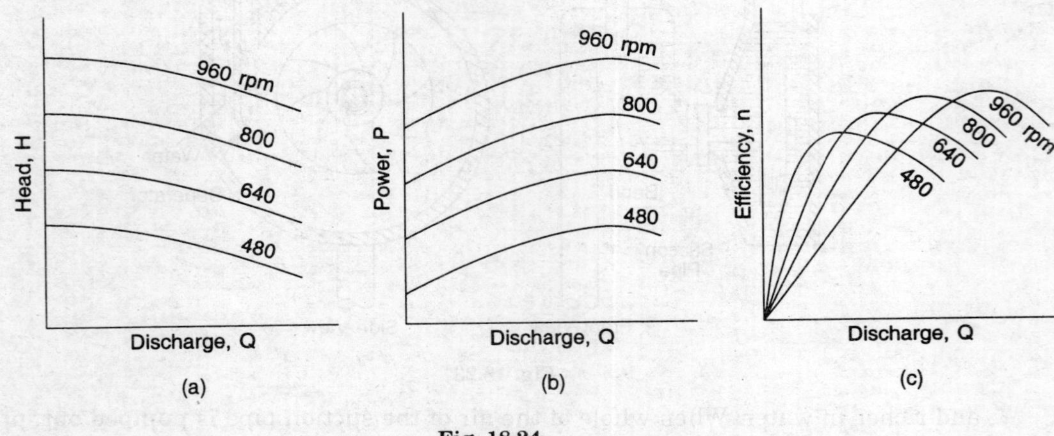

Fig. 18.24

- **P vs. Q curves :** As the discharge increases, the power input increases (Fig. 18.24b). The curves do not pass through the origin, as some power is necessary to overcome the mechanical losses.
- **η vs. Q curves :** The curves pass through origin as efficiency is zero when there is no discharge (Fig. 18.24c).

2. Operating Characteristics

In practice, a pump usually operates at the designed speed. The particular set of main characterstic curves, which corresponds to the designed speed, is known as operating characteristics. Values of Q and H corresponding to maximum efficiency are the normal or designed values (Fig. 18.25). Head that corresponds to zero discharge is called **shut-off head.**

3. Constant-Head and Constant-Discharge Characteristics

These curves are helpful if the pump used is a variable-speed pump. The following curves may be plotted (Fig. 18.26) :

- **Q. vs. N curves :** These curves are drawn by keeping the head constant and varying the speed and the discharge.

$$u = K_u \sqrt{2gH} \qquad \text{or} \qquad u \propto \sqrt{H}$$

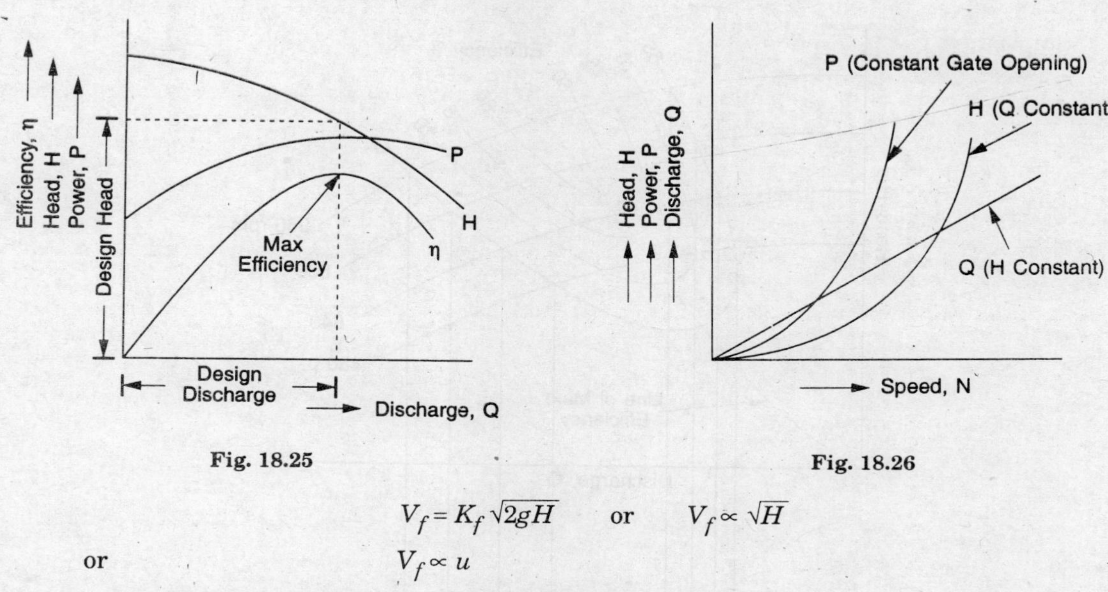

Fig. 18.25

Fig. 18.26

$$V_f = K_f \sqrt{2gH} \quad \text{or} \quad V_f \propto \sqrt{H}$$

or $$V_f \propto u$$

$$Q = k \cdot \pi DBV_f \propto DBV_f \propto D^2 \cdot V_f \propto D^2 \cdot u \propto D^2 \cdot ND \propto ND^3$$

Thus Q increases linearly with speed.

- **H vs. N curves :** The variation of head with speed may be plotted by keeping the discharge constant.

$$u \propto \sqrt{H}$$

$$H \propto u^2 \propto N^2D^2$$

Thus N vs. H is a parabolic curve.

- **P vs. N curves :** By keeping the gate opening constant, P vs. N curves may be plotted.

$$P = wQH/\eta_0 \propto QH \propto ND^3 \cdot N^2D^2 \propto N^3D^5$$

Thus variation of P with N is cubic.

4. Constant Efficiency Curves

These curves are also called Muschel curves. These can be drawn from H vs. Q and η vs. Q curves. Corresponding to a particular value of efficiency, two values of discharge are obtained (Fig. 18.27). On transferring these to H vs. Q plot, constant efficiency curves are obtained. On joining the peak points of all these curves, the curve of best efficiency is obtained. The constant efficiency curves help in knowing the range of operation of a pump for a particular efficiency.

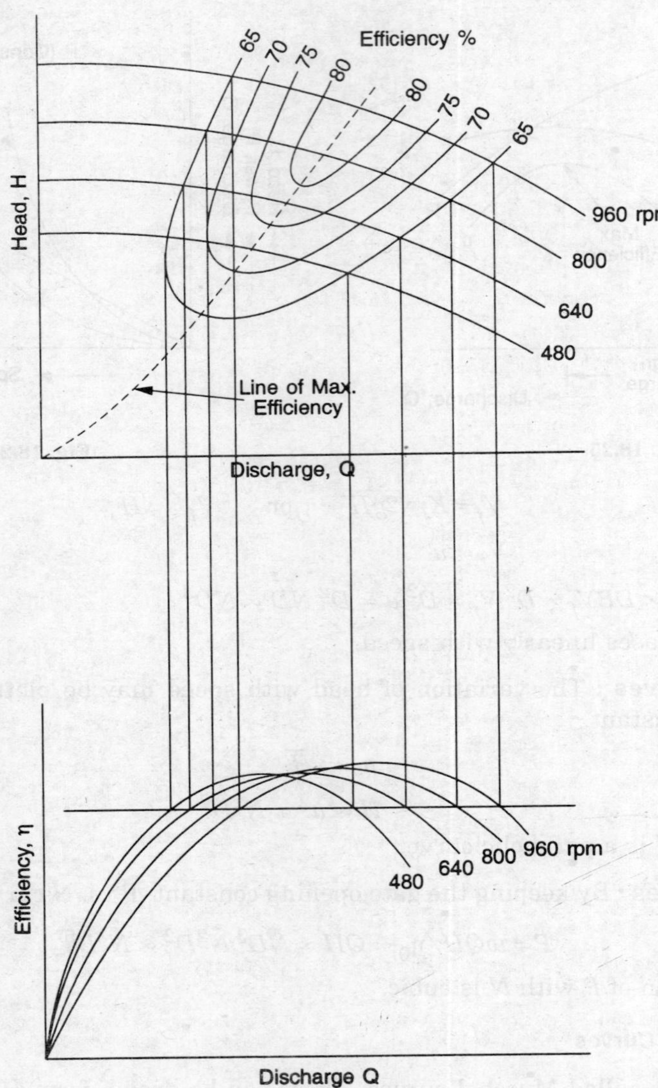

Fig. 18.27

18.20 CLASSIFICATION OF RECIPROCATING PUMPS

There are reciprocating pumps which raise the liquid by suction only and are called **suction pumps.** However, these pumps raise the liquid through small heights only since the maximum height to which the liquid is raised cannot be higher than barometric reading *i.e.* 10 m of water. Other type of reciprocating pumps lift the liquid by means of pressure and are known as **force pumps**. These can be used to generate very high pressures. Reciprocating pumps usually mean force pumps.

Reciprocating pumps are mainly classified according to use of piston sides and according to number of cylinders.

According to Use of Piston Sides

A pump may be single acting or double acting.

Single-acting reciprocating pump : If the liquid is filled only on one side of the piston and there is only one suction and one delivery pipe, it is known as a single-acting reciprocating pump.

Double-acting reciprocating pump : If there are two suction and two delivery pipes and the liquid is received on both sides of the piston in the cylinder and is delivered into the respective delivery pipes, it is called a double-acting reciprocating pump (Fig. 18.28). When one side of the piston becomes a suction side, the other acts as a pressure side. A second arrangement for a double-acting reciprocating pump is also shown in Fig 18.29. In this arrangement, there is only one suction pipe and one delivery pipe. Each pipe is connected to the two ends of the cylinder.

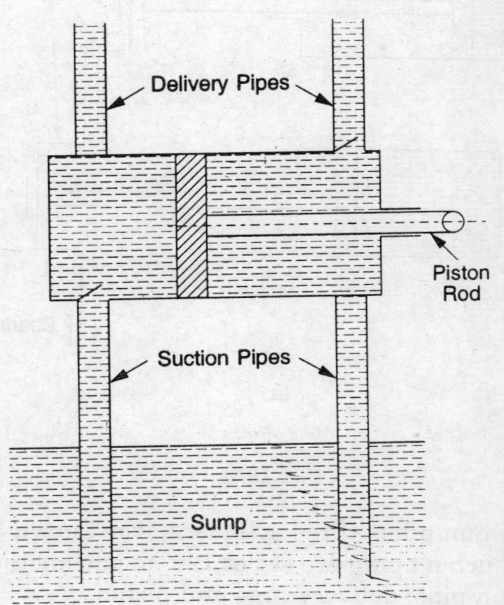

Fig. 18.28

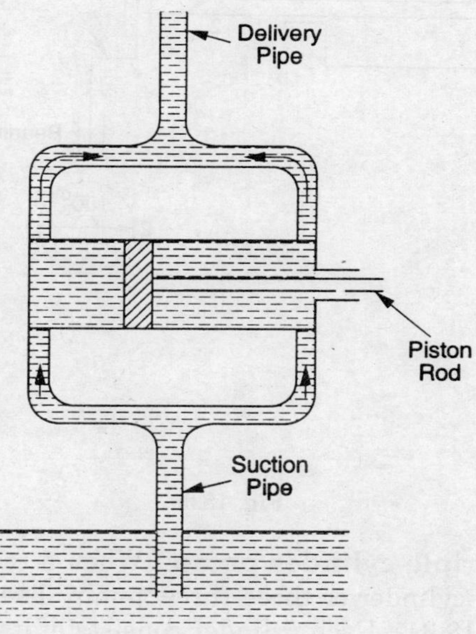

Fig. 18.29

According to Number of Cylinders

Reciprocating pumps with more than one cylinder are known as multi-cylinder pumps.

Single-cylinder pump : A reciprocating pump can be either single acting or double acting.

Double-cylinder pump : If two cylinders are connected to the same shaft, it is called a **double-cylinder or two-throw pump**. Each cylinder consists of its own suction as well as delivery pipe. Cranks set at 180° to each other are connected to the two pistons (Fig. 18.30). When there is suction stroke in one pump, it is delivery stroke in the other. Thus, liquid is delivered to the delivery pipe during each stroke of the piston.

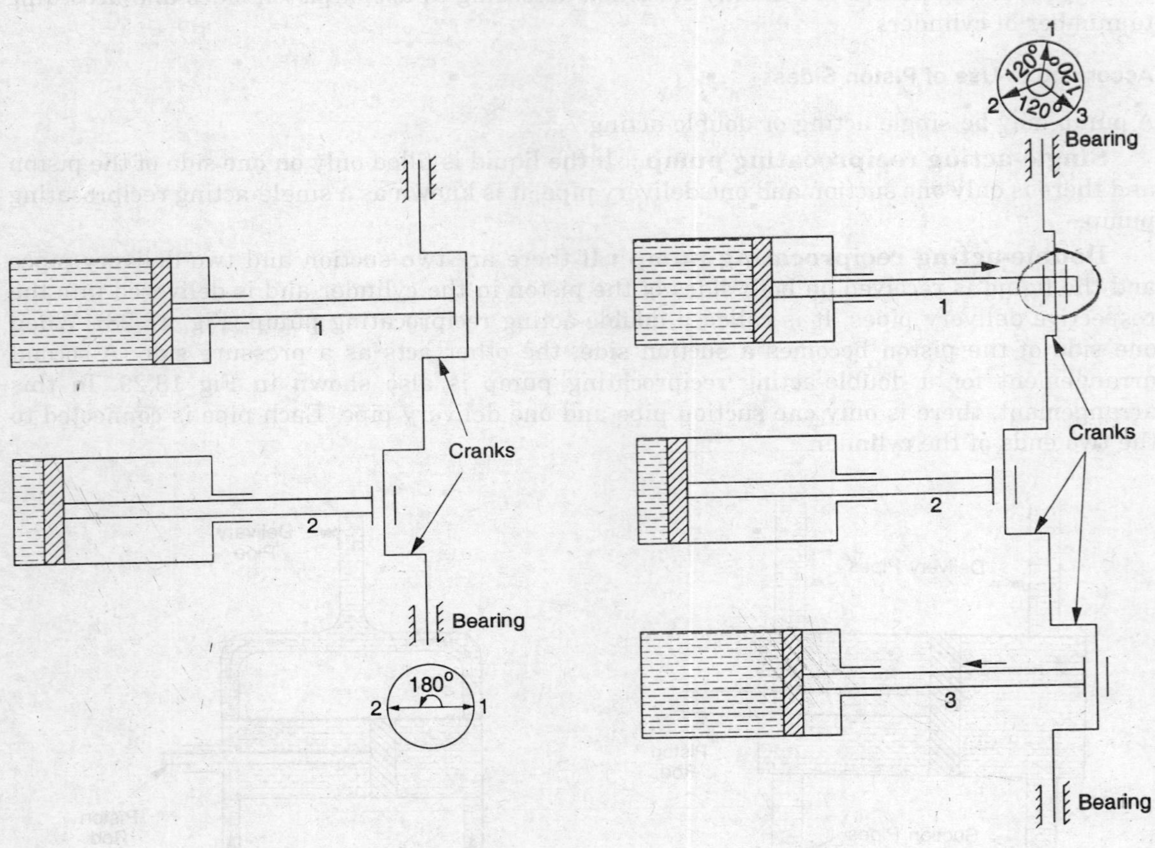

Fig. 18.30

Fig. 18.31

Triple-cylinder pump : When a reciprocating pump has three cylinders, it is known as triple-cylinder or three-throw pump. The cranks of such a pump are set at 120° to one another (Fig. 18.31). Each cylinder consists of its own suction pipe, delivery pipe and piston.

There can also be four-cylinder and five-cylinder pumps.

18.21 COMPONENTS OF A RECIPROCATING PUMP

The components of a reciprocating pump are piston or plunger, cylinder, slider crank arrangement consisting of crank, connecting rod and piston rod, suction pipe and delivery pipe.

The rotary motion of the crank obtained from a motor is transformed into forward and backward motion of the plunger or piston inside the cylinder with the help of connecting and piston rods (Fig. 18.32). During one cylce of the crank, two strokes of the piston, one forward and one backward are completed. The cylinder is connected to the sump by the suction pipe

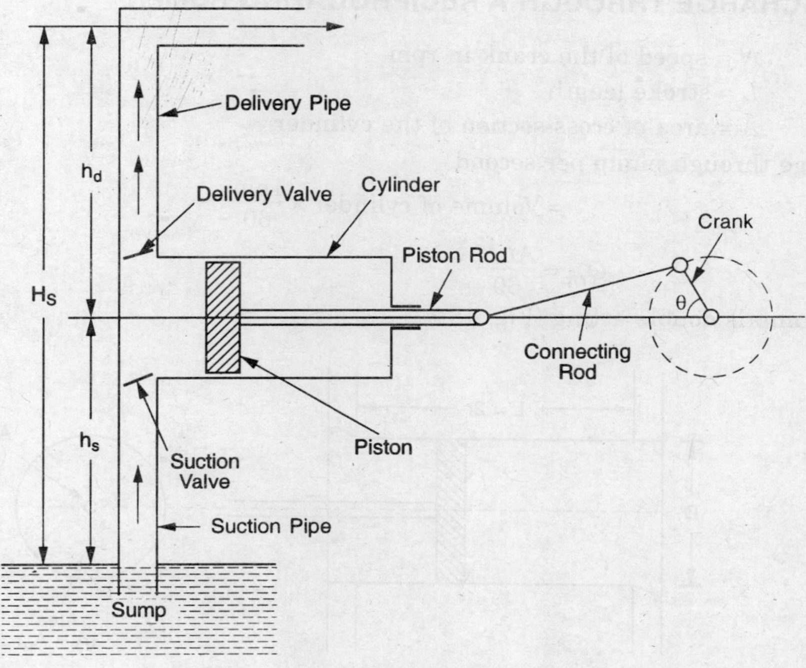

Fig. 18.32

and to the delivery tank by the delivery pipe. Cylinder ends of the pipes are provided with non-return valves which allow the liquid to pass only in one direction.

18.22 WORKING OF A RECIPROCATING PUMP

First assume that the piston slides from left to right. This creates a suction pressure in the cylinder and is called the suction stroke. During this period the suction valve is open and delivery valve is closed. In case the pump is started for the first time or after a long period, air from the suction pipe is sucked during the suction stroke. Liquid rises in the suction pipe by a small height due to the atmospheric pressure on the sump liquid. When the piston moves from right to left, it is known as delivery stroke. During this period, the suction valve is closed and the delivery valve is open. Air in the cylinder is pushed out through the delivery pipe by the thrust of the piston.

When whole of the air of suction pipe is taken out, the liquid from the sump enters into the cylinder. During the delivery stroke, it is pushed into the delivery pipe. Soon, the liquid starts coming out of the delivery pipe intermittently *i.e.* during the delivery strokes only. Note that the theoretical height of suction pipe cannot be greater than the atmospheric pressure head on the sump.

18.23 DISCHARGE THROUGH A RECIPROCATING PUMP

Let N = speed of the crank in rpm

L = stroke length

A = area of cross-section of the cylinder

Discharge through pump per second

$$= \text{Volume of cylinder} \times \frac{\text{rpm}}{60}$$

$$Q_{th} = \frac{ALN}{60} \qquad \text{...(18.20)}$$

If the pump is double acting (Fig. 18.33),

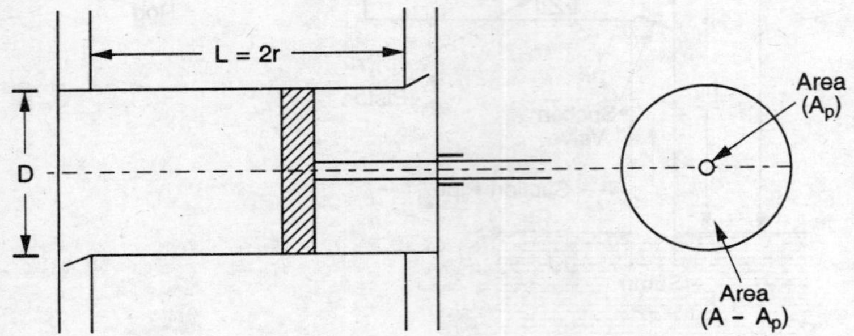

Fig. 18.33

$$Q_{th} = \frac{ALN}{60} + \frac{(A - A_p)\,LN}{60}$$

$$= \frac{(2A - A_p)\,LN}{60} \qquad \text{...(18.21)}$$

where A_p is the cross-sectional area of the piston rod. However, if the area of the piston rod is neglected, the discharge is twice than that of a single-acting reciprocating pump. Due to leakage losses and time delay in closing the valves, the actual discharge is generally less than the theoretical discharge.

18.24 SLIP

Slip is defined as the difference between the theoretical and the actual discharge.

Slip = Theoretical discharge – Actual discharge = $Q_{th} - Q_a$

Slip is also expressed in terms of percentage of the theoretical discharge.

$$\text{Percentage slip} = \frac{Q_{th} - Q_a}{Q_{th}} = \left(1 - \frac{Q_a}{Q_{th}}\right) \times 100 = (1 - C_d) \times 100 \qquad \text{...(18.22)}$$

where C_d is the ratio of the actual discharge to the theoretical discharge and is known as **coefficient of discharge**. The value of coefficient of discharge expressed in percentage is known as **volumetric efficiency** of the pump. Its value lies between 95-98%. Percentage slip is about 2% for pumps in good conditions.

Negative Slip

Sometimes, it is observed that the actual discharge is more than the theoretical discharge. This can happen if the suction pipe of the reciprocating pump is long, the delivery pipe is short and the pump runs at high speeds. In such cases, the inertia force in the suction pipe becomes large as compared to the pressure force above the delivery valve. This forces the delivery valve to open before the piston has completed its suction stroke. Some water is pushed into the delivery pipe even before the delivery stroke is actually started. This results in the actual discharge more than the theoretical discharge. The coefficient of discharge becomes more than one and the slip negative.

18.25 POWER INPUT OF A RECIPROCATING PUMP

In a single-acting reciprocating pump,

Let h_s = suction head

h_d = delivery head

Then total static head, $H_{st} = h_s + h_d$

Theoretical work done by the pump or the power input

$$= wQ_{th} H_{st}$$

$$= W\left(\frac{ALN}{60}\right)(h_s + h_d)$$

Owing to leakage and frictional losses, actual power input is more than the theoretical power. If η is the efficiency of the pump, then

Actual power input to the pump

$$= \frac{1}{\eta} W\left(\frac{ALN}{60}\right)(h_s = h_d) \quad \text{... (18.23)}$$

Example 18.13 The discharge of water through a single- acting reciprocating pump is 27 l/s and the head is 15 m. The piston diameter is 220 mm and the stroke 500 mm. The speed of the pump is 90 rpm. Find

(a) theoretical dischrge

(b) slip

(c) percentage slip

(d) coefficient of discharge

(e) power input.

Solution $D = 0{\cdot}22$ m $N = 90$ rpm

$Q_a = 0{\cdot}027 \text{ m}^3/\text{s}$ $L = 0{\cdot}5$ m

$H_{st} = 15$ m

(a) $Q_{th} = \dfrac{ALN}{60} = \left(\dfrac{\pi}{4}D^2\right)\dfrac{LN}{60} = \left(\dfrac{\pi}{4}\times 0{\cdot}22^2\right)\dfrac{0{\cdot}5\times 90}{60} = 0{\cdot}0285 \text{ m}^3/\text{s}$

(b) Slip $= Q_{th} - Q_a = 0{\cdot}0285 - 0{\cdot}027 = 0{\cdot}0015 \text{ m}^3/\text{s}$

(c) Percentange slip $= \dfrac{Q_{th} - Q_a}{Q_{th}} \times 100 = \dfrac{0{\cdot}0285 - 0{\cdot}027}{0{\cdot}0285} \times 100 = 5{\cdot}26\%$

(d) $C_d = \dfrac{Q_a}{Q_{th}} = \dfrac{0{\cdot}027}{0{\cdot}0285} = 0{\cdot}947$

(e) Power input $= wQ_{th}H_{st} = 9{\cdot}81 \times 0{\cdot}0285 \times 15 = 4{\cdot}19$ kW

Example 18.14 A triple-cylinder pump raises the water level by 100 m and the discharge is 100 l/s. The diameter of the piston is 250 mm and the stroke 600 mm. The velocity of water in the delivery pipe is 1·4 m/s. The friction losses amount to 2 m in the suction pipe and 18 m in the delivery pipe. Taking efficiency of the pump as 90% and slip 2%. find the speed and the power input of the pump.

Solution $Q_n = 0{\cdot}1\ \text{m}^3/\text{s}$ $\qquad H_{st} = 100$ m

$L = 0{\cdot}6$ m $\qquad D = 0{\cdot}25$ m

$H_{fs} = 2$ m $\qquad H_{fs} = 18$ m

$V_d = 1{\cdot}4$ m/s $\qquad s = 0{\cdot}02$

$\eta_h = 0{\cdot}9$

$$\text{Percentage slip} = \frac{Q_{th} - Q_a}{Q_{th}} \times 100$$

$$2 = \left(1 - \frac{Q_a}{Q_{th}}\right) \times 100$$

$$Q_{th} = \frac{Q_a}{0{\cdot}98}$$

$$Q_{th} = \frac{3ALN}{60}$$

$$\frac{Q_a}{0{\cdot}98} = \frac{1}{20} \times \frac{\pi}{4} \times D^2 \times L \times N$$

$$\frac{0{\cdot}1}{0{\cdot}98} = \frac{1}{20} \times \frac{\pi}{4} \times 0{\cdot}25^2 \times 0{\cdot}6 \times N$$

$$N = 69{\cdot}29 \text{ rpm}$$

Total head generated,

$$H = H_{st} + H_{fs} + h_{fd} + \frac{V_d^{\,2}}{2g}$$

$$= 100 + 2 + 18 + 0{\cdot}1 = 120{\cdot}1 \text{ m}$$

Power input

$$= \frac{1}{\eta_h} \cdot wQ_{th}H = \frac{1}{0{\cdot}9} \times 9{\cdot}81 \times \frac{0{\cdot}1}{0{\cdot}98} \times 120{\cdot}1 = 133{\cdot}5 \text{ kW}$$

18.26 EFFECT OF PISTON ACCELERATION

During a stroke, the velocity of the piston does not remain constant. Initially, the piston is stationary at one end and so the velocity is zero. During the first half of stroke, the velocity increases and is maximum at the midpoint. During the latter half of stroke, the velocity decreases and is zero at the end point. If the ratio of the lengths of the connecting rod to crank is large and the rotation of the crank is uniform, the piston approximately executes simple harmonic motion. Since the liquid is physically displaced into the delivery pipe, the velocity of liquid in the suction and delivery pipes is also not uniform. Thus accelerating or retarding heads act on the liquid flowing through the pipes.

Let x = displacement of piston in time 't'

θ = angle rotated by crank in time 't'

r = length of the crank

ω = Angular velocity of the crank

A = cross-sectional area of the cylinder

a = cross-sectional area of the pipe

V = velocity of the piston

v = velocity of the liquid in pipe

l = length of the pipe

f = acceleration of liquid in the pipe

ρ = mass density of the liquid

w = specific weight of the liquid

Displacement of piston (Fig. 18.34),

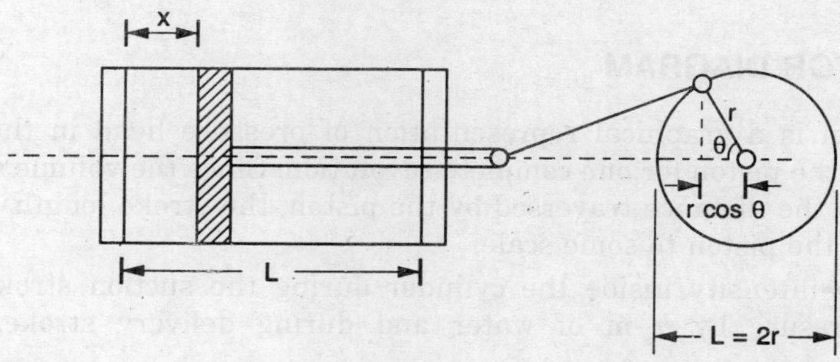

Fig. 18.34

$$x = r - r\cos\theta = r - r\cos\omega t \qquad \ldots(18.24)$$

Velocity of piston,

$$V = \frac{dx}{dt} = 0 - r\,(-\sin\omega t)\,\omega = \omega r\sin\omega t \qquad \ldots(18.25)$$

Applying continuity equation,

$$a \cdot v = A \cdot V$$

or

$$V = \frac{A}{a}\,V = \frac{A}{a}\,\omega r\sin\omega t \qquad \ldots(18.26)$$

Acceleration of liquid in the pipe,

$$f = \frac{dV}{dt} = \frac{d}{dt}\left(\frac{A}{a}\omega r \sin \omega t\right) = \frac{A}{a}\omega^2 r \cos \omega t \qquad ...(18.27)$$

Force to accelerate the liquid in the pipe,

$$F = \text{mass of liquid} \times \text{acceleration}$$

$$= \rho a 1 \times \frac{A}{a}\omega^2 r \cos \omega t$$

Intensity of pressure due to accelerating liquid,

$$p = \frac{F}{a} = \rho 1 \frac{A}{a}\omega^2 r \cos \omega t$$

Accelerating head,

$$h_a = \frac{P}{w} = \frac{P}{\rho g} = \frac{1}{g}\frac{A}{a}\omega^2 r \cos \omega t = \frac{1}{g}\frac{A}{a}\omega^2 r \cos \theta \qquad ...(18.28)$$

Subscript '*s*' and '*d*' are used to denote the quantities on the suction and on the delivery sides respectively.

At the beginning of stroke, $\theta = 0°$

$$h_a = \frac{1}{g} \times \frac{A}{a}\omega^2 r$$

At the middle of stroke, $\theta = 90°$

$$h_a = 0$$

At the end of stroke, $\theta = 180°$

$$h_a = -\frac{1}{g} \times \frac{A}{a}\omega^2 r$$

18.27 INDICATOR DIAGRAM

Indicator diagram is a graphical representation of pressure head in the cylinder and the volume swept by the piston for one complete revolution. Since the volume swept by the piston is proportional to the distance traversed by the piston, the stroke length also represents the volume swept by the piston to some scale.

The pressure intensity inside the cylinder during the suction stroke is less than the atmospheric prerssure by h_s m of water and during delivery stroke, higher than the atmospheric presure by h_d m of water.

In Fig. 18.35.

- Stroke length is represented on *x*-axis and pressure on the *y*-axis.
- Line *ab* represents atmospheric pressure head
- Line *cd* represents pressure head inside the cylinder during suction stroke
- Line *ef* represents pressure head inside the cylinder during delivery stroke.

Work done by the pump per second

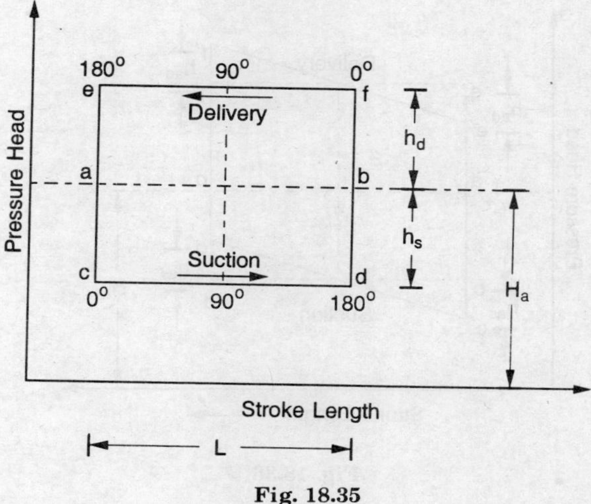

Fig. 18.35

$$W\left(\frac{ALN}{60}\right)(h_s + h_d) = K{\cdot}L\ (h_s + h_d) \qquad \text{...(18.29)}$$

where $$K = \frac{wAN}{60} = \text{a constant}$$

or work done $\propto L\ (h_s + h_d) \propto$ Area of the indicator diagram.

The area *cdfe* in the indicator diagram indicates work done to some scale. Such a diagram can be obtained by placing an indicator card on the cylinder; therefore, it is known as **indicator diagram.**

Effect of Accelerating Head

In the previous section, it was observed that during the first half of stroke, an accelerating head acts on the piston. This means the suction on the liquid in the cylinder is further increased by an amount equal to the head. During the latter half, there is decelerating head which means the suction is reduced by the amount equal to the head.

The change of the suction pressure head during the suction stroke due to accelerating head is shown in Fig. 18.36. The base of the diagram is shifted from cd to $c'd'$. If h_{as} is the maximum value of the accelerating head

$$h_{as} = \frac{l_s}{g} \times \frac{A}{a_s}\,\omega^2 r$$

It is positive at the beginning of stroke and negative at the end of stroke. As the area *c′d′ba* is equal to area *cdba*, the net work done during suction stroke remains the same and the accelerating head has no effect on the work done on the piston.

During the delivery stroke also there is accelerating head during the first half and decelerating head during the latter half movement of the piston. Thus there is increase in the pressure head during the first half and decrease in the pressure head during the latter half.

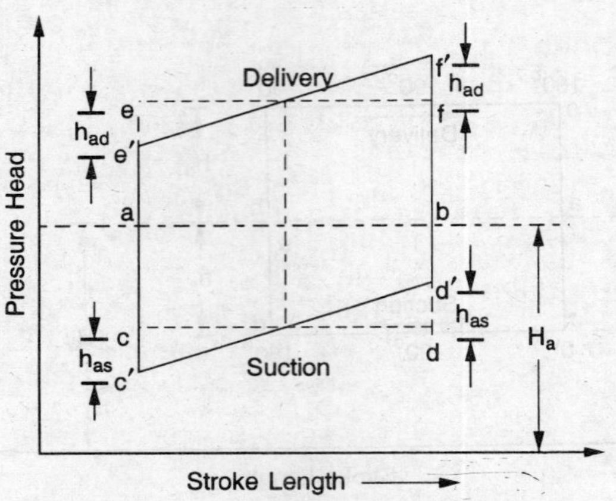

Fig. 18.36

The top of the diagram is shifted from *ef* to *e′ f′*. If h_{ad} is the maximum value of the accelerating head during the delivery stroke

$$h_{ad} = \frac{l_d}{g} \times \frac{A}{a_d}\,\omega^2 r$$

As the area *abf′ e′* is equal. to area *abfe*, there is no change in the net work done on the piston during the delivery stroke.

18.28 EFFECT OF PIPE FRICTION

Let f = coefficient of friction of the suction or delivery pipe

v = velocity of water in the pipe

d = diameter of the pipe

l = length of the pipe

Loss of head in a pipe due to friction

$$h_f = \frac{4flV^2}{2gd} = \frac{4fl}{2gd}\left(\frac{A}{a}\,\omega r \sin\theta\right)^2 \qquad \text{...(18.30)}$$

This shows that the variation of head loss due to friction in the pipe is parabolic with zero values at the beginning and end of stroke and maximum value at the middle of stroke.

At the beginning of stroke $\theta = 0°$, $h_f = 0$

At the middle of stroke $\theta = 90°$,

$$h_f = \frac{4fl}{2gd}\left(\frac{A}{a}\,\omega r \sin\theta\right)^2$$

At the end of stroke $\theta = 180°$, $h_f = 0$

Fig. 18.37. shows the variation of loss of head due to friction. The base of the diameter shifts from *cd* to *cgd* and the top from *ef* to *ehf*. The area of the parabola *cgd* represents the

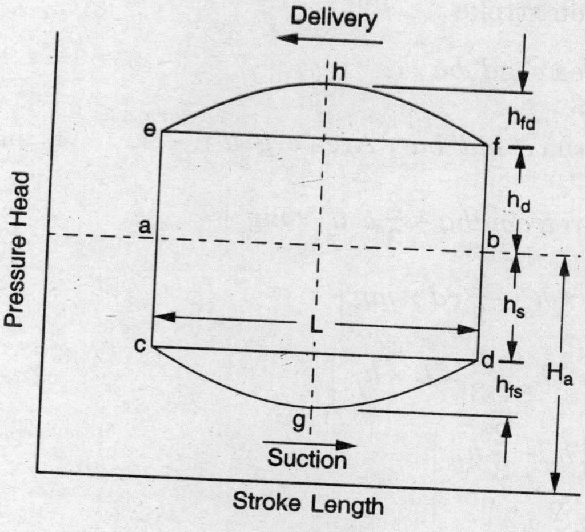

Fig. 18.37

work done against friction in the suction pipe and area *ehf* in the delivery pipe. As area of a parabola equals two thirds of its maximum ordinate,

Area $\qquad cgd = L \times \frac{2}{3} h_{fs}$

Area $\qquad ehf = L \times \frac{2}{3} h_{fd}$

Combined effect of the accelerating head and pipe friction on indicator diagram is shown in Fig. 18.38.

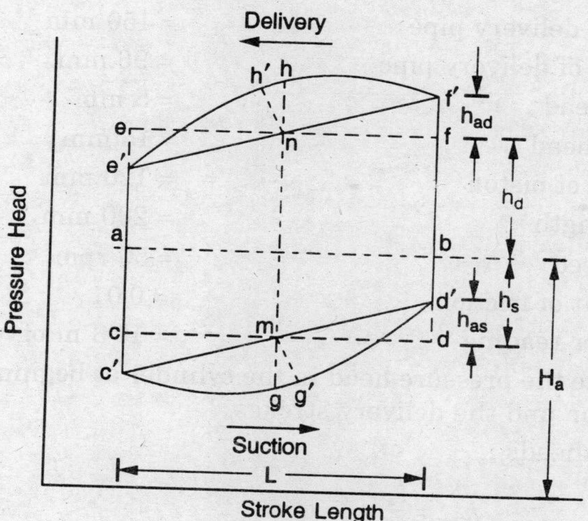

Fig. 18.38

Work done during suction stroke

$$= \frac{wAN}{60} \times \text{Area } c'\,gd'\,ba$$

$$= \frac{wAN}{60} \times (\text{Area } c'\,md'\,ba + \text{Area } c'\,g'\,d')$$

$$= \frac{wAN}{60} \times \left(\text{Area } cmdba + \frac{2}{3}c'\,d' \times mg'\right)$$

$$= \frac{wAN}{60} \times \left(L \times h_s + \frac{2}{3}cd \times mg\right)$$

$$= \frac{wAN}{60} \times \left(L \times h_s + \frac{2}{3} \times L \times h_{fs}\right)$$

$$= \frac{wAN}{60} \times L\left(h_s + \frac{2}{3}h_{fs}\right) \quad \text{...(18.31)}$$

$$= wQ_{th}\left(h_s + \frac{2}{3}h_{fs}\right) \quad \text{...(18.32)}$$

Similarly, work done during delivery stroke

$$= wQ_{th}\left(h_d + \frac{2}{3}h_{fd}\right)$$

Total work done

$$= wQ_{th}\left(h_s + h_d + \frac{2}{3}h_{fs} + \frac{2}{3}h_{fd}\right) \quad \text{...(18.33)}$$

Example 18.15 Following data relate to a single-acting reciprocating pump :

Length of suction pipe	= 75 m*m*
Diameter of suction pipe	= 96 mm
Length of delivery pipe	= 150 mm
Diameter of delivery pipe	= 96 mm
Suction head	= 5 mm
Delivery head	= 15 mm
Diameter of piston	= 120 mm
Stroke length	= 200 mm
Pump speed	= 50 rpm
Coefficient of friction	= 0·01
Barometer reading	= 10·3 m of water

Determine the pressure head in the cylinder at beginning, middle and end of the suction and the delivery strokes.

Solution Accelearating heads :

$$h_{as} = \frac{l_s}{g} \times \frac{A}{a_s}\,\omega^2 r$$

$$= \frac{l_s}{g} \times \left(\frac{D}{d_s}\right)^2 \left(\frac{2\pi N}{60}\right)^2 r$$

$$= \frac{7.5}{9.81} \times \left(\frac{0.12}{0.096}\right)^2 \left(\frac{2\pi \times 50}{60}\right)^2 \times 0.1$$

$$= 3.275 \text{ m}$$

$$h_{ad} = \frac{l_d}{g} \times \frac{A}{a_d} \omega^2 r$$

$$= \frac{l_d}{g} \times \left(\frac{D}{d_d}\right)^2 \left(\frac{2\pi N}{60}\right)^2 r$$

$$= \frac{25}{9.81} \times \left(\frac{0.12}{0.096}\right)^2 \left(\frac{2\pi \times 50}{60}\right)^2 \times 0.1$$

$$= 10.92 \text{ m}$$

Friction heads :

$$h_f = \frac{4fl_s}{2gd_s} \left(\frac{A}{a_s} \omega r\right)^2$$

$$= \frac{4fl_s}{2gd_s} \left[\left(\frac{D}{d_s}\right)^2 \times \frac{2\pi N}{60} \times r\right]^2$$

$$= \frac{4 \times 0.01 \times 7.5}{2 \times 9.81 \times 0.08} \left[\left(\frac{0.12}{0.096}\right)^2 \times \frac{2\pi \times 50}{60} \times 0.1\right]^2$$

$$= 0.128 \text{ m}$$

$$h_f = \frac{4fl_d}{2gd_d} \left[\left(\frac{D}{d_d}\right)^2 \times \frac{2\pi N}{60} \times r\right]^2$$

$$= \frac{4 \times 0.01 \times 25}{2 \times 9.81 \times 0.08} \left[\left(\frac{0.12}{0.096}\right)^2 \times \frac{2\pi \times 50}{60} \times 0.1\right]^2$$

$$= 0.426 \text{ m}$$

Refer Fig. 18.38.

Suction stroke

At the beginning of stroke

Pressure head $= H_a - h_s - h_{as} = 10.3 - 5 - 3.275 = 2.025$ m absolute

At the middle of stroke,

Pressure head$= H_a - h_s - h_{fs} = 10.3 - 5 - 0.128 = 5.172$ m absolute

At the end of stroke,

Pressure head $= H_a - h_s + h_{as} = 10.3 - 5 + 3.275 = 8.575$ m absolute

Delivery stroke

At the beginning of stroke,

Pressure head $= H_a + h_d + h_{ad} = 10.3 + 15 + 10.92 = 36.22$ m absolute

At the middle of stroke,

Pressure head $= H_a + h_d + h_{fd} = 10{\cdot}3 + 15 + 0{\cdot}426 = 25{\cdot}726$ m absolute

At the end of stroke,

Pressure head $= H_a + h_d - h_{ad} = 10{\cdot}3 + 15 - 10{\cdot}92 = 14{\cdot}38$ m absolute

18.29 MAXIMUM SPEED OF A RECIPROCATING PUMP

The maximum speed of a reciprocating pump depends upon the phenomenon of cavitation or flow separation. In case the suction pressure in the cylinder falls below the vapour pressure of the liquid at any instant, the liquid evaporates and the dissolved gases are given out. Thus there is separation of flow resulting in intermittent supply of liquid. The pressure at which the separation takes place is called **separation pressure**. For water, the value of the limiting separation pressure head is 2·6 m absolute or (10·3 – 2·6) = 7·7 m of water below atmospheric pressure head.

During Suction Stroke

The minimum pressure head during suction stroke occurs at the beginning of stroke (Fig. 18.38). Therefore, if the suction pressure falls below the separation pressure during the suction stroke, it can separate at point c'. If H_a is the atmospheric pressure head,

Absolute pressure head at $c' = H_a - h_s - h_{as}$

This value must not fall below the separation pressure head H_{sep}.

$$H_a - h_s - h_{as} > H_{sep} \qquad \ldots(18.34)$$

$$H_a - h_s - \frac{l_s}{g} \times \frac{A}{a_s}\,\omega^2 r > H_{sep}$$

In the limit, $\dfrac{l_s}{g} \times \dfrac{A}{a_s}\left(\dfrac{2\pi N}{60}\right)^2 r = H_a - h_s - H_{sep}$ ...(18.35)

From this equation, the maximum value of the speed can be calculated.

During Delivery Stroke

The minimum pressure head exists at the end of the delivery stroke *i.e.* at e'.

$$H_a + h_d - h_{ad} > H_{sep} \qquad \ldots(18.36)$$

$$H_a + h_d - \frac{l_d}{g} \times \frac{A}{a_d}\,\omega^2 r > H_{sep}$$

In the limit, $\dfrac{l_d}{g} \times \dfrac{A}{a_d}\left(\dfrac{2\pi N}{60}\right)^2 r = H_a + h_d - H_{sep}$ ...(18.37)

From this equation, the maximum value of the speed is calculated. The chances of separation of flow during the delivery stroke are not high. However, if the length of the delivery pipe l_d is very large and the delivery head h_d is small, the absolute pressure at e' may fall below the spearation pressure. This can happen if the delivery pipe is laid in an inclined position or a large protion of the pipe is horizontal.

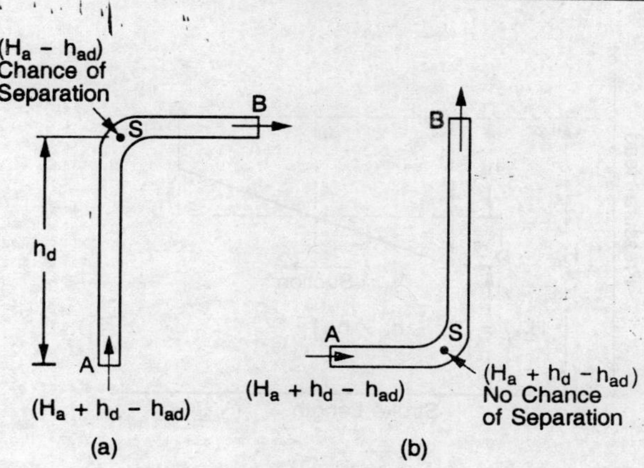

Fig. 18.39

Consider two possibilities of laying down the delivery pipe as shown in Fig. 18.39. Chances of separation of flow lie at the bends.

- In arrangement (a), the delivery pipe is first vertical and then horizontal.

  Absolute pressure head at the end of the delivery stroke = $H_a + h_d - h_{ad}$

  This is the pressure head at the inlet of the delivery pipe.

  At the bend (point s), the pressure head = $(H_a + h_d - h_{ad}) - h_d = H_a - h_{ad}$

- In arrangement (b), the delivery pipe is first horizontal and then vertical.

  At the bend (point s), the pressure head = $H_a + h_d - h_{ad}$

Out of these two arrangements, it is clear that the pressure head at the bend is lower in the first case. Therefore, there are more chances of separating the flow in the first case. Thus this arrangement of laying the delivery pipes should be avoided.

Example 18.16 The suction pipe of a single-acting reciprocating pump is 6·5 m long and 70 mm in diameter. The stroke length is 300 mm and the piston diameter 100 mm. The pump is located at a height of 5 m from the sump water level. The separation of water occurs when the absolute pressure head falls below 2·5 m of water. Determine the safe maximum speed of the pump if the atmospheric pressure is 10·3 m of water.

Solution $l_s = 6{\cdot}5$ m $\qquad d_s = 0{\cdot}07$ m

$D = 0{\cdot}1$ m $\qquad r = L/2 = 0{\cdot}3/2 = 0{\cdot}15$ m

$h_s = 5$ m $\qquad H_{sep} = 2{\cdot}5$ m

$H_a = 10{\cdot}3$ m

$$\frac{l_s}{g} \times \frac{A}{a_s} \left(\frac{2\pi N}{60}\right)^2 r = H_a - h_s - H_{sep}$$

$$\frac{6{\cdot}5}{9{\cdot}81} \times \left(\frac{0{\cdot}1}{0{\cdot}07}\right)^2 \left(\frac{2\pi N}{60}\right)^2 \times 0{\cdot}15 = 10{\cdot}3 - 5 - 2{\cdot}5$$

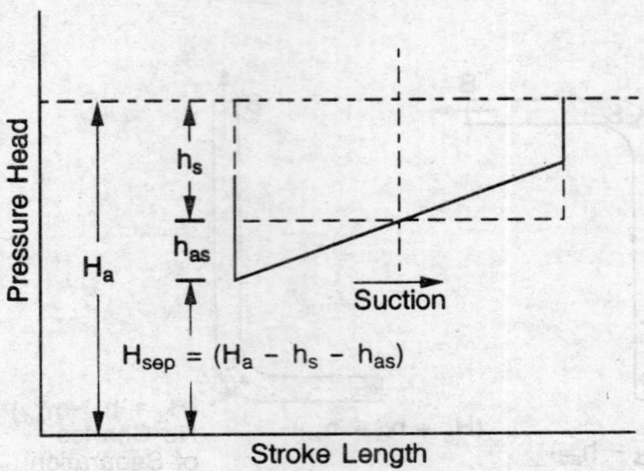

Fig. 18.40

$$= 2.8$$

$$N = 35.48 \text{ rpm}$$

Example 18.17 A single-acting reciprocating pump has its sump level 4 m below and the delivery tank 16 m above the centre line of the pump. The diameter and the length of the suction pipe are 50 mm and 6 m and that of delivery pipe 40 mm and 18 m respectively. The stroke of the piston is 220 mm and the diameter of the piston 120 mm. If the separation of water occurs at 7.8 m below the atmospheric pressure head, what will be the maximum speed at which the pump can be operated without separation?

Solution $l_s = 6$ m $\quad l_d = 18$ m

$h_s = 4$ m $\quad h_d = 16$ m

$d_s = 0.05$ m $\quad d_d = 0.04$ m

$L = 0.22$ m $\quad D = 0.12$ m

$r = L/2 = 0.11$ m $\quad H_{sep} = (H_a - 7.8)$ m absolute of water

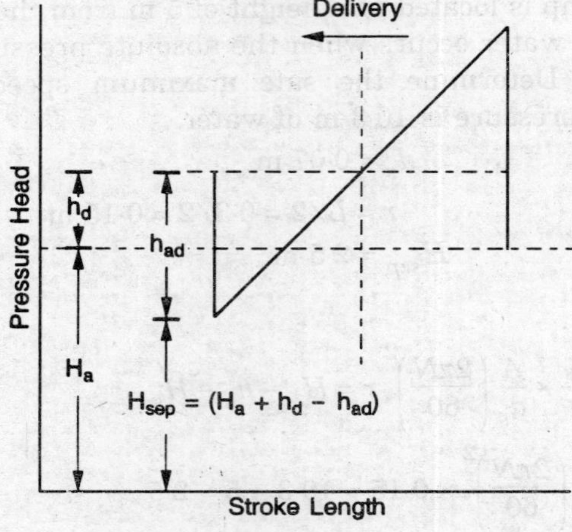

Fig. 18.41

During suction stroke,

$$\frac{l_s}{g} \times \frac{A}{a_s} \left(\frac{2\pi N}{60}\right)^2 r = H_a - h_s - H_{sep}$$

$$\frac{6}{9.81} \times \left(\frac{0.12}{0.05}\right)^2 \left(\frac{2\pi N}{60}\right)^2 \times 0.11 = H_a - 4 - (H_a - 7.8)$$

$$= 3.8$$

$$N = 29.9 \text{ rpm}$$

During delivery stroke,

$$\frac{l_d}{g} \times \frac{A}{a_d} \left(\frac{2\pi N}{60}\right)^2 r = H_a + h_d - H_{sep}$$

$$\frac{18}{9.81} \times \left(\frac{0.12}{0.04}\right)^2 \left(\frac{2\pi N}{60}\right)^2 \times 0.11 = H_a + 16 - (H_a - 7.8)$$

$$= 23.8$$

$$N = 34.57 \text{ rpm}$$

The maximum permissible speed of the pump is the lower of the two *i.e.* 28·73 rpm.

Example 18.18 A single-acting reciprocating pump delivers water at a height of 20 m through a delivery pipe 40 m long and 140 mm in diameter. The stroke and diameter of the piston are is 440 mm and 250 mm respectively. Determine the maximum permissible speed of the pump to run without cavitation to occur on the delivery side if

(a) the pipe runs first horizontally and then vertically upwards

(b) the pipe rises first vertically and then runs horizontally.

Assume the atmospheric pressure head to be 10·3 m of water and the cavitation to occur at 2·5 m of water absolute.

Solution. $D = 0.25$ m $\quad L = 0.44$ m

$l_d = 40$ m $\quad d_d = 0.14$ m

$h_d = 20$ m $\quad r = L/2 = 0.22$ m

$H_{sep} = 2.5$ m

(a) The absolute pressure head at the bend $= H_a + h_d - h_{ad}$

It must be equal to or greater than separation pressure head. At the limit,

$$H_a + h_d - h_{ad} = H_{sep}$$

$$10.3 + 20 - h_{ad} = 2.5$$

or

$$h_{ad} = 27.8 \text{ m}$$

$$\frac{l_d}{g} \times \frac{A}{a_d} \left(\frac{2\pi N}{60}\right)^2 r = 27.8$$

$$\frac{40}{9{\cdot}81} \times \left(\frac{0{\cdot}25}{0{\cdot}14}\right)^2 \left(\frac{2\pi N}{60}\right)^2 \times 0{\cdot}22 = 27{\cdot}8$$

$$N = 29{\cdot}77 \text{ rpm}$$

(b) The absolute pressure head at the bend = $H_a - h_{ad}$

It must be equal to or greater than separation pressure head. At the limit,

$$H_a - h_{ad} = H_{sep}$$

$$10{\cdot}3 - h_{ad} = 2{\cdot}5$$

or $$h_{ad} = 7{\cdot}8 \text{ m}$$

$$\frac{l_d}{g} \times \frac{A}{a_d}\left(\frac{2\pi N}{60}\right)^2 r = 7{\cdot}8$$

$$\frac{40}{9{\cdot}81} \times \left(\frac{0{\cdot}25}{0{\cdot}14}\right)^2 \left(\frac{2\pi N}{60}\right)^2 \times 0{\cdot}22 = 7{\cdot}8$$

$$N = 15{\cdot}77 \text{ rpm}$$

18.30 MULTI-CYLINDER PUMPS

In single-acting reciprocating pumps, the flow is obtained during the delivery strokes only. Moreover, the flow is also not uniform. This is because the velocity in the delivery pipe is not uniform during the stroke. Fig. 18.42 shows the rate of flow *vs*. crank angle curve for this type of pump.

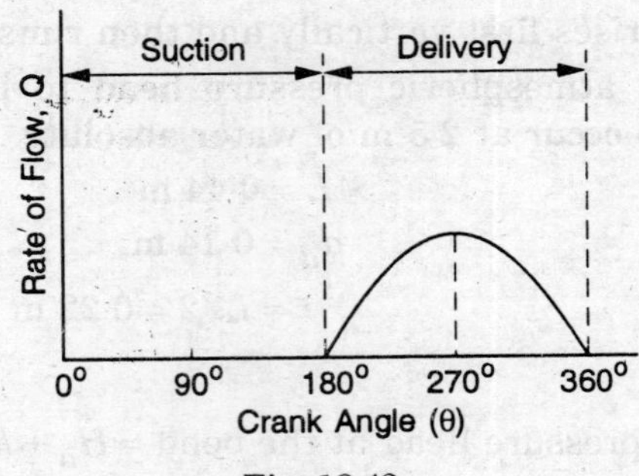

Fig. 18.42

In double-acting reciprocating pumps, the flow is available during both the strokes, though it is not uniform. For multi-cylinder pumps, discharge vs. crank angle curves are shown in Fig. 18.43.

- A double-cylinder pump or two-throw pump with cranks at a phase difference of 180° has the same Q *vs*. θ curve as a double-acting pump (Fig. 18.43a).
- Triple-cylinder pump or three-throw pump with cranks at a phase difference of 120° has Q *vs*. θ curve as shown in Fig. 18.43b.

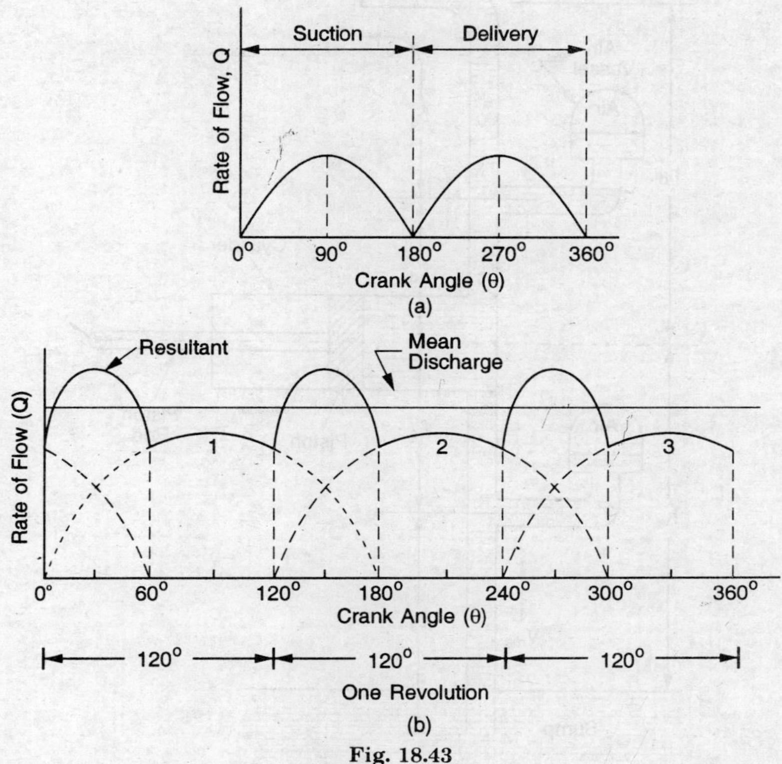

Fig. 18.43

Similarly, the curves for four or five cylinder pumps may be drawn. It is observed that the amount of fluctuations of the resultant flow is decreased and that the resultant flow becomes closer to the mean flow as the number of cylinders is increased.

18.31 AIR VESSELS

In the last section, it was observed that the supply of liquid was not uniform in case of a single-acting reciprocating pump. It can be made uniform by installing multi-cylinder pumps. The purpose is also solved if an air vessel of sufficient capacity is fitted on the delivery pipe as near to the pump as possible (Fig. 18.44). The air vessel consists of compressed air which takes up the fluctuations. It is further compressed when the liquid flows into the vessel and expands when the liquid flows out.

From the Q *vs.* θ curve of a single-acting reciprocating pump, it is seen that the discharge into the delivery pipe is higher than the average during the middle of the delivery stroke (Fig. 18.42). This extra volume of liquid goes into the air vessel and is stored there momentarily though the average supply of liquid in the delivery pipe is maintained. When the discharge is lower than the average in the delivery pipe or during the suction stroke when there is no flow at all, the compressed air of the air vessel pushes the liquid into the delivery pipe and tries to maintain the average flow. In this way, the rate of flow in the delivery pipe beyond the air

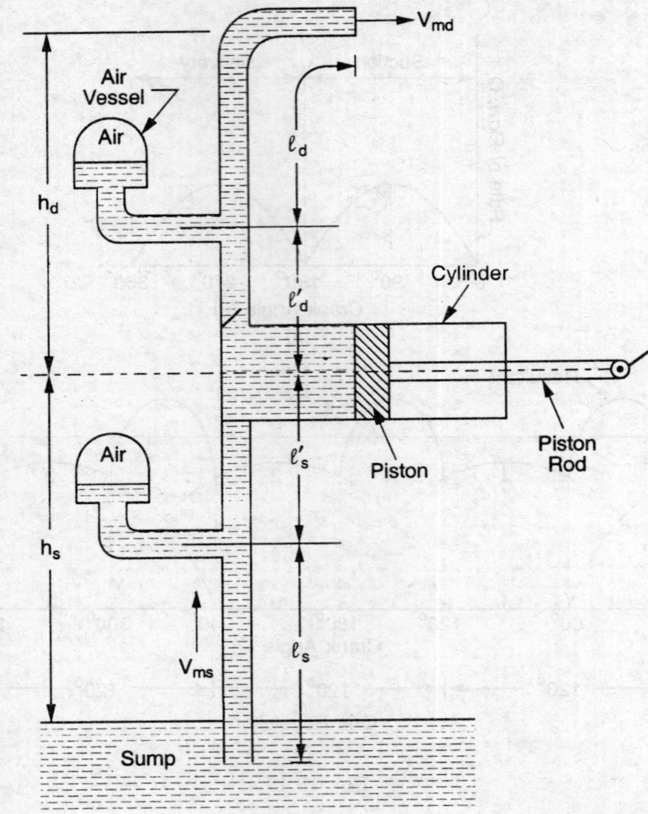

Fig. 18.44

vessel remains constant during the whole period of working. An air vessel of 6 to 7 times the volume of the cylinder gives good results. Thus the function of an air vessel is similar to that of a flywheel of an engine.

An air vessel fitted to the suction side of the pump smoothens the flow in the suction pipe. In the beginning or the end of the suction stroke, the supply needed by the cylinder is less than the average or during the delivery stroke when no supply is required supply is made into the air vessel. During the middle of the suction stroke, when the cylinder needs more discharge, the air vessel does the extra supply whereas the suction pipe makes the average supply. In this way the discharge and the velocity in the suction pipe is constant throughout.

Air vessels may also be fixed to the suction and delivery pipes of double-acting reciprocating pumps.

Rate of Flow In and Out of Air Vessel

In case of single-acting reciprocating pump, mean velocity of flow in the pipe is given by

$$V_m = \frac{\text{Discharge}}{\text{Area}}$$

$$= \frac{ALN}{60a}$$

$$= \frac{A}{60a} \times 2r \times \frac{60\omega}{2\pi} \qquad \left(\because \omega = \frac{2\pi N}{60} \text{ and } L = 2r \right)$$

$$= \frac{A}{a}\frac{\omega r}{\pi} \qquad \text{...(i)}$$

Instantaneous velocity in the pipe

$$v = \frac{A}{a} V = \frac{A}{a} \omega r \sin\theta$$

or

$$v_{max} = \frac{A}{a} \omega r \qquad \text{...(ii)}$$

Dividing (i) by (ii),

$$\frac{v_m}{v_{max}} = \frac{1}{\pi} \qquad \text{...(18.38)}$$

In case of a double-acting pump,

$$\frac{V_m}{V_{max}} = \frac{2}{\pi} \qquad \text{...(18.39)}$$

Now rate of flow from the cylinder into the pipe at any instant

= Area of piston × Velocity of piston

$= A \times \omega\, r \sin\theta$

Rate of flow in the delivery pipe beyond the air vessel

= Area of pipe × Mean velocity of flow

$$= a \times \frac{A}{a}\frac{\omega r}{\pi} = \frac{A\omega r}{\pi}$$

Rate of flow into the air vessel

$$= A\omega r \sin\theta - \frac{A\omega r}{\pi} = A\omega r \left(\sin\theta - \frac{1}{\pi} \right).$$

The positive value of this equation means that the flow is into the air vessel and negative value means it is out of the vessel. If an air vessel is fitted on the suction side, it is opposite.

There is no flow into or out of the air vessel when

$$A\omega r \left(\sin\theta - \frac{1}{\pi} \right) = 0 \quad \text{or} \quad \sin\theta = \frac{1}{\pi}$$

or $\theta = 18{\cdot}56°$ or $161{\cdot}44°$

In case of double-acting reciprocating pump, the rate of flow into the pipe at any instant remains the same. However, the mean discharge is double of the previous value. Therefore,

Rate of flow into the air vessel

$$= A\omega r \sin\theta - \frac{2A\omega r}{\pi} = A\omega r \left(\sin\theta - \frac{2}{\pi} \right) \qquad \text{...(18.40)}$$

There is no flow from or into the air vessel when

$$\sin\theta - \frac{1}{\pi} = 0 \quad \text{or} \quad \sin\theta = \frac{2}{\pi}$$

or $\theta = 39{\cdot}54°$, or $140{\cdot}46°$

Total Work Done with Air Vessels

Let l_s = length of suction pipe below air vessel

l_s' = length of suction pipe between cylinder and air vessel

l_d = length of delivery pipe beyond air vessel

l_d' = length of delivery pipe between cylinder and air vessel

h_{vs} = mean velocity head in suction pipe below air vessel

h_{vd} = mean velocity head in delivery pipe beyond air vessel

h_{as}' = accelerating head in the suction pipe

h_{ad}' = accelerating head in the delivery pipe

h_{fs} = friction head loss in the suction pipe below air vessel

h_{fd} = friction head loss in the delivery pipe beyond air vessel

h_{fs}' = friction head loss in suction pipe between cylinder and air vessel

h_{fd}' = friction head loss in delivery pipe between cylinder and air vessel

V_{ms} = mean velocity in suction pipe below air vessel

V_{md} = mean velocity in delivery pipe beyond air vessel

Total suction pressure head during suction stroke

= Static suction head

+ Accelerating head between cylinder and air vessel

+ Friction head loss between cylinder and air vessel

+ Uniform frictional head loss below air vessel

+ mean velocity head

$$H_s = h_s + h_{as}' + h_{fs}' + h_{fs} + h_{vs}$$

$$= h_s + \frac{l_s'}{g} \times \frac{A}{a_s}\,\omega^2 r \cos\theta + \frac{4fl_s'}{2gd_s}\left(\frac{A}{a_s}\,\omega r \sin\theta\right)^2 + \frac{4fl_d \cdot V_{ms}^2}{2gd_d} + \frac{V_{ms}^2}{2g} \quad \ldots(18.41)$$

If the length l_s' is negligible, second and third terms are omitted. Sometimes, velocity head is also ignored and thus

$$H_s = h_s + h_{fs}$$

Total pressure head during the delivery stroke

= Static delivery head

+ Accelerating head between cylinder and air vessel

+ Friction head loss between cylinder and air vessel

+ Uniform frictional head loss beyond air vessel

+ Mean velocity head

$$H_s = h_d + h_{ad}' + h_{fd}' + h_{fd} + h_{vd}$$

$$= h_d + \frac{l_d'}{g} \times \frac{A}{a_d}\,\omega^2 r \cos\theta + \frac{4fl_d'}{2gd_d}\left(\frac{A}{a_d}\,\omega r \sin\theta\right)^2 + \frac{4fl_d \cdot V_{md}^2}{2gd_d} + \frac{V_{md}^2}{2g} \quad \ldots(18.42)$$

If the length l_d' and the velocity head are negligible,

$$H_d = h_d + h_{fd}$$

The modified indicator diagram incorporating the effect of air vessels is shown in Fig. 18.45. In case small quantities are neglected, the shape of the indicator diagram becomes rectangular.

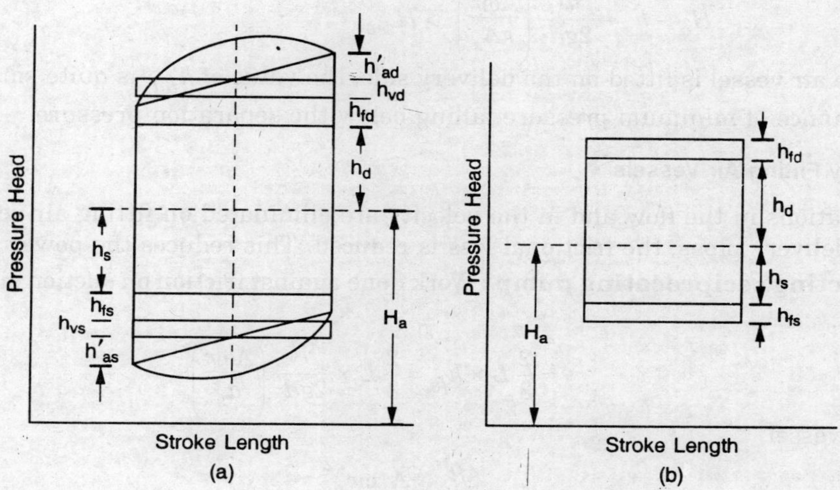

Fig. 18.45

Total work done

$$= \frac{wAN}{60} \text{ (Area of indicator diagram)}$$

$$= \frac{wAN}{60}\left[\begin{array}{l} L\,(h_s + h_d) + L\cdot h_{fs} + L\cdot h_{fd} + \frac{2}{3}\,Lh_{fs}' + \frac{2}{3}\,Lh_{fd}' \\ + Lh_{vs} + Lh_{vd} \end{array}\right]$$

or total power input

$$= wQ_{th}\left[h_s + h_d + h_{fs} + h_{fd} + \frac{2}{3}\,h_{fs}' + \frac{2}{3}\,h_{fd}' + h_{vs} + h_{vd}\right] \quad ...(18.43)$$

If small quantities are neglected,

$$\text{Work done} = wQ_{th}\,[h_s + h_d + h_{fs} + h_{fd}] \quad ...(18.44)$$

Maximum Speed with Air Vessel

During suction stroke, the absolute pressure at the beginning of stroke (θ = 0°) should not fall below the separation pressure head.

$$H_a - H_s > H_{sep}$$

$$H_a - h_s - \frac{l_s'}{g} \times \frac{A}{a_s}\,\omega^2\,r - \frac{4fl_s\cdot V_{ms}^{\;2}}{2gd_s} - \frac{V_{ms}^{\;2}}{2g} > H_{sep}$$

$$H_a - h_s - \frac{l_s'}{g} \times \frac{A}{a_s} \omega^2 r - \frac{4fl_s}{2gd_s}\left(\frac{A\omega r}{\pi a_s}\right)^2 - \frac{1}{2g}\left(\frac{A\omega r}{\pi a_s}\right)^2 > H_{sep} \quad ...(18.45)$$

From the above equation the value of ω can be calculated.

If small quantities are neglected,

$$H_a - h_s - \frac{4fl_s}{2gd_s}\left(\frac{A\omega r}{\pi A_s}\right)^2 > H_{sep} \quad ...(18.46)$$

When the air vessel is fitted on the delivery side, the value of h_{ad}' is quite small and thus there is no chance of minimum pressure falling below the separation pressure.

Work Saved by Fitting Air Vessels

As the fluctuations in the flow and in the velocity are eliminated on fitting air vessels on the suction and delivery pipes, the frictional loss is reduced. This reduces the power input.

Single-acting reciprocating pump : Work done against friction on suction side : Without air vessel

$$= \frac{2}{3} L \times h_{fs} = \frac{2}{3} L \times \frac{4fl_s}{2gd_s}\left(\frac{A\omega r}{a_s}\right)^2$$

With air vessel

$$= L \times \frac{4fl_s}{2gd_s}\left(\frac{A}{a_s}\frac{\omega r}{\pi}\right)^2$$

Work saved

$$= \frac{4fl_s L}{2gd_s}\left(\frac{A}{a_s}\omega r\right)^2\left(\frac{2}{3} - \frac{1}{\pi^2}\right)$$

Precentage of work saved

$$= \frac{\frac{2}{3} - \frac{1}{\pi^2}}{\frac{2}{3}} \times 100 = 84{\cdot}8\%$$

When an air vessel is fitted on the delivery side, same percentage of work is saved. Thus overall saving is 84·8% of the work done against friction.

Double-acting reciprocating pump : Work done against friction without air vessel on suction or delivery side is as before.

For a double-acting pump, the mean velocity is

$$V_m = \frac{2A}{a}\frac{\omega r}{\pi}$$

Work done against friction with air vessel

$$= L \times \frac{4fl}{2gd}\left(\frac{2A}{a}\frac{\omega r}{\pi}\right)^2$$

Work saved

$$= \frac{4fl_s L}{2gd_s}\left(\frac{A}{a_s}\omega r\right)^2\left(\frac{2}{4}-\frac{4}{\pi^2}\right)$$

Percentage of work saved

$$= \frac{\frac{2}{3}-\frac{4}{\pi^2}}{\frac{2}{3}} \times 100 = 39{\cdot}2\%$$

Example 18.19 The bore diameter of a double-acting reciprocating pump is 200 mm and the stroke length 360 mm. The speed of the pump is 140 rpm. The diameter of the suction pipe is 80 mm and it is fitted with an air vessel. Find the rate of flow into or from the air vessel when the crank makes angles of 30°, 60°, 90° and 150° with the inner dead centre. Assume that the piston moves with simple harmonic motion.

Solution $D = 0{\cdot}2$ m $\quad N = 140$ rpm

$L = 0{\cdot}36$ m $\quad r = L/2 = 0{\cdot}18$ m

The air vessel is fixed on the suction side.

Thus rate of flow from the air vessel

$$= A\omega r\left(\sin\theta - \frac{2}{\pi}\right)$$

$$= \frac{\pi}{4}D_s^{\,2} \times \frac{2\pi N}{60} \times r\left(\sin\theta - \frac{2}{\pi}\right)$$

$$= \frac{\pi}{4} \times 0{\cdot}18^2 \times \frac{2\pi \times 140}{60} \times 0{\cdot}2\left(\sin\theta - \frac{2}{\pi}\right)$$

$$= 0{\cdot}0746\left(\sin\theta - \frac{2}{\pi}\right)$$

When $\theta = 30°$, $\quad Q = 0{\cdot}0746\left(\sin 30° - \frac{2}{\pi}\right) = -\,0{\cdot}0102 \text{ m}^3/\text{s}$

ie $\quad 0{\cdot}0102 \text{ m}^3/\text{s}$ into the air vessel

When $\theta = 60°$, $\quad Q = 0{\cdot}0746\left(\sin 60° - \frac{2}{\pi}\right) = 0{\cdot}0171 \text{ m}^3/\text{s}$

ie $\quad 0{\cdot}0171 \text{ m}^3/\text{s}$ from the air vessel

When $\theta = 90°$, $\quad Q = 0{\cdot}0746\left(\sin 90° - \frac{2}{\pi}\right) = 0{\cdot}0271 \text{ m}^3/\text{s}$

ie $\quad 0{\cdot}0271 \text{ m}^3/\text{s}$ from the air vessel

When $\theta = 150°$, $\quad Q = 0{\cdot}0746\left(\sin 150° - \frac{2}{\pi}\right) = -\,0{\cdot}0102 \text{ m}^3/\text{s}$

ie $\quad 0{\cdot}0102 \text{ m}^3/\text{s}$ into the air vessel

Example 18.20 A double-acting reciprocating pump is fitted with an air vessel on the suction side close to the pump. The suction lift of the pump is 4·5 m. The length and the diameter of the suction pipe are 7·5 m and 80 mm respectively. The stroke of the piston and its diameter both are 200 mm each. The coefficient of friction is 0·01. The atmospheric pressure head is 10·3 m and the separation pressure head 2·5 m of water absolute. Determine the speed at which the separation commences. Also find the maximum permissible speed without the air vessel.

Solution $L = 0{\cdot}2$ m $\quad D = 0{\cdot}2$ m

$l_s = 7{\cdot}5$ m $\quad h_s = 4{\cdot}5$ m

$d_s = 0{\cdot}08$ m $\quad r = L/2 = 0{\cdot}1$ m

$l_s' = 0$ $\quad f = 0{\cdot}01$

$H_a = 10{\cdot}3$ m $\quad H_{sep} = 2{\cdot}5$ m

The maximum speed when air vessel is used is given by

$$H_a - h_s - \frac{l_s'}{g} \times \frac{A}{a_s}\,\omega^2 r - \frac{4fl_s}{2gd_s}\left(\frac{A\omega r}{\pi a_s}\right)^2 - \frac{1}{2g}\left(\frac{A\omega r}{\pi a_s}\right)^2 = H_{sep}$$

$$10{\cdot}3 - 4{\cdot}5 - 0 - \frac{1}{2g}\left(\frac{A\omega r}{\pi a_s}\right)^2\left(\frac{4fl_s}{d_s} + 1\right) = 2{\cdot}5$$

$$\frac{1}{2g}\left(\frac{A\omega r}{\pi a_s}\right)^2\left(\frac{4fl_s}{d_s} + 1\right) = 3{\cdot}3$$

$$\frac{1}{2 \times 9{\cdot}81}\left(\frac{A\omega r}{\pi a_s}\right)^2\left(\frac{4 \times 0{\cdot}01 \times 7{\cdot}5}{0{\cdot}08} + 1\right) = 3{\cdot}3$$

$$\frac{A\omega r}{\pi a_s} = 3{\cdot}692$$

$$\frac{\omega r}{\pi}\left(\frac{D}{d_s}\right)^2 = 3{\cdot}692$$

$$\frac{\omega \times 0{\cdot}1}{\pi}\left(\frac{0{\cdot}2}{0{\cdot}08}\right)^2 = 3{\cdot}692$$

$$\omega = 9{\cdot}279$$

$$\frac{2\pi N}{60} = 9{\cdot}279$$

$$N = 88{\cdot}61 \text{ rpm}$$

When no air vessel is used,

$$\frac{l_s}{g} \times \frac{A}{a_s}\left(\frac{2\pi N}{60}\right)^2 r = H_a - h_s - H_{sep}$$

$$\frac{7{\cdot}5}{9{\cdot}81} \times \left(\frac{0{\cdot}2}{0{\cdot}08}\right)^2\left(\frac{2\pi N}{60}\right)^2 \times 0{\cdot}1 = 10{\cdot}3 - 4{\cdot}5 - 2{\cdot}5$$

$$= 3{\cdot}3$$

$$N = 25{\cdot}1 \text{ rpm}$$

It may be observed that the maximum permissible speed is more when the suction pipe is fitted with an air vessel.

18.32 COMPARISON OF CENTRIFUGAL AND RECIPROCATING PUMPS

The following table compares the centrifugal and reciprocating pumps :

S.No.	Centrifugal Pump	Reciprocating Pump
1.	Steady and uniform flow.	Intermittent and pulsating flow.
2.	It can be used for viscous fluids e.g. oils, paper pulp, muddy water etc.	Only pure water or less viscous fluids are used otherwise valves give frequent trouble.
3.	Low initial cost.	High initial cost.
4.	For low heads and large discharge.	For high heads and low discharge
5.	Low maintenance cost. Only periodic check up is sufficient	High maintenance cost. Frequent replacement of parts needed.
6.	It can run at high speeds. It is directly coupled to electric motors.	Low speed due to separation problem. Belt drive is necessary.
7.	Uniform torque.	Non uniform torque.
8.	Less floor area is needed. Compact in design.	Needs 5-6 times area than for a centrifugal pump.
9.	Simple construction. Number of spare parts required is less.	Complicated construction. Number of spare parts required is more.
10.	Efficiency of low head pumps is quite high.	Energy losses are high. The efficiency of low head pumps is about 40%.

PROBLEMS

1. How is the flow in a centrifugal pump different from that in a turbine? Which has higher efficiency?
2. Why is a reciprocating pump called positive displacement pump?
3. How do you classify centrifugal pumps?
4. Why is it necessary to convert a part of the kinetic energy into pressure energy at the outlet of a centrifugal pump?
5. Discuss different types of casings for centrifugal pumps?
6. Differentiate between different types of impellers on the basis of specific speed.
7. What is priming? Why is it necessary to prime a pump before starting it?
8. What are total and manometric heads of centrifugal pumps? What is the difference between them?

9. What is Euler head? Derive an expression for energy conversion in the impeller of a centrifugal pump.
10. In what way the finite number of vanes affects the Euler head of a centrifugal pump? What is vane coefficient?
11. Why is the impeller of a centrifugal pump has backward curved vanes at the outlet?
12. Discuss various types of efficiencies of a centrifugal pump. What is the difference between hydraulic and manometric efficiency?
13. What do you mean by starting speed of a centrifugal pump?
14. What is the net positive suction head? What is its importance?
15. How does the cavitation limit the suction lift of a centrifugal pump?
16. Why should the suction height of a centrifugal pump not exceed a certain limit?
17. Define the terms : speed ratio, flow ratio and diameter ratio of a centrifugal pump.
18. Find an expression for the head loss when the flow through a centrifugal pump is varied from the normal value?
19. What are multistage pumps? Discuss the variation of pressure and velocity in successive stages in multistage pumps.
20. Describe various types of priming devices.
21. Define specific speed of a centrifugal pump. Derive an expression for the same.
22. Plot and discuss various characteristic curves of a centrifugal pump.
23. How do you classify the reciprocating pumps?
24. What are main components of a reciprocating pump? Describe the working of a reciprocating pump?
25. Define slip and coefficient of discharge in case of reciprocating pumps.
26. What is meant by a negative slip in reciprocating pumps? How is it possible?
27. What is an indicator diagram of a reciprocating pump? What is its importance?
28. What is accelerating head in case of reciprocating pumps? How does it affect the indicator diagram?
29. Deduce an expression for the maximum speed of a reciprocating pump if separation of the flow is to be avoided.
30. What is the utility of multistage reciprocating pumps?
31. What is an air vessel? Why is it used in reciprocating pumps?
32. Deduce a relation between the mean velocity and the maximum velocity of flow in the pipes of a (a) single-acting reciprocating pump (b) double-acting reciprocating pump.
33. Derive an expression for the rate of flow from and into an air vessel fitted to the suction side of a double-acting reciprocating pump.
34. Find an expression for the maximum speed of a reciprocating pump fitted with air vessels.
35. In what way the air vessels affect the pressure head in the cylinder of a reciprocating pump?
36. List the advantages and disadvantages of centrifugal pumps over the reciprocating pumps.

37. The impeller of a centrifugal pump rotates at 1000 rpm. It generates 20 m of head while delivering 1·2 m^3 of water per second. Find the type of impeller being used for such a pump. (mixed or axial-flow impeller)

38. A centrifugal pump delivers 0·25 m^3 of water per second. The head developed is 20 m and the speed of rotation 1450 rpm. Velocity of flow at the outlet is 2·5 m/s. The vanes are curved back at an angle of 25°. The hydraulic efficiency is 80%. Find the diameter and width of the impeller at the outlet. (245 mm, 130 mm)

39. The impeller of a centrifugal pump delivers 0·4 m^3 of water per second while rotating at 1000 rpm. The head developed is 40 m. The diameter and the width of the impeller at the outlet are 500 mm and 75 mm respectively. Assuming an efficiency of 75%, determine the vane angles at the outlet of impeller. (28·74°)

40. The diameter and the width of impeller of a centrifugal pump are 300 mm and 50 mm respectively. The impeller rotates at 1400 rpm and the power input is 30 kW. The vanes are curved back at an angle of 45°. Determine the discharge through the pump. (0·065 m^3/s)

41. The diameters at the inlet and the outlet of impeller of a centrifugal pump are 200 mm and 400 mm respectively. The pump raises water to a height of 25 *m*. Determine the minimum starting speed of the pump to deliver water. (1221 rpm)

42. A centrifugal pump raises water to a height of 50 m through a pipe 130 m long. The discharge is 100 litres per second. The impeller rotates at 1400 rpm and the power input is 30 kW. The vane angles are set backwards at an angle of 45°. Determine the discharge through the pump. (0·065 m^3/s)

43. The inside diameter of a centrifugal pump is half that of the outside diameter. The impeller rotates at 1000 rpm and the head developed is 30 m. Determine the least diameter of the impeller which starts just delivering the water. Assume hydraulic efficiency as 80%. (600 mm)

44. The hydraulic efficiency of a radial flow centrifugal pump is 80%. The pump develops a head of 37·5 m and delivers 45 litres of water per second while its impeller rotates at 1450 rpm. The diameter at the outlet is 300 mm and the flow velocity at the outlet 6·6 m/s. Find the velocity of whirl and the vane angles at the outlet. (20·16 m/s, 68·13°)

45. The diameter of the cylinder of a single-acting reciprocating pump is 150 mm and the stroke length 300 mm. The height of water raised is 20 m. The crank speed is 60 rpm. Find the theoretical power input and the theoretical discharge. Also find the percentage slip if the actual discharge is 0·005 m^3/s. (1·04 kW, 0·0053 m^3/s, 5·66%)

46. A three-throw pump delivers 150 litres of water per second at a head of 80 m. The stroke length is 450 mm and the diameter of the piston 200 mm. The velocity of water in the delivery pipe is 1·5 m/s. The total friction losses in the pipes amount to 15 m of head, the slip is 2·5% and the hydraulic efficiency 88%. Determine the speed of the pump and the power input.

47. The discharge of water through a double-acting reciprocating pump is 900 litres per second. The pump runs at 45 rpm. The diameter and the stroke of the piston are 200 mm and 300 mm respectively. Find the percentage slip. (6·2% negative)

48. The suction height of a single-acting reciprocating pump is 4 m. The diameter and the stroke of the piston are 80 mm and 180 mm respectively . The suction pipe is 50 mm in diameter and 6 m long. If the separation occurs at 2·5 m of water absolute and the atmospheric pressure head 10·3 m of water, determine the maximum speed at which the pump can run without separation. (49·6 rpm)

49. The diameter and the stroke of the piston of a single- acting reciprocating pump are 100 mm and 300 mm respectively. The head raised is 20 m above the centre line of the pump. The length and the diameter of the delivery pipe are 25 m and 50 mm respectively. Assuming that the separation takes place at 2·5 m of water absolute and the atmospheric pressure head 10·3 m of water absolute, find the maximum speed of the pump to run without separation of flow during the delivery stroke. (40·7 rpm)

50. A single-acting reciprocating pump runs at 100 rpm. The bore diameter of the cylinder is 150 mm and the stroke of the piston 320 mm. The suction pipe is 60 mm in diameter and is fitted with an air vessel. Determine the rate of flow from or into the air vessel when the crank makes angles of 15°, 30°, 90°, 180° and 240° with the inner dead centre. (0·00176 m^3/s into, 0·00539 m^3/s from, 0·0202 m^3/s from, 0·00944 m^3/s into, 0·00351 m^3/s into)

51. The water tank of a single-acting reciprocating pump is located 15 m above th pump level and at a distance of 30 m horizontally from it. The diameter and the stroke of the piston are 250 mm and 450 mm respectively. The delivery pipe is 100 mm in diameter. The separation occurs at 2·35 m of water absolute and the barometric pressure is 10·3 m of water. Determine the maximum permissible speed of the pump if (a) the pipe is vertical from the pump and then horizontal upto the tank and (b) the pipe runs horizontally from the pump and then vertically to the tank. (13·42 rpm, 18·04 rpm)

Chapter 19

MISCELLANEOUS FLUID MACHINES

19.1 INTRODUCTION

There are some fluid machines which are based on hydraulic systems. A **hydraulic system** is an arrangement in which force and energy are transmitted through incompressible fluids. Hydraulic systems are of two types :

1. Hydrostatic system
2. Hydrokinematic system

In a **hydrostatic system**, the transmission of force and energy is solely by the hydraulic pressure. It consists of three elements :

- Pumping unit such as rotary or reciporcating pump to act as a power source to develop the hydraulic pressure from mechanical work.
- Transmission line or passage through which power or pressure energy can be transmitted from the place of production to the place of necessity.
- Hydraulic motor which can reconvert the pressure energy into mechanical work.

Examples of hydrostatic systems are hydraulic press, hydraulic crane and hydraulic lift etc.

In a **hydrokinematic system**, the transmission of power is obtained through change in the velocity and direction of fluid flow. There is negligible change in the pressure of the fluid. It has two elements :

- Pump-impeller driven by the driving shaft
- Turbine-runner to run the driven shaft.

Examples of the hydrokinematic systems are hydraulic coupling, hydraulic torque converter etc. There is circulation of oil from the pump-impeller to the turbine runner which transmits the power.

19.2 HYDRAULIC ACCUMULATOR

A hydraulic accumulator is a storehouse which stores the pressure energy of a fluid when it is not needed and supplies the same when it is required. The action is similar to that of a flywheel of a reciprocating engine. The fluid under pressure may be required in many hydraulic machines such as hyraulic crane, hydraulic lift etc. In these machines energy is required only when the load is being raised and practically no energy is required to lower the load.

Pressure energy of a fluid is obtained with the help of a pump which works continuously. Fluid under pressure is supplied to the accumulator all the time through a pipe. When a machine does not need the pressurised oil, it goes to the accumulator and is stored there. When there is necessity of the same for use in machine, it goes directly to the machine along with the stored fluid to meet the sudden demand of the fluid. In the absence of an accumulator, the pump would have taken a long period to supply the same. There is also a maximum limit of the accumulator to store the pressurised oil. Any oil in excess of that goes back to the pump through a bypass valve. As the sudden demand of the pressurised oil is met from the accumulator, a low capacity pump usually serves the purpose.

A hydraulic accumulator consists of a cylinder and a loaded ram that can move in the cylinder (Fig. 19.1). Two pipes are provided at the bottom of the cylinder, one for the incoming

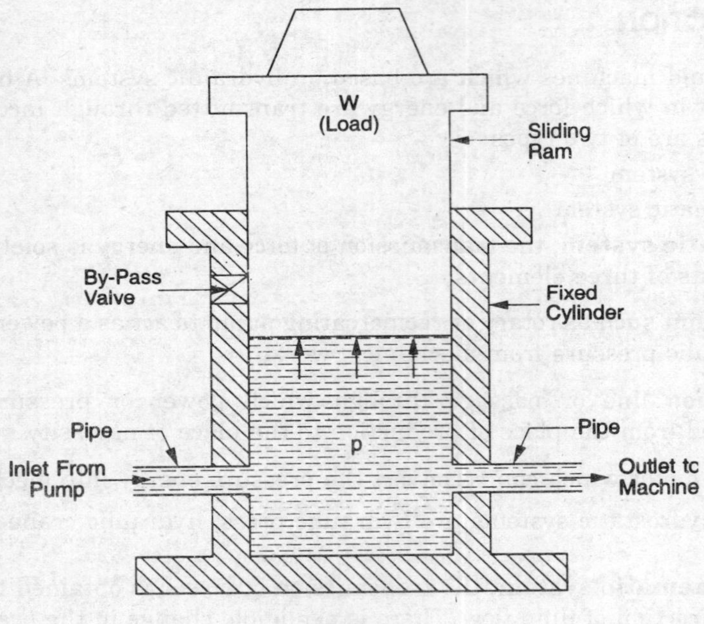

Fig. 19.1

pressurised oil and the other for the supply to the concerned machine. The ram can be raised upto a certain height. Uniform pressure prevails throughout the inlet pipe, in the accumulator and in the outlet pipe.

Let p = pressure intensity of the fluid

L = maximum lift of the ram

A = area of the ram

W = weight of the ram

Then, weight of the ram = Pressure intensity of fluid × Area

$$W = p \cdot A$$

Work done in lifting the ram = $W \cdot L = p \cdot A \cdot L$ ...(19.1)

This is equal to the energy stored in the accumulator and is known as the capicity of the accumulator.

Differential Accumulator

If fluid under high pressure is to be stored in a simple accumulator, heavy weights are needed to do the same. However, in a differential accumulator, comparatively small loads are required to store the fluid at the same pressure. A differential accumulator has a fixed ram consisting of a lower portion of larger diameter made by enclosing the ram in a brass bush or sleeve (Fig. 19.2). A sliding cylinder surrounds the fixed ram. The lower portion has a vertical passage at the axis through which the fluid under pressure can enter into the clyinder. The high-pressure fluid when needed to be used in a machine also pass through the same passage. The internal annular area of the cylinder is equal to the cross-sectional area of the brass bush. To load the cylinder, weights are put on the projected collar at the lower portion of the cylinder.

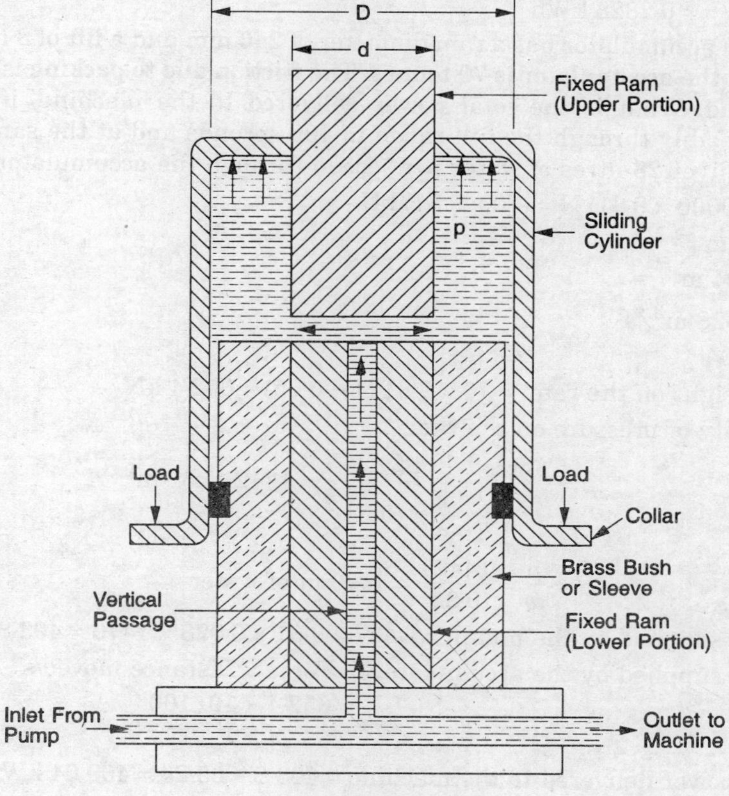

Fig. 19.2

Load on the cylinder, $W = (D^2 - d^2) \cdot p$

Where D and d are the external and internal diameters of the ram respectively. For a fixed load W, pressure p is increased if difference in the external and internal diameters is reduced.

Capacity of the accumulator $= (D^2 - d^2) \cdot p \cdot L$ ... (19.2)

Capacity of an accumulator may be expressed in terms of volume of the accumulator as well as in terms of energy stored *i.e.* either in litres or in kWh.

Example 19.1 An accumulator has a ram diameter 180 mm and a lift of 10 m. It is supplied with water at a pressure of 4000 kN/m^2. Determine the total weight on the ram and the capacity of the accumulator in kWh.

Solution $L = 10$ m $\qquad d = 0{\cdot}18$ m

$p = 4000$ kN/m^3

Total weight on the ram, $W = A \cdot p = \frac{\pi}{4} \times d^2 \times p = \frac{\pi}{4} \times (0{\cdot}18)^2 \times 4000 = 101{\cdot}8$ kN

Capacity of the accumulator $= W \cdot L = 101{\cdot}8 \times 10 = 1018$ kN·m

Capacity of the accumulator is expressed in kWh as follows :

$$\text{Capacity} = 1018 \text{ kN·m} = 1018 \frac{\text{kN·m}}{\text{s}} \cdot s = 1018 \text{ kWs} = \frac{1018}{3600} \text{ kWh}$$

$$= 0{\cdot}2828 \text{ kWh}$$

Example 19.2 An accumulator has a ram diameter of 240 mm and a lift of 8 m. Total weight on the accumulator is 70 tonnes. The friction due to packing is 5% of the total load. What is the total power delivered to the machine, if the ram falls steadily through the full range in 100 seconds and at the same time pumps deliver 28 litres of water per second through the accumulator.

Solution $W = (70000 \times 9{\cdot}81)$ N $= (70 \times 9{\cdot}81)$ kN

$L = 10$ m

$D = 0{\cdot}24$ m

$Q = 0{\cdot}028$ m^3/s

$T = 100$ s

Net weight on the ram $= 70 \times 9{\cdot}81\,(1 - 0{\cdot}05) = 652{\cdot}4$ kN

Intensity of pressure on the ram,

$$p = \frac{\text{Force}}{\text{Area}} = \frac{652{\cdot}4}{\frac{\pi}{4} \times (0{\cdot}24)^2} = 14421 \text{ kN/m}^2$$

or $$H = \frac{p}{w} = \frac{14421}{9{\cdot}81} = 1470 \text{ m of water}$$

Power supplied by the pump $= wQH = 9{\cdot}81 \times 0{\cdot}028 \times 1470 = 403{\cdot}8$ kW

Power supplied by the accumulator = Force × Distance moved/s

$= 652{\cdot}4 \times 10/100$

$= 65{\cdot}24$ kW

Total power delivered to the machine $= 403{\cdot}8 + 65{\cdot}24 = 469{\cdot}04$ kW

Example 19.3 The diameter of the ram of a hydraulic accumulator is 360 mm and the total load 50 tonnes. Frictional resistance for the movement of the ram is 5% of the total load. Determine the intensity of water when the ram moves down with a uniform velocity in the downward direction. How does the result differ when the ram moves up ?

Solution $W = (50000 \times 9{\cdot}81)\ \text{N} = (50 \times 9{\cdot}81)\ \text{kN}$ $\qquad W_f = 0{\cdot}05\ \text{W}$

$d = 0{\cdot}36$ m

$$A = \frac{\pi}{4} \times (0{\cdot}36)^2 = 0{\cdot}1018\ \text{m}^2$$

When the ram moves down, the frictional resistance acts upwards.

Net weight on the ram $= W - W_f = 50 \times 9{\cdot}81\ (1 - 0{\cdot}05) = 466$ kN

Pressure intensity,

$$P = \frac{\text{Net weight}}{\text{Area}} = \frac{466}{0{\cdot}1018} = 4578{\cdot}2\ \text{kN/m}^2$$

When the ram moves up, the frictional resistance acts downwards.

Net weight on the ram $= W - W_f = 50 \times 9{\cdot}81\ (1 + 0{\cdot}05) = 515$ kN

$$P = \frac{\text{Net weight}}{\text{Area}} = \frac{515}{0{\cdot}1018} = 5058{\cdot}9\ \text{kN/m}^2$$

Example 19.4 The diameters of two parts of the ram of a differential accumulator are 200 mm and 180 mm. The stroke length is 1·2 m. The pressure intensity of water is 8000 kN/m^2 when the load is either at rest at the upper end or moves with uniform velocity. Determine the weight of the loaded cylinder and the amount of energy stored in the accumulator. What will be the diameter of the ram of an ordinary accumulator to move the same load under same pressure of water ?

Solution $D = 0{\cdot}2$ m $\qquad d = 0{\cdot}18$ m

$p = 8000\ \text{kN/m}^2$ $\qquad L = 1{\cdot}2$ m

$$A = \frac{\pi}{4} \times (0{\cdot}2^2 - 0{\cdot}18^2) = 0{\cdot}00597\ \text{m}^2$$

Weight of the loaded cylinder $= p{\cdot}A = 8000 \times 0{\cdot}00597 = 47{\cdot}75$ kN

Energy stored in the accumulator = Capacity of the accumulator

$$= p{\cdot}A{\cdot}L$$
$$= 8000 \times 0{\cdot}00597 \times 1{\cdot}2$$
$$= 57{\cdot}3\ \text{kN{\cdot}m}$$

Let d' be the diamter of the ordinary accumulator,

$$\text{Load} = p{\cdot}A$$

$$p \times \frac{\pi}{4}(D^2 - d^2) = p \times \frac{\pi}{4} d'^2$$

$$D^2 - d^2 = d'^2$$

$$0{\cdot}2^2 - 0{\cdot}18^2 = d'^2$$

$$d'^2 = 0{\cdot}0076$$

$$d' = 0{\cdot}0872\ \text{m} \quad 87{\cdot}2\ \text{mm}$$

19.3 HYDRAULIC INTENSIFIER

A hydraulic accumulator is a machine which stores a fluid under pressure and supplies it at the same pressure. A hydraulic intensifier is a machine in which the pressure of the fluid is also increased. A large quantity of low-pressure fluid is needed to accomplish this as a large quantity of fluid goes waste during the operation.

A hydraulic intensifier consists of two cylinders : one fixed and the other movable as shown in Fig. 19.3. There is a fixed ram and the movable cylinder slides between the fixed ram and the fixed cylinder. The fixed cylinder has two valves 1 and 2. The valve 1 allows the low-pressure fluid to enter into the fixed cylinder whereas the valve 2 is used to take it out. The sliding cylinder also consists of two valves 3 and 4. Valve 3 allows the low-pressure fluid to enter into the sliding cylinder and valve 4 is used to take out high-pressure fluid from the sliding cylinder through a vertical passage in the fixed ram. The high-pressure fluid can be used in a machine to do some useful work.

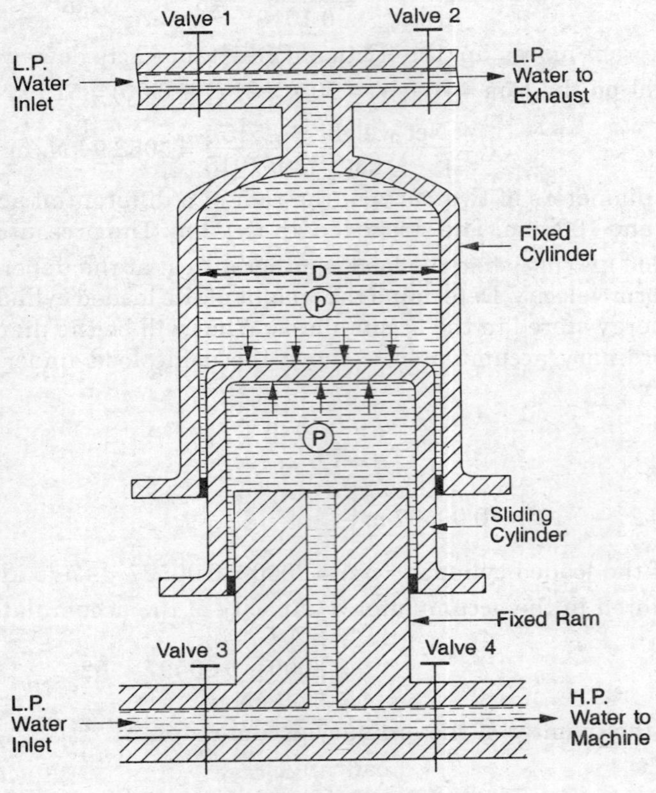

Fig. 19.3

First, assume the sliding cylinder at the bottom position and the fixed cylinder full of low-pressure fluid. Now, if low-pressure fluid is admitted through valve 3 keeping the valves 1 and 4 closed, the sliding cylinder moves up and the low-pressure fluid goes out through valve 2.

At any instant,

Let p = intensity of pressure in the fixed cylinder

P = intensity of pressure in the sliding cylinder

D = internal diameter of the fixed cylinder

= external diameter of the sliding cylinder

d = internal diameter of the sliding cylinder

= diameter of the ram

For equilibrium of the sliding cylinder,

Net upward force = Net downward force

$$\frac{\pi}{4} d^2 \times \rho = \frac{\pi}{4} D^2 \times p$$

$$\rho = \left(\frac{D}{d}\right)^2 \cdot p \qquad \text{...(19.3)}$$

Thus the pressure in the sliding cylinder is higher than that in the fixed cylinder and depends upon the square of the ratio of the outer diameter of the sliding cylinder and the diameter of the ram. If the valves 2 and 3 are closed and low-pressure fluid is supplied through valve 1, the supply of high-pressure fluid can be obtained through valve 4.

Example 19.5 A hydraulic intensifier has diameters of the sliding cylinder and the ram as 800 mm and 180 mm respectively. The pressure on the high-pressure side is 16 MN/m^2. The friction loss at each of the packing is 5% of the total force. Determine the pressure of water on low-pressure side.

Solution $D = 0{\cdot}8$ m $\qquad d = 0{\cdot}18$ m

$P = 16000$ kN/m^2 $\qquad W_f = 0{\cdot}05$ W at each packing

Loss at the packing between fixed and sliding cylinder = 0·05 W

Net force on the sliding cylinder, $\qquad W' = (1 - 0{\cdot}05)\ \text{W} = 0{\cdot}95\ \text{W}$

Loss at the packing between fixed and sliding cylinder = 0·05 W′

$$\text{Net force on the sliding cylinder} = (1 - 0{\cdot}05)\ \text{W}'$$
$$= 0{\cdot}95 \times 0{\cdot}95\ \text{W}$$
$$= 0{\cdot}95^2 \cdot p \cdot A$$

For the equilibrium of the sliding cylinder,

Net upward force = Net downward force

$$P \cdot a = (0{\cdot}95)^2\, p \cdot A$$

$$16000 \times \frac{\pi}{4} \times (0{\cdot}18)^2 = (0{\cdot}95)^2 \times p \times \frac{\pi}{4} \times (0{\cdot}8)^2$$

$$p = 897{\cdot}5\ \text{kN/m}^2$$

Example 19.6 A hydraulic intensifier is used to increase the pressure of water from 4·5 MN/m^2 to 20 MN/m^2. The stroke of the intensifier is 1·5 m and the capacity is 30 litres. Find the diameters of the fixed ram and the sliding cylinder of the intensifier.

Solution $P = 20000$ kN/m^2 $p = 4500$ kN/m^2

Stroke = 1·5 m Capacity = 0·03 m^3

Capacity of the intensifier = Volume of the sliding cylinder
= Area of the ram × Stroke length

$$0{\cdot}03 = \frac{\pi}{4} d^2 \times 1{\cdot}5$$

$$d = 0{\cdot}1596 \text{ m} \quad \text{or} \quad 159{\cdot}6 \text{ mm}$$

Considering the equilibrium of the sliding cylinder,

Net upward force = Net downward force

$$\frac{\pi}{4} d^2 \times P = \frac{\pi}{4} D^2 \times p$$

$$(0{\cdot}1596)^2 \times 20000 = D^2 \times 4500$$

$$D = 0{\cdot}336 \text{ m} \quad \text{or} \quad 336 \text{ mm}$$

19.4 HYDRAULIC PRESS

A hydraulic press is used to raise heavy loads with the application of small forces. Its working is based on Pascal's law *i.e.* the intensity of pressure exerted by a static fluid is same in all directions. As shown in Fig. 19.4, it consists of two cylinders of different diameters joined by a pipe. The cylinder with the larger diameter has a sliding ram with a platform to carry the

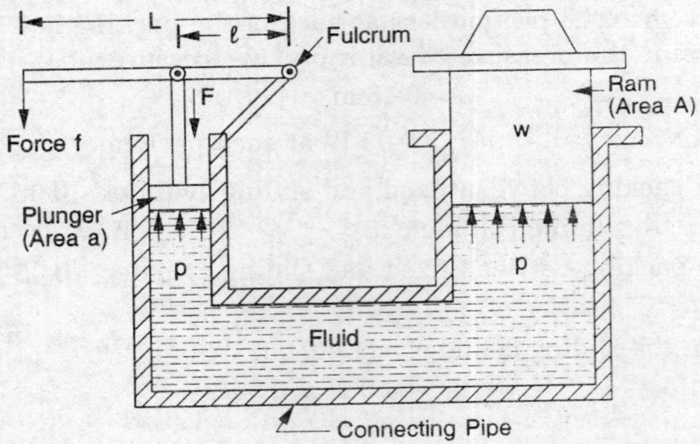

Fig. 19.4

loads to be raised or lowered, whereas the cylinder with the small diameter has plunger that moves upwards or downwards when a force is applied to it through a lever system. On applying a small force F to the plunger in the downward direction, a pressure p is exerted on the fluid which is transmitted equally in all directions. Thus

$$p = \frac{F}{a}$$

where a is the area of cross-section of the plunger. The same intensity of pressure is exerted on the ram also. If W is the load to be lifted and A the area of cross-section of the ram, then

$$p = \frac{W}{A} = \frac{F}{a}$$

Thus, $$W = \frac{A}{a} \times F \quad \text{or} \quad F = W\frac{a}{A} \qquad \text{...(19.4)}$$

As A is greater than a, the load raised is always greater than the force F. Also, the mechanical advantage

$$M{\cdot}A{\cdot} = \frac{W}{F}$$

If a lever of length L is used to apply the force, then the force applied at the end of the lever arm is given by

$$f \times L = F \times l$$

where l is the distance of the fulcrum from the pivot of the force arm.

or $$f = \frac{Fl}{L} = W{\cdot}\frac{a}{A}{\cdot}\frac{l}{L} \qquad \text{...(19.5)}$$

Example 19.7 The diameters of the ram and plunger of a hydraulic press are 144 mm and 18 mm respectively. A weight of 1000 kg is lifted through a distance of 600 mm in 16 minutes. The stroke of the plunger is 220 mm. Determine

(i) the force required on the plunger

(ii) the number of strokes of the plunger

(iii) the power requirement of the plunger

Solution $W = (1000 \times 9{\cdot}81)$ N $= 9{\cdot}81$ kN $\qquad T = 16 \times 60 = 960$ s

$D = 0{\cdot}144$ m $\qquad d = 0{\cdot}018$ m

$L = 0{\cdot}6$ m $\qquad x = 0{\cdot}22$ m

(i) The force F applied on the plunger is given by

$$F \times A = W \times a$$

$$F = W{\cdot}\left(\frac{a}{A}\right) = W{\cdot}\left(\frac{d}{D}\right)^2 = 9{\cdot}81 \times \left(\frac{0{\cdot}018}{0{\cdot}144}\right)^2 = 1{\cdot}533 \text{ kN}$$

(ii) Volume displaced/stroke × Number of strokes = Total volume displaced Plunger area × Plunger stroke × Number of strokes = Ram area × Ram stroke

$$a{\cdot}x{\cdot}n = A{\cdot}L$$

$$n = \frac{A}{a} \times \frac{L}{x} = \left(\frac{D}{d}\right)^2 \times \frac{L}{x} = \left(\frac{0{\cdot}144}{0{\cdot}018}\right)^2 \times \frac{0{\cdot}6}{0{\cdot}22} = 174{\cdot}5 = 175 \text{ strokes}$$

(iii) Power input to the plunger = Work done by the plunger/s

= Weight lifted × Distance moved/s

$$= 9{\cdot}81 \times \frac{0{\cdot}6}{16 \times 60}$$

$= 0{\cdot}00613$ kW or $6{\cdot}13$ W

19.5 HYDRAULIC LIFT

A hydraulic lift is used to carry the goods or passengers from one floor to another in a multi-storied building. It can be either a direct acting or a suspended type of lift. In a direct acting lift, a ram slides inside a fixed cylinder. On underside of the ram, a fluid under pressure

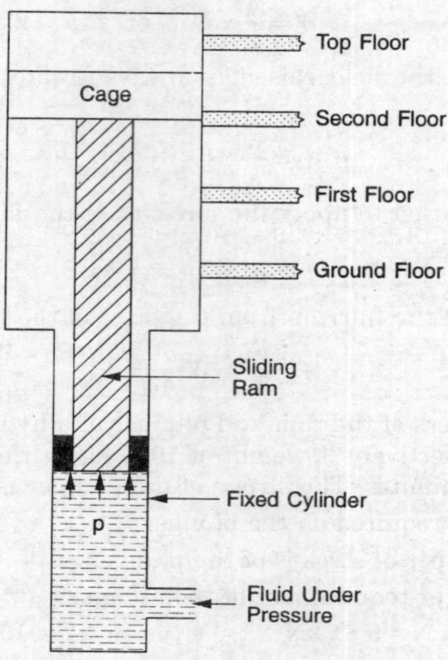

Fig. 19.5

is introduced to lift it and is taken out to lower it. On the upper side a cage is used to carry the load or passengers to different floors (Fig. 19.5).

In suspended type of lifts, a jigger arrangement is provided. A **hydraulic jigger** is a device used to multiply the stroke of a ram. It consists of a fixed cylinder in which a ram can slide as shown in Fig. 19.6. A number of pulleys are fixed to the head of the cylinder and an equal number to the free end of the ram. A wire rope, the one end of which is fixed to the cylinder body is wound around the pulleys on the ram and the cylinder. The free end of the rope is tied to the load which is to be lifted through some guide pulleys. The arrangement magnifies the stroke of the ram. When a fluid under pressure is admitted into the cylinder, the ram moves out increasing the distance between the fixed and the movable pulleys. This shortens the free length of the rope. Similarly, when the fluid is taken out, the tension in the rope moves the ram into the cylinder. This increases the free length of the rope. If n is the number of pulleys on each end of the jigger arrangement, the distance moved by the free end of the rope is equal to n times the stroke of the ram or the travel of the ram. The jigger may be used in a horizontal or in a vertical position.

The free end of the hydraulic jigger is attached to a suspended cage. When the cage is to be lifted, the fluid is admitted into the cylinder which decreases the free length of the rope. Similarly, if the cage is to be lowered the fluid is taken out of the cylinder, thereby moving the ram into the cylinder and increasing the free length of the rope.

These days, the hydraulic lifts have been superseded by electric lifts except where there is chance of fire or explosion.

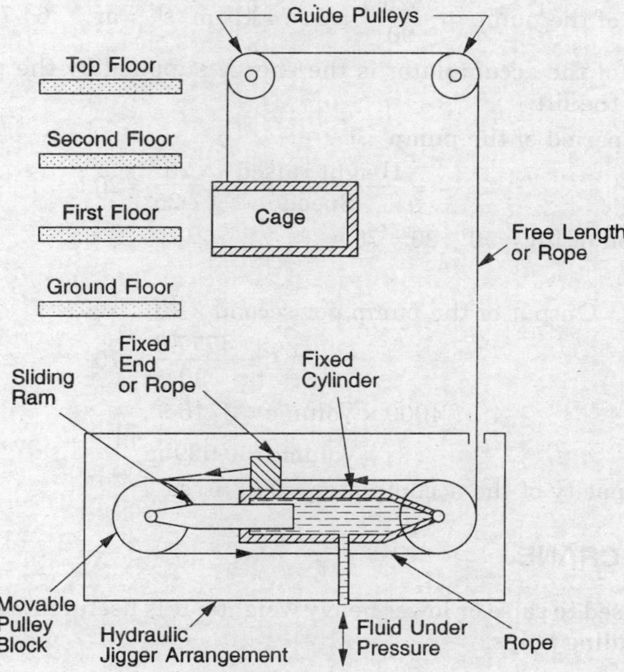

Fig. 19.6

Example 19.8 A hydraulic lift raises a load of 15 tonnes through a height of 25 m once in every 1·5 mintues. An accumulator charged by a pump supplies water to the lift at a pressure of 4000 kN/m^2. The lift moves with a velocity of 1·25 m/s. If the efficiency of the pump is 82% and of the lift 77%, determine the power required to drive the pump and the minimum capacity of the accumulator in litres. Neglect all frictional losses.

Solution $p = 4000 \text{ kN/m}^2$ $\quad W = (15 \times 1000 \times 9{\cdot}81) = (15 \times 9{\cdot}81) \text{ kN}$

$L = 25 \text{ m}$ $\quad V = 1{\cdot}25 \text{ m/s}$

$\eta_p = 0{\cdot}82$ $\quad \eta_L = 0{\cdot}77$

$T = 90 \text{ s}$

$$\text{Output of the lift} = \text{Weight lifted} \times \text{Height raised} = (15 \times 9{\cdot}81) \times 25 = 3678{\cdot}8 \text{ kN·m}$$

$$\text{Input of the lift} = \frac{\text{Output of the lift}}{\text{Efficiency of the lift}} = \frac{3678{\cdot}8}{0{\cdot}77} = 4777{\cdot}6 \text{ kN·m}$$

This is also the output of the pump.

$$\text{Input of the pump} = \frac{\text{Output of the pump}}{\text{Efficiency of the pump}} = \frac{4777{\cdot}8}{0{\cdot}82} = 5826{\cdot}3 \text{ kN·m}$$

$$\text{or power of the pump} = \frac{5826{\cdot}3}{90} = 64{\cdot}74 \text{ kN·m/s} \quad \text{or} \quad 64{\cdot}74 \text{ kW}$$

Capacity of the accumulator is the energy supplied by the pump during the idle period of the lift.

Working period of the pump

$$= \frac{\text{Height raised}}{\text{Speed}} = \frac{25}{1{\cdot}25} = 20 \text{ s}$$

Idle period of lift = 90 – 20 = 70 s

Therefore,

Capacity = Output of the pump per second × 70

$$p{\cdot}A{\cdot}L = \frac{4777{\cdot}6}{90} \times 70$$

$$4000 \times \text{Volume} = 3715{\cdot}9$$

$$\text{Volume} = 0{\cdot}929 \text{ m}^3$$

or the capacity of the accumulator = 929 litres

19.6 HYDRAULIC CRANE

A hydraulic crane is used to raise or lower heavy weights. It is useful in warehouses, workshops or in loading or unloading ships.

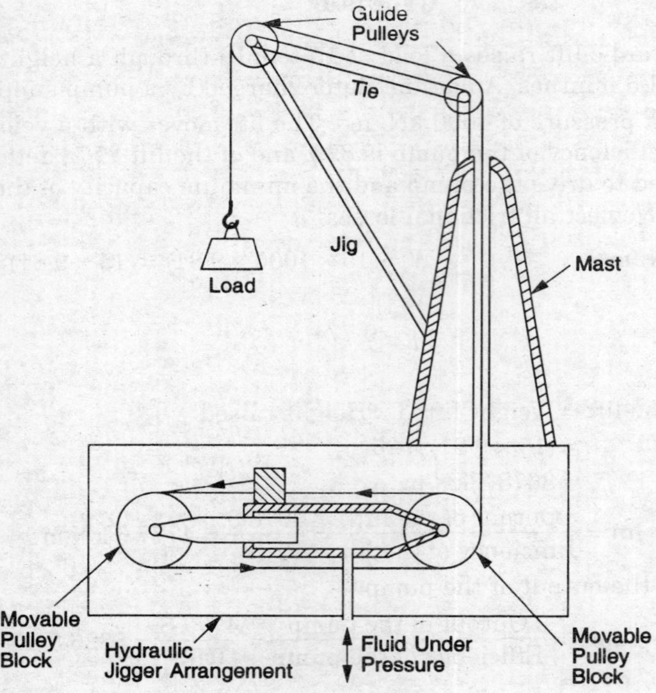

Fig. 19.7

The function of a hydraulic crane is similar to that of a hydraulic lift. It consists of a mast with a jib and tie attached to it as shown in Fig. 19.7. The free end of the wire rope from the jigger arrangement is taken around a guide pulley fixed to the jib and tie. A load can be attached to the free end with the help of a hook. The mast along with the jib and tie can rotate about a vertical axis. This facilitates the load to be carried to any desired place. As discussed in the previous section, the fluid is admitted into the cylinder if the load is to be lifted or it is taken out if load is to be lowered.

Example 19.9 A hydraulic crane lifts a weight of 1500 kg through a height of 10·5 m once in every two minutes. Water pressure is 4500 kN/m^2 and the speed of the lift 0·3 m/s. The efficiency of the crane is 72%. An accumulator charged by a pump supplies water to the crane. Find the capacity of the cylinder of the jigger in litres, power input of the pump and the capacity of the accumulator in litres.

Solution $p = 4500$ kN/m^2 $\quad V = 0{\cdot}3$ m/s

$L = 10{\cdot}5$ m $\quad \eta = 0{\cdot}75$

$T = 120$ s $\quad W = (1500 \times 9{\cdot}81)$ N $= 1{\cdot}5$ kN

$$\text{Output of the crane} = \text{Weight lifted} \times \text{Height raised}$$
$$= (1{\cdot}5 \times 9{\cdot}81) \times 10{\cdot}5$$
$$= 154{\cdot}5 \text{ kN}{\cdot}\text{m}$$

$$\text{Input of the crane} = \frac{\text{Output of the lift}}{\text{Eficiency of the lift}} = \frac{154{\cdot}5}{0{\cdot}72} = 214{\cdot}6 \text{ kN}{\cdot}\text{m}$$

This is also the output of the pump.

$$\text{or power of the pump} = \frac{214{\cdot}6}{120} = 1{\cdot}788 \text{ kN}{\cdot}\text{m/s} \quad \text{or} \quad 1{\cdot}788 \text{ kW}$$

$$\text{Input of the crane} = \text{Work done on the ram of jigger}$$
$$= \text{Force} \times \text{Stroke}$$
$$= p{\cdot}A \times \text{Stroke}$$
$$= p{\cdot} \text{ Volume of cylinder}$$
$$214{\cdot}6 = 4500 \times \text{Volume}$$
$$\text{Volume} = 0{\cdot}0477 \text{ m}^3 \quad \text{or} \quad 47{\cdot}7 \text{ litres}$$

Capacity of the accumulator is the energy supplied by the pump during the idle period of the lift.

Working period of the pump

$$= \frac{\text{Height raised}}{\text{Speed}} = \frac{10{\cdot}5}{0{\cdot}3} = 35 \text{ s}$$

Idle period of lift = 120 − 35 = 85 s

Therefore,

$$\text{Capacity} = \text{Output of the pump per second} \times 85$$
$$p{\cdot}A{\cdot}L = 1{\cdot}788 \times 85$$
$$4500 \times \text{Volume} = 151{\cdot}98$$
$$\text{Volume} = 0{\cdot}0338 \text{ m}^3$$

or the capacity of the accumulator = 33·8 litres

19.7 HYDRAULIC COUPLING

A hydraulic coupling is used to transmit torque from one shaft to another through a fluid medium. The torque on the two shafts remains the same. Its action is similar to that of a clutch of an automobile.

There are two main components of a hydraulic coupling : pump impeller and turbine runner. These are attached to the driving and the driven shafts respectively (Fig. 19.8). Both,

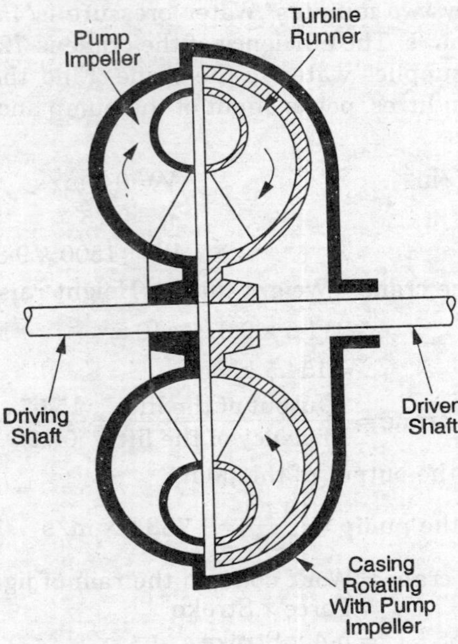

Fig. 19.8

the impeller and the runner are enclosed in a common casing filled with a fluid. The fluid used usually is an oil due to its lubricating power and stability. There is no solid or direct contact between the driving and the driven shafts. During its operation, there is circulation of oil through the turbine rotor and the pump impeller. The energy is received in the pump impeller and is transmitted to the turbine runner.

The action of a hydraulic coupling is similar to the operation of two electric fans kept face to face. If one is run by electric power, the blades of the other fan also start moving due to the circulation of air. In a hydraulic coupling, as the pump starts moving, the oil in the coupling is thrown outwards due to the centrifugal action, thus gaining in torque. When the fluid enters the outer edge of the turbine runner, torque is transferred to the runner which starts rotating. As the algebraic sum of torques is to be zero for the closed circuit,

$$T_P + T_T = 0$$

where T_P and T_T are the torques on the pump impeller and the turbine runner respectively. The direction of one torque is opposite to the other. As the speed of the driven shaft increases, the centrifugal action of the runner starts resisting the flow of the fluid and thus reducing the amount of fluid pumped. In case the speed of both the shafts becomes equal, there can be no pumping of the fluid and thus no torque transmission. Therefore, for normal steady operation, there must exist a difference in the speeds to transmit the torque. Slip of a hydraulic coupling is defined as

$$S = \frac{N_1 - N_2}{N_1} = 1 - \frac{N_2}{N_1} \quad \ldots(19.6)$$

where N_1 and N_2 are the speeds of the pump impeller and the turbine runner respectively. In practice, the value of slip is usually 2 to 4 percent. If ω_1 and ω_2 are the angular velocities of the impeller and the turbine respectively, then the efficiency of the coupling is given by,

$$\eta = \frac{\text{Output}}{\text{Input}} = \frac{T_2 \cdot \omega_2}{T_1 \cdot \omega_1}$$

If friction is neglected in the impeller and runner passages, $T_1 = T_2 = T$ and therefore,

$$\eta = \frac{\omega_2}{\omega_1} = \frac{N_2}{N_1} = 1 - S \quad \ldots(19.7)$$

Power loss in the coupling = Input – Output = $T(\omega_1 - \omega_2)$

The efficiency given by the above equation is for an ideal coupling. However, due to friction losses the actual efficiency is reduced. In practice, the efficiency varies between 95 to 97 percent. Fig. 19.9 shows the speed vs. efficiency curve. The curve starts from zero, increases uniformly upto about 95 percent and then drops to zero. The reason being that when the speed of the runner approaches the speed of the impeller, the flow of oil reduces as discussed earlier. As a result, the torque transmitted approaches zero and so the efficiency.

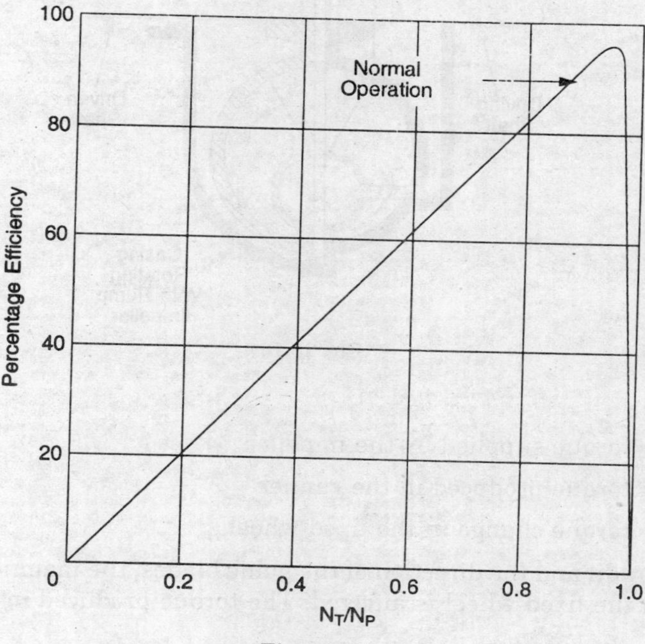

Fig. 19.9

A great advantage of this type of coupling is that the pump impeller can rotate even though it may not be able to rotate the loaded turbine runner. The impeller can first gain the full speed and torque, before taking up the load on the turbine runner to make it rotate. This makes it possible to connect the electric motors to direct loads. Also, a hydraulic coupling acts as a vibration absorber and there is smoothness of operation.

19.8 TORQUE CONVERTER

The function of a torque converter is similar to that of a hydraulic coupling. However, in its steady operation a torque converter is capable to multiply or divide the torque with the help of stationary blades. Its operation is analogous to that of a gearbox in an automobile or an electric transformer.

In addition to a pump impeller and a turbine runner, a torque converter has a fixed wheel with guide blades (Fig. 19.10). For a closed circuit, the algebraic sum of torques must be zero,

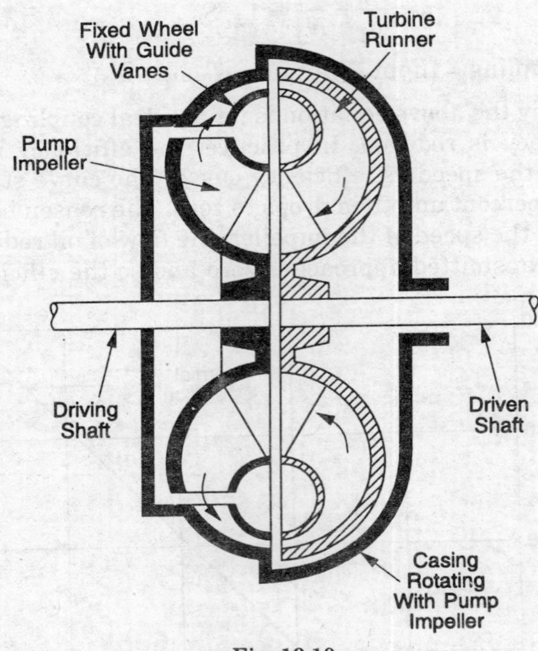

Fig. 19.10

or $$T_1 + T_2 + T_3 \qquad \text{...(19.8)}$$

where T_1 = torque supplied by the impeller

T_2 = torque produced in the runner

T_3 = torque change in the fixed wheel.

By varying the radii and the direction of the guide blades, the magnitude and the direction of torque change in the fixed wheel is affected. The torque produced in the turbine runner is multiplied or divided accordingly. No work is done on the fixed wheel as the torque absorbed by it is transmitted to the earth. If losses are neglected,

Input to the pump = Output of the turbine

$$T_1 \cdot \omega_1 = T_2 \cdot \omega_2$$

$$\frac{\omega_2}{\omega_1} = \frac{N_2}{N_1} = \frac{T_1}{T_2}$$

This shows that the speed of the runner is increased or decreased inversely in the ratio of torques. It is possible to obtain a gear ratio of 5 to 1 with an efficiency of 90 percent. Efficiency of a torque converter is given by

$$\eta = \frac{\text{Output}}{\text{Input}} = \frac{T_2 \cdot \omega_2}{T_1 \cdot \omega_1}$$

$$= 1 \qquad \text{if losses are neglected.}$$

Figure 19.11 shows the characteristic curve of a torque converter. It is observed that a torque converter gives higher efficiency at low speed ratios as compared to a fluid coupling. Due to this reason, it is preferred over for low speed ratios. There is decrease in efficiency when the speed ratio increases from 0·5 to 0·6. However, a combination of a torque converter and a fluid coupling can be used for efficient performance over the full range of speed ratio. In such a machine, as soon as the output torque becomes equal to the input torque, the fixed guide wheel starts rotating and then the combination starts working as a hydraulic coupling.

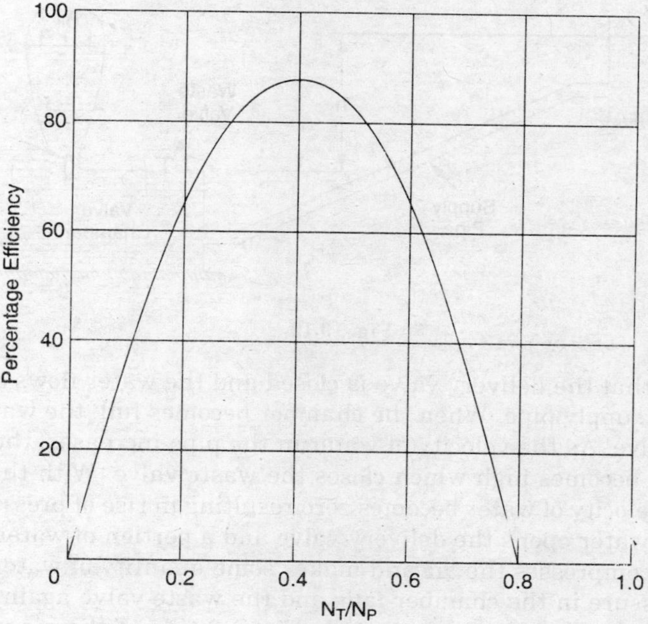

Fig. 19.11

19.9 HYDRAULIC RAM

A hydraulic ram is a pump which is used to raise small quantity of water to a higher level if there is a large quantity of water available at a lower level. Its input is the potential energy of the surplus water and no external power is used to operate it. Its function is based on the principle of water hammer *ie* when a flowing fluid is brought to rest suddenly, there is a rise in pressure which may be utilised to raise a portion of water to a higher level.

A hydraulic ram has a valve chamber fitted with two valves, a waste valve and a delivery valve. Both the valves are of non-return type. The delivery valve opens into an air vessel which contains compressed air and has a delivery pipe attached to it. At the top of the delivery pipe, a tank is connected to receive the water. The available water source is connected to the valve chamber through a pipe (Fig. 19.12).

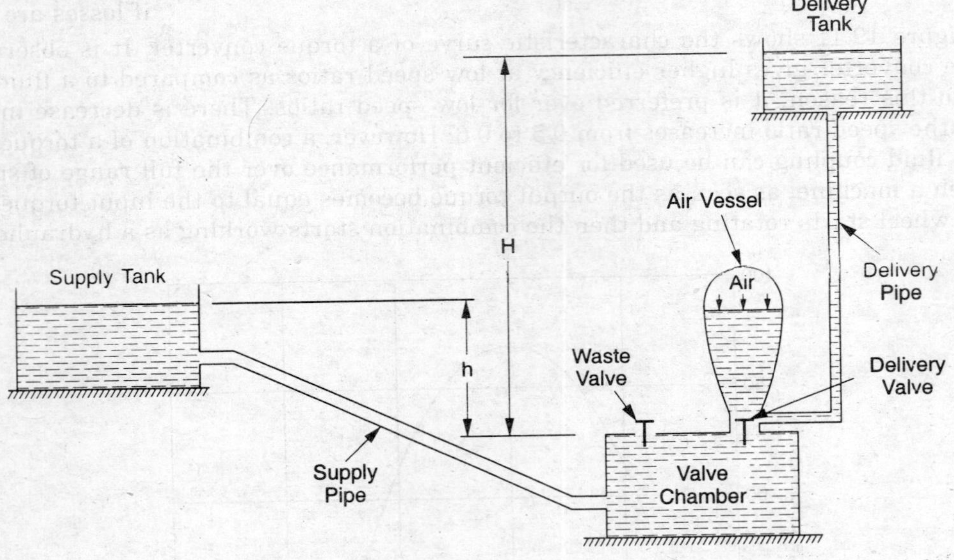

Fig. 19.12

Initially, assume that the delivery valve is closed and the water flows down into the valve chamber through the supply pipe. When the chamber becomes full, the water starts going out through the waste valve. As the velocity of water in the pipe increases, the dynamic pressure in the valve chamber becomes high which closes the waste valve. With the sudden closure of the waste valve, the velocity of water becomes zero resulting in rise of pressure in the chamber. The high pressure of water opens the delivery valve and a portion of water enters into the air vessel which further compresses the air and makes some quantity of water to rise to a higher level. Slowly, the pressure in the chamber falls and the waste valve again opens allowing the water to flow through it. The cycle is repeated. The function of the air vessel is to act as a storage and regulate the flow of water by pushing some water into the delivery pipe even when th delivery valve is closed. Thus the air vessel tends to minimise the fluctuations in the delivery of water.

Let q = quantity of water raised per second
Q = quantity of water into the chamber per second
H = height of water raised above the chamber
h = height of water in the supply tank above the chamber

Energy supplied to the ram = $w \cdot Q \cdot h$

Energy delivered to water by the ram = $w \cdot q \cdot H$

Efficiency,

$$\eta = \frac{\text{energy delivered}}{\text{Energy supplied}} = \frac{w \cdot q \cdot H}{w \cdot Q \cdot h} = \frac{qH}{Qh} \quad \text{...(19.9)}$$

This is known as D' Aubuisson's efficiency. This assumes the datum at the chamber level. However, if the initial water level is taken as datum, then the actual height raised is $(H - h)$.

Energy supplied to the ram = $w \cdot (Q - q) \cdot h$

Energy delivered to water by the ram = $w \cdot q \cdot (H - h)$

Efficiency,

$$\eta = \frac{w \cdot q \cdot (H - h)}{w \cdot (Q - q) \cdot h} = \frac{q\,(H - h)}{(Q - q)h} \quad \text{...(19.10)}$$

This is known as Rankine's efficiency. Its value is always less than D' Aubuisson's efficiency.

Salient features of a hydraulic ram are

- Simplicity
- No electric power consumption
- Absence of moving parts
- Low repair costs
- Negligible running costs.

A hydraulic ram is very useful in hilly areas where large quantity of water flows at lower levels and is required to be raised to higher levels for irrigation purposes.

Example 19.10 A hydraulic ram uses 20 litres of water per second from a height of 2·9 m and raises 2 litres of water to a height of 18 m. The delivery pipe is 80 m long and 50 mm in diameter. The coefficient of friction is 0·01. Determine the efficiency of the ram.

Solution $Q = 0{\cdot}02\ \text{m}^3/\text{s}$ $\qquad q = 0{\cdot}002\ \text{m}^3/\text{s}$

$H' = 18$ m $\qquad h = 2{\cdot}9$ m

$L = 80$ m $\qquad d = 0{\cdot}05$ m

$f = 0{\cdot}01$

$$V = \frac{q}{a} = \frac{q}{\frac{\pi}{4} \times d^2} = \frac{0{\cdot}002}{\frac{\pi}{4} \times 0{\cdot}05^2} = 1{\cdot}0186\ \text{m/s}$$

$$h_f = \frac{4fLV^2}{2gd} = \frac{4 \times 0{\cdot}01 \times 80 \times 1{\cdot}0186^2}{2 \times 9{\cdot}81 \times 0{\cdot}05} = 3{\cdot}384\ \text{m}$$

Head loss due to friction in the delivery pipe

Total head raised = 18 + 3·384 = 21·384 m

D′ Aubuisson's efficiency

$$= \frac{q \times H}{Q \times h} = \frac{0{\cdot}002 \times 21{\cdot}384}{0{\cdot}02 \times 2{\cdot}9} = 0{\cdot}737 \quad \text{or} \quad 73{\cdot}7\%$$

Rankine's efficiency

$$= \frac{q \times (H - h)}{(Q - q) \times h} = \frac{0{\cdot}002 \times (21{\cdot}384 - 2{\cdot}9)}{(0{\cdot}02 - 0{\cdot}002) \times 2{\cdot}9} = 0{\cdot}708 \quad \text{or} \quad 70{\cdot}8\%$$

19.10 JET PUMP

A jet pump is used to raise the suction limit of a radial or centrifugal pump by about 6 m beyond the normal value. However, it needs a supply of high-pressure water or stream for its function which may be taken from the delivery end of the pump or from other sources. A conventional radial-flow pump is provided with a nozzle at the suction end (Fig. 19.13). As the pump is started, a portion of the high-pressure water from the delivery side of the pump is diverted to the nozzle which is enclosed in a constricted passage. Due to the high velocity of the jet, the water around the jet is also pushed up and a local pressure drop occurs. Owing to

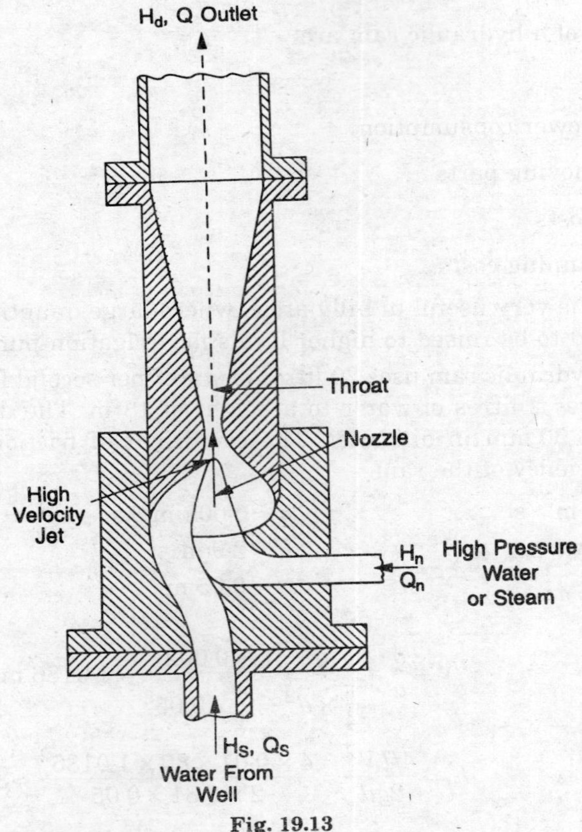

Fig. 19.13

this low pressure, water enters the suction pipe from the well through a strainer and a non-return valve, and is raised upto the suction reach of the pump. As the water is raised in the suction pipe, more water reaches to take its place.

Let Q_s = sucked quantity of water through suction pipe

Q_n = discharge through the nozzle

Q = total discharge through nozzle = $Q_s + Q_n$

H_s = suction head

H_d = delivery head

H_n = supply head to the nozzle

Then, efficiency of the jet pump,

$$\eta = \frac{Q_s\,(H_s + H_d)}{Q_n\,(H_n - H_d)}$$

Jet pumps usually find their use in mines, for pumping oil and to feed steam boilers. In case of steam boilers, steam from the boiler passes through the nozzle of the jet pump.

19.11 AIRLIFT PUMP

An airlift pump is used to lift water from a well by the use of compressed air. It is a simple

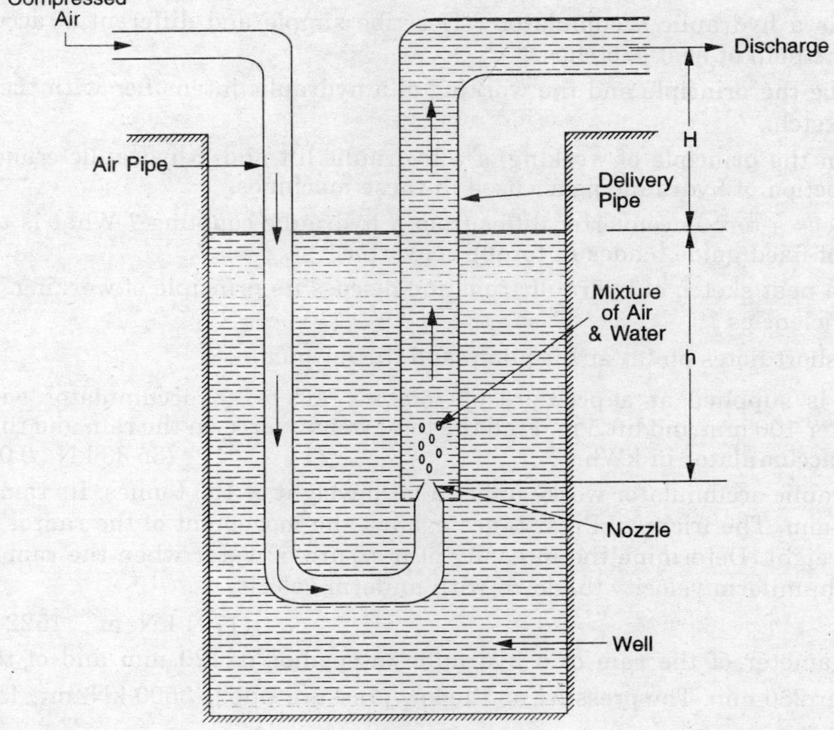

Fig. 19.14

device which consists of an air pipe with a nozzle at one end. A discharge pipe encloses the nozzle. The nozzle is at a depth of h m below the water surface in the well. (Fig. 19.14).

As the compressed air passed through the nozzle, it mixes with water in the delivery pipe. The mixing of air with water reduces its density. The pressure of the mixture column in the delivery pipe becomes less than the pressure of the pure water in the well. This means a longer column of air-water mixture is required to balance against a small column of pure water. This starts a flow of air-water mixture through the delivery pipe if the supply of compressed air is continued. It is observed that the best results are obtained if the ratio h/H is kept between 1 and 4. Its main advantages are that

- it has no moving parts which can wear out
- it can be used to carry solids suspended in water without any damage
- it can be used to drain out water in mines where compressed units are already available.
- It has a low efficiency of the order of 20-30 percent.

PROBLEMS

1. What is a hydraulic system ? What are its types ?
2. What is a hydraulic accumulator ? Describe simple and differential accumulators with the help of neat sketches.
3. Describe the principle and the working of a hydraulic intensifier with the help of a neat sketch.
4. Explain the principle of working of a hydraulic lift and a hydraulic crane. Discuss the function of hydraulic jigger used in these machines.
5. How does a torque converter differ from a hydraulic coupling ? What is the importance of fixed guide blades in torque converter ?
6. Draw a neat sketch of hydraulic ram and discuss its principle of working. Define its two efficiencies ?
7. Write short notes on (a) airlift pump and (b) Jet pump.
8. Water is supplied at a pressure of 4500 kN/m^2 to an accumulator with a ram diameter 100 mm and lift 5 m. Find the total weight force on the ram and the capacity of the accumulator in kWh. (35·33 kN , 0·0049 kWh)
9. A hydraulic accumulator works under a total weight of 105 tonnes. Its ram diameter is 900 mm. The fricitional resistance against the movement of the ram is 6% of the total weight. Determine the intensity of pressure of water when the ram moves (a) up with uniform velocity (b) down with uniform velocity.
(1701 kN/m^2 , 1522·8 kN/m^2)
10. The diameter of the ram of a hydraulic intensifier is 120 mm and of the sliding cylinder 250 mm. The pressure on the low-pressure side is 3500 kN/m^2. The friction

loss at each of the packing is 5% of the total force. Determine the pressure on the high-pressure side. (13·71 MN/m^2)

11. The diameters of the ram and the plunger of a hydraulic press are 220 mm and 30 mm respectively. The stroke of the plunger is 100 mm. The press is used to lift a weight of 3500 kg. Find the force required at the plunger and the work done during 95 strokes of the plunger. (638·5 N , 6066 J)
12. A hydraulic crane used to lift a load of 3 tonnes through a height of 7·5 m is supplied 45 litres of water at a pressure of 3200 kN/m^2. Find the efficiency of the crane. (65·25%)
13. A hydraulic ram uses 800 litres of water per minute and raises it to a height of 20 m. The quantity of water raised is 75 litres per minute. The water is available at a height of 3·5 m. through a delivery pipe 60 m long and 40 mm in diameter. The coefficient of discharge is 0·011. Determine the efficiency of the ram. (*D′* Aubuisson's η = 62·56% , Rankine's η = 58·7%)

INDEX